Hydraulic Systems for Mobile Equipment

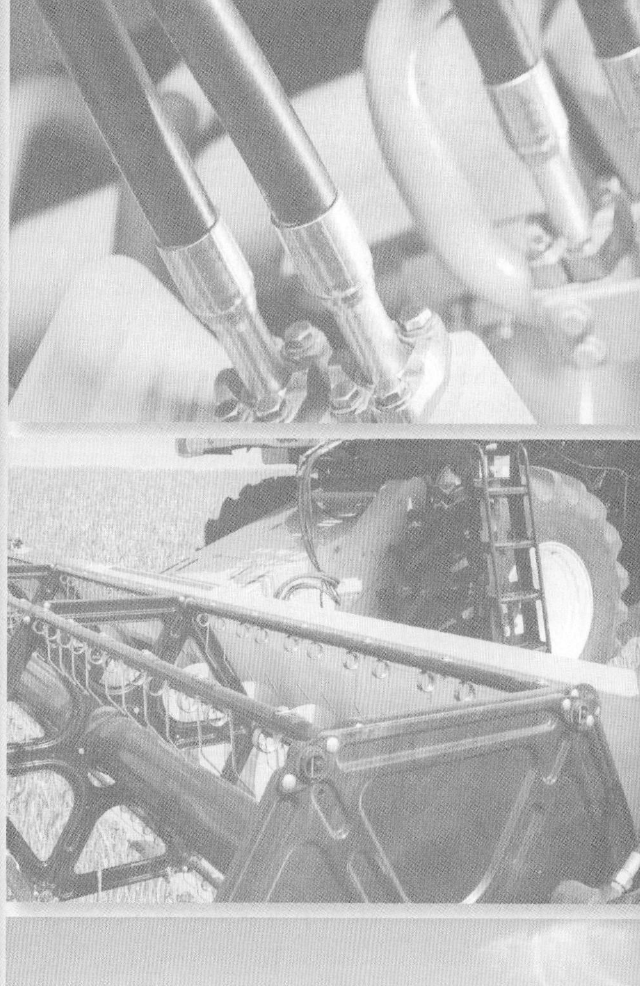

by

Timothy W. Dell, Ph.D.

Publisher
The Goodheart-Willcox Company, Inc.
Tinley Park, IL
www.g-w.com

Manufactured in the United States of America.

Library of Congress Catalog Card Number 2015019166

ISBN 978-1-63126-414-6

1 2 3 4 5 6 7 8 9 – 17 – 20 19 18 17 16 15

Cover image:
Frank L. Junior/Shutterstock.com
Dmitry Kalinovsky/Shutterstock.com
aarrows/Shutterstock.com
Andrey N. Bannov/Shutterstock.com

Library of Congress Cataloging-in-Publication Data
Dell, Tim
Hydraulic systems for mobile equipment / by Tim Dell. -- First edition.
p. cm.
Includes index.
ISBN 978-1-63126-414-6
1. Motor vehicles--Hydraulic equipment. 2. Fluid power technology--Equipment and supplies. I. Title.

TL275.D39 2017
620.1'06--dc23

2015019166

Preface

Hydraulic Systems for Mobile Equipment is intended to educate students in heavy equipment and heavy truck programs. Although the text has a primary emphasis on agricultural and construction machinery, it can empower students working in any related field of hydraulics. To this end, it teaches and is correlated to the competencies of both AED *Hydraulics/Hydrostatics Standards* and the Hydraulics section of the NATEF *Medium/Heavy Duty Trucks Task List*. The scope and approach of the book make it appropriate for all students, whether they are pursuing a certificate, associate's degree, bachelor's degree, or a master's degree.

This textbook includes traditional hydraulic content such as fluid power principles, pumps, motors, safety, valves, filtration, accumulators, plumbing, reservoirs, coolers, and fluids. In addition, the book is especially unique in providing the fundamental explanation of the most common types of mobile hydraulic control systems, specifically open center, pressure compensating, pre-spool load sensing pressure compensating, post-spool compensation (flow sharing), negative flow control, and positive flow control.

The text provides fundamental instruction on hydrostatic transmissions with the goal providing students true comprehension of the systems. This will free the students from relying solely on discrete (or limited) troubleshooting charts. Students will also gain instruction in hydraulic diagnostic principles such as how to tap into the system, what tools are available, how to properly use those tools, how to perform the tests safely, and what to look for when performing the tests.

Hydraulic Systems for Mobile Equipment provides students the necessary foundational building blocks, equipping them for a bright future in hydraulic technology. The book includes hundreds of images comprised of multicolor line art, cross-sectional drawings, photographs, and 3D renderings. The book includes practice questions, cases studies, and helpful techniques for diagnosing hydraulic systems.

As society transitions from printed textbooks to digital learning tools, it is important to have the content you need in the format you need it in. The **Hydraulic Systems for Mobile Equipment** textbook is available in both print and digital formats, and is supported with professionally developed, user-friendly online resources. These resources include electronic study tools and interactive activities designed to prepare you for success in your studies and career.

Congratulations on your study of hydraulics technology, and I encourage you to take advantage of all the resources available as part of the **Hydraulic Systems for Mobile Equipment** learning solution.

Tim Dell

About the Author

Timothy W. Dell is an Associate Professor of Automotive Technology at Pittsburg State University. Dr. Dell received his doctoral degree in curriculum and instruction from Kansas State University, a master of science degree in technology education from Pittsburg State University, and a bachelor of science degree in automotive technology with an emphasis in diesel and heavy equipment from Pittsburg State University. He began his career working for Case IH in their Technical Service Group, specializing in combine diagnostics. Dr. Dell has served on John Deere's Agricultural National Service Training Advisory Board and has been the advisor of Pittsburg State University's Caterpillar ThinkBIGGER four-year degree. He currently teaches Automotive Electricity and Electronic Systems, Automotive Automatic Transmissions, Fluid Power, Advanced Hydraulic Systems and Off Highway Systems. Dr. Dell has served as the automotive department chair for four years, but returned to the classroom full time to pursue his passion for curriculum development and teaching.

Reviewers

The author and publisher wish to thank the following industry and teaching professionals for their valuable input into the development of **Hydraulic Systems for Mobile Equipment**.

James Aakre
Associate Professor
North Dakota State College of Science
Underwood, MN

David Christen
Instructor
University of Northwest Ohio
Delphos, OH

Cole Eddy
Instructor
Lincoln College of Technology—Nashville
Nashville, TN

Nick Deftereos
Instructor
Reedley College
Reedley, CA

Ed Frederick
Instructor
State Technical College of Missouri
Linn, MO

Paul Losh
Instructor
Lincoln College of Technology—Nashville
Nashville, TN

Dennis Massingham
Assistant Professor
University of Alaska, Anchorage
Anchorage, AK

Kent McCleary
Instructor
University of Northwest Ohio
Lima, OH

Chris Thompson
Instructor
Alexandria Technical & Community College
Alexandria, MN

Dick Weber
Instructor
Minnesota State Community and Technical College
Moorhead, MN

Austin Williams
Assistant Professor
Ferris State University
Big Rapids, MI

Bruce Wright
John Deere Tech Program Coordinator
SUNY Cobleskill
Cobleskill, NY

Acknowledgments

This book also would not have been possible without the generous support of many individuals. I would like to thank my family for their patience and encouragement, my students for their valuable input and assistance, and all of the reviewers for their expertise in reviewing the chapters. In addition, I would like to thank the following organizations and individuals for assisting me with their technical expertise during the development of this textbook:

AGCO—Matthew Keller

AED—Steve Johnson

Artic Fox (Phillips & Temro)—Karen Bach, David Mundahl

B&B Hydraulics—Dennis Rayl, Bill Speakman

Bell Farms—Bryan Bell

Caterpillar—Simon Bishop

CNH—Kelly Burgess, Ted Polzer, David North, Rick Cermak, Pete Steiner, Mike Wetzel, Bruce Anderson, Russel Schuchaskie (retired)

Ft. Scott Community College—Chauncey Pennington

Foley Caterpillar—Jarrod Haas, Shannon Dudley

Husco International—Joe Pfaff

John Deere—Glen Oetken, Roger Jenkins, Brian Kopecko, Aaron Vancil, David Rodts, Luke Kurth, William Loch, Alex Anhalt

Industrial Sealing & Lubrication—David Consiglio

O'Malley Implement Company—Jack Dent, Lloyd Bandy, Cleo Jones, Don Workman, Brad Bohnenblust

Pittsburg State University—Bob Schroer, Ken Gordon

San Joaquin Delta College—Rich Dettloff

Ultra Clean Technologies—Bruce Riley

G-W Integrated Learning Solution

Together, We Build Careers

At Goodheart-Willcox, we take our mission seriously. Since 1921, G-W has been serving the career and technical education (CTE) community. Our employee-owners are driven to deliver exceptional learning solutions to CTE students to help prepare them for careers. Our authors and subject matter experts have years of experience in the classroom and industry. We combine their wisdom with our expertise to create content and tools to help students achieve success. Our products start with theory and applied content based upon a strong foundation of accepted standards and curriculum. To that base, we add student-focused learning features and tools designed to help students make connections between knowledge and skills. G-W recognizes the crucial role instructors play in preparing students for careers. We support educators' efforts by providing time-saving tools that help them plan, present, assess, and engage students with traditional and digital activities and assets. We provide an entire program of learning in a variety of print, digital, and online formats, including economic bundles, allowing educators to select the right mix for their classroom.

Student-Focused Curated Content

Goodheart-Willcox believes that student-focused content should be built from standards and/or accepted curriculum coverage. Standards from the *Associated Equipment Distributors (AED)* and the *National Automotive Technicians Education Foundation (NATEF)* were used as a foundation in this text. **Hydraulic Systems for Mobile Equipment** also uses a building block approach with attention devoted to a logical teaching progression that helps students build upon their learning. We call on industry experts and teachers from across the country to review and comment on our content, presentation, and pedagogy. Finally, in our refinement of curated content, our editors are immersed in content checking, securing and sometimes creating figures that convey key information, and revising language and pedagogy.

The AED and NATEF Connections

Goodheart-Willcox is pleased to support the AED and NATEF by correlating **Hydraulic Systems for Mobile Equipment** to both the *AED Standards for Construction Equipment* and the *NATEF Medium/Heavy Duty Truck* task list. These standards were created in concert with industry and subject matter experts to match real-world job skills and marketplace demands.

Features of the Textbook

Features are student-focused learning tools designed to help you get the most out of your studies. This visual guide highlights the features designed for the textbook.

Learning Objectives clearly identify the knowledge and skills to be obtained when the chapter is completed.

Introduction provides an overview and preview of the chapter content.

Warnings alert you to potentially dangerous materials and practices.

Cautions alert you to practices that could potentially damage equipment or instruments.

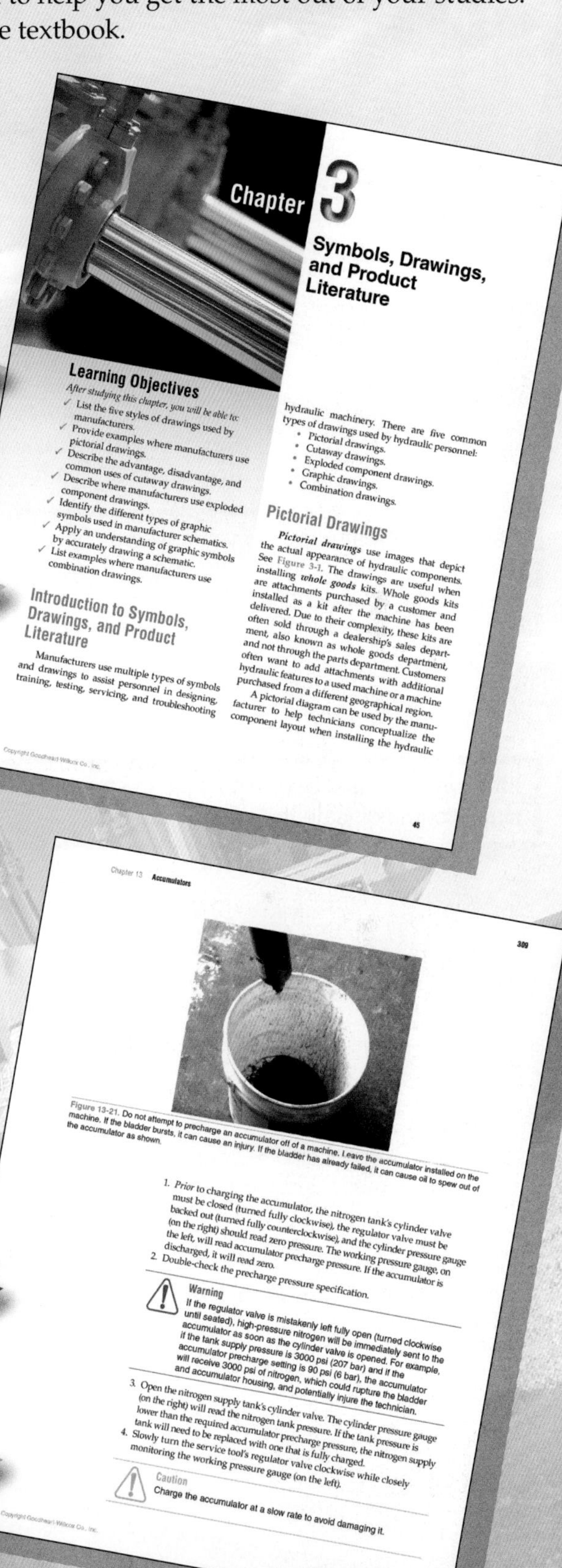

Chapter 3

Symbols, Drawings, and Product Literature

Learning Objectives

After studying this chapter, you will be able to:

- ✓ List the five styles of drawings used by manufacturers.
- ✓ Provide examples where manufacturers use pictorial drawings.
- ✓ Describe the advantage, disadvantage, and common uses of cutaway drawings.
- ✓ Describe where manufacturers use exploded component drawings.
- ✓ Identify the different types of graphic symbols used in manufacturer schematics.
- ✓ Apply an understanding of graphic symbols by accurately drawing a schematic.
- ✓ List examples where manufacturers use combination drawings.

Introduction to Symbols, Drawings, and Product Literature

Manufacturers use multiple types of symbols and drawings to assist personnel in designing, training, testing, servicing, and troubleshooting hydraulic machinery. There are five common types of drawings used by hydraulic personnel:

- Pictorial drawings.
- Cutaway drawings.
- Exploded component drawings.
- Graphic drawings.
- Combination drawings.

Pictorial Drawings

Pictorial drawings use images that depict the actual appearance of hydraulic components. See Figure 3-1. The drawings are useful when installing ***whole goods*** kits. Whole goods kits are attachments purchased by a customer and installed as a kit after the machine has been delivered. Due to their complexity, these kits are often sold through a dealership's sales department, also known as whole goods department, and not through the parts department. Customers often want to add attachments with additional hydraulic features to a used machine or a machine purchased from a different geographical region.

A pictorial diagram can be used by the manufacturer to help technicians conceptualize the component layout when installing the hydraulic

Copyright Goodheart-Willcox Co., Inc.

45

Chapter 13 Accumulators

309

Figure 13-21. Do not attempt to precharge an accumulator off of a machine. Leave the accumulator installed on the machine. If the bladder bursts, it can cause an injury. If the bladder has already failed, it can cause oil to spew out of the accumulator as shown.

1. *Prior* to charging the accumulator, the nitrogen tank's cylinder valve must be closed (turned fully clockwise), the regulator valve must be backed out (turned fully counterclockwise), and the cylinder pressure gauge (on the right) should read zero pressure. The working pressure gauge, on the left, will read accumulator precharge pressure. If the accumulator is discharged, it will read zero.
2. Double-check the precharge pressure specification.

Warning

If the regulator valve is mistakenly left fully open (turned clockwise until seated), high-pressure nitrogen will be immediately sent to the accumulator as soon as the cylinder valve is opened. For example, if the tank supply pressure is 3000 psi (207 bar) and if the accumulator precharge setting is 90 psi (6 bar), the accumulator will receive 3000 psi of nitrogen, which could rupture the bladder and accumulator housing, and potentially injure the technician.

3. Open the nitrogen supply tank's cylinder valve. The cylinder pressure gauge (on the right) will read the nitrogen tank pressure. If the tank pressure is lower than the required accumulator precharge pressure, the nitrogen supply tank will need to be replaced with one that is fully charged.
4. Slowly turn the service tool's regulator valve clockwise while closely monitoring the working pressure gauge (on the left).

Caution

Charge the accumulator at a slow rate to avoid damaging it.

Copyright Goodheart-Willcox Co., Inc.

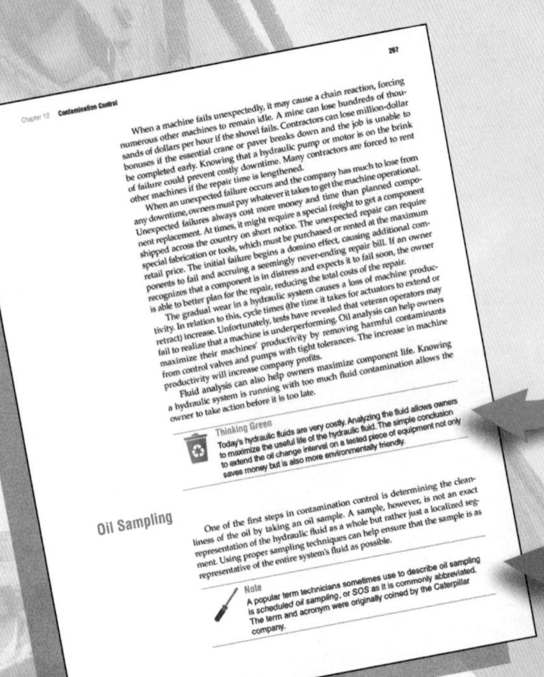

Chapter 12 Contamination Control 267

When a machine fails unexpectedly, it may cause a chain reaction, forcing numerous other machines to remain idle. A mine can lose hundreds of thousands of dollars per hour if the shovel fails. Contractors can lose million-dollar bonuses if the essential crane or paver breaks down and the job is unable to be completed early. Knowing that a hydraulic pump or motor is on the brink of failure could prevent costly downtime. Many contractors are forced to rent other machines if the repair time is lengthened.

When an unexpected failure occurs and the company has much to lose from any downtime, owners must pay whatever it takes to get the machine operational. Unexpected failures always cost more money and time than planned component replacement. At times, it might require a special freight to get a component shipped across the country on short notice. The unexpected repair can require special fabrication or tools, which must be purchased or rented at the maximum retail price. The initial failure begins a domino effect, causing additional components to fail and accruing a seemingly never-ending repair bill. If an owner recognizes that a component is in distress and expects it to fail soon, the owner is able to better plan for the repair, reducing the total costs of the repair.

The gradual wear in a hydraulic system causes a loss of machine productivity. In relation to this, cycle times (the time it takes for actuators to extend or retract) increase. Unfortunately, tests have revealed that veteran operators may fail to realize that a machine is underperforming. Oil analysis can help owners maximize their machines' productivity by removing harmful contaminants from control valves and pumps with tight tolerances. The increase in machine productivity will increase company profits.

Fluid analysis can also help owners maximize component life. Knowing a hydraulic system is running with too much fluid contamination allows the owner to take action before it is too late.

Thinking Green

Today's hydraulic fluids are very costly. Analyzing the fluid allows owners to maximize the useful life of the hydraulic fluid. The simple conclusion to extend the oil change interval on a tested piece of equipment not only saves money but is also more environmentally friendly.

Oil Sampling

One of the first steps in contamination control is determining the cleanliness of the oil by taking an oil sample. A sample, however, is not an exact representation of the hydraulic fluid as a whole but rather just a localized segment. Using proper sampling techniques can help ensure that the sample is as representative of the entire system's fluid as possible.

Note

A popular term technicians sometimes use to describe oil sampling is scheduled oil sampling, or SOS as it is commonly abbreviated. The term and acronym were originally coined by the Caterpillar company.

Copyright Goodheart-Willcox Co., Inc.

Thinking Green notes highlight key items related to sustainability, energy efficiency, and environmental issues.

Note features provide you with advice and guidance that is especially applicable on-the-job.

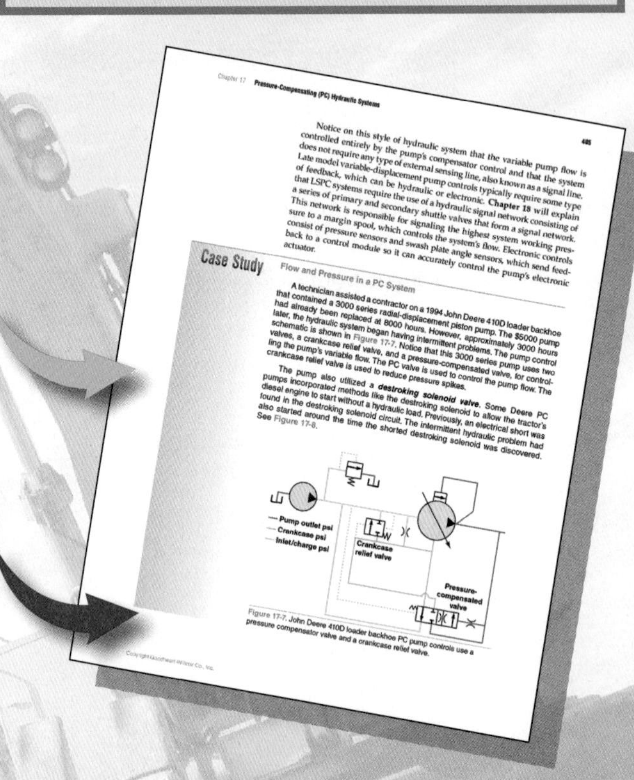

Chapter 17 Pressure-Compensating (PC) Hydraulic Systems 405

Notice on this style of hydraulic system that the variable pump flow is controlled entirely by the pump's compensator control and that the system does not require any type of external sensing line, also known as a signal line. Late model variable-displacement pump controls typically require some type of feedback, which can be hydraulic or electronic. Chapter 18 will explain that LSPC systems require the use of a hydraulic signal network consisting of a series of primary and secondary shuttle valves that form a signal network. This network is responsible for signaling the highest system working pressure to a margin spool, which controls the system's flow. Electronic controls consist of pressure sensors and swash plate angle sensors, which send feedback to a control module so it can accurately control the pump's electronic actuator.

Case Study

Flow and Pressure in a PC System

A technician assisted a contractor on a 1994 John Deere 410D loader backhoe that contained a 3000 series radial-displacement piston pump. The $5000 pump had already been replaced at 8000 hours. However, approximately 3000 hours later, the hydraulic system began having intermittent problems. The pump control schematic is shown in Figure 17-7. Notice that this 3000 series pump uses two valves, a crankcase relief valve, and a pressure-compensated valve, for controlling the pump's variable flow. The PC valve is used to control the pump flow. The crankcase relief valve is used to reduce pressure spikes.

The pump also utilized a **destroking solenoid valve**. Some Deere PC pumps incorporated methods like the destroking solenoid to allow the tractor's diesel engine to start without a hydraulic load. Previously, an electrical short was found in the destroking solenoid circuit. The intermittent hydraulic problem had also started around the time the shorted destroking solenoid was discovered. See Figure 17-8.

Figure 17-7. John Deere 410D loader backhoe PC pump controls use a pressure compensator valve and a crankcase relief valve.

Copyright Goodheart-Willcox Co., Inc.

Case Study features help you develop critical thinking and diagnostic and troubleshooting skills needed in the workplace today.

Illustrations have been designed to clearly and simply communicate the specific topic.

176 Hydraulic Systems for Mobile Equipment

Summary

- ✓ Pressure-relief valves have two designs: direct acting and pilot operated.
- ✓ Pressure-relief valves can be used for main system protection or for trapped oil protection.
- ✓ Relief valves located between the DCV spool and actuator provide protection from external loads but can cause the cylinder to drift.
- ✓ Most adjustable relief valves use a jam nut and an adjustable screw that threads into the valve.
- ✓ Unloading valves are used to unload a hydraulic pump, enabling the system to operate at a low pressure.
- ✓ Pressure-sequence valves are used on agricultural planters causing the planter to rise before it folds the marker.
- ✓ PRVs are normally open and sense the outlet pressure.
- ✓ PRVs will maintain a constant lower pressure, assuming the inlet pressure is higher than the PRV spring pressure.
- ✓ PRVs are used to set a machine's pilot pressure.
- ✓ Lift checks, pilot-operated check valves, and counterbalance valves are used to hold actuators.
- ✓ Lift checks do not use pilot oil.
- ✓ Pilot-operated check valves are used to hold a double-acting cylinder from drifting.
- ✓ Pilot-operated check valves can be located inside the DCV block or at the cylinder.
- ✓ When the pilot-operated check valve is located at the cylinder, it holds the actuator stationary in the event of a hydraulic hose failure.
- ✓ Counterbalance valves are used for load holding in overrunning applications.

Technical Terms

counterbalance valves
cracking pressure
cylinder port relief valves
dead-headed
direct-acting relief valve
full-flow pressure
lift check valves
load-holding valve
pilot-operated relief valve
pressure override
pressure-reducing valve (PRV)
pressure-relief valve
pressure-sequence valve
thermal relief valve
unloading valve

Review Questions

Answer the following questions using the information provided in this chapter.

1. Pressure-relief valve designs are ____.
   A. normally open
   B. normally closed
   C. partially obstructed
   D. None of the above.
2. Pressure-reducing valve designs are ____.
   A. normally open
   B. normally closed
   C. partially obstructed
   D. None of the above.
3. A pressure-relief valve senses ____.
   A. the valve's inlet oil pressure
   B. the valve's outlet oil pressure
   C. case pressure
   D. return pressure
4. A pressure-reducing valve senses ____.
   A. the valve's inlet oil pressure
   B. the valve's outlet oil pressure
   C. system pressure
   D. case pressure
5. Which of the following valves is designed to maintain a constant low pressure, such as pilot oil pressure?
   A. Unloading valve.
   B. Pressure-relief valve.
   C. Pressure-reducing valve.
   D. Pressure-sequence valve.

Copyright Goodheart-Willcox Co., Inc.

Summary feature provides an additional review tool for you and reinforces key learning objectives.

Technical Terms list the key terms to be learned in the chapter.

Review Questions allow you to demonstrate knowledge, identification, and comprehension of chapter material.

Student Resources

Textbook

The **Hydraulic Systems for Mobile Equipment** textbook provides an exciting, full-color, and highly illustrated learning resource. The textbook is available in print or online versions.

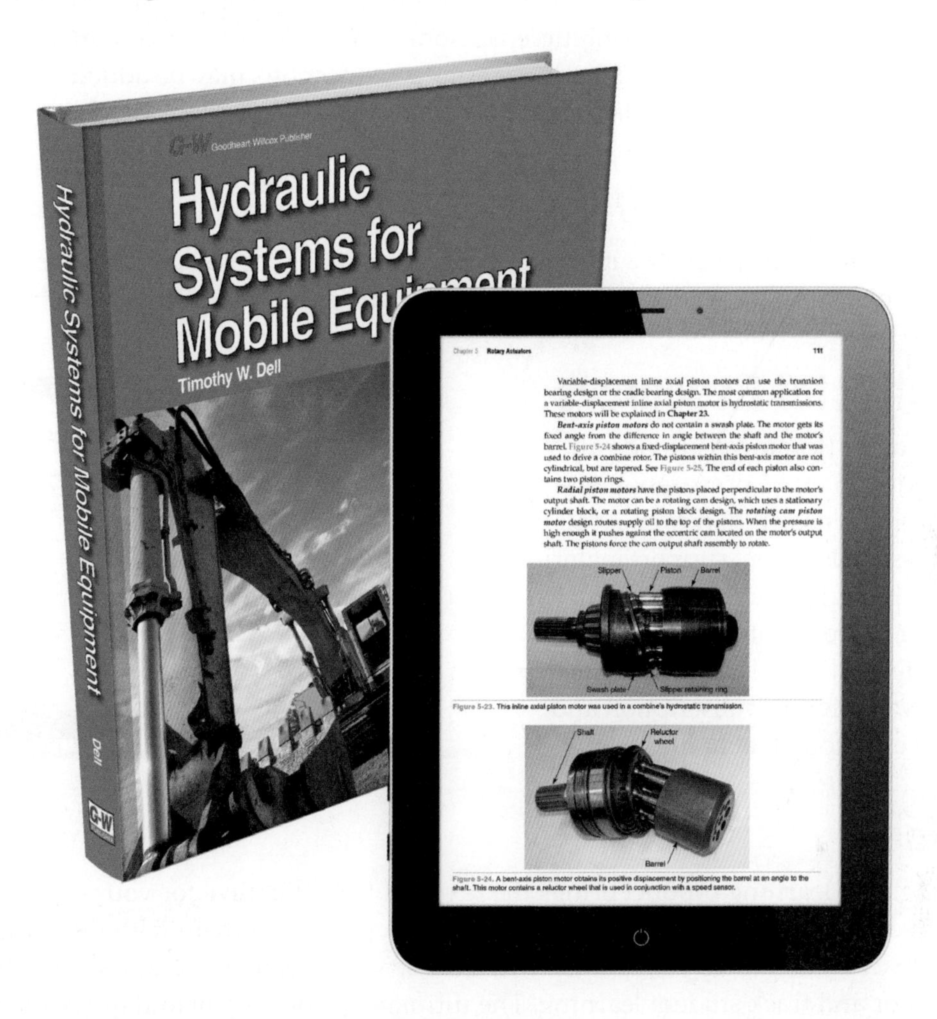

Instructor Resources

Instructor resources provide information and tools to support teaching, grading, and planning; course administration; class presentations; and assessment.

Instructor's Presentations for PowerPoint®

Help teach and visually reinforce key concepts with prepared lectures. These presentations are designed to allow for customization to meet daily teaching needs. They include objectives, outlines, and images from the textbook.

ExamView® Assessment Suite

Quickly and easily prepare, print, and administer tests with the ExamView® Assessment Suite. With hundreds of questions in the test bank corresponding to each chapter, you can choose which questions to include in each test, create multiple versions of a single test, and automatically generate answer keys. Existing questions may be modified and new questions may be added. You can prepare pretests, formative, and summative tests easily with the ExamView® Assessment Suite.

Instructor's Resource CD

One resource provides instructors with time-saving preparation tools such as answer keys; lesson plans; correlation charts to [standards, certification, competencies, accreditation]; and other teaching aids.

Online Instructor Resources

Online Instructor Resources provide all the support needed to make preparation and classroom instruction easier than ever. Available in one accessible location, support materials include Answer Keys, Lesson Plans, Instructor Presentations for PowerPoint®, ExamView® Assessment Suite, and more! Online Instructor Resources are available as a subscription and can be accessed at school or at home.

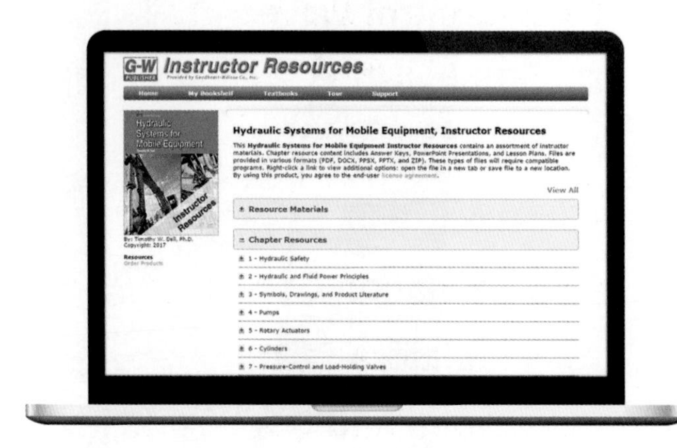

G-W Online

This exciting new learning product brings the text and learning alive for your students! The textbook's interactive learning activities engage your students, providing them with dynamic visuals and immediate feedback. It enhances your course with course management and assessment tools that accurately monitor and track student learning. The ultimate in convenient and quick grading, G-W Online allows you to spend more time teaching and less on administration.

Brief Contents

Contents

Chapter 1
Hydraulic Safety 1

Chapter 2
Hydraulic and Fluid Power Principles 19

Chapter 3
Symbols, Drawings, and Product Literature 45

Chapter 4
Pumps 65

Chapter 5
Rotary Actuators 95

Chapter 6
Cylinders 125

Chapter 7

Pressure-Control and Load-Holding Valves 157

Chapter 8

Flow-Control Valves 179

Chapter 9

Directional-Control Valves 195

Chapter 10

Fluids 225

Chapter 15

Plumbing 341

Chapter 16

Open-Center Hydraulic Systems 379

Chapter 17

Pressure-Compensating (PC) Hydraulic Systems 399

Chapter 18

Load-Sensing Pressure-Compensating (LSPC) Hydraulic Systems 417

Chapter 19

Chapter 20

Chapter 21

Hydraulic Test Equipment 503

Chapter 22

Hydraulic Troubleshooting Principles 535

Chapter 23

Hydrostatic Drives 553

Chapter 24

Hydrostatic Drive Service and Diagnostics 603

Chapter 25

Steering 627

Feature Contents

Case Studies

Chapter 1

Hydraulic Safety

Learning Objectives

After studying this chapter, you will be able to:

- ✓ List PPE for working on hydraulic systems.
- ✓ Demonstrate emergency preparedness by being able to list and describe proper use of emergency equipment.
- ✓ Explain multiple fluid hazards.
- ✓ List five critical pieces of information to provide a surgeon in the event of a fluid injection injury.
- ✓ Identify unsafe practices for working with hydraulic systems.
- ✓ List steps for properly managing a hydraulic system's surrounding environment.

Introduction to Safety

Hydraulic systems pose serious risks to personnel. It takes very little pressure or flow to injure a person or even cause a fatality. Numerous service manuals specify unsafe test procedures. As a result, it is possible for veteran supervisors to unknowingly recommend unsafe practices. Even if a technician has been properly educated on safe practices, it does not mean that the technician will follow those practices.

Technicians, operators, and customers frequently take shortcuts that endanger themselves and others. The old saying is that hindsight is 20/20. If a person only knew when something was going to cause harm, they would have conscientiously taken preventative measures. Many personnel work with hydraulic machinery without a healthy respect of the potential risks and with little expectation that something can go wrong. As a result, they can become complacent, rush through procedures, and take shortcuts, putting themselves and others at serious risk.

A technician must receive proper training before servicing, diagnosing, or repairing hydraulic systems. It is critical to receive instruction on all the risks associated with hydraulic systems, how to respond to an injury, safe practices for working on hydraulic systems, and the location of several key items.

Personal Protective Equipment (PPE)

The ***Occupational Safety and Health Administration (OSHA)*** is the United States federal agency that is responsible for ensuring that employees have a safe work environment. OSHA's regulations carry the force of law, and companies can be fined for failing to follow them. Hydraulic technicians working at mining sites must follow a stricter set of rules as governed by ***Mine Safety and Health Administration (MSHA)***, which is the United States federal agency responsible for ensuring mine site safety.

OSHA regulations require employers to provide workers with ***personal protective equipment (PPE)***, consisting of equipment and clothing that is designed to protect the employees from potential injuries or illnesses. Common PPE includes eye protection, gloves, hard hats, boots, and hearing protection. Construction sites can require safety vests. The employer is also responsible for training the workers for proper use of PPE. See Figure 1-1.

Eye Protection

Eye protection is one of the most important pieces of PPE to be worn by hydraulic technicians. Safety glasses should be equipped with side shields. Hydraulic systems commonly run at 3000–7000 psi (207–482 bar) with flow well over 100 gpm (378 lpm). If a technician's eyes are not protected by side shields, they are susceptible to being injured by fluid being sprayed from multiple angles.

Figure 1-1. Example of PPE: boots, gloves, hard hat, vest, and eye protection.

Clothing

PPE includes task-appropriate clothing. Many companies require their technicians to wear long pants, safety vests, and prohibit long hair and any loose-fitting clothing such as neckties, as these could become caught in operating components. Some technicians wear coveralls. Technicians may choose to wear nitrile gloves to avoid prolonged exposure to petroleum-based oils while they are rebuilding hydraulic components. However, gloves do not protect personnel from hydraulic fluid injection. Fluid injection injuries will be discussed later in this chapter.

Warning

Technicians should avoid listening to music on headphones in order to concentrate 100% of their attention to the job.

Hearing Protection

Diesel-powered off-highway equipment can produce harmful noise, which can cause hearing loss with prolonged exposure. Noise is one of the negative attributes of hydraulic systems. For example, if three different machines equipped with a manual transmission, an electric drive transmission, and a hydrostatic drive transmission were compared, the hydrostatic drive machine would generate considerably more noise than the others when the machine is driving under a heavy load.

Hearing protection is required by OSHA when the workplace noise level reaches certain levels. The maximum permissible noise level without hearing protection ranges from 85 decibels to 140 decibels, depending on the frequency of the noise. Noise frequencies can range from 100 cycles per second up to 8000 cycles per second. Many employers adopt more stringent safety guidelines, requiring employees to wear hearing protection at lower noise levels. Hearing protection can consist of earplugs or earmuffs.

Hard Hats

Hydraulic technicians are frequently tasked with travelling to construction job sites and mine sites. Both OSHA and MSHA require hard hats to be worn on the job site to provide protection from falling objects. Technicians have been kicked off job sites for failing to follow OSHA and MSHA regulations. The hard hats must meet the appropriate OSHA or MSHA regulation, including a specifically colored hat to easily identify new employees or inexperienced employees.

Foot Protection

Most companies require their technicians to wear boots on the job for foot protection. OSHA regulations specify that employee's feet be protected from falling objects, rolling objects, and sole piercings. MSHA's protective footwear regulation does not specifically require steel-toed boots, however most mine sites require their employees to wear steel-toed boots.

Emergency Preparedness

Due to the sheer size of the machines, construction equipment technicians and agricultural equipment technicians work in an environment that inherently has risks. Technicians must be prepared for emergencies. Before stepping into the shop, technicians should know where safety equipment is located and how to use it. Technicians who disregard this rule are placing themselves and those around them at risk.

For example, one instructor reported that his students had experienced a fire while working in the shop. The instructor learned about the incident after a prepared student had quickly extinguished the fire. The student formerly worked aboard a submarine in the U.S. Navy. Starting in boot camp, sailors learn that they have two jobs. The first job is a firefighter. The other job, such as electronics technician, welder, or avionics technician, truly is secondary to firefighting. This sailor had been properly educated to know where the fire extinguishing equipment was located and how to properly use it. A person living 300 feet below the water does not have the luxury of calling the fire department. Just like sailors working in remote locations, hydraulic technicians can be tasked with working in remote locations and must be prepared for the worst case scenario with little or no help from others.

Fire Suppression

Fire extinguishers are rated from class A to class D based on the type of combustible material. Class A extinguishers are rated for combustible solids such as wood, paper, or cardboard. Class B fire extinguishers are used to combat fires fueled by combustible gases, oils, and greases. Class C extinguishers are designed to extinguish electrical fires. Class D extinguishers are used to fight fires caused by combustible metals. Construction and agricultural technicians use class A, B, C fire extinguishers. See **Figure 1-2**.

Hydraulic technicians working in a manufacturing environment, not only need to know where the fire extinguishers are located, but know the location of the fire exits and fire alarm switches, **Figure 1-3**. The safest companies drill safe practices into all their employees. Recalling the location of the extinguishers, exits, and fire alarms becomes second nature, just like breathing.

The most expensive off-highway machines commonly use onboard fire-suppression systems. Extra care must be taken when servicing machines with onboard suppression systems. These systems use inert gases such as nitrogen and chemical agents for the purpose of smothering the fire. The systems can be manually actuated by an operator or can be automatically triggered due to the heating of a sensor or sensing wire. If a technician inadvertently deploys the system, not only will it cause a considerable expense and mess, it could injure or even cause a fatality. The deployment of the nitrogen gas and chemical agents would virtually eliminate a technician's oxygen if he or she was working in a confined area, such as an engine compartment, when the suppression system was activated.

A—Fire Classifications					
Pressurized Water	**Carbon Dioxide (CO_2)**	**Dry Chemical**	**Dry Chemical (Granular)**	**Dry Chemical or Wet Chemical**	**Multi-Purpose Dry Chemical**
A	B	C	D	K	A B C
Ordinary combustibles (wood, paper, or textiles)	**Flammable liquids** (grease, gasoline, oils, and paints)	**Electrical equipment** (wiring, computers, and any other energized electrical devices)	**Combustible metals** (magnesium, potassium, titanium, and sodium)	**Kitchen fires** (grease fires in commercial kitchens)	Labeled for use on ordinary combustibles, flammable liquids, and electrical equipment fires

B—Fire Extinguisher Classifications					
A	B	C	D	K	A B C
Direct stream at base of flame.	Direct discharge as close to fire as possible, first at edge of flames and gradually forward and upward.	Direct stream at base of flames. Use rapid left-to-right motion toward flames.	Smother flames by scooping granular material from bucket onto burning metal.	Direct stream at base of flames. Use rapid left-to-right motion toward flames.	Direct stream at base of flames. Use rapid left-to-right motion toward flames.

Figure 1-2. Fire safety charts. A—Fire classifications. B—Fire extinguisher classifications.

Figure 1-3. Fire extinguishers and fire alarms are often located near an entryway. The fire extinguisher shown here is rated for class A, B, and C fires.

Safety Data Sheets (SDS)

Technicians work with a wide variety of chemicals and products, such as hydraulic oils, greases, engine oils, and cleaners. ***Safety data sheets (SDS)*** provide end users important information regarding products. Employees must know the location of the data sheets and be able to quickly access that information so they are prepared for emergencies. See Figure 1-4. The data sheets provide 16 different categories of information:

1. Product and company identification.
2. Hazards identification.
3. Composition information on ingredients.
4. First aid measures.
5. Firefighting measures.
6. Accidental release measures.
7. Handling and storage.
8. Exposure controls and personal protection.
9. Physical and chemical properties.
10. Stability and reactivity.
11. Toxicological information.
12. Ecological information.

Figure 1-4. An unobstructed view of the location of safety data sheets in a shop.

13. Disposal considerations.
14. Transport information.
15. Regulatory information.
16. Other information.

Exposure

While there is usually no immediate serious risks for skin exposure to hydraulic oil, irritation can occur. As previously mentioned, some technicians wear nitrile gloves to reduce the effects of long-term exposure to hydraulic oils.

Ingestion

Although swallowing hydraulic fluid might be one of the rarest risks associated with hydraulic systems, it is possible, especially if a hose fails and sprays a technician in the face. In addition to rinsing the oil from the patient's mouth, some hydraulic oil safety data sheets state to *not* induce vomiting, but to seek medical attention. Follow the SDS's instructions.

Inhalation

Oil can become vaporized due to a fire or due to oil being sprayed. Both examples are harmful to a human's lungs. Seek medical attention in the event of oil inhalation.

Other Information

The SDS can also contain important information such as a phone number to call in case an accident has occurred involving the product.

First Aid Kit and Eyewash Station

Knowing the locations of a first aid kit and an eyewash station will also assist a technician in being prepared for an emergency. Technicians working in a new environment should familiarize themselves with the new workplace in order to be prepared. See Figure 1-5.

Fluid Hazards

Hydraulic technicians are at risk of injuries anytime they work with pressurized hydraulic fluids. Two of those risks are burns and ***fluid injection***. Fluid injection occurs when pressurized fluid penetrates the skin. Untrained technicians and lackadaisical technicians are at risk of receiving serious burns, having a limb amputated, or worse yet, losing their life.

Fluid-Injection Injuries

Off-highway hydraulic systems operate at high system pressures, sometimes two to three times higher than pressures in a manufacturing setting. A common implement high-pressure relief ranges from 3000 to 4000 psi (206 to 275 bar). Some hydrostatic drive pressures can exceed 7000 psi (482 bar). As noted in a 1983 article by Dr. A.R. Scott in the *Journal of Society of Occupational Medicine,* it only takes a pressure of 7 atmosphere (or 100 psi) for fluid to puncture the skin. The article reports that a high-pressure oil leak can cause fluid to spray at a velocity of 300 meters per second (671 miles per hour).

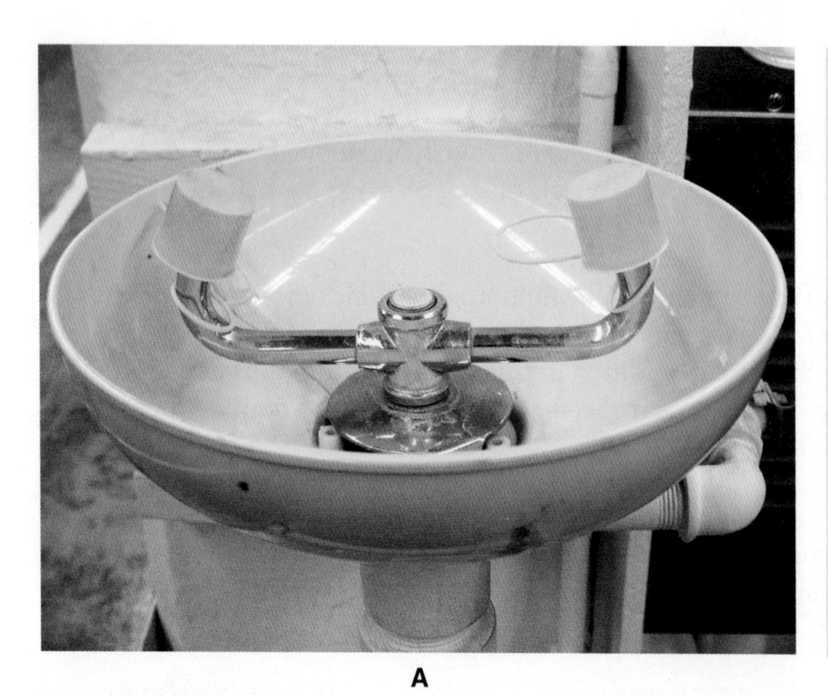
A

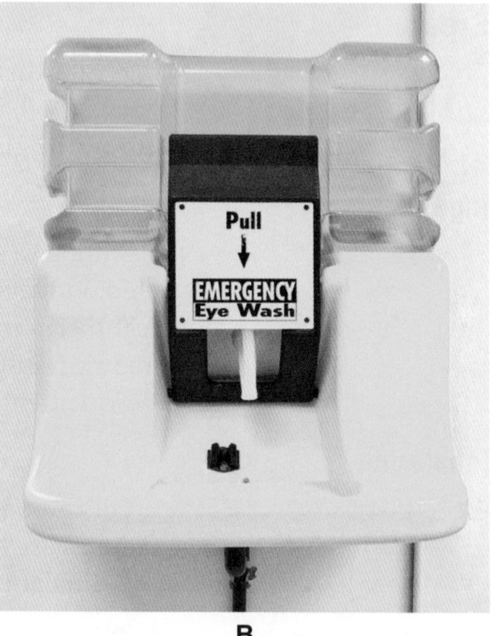

B

Figure 1-5. Two types of eyewash stations. A—This station is tied into the building plumbing. B—This station is self-contained and can be installed where there is no plumbing.

Although fluid-injection injuries are rare, the actual injection injury itself often consists of a very small pinhole to the skin, so small that it is easily overlooked. Some patients have reported that the injury did not initially cause intense pain, leading many patients to delay seeking medical attention. If medical attention is delayed, swelling and pain will increase.

Patients need to quickly seek the care of a surgeon. The two immediate treatments consist of surgical decompression and debridement, which is the removal of damaged tissue. The severity of the injury is affected by multiple factors: the quantity of fluid injected, the pressure and velocity of the injected fluid, the toxicity of the oil, and the delayed amount of time before medical attention is obtained.

Patients with injection injuries can mistake the cause of their injury from something as small as a nick to the hand. This oversight allows the hydraulic fluid to cause further damage to the skin, which can lead to gangrene if not properly treated. 40% of all fluid injection injuries result in some form of amputation, and amputation is required nearly 100% of the time if the patient does not receive fast medical care. Injuries resulting from system pressures of 7000 psi (482 bar) and higher result in amputation 100% of the time. Even for the most fortunate patients that are able to avoid amputation, nearly all of them are unable to have their fingers or hand restored to the pre-injury condition.

For these reasons hydraulic personnel should never use their hands for trying to locate hard to find hydraulic leaks. Some manufacturers recommend using cardboard for locating a hydraulic pinhole leak. The most common cause for injection injuries is a ruptured hose. Therefore technicians should avoid handling pressurized fluid conductors, because they are unable to know when a hose might burst.

Note

When machines are being built on the assembly line, many manufacturers place dye in the oil and use a black light for diagnosing hard-to-find leaks.

Because of the rarity of fluid injections, it is necessary for technicians to be prepared in the event of an injury. The International Fluid Power Society (IFPS) provides its members a reminder card that can be carried on their person to remind them of the five things to share with the emergency room doctor in the event of a fluid-injection injury:

- Type of fluid.
- Quantity of fluid injected.
- The fluid pressure.
- How far the fluid injury has spread.
- The amount of time since the injury.

Technicians should also have fast and easy access to the fluid's SDS so the data sheet can be provided to the surgeon as well.

Case Study

Fluid Injection

A technician was working on the header float system of a self-propelled windrower. The system had a relief pressure of 2100 psi (145 bar), but a pressure sensor was reading 3800 psi (262 bar). To determine if the pressure sensor was malfunctioning, a diagnostic test port was going to be installed in the circuit to directly measure the circuit's pressure.

The technician shut off the machine and followed the service manual's procedure for depleting the pressure in the circuit. Note that the circuit did have an accumulator. The circuit was bled by manually pressing a bypass valve multiple times.

A wrench was used to crack the fitting on a 1/4″ hydraulic hose. Approximately a half gallon of oil leaked from the cracked hydraulic line. After oil quit draining from the hose, the technician began to remove the hose by hand. Keep in mind that the oil had quit draining and the attached hose end was quite loose, with no tension on the fitting.

The technician used his hand to back off the remaining threads on the loose fitting, and this is when things went awry. A tremendous amount of fluid under high pressure blew out of the hose, injecting the technician's fingers. One finger had approximately a half square inch of skin removed by the force of the hydraulic fluid. An inch-long blister immediately formed on his middle finger, and the technician was completely covered in oil.

He covered his bleeding fingers and travelled 40 minutes to the hospital. He chose the hospital that was 40 minutes away because it was a little larger facility. He assumed the doctors at the larger facility would have more experience with this type of injury.

When he arrived at the hospital, oil was still oozing out of his fingers. The doctor soaked his hand and treated the wound as a common hand injury. The technician was unsettled by the lack of concern shown by the doctor. The technician attempted to give the doctor the 1-800 number of the OEM's 24-hour medical hotline so she could consult with them regarding the injury. The emergency room doctor advised the technician that she had gone to medical school, and that the problem was just a common hand injury. After soaking the fingers and wrapping them, the doctor sent the technician home, stating that he might feel some tingling, numbness, and soreness.

After leaving the hospital, the technician still felt a little uneasy about the course of treatment he had received. He called the OEM's 24-hour medical hotline, using the phone number listed in the front of the manuals. The hotline advised the technician regarding hydraulic fluid injuries, what symptoms might occur, and what information to provide the medical personnel, which included information from the oil's safety data sheet.

Approximately two hours after he left the emergency room, his finger and the blister on it began to swell quite large, his fingers tingled, his arm went numb, and he suddenly felt as if he had the flu, causing him to vomit violently.

The technician called the OEM's hotline again, and was advised to go to a different hospital. The hotline staff called and talked to the emergency room's physician's assistant. This medical team was much more concerned than the previous hospital. After discussing the injury with the OEM's medical hotline, they called the state's university hospital and consulted with a hand surgeon.

The medical staff was unable to find a puncture wound on the fingers. They took an X-ray of the hand to investigate the extent of the damage. They lanced the blister on the finger and drained four cubic centimeters of oil from the technician's finger. The hand immediately began to feel better. They brushed the wound to clean away the remaining hydraulic oil. Unfortunately, the medical staff was advised to not administer local anesthesia because it would interfere with the treatment.

Burns

During a hot summer day, hydraulic operating temperatures can exceed 200° F (93° C). Malfunctioning hydraulic systems can overheat causing the oil temperatures to exceed 300° F (149° C). In the event of a hose failure, a technician can receive serious burns.

Burns are categorized as first degree, second degree, and third degree depending upon their severity. The least severe burn is a first degree burn, which affects the outer layer of the skin, called the epidermis. The burn can be the result of exposure to the sun, scalding hot water, chemicals, or even a fire. First degree burns cause patients pain, exhibit redness of the skin, and cause swelling. These burns are often treated at home without the assistance of a doctor. However, some patients do seek medical attention, especially due to the possibility of infection. One of the most common treatments is running cool water on the burn for ten minutes, or until the pain subsides.

A second degree burn affects the two outer layers of the skin the epidermis and dermis. The burn will look red, splotchy and blistered and can even result in disfigurement. Close to one-fifth of second degree burns later become infected. Patients must seek medical attention.

Third degree burns are the most serious types of burns. A third degree burn affects the deepest skin, penetrating through the first two layers of the skin, the epidermis and dermis, and reaching the inner hypodermis layer. The patient's skin is usually charred black or dry and white. The burn frequently leads to disfigurement and potentially death. As with fluid-injection injuries, patients with third degree burns must seek immediate medical attention. If the skin is not broken, cool water or a cool moist towel can be applied to the burn while the patient is being transported to the doctor. Do *not* apply an ointment or ice.

Warning

In the event of a fire, remember to "stop, drop, and roll." If helping someone on fire, use a blanket to smother the fire and call 911.

Combustion

Although hydraulic oil is not considered highly volatile, it can ignite if it is heated to its flash point. Typical hydraulic oil flash points range from 338°F (170°C) to 590°F (310°C). Unfortunately, technicians have lost their life due to machine fires. A diesel engine's exhaust, especially the turbocharger, is a source of heat that can cause oil from a ruptured hose to quickly ignite.

Figure 1-6. Typical tag and lock used for lock-out, tag-out procedures.

Safe Practices for Working with Hydraulic Systems

Safe employees are more profitable to their company. When an accident occurs, not only does it cause potential damage to the machine and equipment, but it can put the technician out of work, costing employers thousands of dollars in worker compensation. Also, the machine availability can be reduced while the technician is recovering away from work. For these reasons, many employers offer safety bonuses when employees or shops work consecutive weeks or months without a reportable incident or injury.

In addition to hazards associated with fluid-injection injuries and burns, technicians face other serious risks while working on hydraulic systems. Technicians routinely complete unsafe tasks, and unfortunately do not realize that the methods they were taught are unsafe. Numerous service manuals specifically direct technicians to perform unsafe tasks. As a result, hydraulic technicians must be trained to safely diagnose and repair hydraulic systems.

Before Starting a Job

Before a technician ever starts a job, he or she must take action so that the work site is safe. One manufacturer recommends that the machine is lowered to the ground, the engine is shut off, and the ignition key is removed. To promote safety, it is often best to disable the machine before working on it. Most manufacturers recommend disconnecting the battery's negative cable before completing any substantial work on the machine. This practice ensures that someone will not crank the engine while a technician's arm is wrapped around a shaft. The practice also prevents machine and equipment damage.

For example, disabling the machine might prevent a pump failure by not allowing an engine to crank while the reservoir is empty. A technician should also install a note in the cab that clearly states "Do not operate." The tag should include the name of the technician, the date, and time.

Lock-Out, Tag-Out

If the technician has been tasked with diagnosing or repairing a hydraulic system in a manufacturing environment, the technician will have to disable the machine by following lock-out, tag-out procedures. The machine's power must be shut off and a lock must be installed to prevent power from being restored to the machine. After the required service is completed, the lock and tag are removed. See **Figure 1-6**.

Safety Shields and Guards

Safety guards and panels are purposely installed on machines by manufacturers to protect personnel. Unfortunately, people are injured and die each year due to working carelessly around machines while the safety guards are removed. The agricultural industry, construction industry, and mining industry have all placed tremendous amounts of energy into ensuring that technicians are safeguarded from rotating shafts, belts, gears, pumps, and motors. The evening television news and morning newspapers are an unfortunate place to be reminded about the consequences of not working safely around machines or disregarding safety shields and guards. Many technicians personally know someone who has been injured while working on a machine without safety guards or shields. Do not allow yourself to become a statistic by taking unnecessary shortcuts.

Avoid Heat near Hydraulic Components and Conductors

Technicians frequently use tools that discharge high amounts of heat, such as welders, plasma cutters, and torches. All forms of heat must be kept away from hydraulic cylinders, hoses, steel lines, accumulators, and other hydraulic components.

Know the System

Hydraulic systems can impose serious risks. Prior to working on any hydraulic system, it is important to first gain an understanding of the system. Do not attempt to service or repair a system without first studying the system. Otherwise, machine damage, personal injury, or a fatal accident can occur.

Case Study

Importance of PPE

A customer requests some assistance with a tractor. The tractor's transmission clutch was replaced, and after the new clutch was installed, the hydraulic three-point hitch began malfunctioning. The hydraulic hitch is now jerky, sluggish and sometimes will not lift a bale of hay. The technician has no familiarity with the tractor. If the technician is limited to taking only one item to the tractor, what should the one-and-only item be?

Inexperienced technicians often recommend taking a service manual, a pressure gauge, a flowmeter, a bucket of oil, or even an experienced technician. However, it is surprising that inexperienced technicians very seldom mention the single most important item to bring, *safety glasses!* In this real case scenario, the hydraulic system spewed oil in the face of the technician. Fortunately the one-and-only item that was brought to the tractor was the pair of safety glasses that technician was wearing. Technicians must work with the expectancy that the hydraulic system could fail and must be prepared for when that failure occurs.

Hydraulic Environmental Management

Hydraulic systems not only pose risks to personnel, the hydraulic fluids also pose risks to the environment. Many systems use petroleum-based fluids that can harm soils and waterways. An oil spill as little as one gallon can contaminate more than a million gallons of water, giving it far-reaching effects. An oil leak or an oil spill can require specific action. Technicians must be prepared and equipped to respond to oil leaks and oil spills.

In addition, many off-highway machines require large oil capacities, which forces shops and service trucks to store large quantities of oil. Technicians must be familiar with containment methods for storing oil and the proper disposal of hydraulic oils.

Reporting an Oil Spill to the Authorities

When an oil spill occurs, a technician can be tasked with reporting the event to multiple federal, state, or local agencies. Different guidelines stipulate if an oil spill requires notifying authorities. One guideline is labeled ***reportable quantity (RQ)***, which is the minimum quantity of oil that requires someone to file a report.

Federal

The ***Environmental Protection Agency (EPA)*** is the United States' federal agency that is responsible for protecting the nation's land and bodies of water. The EPA's website states that a report is necessary when the oil spill equates to a "harmful quantity." A spill is considered to be a harmful quantity if it meets any of the following conditions:

- Exceeds the state RQ.
- Leaves an oil sheen on the water's surface.
- Causes sludge or oil emulsion below the water surface.

When an RQ has occurred, the EPA states that the "person in charge" must notify the ***National Response Center (NRC)***. The NRC can be contacted by phone or through links in the EPA and U.S. Coast Guard websites.

According to the 1990 Oil Pollution Act (OPA), a responsible party who knowingly fails to report an oil spill is subject to a fine of up to $250,000 and up to five years imprisonment. If a person is found guilty of a violation, the prison term can be increased to 15 years. An organization can be fined up to $500,000.

State

The Transportation Environmental Resource Center (TERC) provides a web page that contains a matrix that includes information for: each state's RQ, a phone number, and a web link for all 50 of the states. Some states have adopted the federal EPA guidelines, while other states have more stringent RQs than the EPA, for example requiring a report for "any size of oil spill." A common state RQ is 25 gallons of oil. For more information regarding the RQ and the contact information, visit the TERC website.

Local

Some technicians will experience a job site that has extreme RQ standards by a local agency. The local agency could be a city, county agency, or the owner of the job site. In extreme situations, technicians can be required to report as little as one drop of oil that has been spilled, and the course of action taken after the spill.

Fluid Containment

Some mobile hydraulic technicians need to be aware of the ***Spill Prevention, Control, and Countermeasure (SPCC) Act*** which specifies containment guidelines if an operation contains above ground oil storage exceeding 1320 gallons. The SPCC Act stipulates special regulations for these larger operations. For the facilities that must comply with the SPCC Act, the EPA 40 CFR 264.175 states that the containment system must be able to contain 10% of the stored oil. When determining if a facility exceeds the SPCC rule, entities only have to count storage tanks that are 55 gallons or larger. However, the agency must count the volume of the tanks, rather than the quantity of oil currently being stored on site.

Although most facilities will not exceed 1320 gallons of stored oil, containment storage devices provide a safe measure for containing oil in the event of spills and can also help maximize storage space inside shops. See **Figure 1-7**.

Most fluid manufacturers recommend storing 55 gallon drums of oil on their side, especially if the drums are stored outside. This will prevent water and dirt from migrating into the drums. If the drums must be temporarily stored in a vertical position, cover the drum with a drum cover, which can be purchased through an industrial supply company.

Disposal

If used oil is dumped, it has long-lasting effects, such as damaging water supplies and contaminating soils. The American Petroleum Institute (API) estimates that personnel mismanage the disposal of 200 million gallons of used oil every year. Simply recycling used oil not only helps prevent pollution to the environment, but it also improves the sustainability of natural resources. For example, if 1 gallon of used oil is recycled, it can harness enough energy to power a home's electrical power for 12 hours.

Figure 1-7. Example of an oil containment system for 55-gallon drums of oil.

Thinking Green

Re-refining of used oil requires only 1/3 of the energy required to produce the same quantity of lubricating oil from crude oil.

Today's maintenance practices are resulting in a decreased need to dispose of large quantities of hydraulic oil. Many shops are using filter carts to ensure that their machines are operating with extremely clean oil when compared to the common maintenance practices that took place just a few years ago. As a result of these contamination control practices, many machines have their oil filtered repeatedly until the fluid has reached the appropriate cleanliness level. The oil will only need to be changed once the fluid's properties have begun to deteriorate. **Chapters 11** and **12** will detail filtration and contamination control practices.

In the event that a technician must change the machine's hydraulic oil, the fluid must be properly disposed of or, preferably, recycled. Proper disposal or recycling of hydraulic oil is the same as for used engine oil. Many waste management firms are readily available to recycle a shop's waste oil. Most of waste management companies will actually pay shops for the opportunity to remove their used oil. Prices vary depending on the region.

Summary

- ✓ Hydraulic systems pose risks to personnel and the environment.
- ✓ It takes little pressure or flow to seriously injure a person or cause a fatality.
- ✓ Technicians must apply known proven safe practices for disassembling, testing, and repairing hydraulic systems.
- ✓ Technicians should never use any part of their body to locate a hydraulic leak. Nor should technicians handle pressurized fluid conductors such as hoses or components.
- ✓ Prior to servicing or making repairs, a technician must depressurize the system.
- ✓ Pressure gauges and flowmeters must be used for measuring pressure and flow.
- ✓ Technicians must also be properly educated for managing hydraulic oils and know the potential effects that hydraulic oils have on the environment.
- ✓ Technicians need to know what the RQ is for their job site and who should be contacted when a spill reaches the level of the RQ.
- ✓ When the oil has reached the end of its useful life, technicians must contact a local waste management agency to pick up the used oil so it can be recycled.

Technical Terms

Environmental Protection Agency (EPA)
fluid injection
Mine Safety and Health Administration (MSHA)
National Response Center (NRC)
Occupational Safety and Health Administration (OSHA)
personal protective equipment (PPE)
reportable quantity (RQ)
safety data sheets (SDS)
Spill Prevention, Control, and Countermeasure (SPCC) Act

Review Questions

Answer the following questions using the information provided in this chapter.

1. OSHA stands for?
   A. Occupational Safety and Health Association
   B. Occupational Standard and Health Association
   C. Occupational Safety and Health Administration
   D. Occupational Safety and Helping Association
2. What does PPE stand for?
   A. Personnel protection extension.
   B. Personal protective equipment.
   C. Personnel protection exam.
   D. Personal protective essentials.
3. A technician is going to work on a hydraulic system. If the technician was limited to take only one item, which item should the technician take to the machine?
   A. Service manual.
   B. Safety glasses.
   C. Flowmeter.
   D. Pressure gauge.
4. A technician has been tasked with diagnosing a hydraulic system. In order to be prepared for emergencies, a technician must know the location of all of the following, *EXCEPT*:
   A. the applicable SDS.
   B. a pressure gauge.
   C. a fire extinguisher.
   D. the exit.

5. Which class of fire extinguisher is used to extinguish oil or grease fires?
   A. Class A.
   B. Class B.
   C. Class C.
   D. Class D.
6. Which class of fire extinguisher is used to extinguish an electrical fire?
   A. Class A.
   B. Class B.
   C. Class C.
   D. Class D.
7. A technician's arm breaks out in a rash after rebuilding a hydraulic cylinder. Where should the technician look for information related to this problem?
   A. Service manual.
   B. Operator's manual.
   C. Training manual.
   D. SDS.
8. What is the minimum oil pressure that can cause oil to penetrate a human's skin?
   A. 50 psi.
   B. 100 psi.
   C. 500 psi.
   D. 1000 psi.
9. In the event of a fluid injection in a technician's hand, all of the following will affect the chances of potential amputation, *EXCEPT*:
   A. delay before receiving medical attention.
   B. quantity of fluid.
   C. the type of fluid.
   D. weather conditions.
10. Which of the following burns has a greater chance of causing disfigurement and potential death?
    A. First degree.
    B. Second degree.
    C. Third degree.
    D. All classes of burns pose the same degree of risk.
11. A hydraulic system operating at high system pressure can spray fluid up to?
    A. 60 mph.
    B. 300 mph.
    C. 670 mph.
    D. 2000 meters per second.
12. According to some manufacturers, when checking for high-pressure hydraulic leaks, a technician should use _____.
    A. a piece of cardboard
    B. latex gloves
    C. bare hands
    D. rubber gloves
13. The most common cause of fluid injection injuries is _____.
    A. a burst hose
    B. pump cavitation
    C. engine overspeed
    D. pump aeration
14. Which burn affects only the epidermis?
    A. First degree.
    B. Second degree.
    C. Third degree.
    D. All of the above.
15. What is the minimum amount of stored oil that requires facilities to meet the SPCC act?
    A. 55 gallons.
    B. 220 gallons.
    C. 440 gallons.
    D. 1320 gallons.
16. When determining if an agency must comply with the SPCC act, what is the size of oil containers that must be counted in the assessment?
    A. 5-gallon containers and larger.
    B. 55-gallon containers and larger.
    C. 100-gallon containers and larger.
    D. 250-gallon containers and larger.

Chapter 2

Hydraulic and Fluid Power Principles

Learning Objectives

After studying this chapter, you will be able to:

- ✓ List the fundamental advantages and disadvantages of hydraulic systems.
- ✓ Explain differences between hydrodynamic and hydrostatic drives.
- ✓ Apply Pascal's law by solving for pressure, force, or area.
- ✓ Describe hydrodynamic fundamental principles.
- ✓ Explain how energy is transformed in hydraulic systems.
- ✓ Compute horsepower requirements.

Introduction to Fluid Power

The term "fluid power" refers using fluid, in the form of either a gas or a liquid, for the purpose of performing work, such as extending a cylinder to lift a heavy load. If the fluid medium is a gas, the field of study is called ***pneumatics***. It is called ***hydraulics*** if the fluid medium is a liquid. Although some off-highway machinery such as construction and mining equipment might use pneumatic systems, this text will focus on the heart of off-highway fluid power systems, which is the hydraulic system.

The word "hydraulic" is derived from an ancient Greek term meaning water. For centuries, water-powered systems, such as early aqueducts and waterwheels, have been used to improve civilization. See **Figure 2-1**.

Advantages of Hydraulic Systems

Today's hydraulic systems use oil instead of water because oil provides better lubrication than water. Hydraulic oils, like all liquids, will conform to the shape of their container, **Figure 2-2**, and are practically incompressible, compressing only 0.5% at 1000 psi of pressure. The compressibility of oil is not linear. For example, if the fluid pressure was tripled to 3000 psi, the oil would compress to approximately 1.2%, rather than 1.5%. ***Bulk modulus*** is the term used to describe the compressibility of oil. Fluid properties will be detailed in **Chapter 10**. This rigidity of liquid provides advantages over gas, such as precision and predictability, especially when electronics are used to operate the control valves. Hydraulic systems can provide operators an element of safety when compared to the less-predictable, spongy, pneumatic systems.

Figure 2-1. A waterwheel is one of the earliest examples of using liquid to power a fluid drive.

The rigidity of the liquid allows for a large multiplier effect, enabling cylinders to lift thousands of pounds of weight with a much smaller input force. Hydraulic systems do not require the use of a lubricator, because the fluid itself inherently provides good lubrication.

Hydraulic components offer the advantage of *compact* size. For example, a 10-horsepower electric motor can be nearly three times the size of a 10-horsepower hydraulic motor. When compared to gearing, shafts, belts, or chains, hydraulic power systems provide engineers with a *simple* means for transmitting power. The engineers use the *adaptability* of hydraulic components by placing them in unique places without having the constraints or limitations that exist with belts and chains. For example, years ago, an agricultural combine used dozens of belts and chains. Today, by incorporating additional hydraulic pumps, motors, cylinders, and hydraulic clutches, manufacturers can build machines using far fewer belts and chains for driving augers, drums, and shafts.

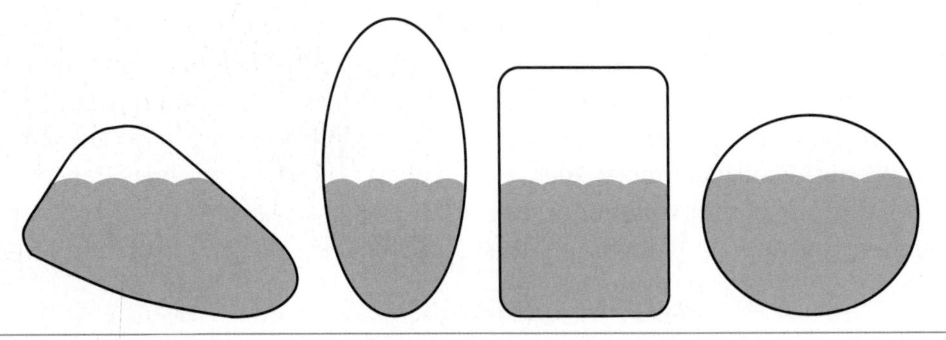

Figure 2-2. Liquids conform to the shape of their container.

The hydraulic controls also provide fast and easy adjustments, enabling the operator to vary the speeds and component positions while the machine is in operation. One example of this ***variable adjustment*** is a reel on a combine header. See Figure 2-3. Operators can hydraulically raise and lower the reel, push it forward, pull it back, and vary its speed. Years ago, the operator had to make all of those adjustments by mechanical means with sprockets and threaded adjustment rods. In addition, today, the reel speed can be varied and controlled in proportion to the machine's ground speed. As the machine speed increases, so does the reel speed.

When electronics are used to control the hydraulic system, the machines offer their greatest advantage, which is increased productivity. An advanced late-model hydraulic machine, with its incredible speed and agility, can outperform older hydraulic machines. Consider the combine header illustrated in Figure 2-3. If the machine is equipped with automatic header height control and automatic lateral tilt control, the header can gather the crop at great speeds and efficiency. The header can automatically sense the ground's contour and elevation by sending variable voltage signals back to an electronic control module (ECM). An operator can adjust the sensitivity of the potentiometers and the reaction time of the ECM outputs in order to dial the machine's performance to a near perfect operation. When set, adjusted, and calibrated correctly, the operator only needs to steer. In fact, with today's global positioning satellite (GPS) systems, many machines no longer rely on their operators for guidance. Autonomous vehicles allow companies to operate machines remotely without having to place an operator in the machine, which provide longer and safer hours of operation by eliminating the risk of operator fatigue.

Figure 2-3. A combine header reel has five hydraulic adjustments: raise, lower, fore, aft, and rpm.

Hydraulic systems provide overload protection with the use of relief valves, as compared to a machine with a manual transmission that can only rely on wheel slip before overloading the machine's powertrain. Hydraulic controls also allow an operator to operate at a constant force, torque, or speed. One example of a constant speed or constant torque application is controlling the speed of a concrete truck's drum.

Disadvantages of Hydraulic Systems

Along with all of the advantages, a hydraulic system has disadvantages. Noise, as mentioned in chapter 1, is a negative attribute of hydraulic systems. As the load increases, the pressure increase causes hydraulic noise. A ***hydrostatic transmission*** is a hydraulic pump and motor that is used to change a machine's speed and direction. These transmissions are notorious for generating lots of noise due to their high operating pressures. Some manufacturers add a noise suppression device known as an ***attenuator***. The attenuator acts like a hydraulic muffler. Because it receives high pressure oil, it consists of a heavy steel shell and tubes. It is designed to reduce the fluid pulsations that normally cause noise to resonate. See Figure 2-4.

In order to achieve high operating pressures, a hydraulic system must have extremely tight tolerances or clearances between moving components, such as a spool valve within a bore of a housing. These tight tolerances cause the system to be sensitive to contamination. Some manufacturers specify that brand new oil must first be filtered before it is used to fill a reservoir. If any particles, from normal wear or any outside dirt source, are left in the system, the hydraulic component's life will be reduced. **Chapters 11** and **12** focus on oil filtration and oil contamination control.

Hydraulic systems are harmed anytime air infiltrates the system. Oil aeration can be caused by leaking components or poor machine design. Anytime a fluid becomes aerated, the machine operation loses its predictability and precision, placing personnel at risk. The aeration can cause a pump to quickly fail. Aeration also causes overheating and an increase in noise.

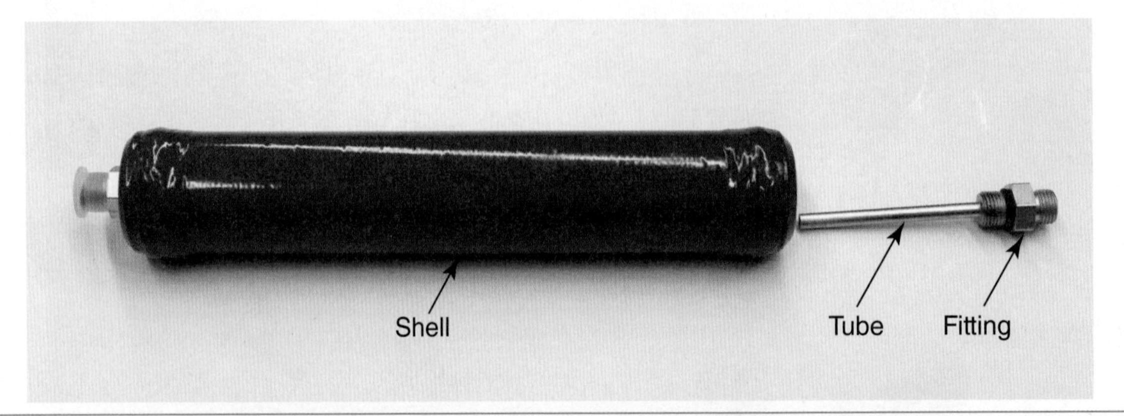

Figure 2-4. A hydraulic attenuator acts like a muffler to reduce hydraulic noise. This attenuator was placed in series between a combine's hydrostatic transmission's foot-and-inch relief valve and the hydraulic motor. Located in each end of the attenuator is a fitting that has a specially designed tube. The tubes are tuned to reduce noise that occurs when the operator actuates the foot-and-inch valve. **Chapter 23** will explain hydrostatic transmissions and foot-and-inch valves.

Hydraulic systems are sensitive to elevated temperatures. The expected life of the fluid is cut in half for every 18°F that the operating oil temperature exceeds the recommended operating temperature. Increased temperature can be caused by worn components, fluid aeration, internally restricted coolers, externally plugged coolers, or poor machine circulation design.

Petroleum-based fluids are the primary fluid medium used in off-highway hydraulic systems. As mentioned earlier in this chapter, oil spills can be environmentally harmful. As a result, a hydraulic system with a leak has a greater negative impact on the environment than a pneumatic system with a ruptured hose; venting compressed air will not harm the environment. Pneumatic engineers can vent air directly at the control valve, eliminating the need to return the air back to a reservoir, which is necessary with hydraulic systems.

Although hydraulic systems provide increased machine productivity, that advantage comes at the cost of mechanical inefficiency. For example, some mechanical drives can exhibit very high mechanical efficiencies, approaching close to 99%. However, due to fluid losses across clearances, hydraulic systems will have less mechanical efficiency requiring more fuel to perform the same amount of work performed by a traditional mechanical drive.

Hydrodynamic versus Hydrostatic

Fluid power systems can be further classified into two different types of fluid drives: ***hydrodynamic*** (fluids in motion) and ***hydrostatic*** (fluids at rest or under pressure). Both systems use fluid as the critical medium for transmitting power. Without fluid, the drives will be inoperable.

Hydrodynamic drives operate at significantly lower pressure than hydrostatic systems, such as 60 psi (4 bar). Examples of hydrodynamic drives are fluid couplings, torque converters, torque dividers, hydro-powered generators, and home sump pumps. Hydrostatic drive pressures can exceed 7000 psi (482 bar). Examples of hydrostatic drives are skid steer and combine transmissions. Both hydrodynamic and hydrostatic fluid drives will be further explained in **Chapter 23**.

Hydrostatic Fundamental Principles

Fluid power uses several key principles that were developed centuries ago. First, it is important to understand a few key terms.

- ***Force***—Occurs when a push or pull, measured in Newtons or pounds, is applied to an object.
- ***Pressure***—Equals a force distributed over a given unit of area. It is measured in pounds per square inch (psi), bar, or kilopascals (kPa).

Pascal's Law

Blaise Pascal is the 17th century (1623–1662) French mathematician, philosopher, and inventor who is responsible for theorizing the most widely known principle used in hydraulic systems. ***Pascal's law*** stipulates when a force is placed upon a container full of liquid, that a pressure will be developed and that the pressure will act equally in all directions, Figure 2-5. It is the generation of this pressure that provides the foundation for today's hydraulic systems.

Pascal's law can be expressed using an equation of three variables: force, pressure, and area. The variables are placed in a triangle, making it easier for a technician to solve for the missing variable. See Figure 2-6.

Before applying Pascal's law it is important to know how to determine the area of a circle, because cylinder bores and pistons are specified by diameter, not area. Two different formulas can be used for calculating the cross-sectional area of a cylinder or its piston:

$$\text{Area} = D^2 \times 0.7854$$

where D = the diameter of the cylinder

$$\text{Area} = \pi r^2$$

The second formula, πr^2, is the most popular equation for finding the area of a circle. However, the first formula, $D^2 \times 0.7854$, eliminates the step of computing a cylinder's radius. Although it may appear that finding the radius only requires dividing the diameter in half, technicians will not always be able to

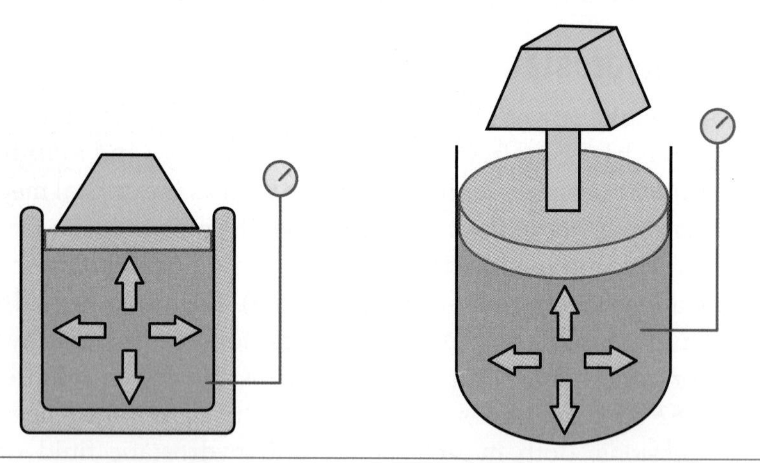

Figure 2-5. Pascal's law stipulates that a container filled with fluid will exert pressure equally in all directions when a force is applied to that confined body of fluid.

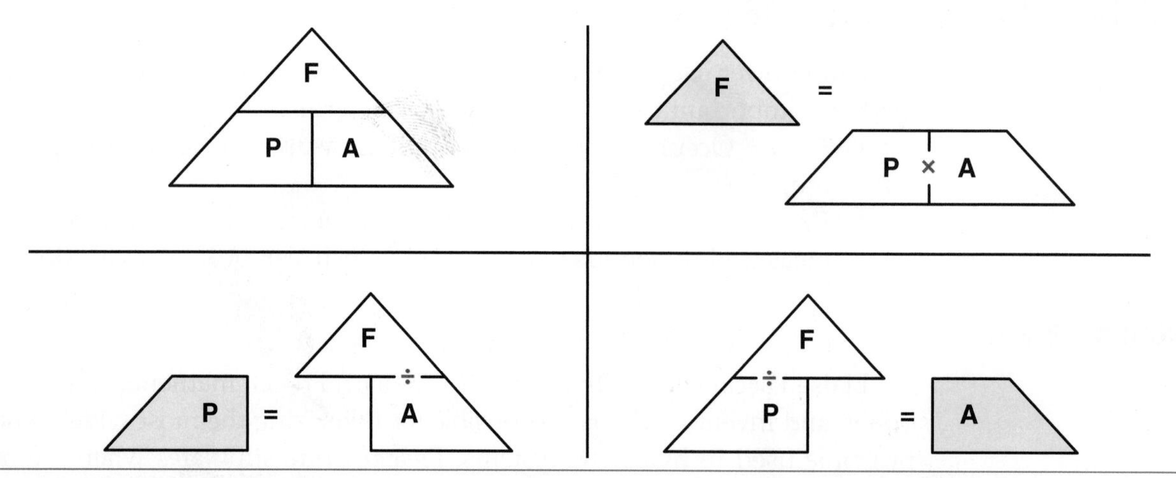

Figure 2-6. When any of two variables are known (force, area, or pressure), the third variable can be found using the following equations: (F = P × A), or (P = F/A), or (A = F/P).

easily compute this value in their head. For example, when a cylinder's piston diameter equals 1.25 inches, the radius equals 1.25/2 = 0.625 inches, which is an additional calculation. For this reason, this text will always use the first formula, $D^2 \times 0.7854$. See **Figure 2-7**. When the diameter is given in inches, the area's unit of measurement will be square inches (in^2). If the diameter is provided in millimeters the area's unit of measurement will be square millimeters (mm^2).

Note

For this discussion, the cross-sectional areas of the pistons are considered to be identical to the cross-sectional areas of the cylinders that house them.

A simplified hydraulic system contains a small-diameter input cylinder and a large-diameter output cylinder. See **Figure 2-8**. When a small force is applied to the input cylinder's piston, a potential pressure is developed. Pascal's law stipulates that the pressure acts equally in all directions. When the pressure acts on a cylinder that has a larger surface area, it enables the piston in the large cylinder (also known as the output cylinder) to move, exerting a larger output force.

Note

Pressure will only develop in the system if a load is placed on output cylinder. If a load is not placed on the output cylinder, the output cylinder's piston will still move. However, the pressure developed in the cylinder will be only the minimum needed to move the output piston.

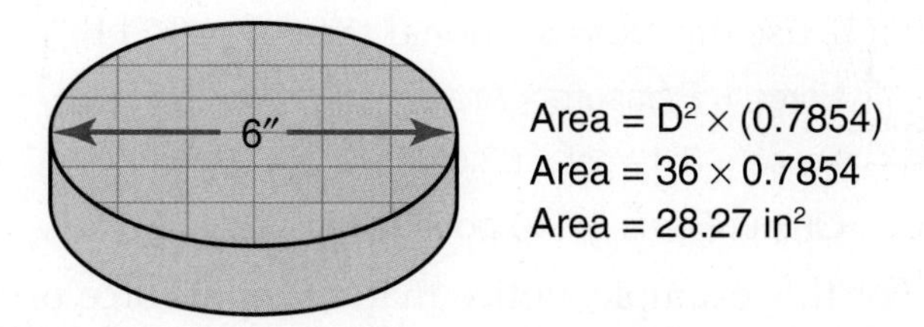

Figure 2-7. Finding the area of a circle requires squaring the diameter (multiply the diameter times itself) and multiplying that number times 0.7854.

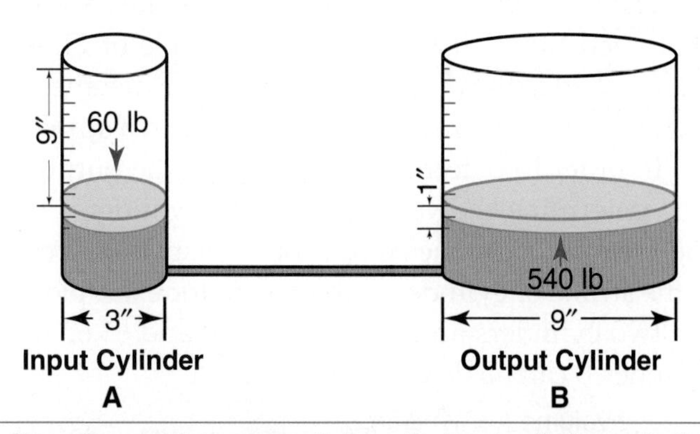

Figure 2-8. A simplified hydraulic system contains a small diameter input cylinder and a large diameter output cylinder. If the output cylinder area is nine times larger than the input cylinder, the input force will be multiplied nine times.

To compute the output force, several things must first be determined. The cylinder diameters must be converted to cross-sectional areas of the pistons.

Cross-sectional area of piston A

Area = $D^2 \times 0.7854$

Area = $3^2 \times 0.7854$

Area = 9×0.7854

Area of Piston A = 7.07 square inches

An output force cannot be determined unless an input force has been provided. For this example, 60 pounds of force will be applied to the input cylinder's piston (piston A).

When determining the potential system pressure, divide the input force by the cross-sectional area of the input cylinder's piston.

Pressure = Force (A)/Area (A)

Pressure = 60 pounds/7.07 in^2

Potential Pressure = 8.49 psi

Next determine the cross-sectional area of piston B.

Cross-sectional area of piston B

Area = $D^2 \times 0.7854$

Area = $9^2 \times 0.7854$

Area = 81×0.7854

Area of Piston B = 63.62 square inches

The potential output force can now be determined by multiplying the pressure times the piston's cross-sectional area. Do not be confused regarding which piston area should be used. When computing the potential force of piston B, use the cross-sectional area of piston B.

Force = Pressure × Area

Force = 8.49 psi × 63.62 in^2

Output Force = 540 pounds

For this example, notice that an input force of 60 pounds was multiplied nine times to 540 pounds. This multiplication effect is the result of the output cylinder's cross-sectional area being nine times the size of the input cylinder's cross-sectional area.

The multiplication of force comes at the cost of a reduced output stroke, or shortened travel distance. In the example of Figure 2-8, if the input cylinder was pushed a distance of nine inches, the distance of the output cylinder stroke would be reduced nine times, resulting in a stroke of one inch.

In order to determine the travel of the output cylinder's stroke, you must first determine the volumes of both cylinders. The volume displaced at cylinder A will equal the volume of oil displaced at cylinder B, due to Pascal's law. If the stroke of cylinder A is known, for example nine inches, and the areas of the two cylinders are known (7.07 in^2 and 63.62 in^2), then the output stroke of cylinder B can be determined.

Volume A = Volume B

Stroke A × Area A = Stroke B × Area B

$9'' \times 7.07 \text{ in}^2 = \text{Stroke B} \times 63.62 \text{ in}^2$

$63.62 \text{ in}^3 = \text{Stroke B} \times 63.62 \text{ in}^2$

Stroke B = 1 inch

Consider the relationship between force, area, and pressure in **Figure 2-8**. If a larger output force is desired, three different changes could be made. Those changes center on increasing the pressure or increasing the area.

- Reduce the area of the input cylinder (resulting in an increase in system pressure).
- Apply a larger force to the input cylinder (resulting in an increase in system pressure).
- Increase the output cylinder's area (applying system pressure to a larger area).

To summarize the effects of a simple hydraulic system

- A small input cylinder and a small input force can generate a high output force.
- The ability to generate tremendous output forces comes at the cost of a reduced output stroke.
- Force = Pressure × Area

Case Study

Force and Pressure

The terms "force" and "pressure" are sometimes easily confused for one another. Consider two siblings that are upset with each other. The sister is wearing high heel shoes and places all of her weight on the top of her brother's foot. How much pressure is the high heel placing on top of the brother's foot? If the sister weighs 125 pounds and if the area of the heel is 0.25 square inches, then the pressure equals:

Pressure = Force/Area

Pressure = 125/0.25

Pressure = 500 psi

Notice that the brother is not feeling the effect of the sister's 125 pounds of force (weight). The brother is experiencing the effect of pressure, which can be defined as a "force for a given unit of area."

Common Triangles

The triangle in **Figure 2-6** works the same as two other important triangles found in the field of electricity and electronics, Ohm's law triangle and the power equation triangle. See **Figure 2-9**. As previously discussed, if any two variables are known, the third variable can be easily found.

Another key and unique feature to the triangles is that if they are inverted, they exhibit inverse relationships. See **Figure 2-10**. For a fixed amount of force, the area and pressure are inversely proportional. The same holds true with Ohm's law; for a fixed amount of voltage, the amperage and ohms are inversely proportional. Again, the same holds true with Watt's law; for a fixed amount of wattage, the voltage and amperage are inversely proportional.

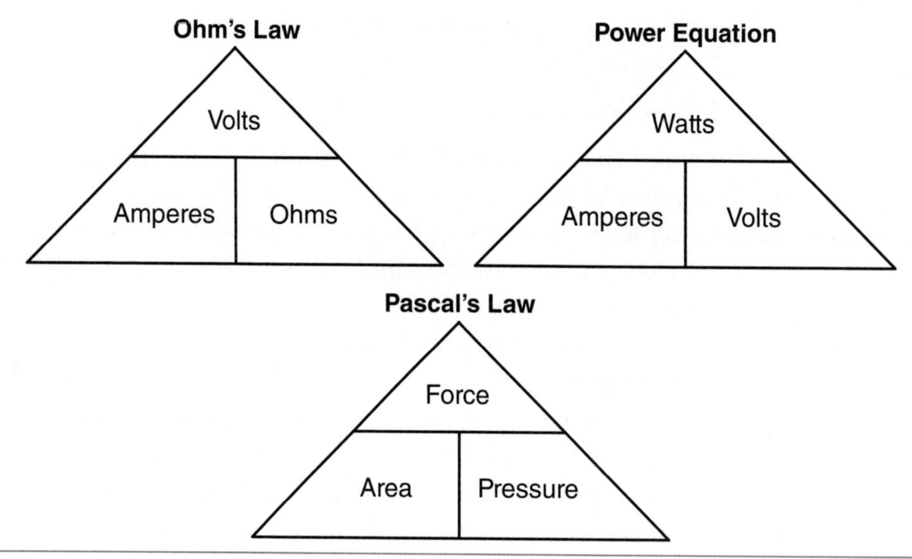

Figure 2-9. The three triangles: Pascal's law, Ohm's law, and the power equation, all have the same relationship. If any two variables within a single triangle are known then the third variable can be algebraically solved.

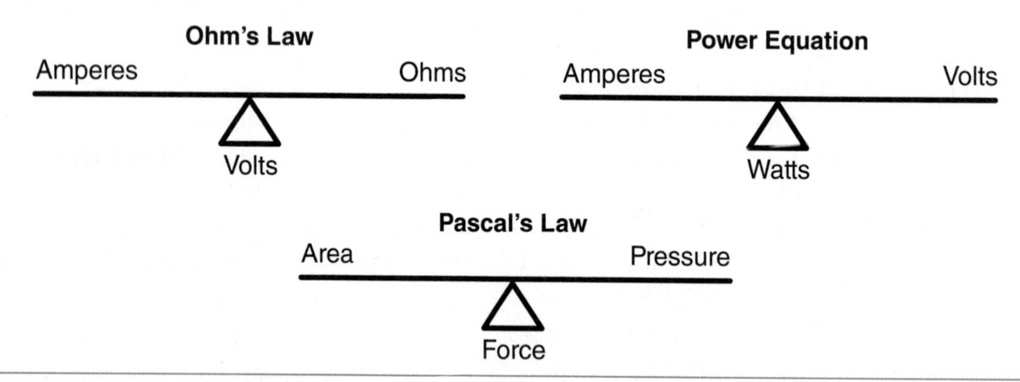

Figure 2-10. When one value in a triangle (Pascal's law, Ohm's law, or the power equation) is held constant, the other two values have an inverse relationship. For a fixed amount of force, the area is inversely proportional to pressure. For a fixed amount of voltage, the resistance is inversely proportional to amperage. For a fixed amount of wattage, the voltage is inversely proportional to amperage.

Hydrodynamic Fundamental Principles

The inverse relationship of pressure and area in the simplified hydraulic system found in Figure 2-8 relates similarly to hydraulic fundamental principles that are based on moving fluid, specifically a fixed amount of flow. ***Flow*** can be defined as a quantity of fluid moving past a given point during a specified time period. The two most common units of measurement of flow are ***gallons per minute (gpm)*** and ***liters per minute (lpm)***.

Continuity Equation

The continuity equation, also known as the principle of mass conservation, stipulates that for a fixed amount of flow, the fluid velocity is inversely proportional to the cross-sectional area of the fluid passage. See Figure 2-11. Velocity is defined as the speed of the fluid moving in a hydraulic system.

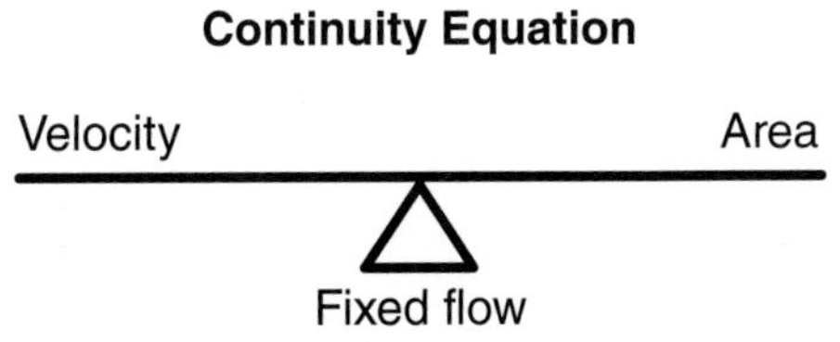

Continuity Equation

Velocity Area

Fixed flow

Figure 2-11. The continuity equation stipulates that if a conductor has a fixed amount of flow, the velocity will be inversely proportional to the conductor's cross-sectional area.

The inverse relationship of velocity and area applies to an operational hydraulic system that contains a ***fixed displacement pump***, a pump that produces a constant flow for a fixed pump speed. As previously stated, if a larger output force is desired, the output cylinder's area can be enlarged. However, when factoring in the continuity equation, the larger area results in slower fluid velocities, which translates into a slower cylinder speed.

Figure 2-12 is a two-dimensional illustration for explaining the continuity equation. The left side of the pipe is sized to hold seven droplets of oil side-by-side. On the right side, the pipe converges to a size that holds only three droplets of oil side-by-side. If the oil was flowing through the pipe at a rate of seven droplets per second, also known as a fixed amount of flow, the velocity would have to speed up when the oil reaches the smaller diameter pipe. The velocity would have to more than double in order to maintain the fixed amount of flow, in this case, seven droplets of oil per second.

Now, imagine the same pipe in three dimensions. Instead of seven droplets side by side, there is room for about 39 droplets in the wide cross-section of the pipe (7 × 7 × .7854). There is room for only about seven droplets in the narrow section of the pipe (3 × 3 × .7854). As a result, the droplets in the narrow section of pipe would need to move more than five times as fast as the droplets in the wide section in order to maintain the same rate of flow.

The formula Q = A × V × 3.12 can be used for computing the flow, velocity, or area in a fluid pipe. Note the converging pipe in Figure 2-13. The left side of the pipe has a diameter of two inches and the right side of the pipe has a diameter of one inch. For this example, if the fluid was flowing at a fixed rate of 40 gpm, what would happen to the velocity when the fluid reached the portion of the pipe with a one-inch diameter?

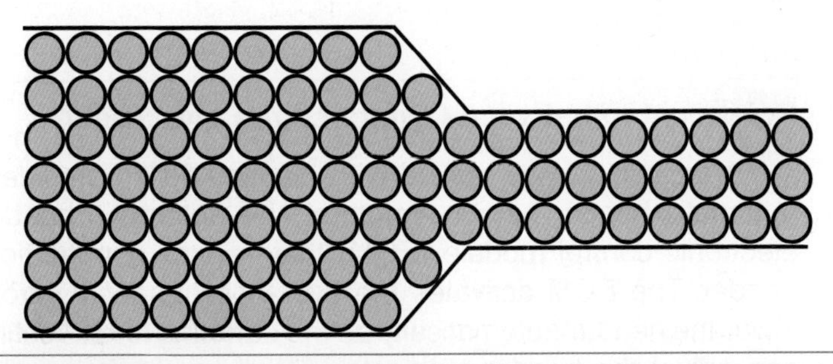

Figure 2-12. A two-dimensional illustration of a converging pipe illustrates that, when a system has a fixed amount of flow, the seven droplets of oil must speed up when they reach the three-droplet-sized pipe.

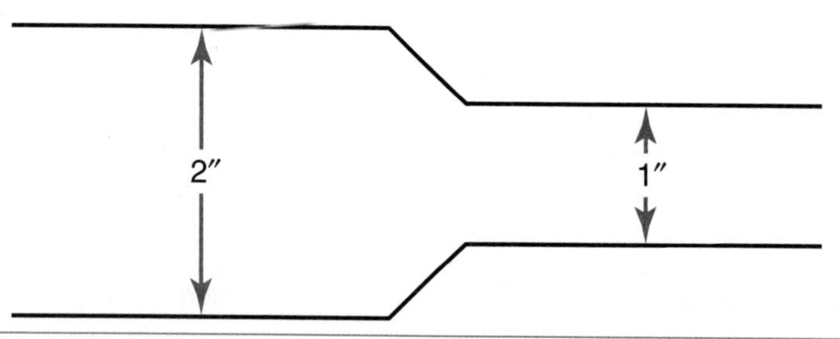

Figure 2-13. When a conductor's diameter is reduced by half, the velocity will increase four times.

Velocity in the two-inch pipe

$Q = \text{Area} \times \text{Velocity} \times 3.12$

$40 = (2 \times 2 \times 0.7854) \times \text{Velocity} \times 3.12$

$40 = 3.14\ \text{in}^2 \times \text{Velocity} \times 3.12$

$40 = 9.80 \times \text{Velocity}$

Velocity = 4.08 feet per second

Velocity in the one-inch pipe

$Q = \text{Area} \times \text{Velocity} \times 3.12$

$40 = (1 \times 1 \times 0.7854) \times \text{Velocity} \times 3.12$

$40 = 0.79\ \text{in}^2 \times \text{Velocity} \times 3.12$

$40 = 2.46 \times \text{Velocity}$

Velocity = 16.26 feet per second

The math reveals two interesting points. Reducing the diameter of a pipe by half will increase the velocity four times. Doubling the diameter reduces the velocity four times.

Nomograph

A ***nomograph*** is a handy tool that allows a technician to quickly find one variable as long as the other two variables are known. Whether a technician is troubleshooting a system, helping a customer build a new hydraulic system from scratch, or simply trying to determine the effect of a larger or smaller diameter hose, a nomograph can enable a technician to quickly find a value without having to make any computations. See Figure 2-14.

Notice that this nomograph contains three variables and that these variables are the three key components found in the continuity equation:

- Flow.
- Area.
- Velocity.

Case Study

Relationship between Cylinder Area and Speed

A technician was tasked with diagnosing a combine with a slow-responding automatic header height symptom. The automatic header height consists of an electronic control module (ECM) that senses a potentiometer on the combine's header. The ECM activates the hydraulic header control solenoids to raise and lower the header automatically as the combine is harvesting the crop. The operator can adjust the header-raise-rate potentiometer, header-lower-rate potentiometer, and the header sensitivity potentiometer to fine-tune the ECM's reaction time.

The technician had completed a thorough assessment of the inputs and outputs to the ECM and found that everything was operating correctly. Unfortunately, the automatic header height was still too sluggish for the customer, requiring the machine speed to be slowed during harvesting. The technician enlisted the help of the manufacturer's hotline. After spending lots of time and energy diagnosing the machine's electronic controls, it was later determined that the customer added two additional header lift cylinders so that the machine would have the capacity to lift a heavier header. The combine's fixed displacement pump remained the same. However, the addition of two more lift cylinders caused an increase in area, resulting in a drop in fluid velocity, greatly slowing the header lift, which made the customer and technician think that the ECM was unresponsive. The customer would need to use a smaller header and remove the extra lift cylinders in order to increase the header reaction time. Later in the chapter, it will be explained that adding a larger pump is not a good solution because it would require more engine horsepower.

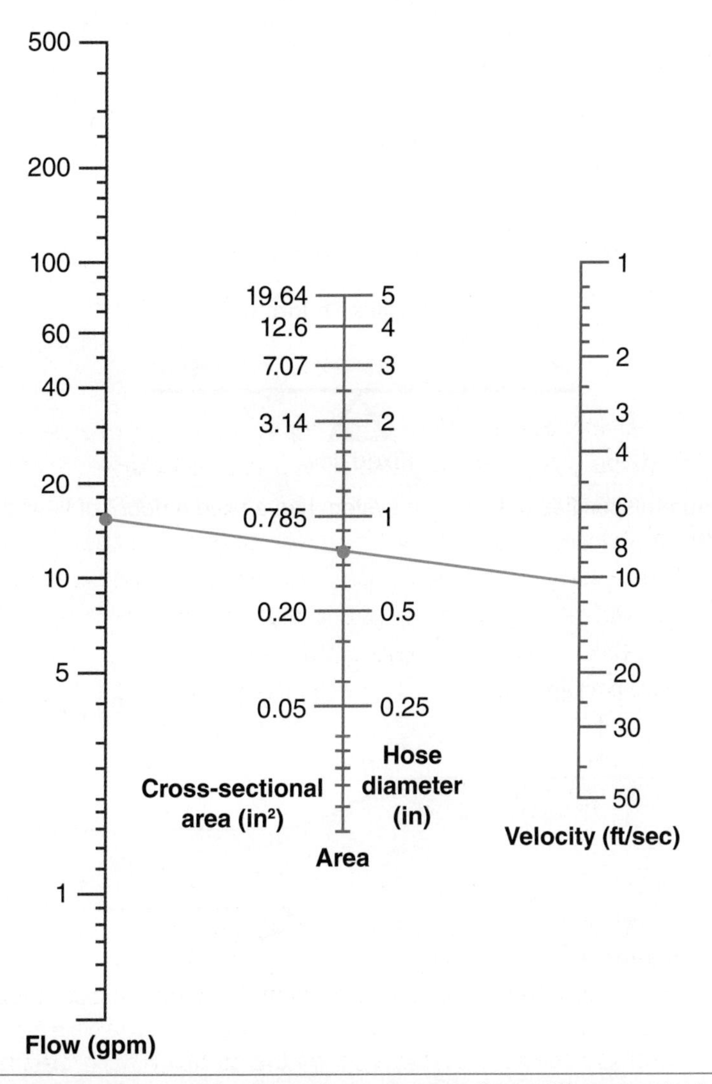

Figure 2-14. A nomograph contains three vertical lines that represent three different variables, for example, flow, area, and velocity. If two variables are known and plotted on the graph, the third variable can be found by drawing a line across the three vertical lines connecting the plotted values. This graph has been simplified.

Bernoulli's Theorem

Another important fluid power principle that is similar to the continuity equation is Bernoulli's theorem. Daniel Bernoulli was an 18th century Swiss mathematician and physicist. His theorem specifies that when a hydraulic system contains a fixed amount of flow, the fluid velocity is inversely proportional to pressure. When viewing Figure 2-15, notice that the relationship looks identical to the equation of continuity, except the term "area" is replaced with the word "pressure."

An example of a fixed amount of flow is 10 gpm. For a hydraulic system that is flowing a constant 10 gpm, if the pressure is high, the velocity will be low. If the velocity is high, the pressure will be low. Figure 2-16 illustrates another critical concept of hydraulic systems related to Bernoulli's theorem. If a conductor is flowing oil and a restriction exists, the pressure will drop after the restriction. Keep in mind that the system must have flow. Notice in Figure 2-17 that once the pipe is capped, the pressures equalize.

The concept of pressure dropping after the load, such as an orifice, in a hydraulic system with a fixed amount of flow is similar to voltage drops in an electrical series circuit. See Figure 2-18. In a series electrical circuit with flowing current, the voltage (pressure) will drop after the load, and the current flow will remain constant. The electrical power source is similar to a fixed displacement hydraulic pump flowing a constant flow of fluid.

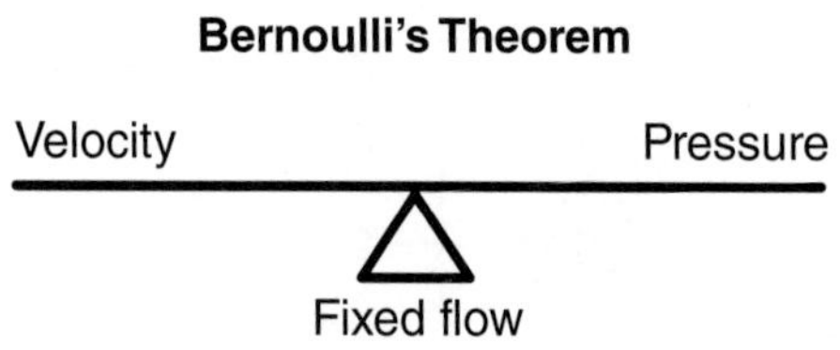

Figure 2-15. Bernoulli's theorem specifies that when a system has a fixed amount of flow, the area will be inversely proportional to the pressure.

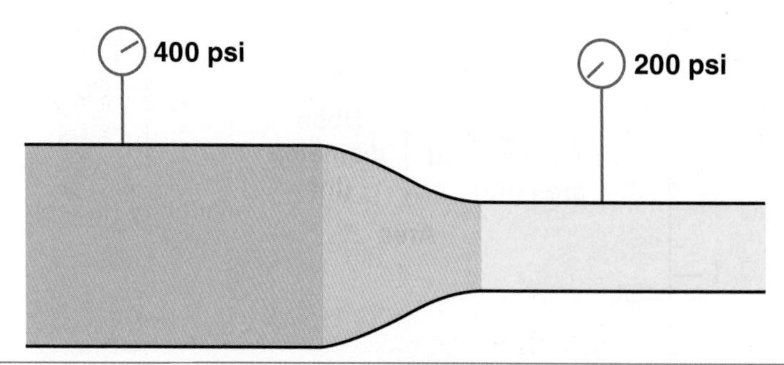

Figure 2-16. A converging pipe illustrates Bernoulli's theorem. Notice that the pressure drops after the conductor's diameter has been reduced.

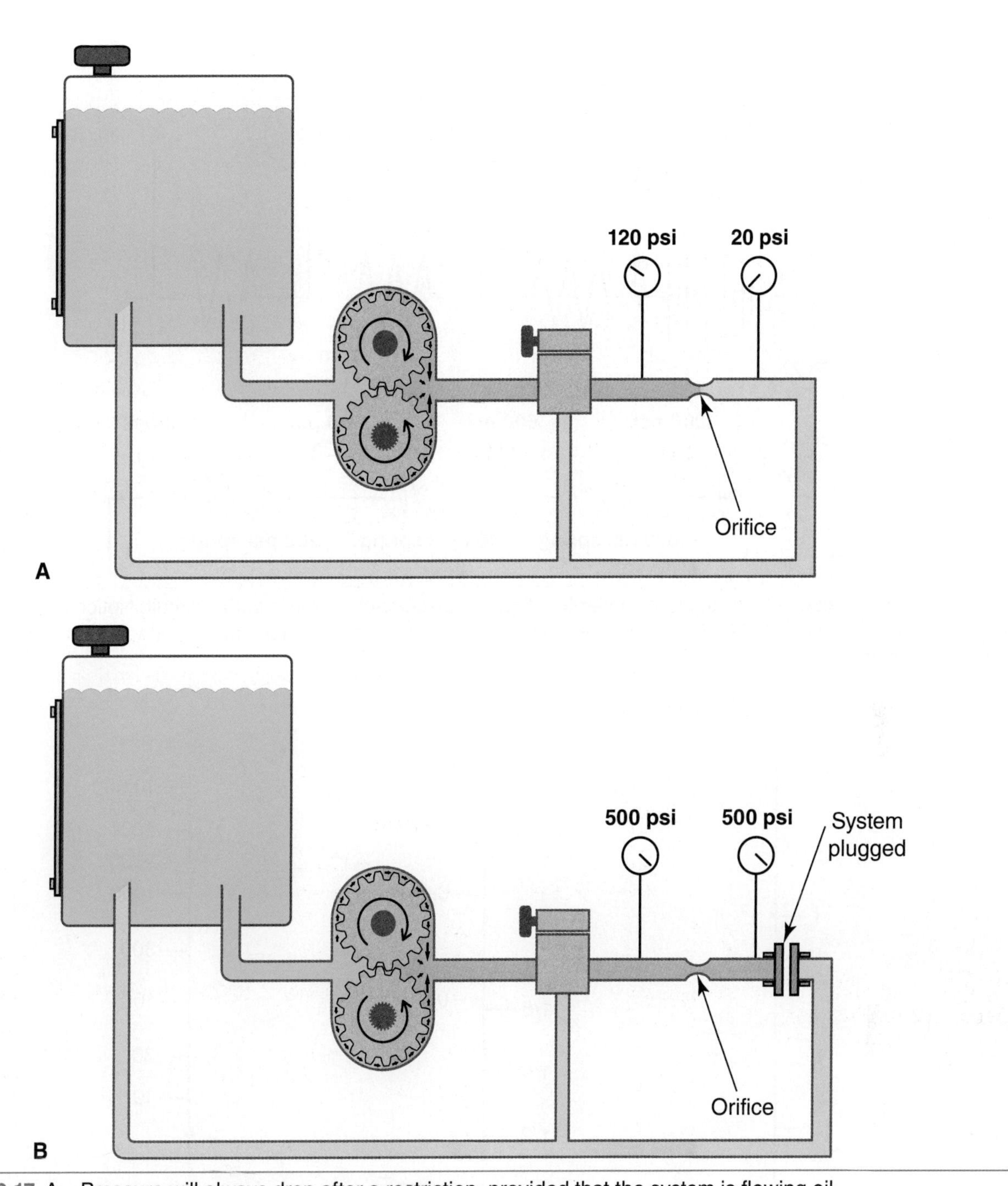

Figure 2-17. A—Pressure will always drop after a restriction, provided that the system is flowing oil. B—If the conductor is plugged, notice that the pressures equalize.

Nomograph

A nomograph for Bernoulli's theorem is similar to the continuity equation nomograph. Notice that the graph in **Figure 2-19** contains the three key factors of Bernoulli's theorem: flow, area, and pressure drop. If a technician was attempting to determine the effect of a different sized orifice, the nomograph would save the effort of having to complete several computations.

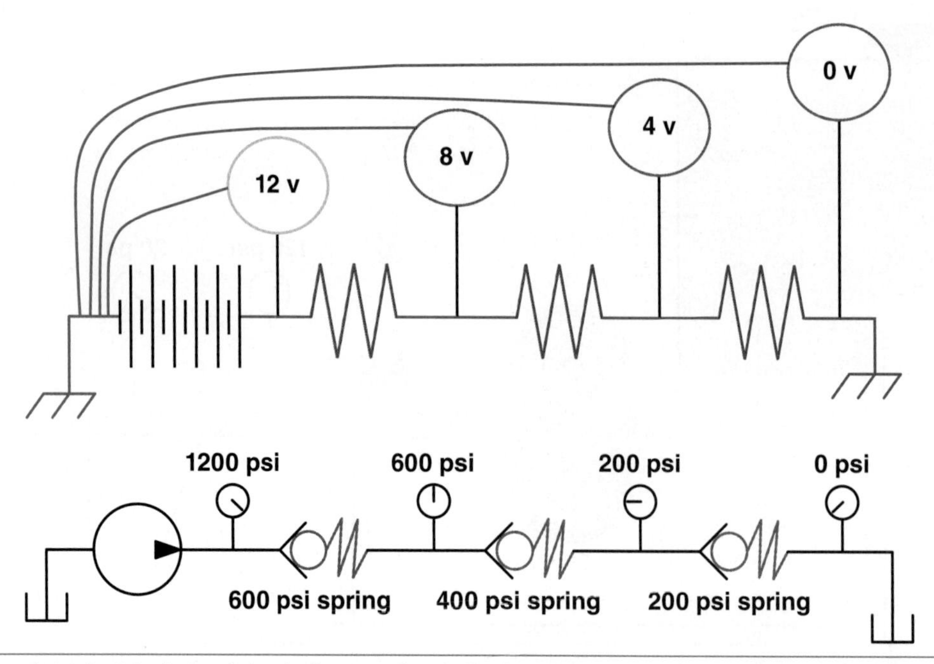

Figure 2-18. A series electrical circuit is similar to a fixed-displacement hydraulic series circuit. Notice in both the electrical and hydraulic circuit that the pressure will drop after each load. Also note that the flow in both circuits remains the same throughout the entire circuit.

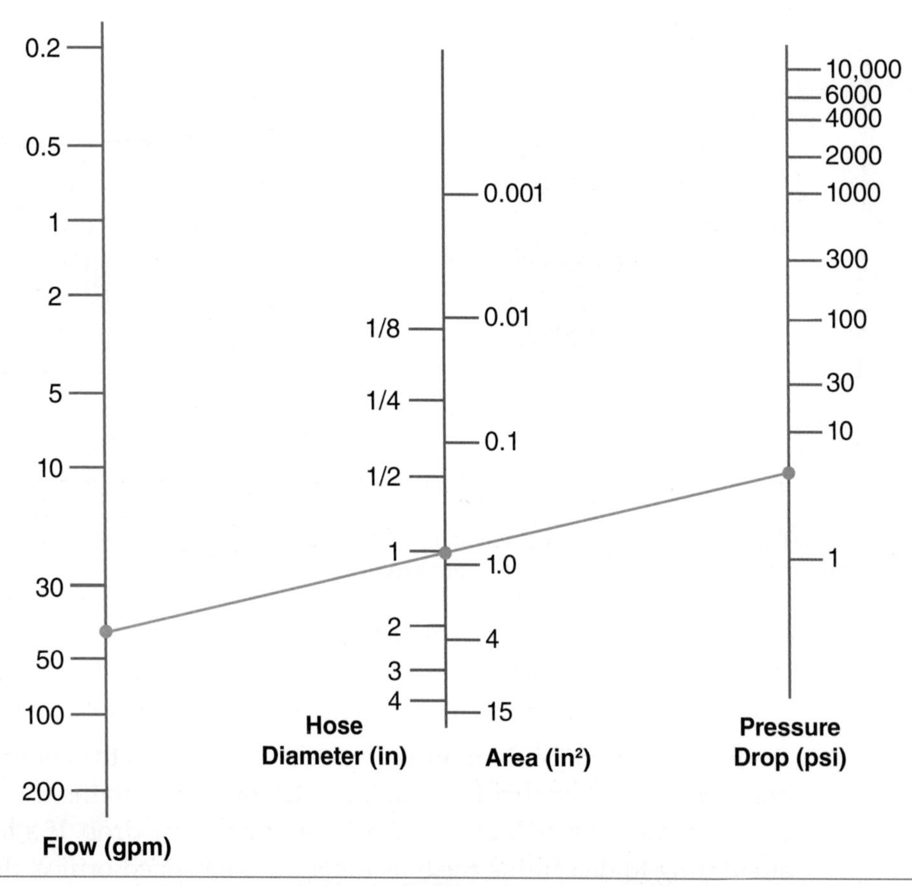

Figure 2-19. This nomograph lists three variables: flow, area, and pressure drop. Knowing the system flow and size of the conductor allows a technician to determine the pressure drop. This nomograph has been simplified.

Case Study

Slowing Cylinder Action

An older combine has a header that rises too fast. The machine has no provision for slowing the header raise rate. The customer visits with the local dealer and they decide that they have four choices for solving this problem:

- Installation of an accumulator.
- Installation of a smaller pump.
- Slow the pump's rpm.
- Add larger lift cylinders.

Among those four choices, which choice is the best solution? The accumulator kit uses a switch on the feeder house to actuate the accumulator valve when the header is raised. This solution provides only a slight, momentary pause during lift and would not be the best solution. Installing a smaller pump is a possibility, but it seems like a costly decision that provides no additional benefit other than slowing the system's flow. Slowing the pump's rpm might be quite difficult depending on the mechanism that is used to drive the pump. If the pump is driven by a belt, a pulley change is a potential solution, but it provides no additional benefit aside from slowing the system's flow. Adding additional lift cylinders or replacing the original lift cylinders with larger-diameter cylinders would increase the area, and would result in a slower lift speed. Bernoulli's theorem also explains that the decreased velocity would result in a pressure drop, which would extend the operational life of the system. In addition, the larger lift cylinders would enable the combine to lift heavier headers.

Law of Conservation of Energy

Both the continuity equation and Bernoulli's theorem are derived from the ***law of conservation of energy***, which states that energy cannot be created or destroyed, but can only be transformed from one form to another. A 5-gallon can of diesel is transformed into heat energy inside a diesel engine. The diesel engine transforms the heat energy into rotational mechanical energy. The diesel engine then drives a hydraulic pump, which transforms the mechanical rotational energy of the engine's crankshaft into hydraulic fluid energy for the purpose of extending hydraulic cylinders and rotating hydraulic motors. See **Figure 2-20**.

In a hydraulic system, the system's total energy equals the sum of the kinetic energy plus the potential energy. Kinetic energy results from fluid flow and is comprised of the weight of the fluid and the velocity of the fluid. Potential energy is stored energy, which results from elevating the fluid or pressurizing the fluid.

Pressure energy occurs when the fluid flow meets resistance or a load. This energy can be stored in an ***accumulator***, a vessel used to store fluid pressure, and can be released when system pressure drops. Pressure can be measured in units of pounds per square inch (psi), bar, and kilopascals (kPa).

When fluid is elevated, it is measured in head-feet, head-meters, or pressure. Pump manufacturers prefer that the machine's reservoir supply a positive pressure head to eliminate cavitation. One of the easiest ways to ensure a positive pump pressure head is to elevate the reservoir above the pump's inlet.

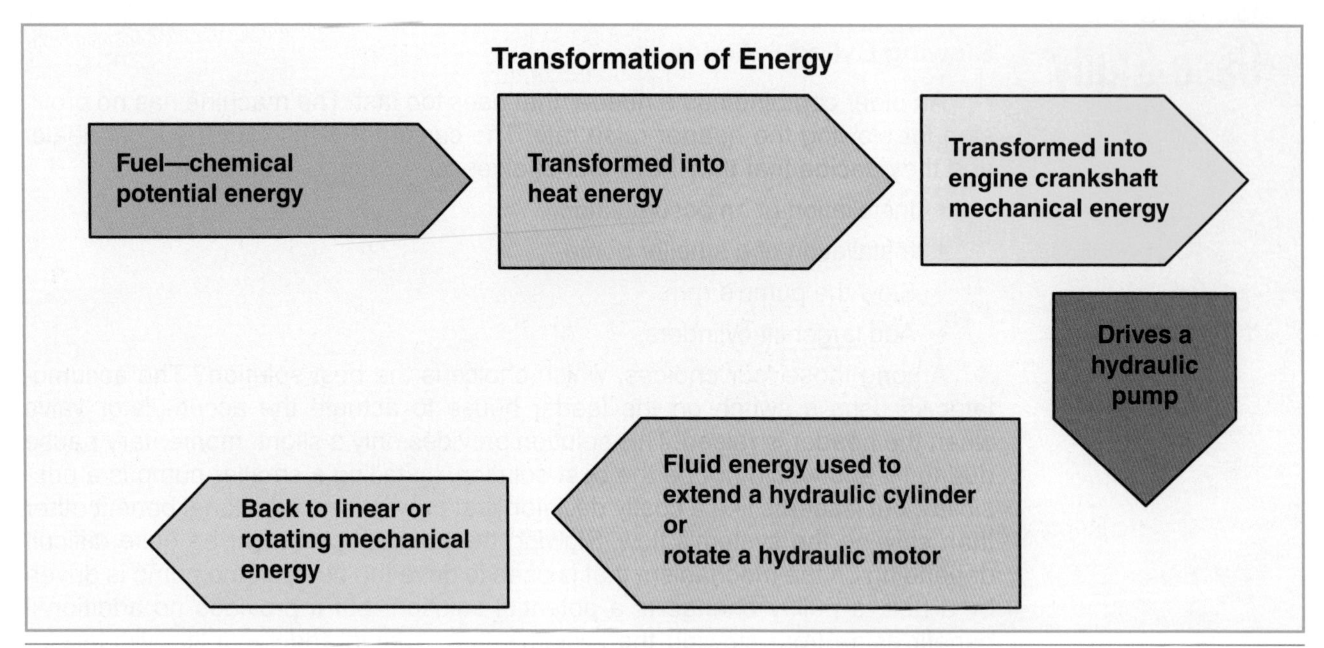

Figure 2-20. Energy is transformed multiple times in order to perform work. This flow chart depicts the transformation of energies in an off-highway machine.

The potential of an elevated fluid can be determined by using the following formula: height (in ft) × Sg (specific gravity) × 0.433. The last two factors both relate to the weight of water. The ***specific gravity (Sg)*** is the ratio of the fluid's weight in relation to the weight of water. For example, many petroleum-based oils are lighter than water. An example of a petroleum-based oil's specific gravity is 0.85, meaning that the oil is 15% lighter than water. This explains why most oils will not mix with water and will float on the top of water. A range of specific gravity values for petroleum oils is 0.75 to 0.90. Synthetic oils can actually weigh more than water and have a specific gravity ranging from 1.02 to 1.20.

The formula's 0.433 factor also relates to the weight of water. This value equals the amount of pressure that is exerted at the base of one foot of water. The easiest way to solve for the pressure of a foot of water is to use a cubic foot container. A cubic foot of water weighs 62.4 pounds. The area at the base of a cubic foot of water equals 144 square inches. See Figure 2-21. When solving for pressure, divide force (62.4 pounds) by the area (144 square inches), which equals a pressure of 0.433 psi.

Figure 2-22 illustrates the potential found at the base of a water tower. If the 60-foot tower was filled with oil that had a specific gravity of 0.85, at the base elevation the fluid would have a pressure of 22 psi.

Note

The effect of the weight of a fluid and the height of fluid is precisely why it is difficult to explore the depths of the ocean. The further down a vessel is submerged in water, the higher the pressure that is exerted on the vessel due to the weight of the fluid. One mile below the ocean's surface, water pressure will be (5280 feet × 0.433) = 2286 psi. The bottom of the ocean is 6.85 miles deep, and a vessel at that depth must be able to withstand a pressure of 15,674 psi.

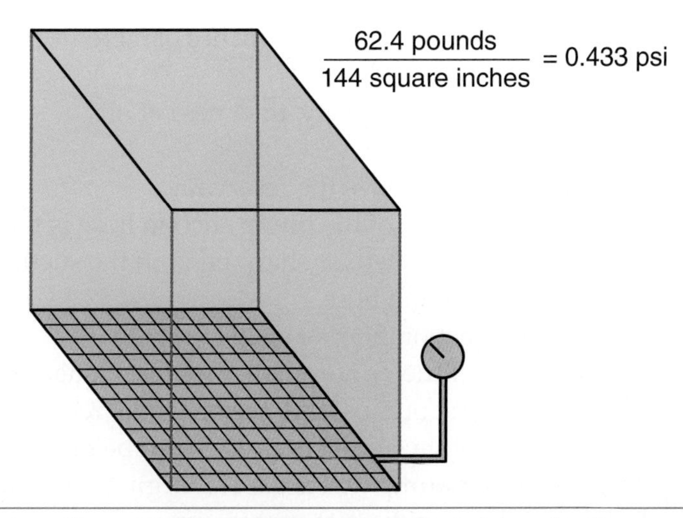

Figure 2-21. One foot of water will exert 0.433 psi at the base of the column of water.

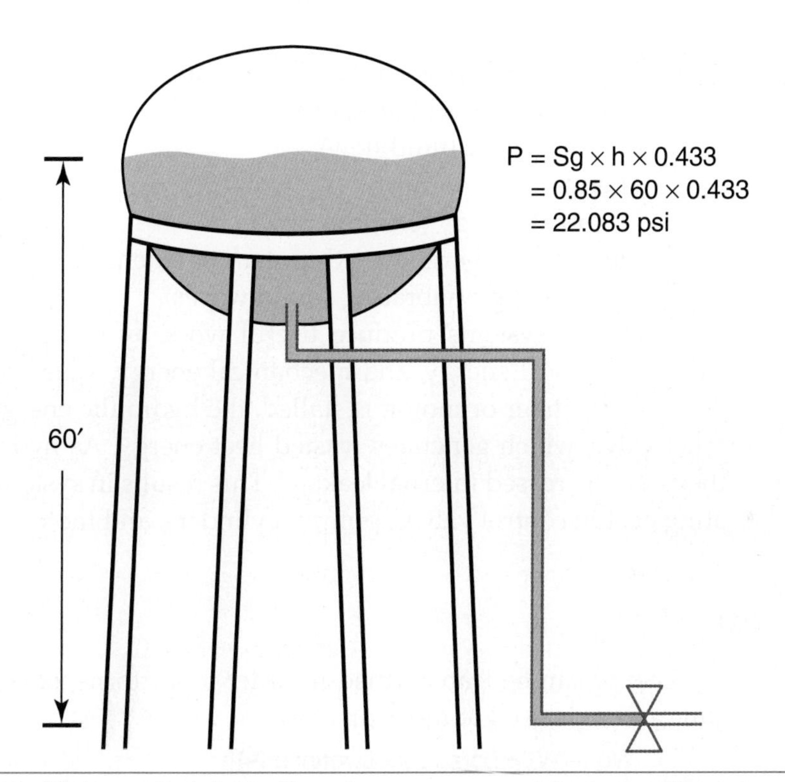

Figure 2-22. The weight of fluid and its height will produce a pressure at the base of the column of fluid.

Pump Inlet

A reservoir that is elevated above a pump's inlet uses the weight of the fluid and the height of the fluid to provide a good supply of fluid to the pump's inlet. Most pump manufacturers provide a maximum pump inlet specification, which is a value that measures the amount of vacuum or restriction that exists at the pump's inlet, also known as a pump's suction port. An example of a max pump inlet specification is 7 to 10 inches of mercury (in Hg). Several factors can cause a restricted pump inlet:

- Too high of oil viscosity, thickness of oil.
- Oil not warmed.
- Pump located above the reservoir.
- Inside diameter of the pump suction hose is too small.
- Kinked, obstructed, or sharp bend in the suction hose or tube.
- Incorrect suction hose.
- Plugged suction filter or suction screen.

A pump with little or no vacuum will last longer than a pump with a high inlet vacuum. Some pumps have the benefit of receiving a positive pressurized supply of oil. Positive supply pressures can be obtained with pressurized reservoirs or another pump that is supercharging the pump's inlet. Later chapters will discuss filtration, how to measure pump inlet pressure, hydrostat charge pumps, and pressurized reservoirs in further detail.

Forms of Energy

Energy can exist in several different states.

- Static energy (fluid pressure).
- Kinetic energy (fluid flow).
- Thermal energy (heat).
- Electrical energy (electron flow).
- Mechanical energy (rotating shaft or extending cylinder).
- Sound energy (vibrating sound waves).

Hydraulic systems produce useful work by using static energy, kinetic energy, electrical energy, and mechanical energy. When a hydraulic actuator, such as a cylinder or motor, is stalled, the hydraulic energy is dumped over a relief valve, which generates wasted heat energy. As hydraulic systems wear, they have increased internal leakage. This results in system inefficiency, generating heat in control valves, pumps, cylinders, and motors.

Work and Power

Energy can be transformed in order to perform ***work***, which occurs anytime a force is used to move an object.

$$\text{Work (W)} = \text{Force (f)} \times \text{Distance (s)}$$

An example of work being performed is a hydraulic jack lifting a tractor. The jack is using hydraulic pressure acting on a cylinder to exert a force to lift a tractor in the air a certain distance. Since liquids are practically incompressible, and because a developed pressure will act equally in all directions, hydraulic systems enable personnel to complete amazing tasks, such as a 150 pound adult lifting a 40,000 pound tractor by simply pumping a small bottle jack handle.

Note that the variable of time is not included in the work equation. When the element of time is added to the work equation, it becomes the power equation, a measure of how fast work can be completed. In many fields, including hydraulics, power is commonly measured in horsepower.

James Watt was an 18th century Scottish mechanical engineer that is credited with revolutionizing the steam engine. He needed a way to accurately compare the power of a steam engine and a horse. Mr. Watt conducted various experiments in which he discovered that a work horse could perform work at a rate of 22,000 ft-lb per minute. To ensure that his steam engine could outperform even stronger horses, he made sure his "one horsepower" steam engine could lift 33,000 ft-lb/min, which was 50% more than the horse in his experiment. As a result, a measurement of ***mechanical horsepower*** had been developed to measure how fast work can be performed.

1 hp = 33,000 ft-lb/min

This mechanical horsepower value can be converted to a rate 'per second', by dividing 60 seconds into the 33,000 ft-lb/min.

1 hp = 33,000/60

1 hp = 550 ft-lb/second

Mechanical horsepower can be computed by multiplying torque times speed.

Mechanical Horsepower = Torque (ft-lbs) × Speed (rpm)/5252

For a fixed amount of horsepower, torque is inversely proportional to speed. See Figure 2-23. Many customers and machine operators misunderstand that their machine can somehow magically perform more work faster if the gearing is simply changed. They fail to understand that as long as the engine's horsepower remains fixed, that an increase in output speed results in a reduction in output torque. For example, the 2100 and 2300 series Case IH combines had two different mechanical transmissions, one for flat terrain offering faster travel speeds and one for hilly terrain providing higher torque. A customer operating an incorrectly equipped transmission will condemn the engine or hydrostatic transmission, not realizing the machine is performing as designed, for either higher travel speeds or higher torque multiplication. **Chapters 23** and **24** will explain hydrostatic transmissions in detail.

Hydraulic horsepower is the product of pressure and flow and can be computed using different formulas:

Hydraulic Horsepower = Flow (gpm) × Pressure (psi)/1714

or

Hydraulic Horsepower = Flow (gpm) × Pressure (psi) × 0.000583

or

Hydraulic Horsepower = Flow (lpm) × Pressure (bar)/600

Notice that hydraulic flow is what brings the time element into hydraulic horsepower equation. As an example, compute how much horsepower is required to drive a hydraulic pump with a 3.5 cubic inch displacement that is rotating at 1000 rpm and operating at 2000 psi.

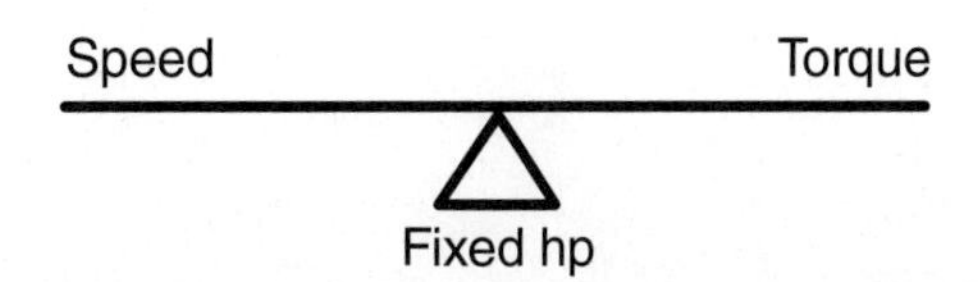

Figure 2-23. For a fixed amount of mechanical horsepower, speed is inversely proportional to torque.

The flow rate must first be computed:

3.5 in^3 per revolution × 1000 rpm=3500 in^3 per minute

Notice that both horsepower equations use 'gallons per minute', therefore the 3500 in^3 per minute must be converted into gpm. The volume of one gallon contains 231 cubic inches. The flow rate of 3500 in^3 per minute, must be divided by 231, which equals 15.15 gpm. Horsepower can now be computed:

(15.15 gpm × 2000 psi)/1714 = 17.68 hp

One more computation is required to determine the input horsepower necessary to drive the hydraulic pump and that factor is hydraulic system efficiency. Assuming the hydraulic system is operating at 90% efficiency, it is necessary to divide the 17.68 hp by 0.90. This determines that the machine must have 19.64 engine hp to operate the hydraulic system. Equally important is the understanding that if the operator desires more flow or more pressure, the engine would have to be capable of delivering more horsepower.

Although electrically driven hydraulic pumps in off-highway machinery are rare, they do exist. For example, a Caterpillar M-series motor grader uses an electrically driven pump as a backup for steering.

If an electric motor is used to drive the pump, it is necessary to determine how much electrical horsepower is required to drive the pump. One mechanical (imperial) horsepower equals 746 watts. Using the same 19.64 hp and multiplying it times 746, equals 14,651 watts (14.65 kilowatts).

A rule of thumb is that one mechanical horsepower is the equivalent of three fourths of a kilowatt. Stated another way, if someone purchased a 7.5 kilowatt electric motor, how much mechanical horsepower would it be capable of delivering?

7500 watts/746 = 10 mechanical horsepower

Conversion Table

Pressure	1 bar = 14.5 psi
	1 bar = 100 kPa
	1 psi = 6.89 kPa
	psi = 0.0689 bar
Force	1 pound = 1 newton
Flow	1 gpm = 3.785 lpm
	1 lpm = 0.264 gpm
Volume	1 gallon = 231 cubic inches
Power	1 hp = 745.7 watts
	1 hp = 42.4 Btu/minute

Summary

- ✓ Liquids are practically incompressible and take the shape of their container.
- ✓ Hydraulic systems have many advantages: large multiplier of force; self-lubricating; compact; adaptable; simple; variable; productive; provide overload protection; and are capable of constant force, torque, or speed.
- ✓ Hydraulic systems have several disadvantages: noisy; sensitive to heat, contamination, and air; less mechanically efficient; potentially harmful to the environment.
- ✓ Hydrodynamic fluid drives operate on the principle of fluids in motion and have lower pressures.
- ✓ Hydrostatic fluid drives operate on the principle of fluids under pressure and can have very high system pressures.
- ✓ Pascal's law explains that a force applied to a confined body of fluid will exert a pressure equally in all directions.
- ✓ Increasing the area of an output cylinder increases force and decreases speed.
- ✓ The continuity equation explains that if a conductor has a fixed amount of flow, the fluid velocity is inversely proportional to the cross-sectional area of the conductor.
- ✓ Bernoulli's theorem explains that if a conductor has a fixed amount of flow, the pressure is inversely proportional to the velocity.
- ✓ Pressure will drop after an orifice, as long as the system has flow.
- ✓ The law of conservation of energy stipulates that energy is not created or destroyed, but is transformed from one form to another.
- ✓ The pressure at a pump's inlet is affected by the weight of the fluid and the height of the fluid.
- ✓ A pump's inlet can be inhibited by a fluid that is too thick, a reservoir that is too low, a suction hose that is too small, an obstructed suction hose, or a plugged suction filter or strainer.
- ✓ Work is performed when a force is used to move an object.
- ✓ Power is a measure of how fast work can be performed.
- ✓ For a fixed amount of horsepower, speed is inversely proportional to torque.

Technical Terms

accumulator
adaptability
attenuator
bulk modulus
fixed displacement pump
flow
force
gallons per minute (gpm)
hydraulic horsepower
hydraulics
hydrodynamic
hydrostatic
hydrostatic transmission
law of conservation of energy
liters per minute (lpm)
mechanical horsepower
nomograph
Pascal's law
pneumatics
pressure
specific gravity (Sg)
variable adjustment
work

Formulas and Equations

- ✓ Pressure = Force/Area
- ✓ Force = Pressure × Area
- ✓ Pump flow rate = Pump displacement × rpm
- ✓ One US gallon = 231 in^3
- ✓ One mechanical hp = 550 ft lbs/s, 33,000 ft-lbs/min, or 746 watts.
- ✓ Hydraulic horsepower = (gpm × psi)/1714

Review Questions

Answer the following questions using the information provided in this chapter.

1. All of the following are considered hydraulic system advantages, *EXCEPT*:
   A. mechanically efficient.
   B. productive.
   C. provides overload protection.
   D. adjustable.
2. Which of the following is *not* an advantage of a hydraulic system?
   A. Compact components.
   B. Provides overload protection.
   C. Insensitivity to temperature.
   D. Increased productivity.
3. Which of the following is *not* an advantage of a pneumatic system?
   A. Cheap to build.
   B. Its medium (air) is readily available.
   C. Rigid and responsive.
   D. Environmentally friendly.
4. Which of the following is *not* an advantage of a pneumatic system?
   A. Self-lubricating.
   B. Economical.
   C. Easily exhausted.
   D. The medium (air) is readily available.
5. All of the following are attributes of hydrostatics, *EXCEPT*:
   A. fluids in motion.
   B. higher pressures.
   C. fluid drive.
   D. fluids at rest.
6. Liquid will compress how much at 1000 psi?
   A. 0.5%.
   B. 1%.
   C. 1.5%.
   D. 2%.
7. Who is credited with developing theory that a force applied to a confined fluid will exert a pressure equally in all directions?
   A. Joseph Bramha.
   B. Blaise Pascal.
   C. Daniel Bernoulli.
   D. James Watt.
8. All of the following are units of measurement for pressure, *EXCEPT*:
   A. pounds per square inch.
   B. Newton-meters.
   C. bar.
   D. kiloPascals.
9. The units of measurement for pressure are ______.
   A. N and lb
   B. kPa and bar
   C. gpm and lpm
   D. fps and mps
10. Force equals ______.
    A. P × A
    B. P/A
    C. A/P
    D. None of the above.
11. A cylinder has a 5-inch diameter. What is the cylinder's area?
    A. 7.85 square inches.
    B. 15.7 square inches.
    C. 19.64 square inches.
    D. 189.6 square inches.

12. Hydraulic flow is measured in what units of measurement?
    A. in^3 or cm^3.
    B. psi or bar or kPa.
    C. gpm or lpm.
    D. lbs or N.
13. Pump displacement is measured in what units of measurement?
    A. in^3 or cm^3.
    B. psi or bar or kPa.
    C. gpm or lpm.
    D. lbs or N.
14. The continuity equation is based on a _____.
    A. fixed amount of pressure
    B. fixed amount of flow
    C. fixed amount of load
    D. fixed amount of resistance
15. According to the continuity equation, when a pipe's diameter is decreased by half, _____.
    A. velocity is multiplied by a factor of two
    B. velocity is multiplied by a factor of four
    C. velocity reduced by a factor of two
    D. velocity is reduced by a factor of four
16. A technician is asked to determine the size of a conductor using a nomograph based on the continuity equation. Two variables are provided: gallons per minute (gpm) and feet per second (fps). Which of the following will be obtained from the nomograph?
    A. psi, kPA, or bar.
    B. rpm.
    C. in or in^2.
    D. lbs or N.
17. A technician is troubleshooting an overheating hydraulic system. The service manual states that the cooler should flow 40 lpm when it is unrestricted. The technician's flowmeter only reads gpm. What should the flowmeter read if the cooler is unrestricted?
    A. 10.5 gpm.
    B. 11.5 gpm.
    C. 12.5 gpm.
    D. 13.5 gpm.
18. To summarize Bernoulli's theorem, if a fluid is moving in a steady state, _____.
    A. as the velocity increases, the pressure will not change
    B. as the velocity increases, the pressure will decrease
    C. as the velocity increases, the pressure will increase
    D. as the velocity increases, the area will increase
19. In a hydraulic system, if there is no flow, _____.
    A. pressure drop will fluctuate
    B. pressure drop will be low
    C. pressure drop will be high
    D. there will be no pressure drop
20. A concrete mixer truck has high inlet vacuum at the hydrostatic pump's inlet. Technician A states that the mixer may have the wrong type of suction oil filter installed. Technician B states that the suction hose may be bent or kinked. Who is correct?
    A. Technician A.
    B. Technician B.
    C. Both A and B.
    D. Neither A nor B.
21. If a hydraulic system is flowing a fixed amount of flow (for example 30 gpm) and if it is trying to raise a 4000 pound weight, which of the following cylinders would require the least amount of pressure?
    A. 2-inch diameter cylinder.
    B. 3-inch diameter cylinder.
    C. 4-inch diameter cylinder.
    D. 5-inch diameter cylinder.
22. All of the following are correct regarding the law of conservation of energy, *EXCEPT*:
    A. energy cannot be created.
    B. energy cannot be destroyed.
    C. energy cannot be changed from one form to another.
    D. None of the above.

23. Pressure and elevation are both forms of _____ energy.
    A. kinetic
    B. hydrodynamic
    C. hydrostatic
    D. potential
24. Work is best described as _____.
    A. any application of force
    B. force used to move an object
    C. force used to move an object a certain distance in a fixed amount of time
    D. force used to move an object a certain distance using a fixed amount of pressure
25. Who was responsible for developing a method for measuring horsepower?
    A. Blaise Pascal.
    B. Joseph Bramha.
    C. James Watt.
    D. Daniel Bernoulli.
26. One mechanical horsepower equals _____ watts.
    A. 546
    B. 746
    C. 846
    D. 946
27. Pump displacement times pump rpm equals _____.
    A. pressure
    B. stroke
    C. volume
    D. flow rate
28. One horsepower equals _____.
    A. 550 ft-lbs per second
    B. 650 ft-lbs per second
    C. 750 ft-lbs per second
    D. 850 ft-lbs per second
29. For a fixed amount of horsepower, _____.
    A. if the speed is increased, the torque will increase
    B. if the torque is decreased, the speed will decrease
    C. if the torque is increased, the speed will decrease
    D. torque must equal speed
30. Hydraulic horsepower is a function of _____.
    A. velocity and flow
    B. flow and turbulence
    C. restriction and pressure
    D. pressure and flow

Chapter 3

Symbols, Drawings, and Product Literature

Learning Objectives

After studying this chapter, you will be able to:

- ✓ List the five styles of drawings used by manufacturers.
- ✓ Provide examples where manufacturers use pictorial drawings.
- ✓ Describe the advantage, disadvantage, and common uses of cutaway drawings.
- ✓ Describe where manufacturers use exploded component drawings.
- ✓ Identify the different types of graphic symbols used in manufacturer schematics.
- ✓ Apply an understanding of graphic symbols by accurately drawing a schematic.
- ✓ List examples where manufacturers use combination drawings.

Introduction to Symbols, Drawings, and Product Literature

Manufacturers use multiple types of symbols and drawings to assist personnel in designing, training, testing, servicing, and troubleshooting hydraulic machinery. There are five common types of drawings used by hydraulic personnel:

- Pictorial drawings.
- Cutaway drawings.
- Exploded component drawings.
- Graphic drawings.
- Combination drawings.

Pictorial Drawings

Pictorial drawings use images that depict the actual appearance of hydraulic components. See Figure 3-1. The drawings are useful when installing ***whole goods*** kits. Whole goods kits are attachments purchased by a customer and installed as a kit after the machine has been delivered. Due to their complexity, these kits are often sold through a dealership's sales department, also known as whole goods department, and not through the parts department. Customers often want to add attachments with additional hydraulic features to a used machine or a machine purchased from a different geographical region.

A pictorial diagram can be used by the manufacturer to help technicians conceptualize the component layout when installing the hydraulic

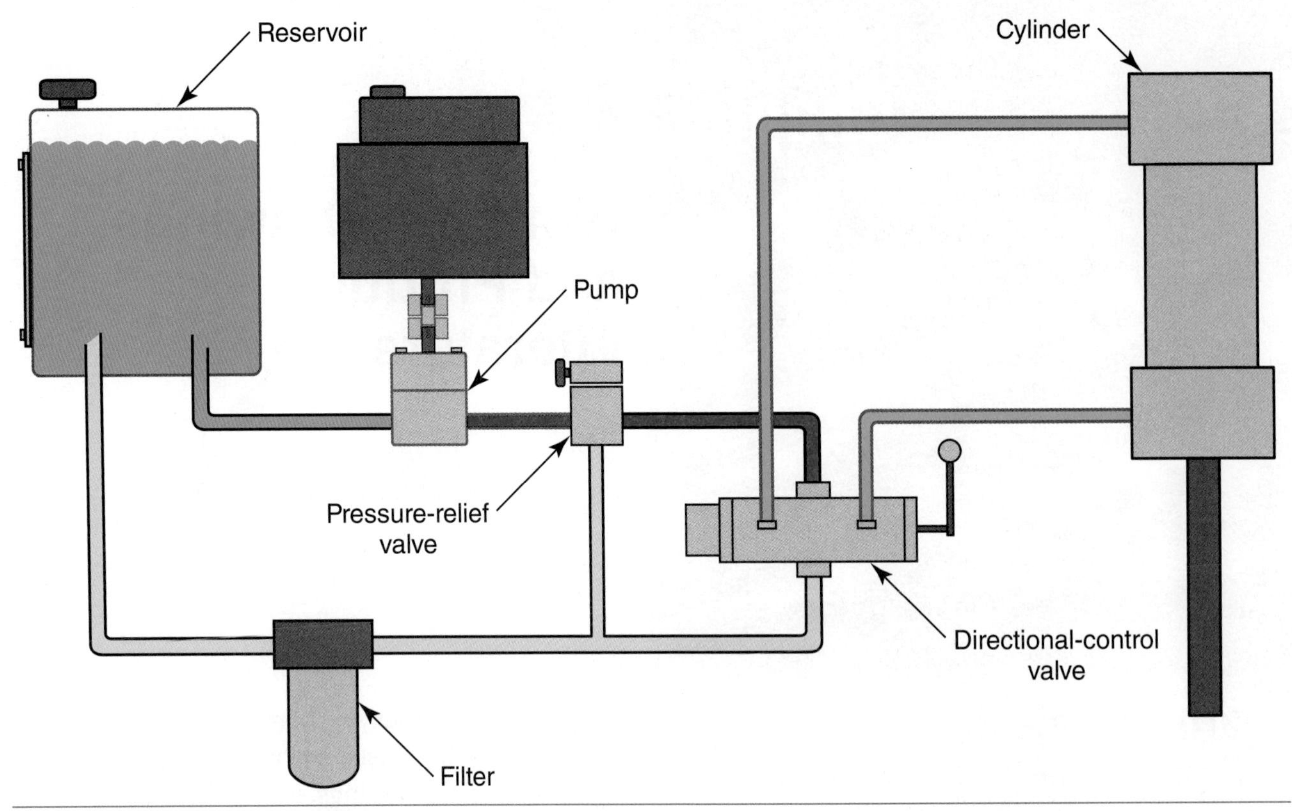

Figure 3-1. A pictorial diagram contains images that resemble the appearance of actual hydraulic components. The drawings are useful to technicians when installing kits.

attachment. Examples of whole good hydraulic attachments are: an automatic header tilt for a combine, a rear power guide axle for a combine, auxiliary hydraulics for a loader, a hydraulic grapple for a loader, or a thumb for backhoe.

Pictorial drawings can also be used for service training, service bulletins, or service letters. ***Service bulletins*** are documents used by manufacturers to distribute technical-update information to dealerships after a product has been released. This additional information is often related to a fix or an update that is not documented in the service manual. ***Service letters*** are the same as service bulletins except they contain much more sensitive information and are sent only to the manufacturer's field representatives so that they can be made aware of short-term solutions or fixes that are not ready to be distributed throughout the dealership network.

Cutaway Drawings

A ***cutaway drawing*** consists of a component that has been virtually sliced in half, or cutaway, for the purpose of viewing its internal parts. See Figure 3-2. Cutaway drawings are limited. The drawing is a single image of a two-dimensional slice of a component and does not depict other parts that are located above or below the slicing plane. When using the drawings for diagnostics, it is easy to make the mistake of forgetting that the component, such as a control valve, has other parts not illustrated in the drawing.

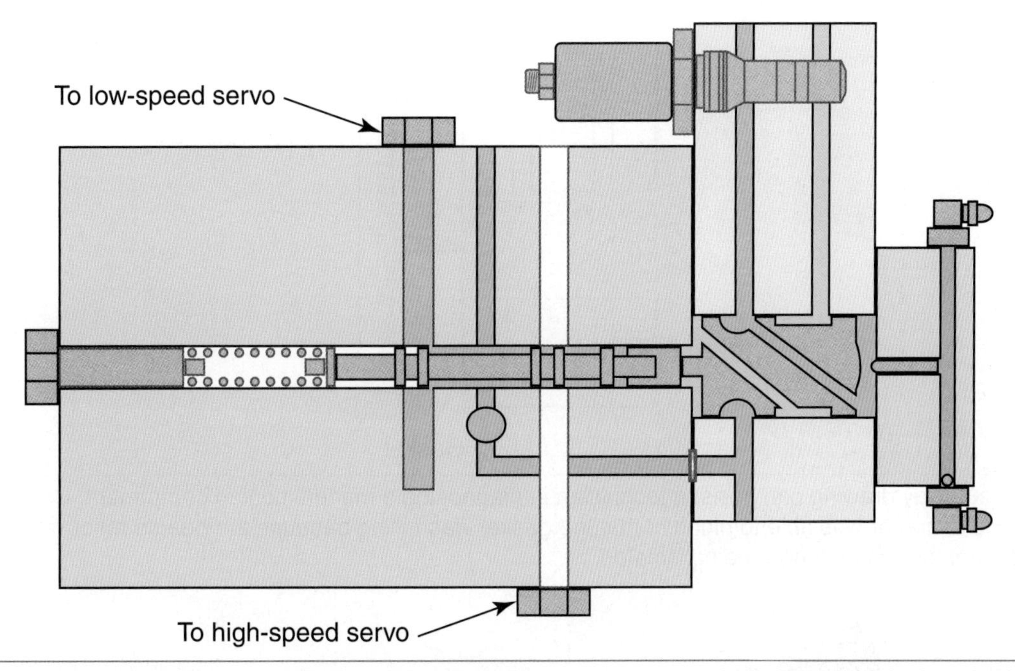

Figure 3-2. A cutaway drawing is a two-dimensional illustration used by manufacturers to depict how intricate parts of a component are positioned and located within the component.

Cutaways are commonly used in training manuals to explain how the parts are supposed to work within the component. They are also used in service manuals and service bulletins.

Case Study

Role of Drawings in Diagnosis

A technician called the manufacturer's hotline regarding a combine with a two-speed hydrostatic motor. The motor's control valve was blowing a gasket each time the operator attempted to propel the machine. A veteran troubleshooter was unable to find any information in the database or service manual. The troubleshooter found a cutaway drawing of the control valve in a training manual. After studying that single drawing for 30 minutes, he was able to call the technician back with a solution to the problem. See Figure 3-3. The needle roller was missing, and as a result the high pressure was no longer isolated from the lower pressure.

Exploded Component Drawings

An ***exploded component drawing*** is most commonly used by the parts department. The drawings illustrate each of the individual parts that are located inside of a component or assembly. See Figure 3-4.

It can be difficult to locate the exact part that needs to be ordered in complex exploded views. Service manuals frequently use other styles of drawings, and those drawings do not easily help an individual find a part in an exploded drawing. A veteran parts person can greatly assist in navigating the exploded component drawings to find the necessary components. Exploded component drawings can also be used in whole good attachment kits and service bulletins.

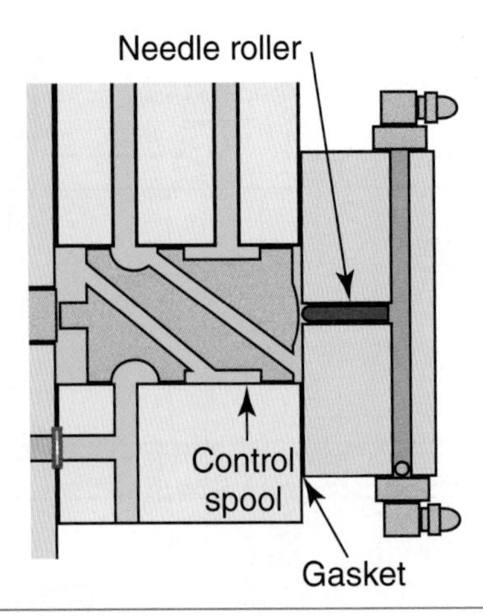

Figure 3-3. A cutaway drawing can assist a technician in diagnosing a malfunctioning component. Using this drawing, a troubleshooter was able to pinpoint that the gasket was failing because the needle roller was missing and high-pressure oil was no longer being isolated.

Figure 3-4. Exploded view drawings illustrate each of the individual parts that are located within a component.

Graphic Drawings

Graphic drawings are also known as schematics. Just like electrical schematics and mechanical power flow schematics, hydraulic schematics use symbols to depict a complete circuit or system. Graphic drawings have the distinct advantage over cutaway drawings in that they illustrate all of the system components and how they are connected or related to the other components within the system.

Schematics are one of the most important tools in the toolbox because they aid the diagnostic process by enabling a person to:

- Trace circuit flows.
- Determine the location of parts within housings and components.
- Determine if the system is open center or closed center, which will be explained in **Chapter 9**.
- Determine what components receive priority oil, which will be explained in **Chapter 25**.
- Determine what circuits have relief valves.

A common question in an interview for a hotline troubleshooting job is, "How do you read a schematic?" The answer is, "Like a road map." Start at the beginning, which is the reservoir, and trace the flow throughout the entire circuit until you reach the destination, which commonly is an actuator, such as a cylinder or motor. See **Figure 3-5**.

Types of Graphic Symbols

Graphic symbols have been used for decades. Several different governing bodies have evolved over those decades, assisting manufacturers by developing common standards and guidelines. Each of the following organizations has influenced the common hydraulic symbols that are used in today's hydraulic schematics:

- Joint Industry Council (JIC).
- American Standards Associations (ASA).
- United States of America Standards Institute (USASI).
- American National Standards Institute (ANSI).
- International Standards Organization (ISO).

Whether in JIC, ASA, USASI, ANSI, or ISO standards or guidelines, all of the symbols are very similar, having much in common. Today, the most common standard used by manufacturers is ISO.

Reservoirs

Hydraulic schematics commonly have reservoir symbols placed throughout a drawing in the same manner ground symbols are placed throughout an electrical schematic. It is convenient and eliminates the need to draw numerous lines snaking all the way back to a common reservoir.

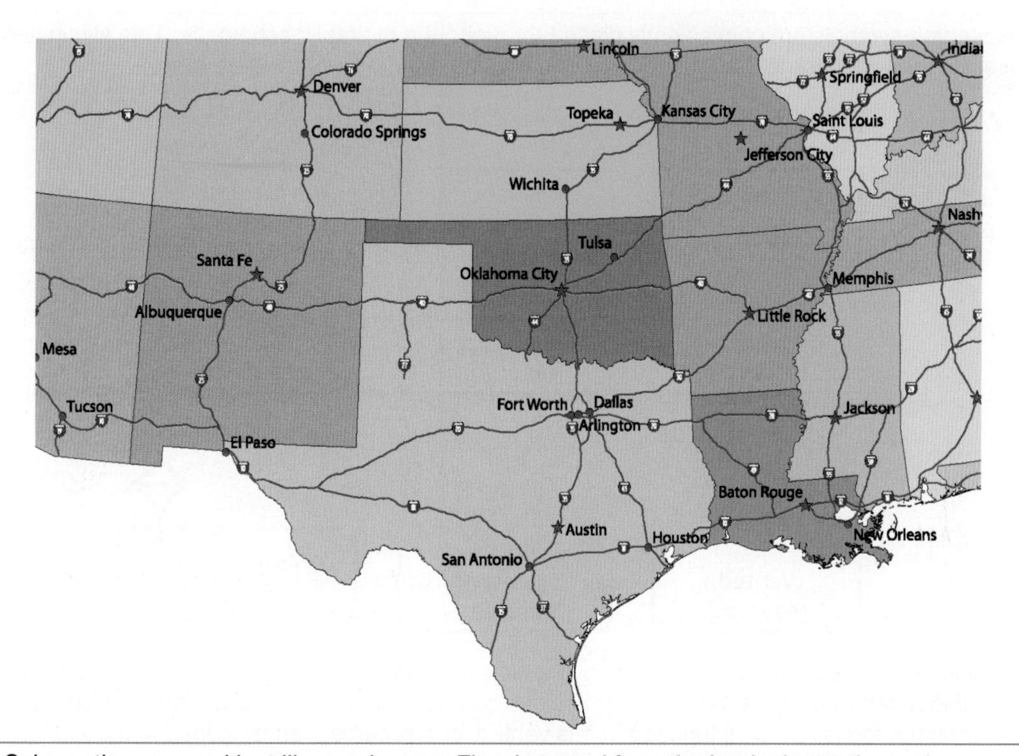

Figure 3-5. Schematics are read just like road maps. Flow is traced from the beginning to the end.

The most common reservoir symbols are found in Figure 3-6. These reservoir symbols depict a vented reservoir, but not everyone uses them that way. Some people will use the symbol for both vented and pressurized reservoirs. The symbol can be placed next to the individual components as needed, similar to a ground symbol in an electrical schematic.

An enclosed shape is used to depict a sealed reservoir, as illustrated in Figure 3-7. When a line is added to the reservoir, the line signifies that the reservoir is pressurized. However, it is similarly easy for a designer to add a breather to the top of the reservoir in order to indicate that the reservoir is vented. **Chapter 14** explains reservoirs in further detail.

Pumps

Figure 3-8 illustrates pump symbols. These symbols are frequently confused with motor symbols. Notice that pump symbols have the triangle pointing out of the circle, which indicates that the component is pushing flow out into the system. If the triangle is empty, it signifies that the component is pumping a gas (pneumatic). For example, an air compressor would be identified with this type of symbol. If the triangle is darkened or filled-in, the component is pumping a liquid (hydraulic).

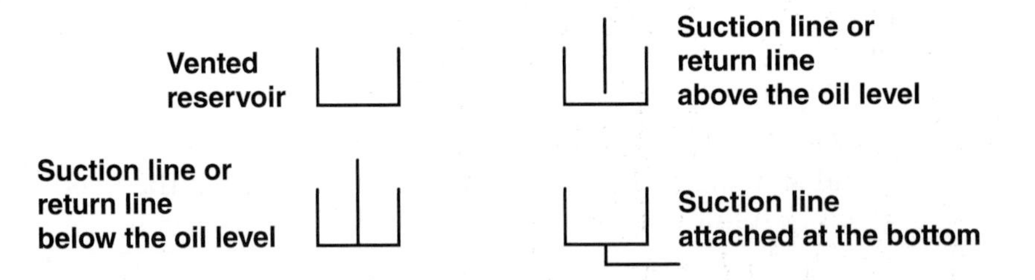

Figure 3-6. Reservoir symbols are conveniently placed throughout a hydraulic schematic. This eliminates the need to retrace hoses all the way back to a common reservoir, making the drawing easier to read.

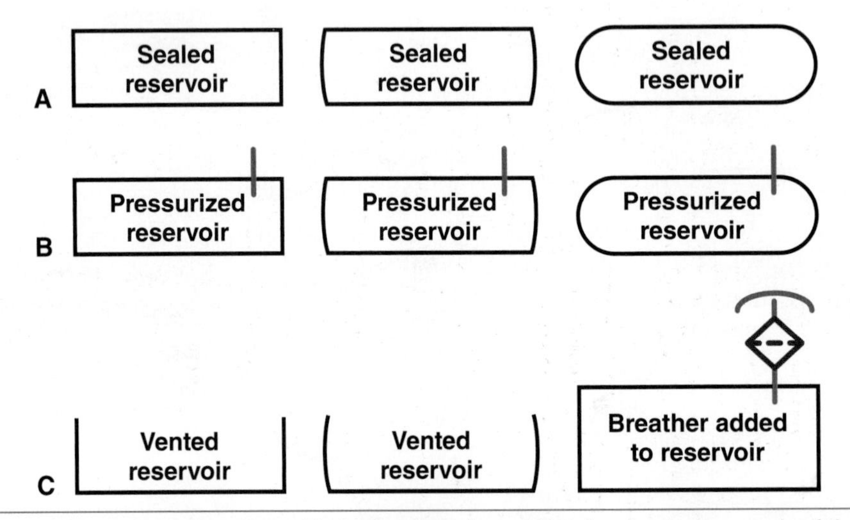

Figure 3-7. Enclosed reservoirs are also placed in hydraulic schematics. A—As drawn, each of the symbols represents a sealed reservoir. B—A short line segment can be added to the symbol to indicate a pressurized reservoir. C—When the top is removed the reservoir is vented, or a breather can be added to the enclosed box to indicate a vented reservoir.

The simplest and most common hydraulic pump is the fixed-displacement unidirectional pump. This pump is commonly used for implement hydraulics. If the circle contains two triangles, the pump is reversible. This is the common symbol for a hydrostatic pump. If the symbol contains an arrow placed diagonally across the circle, the pump's displacement is variable. **Chapter 4** explains pump fundamentals.

Motors

Motor symbols have the triangle pointing into the circle, indicating that it must receive flow in order to rotate. See Figure 3-9. Motors can be unidirectional, bidirectional, fixed-displacement, or variable-displacement. **Chapter 5** explains motor fundamentals.

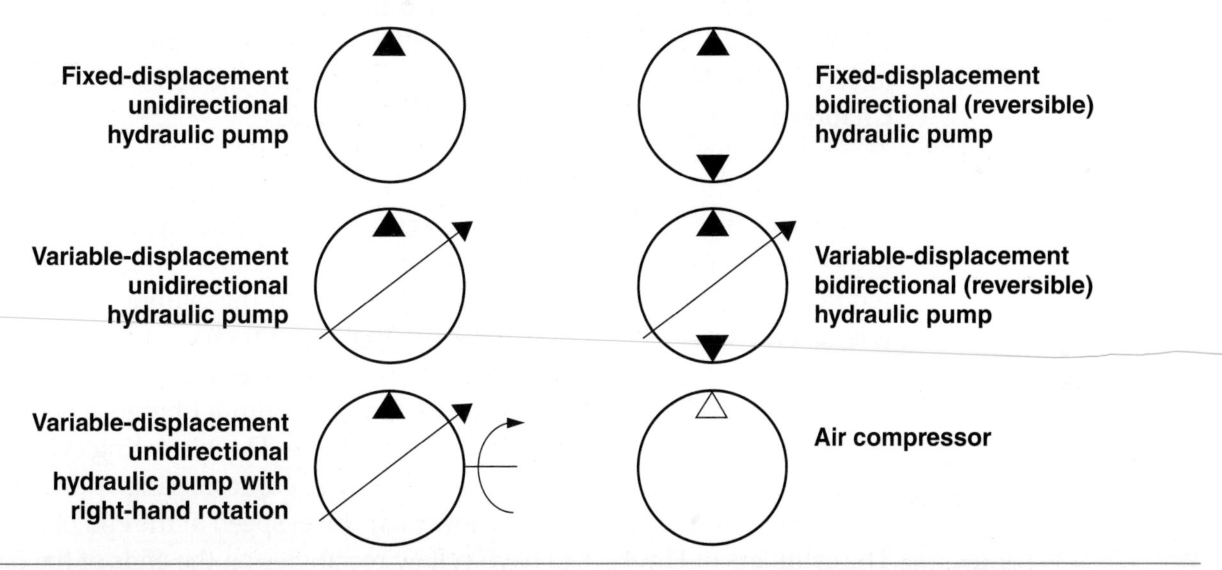

Figure 3-8. Pump symbols have the triangle pointing out of the circle to depict that the component is generating fluid flow. If the triangle is empty, the symbol represents air or gas (pneumatic pump), and if the symbol is filled-in, it represents liquid (hydraulic pump).

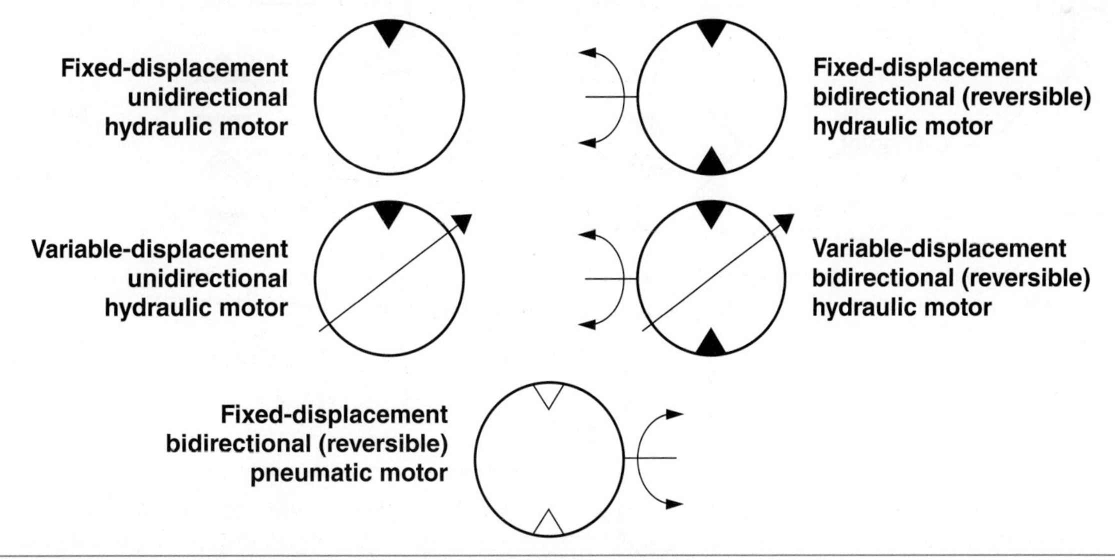

Figure 3-9. Motor symbols have the triangles pointing into the circle to depict that the motor must receive flow in order to rotate. Motors can be fixed-displacement or variable-displacement, and unidirectional or bidirectional.

Coolers

Hydraulic temperature-related devices, such as coolers, heaters, and heat exchangers, are illustrated in Figure 3-10. Notice that the triangles point in the direction of the heat. When the heat needs to be removed or cooled, the arrows point out of the box. When the oil needs to be heated, such as a heater, the triangles point into the center of the box. If the device has triangles pointing inward and outward, it depicts a heat exchanger (also known as a temperature controller). Almost all mobile hydraulic systems require the use of an oil cooler. Coolers will be further explained in **Chapter 14**.

Filters and Strainers

The symbol in Figure 3-11 can be used to depict multiple types of filtration devices. The location of the symbol within the circuit can help a person guess whether the device is being used as a filter, screen, or a strainer. It is impossible to be certain of the type of filtration device, unless the schematic identifies it. **Chapter 11** will further explain filtration devices.

Cylinders

Figure 3-12 illustrates several types of hydraulic cylinder symbols. If the cylinder only has one port or hose, it is considered single-acting, which requires an outside force to retract the cylinder. If the cylinder has two ports, it is considered double-acting, meaning that oil pressure is used to extend the cylinder and oil is also used to retract the cylinder. If a double-acting cylinder only has one rod, it can be labeled a differential cylinder because of the difference in piston areas, rod-side versus non-rod side. Double-acting cylinders can have two rods. Such cylinders are commonly used in steering applications.

Cylinders can be designed to slow their travel speed at the end of the stroke. The cylinders in Figure 3-13 have yellow rectangles on the ends of the piston to represent dampers that cushion the cylinder's end of stroke. The cylinder can be cushioned at the end of extension or at the end of retraction. The cushion speed can be fixed or variable. **Chapter 6** will further explain hydraulic cylinders.

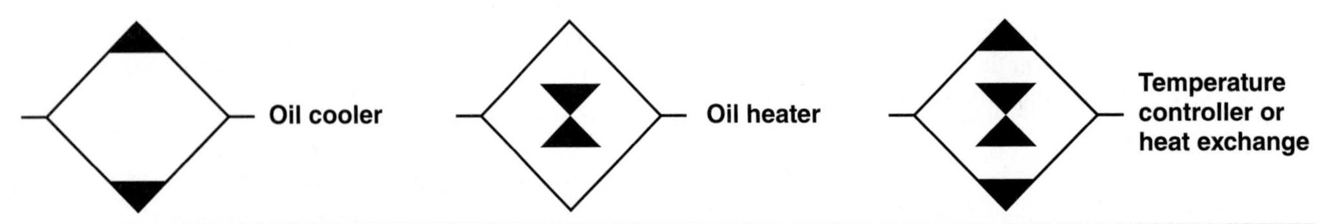

Figure 3-10. Triangles can be used to depict the direction of heat, such as pointing outward to cool the oil, pointing inward to heat the oil, and pointing in both directions to illustrate an exchange of heat.

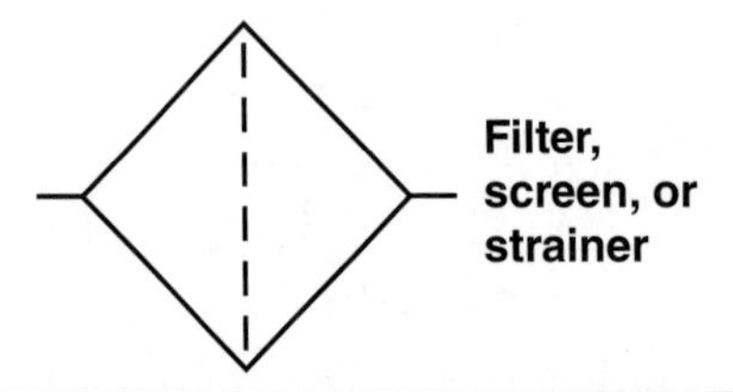

Figure 3-11. Filtration symbols can be used to depict a filter, screen, or a strainer.

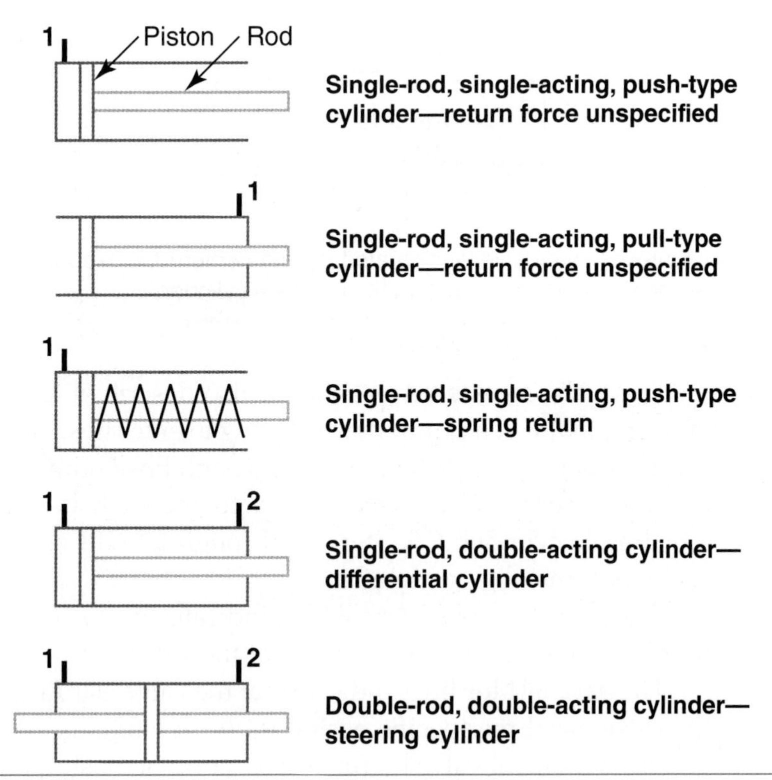

Figure 3-12. Single-acting cylinders are drawn with one hose ported on the cylinder, and double-acting cylinders are drawn with two hoses ported on the cylinder.

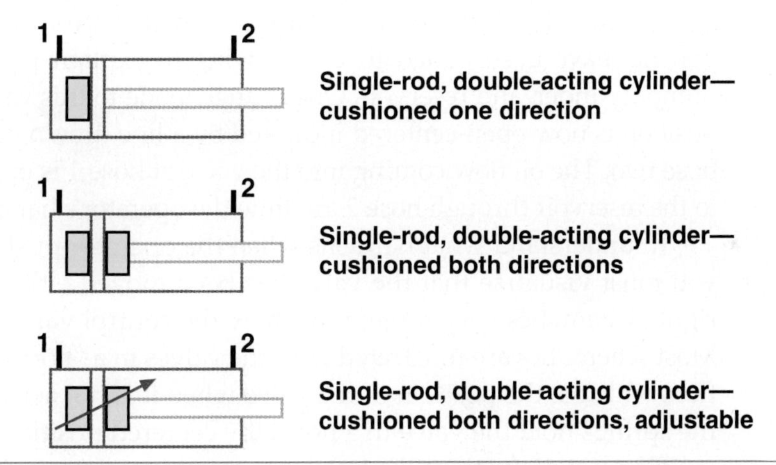

Figure 3-13. Cushions are added to cylinders to slow the travel speed as the pistons approach the end of their travel. They are depicted by the yellow rectangles on the pistons.

Control Valves

Chapters 7, **8**, and **9** will explain pressure-control valves, flow-control valves, and directional-control valves respectively. **Figure 3-14** illustrates a manual or hand-lever–actuated directional-control valve. The valve has three positions, which have been highlighted in green, yellow, and light purple. Position B (yellow) is the neutral position because the cylinder is not activated when the valve is in this position. When the operator's hand is off the lever, the orange springs hold the valve in this position.

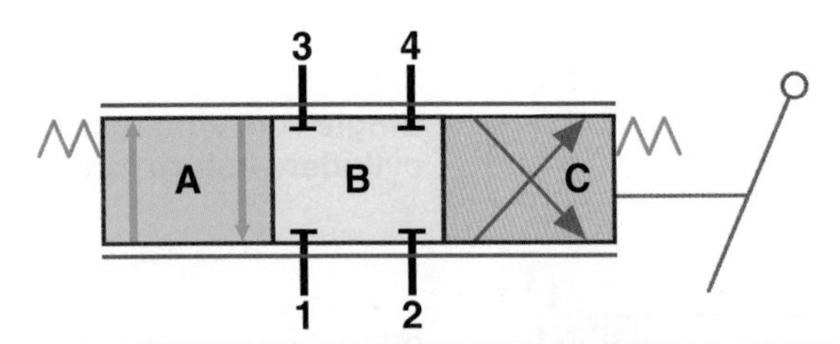

Figure 3-14. A three-position, four-way, manual lever-operated, spring-centered, closed-center, directional-control valve is commonly used to operate a double-acting actuator, such as a cylinder.

This valve's centered position is also described as a closed-center design, meaning that when pump flow is trying to enter hose number one, it is blocked and cannot return to the reservoir through hose number two. Positions A and C are commonly used to control an actuator, such as extending and retracting a hydraulic cylinder. **Chapters 16** through **20** will explain open- and closed-center designs in great detail.

The valve has four ports which indicates it is a "four-way valve." Note the four hoses plumbed into the center of the valve.

The parallel blue lines indicate that the valve is infinitely variable. Although the valve has three specific positions, the operator can infinitely vary the position of the control valve by further opening or closing the valve from the neutral position to either of the outside blocks. The infinitely variable position allows the operator to increase or decrease the flow, sometimes described as feathering.

Figure 3–15 illustrates a similarly designed control valve. A few more components have been added to the drawing to explain the valve's operation: a pump, cylinder, and reservoir. One change made to this valve is that the neutral position is now open-centered indicated by a line drawn between hose one and hose two. The oil flow coming into the valve at hose 1 is open and free to return to the reservoir through hose 2 anytime the operator's hand is off of the lever.

To understand what happens when the operator pushes or pulls the lever, you must visualize that the valve has been moved either to the left or to the right. Schematics very rarely will show the control valve in an actuated state. Most schematics are illustrated with the valves in a "normal" or "neutral" position, which is the position of the valve when it is not activated. In Figure 3-15, the springs hold the valve in a normally centered position.

Figure 3-16 further illustrates how position A (green) of the valve is used to extend the cylinder and how position C (purple) is used to retract the cylinder. Because schematics are usually drawn in the neutral position, you must conceptualize the valve moving. This is the same way an electrical switch is represented in an electrical schematic.

Valves are commonly configured with two, three, or four positions. However, it is possible to have more positions. Figure 3-17 illustrates a two-position, two-way, solenoid-operated valve. In its normal state, it is blocking oil flow, because the spring pressure forces the valve to the right. The solenoid is highlighted in yellow. When the solenoid is energized, the solenoid pushes the valve to the left, allowing oil to flow through the valve. Notice that this valve is not infinitely variable, but has two fixed positions, open and closed.

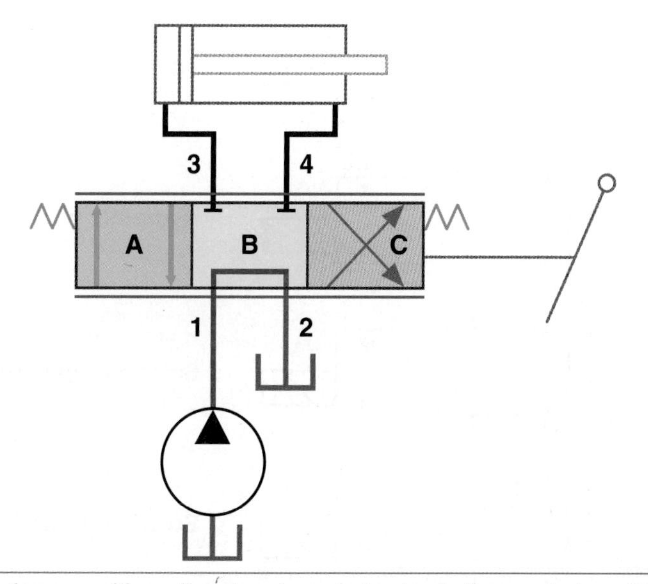

Figure 3-15. An open-center, three-position, directional-control valve in the neutral position.

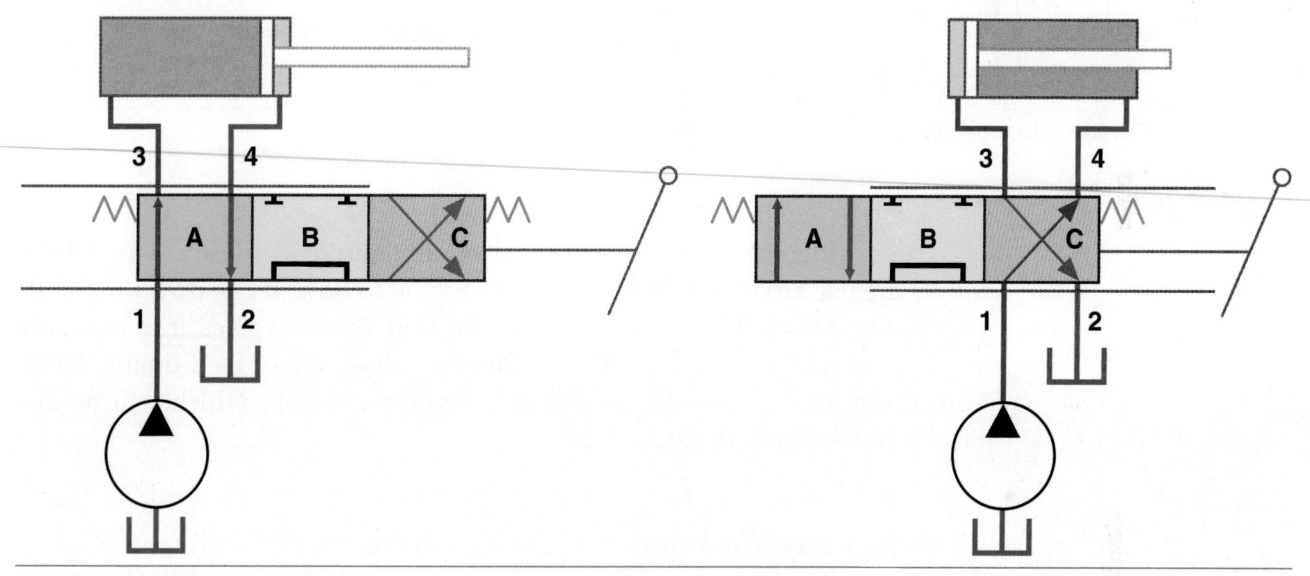

Figure 3-16. A three-position, directional-control valve is used to extend and retract a double-acting hydraulic cylinder.

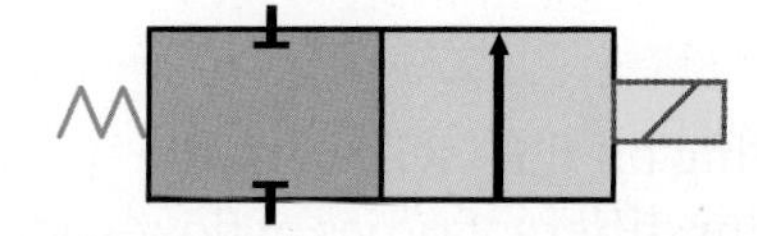

Figure 3-17. A two-position, solenoid-operated control valve will either fully block oil flow or fully allow oil flow.

Figure 3-18 illustrates different methods for actuating a control valve. The valves can be pilot-actuated, by either air pressure or hydraulic pressure. Pilot controls work on the same principle as an electronic transistor, which uses a small amount of electricity to control a larger amount of electricity. Pilot controls will be further explained in **Chapter 20**. Solenoid-operated valves can be variable or non-variable. Lever-operated valves can also be detented so that the spool can be held in a fixed position, allowing the operator to take his or her hand off the lever.

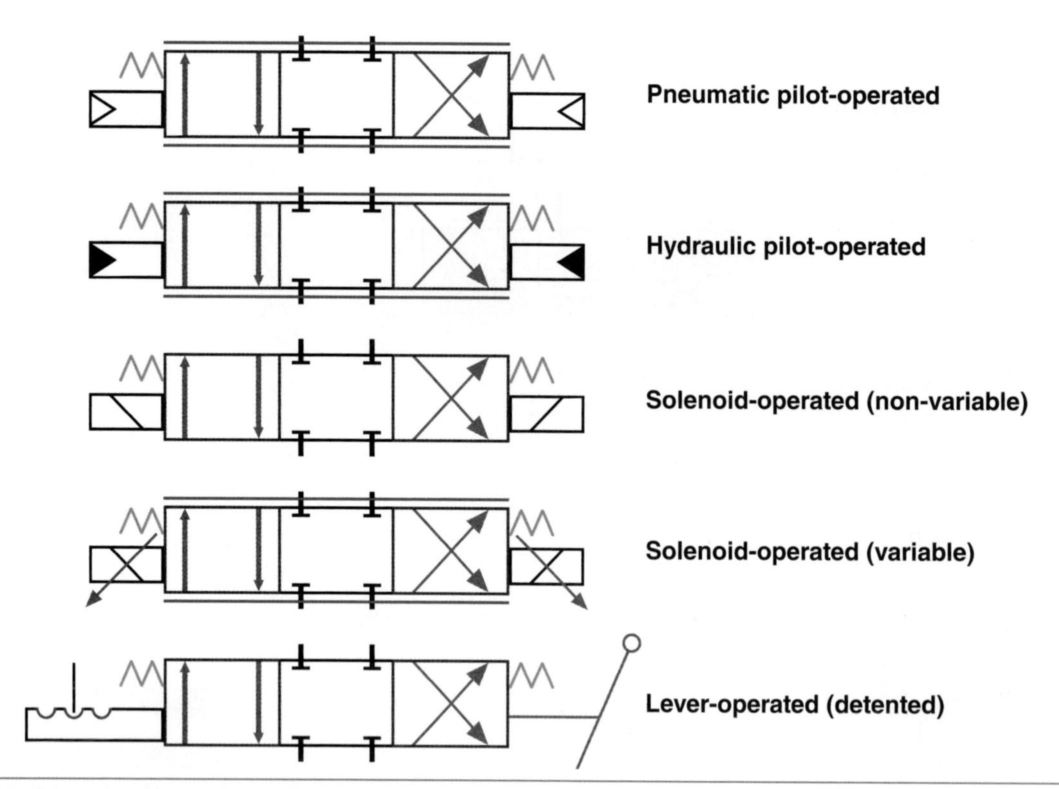

Figure 3-18. Control valves can be pilot-, solenoid-, or lever-operated.

Check Valves

Three different types of check valves are found in Figure 3-19. Check valves are used to direct oil flow by allowing oil to flow in one direction while preventing it from flowing in the other direction. Bypass valves are commonly used in oil filter and cooler circuits. Shuttle valves, also called double check valves, are used in variable displacement pump controls, which will be discussed in **Chapters 18** through **20**.

Orifices

Orifices, shown in Figure 3-20, are some of the most basic oil control valves found in hydraulic systems. These valves are non-pressure-compensated and are used to control flow and pressure.

Flow-Control Valves

Controlling oil flow will vary the speed of hydraulic actuators. Figure 3-21 illustrates four different types of flow-control valves. Notice the valves are commonly two-port and three-port, which signify the number of hoses connected to them. The valves can be adjustable, and might be pressure-compensated or temperature-compensated.

Pressure-Control Valves

Two of the most common styles of pressure-control valves are the pressure-relief valve and the pressure-reducing valve. See Figure 3-22. A pressure-reducing valve is normally open, while a pressure-relief valve is normally closed. The valves can have a fixed spring value or it can be adjustable. **Chapter 7** will explain pressure-control valves.

Figure 3-19. Check valves allow oil to flow in one direction while preventing it from flowing in the other direction.

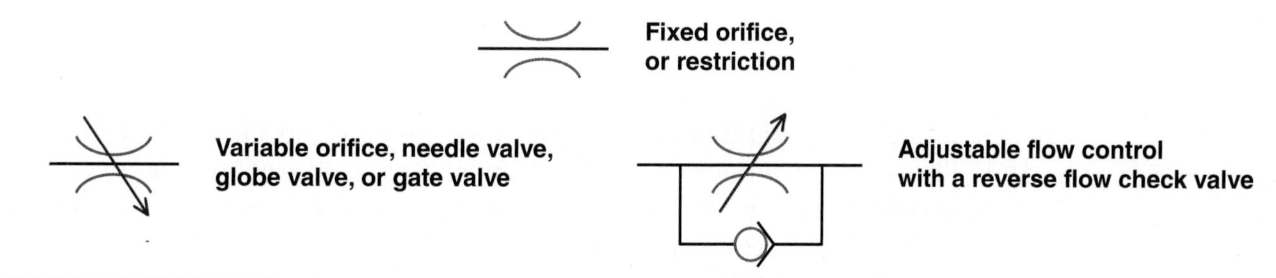

Figure 3-20. Fixed and variable orifices are used to control flow and pressure.

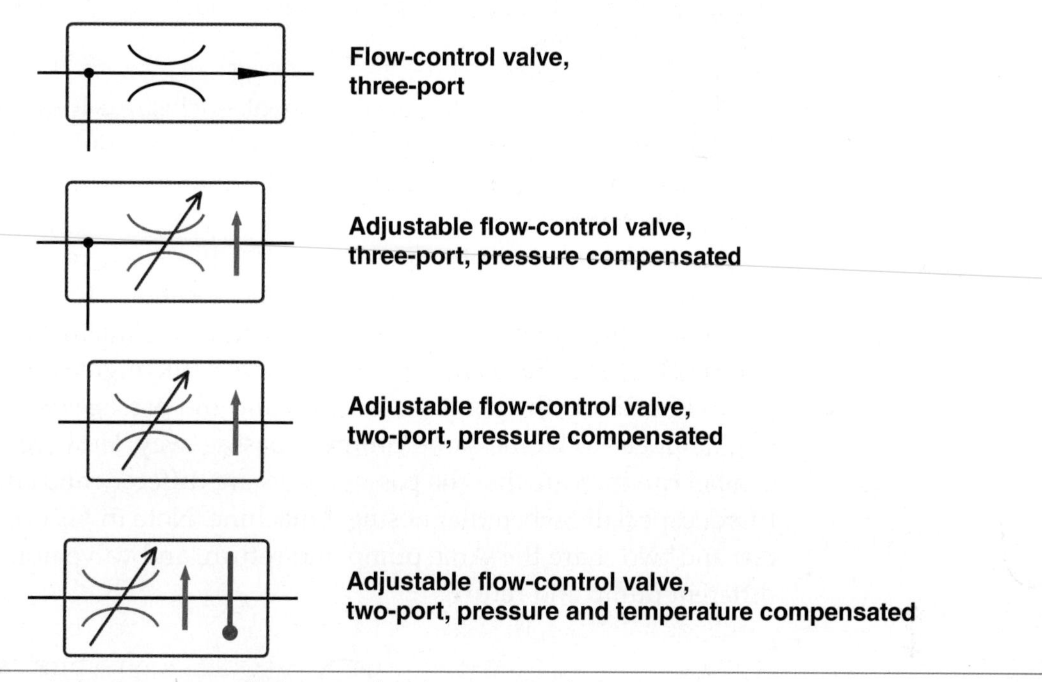

Figure 3-21. Flow control valves vary cylinder speed and motor speed.

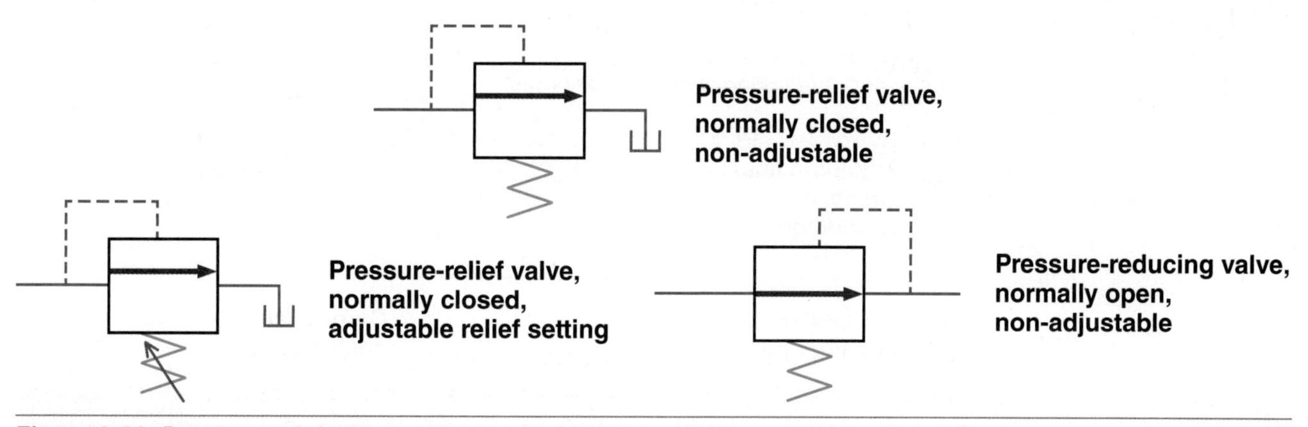

Figure 3-22. Pressure-relief valves and pressure-reducing valves are common types of pressure-control valves used in hydraulic systems.

Accumulators

Accumulators store fluid energy in the form of fluid pressure in the same way a capacitor stores electrical energy in an electrical circuit. Figure 3-23 illustrates the three main styles of accumulators: spring, weighted, and gas. Weighted accumulators are not commonly found in mobile applications. **Chapter 13** will explain accumulators in detail.

Additional Schematic Symbols

Schematics at times can become quite busy and hard to follow. Figure 3-24 illustrates a few more necessary symbols that help explain how to navigate drawings. Some schematics are drawn strictly in black-and-white, which makes it difficult to trace conductors.

Note

If the conductors are connected, a dot is commonly used to join the two lines.

A dashed box indicates a component enclosure. It is common for one housing, such as a pump or a valve, to contain multiple individual parts. A component enclosure is used to illustrate that all of those components are enclosed in one housing.

An "X" at the end of a line indicates a place where a passage has been plugged. Sometimes these ports are good locations for installing pressure gauges. At other times, it is an important indication that the passageway stops, as in a valve bank. See Figure 3-25. The valve bank might contain three or more valves bolted together as one assembly, and the passageways from a distance might appear to be the same common passageway. However, a plugged port symbol can indicate that the passageways are different and should not be confused, especially when diagnosing a machine. Note in Figure 3-25 that valves one and two share the same pump and return, and valve number three uses a different pump and return.

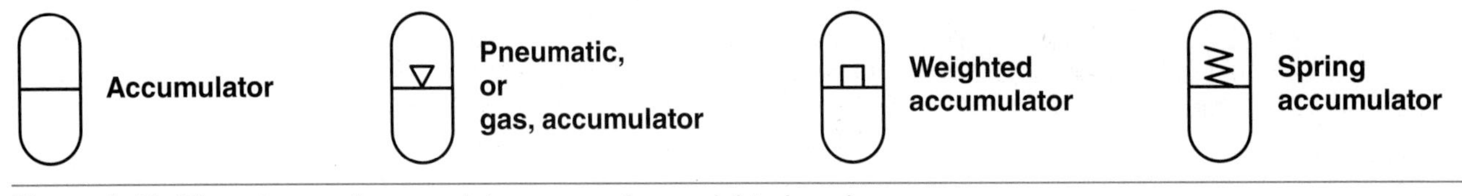

Figure 3-23. Three types of accumulators are spring, weighted, and gas.

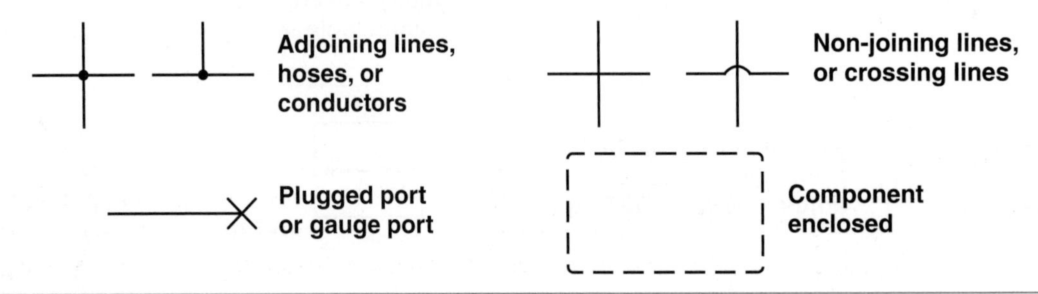

Figure 3-24. Additional symbols that are commonly used in hydraulic schematics are crossing lines, a plugged port, and a component enclosure.

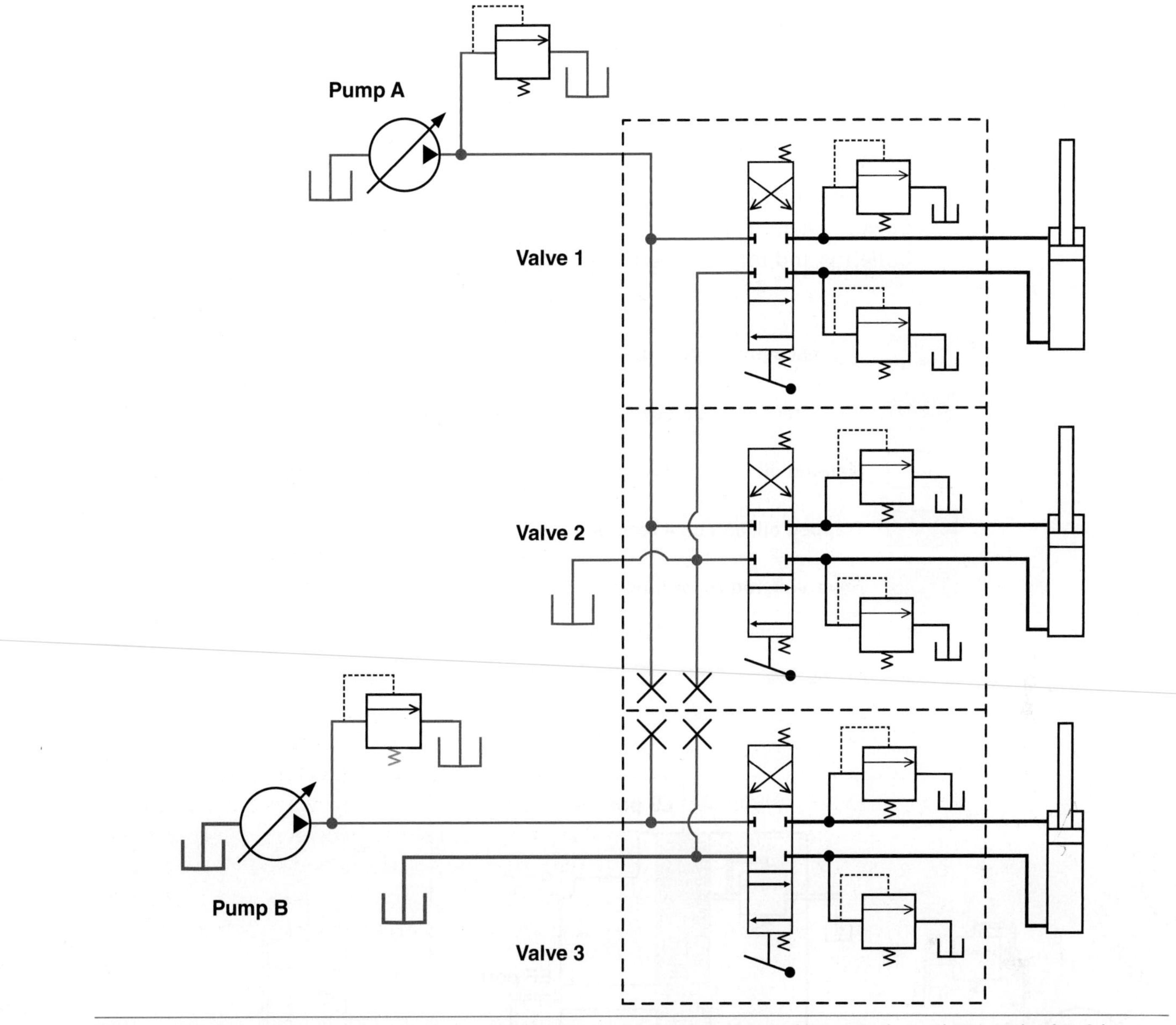

Figure 3-25. Pump A is responsible for supplying oil to valve 1 and valve 2. Also note that valve 1 and valve 2 have their own individual return passageway. The passageways between valve 2 and valve 3 are plugged to isolate valve 3. Pump B is responsible for supplying oil to valve 3, and valve 3 has its own individual return to the reservoir.

Colors

Schematics often have color codes. Not all schematics follow the same color conventions. Red is the color most commonly used to indicate pump outlet, high pressure, or the system's highest pressure. Orange is sometimes used to indicate a pressure lower than the main system high pressure; one example would be charge pump pressure. A charge pump is a lower pressure pump used to supercharge the inlet of another pump. Yellow is sometimes used to depict metered oil, such as servo pressure found in a hydrostatic transmission. Blue is often used for reservoir, drain, or suction pressure; however, green is also sometimes used for reservoir, drain, or suction pressure. Green can also be used to depict trapped oil pressure. See Figure 3-26.

Combination Drawings

Other drawings are a combination of the cutaway, graphic, pictorial, and exploded drawings, and are appropriately named ***combination drawings***. Figure 3-27 illustrates a cutaway drawing incorporated into a traditional schematic.

Combination drawings are used to detail a particular portion of a fluid power system. The two most popular uses of combination drawings are service bulletins and for training technicians on new machines.

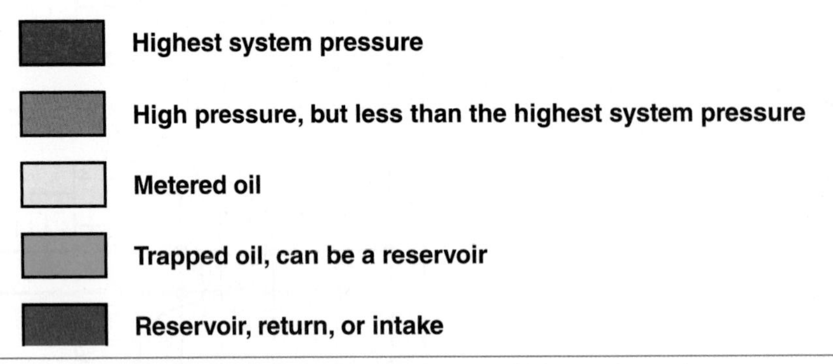

Figure 3-26. Colors are used to depict different levels of pressure within a schematic.

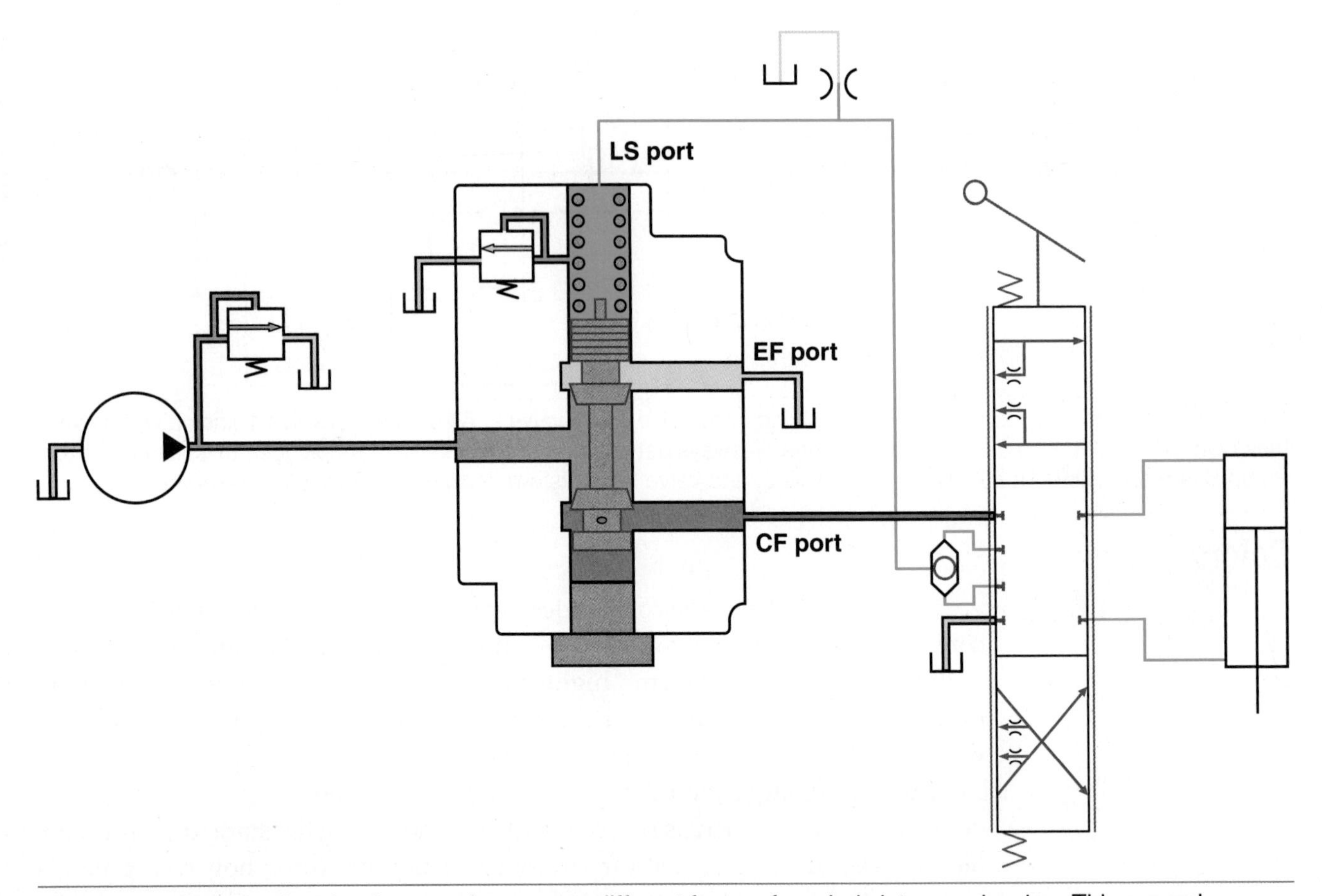

Figure 3-27. Combination drawings can incorporate different forms of symbols into one drawing. This example includes a cutaway drawing and traditional graphic symbols.

Summary

- ✓ Manufacturers use multiple styles of drawings to help service personnel maintain and repair mobile machinery.
- ✓ Pictorial drawings can be helpful for installing hydraulic attachments.
- ✓ Cutaway drawings are two-dimensional, illustrating just one slice of a component.
- ✓ Graphic drawings (schematics) enable technicians to conceptualize the parts that are located in each of the components, allowing technicians to trace flow and determine priorities of oil flow.
- ✓ Combination drawings incorporate more than one style of symbol and are sometimes found in service bulletins and training manuals.

Technical Terms

combination drawings
cutaway drawing
exploded component drawing
graphic drawings
pictorial drawings
service bulletins
service letters
whole goods

Review Questions

Answer the following questions using the information provided in this chapter.

1. Which of the following is the most common style of drawing used by technicians?
   A. Pictorial.
   B. Graphic.
   C. Cutaway.
   D. Exploded view.
2. Which of the following styles of drawings has the disadvantage of being limited to a single two-dimensional view, leaving out other parts within a component?
   A. Pictorial.
   B. Graphic.
   C. Cutaway.
   D. Exploded view.
3. Which of the following is the most common drawing style used by parts personnel?
   A. Pictorial.
   B. Graphic.
   C. Cutaway.
   D. Exploded view.

For Questions 4 through 6, match the description with the appropriate symbol in the following drawing.

4. Non-variable solenoid-controlled valve. _____
5. Pneumatic pilot-actuated valve. _____
6. Hydraulic pilot-actuated valve. _____

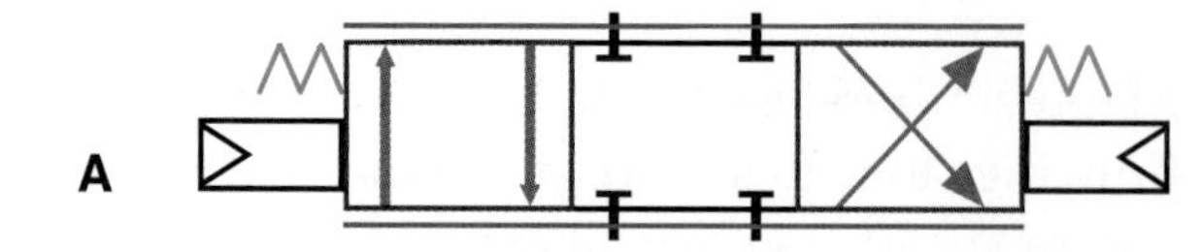

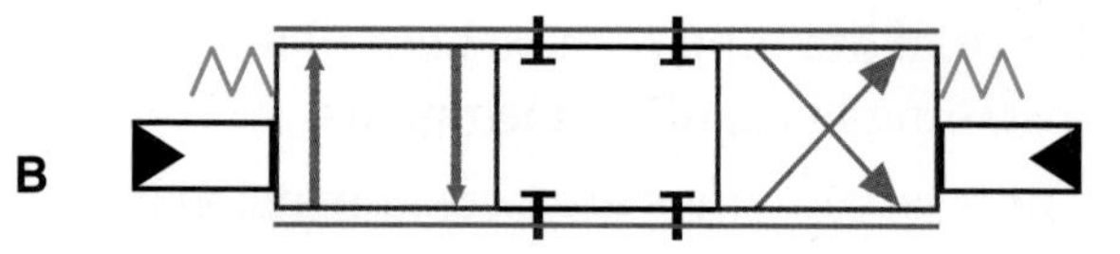

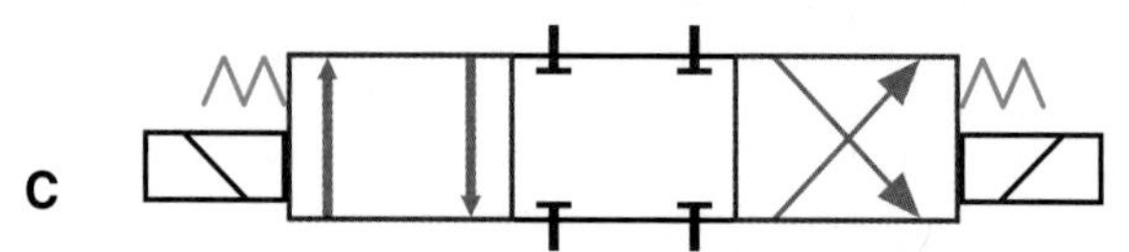

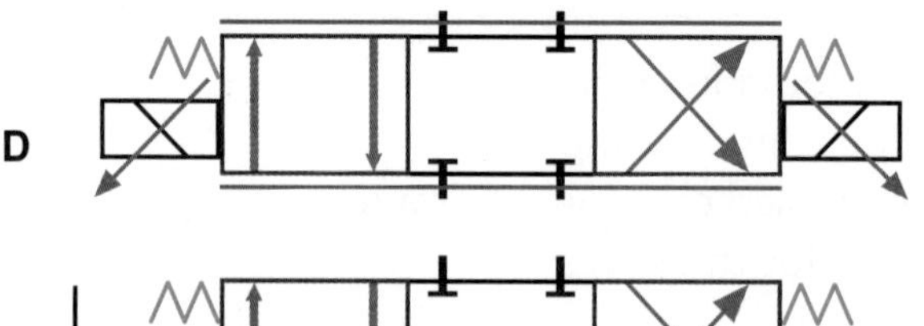

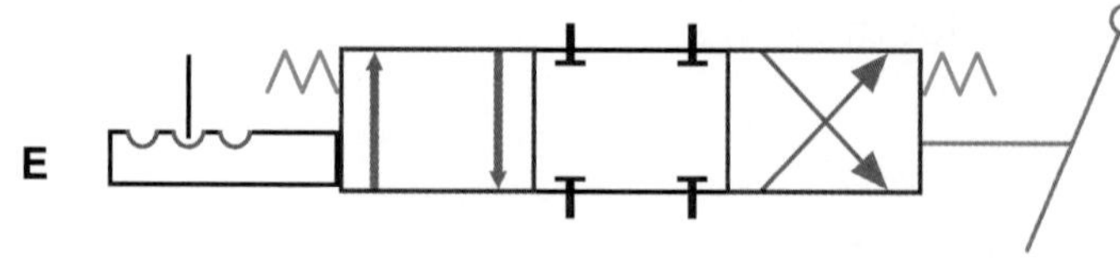

For Questions 7 through 10, match the description with the appropriate symbol in the following drawing.

7. Reversible fixed-displacement hydraulic pump. _____
8. Reversible variable-displacement hydraulic pump with right-hand rotation. _____
9. Unidirectional fixed-displacement hydraulic pump. _____
10. Air compressor. _____

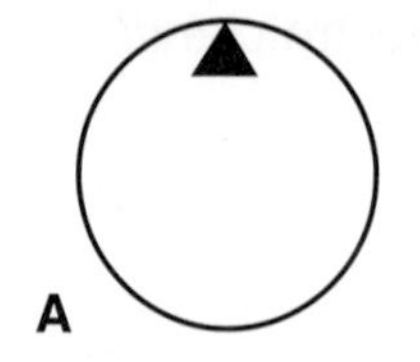

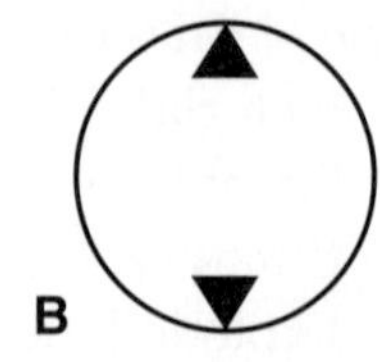

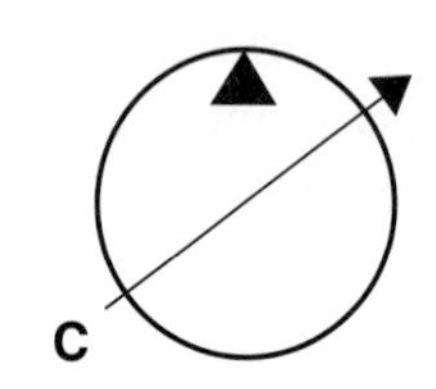

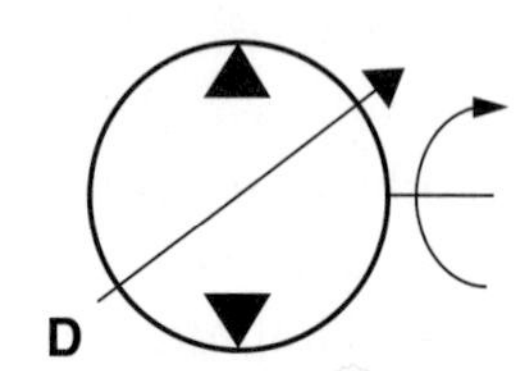

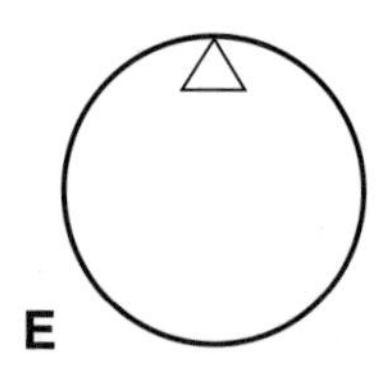

For Questions 11 through 13, match the description with the appropriate symbol in the following drawing.

11. Unidirectional variable-displacement hydraulic motor. _____
12. Unidirectional fixed-displacement hydraulic motor. _____
13. Fixed-displacement reversible pneumatic motor. _____

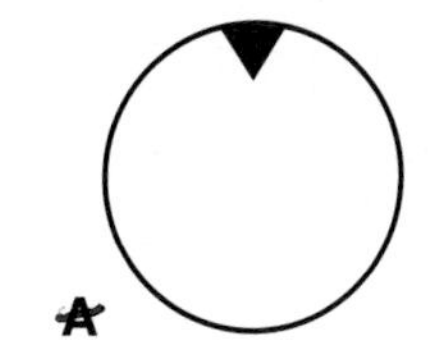

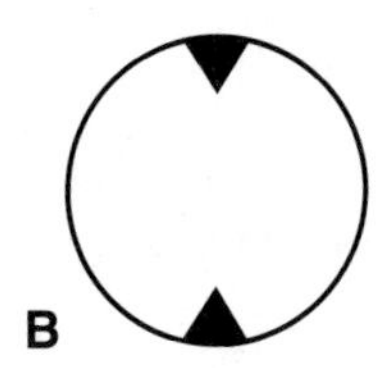

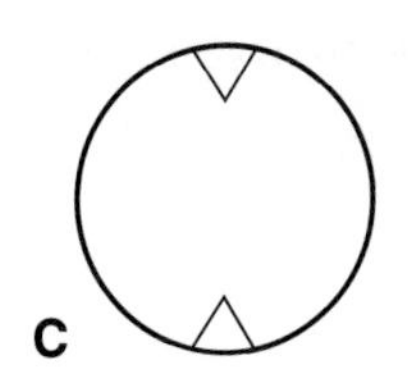

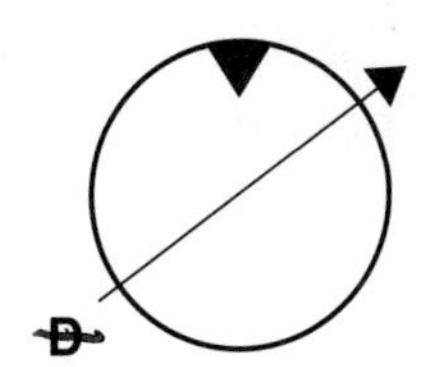

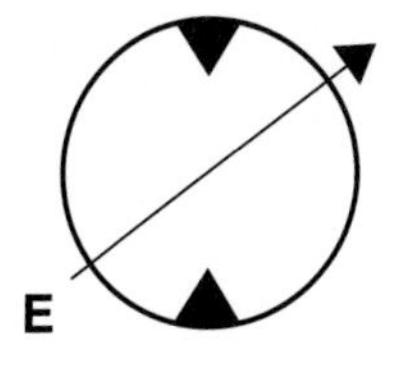

For Questions 14 through 17, match the description with the appropriate symbol in the following drawing.

14. Heat exchanger/temperature controller. _____
15. Oil heater. _____
16. Oil cooler. _____
17. Filter/strainer. _____

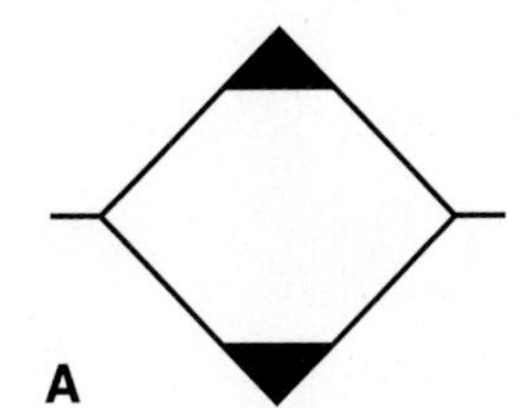

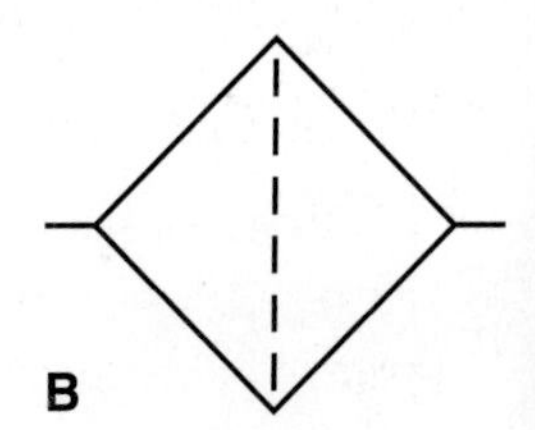

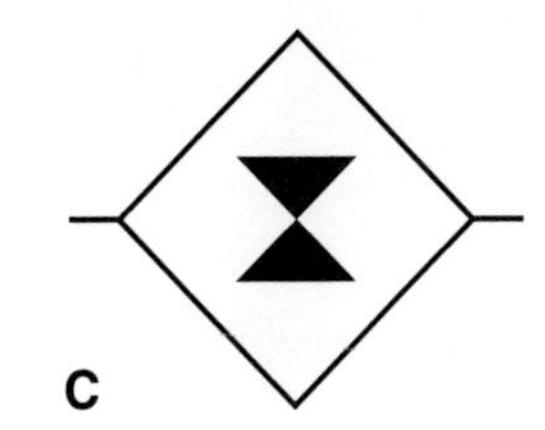

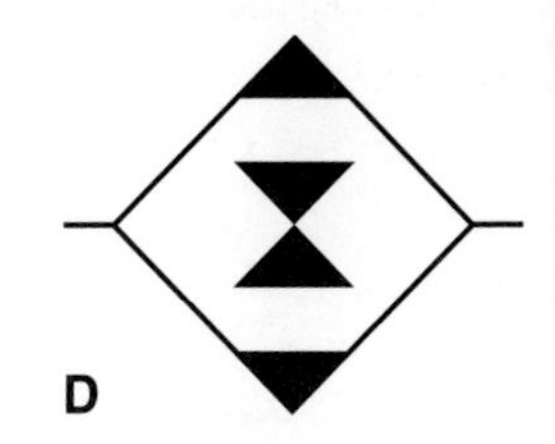

For Questions 18 through 22, match the colors used in schematics with the appropriate descriptions.

18. Signifies the highest system pressure. _____
19. Signifies the second highest system pressure. _____
20. Signifies trapped oil. _____
21. Signifies metered oil. _____
22. Signifies the reservoir, return, or suction pressure. _____

A. Blue.
B. Green.
C. Red.
D. Orange.
E. Yellow.

Chapter 4

Pumps

Learning Objectives

After studying this chapter, you will be able to:

- ✓ Explain foundational principles of hydraulic pumps relating to displacement, rotation, and pump inlets.
- ✓ Identify the differences between external-toothed and internal-toothed gear pumps.
- ✓ List attributes that are unique to gear pumps.
- ✓ Explain the different types of vane pumps.
- ✓ List attributes that are unique to vane pumps.
- ✓ Explain the differences in piston pump designs.
- ✓ List attributes that are unique to piston pumps.
- ✓ List the different attributes that must be considered when choosing a pump.

Introduction to Pumps

A hydraulic pump is the heart of a hydraulic system. The pump's primary purpose is to supply flow. Pressure occurs when the pump's flow meets resistance, usually in the form of a load, such as a hydraulic cylinder lifting a load. The resistance can also be the result of restricting the oil flow, for example by using an orifice.

Pumps can be classified into several different categories:

- Positive and non-positive displacement.
- Fixed displacement and variable displacement.
- Unidirectional and reversible.
- Open loop and closed loop.
- Gear, vane, and piston.

Non-Positive-Displacement Pumps

A ***non-positive-displacement pump*** works on the principle of centrifugal force. It uses an impeller that is driven by an input shaft. The impeller does not have a tight sealing surface within the pump body. As the impeller is driven, it develops flow.

A common application is an internal combustion engine's coolant pump. See **Figure 4-1**. The pump's flow will vary depending on the system pressure, and that is why the pumps are known as non-positive-displacement pumps. These pumps, also known as centrifugal pumps, are incapable of building high pressure if the flow encounters high resistance. As system pressure increases, the pump flow will slip at the impeller, resulting in less flow or potentially no flow. For example, an engine's water pump uses a thermostat to block the coolant's flow until the coolant reaches operating temperature. The pump continues to turn, but there is no flow developed. In contrast, if a positive-displacement pump's flow was blocked, a relief valve would be needed to prevent the pump from seizing or the weak part of the system from bursting. Two other examples of non-positive-displacement pump applications are a household sump pump and a windshield wiper washer pump. However, windshield washer pumps can also be positive-displacement pumps.

Most mobile hydraulic systems do not use non-positive-displacement pumps. In a few rare applications, a non-positive-displacement pump can be used as a ***charge pump***, which supercharges the main pump's inlet with oil flow. However, most mobile charge pumps are positive-displacement pumps.

Positive-Displacement Pumps

A ***positive-displacement pump*** uses tight sealing surfaces, which provides the foundation for the pump to displace a definite volume of oil. The flow varies very little when it encounters pressure. For every cup of oil that comes into the pump's inlet, the majority of that cup of oil will flow out of the pump.

Figure 4-2 is an example of a pump flow and pressure curve that is commonly provided by pump manufacturers. The graph's vertical (Y) axis lists the

Figure 4-1. An internal combustion engine's coolant pump is an example of a non-positive-displacement pump.

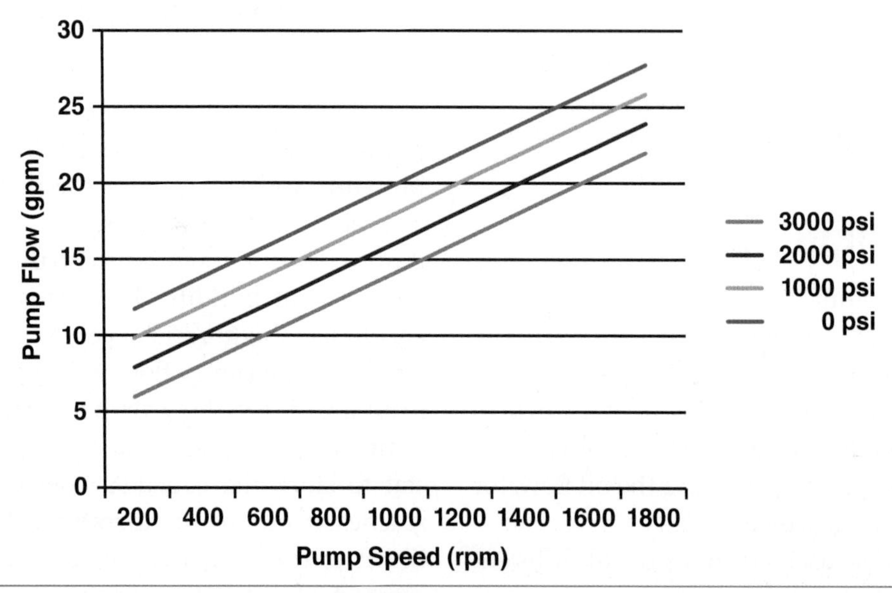

Figure 4-2. Pump suppliers provide machine manufacturers with a graph that illustrates the pump's flow based on system pressures. The vertical (Y) axis lists pump flow in either gpm or lpm. The horizontal (X) axis lists pump rpm.

pump's flow, which can be in units of gpm or lpm. The horizontal (X) axis lists the pump's rpm. The graph illustrates several key points regarding positive-displacement pumps:

- Pump flow is proportional to the pump's rpm.
- Pump rpm has the largest effect on pump flow.
- Pump flow is not greatly affected by pressure.
- Pressure has some effect on pump flow, but the pump's positive displacement is still displacing a definite volume for a given system pressure.
- It takes a substantial amount of system pressure to cause a drop in pump flow.

A non-positive-displacement pump's flow, on the other hand, is greatly affected by pressure. A substantial increase in system pressure can cause a centrifugal pump's flow to drop to zero flow.

Figure 4-2 includes the pump's theoretical output graphed at zero psi. Note that as the system pressure increases, the pump's volumetric efficiency will decrease by a small portion. A pump's ***volumetric efficiency*** is determined by dividing the actual measured pump output by the theoretical pump output.

Actual Flow ÷ Theoretical Flow × 100 = Volumetric Efficiency

The pump's flow drops a little as system pressure increases. This drop in volumetric efficiency is the result of the pump's internal clearances leaking more oil as system pressure increases.

Positive-displacement pumps work on the principle of an expanding volume at the pump's intake and a decreasing volume at the pump's outlet. The pump inlet's expanding volume generates a low-pressure cavity. If the reservoir is vented, the pump's inlet pressure drops to a value less than atmosphere, which causes the higher pressure of the atmosphere to push oil into the pump's inlet. See Figure 4-3.

A positive-displacement pump's inlet works on the same principle as a person drinking a beverage through a straw. When the person sips on the straw, a low pressure is generated inside the straw, causing atmospheric pressure to push the liquid into the straw. See Figure 4-4.

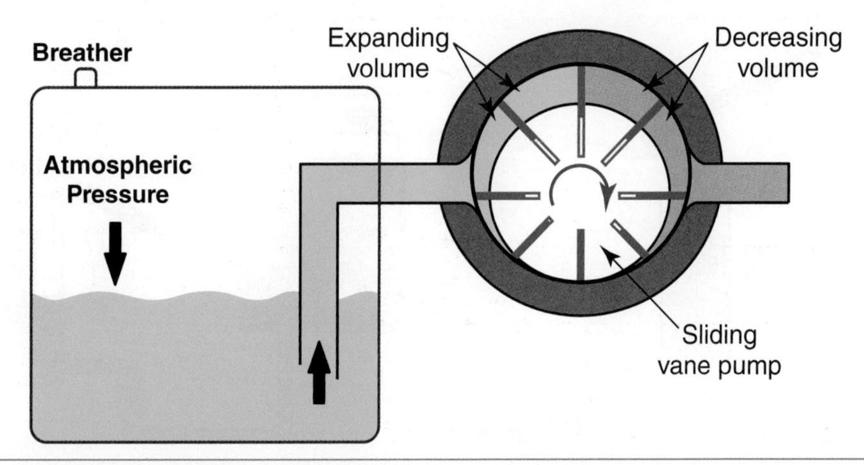

Figure 4-3. A positive-displacement vane pump's intake has an expanding volume that generates a low pressure, allowing atmospheric pressure to push oil into the pump's inlet. The pump's outlet uses a decreasing volume to push the oil into the hydraulic system.

The hand pump in Figure 4-5 illustrates the changing of volumes that takes place within a positive-displacement pump. A ***hand pump*** is a lever-operated hydraulic pump consisting of a piston, reservoir, two check valves, and an actuating lever. When the lever is lifted, it causes the piston to draw oil into the chamber through the inlet check valve. When the lever is pushed down, the piston decreases the volume in the cylinder, causing the pump to push oil out of the pump's outlet valve.

Fixed versus Variable Displacement

Positive-displacement pumps can be further classified into two categories, fixed displacement or variable displacement. A ***fixed-displacement pump*** has no method of varying the pump volume, except for changing the pump speed. The chambers in a fixed-displacement pump have a set volume and will produce a fixed amount of flow for a given pump speed. Keep in mind that most off-highway machines operate at a set engine speed. If the pump is a fixed-displacement design, the machine is producing maximum hydraulic flow, regardless of the machine's need for flow. The excess hydraulic flow is wasting horsepower and energy.

Variable-displacement pumps have the ability to vary the effective volume of their chambers to adjust the pump's flow. For example, if the machine is operating at a set speed of 1800 rpm, and if the machine does not require any

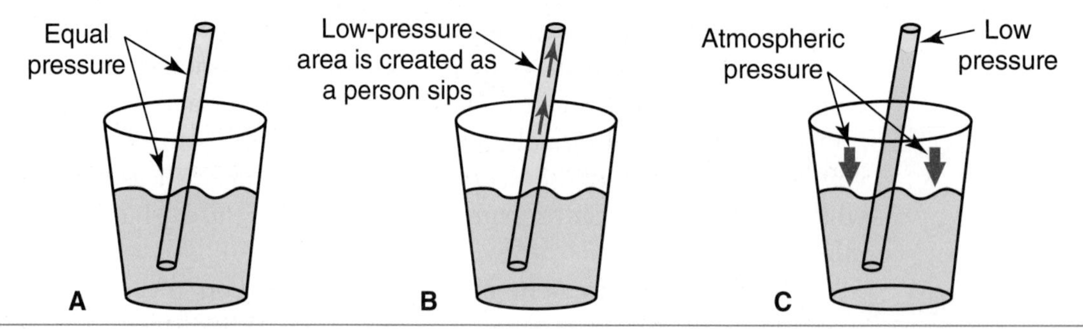

Figure 4-4. Liquid is not sucked out of a container, it is pushed. A—Before the person sips, the pressure in the straw and at the top of the glass are equal. B—A low-pressure area is created inside of the straw as the person sips. C—Atmospheric pressure forces the liquid up through the straw.

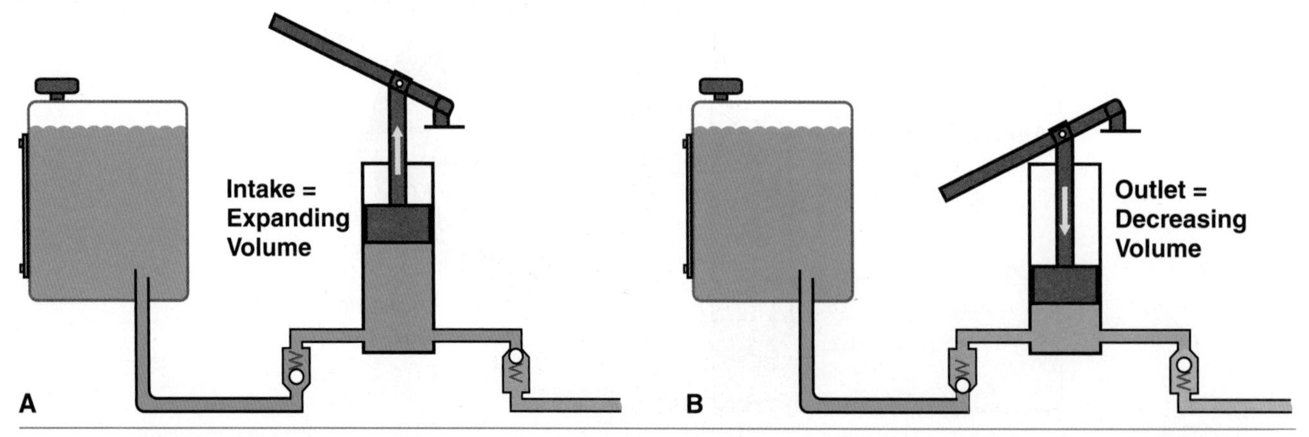

Figure 4-5. A hydraulic hand pump consists of a piston, two check valves, and a manual lever. A—As the piston is pulled up in its bore, it draws oil through the inlet check valve. B—As the piston is pushed down in the bore, it forces oil out of the outlet check valve.

hydraulic flow, the pump's flow can be reduced. Some variable-displacement pumps can have their flow changed to any level between zero flow and maximum flow. This style of pump is commonly used in agricultural equipment.

Other variable-displacement pumps are limited to varying their flow between minimum flow and maximum flow. This style of pump will never operate at zero pump flow, but can be reduced to a minimum flow when the machine does not require flow. An excavator is a machine that commonly uses a variable-displacement pump that varies between minimum flow and maximum flow. **Chapters 17** through **20** will detail the complexities of the controls for a variable-displacement pump's flow.

Unidirectional and Reversible

Pumps can be classified as unidirectional or reversible. A ***unidirectional pump*** is designed to be rotated in only one direction, either clockwise (CW) or counterclockwise (CCW). The pump's direction of rotation is determined by viewing the pump with the shaft pointed toward you. See **Figure 4-6**. Many unidirectional pumps use a large suction port, also known as the intake port, and a small outlet port, also known as the pressure port.

Caution

When replacing a pump, it is important to ensure that the new pump has the same direction of rotation as the last pump. If a pump is driven in the wrong direction, it will quickly fail.

Some unidirectional pumps can be altered to reverse the pump's direction of rotation. On some pumps, it is easy to reverse the pump's flow, and does not require purchasing any new parts. Other pumps require, at a minimum, a new end cap to reverse the pump's rotation. Be sure to follow the pump manufacturer's directions for changing the pump's direction of rotation.

The term "reversible pump" can have two different meanings. The first definition of a ***reversible pump*** is a pump that can be rotated in either direction, CW or CCW, without alterations to the pump. This style of pump provides manufacturers the flexibility of mounting the pump in multiple locations, having to worry about only the correct plumbing, and not the direction of shaft rotation.

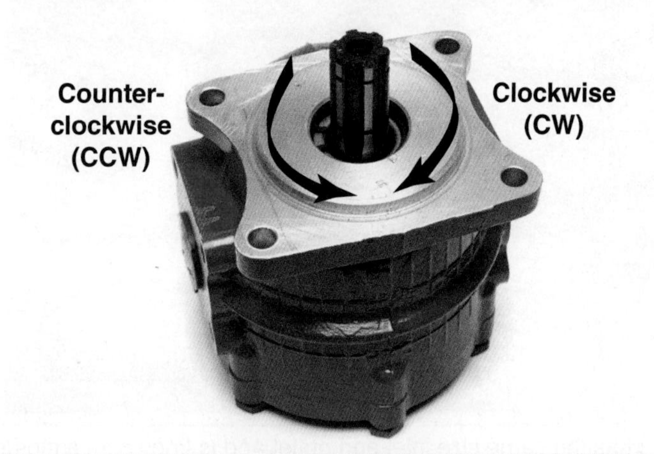

Figure 4-6. Pump rotation is determined by viewing the end of the pump's input shaft.

A reversible pump can also be defined as a hydrostatic transmission pump that is driven in one direction, either CCW or CW, but has the capacity to internally reverse the direction of the pump's flow. These pumps are commonly used for propelling a machine. **Chapters 23** and **24** will cover hydrostatic transmissions.

Open Loop and Closed Loop

As mentioned, some pumps use a large inlet port and a small outlet port, and those pumps are frequently labeled open-loop pumps. An ***open-loop pump*** draws all of its inlet oil directly from the reservoir. Open-loop pumps do not require the use of a charge pump or any other method to assist in supplying oil to the pump's inlet. The primary purpose of a charge pump is to provide a positive supply of oil to a pump's inlet. Gear pumps are the most common style of charge pumps used in off-highway machinery, and they frequently are used to supply a charge of oil to the inlet of a piston pump.

A ***closed-loop pump*** does not need to rely solely upon itself for drawing all of its oil directly from the reservoir. A closed-loop pump uses some type of external source that helps supply a positive charge of oil into the pump's inlet. The two common methods of supplying oil to a closed-loop pump are:

- Directing return oil back to the pump inlet and not the reservoir.
- Using a charge pump to supercharge the main pump's inlet.

Closed-loop pumps frequently use the same size ports for the pump inlet and the pump outlet. See Figure 4-7. Open- and closed-loop pumps will be discussed further in **Chapter 23**.

Gear Pumps

Gear pumps are fixed-displacement pumps. Their pumping chambers cannot be varied. Gear pumps can be classified as internal-toothed or external-toothed pumps. The types of internal-toothed pumps are gerotor and crescent pumps. External-toothed pumps use spur-type, external-toothed gears.

Figure 4-7. The pump on the left has the same size inlet and outlet and is known as a closed-loop pump. The pump on the right is an open-loop pump, which uses a large suction port and a small outlet port.

Gear pumps commonly use two gears: either both external-toothed gears, or one internal-toothed and one external-toothed gear. The drive gear is attached to the pump's input shaft. The other gear is the driven gear. See Figure 4-8.

Gear Pump Attributes

Gear pumps are the most economical and simplest type of pump used in mobile equipment. Although contamination is harmful to all mobile hydraulic systems, gear pumps are the least sensitive to contamination when compared to vane and piston pumps. Gear pumps are the most mechanically efficient, having little loss due to friction or drag. However, gear pumps are the least volumetric efficient, meaning that as pressure increases, a gear pump's internal leakage will lose more flow than a piston or vane pump.

Note

The most economical mobile machines commonly use gear pumps. It is difficult to establish a price threshold where manufacturers quit using gear pumps. However, very few machines that cost more than $150,000 use a gear pump as the primary pump for implement hydraulics.

External-Toothed Gear Pumps

An ***external-toothed gear pump*** uses two external-toothed gears, a drive gear and a driven gear. The displaced oil travels around the outside perimeter of the pump. See Figure 4-9. The expanding volume takes place as the pump's teeth unmesh, which generates a low-pressure cavity. As the teeth of the two gears come back into mesh, it causes a decreasing volume, which results in the pump pushing the oil out of the pump's outlet. The spur gear, external-toothed gear pump is the most popular type of gear pump used in the mobile industry.

Note

Oil flow through an external-toothed gear pump is parallel to the gear faces.

Figure 4-8. An external-toothed gear pump is one of the most popular styles of fixed-displacement pumps in mobile machinery.

External-toothed gear pumps use pressurized side plates to overcome fluid losses. The plates can be called pressure plates, thrust plates, wear plates, or side plates. The pressure plates provide the only means for limiting internal leakage in a gear pump. See Figure 4-10. The pressure plates have a bronze coating to provide a bearing-type surface between the face of the gears and the plates. Seals must be installed in the plates to prevent them from leaking under pressure.

A gear pump's housing is typically made of cast aluminum or cast iron. Interestingly, during the pump manufacturing process, manufacturers commonly install the pump's steel gears tightly into the aluminum housing and run the pump. In effect, this causes the steel gears to machine the bore of the aluminum housing. The gears cut a path in the bore of the housing, called the ***gear track***. This manufacturing process provides a close tolerance fit, enabling the pump to be volumetric efficient. An example of the depth of the gear track is .008″. Figure 4-11 illustrates the gear tracks of a new gear pump.

Internal-Toothed Gear Pumps

Internal gear pumps use two gears, one external-toothed gear as a drive gear and one internal-toothed gear, which is the driven gear. Crescent and gerotor are the two types of internal-toothed gear pumps. The ***crescent internal-toothed gear pump*** uses a crescent spacer to separate the external-toothed drive gear and the internal-toothed driven gear. See Figure 4-12. As the teeth come unmeshed, an expanding volume generates a low pressure to pull oil into the pump's inlet. As the teeth come back into mesh, the pump's volume decreases, forcing oil out of the pump and into the hydraulic system.

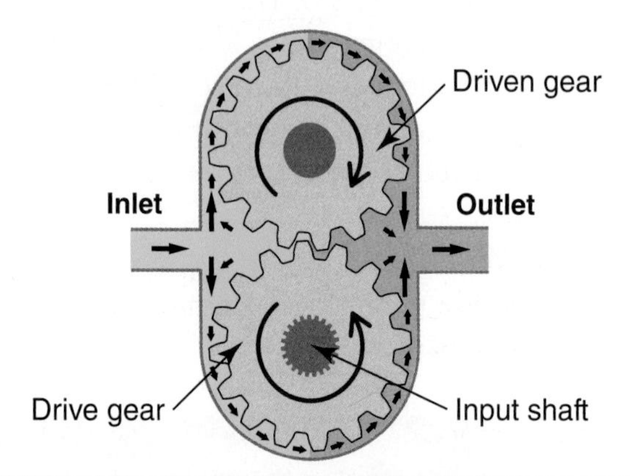

Figure 4-9. Oil flow in an external-toothed gear pump travels around the perimeter of the gears, in the spaces between the teeth.

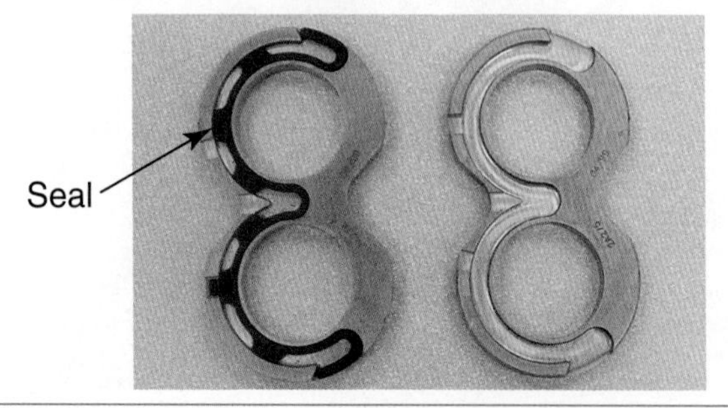

Figure 4-10. Bronze-coated steel side plates are pressurized to limit internal leakage in gear pumps.

A ***gerotor internal-toothed gear pump***, like the crescent gear pump, has a smaller external-toothed drive gear placed inside a larger internal-toothed driven gear. Teeth on the external-toothed gear have lobes rather than spur-shaped teeth. See **Figure 4-13**. As the smaller external-toothed gear is driven, the outer gear with internal teeth will orbit around the external-toothed gear.

Oil flow through an internal-toothed gear pump is perpendicular to the gear faces. The perpendicular oil flow in an internal-toothed gear pump requires the use of port plates, also known as valve plates. The port plate acts like a fixed directional control valve that directs oil in and out of the pump, perpendicular to the face of the gears.

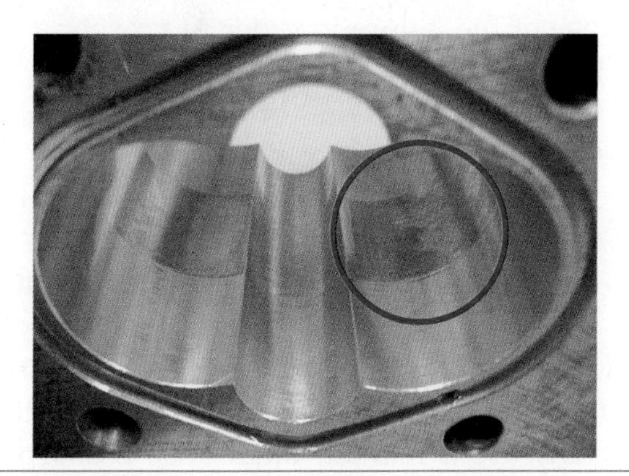

Figure 4-11. Manufacturers commonly break in a new gear pump by allowing the steel gears to machine a gear track inside the aluminum housing. The gear track of a worn pump can have deep grooves.

Figure 4-12. A crescent gear pump uses a crescent spacer to separate the external-toothed gear from the internal-toothed gear.

Figure 4-13. A gerotor gear pump contains a smaller external-toothed gear placed inside a larger internal-toothed driven gear.

Port plates contain kidney-shaped pockets, also known as ports. The ports direct intake oil into the pump and direct flow out of the pump. See Figure 4-14.

External-toothed gear pumps are used more often than internal-toothed gear pumps in mobile hydraulic systems. Internal-toothed gear pumps are commonly used in automotive automatic transmissions and are used in mobile machinery as lube pumps in mechanical transmissions.

Gear Pump Inspection

After a gear pump fails, the pump is usually replaced rather than rebuilt, unless it only needs a bearing or new seals. It is important to determine the root cause of the failure in order to prevent a repeat failure. Some causes of pump failure and their indicators are:

- Over pressurization—broken teeth or shaft.
- Contamination—scoring.
- Aeration or cavitation—pitting.
- Overheating—dark discoloration.
- Improper installation—contamination, aeration, cavitation, or overheating.

Prior to disassembling a gear pump, be sure to mark the housing so that each of the components can be reassembled with the correct orientation. Some technicians use a marker; others use a sharp scribe to mark the housing. Although some technicians simply mark a continuous horizontal line from one end of the pump to the other, it is better to number each of the housing's individual components, especially in housings that contain multiple pumps. See Figure 4-15.

When inspecting a failed gear pump, be sure to check the following items:

- Inspect the bore of the housing for scoring, pitting from cavitation, and wear.
- Inspect the port plate seals for wear, deformation, and erosion.
- Inspect the shaft bearings and bushings for wear and discoloration.
- Inspect the bronze-coated port plates for wear, scoring, or pitting caused by aeration or cavitation.
- Inspect the gears for scoring, wear, cracks, or discoloration.

Figure 4-14. Port plates, also known as valve plates, provide pockets for directing oil into and out of the pump.

Figure 4-15. When disassembling a gear pump, it is important to label each of the individual pump housings to ensure the pump is assembled correctly.

If the pump was misaligned during the manufacturing process, the input shaft's teeth, bearings, or bushings could have failed, requiring a technician to replace those failed components.

Case Drain

Case drain refers to oil leakage inside the pump's case. Case drain leakage will increase as the pump pressure increases and as the pump wears over the course of its useful life. Many pump manufacturers route this heated oil back to the reservoir, or in some instances a cooler, requiring the pump to draw all of its intake oil from the cooled oil located inside the reservoir.

Case drain is usually a low pressure, for example 15 psi (1 bar). If case drain pressure increases too much, it can cause the pump's input shaft seal to leak. Most gear pumps do not route case drain back to the reservoir. Instead, they have the internal leakage routed back to the pump's inlet. Case drain will be discussed further in **Chapters 23** and **24**.

Gear Pump Leakage

Fluid losses in a gear pump can occur in three areas:

- Between the side plates and the side faces of the gears.
- Between the housing's pocket and the gear's teeth.
- Between the two gears' teeth where the gears mesh.

The largest contributor of internal leakage, nearly 65% of the leakage, occurs between the side plates and gear's end face. This is the reason manufacturers use oil pressure to hold the side plates against the gears' end faces.

Load-Sensing Gear Pumps

As mentioned earlier in the chapter, gear pumps are fixed-displacement pumps. Concentric, formerly Haldex, produces a load-sensing gear pump. ***Load-sensing (LS) pumps*** have the ability to vary pump flow based on actuators' working pressures. At first glance, it would appear that a LS gear pump is a variable-displacement pump. However, the LS gear pump is technically a fixed-displacement pump that incorporates an unloading valve that bypasses oil flow at low pressure whenever system flow is not needed. Fixed-displacement pumps used in load-sensing circuits will be further explained in **Chapter 18** of this text.

Vane Pumps

Vane pumps are used in mobile machinery; however, they are not as common as gear pumps and piston pumps. Vane pumps can be classified into different categories:

- Sliding vane, roller vane, articulated vane, or slipper vane.
- Single lobe or dual lobe.
- Fixed-displacement or variable-displacement.

Types of Vane Pumps

The ***sliding-vane pump*** is the type of vane pump commonly used in mobile machinery. It consists of a rotor that is driven by the pump's input shaft, has vanes that slide in and out of the rotor, and a cam-shaped housing that forms the outer edge of the pump chamber. See **Figure 4-16**.

As the input shaft spins the rotor, centrifugal force extends the vanes against the housing, also known as the cam ring. Because the vanes are initially extended, the pump will exhibit the characteristics of a positive-displacement pump. The intake side of the pump will have an expanding volume that generates a low pressure in the pump's inlet, allowing atmospheric pressure to push oil into the pump's suction port. The decreasing volume pushes oil out of the pump's discharge port.

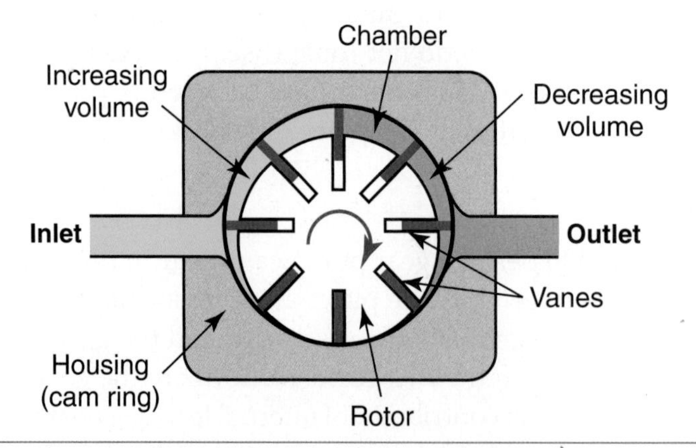

Figure 4-16. A sliding-vane pump has a rotor driven by an input shaft. The vanes slide in and out of the rotor to form the pumping chambers.

See Figure 4-20. The smaller area results in a reduced force, which increases the life of the pump. The intra-vane pumps are also known as high-performance vane pumps.

Cartridges

Some vane pumps can be quickly rebuilt by simply installing a new cartridge assembly. See Figure 4-21. The ***vane pump cartridge*** consists of the cam ring, rotor, vanes, and side plates all assembled in a compact unit. The old vane pump is disassembled by removing the housing's bolts. The old cartridge can be lifted out of the pump housing and input shaft. The new cartridge is inserted on the input shaft and placed inside the housing and assembled. Installing a cartridge can lower the costs for replacing a vane pump.

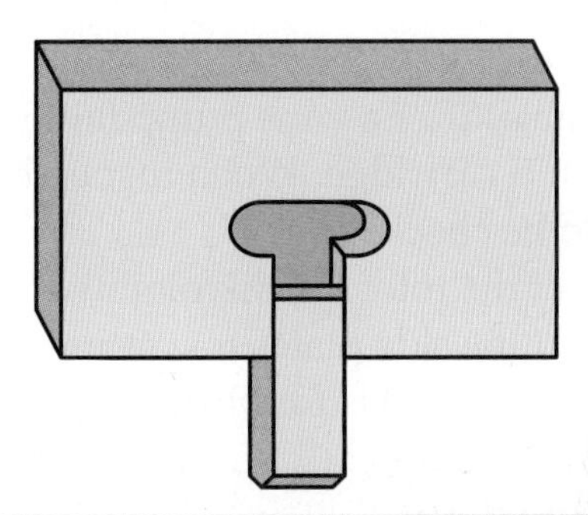

Figure 4-20. An intra-vane pump consists of a rotor that uses a small vane placed inside a larger vane.

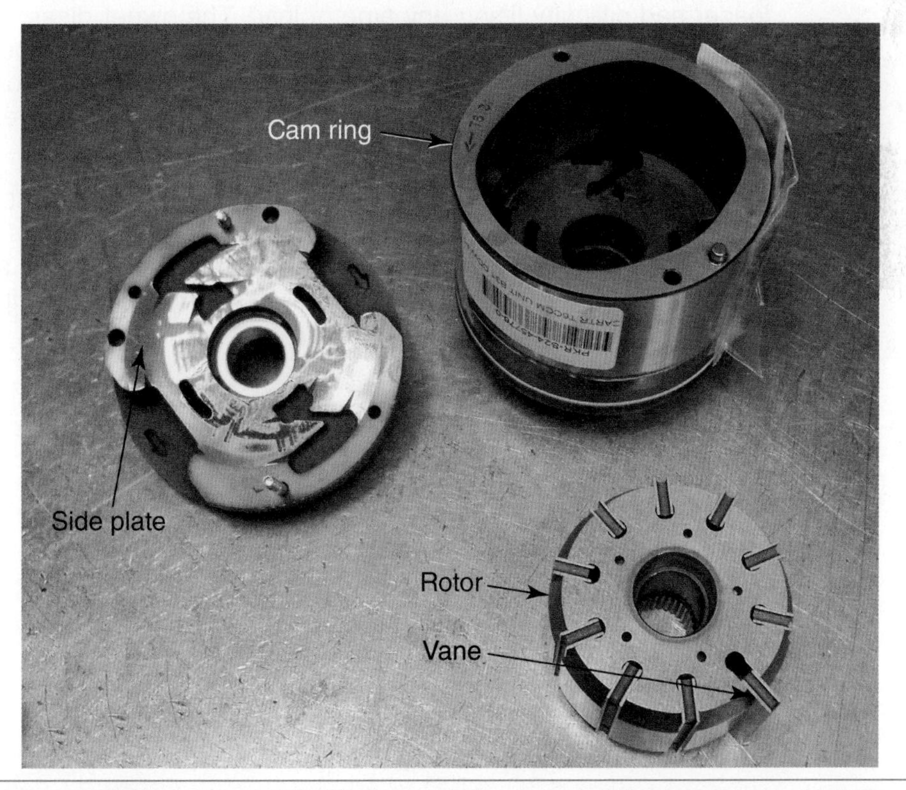

Figure 4-21. Vane pump cartridges contain all of the essential components in a vane pump, including the rotor, cam ring, and vanes. The cartridge allows a technician to quickly rebuild a pump without having to handle intricate components, such as the vanes and rotor.

Variable-Displacement Vane Pumps

One of the most common applications of variable-displacement vane pumps is automotive automatic transmissions. See Figure 4-22. The pump's rotor is located inside of a cam ring that pivots on a pin. A bias spring is used to hold the pump at a maximum displacement. When the hydraulic system's fluid requirements are met, a regulator valve directs oil to a chamber outside the cam ring, where it presses on the outer surface of the cam ring, causing it to pivot.

Note

Vane pumps have greater volumetric efficiency than gear pumps and are less complicated than piston pumps.

Vane Pump Inspection

If applicable, after a vane pump failure, a new cartridge will be installed in place of the old cartridge. When inspecting a failed vane pump, the following items should be checked:

- Inspect the rotor for discoloration and scoring.
- Inspect the side plates for pitting, cavitation, and wear.
- Inspect the vanes for burrs and contamination.

Case Study

Pump Refurbishing

A vane pump that was used for a skid steer's hydraulic system failed. The loader had difficulty lifting any type of load. The owner disassembled the pump and took it to a hydraulic repair shop. The repair shop pointed out to the customer the cavitation on the pump's housing, discoloration on the rotor due to a lack of oil, and the formation of burrs on the vanes. The repair shop recommended that the customer purchase a new vane pump.

The owner took the pump to a friend who was a machinist. He removed the cavitation on the pump housing. The owner polished the vanes with Emery cloth and lubed the vanes and rotor, ensuring that the vanes could slide freely inside the rotor. After the pump was reinstalled, the loader's lift capacity was restored to its full break-out force.

Figure 4-22. A variable-displacement vane pump uses a cam ring that pivots on a pin. The bias spring is used to hold the cam ring in the position that produces maximum displacement. If pump flow exceeds demand, fluid pressure causes the cam ring to overcome the force of the bias spring, reducing displacement.

Piston Pumps

The more a machine costs, the more likely it will use a piston pump for its hydraulic system. A machine that costs more than $150,000 is more likely to use a piston pump than a gear pump.

Piston pumps are positive displacement. Although piston pumps can be fixed displacement or variable displacement, among the mobile machines that use piston pumps, the majority will be variable displacement.

Piston pumps can be categorized as:

- Radial piston pump types.
  - Rotating cam.
  - Rotating piston.
- Axial piston pump types.
  - Inline.
  - Bent-axis.

Radial Piston Pumps

Radial piston pumps have the pistons positioned radially, or perpendicular to the pump's input shaft. See Figure 4-23.

In a ***rotating-cam radial piston pump,*** the pump housing is stationary and the pistons reciprocate in and out of their individual bores, creating an expanding volume and a decreasing volume. Figure 4-24 illustrates a rotating-cam piston pump that has been cutaway. An eccentric cam is mounted on the pump's input shaft and is used to force the pistons into their bores, which pushes oil out of the pump. Springs push the pistons out of their bores against the eccentric cam, which draws intake oil into the pump.

Rotating-cam piston pumps were used by John Deere from 1960 through 1994, in both agricultural equipment and construction equipment. Each piston has its own intake check valve and its own outlet check valve. All of the inlet check valves are located within an intake oil gallery. All of the outlet check valves are also located in a pump discharge oil gallery.

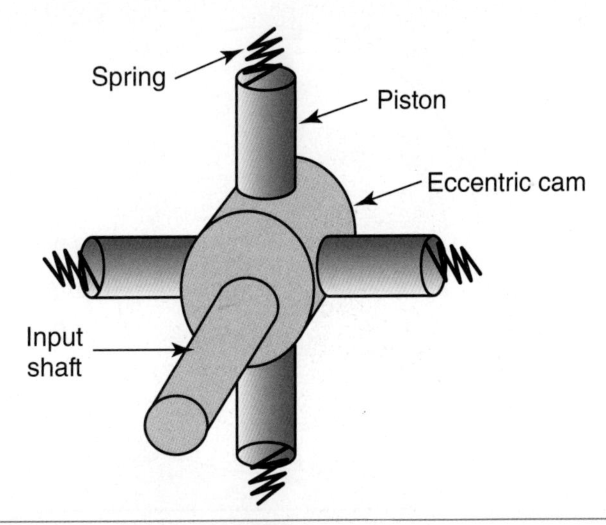

Figure 4-23. Radial piston pumps have the pistons and cylinders arranged perpendicular to the pump's input shaft. This is similar to the way pistons are positioned in an airplane engine.

The John Deere radial displacement piston pumps were variable-displacement pumps used in their pressure-compensated hydraulic systems. A *pressure-compensated hydraulic system* is designed to be used with closed-center DCVs so that the hydraulic system can operate at a high pressure, such as 2500 psi, and have no flow when the DCVs are in a neutral position. **Chapter 17** explains pressure-compensated hydraulic systems in further detail.

Rotating-block radial piston pumps use a block assembly that is driven by the pump's input shaft. See Figure 4-25. Notice that the pistons slide in and out of the rotating piston block in the same manner as vanes slide in and out of a rotor in a sliding vane pump. Rotating-block piston pumps are not commonly found in mobile machinery.

Figure 4-24. A John Deere radial piston pump is a rotating cam design that uses an eccentric cam mounted on the input shaft. The pistons reciprocate in and out of their own bore.

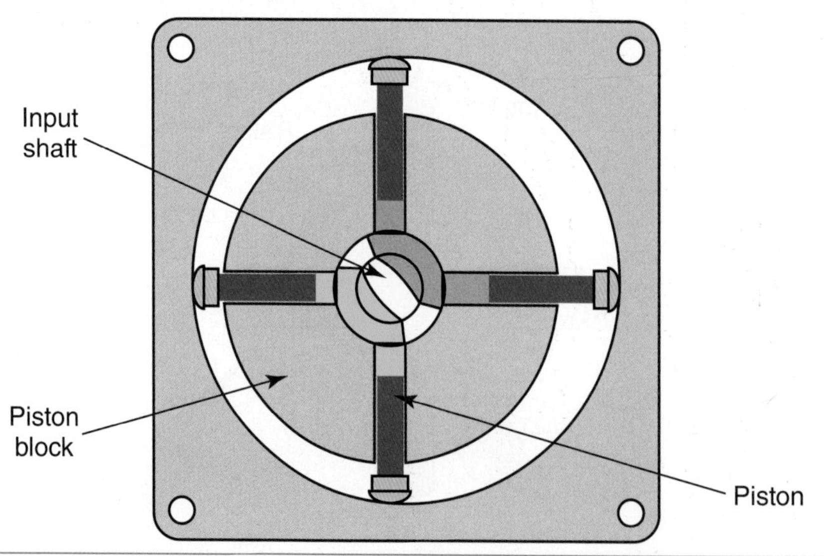

Figure 4-25. Rotating-block radial piston pumps contain a block that is driven by the pump's input shaft. The reciprocating action of the pistons causes oil to be drawn in and pumped out of the pump.

Axial Piston Pumps

Axial piston pumps have been used in mobile machinery since the 1970s and continue to be the most common style of piston pump used in today's advanced mobile machinery. Axial pumps are categorized as inline axial piston pumps, which are commonly found in agricultural equipment and construction equipment, and bent-axis piston pumps, which are found in construction equipment.

An ***inline axial pump*** is a piston pump that has a rotating ***cylinder block***, also known as a barrel, that is splined to the pump's input shaft. As the input shaft rotates the cylinder, the pistons reciprocate in and out of the cylinder, causing the pump to pull fluid into the pump's inlet during 180 degrees of pump rotation and push fluid out of the pump's discharge during the other 180 degrees of pump rotation. Figure 4-26 illustrates a cutaway of an inline axial piston pump.

A ball on one end of the piston fits into a socket on the ***slipper pad***, sometimes known as the shoe. See Figure 4-27. As the pump's cylinder rotates, the slipper pad rides against the swash plate's surface.

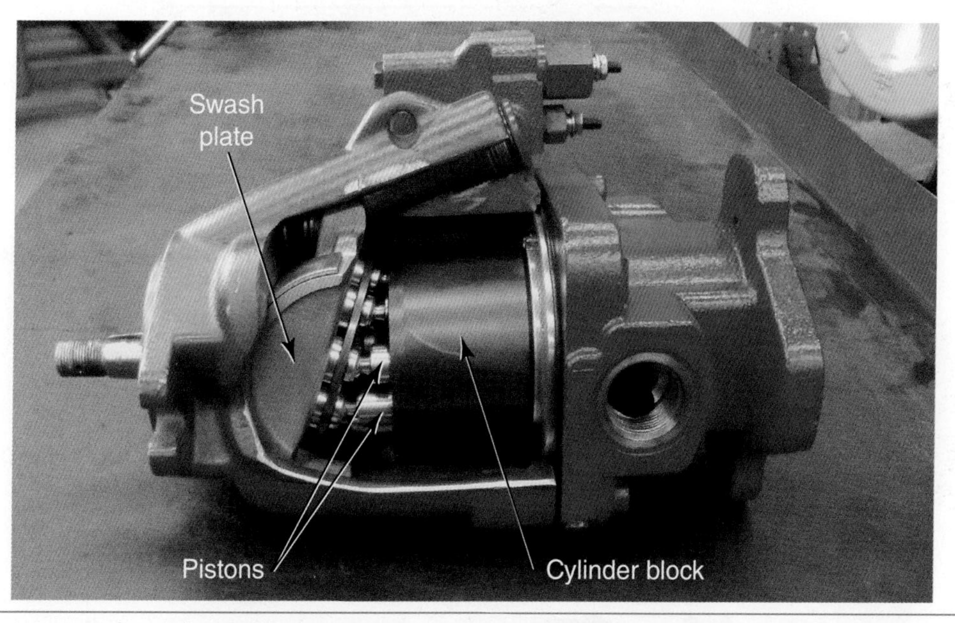

Figure 4-26. A cutaway of an inline axial piston pump reveals a swash plate, which causes pistons to reciprocate in and out of the rotating block as the input shaft rotates.

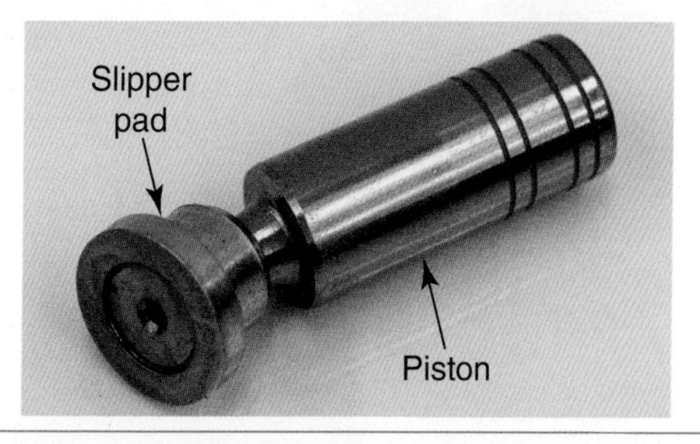

Figure 4-27. The pistons are attached to bronze-coated slipper pads. The slipper pads slide around the swash plate, causing the pistons to reciprocate in and out of their bores.

A fixed-displacement inline axial pump will have a fixed swash plate angle, for example 18 degrees. See Figure 4-28. As the pump's cylinder rotates, the piston's slipper pad is held against the swash plate with a ***slipper retaining ring***, Figure 4-29. The rotation of the cylinder causes the piston to reciprocate in and out of the barrel's bore once every revolution, as the shoe slides up and down the angled swash plate. The slipper pad is bronze-coated. The bronze coating provides the bearing surface as the slipper pad slides around the pump's swash plate.

As the pistons are pumping oil and the system pressure builds, the system's pressure exerts force on the pistons. That force is transmitted to the slipper pads where it causes wear. To offset this force and reduce slipper pad wear, the pistons have a small orifice that allows oil to travel from end to end in the piston. This oil passageway enables the system pressure to act on the end of the slipper, offsetting the load that is exerted on the slipper.

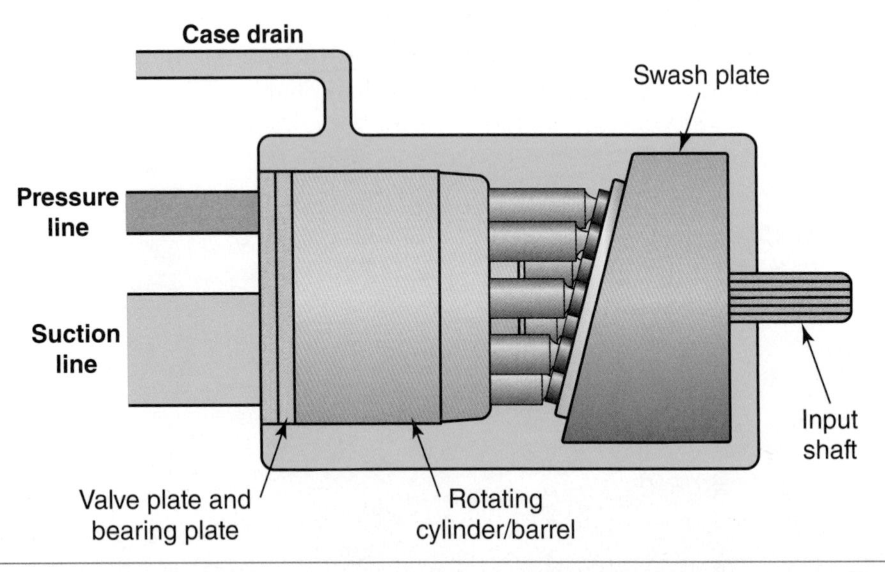

Figure 4-28. A fixed-displacement inline axial piston pump has a swash plate with a fixed angle. The pistons slide up and down the swash plate as the cylinder barrel rotates.

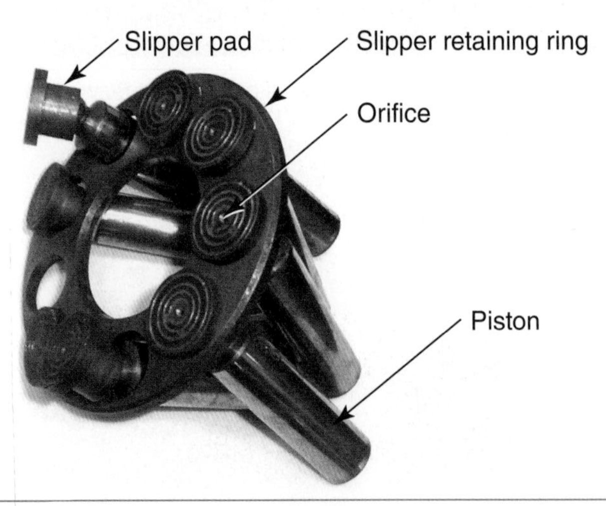

Figure 4-29. A retaining ring holds the slipper pads against the pump's swash plate, locking the piston assemblies in place.

Valve Plates and Bearing Plates

Axial piston pumps often contain one or two plates. If the pump contains two plates, one is the valve plate and the other is the bearing plate. Both are located at the end of the cylinder's barrel. Typically, two plates are used in high-pressure applications such as hydrostatic transmissions.

The ***valve plate*** is pinned to the pump's case, often with one or two dowel pins, to keep the plate stationary. The valve plate is unidirectional, meaning that its design dictates the direction of rotation of the pump's input shaft. The valve plate contains one or two metering slots, which determine the direction of rotation for the pump. The metering slots are often in the shape of a V at the end of the valve plate's pockets. The V-shaped metering slots are primarily used on valve plates inside implement pumps. See Figure 4-30. The metering slots trail in the direction of rotation. If the valve plate contains an equal number of pockets on each side, then it is used in a closed-loop pump.

The valve plates used in open-loop pumps are unique in that the suction side of the plate uses one large port. The pressure side of the valve plate has multiple ports, such as three, four, or five individual ports. See Figure 4-31.

Sometimes the metering slots can be in the shape of a kidney. The kidney-shaped slots are located on valve plates used inside propulsion pumps, such as hydrostatic transmission pumps. See Figure 4-32.

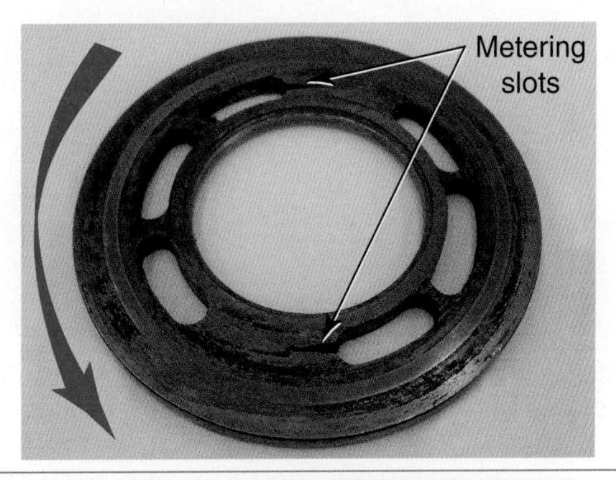

Figure 4-30. This bronze-coated valve plate has two V-shaped metering slots. This plate was used in a closed-loop implement pump. The pump was worn out and replaced. Notice the pitting on the valve plate.

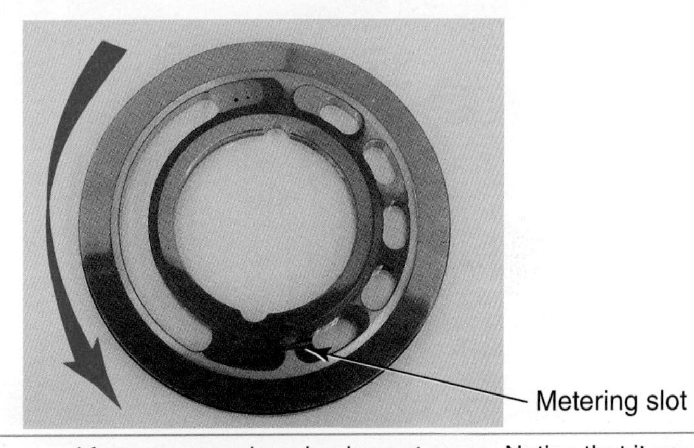

Figure 4-31. This valve plate was removed from an open-loop implement pump. Notice that it uses a single V-shaped metering slot. The large single port is the suction side and the five individual ports direct oil out of the pump's outlet.

Some inline axial piston pumps do not use any type of valve plate or bearing plate. In these designs, the pump's case also serves as the valve plate. See Figure 4-33.

Hydrostatic transmission pumps often contain a second plate called a ***bearing plate***, which is pinned to the end of the barrel and will rotate along with the barrel assembly. In this application, the bearing plate usually is coated in bronze and the valve plate is not coated. Refer to Figure 4-32. The two plates direct oil in and out of the barrel assembly as the barrel and bearing plate assembly rotate past the valve plate. The bearing plate does not contain metering slots. Sometimes the service literature does not distinguish the difference between a bearing plate and a valve plate, but calls both of the plates "valve plates." When a pump does not contain a bearing plate, the end of the barrel serves as the bearing plate.

Note that the pump's barrel, piston slipper assemblies, slipper retaining ring, and bearing plate are all part of an assembly called a ***rotating group***. Most axial piston pumps contain an odd number of pistons, such as seven or nine.

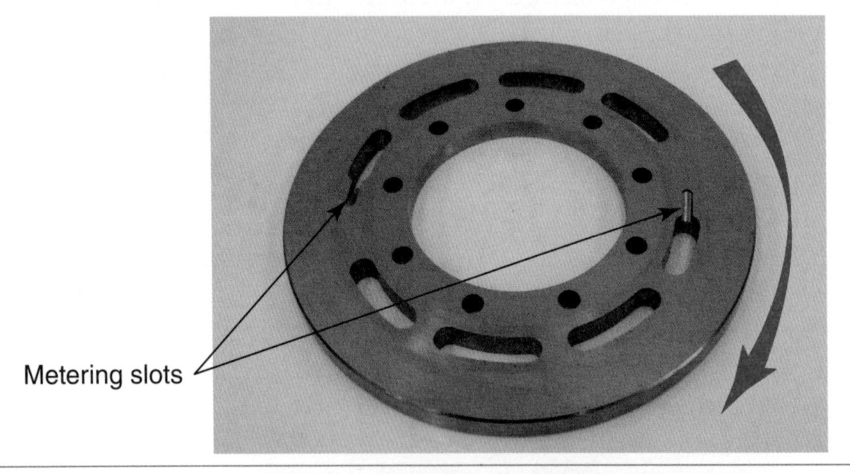

Figure 4-32. The valve plate's metering slots are in the shape of a kidney because the plate is used in a hydrostatic pump. Notice that this plate is not coated with bronze. The pump's bearing plate is coated in bronze.

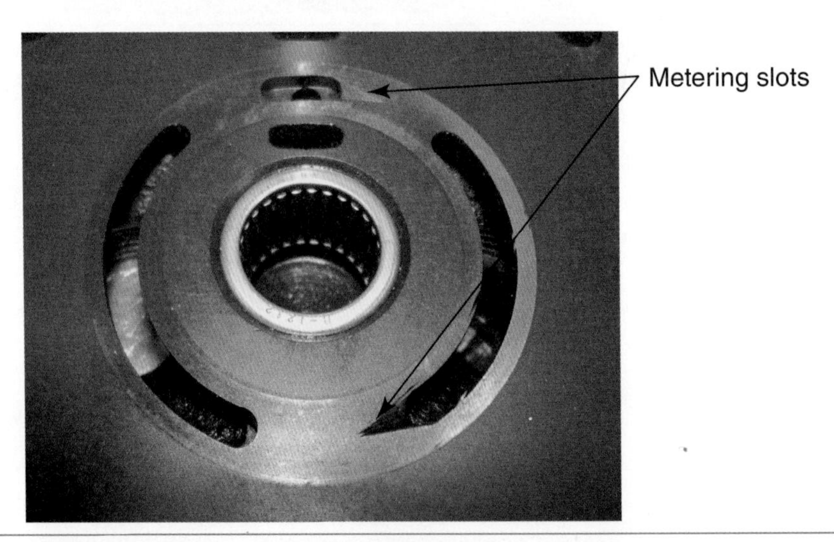

Figure 4-33. This pump contained no individual valve plate or bearing plate. Notice the fixed pump housing contains two V-shaped metering slots similar to those found in valve plates.

Variable-Displacement Inline Axial Pumps

As stated, most inline axial piston pumps are variable displacement. When the pump swash plate is in a straight position, also known as parallel to the rotating group, the pump is not producing flow because the pistons are unable to reciprocate in and out of the cylinder block. This means that an inline axial pump's swash plate must have some type of a positive angle in order for the pump to generate flow. A larger degree of swash plate angle will result in more pump flow.

Pump manufacturers use different types of axial pump frame designs. Refer back to Figure 4-26. That figure illustrates a cutaway of a ***cradle bearing axial piston pump***. This style of pump uses two bearings or bushings in a cradle-shape form, similar to the main bearings in an engine. The cradle bearings allow the swash plate to pivot. This pump frame design is the most common style used in mobile equipment. See Figure 4-34. A bias spring and control piston are responsible for changing the angle of the swash plate.

A ***bias spring*** has the purpose of pushing the swash plate to a maximum angle, also known as turning the pump on. Some pumps use a ***bias piston*** instead of a bias spring for maximizing the pump flow. A ***control piston*** is used to push the swash plate back to a neutral angle, thereby shutting off the pump flow.

Chapters 17 through **20** will elaborate, in detail, on the different types of pump controls used for varying an implement pump's displacement.

A ***trunnion bearing axial piston pump*** uses a swash plate that pivots on two trunnions that are located inside two bearings. The trunnion bearing pumps were the first style of axial piston pumps used in mobile machinery and are becoming less common than cradle bearing pumps. See Figure 4-35.

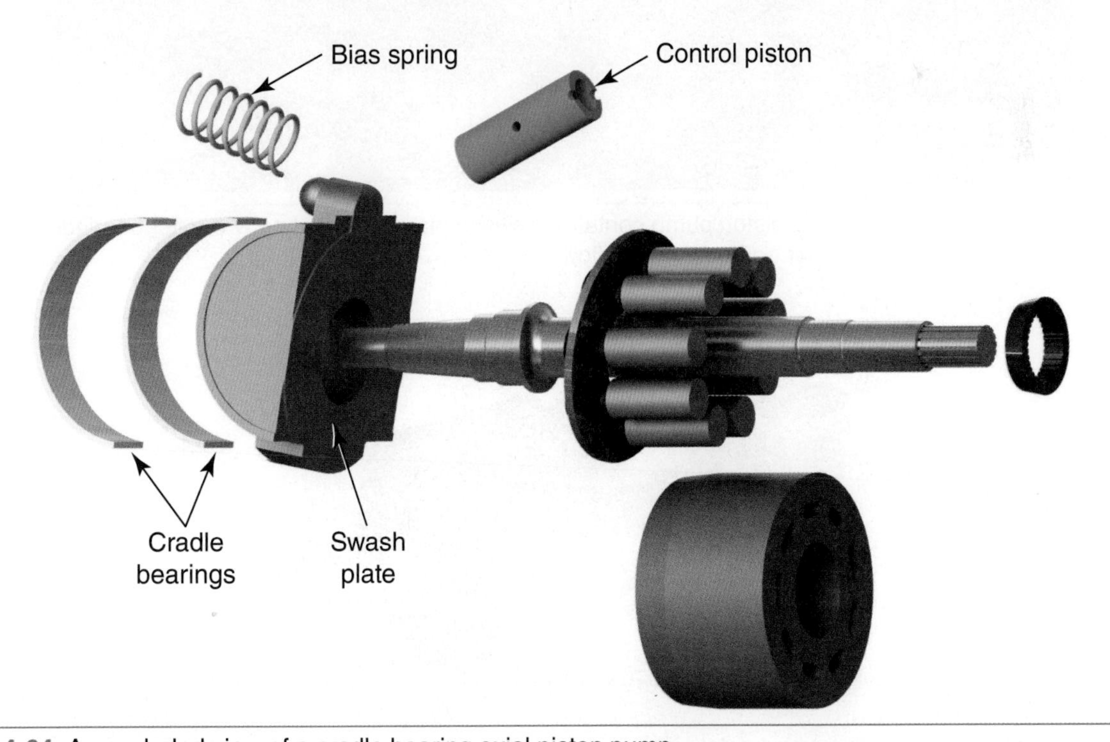

Figure 4-34. An exploded view of a cradle bearing axial piston pump.

Reversible Axial Piston Pumps

Both cradle bearing and trunnion bearing pumps can be used in reversible pump applications. A reversible axial piston pump has the ability to move the pump's swash plate angle in opposite directions. Reversible pumps are used in hydrostatic transmission applications in order to reverse the pump flow and allow a machine to back up. The cradle bearing reversible axial piston pump is the modern style of hydrostatic piston pump used in today's mobile applications. It does not use a bias spring or a control piston. It uses a single servo assembly for stroking the pump forward or reverse. See Figure 4-36.

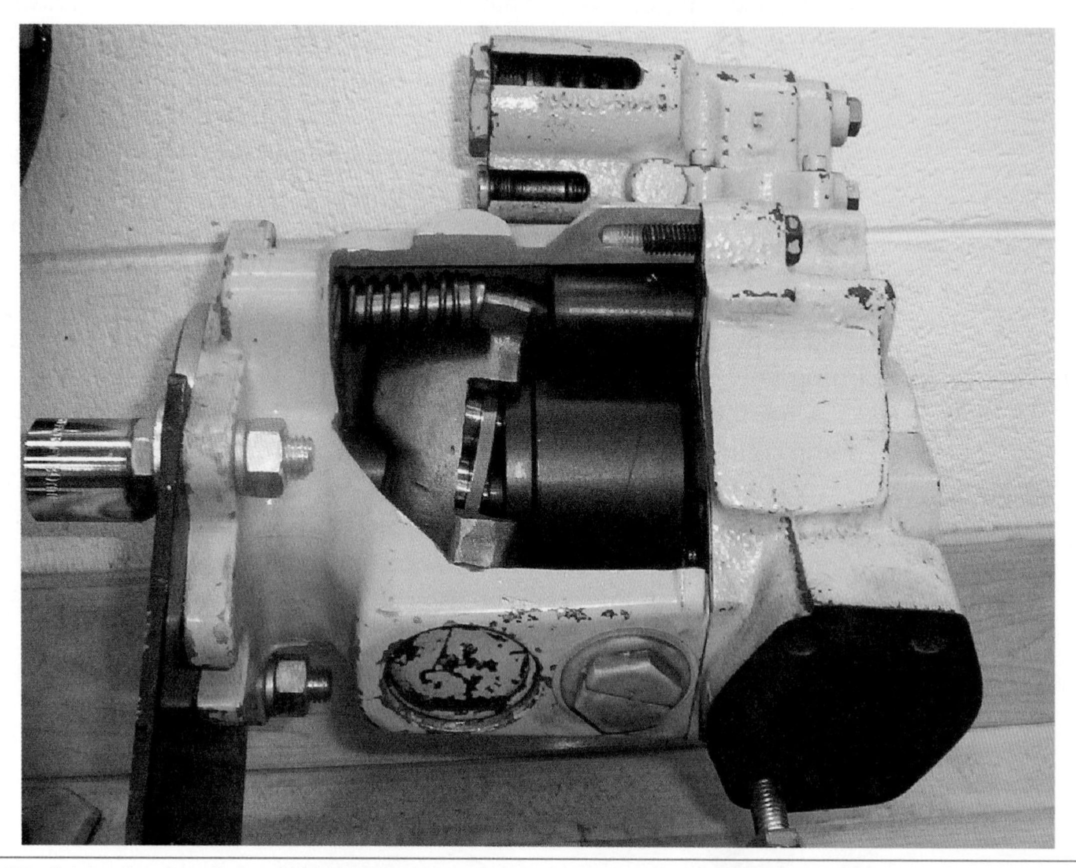

Figure 4-35. A trunnion bearing axial piston pump contains a swash plate mounted on two roller bearings that allow the swash plate to pivot to increase or decrease pump flow.

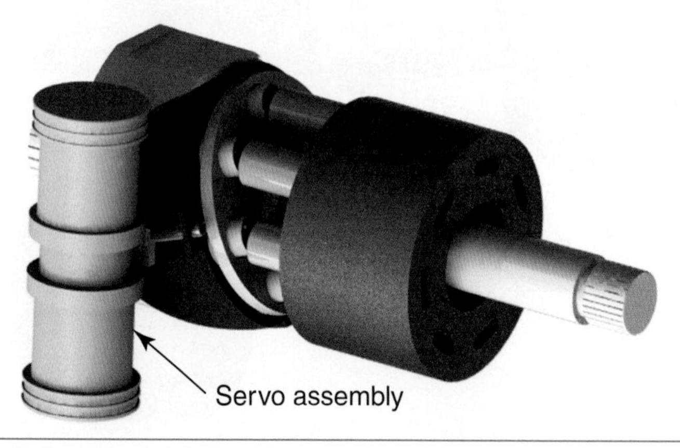

Figure 4-36. A reversible cradle bearing axial piston pump.

The trunnion bearing reversible pump is the oldest design of axial piston pumps used for hydrostatic propulsion. This design as well does not use a bias piston or a control piston. This reversible pump uses two servo pistons for stroking the pump forward or reverse. See Figure 4-37. **Chapter 23** further explains hydrostatic pumps.

Bent-Axis Piston Pumps

The second style of axial piston pump is the bent-axis pump. As its name implies, the rotating group is not inline with the input shaft. This type of pump has an angled housing. See Figure 4-38.

The input shaft is often supported by three sets of bearings, either ball bearings or tapered roller bearings. The input shaft assembly contains a drive flange that couples to the pistons through their ball sockets. As the input shaft rotates, the drive flange pulls and pushes the pistons in and out of the rotating group's barrel.

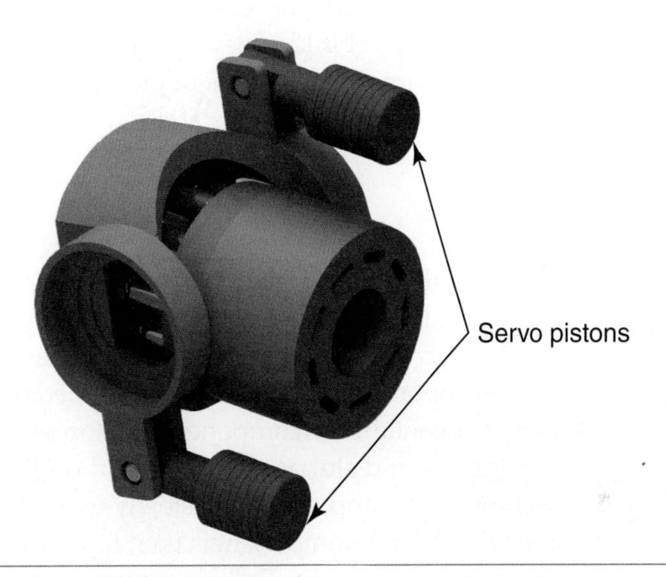

Figure 4-37. A reversible trunnion bearing axial piston pump.

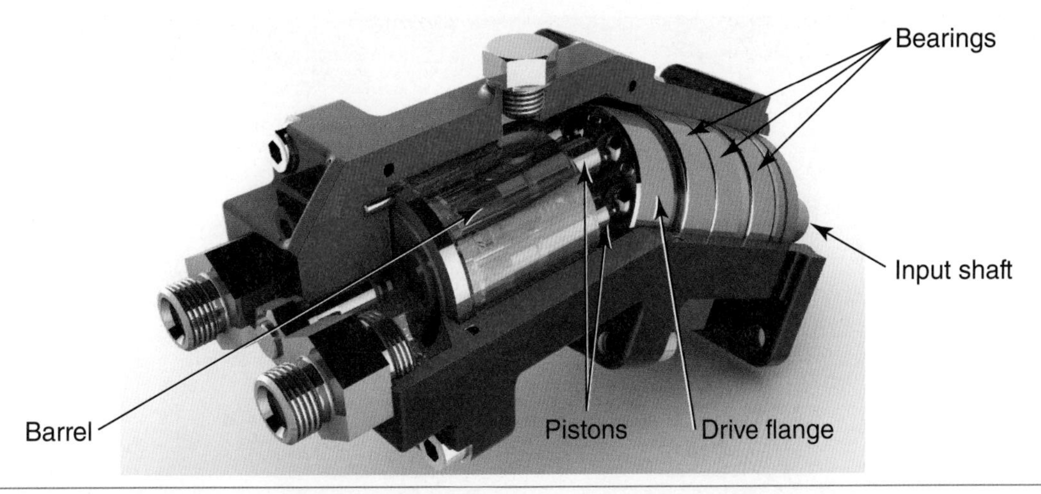

Figure 4-38. Bent-axis pumps are axial piston pumps that do not use a swash plate. Instead, they have a pivoting rotating group that varies the pump's displacement.

Variable-Displacement Bent-Axis Piston Pump

A bent-axis piston pump uses a valve plate that has a partial spherical shape that mates against the rotating barrel. Within a variable-displacement bent-axis piston pump, the valve plate is actuated by a servo piston in order to change the pump's displacement. As the valve plate is actuated, it causes the rotating group to vary its angle. A larger angle will equal more pump flow. Flow will be decreased as the angle decreases and approaches an inline angle with the input shaft.

Bent-axis piston pumps are commonly used in excavator applications. Most excavator variable-displacement bent-axis pumps are designed to vary their flow between a minimum displacement rate and a maximum displacement rate, but rarely will an excavator pump flow be completely shut off. **Chapter 20** will further explain excavator pump applications.

Piston Pump Attributes

Piston pumps are the least mechanically efficient pumps and the most complex pumps to rebuild. They are the least tolerant to contamination, but are the most volumetrically efficient pumps. Due to their high volumetric efficiencies, they are used in hydrostatic transmission applications, which operate at extreme pressure, up to 7000 psi.

Piston Pump Service

Hydraulic pump repair shops specialize in rebuilding piston pumps. The slippers, valve plates, bearing plates, and cylinder blocks all wear during the life of a pump. Repair shops use a ***lapping machine*** to machine the rotating group components. See Figure 4-39. The components are placed inside of steel rings that contain the components on top of the lapping machine.

A compound slurry mix is sprayed on the rotating surfaces of the lapping machine. The components are slowly machined flat as they rotate within each of the rings. Each component is then measured to see if it falls within the limits of service.

Figure 4-39. A lapping machine is used to machine barrels, valve plates, and bearing plates. A compound slurry mix is sprayed on the rotating lapping plates.

Valve plates and bearing plates are also placed under a crystal. See Figure 4-40. A light shines over the crystal and plate. The flatness of the plate is determined by the straightness of the emitted light waves. The flatness must be within extremely small tolerances, otherwise the pump will leak.

Piston Pump Inspection

When inspecting a piston pump, the following items should be checked:

- Inspect the end of the barrel, valve plate, and bearing plate for cavitation (pitting or erosion), discoloration, or scoring from contamination.
- Inspect the bearing and valve plates for straightness.
- Inspect the slipper pads for wear.
- Check the piston-to-bore fit. Note some pump technicians place a thumb over the end of the barrel and use the slipper pad to push and pull the piston out of its bore, checking for vacuum and pressure.

Factors Affecting Pump Selection

When building a new hydraulic system from scratch, there are several factors that need to be considered when choosing a new pump. If the wrong pump is incorrectly matched in the wrong application, the pump can have a sudden failure or, at a minimum, have its life greatly reduced. The following pump attributes should be considered when building a new hydraulic system:

- Pump maximum pressure.
- Pump maximum inlet vacuum.
- Open loop versus closed loop.
- Direction of input shaft rotation.
- Fixed displacement or variable displacement.
- Unidirectional versus reversible.
- Power source (electric or engine-driven).
- Maximum pump rpm.
- Pump displacement.

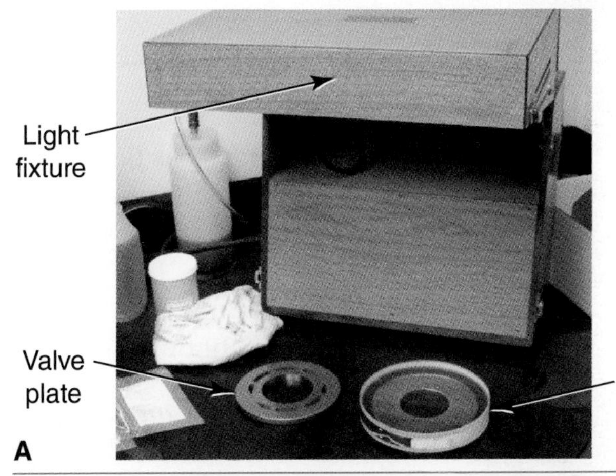

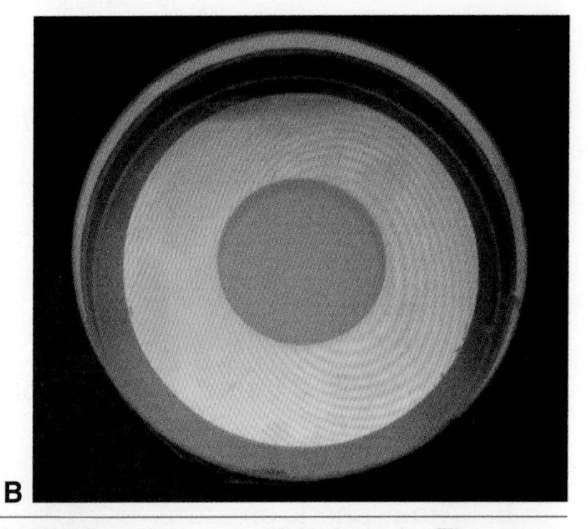

Figure 4-40. A crystal and light is used to determine the true flatness of a bearing plate or a valve plate. A—The setup for inspecting a plate. B—Closeup of a bearing plate under a crystal. Notice the curved lines of reflected light.

Summary

- ✓ Hydraulic pumps can be classified based on different attributes:
  - Positive and non-positive displacement.
  - Fixed displacement and variable displacement.
  - Open loop and closed loop.
- ✓ Gear pumps are mechanically efficient, are the most tolerant to contamination, have fixed displacement, and are the least volumetric efficient.
- ✓ Vane pumps will self-compensate for wear. They can be single-lobe, variable-displacement pumps, or dual-lobe, fixed-displacement pumps.
- ✓ Piston pumps are the most volumetric efficient, the most expensive and complex, and the most susceptible to failure from contamination. They are the least mechanically efficient.

Technical Terms

bearing plate
bias piston
bias spring
case drain
charge pump
closed-loop pump
control piston
cradle bearing axial piston pump
crescent internal-toothed gear pump
cylinder block
external-toothed gear pump
fixed-displacement pump
gear pumps
gear track
gerotor internal-toothed gear pump
hand pump
inline axial pump
intra-vane pump
lapping machine
load-sensing (LS) pumps
non-positive-displacement pump
open-loop pump
positive-displacement pump
reversible pump
rotating-block radial piston pumps
rotating-cam radial piston pump
rotating group
sliding-vane pump
slipper pad
slipper retaining ring
trunnion bearing axial piston pump
unidirectional pump
valve plate
vane pump cartridge
variable-displacement pumps
volumetric efficiency

Review Questions

Answer the following questions using the information provided in this chapter.

1. Which pump uses an impeller?
   A. Piston pump.
   B. Vane pump.
   C. Gear pump.
   D. Non-positive-displacement pump.
2. What pushes oil into a positive-displacement pump's inlet?
   A. Pilot oil.
   B. Atmospheric pressure.
   C. Case pressure.
   D. Signal pressure.
3. Which of the following pumps has tight sealing surfaces and works on the principle of an expanding volume and a decreasing volume?
   A. Positive-displacement pump.
   B. Non-positive-displacement pump
   C. Both A and B.
   D. Neither A nor B.

4. Which of the following pumps requires the use of a port plate?
   A. Crescent.
   B. Spur.
   C. Helical.
   D. Herringbone.
5. Which of the following pumps has the most tolerance for dirt contamination?
   A. Gear pump.
   B. Vane pump.
   C. Piston pump.
   D. All pumps have equal tolerance for contamination.
6. Which of the following has a high volumetric efficiency?
   A. Gear pump.
   B. Vane pump.
   C. Piston pump.
   D. All pumps have equal volumetric efficiency.
7. Which of the following has the best mechanical efficiency?
   A. Gear pump.
   B. Vane pump.
   C. Piston pump.
   D. All pumps have equal mechanical efficiency.
8. Which of the following components is pressurized on an external-toothed gear pump in order to overcome internal leakage?
   A. Port plate.
   B. Side plate.
   C. Drive gear.
   D. Driven gear.
9. Pump manufacturers will break in an external-toothed gear pump, causing which of the following?
   A. Cavitation.
   B. Aeration.
   C. Gear tracks.
   D. Gear galling.
10. During operation, where does the majority of oil travel in an external-toothed gear pump?
    A. Between the two meshing gears.
    B. Around the outside periphery.
    C. Both A and B.
    D. Neither A nor B.
11. Gear pumps are what type of pumps?
    A. Fixed displacement.
    B. Variable displacement.
    C. Both A and B.
    D. Neither A nor B.
12. Vane pumps are what type of pumps?
    A. Fixed displacement.
    B. Variable displacement.
    C. Both A and B.
    D. Neither A nor B.
13. Technician A states that most fixed-displacement vane pumps are balanced. Technician B states that balanced vane pumps have two cam lobes. Who is correct?
    A. Technician A.
    B. Technician B.
    C. Both A and B.
    D. Neither A nor B.
14. What do vane pumps sometimes contain in order to speed the process of rebuilding the pump?
    A. Port plate.
    B. Side plate.
    C. Cam ring.
    D. Cartridge.
15. Which of the following is the most popular type of piston pump found in mobile equipment?
    A. Rotating-cam radial piston pump.
    B. Rotating-piston radial piston pump.
    C. Inline axial piston pump.
    D. Bent-axis piston pump.
16. Which of the following piston pumps has an eccentric mounted to the pump's input shaft?
    A. Rotating-cam radial piston pump.
    B. Rotating-piston radial piston pump.
    C. Inline axial piston pump.
    D. Bent-axis piston pump.
17. Which one of the following types of piston pumps is commonly used in excavators?
    A. Rotating-cam radial piston pump.
    B. Rotating-piston radial piston pump.
    C. Inline axial piston pump.
    D. Bent-axis piston pump.

18. Which of the following styles of pumps uses a swash plate?
    A. Rotating-cam radial piston pump.
    B. Rotating-piston radial piston pump.
    C. Inline axial piston pump.
    D. Bent-axis piston pump.
19. Within an inline axial piston pump, which of the following components is splined to the input shaft?
    A. Slipper plate.
    B. Bearing plate.
    C. Valve plate.
    D. Barrel.
20. All of the following components are part of the rotating group, *EXCEPT*:
    A. barrel.
    B. piston and slipper assembly.
    C. valve plate.
    D. bearing plate.
21. Hydrostatic inline axial piston pumps commonly have how many total plates, including valve plates and bearing plates?
    A. Zero.
    B. One.
    C. Two.
    D. Three.
22. Which of the following plates is commonly coated in bronze, pinned to the barrel, and rotates at pump rpm?
    A. Side plate.
    B. Port plate.
    C. Valve plate.
    D. Bearing plate.
23. Bent-axis piston pumps use valve plates with what type of shape?
    A. Flat.
    B. Partial spherical shape.
    C. Octal shape.
    D. Hex shape.
24. Which component within a bent-axis piston pump is actuated in order to vary the pump's displacement?
    A. Swash plate.
    B. Valve plate.
    C. Slipper retaining ring.
    D. Bearing plate.
25. Which of the following is used to machine the rotating group components?
    A. Lapping machine.
    B. Desktop belt sander.
    C. Hand grinder.
    D. Hand file.
26. In an inline axial piston pump, which one of the following components holds the pistons against the swash plate?
    A. Port plate.
    B. Slipper retaining ring.
    C. Bearing plate.
    D. Valve plate.
27. Inline axial piston pumps typically have how many pistons?
    A. Three.
    B. Four.
    C. Six.
    D. Nine.
28. Which one of the following inline axial piston pump components has the *primary* responsibility to shut off pump flow?
    A. Bias spring.
    B. Control piston.
    C. Relief valve.
    D. Bias piston.
29. Which one of the following inline axial piston pump components has the *primary* responsibility to turn on the pump flow?
    A. Bias spring.
    B. Control piston.
    C. Relief valve.
    D. Slipper retaining ring.
30. Which of the following components within a variable-displacement inline axial piston pump dictates the pump's quantity of flow?
    A. Port plate.
    B. Swash plate.
    C. Valve plate.
    D. Side plate.

Chapter 5

Rotary Actuators

Objectives

After studying this chapter, you will be able to:

- ✓ Explain how motor displacement affects torque, speed, and pressure.
- ✓ Define motor efficiencies.
- ✓ List the different types of gear motors, including their components.
- ✓ List the three different methods for extending vanes inside motors.
- ✓ List the different types of piston motors and explain their operation.
- ✓ Draw and explain examples of motor circuits.
- ✓ Explain the operation of a limited rotary actuator.

Introduction to Rotary Actuators

Hydraulic systems use actuators as output devices to perform useful work. The actuators convert fluid energy into mechanical energy. Actuators can be divided into two groups, linear actuators, which will be covered in **Chapter 6**, and rotary actuators, the topic of this chapter.

Rotary actuators transform fluid energy into rotational mechanical energy. The actuators can transmit energy in three different forms: high-speed rotational energy, high-torque rotational energy, or limited-rotation energy.

- High-speed motor applications commonly drive components, gear boxes, or shafts at high speeds, for example 1000 to 5000 rpm. In order to achieve the high speeds, the motors commonly have smaller displacements.
- High-torque motor applications are used to drive components that have high operating loads, such as a hydrostatic drive system on a dozer. The high-torque motors commonly use larger displacements and potentially high drive pressures, in excess of 6000 psi (414 bar).
- Limited-rotation actuators rotate less than a full revolution, or just a few rotations. The actuators use fluid energy to move an object in an arcing motion, but are often limited to less than 360 degrees.

A ***hydraulic motor*** is a positive-displacement actuator, containing several oil chambers that receive oil flow for the purpose of driving an output shaft. The output shaft will rotate when the oil pressure acting on the motor's chambers is high enough to overcome the load on the output shaft. Because hydraulic motors are positive displacement, they generally rotate at a speed proportional to the input flow. Stated another way, for a given amount of flow, the motor's rpm should be consistent, and change very little based on small changes in pressure.

However, remember that as system pressure increases, a pump's flow rate drops a little. The same can be stated for hydrostatic motors. As the system pressure substantially increases, internal losses in the motor can cause a drop in speed. For this reason, hydrostatic transmission designers use graphs to size the transmission's pump and motor by plotting the motor's output speed based on the pump's input speed, and the system's pressure. See Figure 5-1.

A motor is similar to a pump, except that a pump generates oil flow and the motor is driven by oil flow. The analogy is similar to an alternator and a starter motor. The alternator generates electron flow and the starter motor is driven by electron flow. See Figure 5-2.

Displacement and Direction

A hydraulic motor can be designed as ***unidirectional,*** meaning that the motor is configured to only rotate in one direction, either clockwise or counter-clockwise. See Figure 5-3. Notice that this application only has a need to drive the motor in one direction. The motor is also a ***fixed-displacement motor,*** which will operate at a constant rpm for a given amount of input flow.

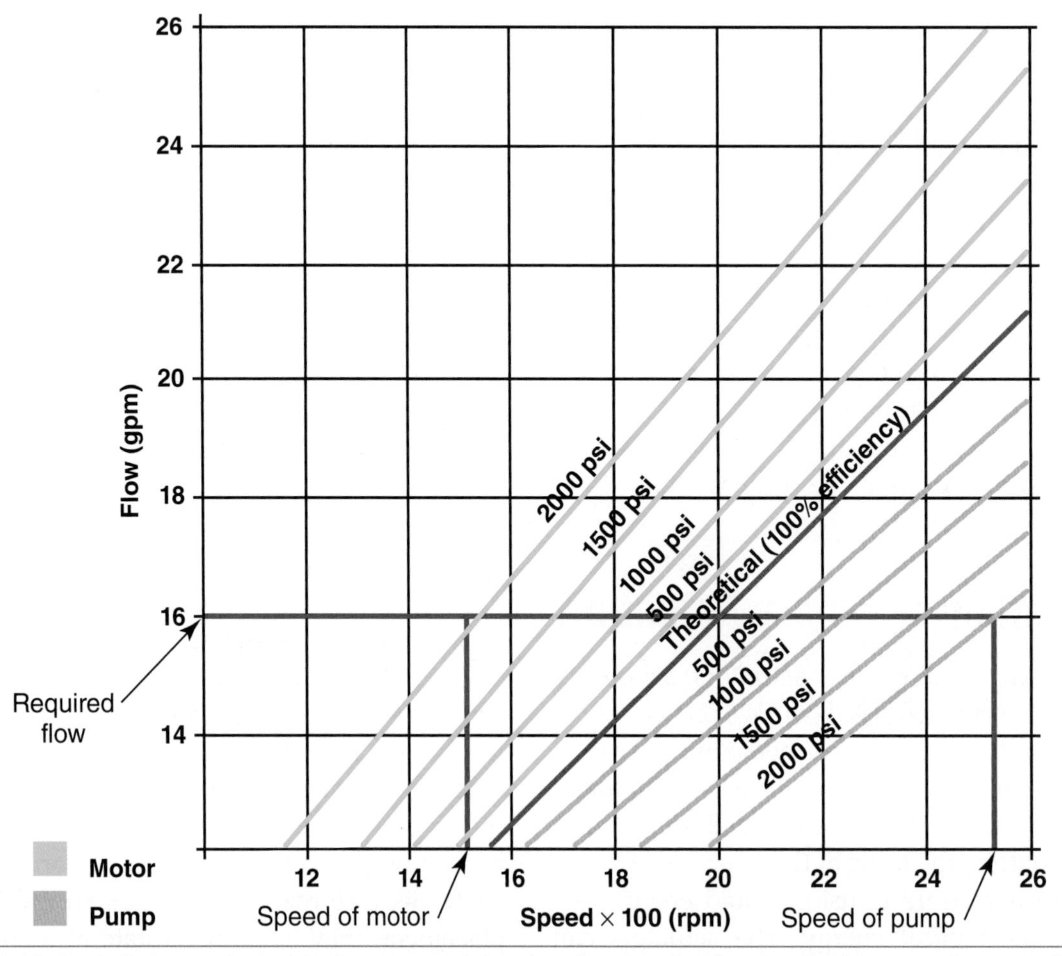

Figure 5-1. Hydrostatic transmission designers can determine the pump and motor displacement for a machine based on the speed the engine is driving the pump, the desired motor output shaft speed, and the maximum system operating pressure. In the example shown, a pump operating at 2500 rpm and producing 16 gpm of flow at 2000 psi would drive the motor at a speed of a little over 1500 rpm.

Other motors are designed to be bidirectional, also known as reversible. This configuration allows the motor's rotation to be reversed by switching the direction of flow at the motor. See Figure 5-4. The following are some examples of reversible motor applications:

- Track drives on excavators.
- Winch drives on dozers.
- Engine cooling fans on motor graders.

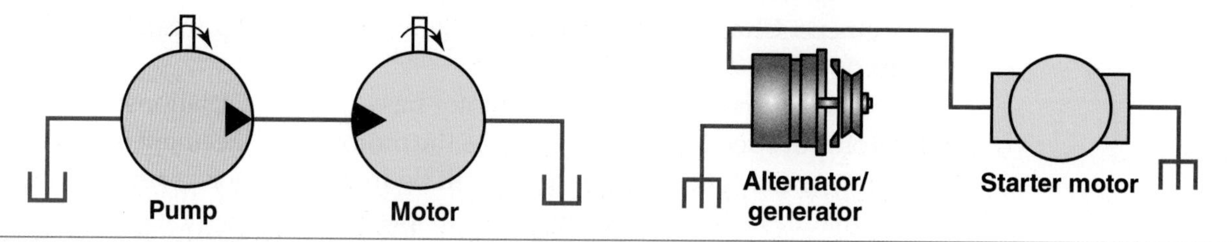

Figure 5-2. A pump is similar to an alternator because they both generate flow. A hydraulic motor is similar to an electric motor, because both will rotate only if they receive flow.

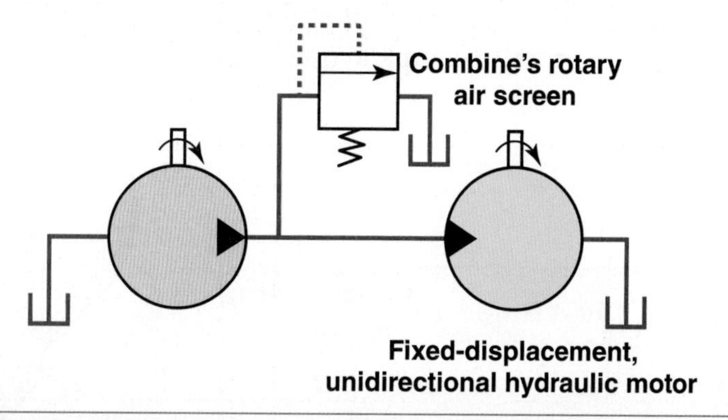

Figure 5-3. An agricultural combine often uses a rotary air screen that provides clean air to the coolers. The screen can be driven by a fixed-displacement, unidirectional hydraulic motor.

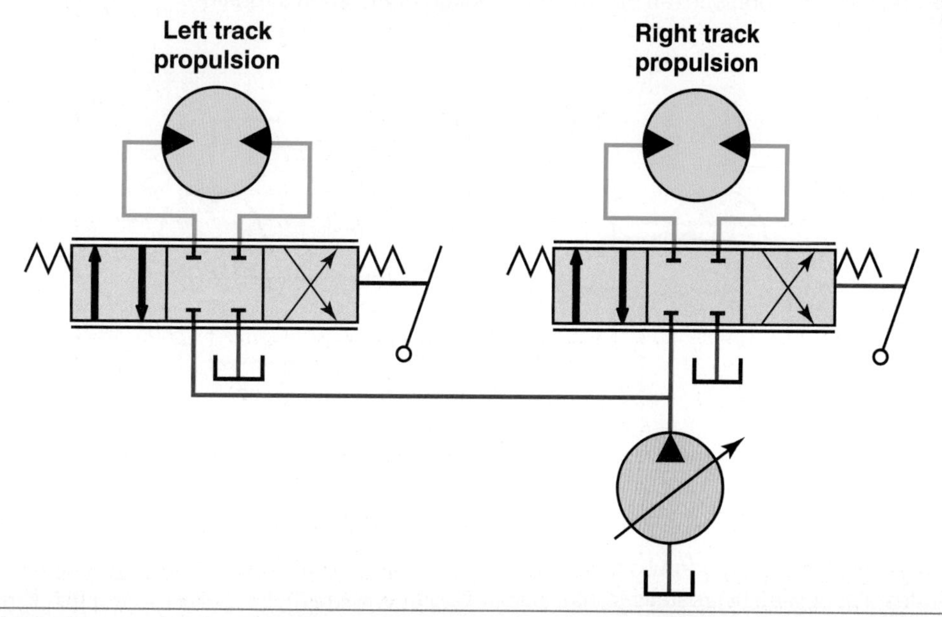

Figure 5-4. A bidirectional hydraulic motor can reverse the motor shaft's rotavtion when the direction of oil flow is switched at the motor. Many excavators use pilot controls to actuate the DCV, which will be covered in **Chapter 20**.

Determining the direction of rotation of a motor is done the same way as determining the direction of a pump. It requires viewing the motor from the end of the motor's shaft. For example, motor manufacturers will specify the rotation based on which port receives the pressurized oil. See **Figure 5-5**.

Variable-displacement motors are commonly used in hydrostatic transmission applications. See **Figure 5-6**. The motor can be designed to have two distinct displacements, for example a high speed and low speed, or the motor can be designed to be infinitely variable between a minimum and maximum displacement. **Chapter 23** and **Chapter 24** will explain hydrostatic transmissions in detail.

The displacement of a hydraulic motor has an inverse relationship with the motor's output speed. See **Figure 5-7**. As the motor's displacement is decreased, the output of the speed of the motor will be increased. For example, consider a motor that has a four cubic inch fixed displacement. The motor must receive four cubic inches of oil for the motor to make one complete revolution. If the motor was replaced with a smaller displacement motor, for example a two cubic inch displacement motor, then it would require only two cubic inches of oil flow in order to complete one revolution.

Port A	Port B	Direction of Rotation
Pressurized		CW
	Pressurized	CWW

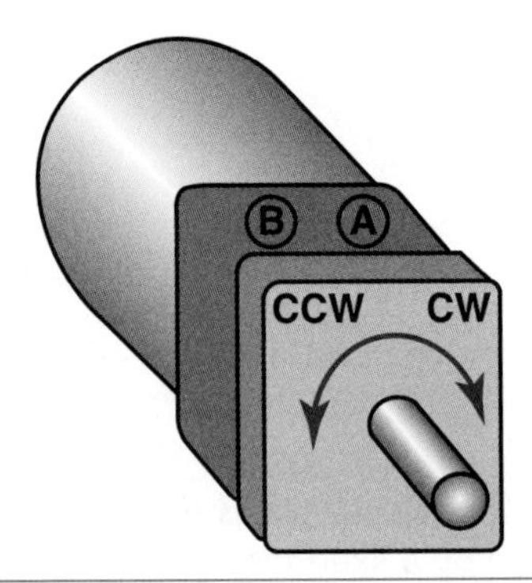

Figure 5-5. A motor's direction of rotation is determined by viewing the end of the motor's shaft. Manufacturers will specify which port receives the pressurized oil and the direction of the shaft's rotation.

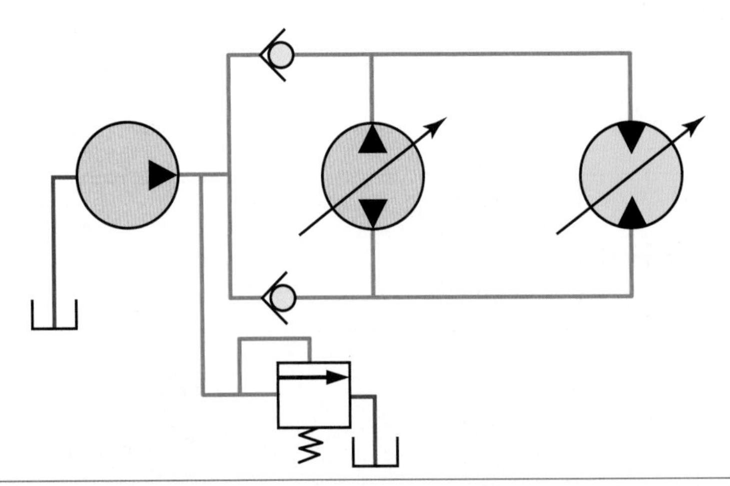

Figure 5-6. Variable-displacement hydraulic motors are used in many hydrostatic transmission applications. This figure illustrates a simplified hydrostatic transmission. Notice that the motor is also reversible. Practically all hydrostatic transmissions use a variable-displacement pump to obtain a variable range of travel speeds. Some transmissions use a variable-displacement motor to obtain a wider range of travel speeds.

If the first motor received a fixed amount of flow, such as four cubic inches per minute, the first motor would make one revolution every minute. If the second motor also received four cubic inches of oil per minute, that motor would rotate twice as fast. It would complete two revolutions per minute, because its displacement is half the size of the first motor. In addition to using a smaller-displacement motor, motor speed can also be increased by increasing the oil flow supplied to the motor.

Torque

A motor's torque has a directly proportional relationship with its displacement. ***Torque*** is the strength of force being applied through an arcing motion, such as a rotating lever. See **Figure 5-8**. For example, if a person applied 30 pounds of force at the end of a one-foot long wrench, 30 foot-pounds of torque would be exerted through the wrench. Torque applied to a lever equals force times the distance between the point where the force is applied and the lever's fulcrum. Torque can be measured in:

- Pound-inches (lb-in)
- Pound-feet (lb-ft)
- Newton meters (N•m)

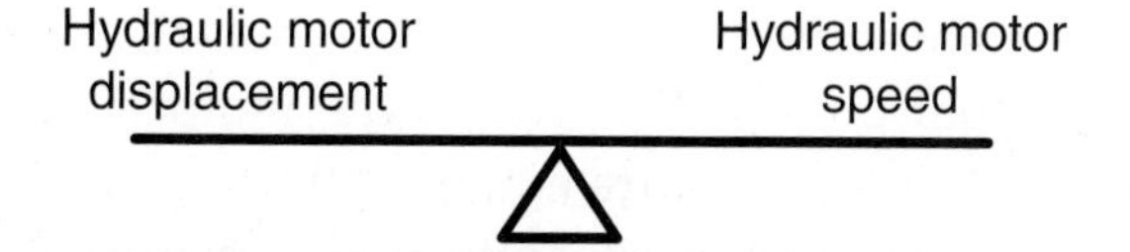

Figure 5-7. A motor's displacement is inversely proportional to its output speed. For a fixed amount of flow, a smaller-displacement motor will rotate faster than a larger-displacement motor.

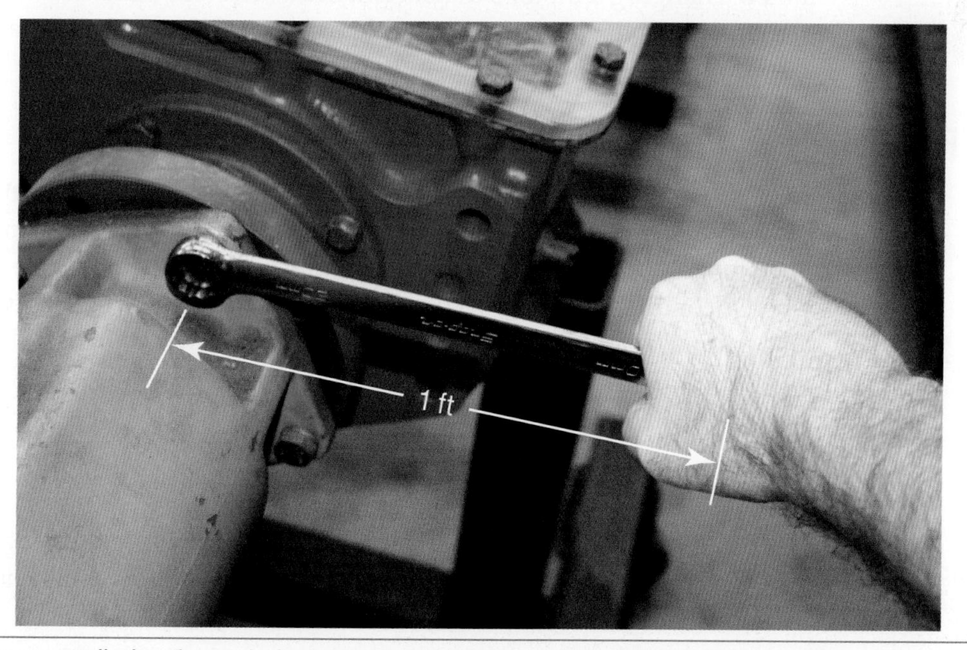

Figure 5-8. Torque applied to the end of a lever equals the force times the length of the lever.

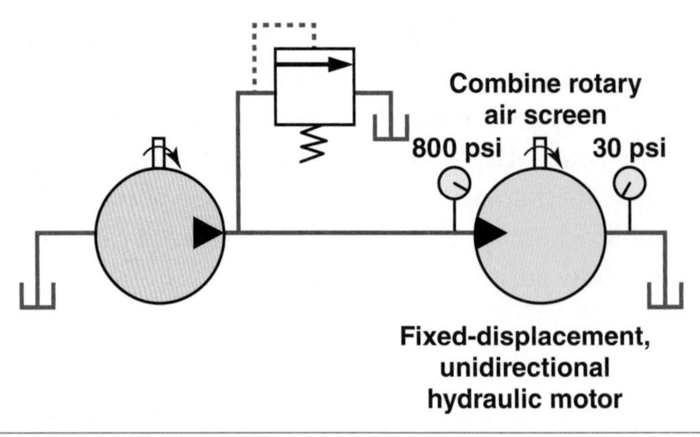

Figure 5-9. The effective pressure acting on a motor is called the differential pressure, which equals the motor's inlet pressure minus the motor's outlet pressure (return pressure).

For any given amount of pressure, as a motor's displacement is increased, the motor's torque will also increase. A larger-displacement motor has a larger effective area, and this larger area multiplied by the pressure equals a larger torque. A motor's torque can be found by multiplying the motor's displacement times the motor's differential pressure, and dividing that product by 2π.

$$\text{Torque} = (\text{Displacement} \times \text{Pressure})/2\pi$$

The motor's ***differential pressure*** is defined as the difference in pressure between the motor's inlet and the motor's outlet. The differential pressure is the actual pressure acting on the motor's fluid chambers. For example, if the motor's inlet pressure equals 800 psi (55 bar) and the return line pressure equals 30 psi (2 bar), the net effective pressure acting on the motor would equal 770 psi (53 bar). See Figure 5-9.

A motor's torque can be increased two different ways: increasing the motor's displacement or increasing the fluid differential pressure acting on the motor. Changing a motor's displacement has two opposite effects. A larger displacement equals more torque and less speed. Conversely a smaller displacement equals decreased torque and increased speed.

Operating Pressure

A motor's displacement not only affects speed and torque, but it influences the operating pressure. For a given load, a larger-displacement motor will operate at a lower pressure than a smaller-displacement motor. For example, when a combine is configured with the optional rear-wheel drive, the operator's manual might recommend running the rear-wheel drive while operating in the field. The net result is more displacement, which lowers the drive pressure and generates less heat.

Speed

The speed of a motor can be calculated by multiplying the flow rate (in gpm) times 231 (the number of cubic inches in a gallon) and dividing that product by the motor's displacement.

$$\text{Speed} = (\text{Flow Rate} \times 231)/\text{Displacement}$$

For example if a 3.0 cubic-inch displacement motor was receiving 50 gpm, the motor speed would equal 3850 rpm:

Speed = (Flow Rate × 231)/Displacement

Speed = (50 gpm × 231)/3.0 in^3

Speed = 3850 rpm

Volumetric Efficiency

The computed speed of 3850 rpm is the theoretical speed, which will drop a little as pressure rises due to internal losses in the motor. A motor's volumetric efficiency can be computed by dividing the actual rpm of the motor by the theoretical rpm and multiplying it by 100.

Volumetric Efficiency = (Actual rpm/Theoretical rpm) × 100%

For example if the motor's actual rpm was 3650, then the motor's volumetric efficiency would be 94.8%:

Volumetric Efficiency = (Actual rpm/Theoretical rpm) × 100%

Volumetric Efficiency = (3650/3850) × 100%

Volumetric Efficiency = 0.948 × 100%

Volumetric Efficiency = 94.8%

Mechanical Efficiency

Mechanical efficiency accounts for energy losses that occur due to friction and drag. A motor's mechanical efficiency is more difficult to determine, because it requires measuring the motor's output with a torque sensor. Torque sensors are typically used only in engineering. The mechanical efficiency is determined by dividing the actual motor's torque by the theoretical torque and multiplying that product times 100.

Mechanical Efficiency = (Actual Torque/Theoretical Torque) × 100%

For example, if a motor's actual output torque was 450 pound feet, and its theoretical torque was computed to be 475 pound feet, the motor's mechanical efficiency would be 94.7%:

Mechanical Efficiency = (Actual Torque/Theoretical Torque) × 100%

Mechanical Efficiency = (450/475) × 100%

Mechanical Efficiency = 0.947 × 100%

Mechanical Efficiency = 94.7%

Overall Efficiency

Overall efficiency accounts for losses due to both volumetric and mechanical inefficiencies. It is calculated by multiplying the volumetric efficiency by the mechanical efficiency and dividing by 100.

Overall Efficiency = (Volumetric Efficiency × Mechanical Efficiency)/100%

Using the previous examples, a motor operating at 94.8% volumetric efficiency and 94.7% mechanical efficiency would have an overall efficiency of 89.8%:

Overall Efficiency = (Volumetric Efficiency × Mechanical Efficiency)/100%

Overall Efficiency = (94.8 × 94.7)/100%

Overall Efficiency = 8977.56/100%

Overall Efficiency = 89.8%

Types of Motors

Hydraulic motors can be categorized into three groups:

- Gear motors.
- Vane motors.
- Piston motors.

All three styles of motors are used in mobile hydraulic systems. All three are positive displacement. As fluid enters the motor, the fluid pressure acts on the motor chamber's effective area. The motor will rotate when the combination of the fluid pressure and motor displacement can overcome the load on the motor's output shaft.

Each of the three different types of motors has advantages and disadvantages when compared to each other.

Gear Motors

Some gear motors are classified as external-toothed gear motors, which contain two external-toothed gears. Other gear motors are classified as internal-toothed gear motors, which contain one internal-toothed gear and an external-toothed gear. ***External-toothed gear motors*** use two spur gears. One gear is called the drive gear, which is coupled to the output shaft. The other external-toothed gear is the idler. See Figure 5-10. As fluid is supplied to the motor's inlet, the pressure causes the gears to rotate. Because the two gears have external teeth, the gears rotate in the opposite direction.

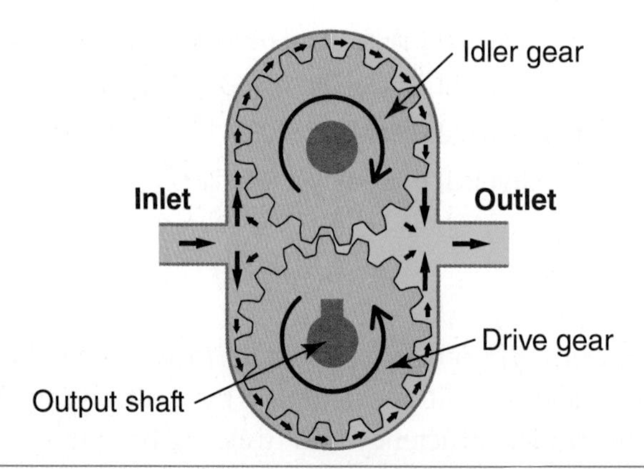

Figure 5-10. An external-toothed gear motor has two spur gears meshed to each other. The drive gear is splined to the motor's output shaft. The idler gear's shaft is enclosed inside the motor's housing and is not used to perform work. The motor is driven by fluid flow.

The torque capacity of an external-toothed gear motor is the result of three factors:

- The differential pressure across the motor.
- The area of one tooth on one gear. See Figure 5-11.
- The diameter of the gear.

The motor's differential pressure will always be acting across three teeth. See Figure 5-12. Two teeth are located on one gear, and one tooth is located on the other gear. The force generated by one tooth on each of the gears will offset each other, which cancels the effect of two teeth. As a result, one tooth remains left to develop the motor's output force. It is the area of this single tooth, the distance of how far that tooth is located from the axis of the shaft, and the differential pressure across that tooth that determines the motor's torque.

Note

The tooth serving as the effective tooth alternates from gear to gear as the motor gears rotate.

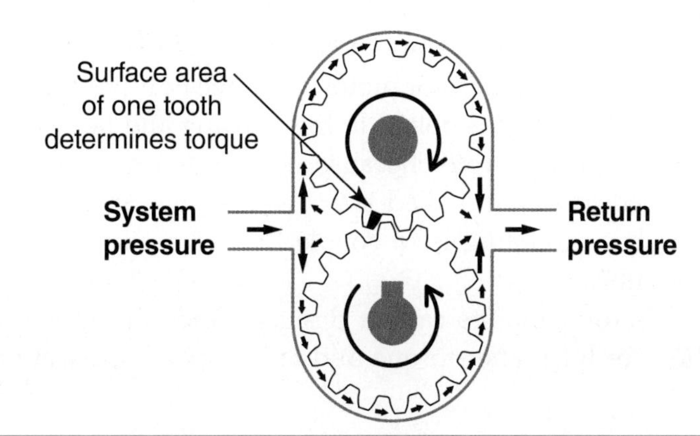

Figure 5-11. The torque of an external-toothed gear motor is dependent on the area of one tooth on one gear.

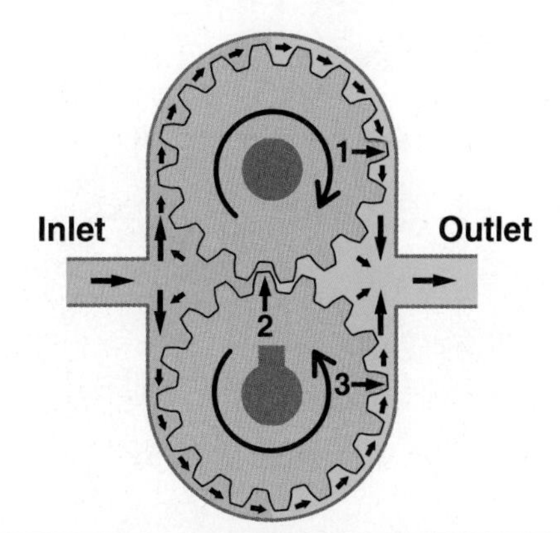

Figure 5-12. The differential pressure acting on an external-toothed gear motor will generate a force across three teeth, two of which oppose each other, which cancels the force generated by two teeth. Notice that the force applied to tooth 2 works in opposition to the force applied to tooth 3. These forces cancel each other out, leaving the force applied to tooth 1 to drive the motor.

Gerotor

Internal-toothed gear motors are commonly configured with a large internal-toothed ring gear that surrounds a smaller external-toothed gear. The ***gerotor motor*** is a popular type of internal-toothed gear motor used in the mobile equipment industry. It contains an internal-toothed ring gear and a star-shaped external-toothed gear. Both the internal- and external-toothed gears have lobe-shaped teeth. See Figure 5-13. The larger internal-toothed gear always has one more tooth than the external-toothed gear. Gerotor motors are sometimes called orbital motors, because the small star-shaped gear orbits inside of the ring gear.

Note that the center axis of the star-shaped external-toothed gear is offset from the axis of the large internal-toothed gear. Due to the offset gears, the motor's internal shaft has to rotate in a pivoting fashion. As the external-toothed gear spins, the shaft rotates and simultaneously orbits around the inside of the ring gear.

A gerotor motor can be designed in one of two types of configurations. The Eaton Charlynn gerotor motor has the internal-toothed ring gear held stationary while the smaller, external-toothed gear rotates inside of the ring gear. Other manufacturers have designed the gerotor motor so that both the internal- and external-toothed gears rotate.

In the Eaton gerotor motor, the large internal-toothed ring gear is called the stator, because it is bolted to the housing and held stationary. The star-shaped gear is called a rotor because fluid pressure causes it to rotate. The rotor is splined to a shaft called a drive coupling. See Figures 5-14. The purpose of the drive coupling is to couple the rotor with the rotary valve output shaft assembly.

Gerotor motors can be designed for high-speed and low-torque applications or for higher-torque and lower-speed applications. See Figure 5-15. For

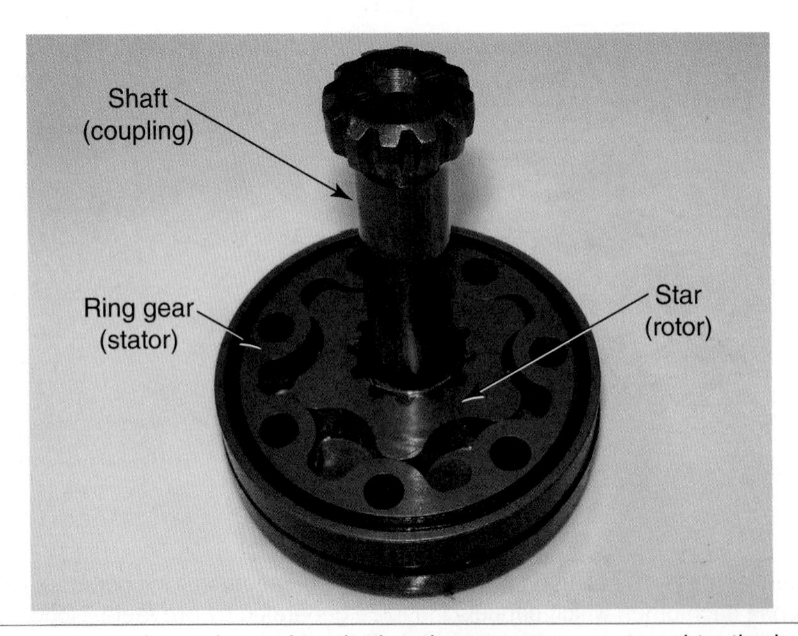

Figure 5-13. A gerotor motor contains an internal-toothed stationary gear, an external-toothed rotating gear, and a drive coupling.

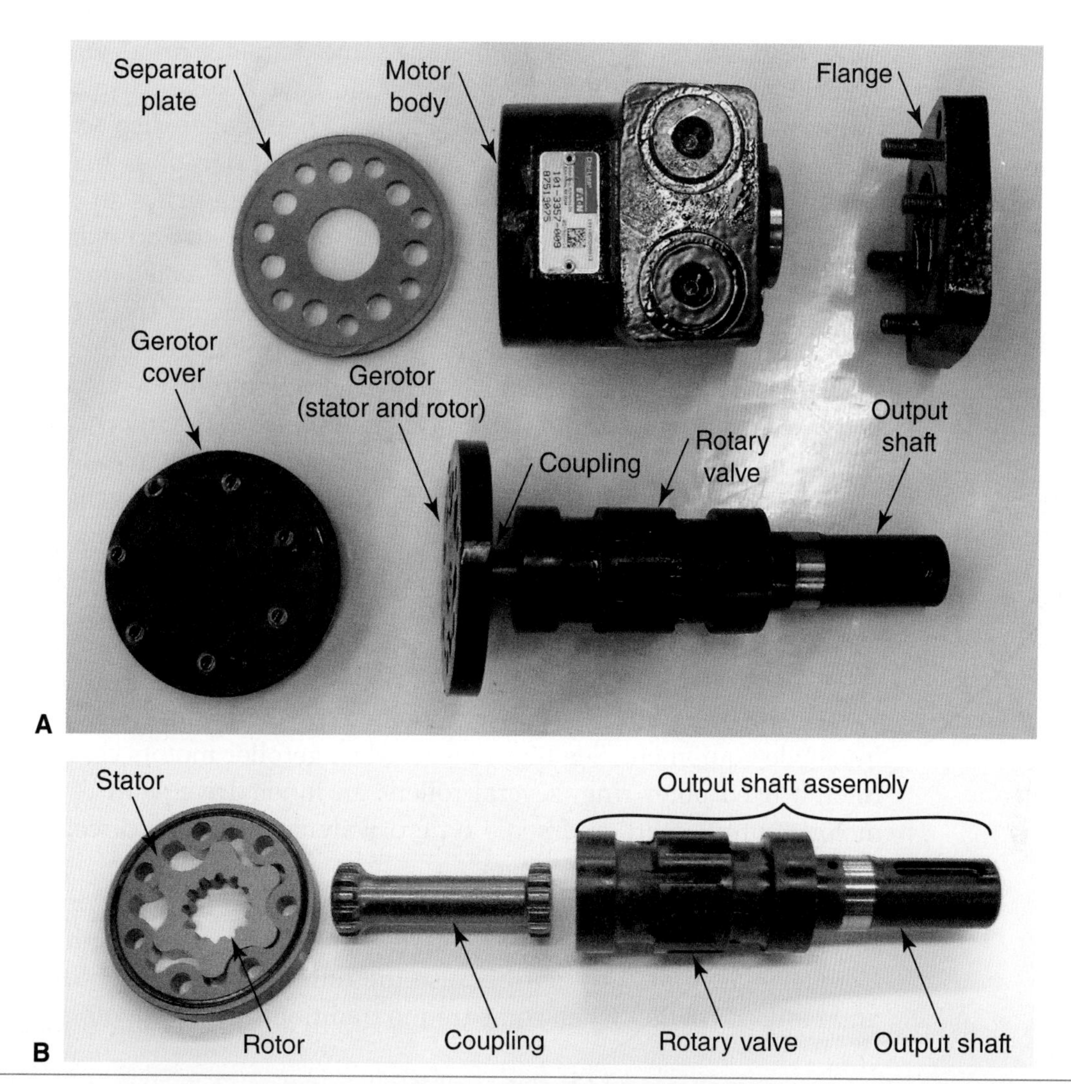

Figure 5-14. Parts of an Eaton Gerotor motor. A—The gerotor and output shaft assembly are installed in the motor body and held in place by the gerotor cover and flange. B—The gerotor consists of a stationary ring gear (stator) and a rotating rotor gear. The rotor gear is splined to a drive coupling, which is splined to the rotary valve output shaft assembly.

Figure 5-15. Gerotor motor sizes. A—Gerotor motors can be designed with thin rotors for higher speeds or with wide rotors for slower speeds or higher torque applications. B—The smaller-displacement motor was used to drive a combine's chaff spreaders. The larger-displacement motor has a bypass relief valve mounted to the motor's housing and was used to drive a combine's rotary air screen. The small-displacement motor has a maximum speed in excess of 3000 rpm. The large-displacement motor is designed to operate at a much slower speed, 200 rpm or less.

example, if two gerotor motors had the same diameter rotors and stators, the motor with a thicker rotor and stator would have a larger displacement, providing slower speeds and higher torque. The smaller-displacement motor would be used for higher-speed and lower-torque applications. In Figure 5-15B, note that the motor with the thicker rotor and stator (top) has an overall longer motor housing than the smaller-displacement, thin rotor and stator assembly.

Within a gerotor motor the fluid pressure is directed to the opening between the rotor and stator. The fluid pressure acts on the rotor causing it to rotate the drive coupling. The drive coupling is splined to the valve output shaft assembly. Within the Eaton Charlynn gerotor motor, the valve and output shaft are an assembly and rotate at the same speed as the rotor.

Within the Eaton gerotor motor, as the rotor spins, the motor's expanding volume orbits along with the rotor, which requires a rotating valve. The rotary valve directs supply oil to motor inlet and directs the motor's outlet oil to the housing's return port. The types of internal valving mechanisms within internal gear motors vary from manufacturer to manufacturer.

Geroller

A ***geroller motor*** is another popular type of internal-gear motor found in mobile equipment. See Figure 5-16. Most geroller motors contain an inner rotor, a fixed outer ring, several rollers, an inner drive shaft coupled to the motor's output shaft, and some type of valving device to direct fluid to the motor's expanding oil chambers.

If the geroller's external-toothed gear has six teeth, the motor will contain seven rollers, which is one more than the number of external teeth on the rotor. Geroller motors are popular because the rollers provide smoother operation and reduced friction when compared to traditional gerotor motors.

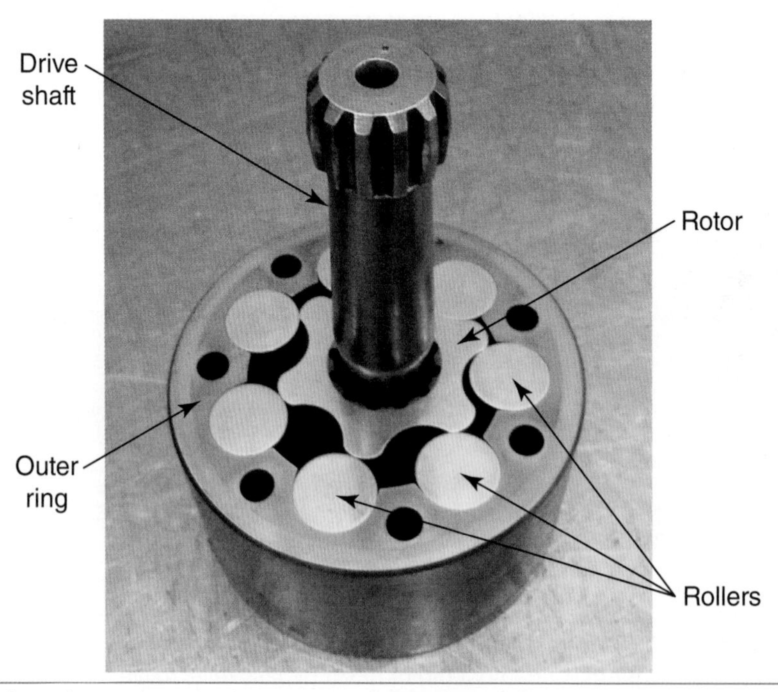

Figure 5-16. A geroller is an internal-toothed gear motor. It has steel rollers between an external-toothed gear and an internal-toothed gear.

Another type of geroller motor contains an inner rotor, several sealing rollers, an orbiting outer ring, several anti-rotation rollers, and an outside locating ring. These motors are also called orbital motors because the middle ring orbits around the inside rotor during motor operation.

Vane Motors

A ***vane motor*** contains a rotor, with sliding vanes, that is splined to the motor's output shaft. As described in **Chapter 4,** vane pumps require the vanes to be fully extended in order to achieve positive displacement, and pumps rely on centrifugal force to initially extend the vanes. Vane motors are different because the rotor will not turn until the vanes extend to provide a positive-displacement chamber, and centrifugal force cannot extend the vanes until the rotor is spinning. For this reason, vane motors must use some other method to extend the vanes prior to operation. Three methods are commonly used for extending the vanes:

- Coil springs.
- Wire spring clips.
- Fluid pressure.

Coil springs can be placed directly underneath the vanes in order to force the vanes against the cam ring. The vane in Figure 5-17 is extended by three coil springs. Figure 5-18 illustrates the vanes located inside the rotor and held against the cam ring.

Wire spring clips perform the same function as the coil springs. They hold the vanes in the extended position, providing a positive-displacement chamber. See Figure 5-19. The clips loop around the rotor assembly in an "S" shaped pattern. A motor with 12 vanes uses six dowel pins for securing the spring clips. The clips place a force on the vanes, causing the vanes to be held against the cam ring. Fluid pressure works in conjunction with the spring clips.

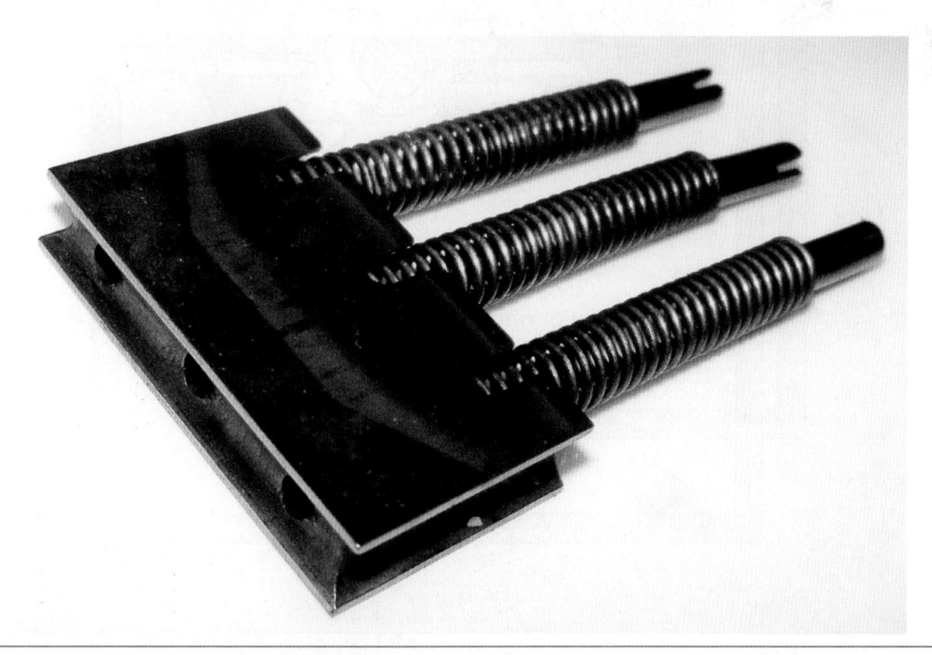

Figure 5-17. Three coil springs are placed under the vane to hold it extended against the motor's cam ring. This vane also contained roll pins inside of the springs.

Figure 5-18. The cross-sectional view of this vane motor shows the coil springs holding the 12 vanes in the extended position. The motor contains 36 coil springs to hold the vanes against the cam ring, although only 12 of the springs are visible from either end of the motor.

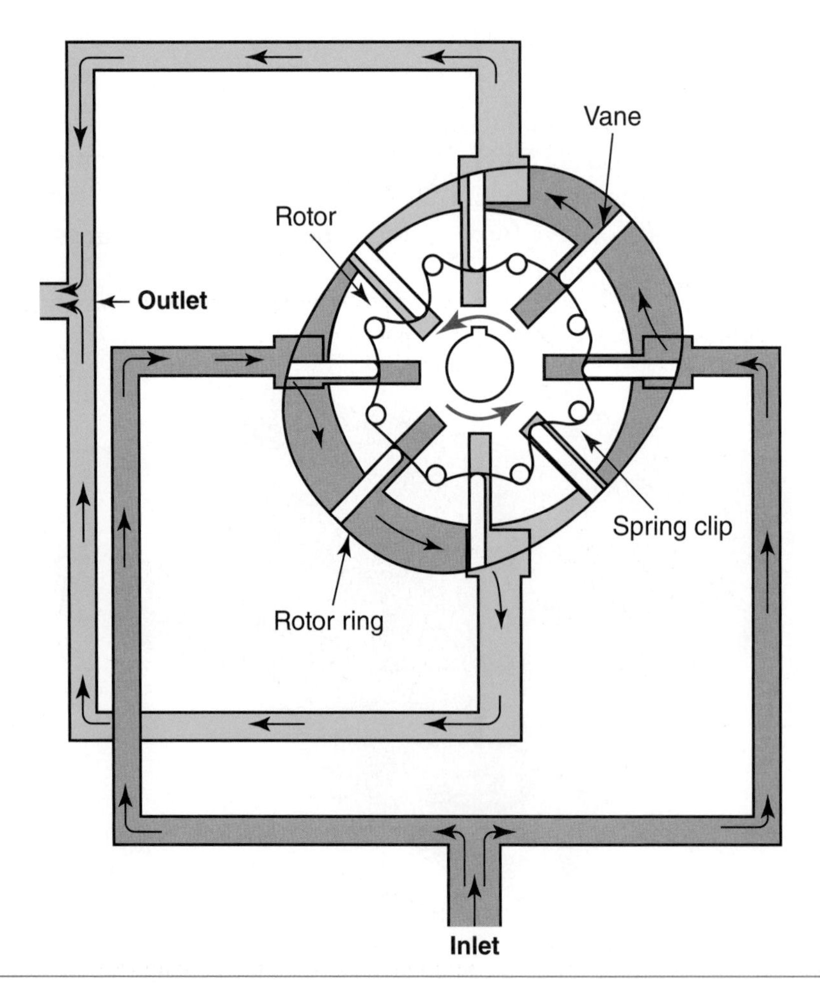

Figure 5-19. Spring clips can be used to hold the vanes against the cam ring.

Fluid pressure is the third method used for extending a motor's vanes. A spring-loaded check valve initially prevents oil from entering the motor's chambers. After fluid has extended the vane, pressure will build, causing the check valve to open, which directs the motor's inlet fluid to the chambers. See Figure 5-20. The motor will then rotate if the differential pressure is high enough to overcome the load on the motor's shaft.

An ***overrunning load*** is a load that causes the motor to continue to freewheel when flow is reduced. If a vane motor drives an overrunning load and uses fluid pressure to extend the vanes, a spring-loaded check valve must be used in the return line. See Figure 5-21. The check valve pressure setting is approximately 90 psi (6 bar) ± 30 psi (2 bar). The valve will create a back pressure high enough to keep the vanes extended while the motor freewheels to a stop.

Vane motors are typically fixed displacement. They are balanced and commonly use a dual-lobe cam ring. The two-lobe design places offsetting loads on the rotor, shaft, and bearings, enabling motor to have a longer service life. Note that Rexroth manufactures a cam ring with four lobes. See Figure 5-22. This motor contains four additional vanes located in the cam ring to ensure a positive seal between each of the four lobes.

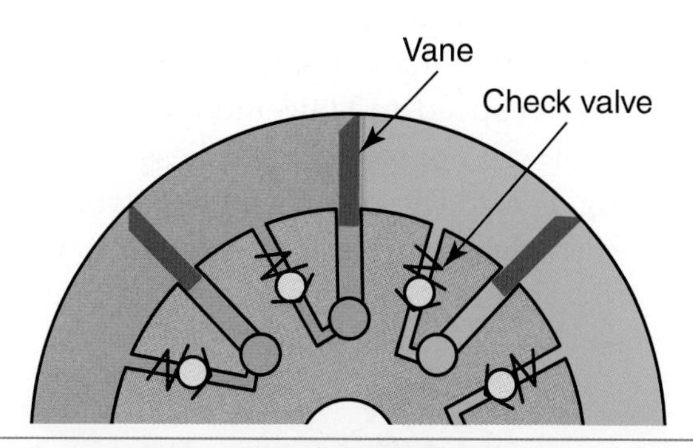

Figure 5-20. Fluid pressure is used to extend a vane motor's vanes. After the vanes have been extended, a check valve opens, allowing fluid pressure to be directed to the motor's chambers.

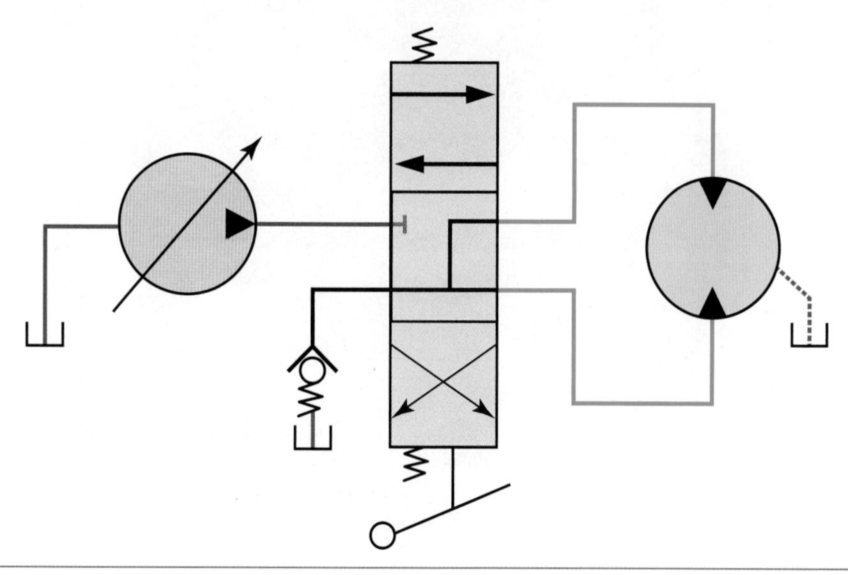

Figure 5-21. Vane motors that use fluid pressure to extend the vanes have a spring-loaded check valve to create back pressure to hold the vanes extended as the motor coasts to a stop.

Piston Motors

Piston motors are commonly used in mobile hydraulic systems and can be fixed displacement or variable displacement. They are used in traditional implement systems and hydrostatic transmissions. Piston motors are categorized as inline axial, bent axis, radial, and cam lobe designs.

Inline axial piston motors contain a rotating group consisting of a barrel, pistons, a slipper retaining ring, and a bearing plate. In addition to the rotating group, the motors contain a swash plate and valve plate. Fixed-displacement piston motors use a stationary swash plate that has a fixed angle. See Figure 5-23.

The motor's valve plate routes oil into the motor's rotating group. The fluid is directed to the end the motor's pistons. When fluid pressure is high enough to overcome the load on the output shaft, the pistons extend, which causes the pistons to slide up the motor's swash plate as the barrel rotates the motor's output shaft. As the motor continues to rotate, the pistons reciprocate back into their individual cavities inside the motor's barrel. As the pistons are pushed back into their bores, they exhaust the motor's outlet oil back to the reservoir.

A larger swash plate angle equals a larger-displacement motor, which requires more oil to make a single revolution. A small swash plate angle will result in a faster speed and lower torque output. If a motor has a neutral swash plate angle, it will not rotate, regardless of the amount of pressure applied to the rotating group. The pistons must be able to slide up and down along the angled swash plate in order for the motor to rotate. Fixed-displacement piston motors are sometimes called single-speed motors, meaning that they deliver a single speed for a given amount of flow.

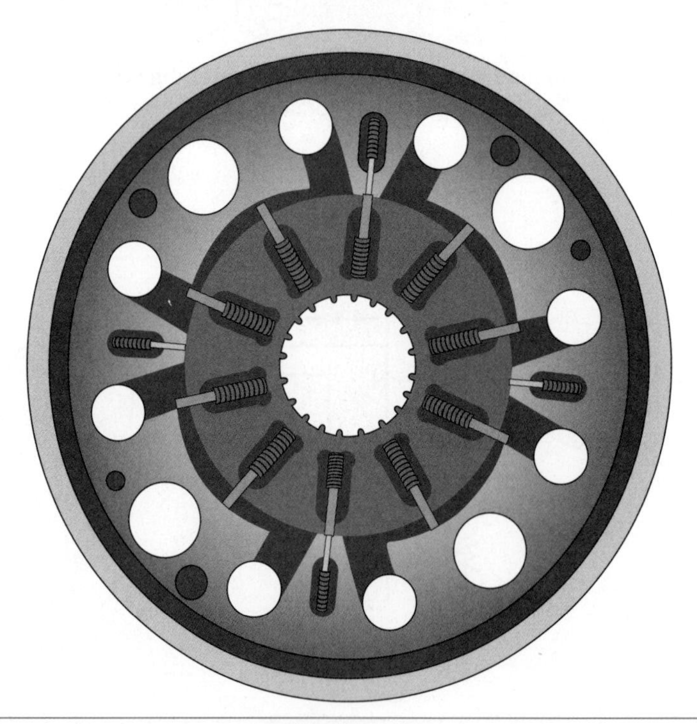

Figure 5-22. Rexroth produces a four-lobe vane motor that contains four vanes in the cam ring to provide a seal between each of the four lobes.

Variable-displacement inline axial piston motors can use the trunnion bearing design or the cradle bearing design. The most common application for a variable-displacement inline axial piston motor is hydrostatic transmissions. These motors will be explained in **Chapter 23**.

Bent-axis piston motors do not contain a swash plate. The motor gets its fixed angle from the difference in angle between the shaft and the motor's barrel. Figure 5-24 shows a fixed-displacement bent-axis piston motor that was used to drive a combine rotor. The pistons within this bent-axis motor are not cylindrical, but are tapered. See Figure 5-25. The end of each piston also contains two piston rings.

Radial piston motors have the pistons placed perpendicular to the motor's output shaft. The motor can be a rotating cam design, which uses a stationary cylinder block, or a rotating piston block design. The ***rotating cam piston motor*** design routes supply oil to the top of the pistons. When the pressure is high enough it pushes against the eccentric cam located on the motor's output shaft. The pistons force the cam output shaft assembly to rotate.

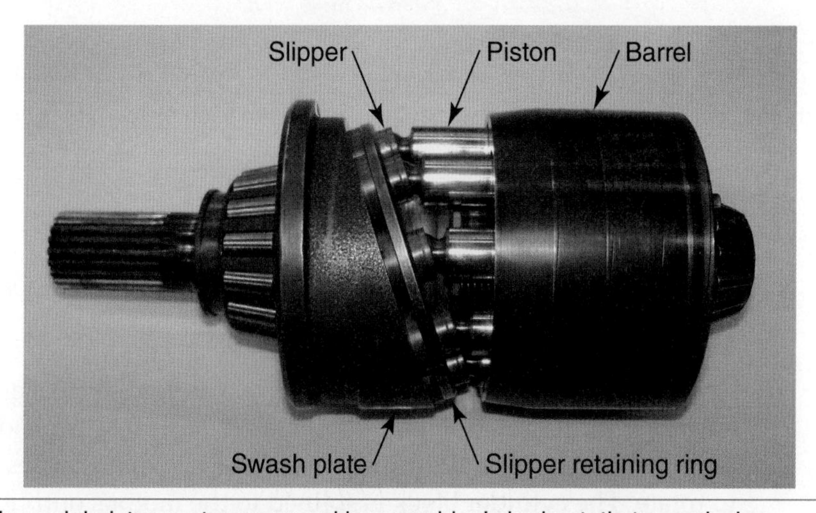

Figure 5-23. This inline axial piston motor was used in a combine's hydrostatic transmission.

Figure 5-24. A bent-axis piston motor obtains its positive displacement by positioning the barrel at an angle to the shaft. This motor contains a reluctor wheel that is used in conjunction with a speed sensor.

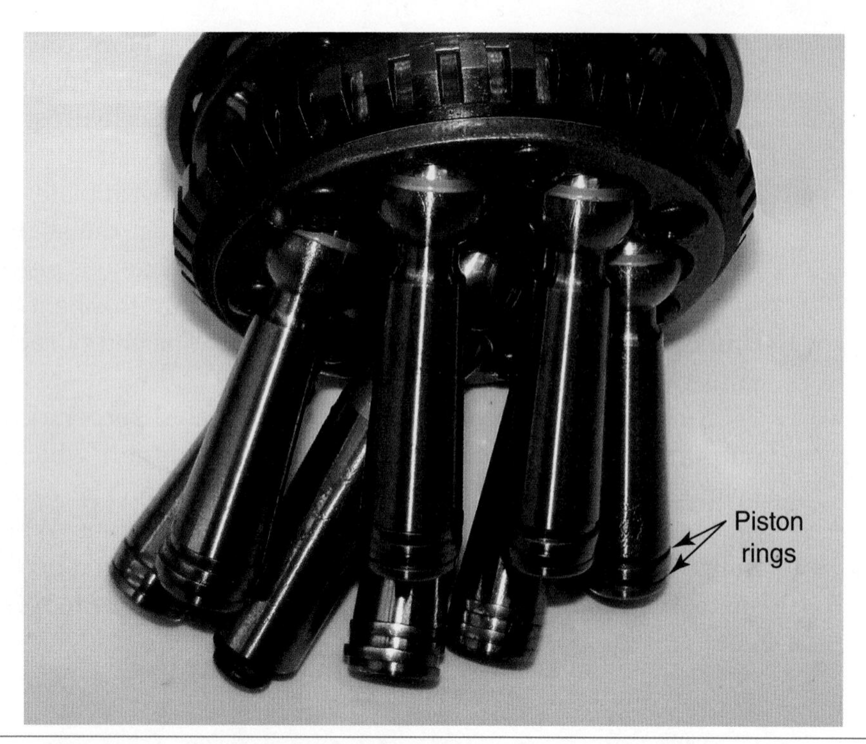

Figure 5-25. This bent-axis piston motor uses tapered pistons. The end of each piston has two piston rings.

Note

Rotating cam piston motors are not commonly found in mobile hydraulic systems.

A ***rotating cylinder block piston motor*** contains several pistons that are located inside a rotating cylinder block. Fluid pressure is directed to the pistons causing them to slide out against a ring, which causes the cylinder block to rotate. The cylinder block is splined to the motor's output shaft.

The ***cam lobe motor*** is the most popular type of radial piston motor used in the mobile hydraulic industry. It is commonly used in hydrostatic transmission applications, for example in skid steers and rear drive axles on combines. It contains a rotating cylinder block, also known as a piston carrier or rotor, with several pistons that reciprocate in and out of the cylinder block. The pistons have a roller, also known as a follower, mounted on the end of the piston assembly. When fluid pressure is directed to the pistons, it causes the pistons to extend. As the pistons extend, they cause the rollers to ride along the inside of a cam ring which forces the cylinder block to rotate. See Figure 5-26. The motor's output shaft or wheel spindle is splined to the cylinder block.

Cam lobe motors commonly are configured with eight, ten, or twelve piston assemblies. A cam lobe motor uses an oil distributor, also known as a manifold, to distribute the oil to the individual pistons in the correct sequence. The distributor contains multiple galleries, also known as fluid passageways. These individual fluid passageways lead to a group of pistons that are timed to power the motor as they receive fluid pressure. The galleries also direct the motor's return oil out of the motor.

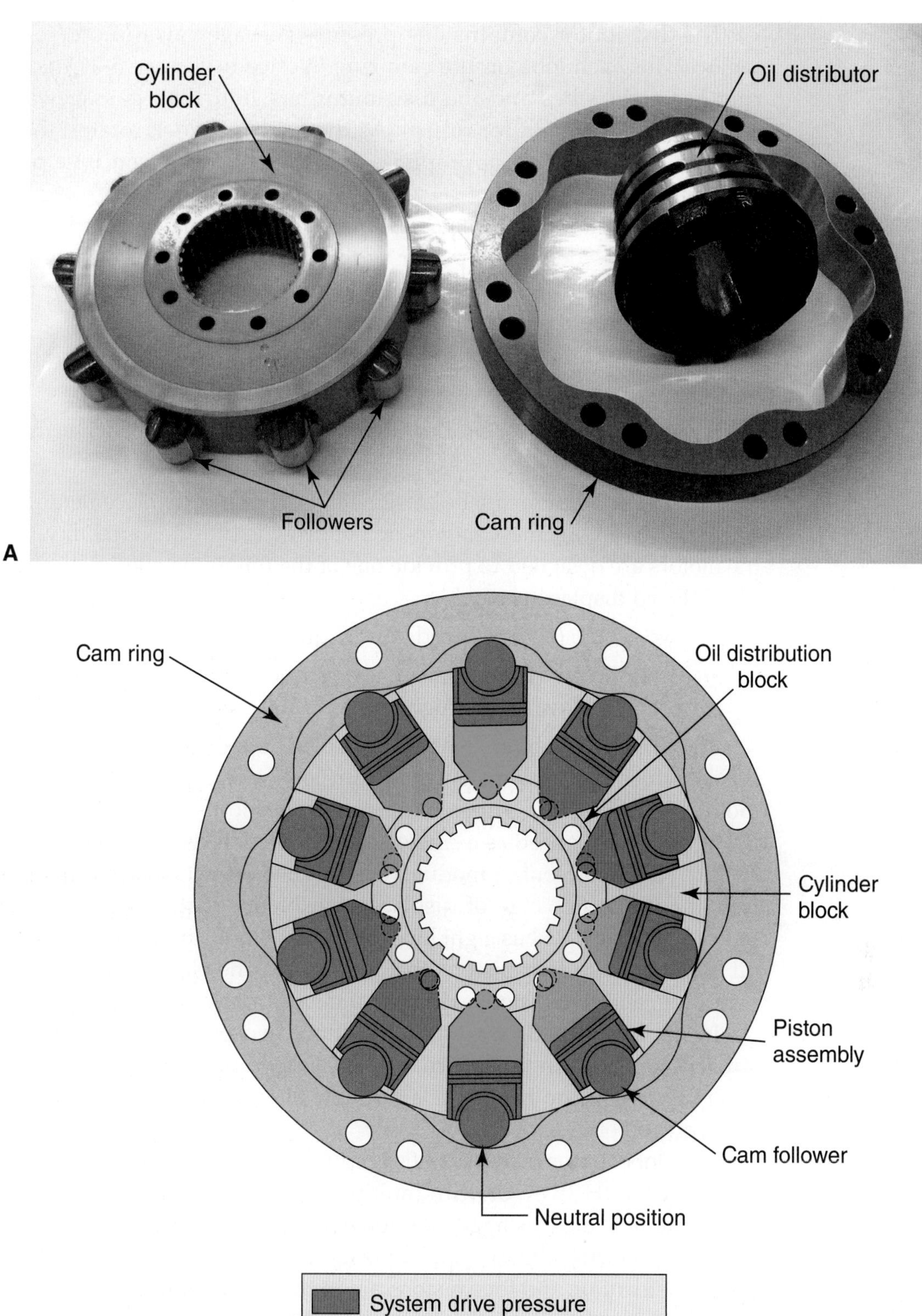

Figure 5-26. A cam lobe motor. A—Components of a cam lobe motor. B—During operation, pistons reciprocate in and out of a rotating cylinder block. Fluid pressure forces the pistons to extend against a cam ring. As the rollers ride along the cam ring, it causes the cylinder block to rotate.

The distributor contains one pressure passageway and one return passageway for each lobe on the cam ring. Notice in Figure 5-26 that the cam ring has eight lobes and the distributor has 16 drilled passageways, eight for pressure and eight for return. As the cylinder block rotates, the distributor redirects oil to the appropriate oil passageways to continue propelling the motor.

As the cylinder block rotates, the pistons will be in different states:

- Driving—delivering power by receiving high pressure oil, which causes the piston followers to follow the expanding cam lobe on the fixed cam ring, resulting in the cylinder block rotating.
- Neutral—piston is located between the power stroke and return stroke.
- Returning—pistons following the decreasing cam lobe. During this state, the piston's oil is routed to the return circuit.

Cam lobe motors can be designed as variable displacement. However, unlike axial motors, these motors do not provide infinite variability. Generally the motors are designed to provide one of the following configurations:

- Fixed displacement.
- Two speed: (1) high speed/low torque, and (2) low speed/high torque.
- Three Speed: (1) high speed/low torque, (2) medium speed/medium torque, (3) low speed/high torque.

The Case IH combine's rear axle can be equipped with a Power Guide Axle (PGA), which contains a right rear cam lobe motor and a left rear cam lobe motor. When the PGA is operational, it provides four-wheel drive. The Case IH PGA can be configured as a single-speed drive or a two-speed drive.

The electronic control module (ECM) is used to actuate a solenoid that controls the motor's change of displacement. Notice that the Case IH cam lobe motor in Figure 5-26 has eight lobes and ten pistons. In the low-speed position, oil is delivered to four pistons that are driving the block, four pistons will be returning, and two pistons will be in a neutral stroke.

In the high-speed, low-torque position, two pistons will be in the drive stroke, six pistons in the return stroke, and two pistons in the neutral stroke. Note that one piston has to cycle through all eight fixed cam lobes in order to make one revolution.

Cam lobe motors can be configured with a single bank (or row) of pistons, like the Case IH PGA, or with multiple banks (or rows) of pistons. The John Deere combine uses a staggered double row of eight pistons totaling 16 pistons per motor. During the low-speed position six pistons are always driving, six pistons returning, and four pistons are in the neutral stroke. In the high-speed position three pistons are driving, nine pistons are returning, and four pistons are on the neutral stroke.

Three-speed cam lobe motors are less common than single-speed or two-speed motors. One manufacturer produces a three-speed motor with 15 lobes and 12 pistons. They operate similarly to the single-speed and two-speed motors, except they offer a medium-speed, medium-torque configuration.

Motor Circuits and Configurations

Hydraulic motors can be designed to operate in several different configurations. Depending on the manufacturer's application, the motor might be designed to coast to a stop or brought to an abrupt stop. Some applications combine two motors into a single housing providing two-speed control.

Tandem Motors

Two motors can be designed into one assembly using a single output shaft for both motors. The unit contains a valve that enables the operator to choose placing the two motors in parallel or series. When the valve is placed in the series mode, it directs the outlet oil of motor A, into the inlet of motor B. See Figure 5-27. The series mode provides high speed and low torque.

When the control valve selects the parallel position, motors A and B receive motor inlet oil simultaneously, which increases the unit's displacement. The parallel position provides low speed and high torque. See Figure 5-28.

Back Pressure

As stated earlier, a hydraulic motor will operate as a result of its differential pressure. Two factors influence the amount of pressure that is required to rotate a motor: back pressure in the motor's return line and the load placed on the motor. For example, if a motor has 60 psi (4 bar) of back pressure in the return line, and if the load on the motor requires 600 psi (41 bar), then the motor's inlet must receive a minimum of 660 psi (45 bar) in order for the motor to rotate. See Figure 5-29.

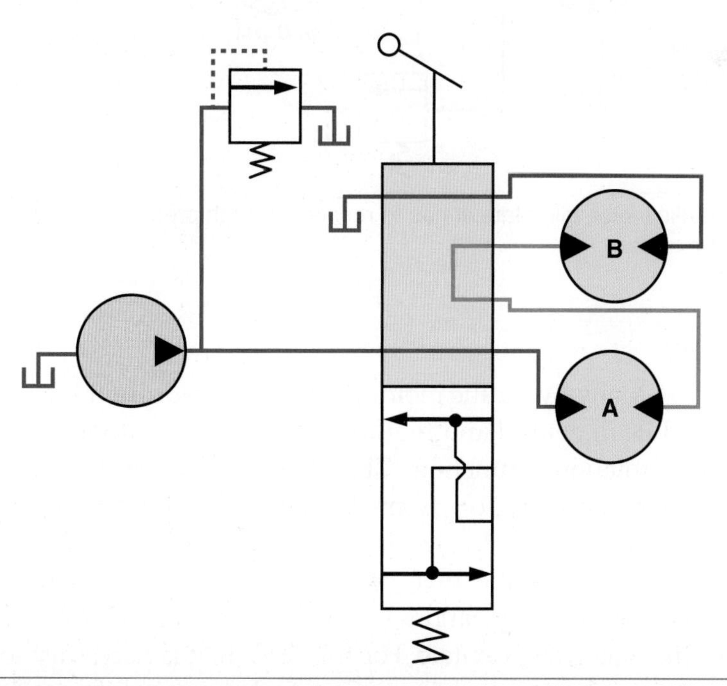

Figure 5-27. A tandem motor can be designed to operate in a series mode. The control valve connects the two motor assemblies in series, providing high speed and low torque.

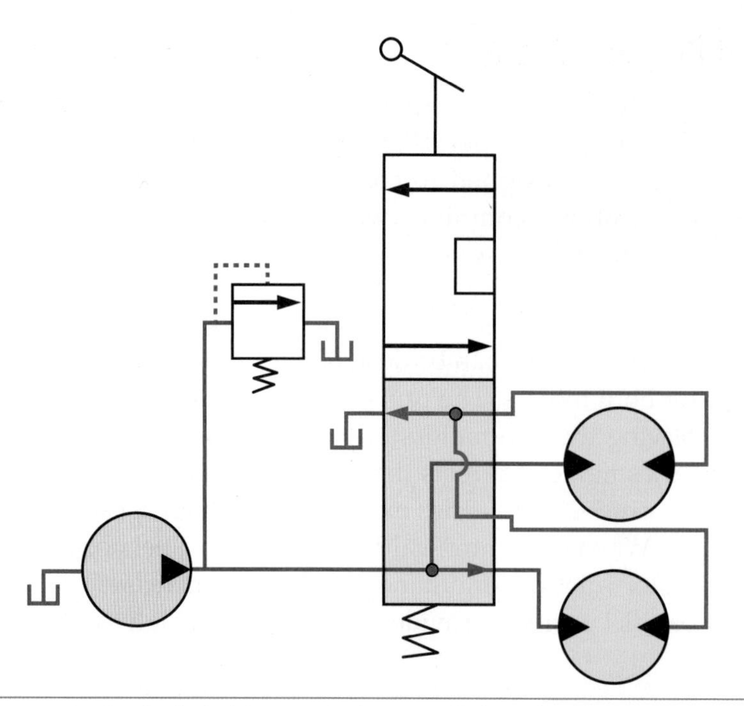

Figure 5-28. A tandem motor can be designed to operate in a parallel mode. The control valve connects the two motor assemblies in parallel allowing the motor to provide low speed and high torque.

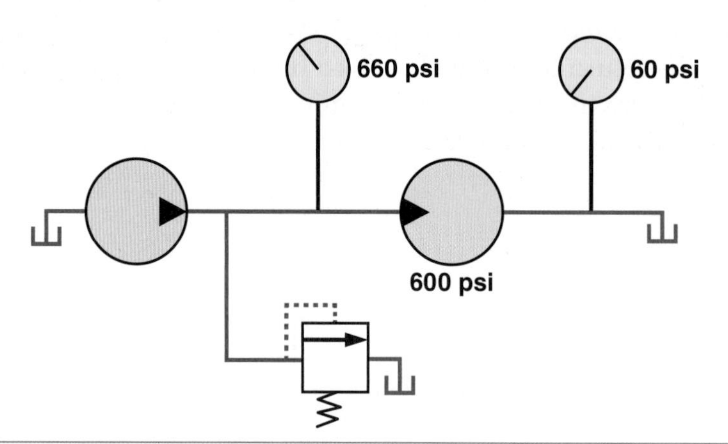

Figure 5-29. Before a hydraulic motor will rotate, it has to receive enough pressure to overcome the return line back pressure and the load on the motor.

Braking a Rotating Motor

Many hydraulic motor applications require the motor to be held stationary while in neutral and require the motor to be stopped quickly after performing its rotational function. These two tasks are accomplished by blocking the motor's return port from the tank and connecting a motor brake valve, also known as a relief valve, in the motor's return. See Figure 5-30.

Keep in mind that as the motor rotates to a stop, the motor's inlet is still generating an expanding volume. If the motor's inlet is blocked at shut down, the motor will cavitate. For this reason, it is necessary to provide a path for the motor's inlet to draw in oil. Counter balance valves are also useful for preventing cavitation in hydraulic motor applications and are explained in **Chapter 7**.

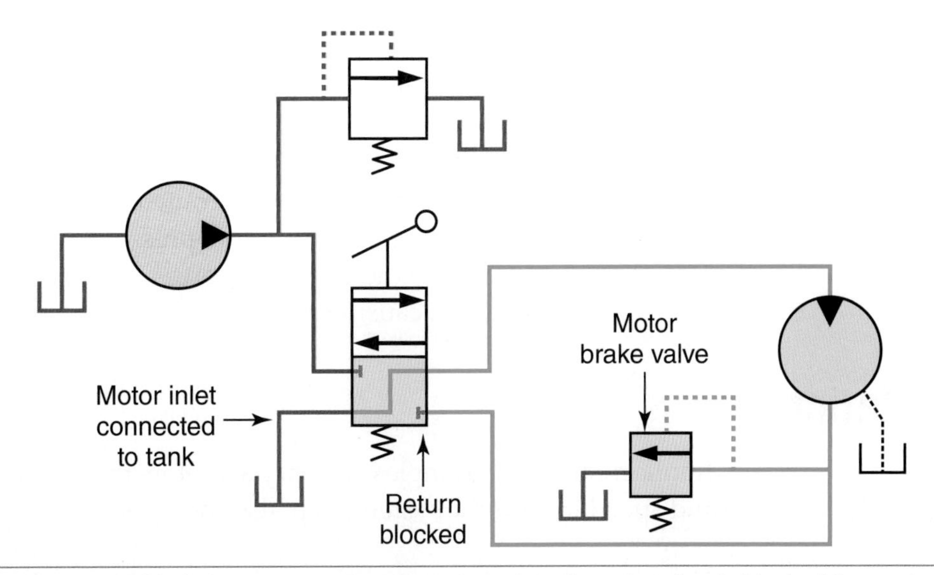

Figure 5-30. In order to quickly halt motor rotation after operation, the return line is blocked from returning to the tank and a relief valve is placed in the motor's return.

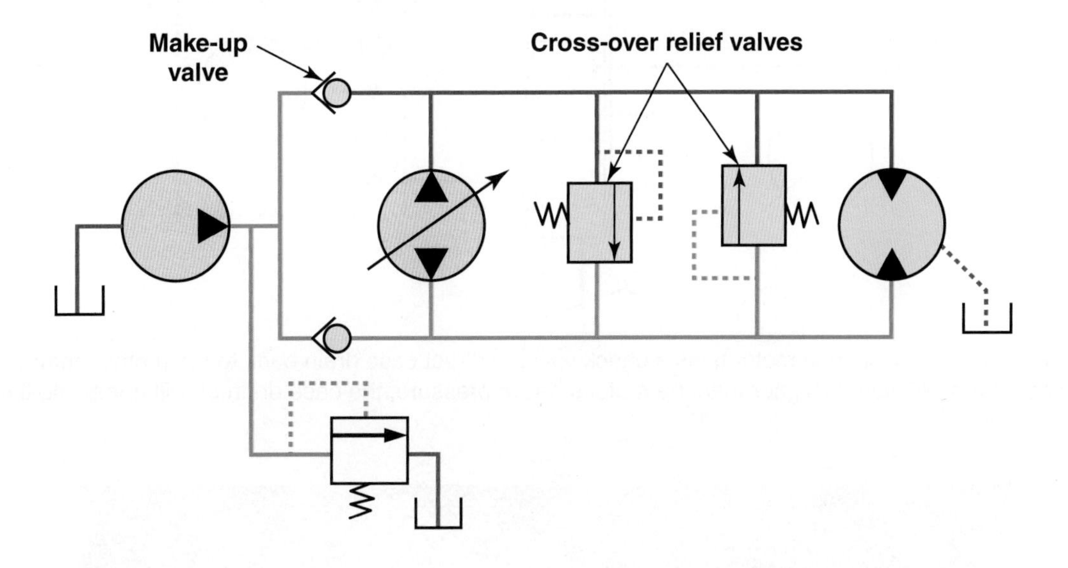

Figure 5-31. Cross-over relief valves dump the motor's return oil back into the motor's inlet.

Cross-Over Reliefs

One of the most popular motor relief applications is the closed-loop hydrostatic transmission that contains two cross-over relief valves. The ***cross-over relief valves*** dump relief oil back into the motor's inlet. See **Figure 5-31**. At first glance, it would appear that the cross-over relief valves would sufficiently prevent motor cavitation. However, hydrostatic transmissions typically are equipped with a separate case drain line that allows motor leakage to return to the reservoir. In this scenario, hydrostatic transmissions use low-pressure check valves to serve the purpose of making up the lost oil. The valves are frequently called ***make-up valves*** and are used to prevent cavitation. Hydrostatic transmissions will be discussed in detail in **Chapter 23**.

Internal Case Drain

Some hydraulic motors do not use a separate case drain line. The internal case drain requires the use of one or two check valves. Two valves are required with reversible motors. See Figure 5-32. Once the case drain pressure is high enough to open one of the check valves, oil is routed to the motor's return. It is critical that these motors operate with a low-pressure return, otherwise a high return pressure will cause the case drain pressure to rise and eventually damage the motor's shaft seal, causing a leak.

Limited Rotation

Hydraulic motors are examples of rotary actuators that produce continuous torque and speed. Hydraulic systems can also incorporate the use of a limited

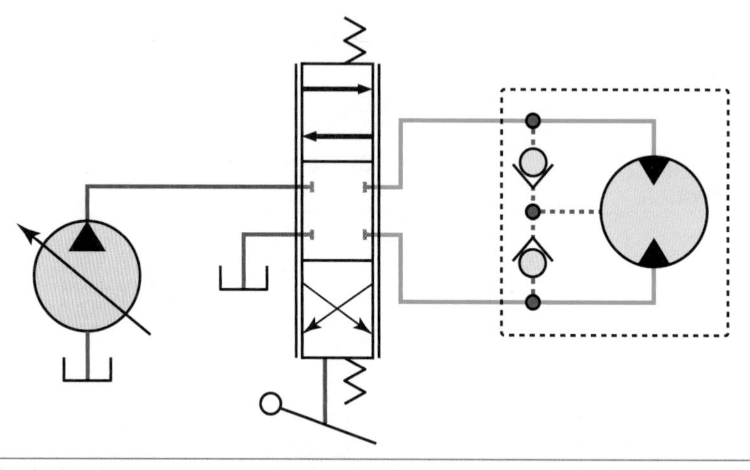

Figure 5-32. An internally drained motor uses a check valve to direct case drain back to the motor's return. Once the case drain pressure is higher than the motor's return pressure, the case drain oil will dump into the motor's return line.

Figure 5-33. Limited rotary actuators are used in mobile machinery to rotate a load within a limited range of motion.

rotary actuator, which will rotate just a few rotations, or less. These actuators use fluid energy to move an object in an arcing motion, but are often limited to less than 360 degrees of rotation.

Helac Corporation manufactures a limited rotary actuator that is used in multiple mobile applications. Two common applications of limited rotational actuators are rotating the basket on the personnel lift, Figure 5-33, and swiveling a backhoe's bucket for forming ditches.

The Helac actuator contains a cylinder that resembles a traditional hydraulic cylinder. The actuator converts a linear moving piston into a limited rotary motion. The actuator contains a helical internal-splined ring, two helical splined shafts, and a linear piston. See Figure 5-34. The centrally located ring is held stationary within the actuator's bore. The ring's internal splines mesh with a central shaft. As the linear piston is actuated, it forces the central shaft to rotate through the internal ring. The central shaft also has internal helical splines that mesh with the output shaft. As the central shaft is forced through the internal ring it causes a limited rotation on the output shaft, resulting in a shaft that rotates from approximately 0 to 180 degrees.

Two other types of limited rotary actuators are oscillators and rack-and-pinion gearsets. Both take linear motion and convert it into a limited rotary motion. A rack-and-pinion gearset is commonly used for steering in automotive applications. Oscillators are commonly used in industrial manufacturing settings.

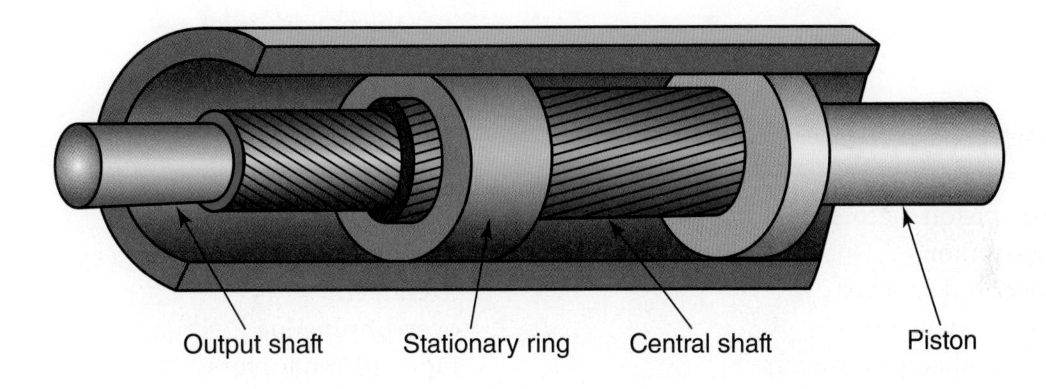

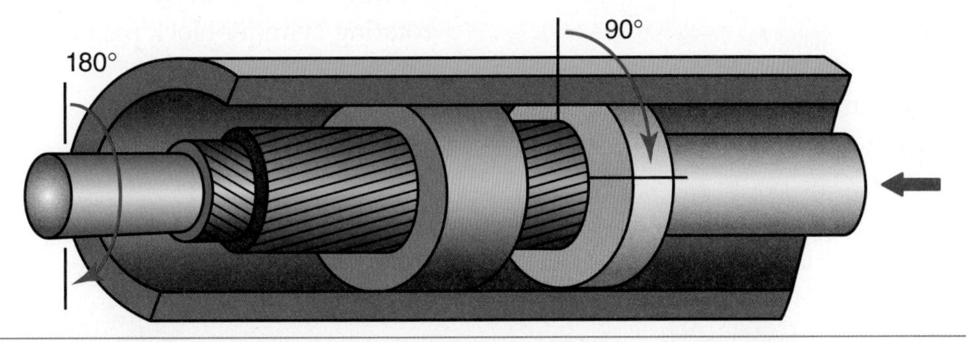

Figure 5-34. A cross-sectional view of a Helac actuator illustrates the actuator contains an internal ring fixed to the barrel, two helically splined shafts and a linear piston. As the linear piston is actuated it causes the output shaft to rotate in a limited direction, clockwise and counterclockwise.

Summary

- ✓ The speed of a motor can be increased by decreasing the motor's displacement or by increasing the flow supplied to the motor.
- ✓ The torque of a hydraulic motor can be increased by increasing the motor's displacement or by increasing the pressure applied to the motor.
- ✓ For a given load, a larger-displacement motor will operate at a lower pressure.
- ✓ Motor direction of rotation is determined by viewing the end of the output shaft.
- ✓ A motor volumetric efficiency is found by dividing the actual motor speed by the theoretical speed.
- ✓ A motor mechanical efficiency is found by dividing the actual motor torque by the theoretical motor torque.
- ✓ Motor speed is calculated by multiplying the flow times 231 and dividing that product by the displacement.
- ✓ The torque of an external-toothed gear motor is dependent on the area of one tooth on one gear.
- ✓ Gerotor motors can be designed so that both the internal- and external-toothed gears rotate, or with a stationary internal-toothed gear and a rotating external-toothed gear.
- ✓ A wider gerotor assembly has a larger displacement.
- ✓ Vane and gear motors are fixed displacement.
- ✓ Piston motors can be fixed or variable displacement.
- ✓ A variable-displacement axial piston motor can be infinitely variable.
- ✓ Swash plates are used only in inline axial piston motors.
- ✓ A greater swash plate angle equals a larger displacement and slower speed.
- ✓ Cam lobe motors are commonly configured as fixed-displacement or two-speed.
- ✓ Make-up valves are low-pressure check valves used to prevent cavitation.
- ✓ Cross-over reliefs are used in hydrostatic transmissions and dump the relief oil back into the motor's inlet.

Technical Terms

bent-axis piston motors
cam lobe motor
cross-over relief valves
differential pressure
external-toothed gear motors
fixed-displacement motor
geroller motor
gerotor motor
hydraulic motor
inline axial piston motors
internal-toothed gear motors
make-up valves
mechanical efficiency
overall efficiency
overrunning load
radial piston motors
rotary actuators
rotating cam piston motor
rotating cylinder block piston motor
torque
unidirectional
vane motor

Review Questions

Answer the following questions using the information provided in this chapter.

1. If a positive-displacement motor is receiving a constant flow of oil, the motor will _____.
   A. rotate faster when receiving a higher operating pressure
   B. will rotate faster when receiving a lower operating pressure
   C. maintain relatively the same speed regardless of the pressure
   D. increase in speed when the flow rate drops
2. A hydraulic circuit is delivering a fixed amount of flow to a hydraulic motor. Which of the following motor displacements will produce the fastest motor speed?
   A. 1.0 in^3/rev.
   B. 2.0 in^3/rev.
   C. 3.0 in^3/rev.
   D. 4.0 in^3/rev.
3. A hydraulic motor is being used to drive a shaft. Which of the following motor displacements will operate at the lowest operating pressure for a given load?
   A. 1.0 in^3/rev.
   B. 2.0 in^3/rev.
   C. 3.0 in^3/rev.
   D. 4.0 in^3/rev.
4. Which of the following motor displacements will produce the highest amount of torque for a given amount of pressure?
   A. 1.0 in^3/rev.
   B. 2.0 in^3 /rev.
   C. 3.0 in^3/rev.
   D. 4.0 in^3/rev.
5. Reversible motors commonly reverse the motor's shaft direction of rotation using what method?
   A. Flipping over a cam ring.
   B. Swapping out the valve plate.
   C. Reversing the direction of oil flow to ports A and B.
   D. Swapping the motor's end housing with another housing.
6. What is the theoretical rpm of a motor that is receiving 30 gpm and has a 2.0 in^3/rev displacement?
   A. 2310 rpm.
   B. 3210 rpm.
   C. 3465 rpm.
   D. 3850 rpm.
7. A motor's theoretical speed has been computed to be 2850 rpm. The actual rpm was measured at 2575 rpm. What is the motor's volumetric efficiency?
   A. 89.3%.
   B. 90.3%.
   C. 95.9%.
   D. 96.1%.
8. A motor's mechanical efficiency is 95% and its volumetric efficiency is 94%. What is the motor's overall efficiency?
   A. 89.3%.
   B. 90.3%.
   C. 95.9%.
   D. 96.1%.
9. A customer wishes to increase a hydraulic drive's torque. Which of the following would not increase torque?
   A. Increase supply of oil flow.
   B. Increase system pressure.
   C. Increase the motor's displacement.
   D. Increase the motor's swash plate angle.
10. A customer wishes to increase the speed of a hydraulic drive. Which of the following would be the best choice for increasing the drive's speed?
    A. Increase system pressure.
    B. Decrease motor's supply of oil flow.
    C. Decrease system pressure.
    D. Decrease the motor's displacement.

11. The torque produced by an external-toothed gear motor is dependent on the area of _____.
    A. one tooth on one gear
    B. two teeth on one gear
    C. two teeth on two gears
    D. four teeth on two gears
12. Within an Eaton Charlynn gerotor motor, which of the following components is always held stationary?
    A. External-toothed gear.
    B. Internal-toothed gear.
    C. Drive coupling.
    D. Output shaft.
13. A technician is looking at the thickness of four different gerotor assemblies with all the same diameters. Which gerotor will provide the fastest motor rpm?
    A. 0.5″ thick gerotor.
    B. 1.0″ thick gerotor.
    C. 1.5″ thick gerotor.
    D. 2.0″ thick gerotor.
14. A geroller motor has an external-toothed gear with six lobes. How many rollers will be located between the two gears?
    A. Five.
    B. Six.
    C. Seven.
    D. Eight.
15. Which of the following cannot be used to initially extend a vane prior to motor operation?
    A. Centrifugal force.
    B. Coiled spring.
    C. Wire spring clip.
    D. Fluid pressure.
16. A vane motor is using fluid pressure to hold the vanes extended. What is typically placed in the motor's return circuit?
    A. 15 psi check valve.
    B. 90 psi check valve.
    C. 150 psi check valve.
    D. 250 psi check valve.
17. A vane motor has what type of displacement?
    A. Fixed.
    B. Variable.
    C. Both A and B.
    D. Neither A nor B.
18. Most vane motors are _____.
    A. unbalanced
    B. balanced
    C. variable displacement
    D. single lobe
19. An axial piston motor with a neutral angle will exhibit what type of motor speeds?
    A. Faster speeds.
    B. Slower speeds.
    C. No speed.
20. Which of the following motors can be designed as a variable-displacement motor?
    A. Gear motor.
    B. Vane motor.
    C. Piston motor.
    D. All of the above.
21. A piston motor with a large swash plate angle will exhibit what characteristic?
    A. Slower speed.
    B. Lower torque.
    C. Both A and B.
    D. Neither A nor B.
22. Which of the following motors provides a variable displacement, but is *not* infinitely variable?
    A. Inline axial.
    B. Bent axis.
    C. Cam lobe.
    D. All of the above.
23. Which of the following cam lobe motor components will *not* rotate?
    A. Cam lobe ring.
    B. Cylinder block.
    C. Output shaft.
    D. All of the above.

24. A cam lobe motor has been designed as a two-speed hydrostatic drive motor. It has eight pistons. How many pistons receive oil in the high-speed position?
    A. Two.
    B. Four.
    C. Six.
    D. Eight.
25. What is used in a cam lobe motor to route oil to the pistons?
    A. Oil distributor/manifold.
    B. Directional control valve.
    C. Solenoid cartridge valve.
    D. Rotary valve.
26. The load on a motor requires 1000 psi (69 bar) of differential pressure. The motor's return circuit is running at 90 psi. How much pressure must be applied to the motor's inlet in order for it to rotate?
    A. 90 psi (6 bar).
    B. 910 psi (63 bar).
    C. 1000 psi (69 bar).
    D. 1090 psi (75 bar).
27. Which of following components is a low-pressure check valve used to prevent cavitation?
    A. Brake valve.
    B. Make-up valve.
    C. Pilot relief valve.
    D. Cross-over relief.
28. Hydraulic motor brakes are typically operational when the directional control valve is in what position?
    A. Center position.
    B. Motor forward operation.
    C. Motor reverse operation.
    D. Both forward and reverse operation.
29. To prevent motor cavitation anytime the motor is braking, the motor inlet should be connected to _____.
    A. the pump outlet
    B. the brake valve
    C. the main relief valve
    D. the tank/reservoir
30. Technician A states that all hydraulic motors use a case drain line/hose. Technician B states that case drain lines route low-pressure oil back to the reservoir. Who is correct?
    A. Technician A.
    B. Technician B.
    C. Both A and B.
    D. Neither A nor B.

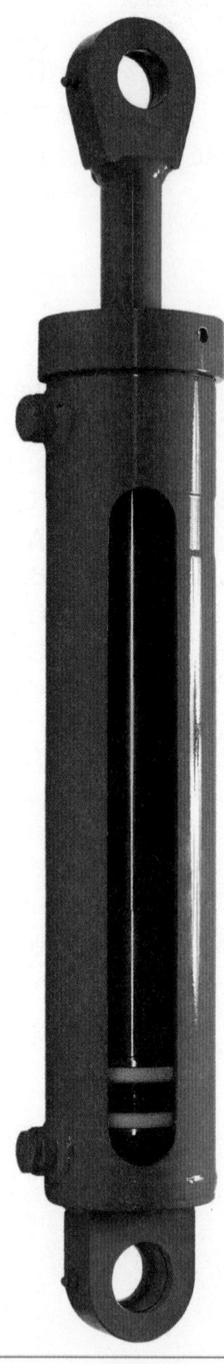

A cutaway of a double-acting hydraulic cylinder is shown here. Hydraulic cylinders are the simplest type of hydraulic actuators and are found on a wide range of equipment.

Chapter 6

Cylinders

Objectives

After studying this chapter, you will be able to:

- ✓ Describe attributes related to the different types of linear actuators.
- ✓ Identify the different components used in hydraulic cylinders.
- ✓ List the different methods manufacturers use to dampen a cylinder's stroke.
- ✓ Explain different methods used for sensing a cylinder's position.
- ✓ Explain the difference between tension and compression cylinder loads.
- ✓ Describe how to compute cylinder speeds.
- ✓ List different types of valves that can be integrated into a cylinder housing.
- ✓ Explain the principle of cylinder regeneration.
- ✓ Identify several items that can cause a cylinder to drift.
- ✓ List the problems that occur when metering-in or metering-out an overrunning load.
- ✓ Explain the different methods used for synchronizing hydraulic cylinders.
- ✓ List unsafe actions to avoid when servicing and diagnosing hydraulic cylinders.

Types of Linear Actuators

Hydraulic cylinders are linear actuators that convert fluid energy into linear mechanical energy. Like hydraulic motors, cylinders are the output devices that perform the heavy lifting for mobile equipment. Examples of work performed by the hydraulic cylinders are a haul truck dumping a heavy payload, or an agricultural tractor lifting and moving hay bales, or a combine swinging an unloading auger.

Mobile equipment requires the use of several different types of hydraulic cylinders:

- Single acting.
- Double acting.
- Single rod.
- Double rod.
- Ram.
- Telescoping.

Single-Acting Cylinders

A ***single-acting cylinder*** requires only one hydraulic hose and is hydraulically actuated in just one direction. The cylinder uses an outside force to return the cylinder back to its original state. The two common methods for returning a single-acting cylinder is a spring or some type of weight, which is the most common. See **Figure 6-1**

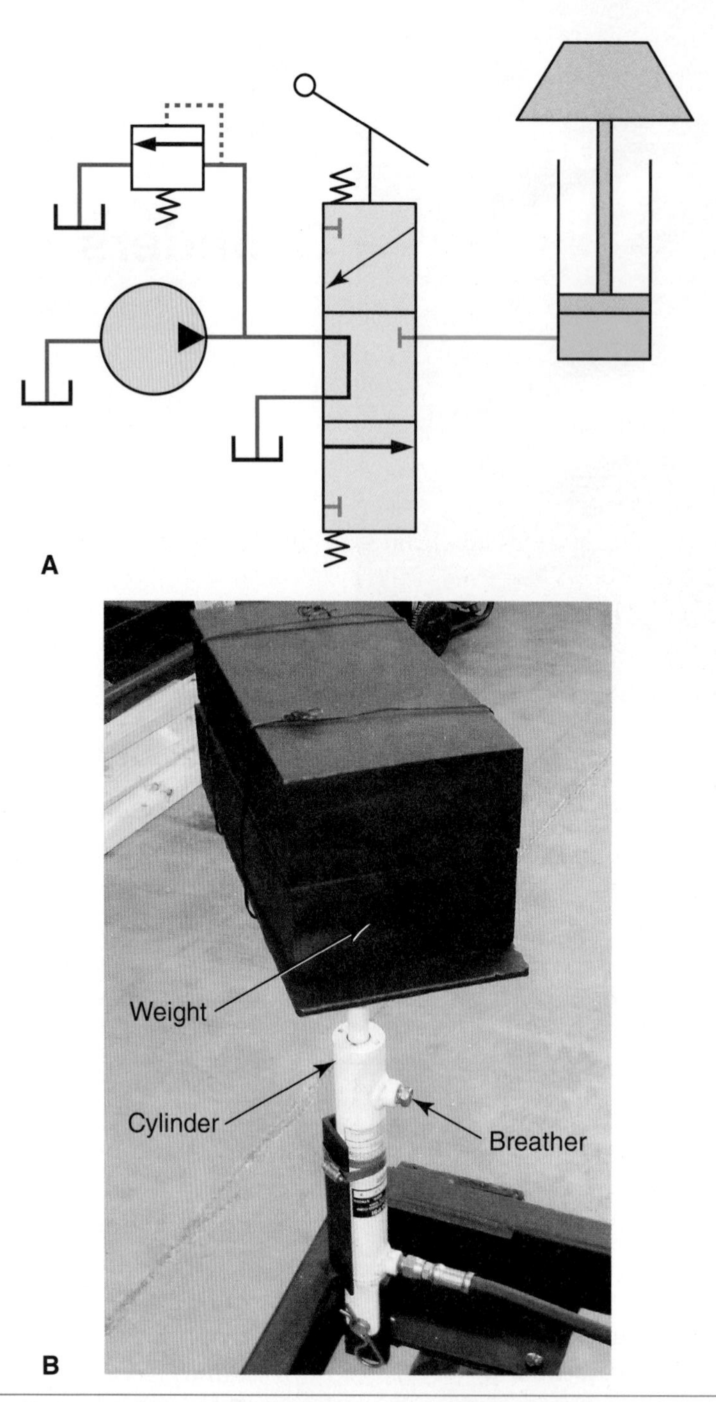

Figure 6-1. Single-acting cylinders. A—The most common single-acting cylinder application uses fluid pressure to extend the cylinder and some type of external force to retract the cylinder. B—This single-acting cylinder is being used on a combine simulator. Notice the cylinder only has one hydraulic hose, and a heavy weight is used to retract the cylinder. The cylinder is emulating the header raise function. The cylinder could be used as a double-acting cylinder, but a breather is installed in the other hydraulic port.

A single-acting cylinder could be designed so that fluid pressure is used only to retract a cylinder, and a weight is used to extend the cylinder. See Figure 6-2. However, this arrangement is rarely found in mobile equipment. This application is also an example of an overrunning load, which creates problems. **Chapter 7** will discuss a pressure control valve that is used to resolve problems with overrunning loads.

Double-Acting Cylinders

A ***double-acting cylinder*** uses fluid pressure to extend the cylinder and fluid pressure to retract the cylinder. The most common type of double-acting cylinder is equipped with a single rod. Another name for this actuator is a ***differential cylinder,*** due to the piston area differences. See Figure 6-3. The rod side of the piston has less effective area because a portion is displaced by the cylinder's rod, resulting in an effective area that resembles a ring. The ring's area is always less than the cap side of the piston.

Another term that is used to frequently describe one side of a cylinder is "head end." The National Fluid Power Society specifies the rod end of the cylinder as the ***head end,*** and the opposite end as the ***cap end,*** or blind end. In contrast, Caterpillar and John Deere's Construction and Forestry (C&F) division frequently label the side that is opposite the rod end as the head end. To avoid confusion resulting from different agencies using the term in different fashions, this text will not use the term "head end." Instead, it will identify the cylinder ends as the rod end and the cap end.

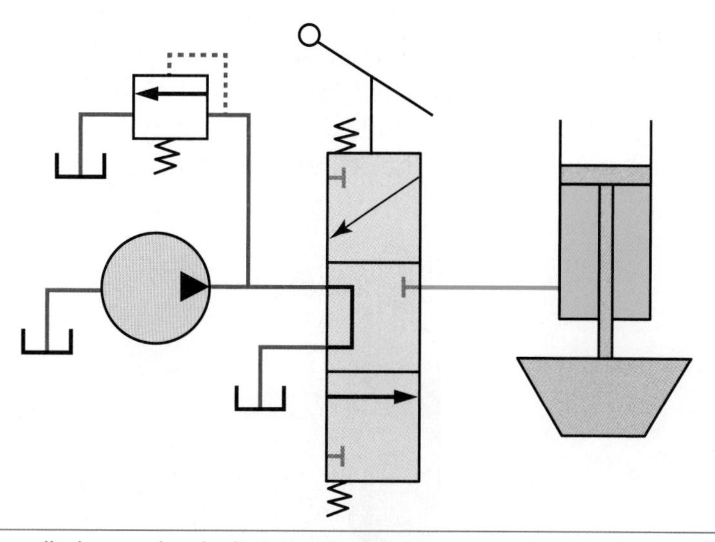

Figure 6-2. A single-acting cylinder can be designed to hydraulically retract the cylinder, with a weight used to extend the cylinder. However, this application is rare.

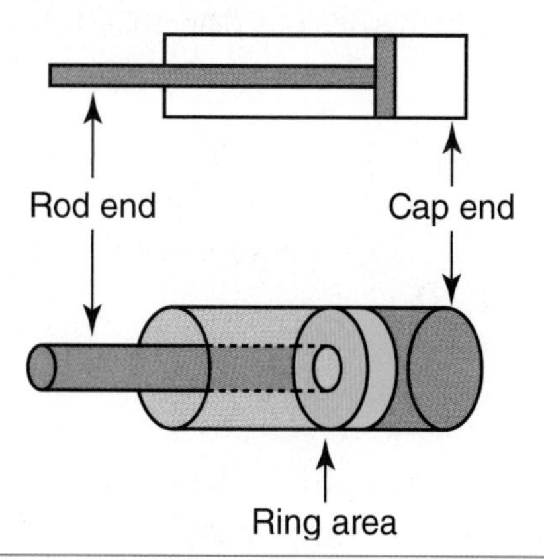

Figure 6-3. A double-acting cylinder uses fluid pressure to extend and to retract a cylinder. The area of the ring side of the cylinder piston (pink) is smaller than the area of the piston's cap side. Note that the cross-sectional area of the rod is subtracted from the total area on the piston's ring side.

The difference in areas between a piston's rod end and cap end affects the cylinder's output force and output speed. See Figure 6-4. If an equal amount of pressure, such as 1000 psi (69 bar), was applied to both sides of a differential cylinder simultaneously, the cylinder would extend. At first glance it might appear that the cylinder would be statically locked. However, as described by Pascal's law (Force = Pressure × Area), the difference in areas causes two different forces.

If the cylinder's piston was three inches in diameter, the cap side area would equal 7.07 in^2. If the rod's diameter was 1.5″, the rod would displace 1.77 in^2 of area from the center of the ring, leaving only 5.3 in^2 of area on the ring. See Figure 6-5. Notice when 1000 psi is applied to both sides of the cylinder at the same time that the net effective force is a factor of the cylinder's rod area, in this case 1.77 in^2.

This principle of different areas within a double-acting cylinder will always result in a stronger extension force than a retraction force for a given system pressure. This principle is also necessary to understand cylinder regeneration, which will be explained later in this chapter.

A similar characteristic exists for cylinder speeds. **Chapter 2** explained that, when given a fixed amount of flow, a smaller cylinder area will result in a faster cylinder speed. A larger cylinder area will result in a slower cylinder speed. As a result, anytime a differential cylinder is used in a system with a fixed amount of flow, the cylinder will always retract faster and extend slower.

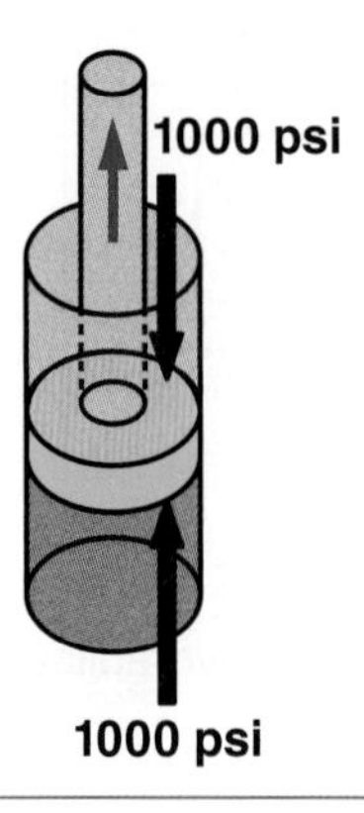

Figure 6-4. If an equal pressure is applied to both sides of a differential cylinder, the cylinder will extend because the larger area will generate a larger force.

Piston Area	3 × 3 × 0.7854	7.07 in^2
Rod Area	1.5 × 1.5 × 0.7854	1.77 in^2
Ring Area	7.07–1.77	5.3 in^2
Extension Force	1000 × 7.07	7070 pounds
Retraction Force	1000 × 5.3	5300 pounds
Net Extension Force	7070–5,300	1770 pounds

Figure 6-5. If a differential cylinder has an equal pressure applied to the rod and cap end simultaneously, the cylinder will extend with a force equal to the rod's area times the system's pressure.

Excavator manufacturers make use of this inherent principle and maximize it on the boom, stick, and bucket controls. For example, when digging in difficult applications that require high break-out forces, the boom, stick, and bucket cylinders will be extending, which provides maximum cylinder force. See Figure 6-6. However, after the excavator has loaded the bucket, high speed is more desirable, and the extra force is not necessary. Therefore, to dump the bucket, to extend the stick, or to extend the boom, the excavator will retract a hydraulic cylinder for the fastest cylinder speeds at a reduced force.

Double-Acting, Double-Rod Cylinders

When manufacturers want a cylinder that delivers the same speed and same force in both directions, a second rod is added to the cylinder. See Figure 6-7. The most common application for double-acting, double-rod cylinders is steering.

Rams

A ***ram*** is a single-acting cylinder that uses a cylinder rod that is the same diameter as the cylinder's piston. Rams are used in applications that require long stroke and must maintain the rigidity of the rod. Some combines use rams to lift the header. See Figure 6-8.

Figure 6-6. Excavators optimize differential cylinders by retracting cylinders when speed is more important than force, and extending cylinders when force is more important than speed. A—Force is important during digging and lifting, so the cylinders are extended. B—Speed is important during dumping, so cylinders are retracted.

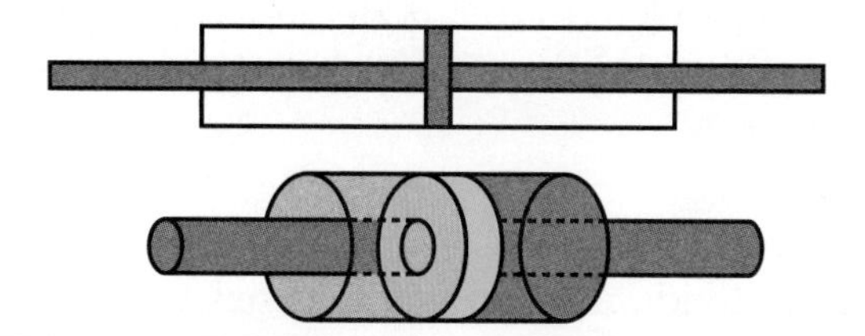

Figure 6-7. A double-rod cylinder is used for some steering applications so that, as the operator steers, the forces and speeds are equal in both directions.

Figure 6-8. Some combines use single-acting rams to raise the feeder house, which lifts the combine's header. Notice that the ram only has one hydraulic hose.

As previously mentioned, single-acting cylinders are retracted using an outside force, something other than fluid pressure. When rams are used to lift a header, the weight of the header must retract the rams. If the feeder house (the housing that feeds crop from the header into the combine and is used for lifting the header) has extra rams added to gain extra lift force, and if the header is removed from the feeder house, the feeder house might be too light to overcome the back pressure in the return circuit. This concern is especially true if the hydraulic system is an open-center system with a fixed-displacement pump. **Chapter 16** will explain open-center hydraulic systems in greater detail. To overcome this problem, some manufacturers use double-acting cylinders in place of single-acting rams. Note that the double-acting cylinders can also improve a cylinder's retraction speed, which can be critical if the machine is operating at fast speeds.

Telescoping Cylinders

A ***telescoping cylinder*** is a linear actuator that contains up to six cylinder tubes that extend to provide a very long cylinder stroke. During retraction, the tubes (also known as sleeves) collapse into a compact cylinder. The cylinder can retract to a length that is 20–40% of the overall extended stroke, compared to the traditional cylinder, which can only be retracted to 50% of the cylinder's overall length.

Components

The individual components consist of a barrel, one to four intermediate stages, and a final stage called a plunger. See Figure 6-9. When retracted, the barrel houses all of the collapsed stages.

Application

The telescoping cylinders can be single acting or double acting. An example of a single-acting telescoping cylinder application is a haul truck's dump bed. See Figure 6-10. A traditional cylinder would not be able to provide the overall stroke length nor the necessary compactness of the retracted cylinder.

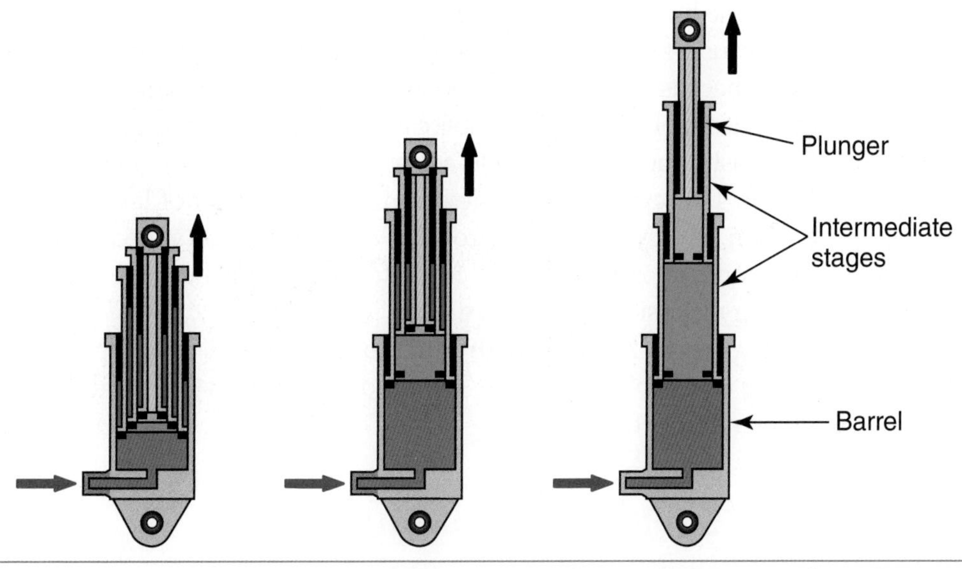

Figure 6-9. A single-acting telescoping cylinder contains a large-diameter barrel, one to four tubes, and a smaller-diameter plunger.

Figure 6-10. Telescoping cylinders are commonly used in dump bed applications, providing very long strokes and very short retraction lengths.

Operation and Inherent Benefit

The cylinder extends in stages. The cylinder's operation has benefits that are inherent to its design. The larger-diameter stages extend first, because they have the most effective area. The smaller stages extend last. During the initial extension the cylinder extends at the slowest speed and with the most force. This operation is beneficial in a dump truck because it provides smooth and safe extension while the bulk of the load begins to move. A smaller-diameter stage will extend next, causing the cylinder to extend faster with less force. This operation is beneficial because the bulk of the load has been dumped, and an increase in speed is more desirable than greater force as the remaining payload is dumped.

Trash trucks use double-acting telescoping cylinders to actuate the compactor. This cylinder provides the inherent benefit of large forces during compaction. The cylinder also provides a fast retraction speed with a reduced force. The compactor needs little force for retraction because the load exists only during extension. See Figure 6-11.

Figure 6-12 shows hydraulic hoses attached to the plunger of a double-acting telescoping cylinder. The fluid passageways are drilled through the cylinder's plunger.

Cylinder Components and Nomenclature

A hydraulic cylinder contains several individual components: piston, gland, rod, barrel, and seals. The cylinder's piston is secured to the cylinder's rod by a nut or a bolt. See Figure 6-13.

Piston Rings and Seals

Pistons have multiple rings and seals installed around their outside circumference. A piston *wear ring* prevents the piston from rubbing the barrel. A piston

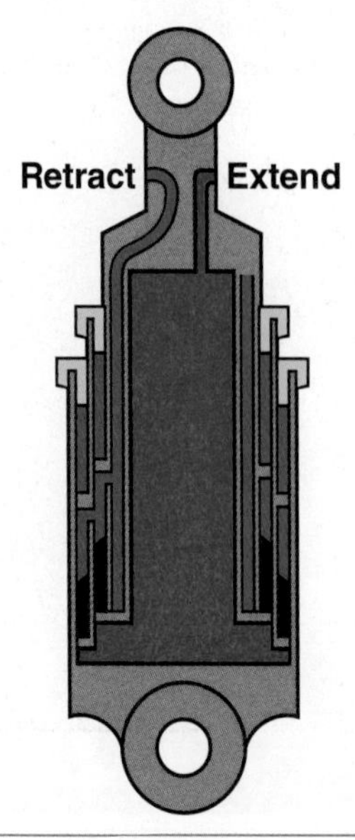

Figure 6-11. Double-acting telescoping cylinders have the oil routed internally through the cylinder. The effective area for retraction is small, resulting in fast retraction speeds with reduced force.

can be configured with one narrow wear ring, one wide wear ring, or two narrow wear rings. Wear rings are commonly made of nylon. See **Figure 6-14**.

A piston seal is installed around the outside circumference of the piston. The piston seal's job is to maintain a fluid-tight seal between the barrel and piston as the piston slides in and out of the barrel. Piston seals vary in composition and design. A few factors that manufacturers must consider when choosing a seal is the compatibility of the oil, fluid temperature, seal longevity, cylinder speed, and pressure. Piston seals designs can be categorized into three groups:

- Metal seals.
- Double-acting seals.
- Lip seals.

Seal design and composition determine the effectiveness, durability, and leakage potential of the piston seal. For example, metal seals will leak the most, but are more durable and longer lasting. Metal seals can be steel, cast iron, or chrome-plated steel.

Figure 6-12. This articulated dump truck uses two double-acting telescoping cylinders to dump the bed. The fluid passageways are drilled through the plunger and stages.

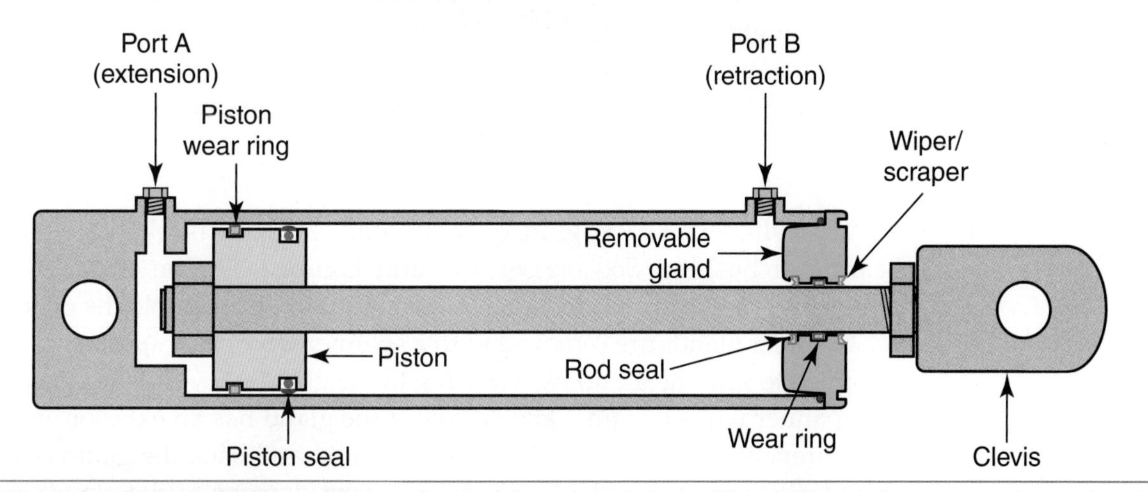

Figure 6-13. A cylinder contains a piston, rod, gland, barrel, and clevis. The piston requires a wear ring and seal. The gland requires a rod seal and a wiper.

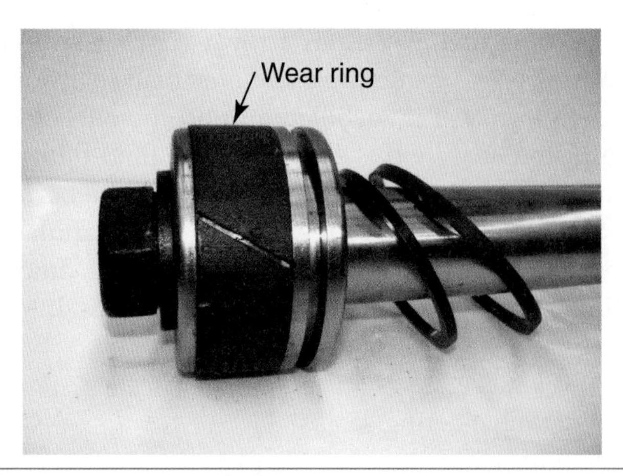

Figure 6-14. This double-acting cylinder piston has a wide wear ring.

A double-acting piston seal holds fluid in both directions. It provides improved sealing over metal seals, but is less durable. O-ring piston seals can be used in double-acting applications. However, the soft rubber O-ring requires the use of a stiff plastic backup ring. The more popular type of double-acting seal uses an outer seal ring made of polytetraflourethylene (PTFE), commonly known as Teflon. The Teflon reduces friction, enabling the seal to easily slide between the barrel and the piston. However, Teflon is fairly rigid and requires the use of a rubber ring underneath the Teflon ring. The rubber ring, sometimes known as an expander, places pressure on the sliding Teflon ring. In Figure 6-15A, the piston seals have been removed. The red ring is the rigid Teflon seal, and the neighboring black ring is the soft rubber ring that holds pressure against the Teflon seal as it rides against the barrel. Figure 6-15B shows the piston with the seal ring installed.

A single-acting piston's ***lip seal***, U-shaped or V-shaped, is designed to hold fluid in one direction. If used in a double-acting cylinder, the piston will require two lip seals. They are less durable than metal seals, but are less likely to leak. Fluid pressure forces the lip to seal against the barrel and piston, providing a snug connection. See Figure 6-16.

Follow the manufacturer's service literature during assembly. The manufacturer might state to coat the seal with petroleum jelly prior to assembly. One service manual recommended heating a piston seal ring prior to installation by placing the ring in 180°–200° Fahrenheit (80°C–90°C) water.

Gland

The cylinder's rod is guided by a ***gland***, which acts like a cylinder rod bearing or bushing. See Figure 6-13 and Figure 6-17. The gland is usually removable, providing the technician the ability to disassemble the cylinder.

Cylinder glands are removed with a spanner wrench. A ***spanner wrench*** is designed to grip the gland by attaching to the gland's key slots or dowel holes. If a spanner wrench is unavailable, and if the gland has an exterior lip, a large pipe wrench can be used to grip the gland's lip and back the gland out of the cylinder. See Figure 6-18. However, to prevent damage to the gland's outside surface, a spanner wrench should be the first choice for removing the gland.

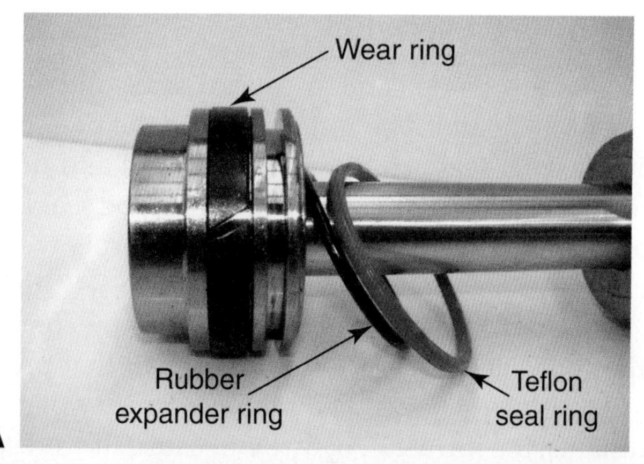

Figure 6-15. Double-acting Teflon seal. A—The piston seals have been removed. The soft black rubber ring rides underneath the rigid red Teflon ring, placing pressure against the Teflon ring. B—When the rings are installed, the black rubber ring is covered by the red seal ring and is unseen.

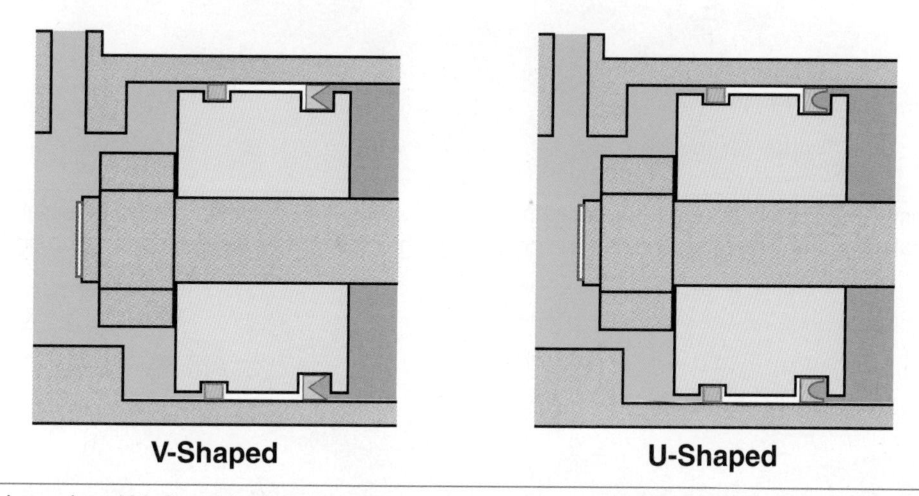

Figure 6-16. U-shaped and V-shaped piston lip seals use fluid pressure to force the lip against the cylinder's barrel.

Note

If a spanner wrench is unavailable, and if a pipe wrench does not fit on the gland, it is possible to fabricate a spanner wrench. A spanner wrench can be made by welding properly spaced dowel pins to a piece of steel.

Rod Rings and Seals

Cylinder glands contain several seals and rings. The ***rod seal*** is located inside of the gland. The rod seal keeps the pressurized oil retained inside the cylinder. This seal is a dynamic-type seal and must hold pressure as the rod moves in and out of the cylinder. A lip-type seal is commonly used in rod seal applications. This single-acting seal performs well at retaining high pressure during cylinder operation. Refer to **Figure 6-17** and **Figure 6-13**.

The gland uses a ***wiper*** to remove any contamination from the surface of the rod. The wiper is the external seal that protrudes out of the cylinder's gland. It is also known as a scraper. If the wiper fails to remove outside contaminants, the intrusion of dirt will cause the rod seal to fail. If a cylinder's rod has oil seepage, the leakage is a direct result of a failed rod seal. The rod seal failure might be the result of a bad wiper. However, a wiper does not have the responsibility to retain the high internal cylinder pressure. Some glands use a dual-purpose seal that includes both a wiper and a rod seal in one ring.

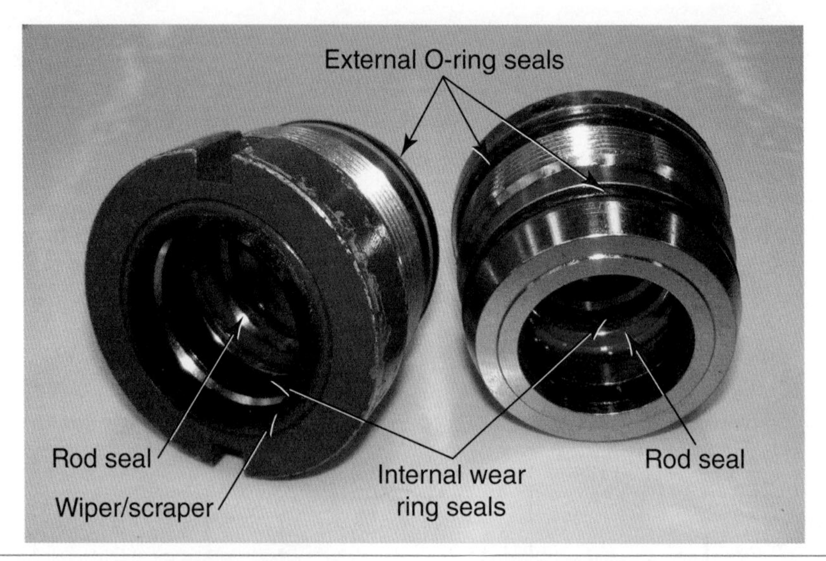

Figure 6-17. Opposite ends of two different cylinder glands are shown here. The glands contain external seals, an internal seal, and an internal wear ring. The red gland has two notches for the placement of a spanner wrench during disassembly.

Figure 6-18. A pipe wrench is sometimes used to remove the cylinder's gland.

A gland contains an internal wear ring that prevents the cylinder rod from rubbing on the gland. An external O-ring seal is also located on the threads of the gland. This static seal prevents leakage between the gland and the barrel.

Barrel

After a gland has been removed, the barrel can be honed. A Flex-Hone® is an attachment placed on a lathe or a drill and is used to machine the inside surface of the barrel. See Figure 6-19. The hone deglazes the surface removing rust and small blemishes. If a cylinder is rebuilt and not honed, it is possible that the piston seals will leak because the barrel's surface is too slick. The hones are offered in eight different material compositions. One of the most popular hones used for hydraulic cylinders is made of silicon-carbide.

Cylinder housings vary in design. Some cylinders cannot be rebuilt, because they are welded together from the factory. A ***tie bolt cylinder*** is designed to be repairable. The housing contains four long bolts that mate the rod end with the cap end. See Figure 6-20. These cylinders are commonly found on agricultural implements. The cylinders can be easier to rebuild and contain an O-ring seal between the cylinder's barrel and the end plates. The cylinders are also known as tie rod cylinders.

Figure 6-19. A Flex-Hone is used to machine the inside surface of a cylinder's barrel. The hone deglazes the cylinder's surface while removing rust and small blemishes.

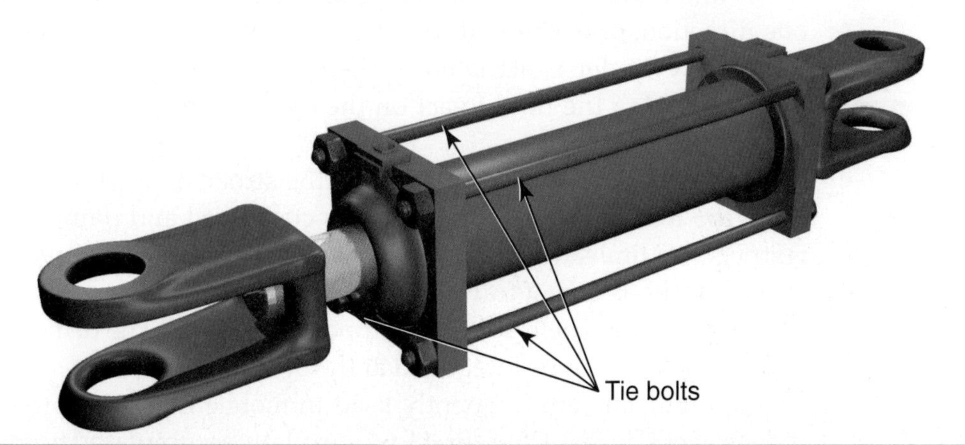

Figure 6-20. A tie bolt cylinder uses four long bolts to clamp the cylinder's end caps to the barrel. The cylinder can be easier to rebuild.

Cylinder Rod

The cylinder's rod must also be considered during a rebuild. If the rod has cracks, nicks or is bent, it will cause leakage. The rod's surface can be rechromed. If repair parts are unavailable because the cylinder is too old, a hydraulic repair shop can fabricate a rod, or even an entire cylinder, using materials they have on hand.

Cylinder Loads, Dampening, and Speeds

As machines are manufactured, the type of cylinder load, potential shocks, and cylinder speeds must all be factored into the machine's design. Loads exerted on cylinders are classified as thrust loading and tension loading. A ***thrust load*** occurs anytime a cylinder must push a load. This is also known as compression loading. A ***tension load*** occurs any time a cylinder pulls a load. It is also known as shear loading.

Cylinder Dampening

Hydraulic cylinders can exert shock loads to machines. Shock loads occur when the cylinder's piston harshly hits the cylinder barrel as it reaches the end of travel. The shock loads jar machine components and linkages causing potential damage. Manufacturers use different techniques to reduce the shock loads placed on the machine:

- Cushioned cylinders.
- Orifices.
- Accumulators.
- Programmable kick-outs.

A cylinder ***cushion*** consists of an internal plunger that blocks off a larger portion of the cylinder's flow as the cylinder reaches the end of travel. See Figure 6-21. When the plunger reaches its cavity, the cylinder's remaining oil must flow through the cushioned oil passage. A threaded adjustment is often placed in the cavity providing some adjustment to the cushion.

Cylinder cushions can be designed as fixed, variable, cushioned in just one direction, or cushioned in both directions. A reverse flow check is used when the cylinder is actuated in the opposite direction. It routes oil to the piston allowing the fluid to act on the entire piston area, not just the area of the plunger.

An orifice can be used to dampen the stroke of a cylinder. The schematic in Figure 6-22 shows a cylinder that is cushioned and dampened. The orifice restricts and limits the rush of fluid into the cylinder's inlet. Only after the cylinder reaches its end of travel will the cylinder's supply pressure reach the full system pressure. The result is a dampened cylinder. The orifice can be located at the DCV or it can be integrated at the cylinder.

Accumulators are frequently used in mobile machinery to dampen a cylinder's shock loads. **Chapter 13** will explain accumulator fundamentals and applications.

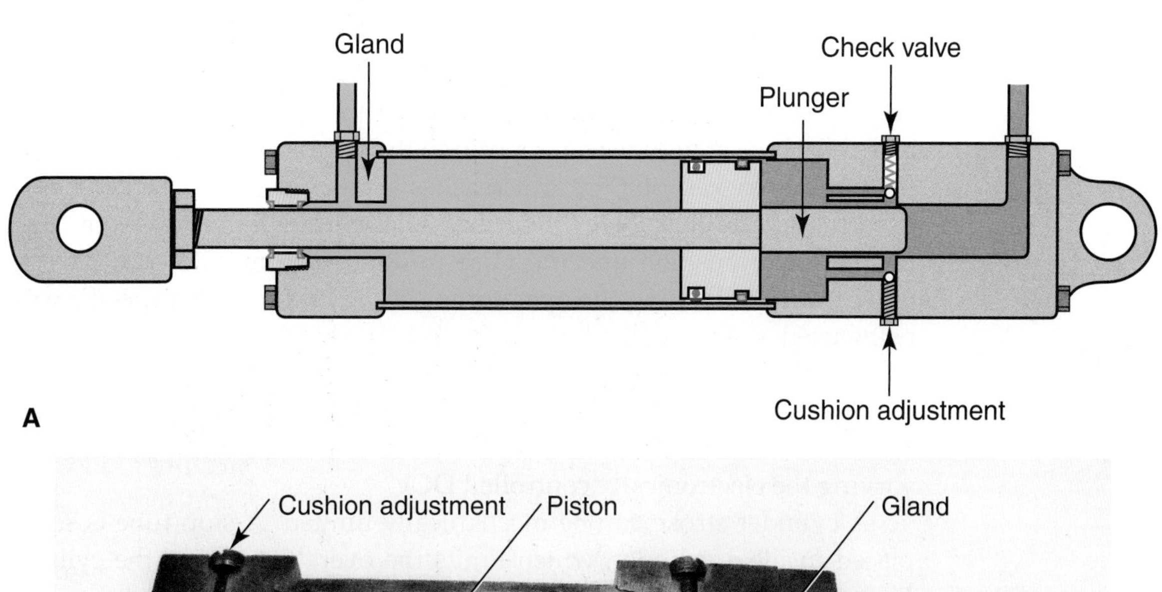

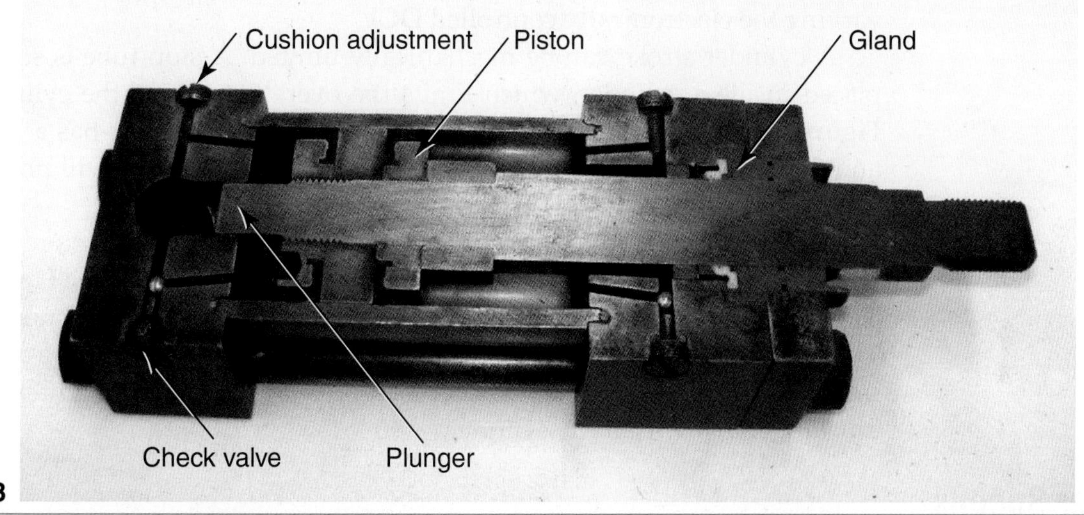

Figure 6-21. A cushioned cylinder consists of a plunger that blocks off the cylinder's oil flow as the cylinder reaches the end of travel, which slows the cylinder as it reaches its end of travel. A—Cross-sectional drawing of a cushioned cylinder. B—Cutaway of a cushioned cylinder.

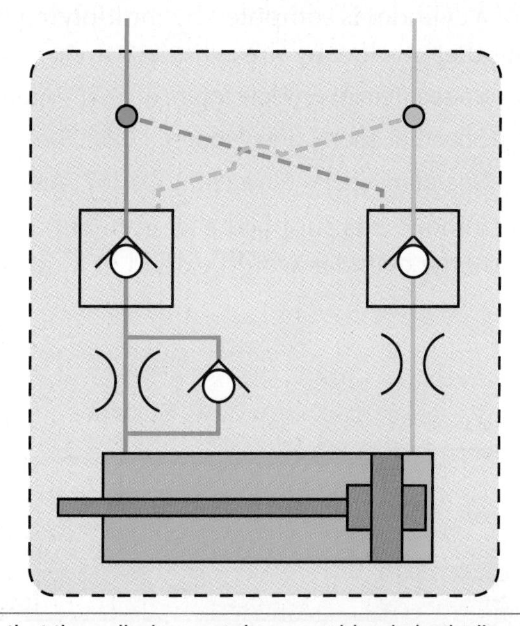

Figure 6-22. The schematic shows that the cylinder contains a cushion in both directions, an orifice in both directions, and pilot-operated check valves in both directions. Pilot-operated check valves will be detailed in **Chapter 7**.

Today's electronic controls can also be used to reduce cylinder shock loads. Construction wheel loaders use ***programmable kick-outs***. The operator accesses the programmable option in the machine's monitor. The operator places the bucket in the desired position then presses the monitor's key to have the bucket "kicked out" once it reaches that set position. Kick-out is a term used to indicate that the cylinder has stopped moving. A kick-out can be set for the lower position and the raise position. The monitor can have several other features allowing the operator to set the kick-out for a specific work tool being used, such as pallet forks or bucket. The kick-out can be set for tilting, or curling, the bucket as well.

The machine's ECM determines the cylinder's position through a sensor. As the cylinder reaches the end of travel, the ECM will slow the cylinder by varying the electronically controlled DCV.

A cylinder stroke can be mechanically limited. A stop tube is sometimes placed inside a cylinder, which limits the overall stroke of the cylinder. See Figure 6-23. Stop tubes are sometimes used when a cylinder has a long rod, and the rod has a tendency to buckle or bend. The stop tube will prevent the rod from fully extending.

Agricultural implements frequently use an external depth stop that consists of an aluminum clamp that fits around the cylinder's rod. See Figure 6-24. These types of depth stops are readily available at farm equipment stores. The clamps are available in a variety of thicknesses and are used to adjust the individual wings on implements. The clamps are easily installed and can be quickly removed from the cylinder.

Cylinder Speeds

A cylinder's speed can be calculated if the cylinder's area is known and the quantity of oil flow into the cylinder is known. The speed is measured in feet per minute (ft/min), feet per second (ft/sec), or meters per second (m/sec). The speed of a cylinder is computed by multiplying the flow rate times a factor, and dividing that product by the cylinder's area:

$$\text{Speed (ft/min)} = (\text{Flow (gpm)} \times 19.25)/\text{Area (in}^2)$$

$$\text{Speed (ft/sec)} = (\text{Flow (gpm)} \times 0.3208)/\text{Area (in}^2)$$

$$\text{Speed (m/sec)} = \text{Flow (lpm)} \times 0.167)/\text{Area (cm}^2)$$

If a cylinder was supplied a 20 gpm of oil, and if the cylinder had an area of 7.07 in^2, the cylinder would extend at a rate of 54 feet per second. A factor

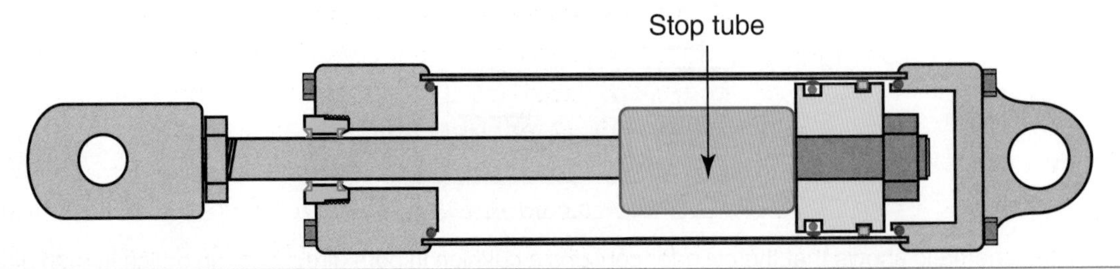

Figure 6-23. Stop tubes can be installed inside a cylinder, limiting the cylinder's overall stroke.

Figure 6-24. External clamps can be used as a cylinder depth stop.

of 19.25 is used. It was derived by dividing the quantity of a gallon (231 in^3) by 12″ (foot), which equals 19.25. The same factor can be converted to solve for feet per second by dividing the 19.25 by 60 seconds, equaling 0.3208.

The same cylinder can be computed in metric units of measurement. A flow rate of 75.7 lpm and an area of 45.6 cm^2 would equal a cylinder speed of 0.277 m/sec.

Cylinder Electronic Sensing

Electronically controlled machines use a variety of sensors to measure a cylinder's position and the load on a cylinder. Sensing the position and/or cylinder load provides inputs for two types of systems: automatic implement controls and automatic guidance systems. Some examples of automatic implement controls are:

- Automatic header height on a combine.
- Load control for an agricultural tractor's three-point hitch.
- Slope control on a motor grader.
- Grade control on an excavator.
- Grade control on a dozer.

Two common technologies used for sensing a cylinder's position are potentiometers and magnetostrictive transducers. A ***potentiometer*** is a three-wire sensor that provides a variable resistance based on the location of the signal wiper. See Figure 6-25. Potentiometers provide an indirect method of measuring a cylinder's position. The sensor is often located on the implement away from the cylinder, as on a combine's feeder for example. As the lift cylinders are actuated, the feeder moves, causing the potentiometer to rotate. This provides the ECM with an indirect input of the position of the feeder cylinders.

A ***magnetostrictive cylinder sensor*** is an internal linear-displacement transducer (LDT) that directly measures a cylinder's position. The cylinder's piston and rod have a passageway drilled through their center. The sensor has a shaft that fits inside the rod's passageway. See Figure 6-26. A magnet is located in the cylinder's piston. As the piston is actuated, a pulse-width modulated signal is transmitted through two wires located inside the sensor's shaft. The magnet moves in relation to the shaft, allowing the sensor to gauge how far the cylinder has been extended. Magnetostrictive cylinders are used for steering cylinders on combines and motor graders. The sensors are used for automatic guidance systems.

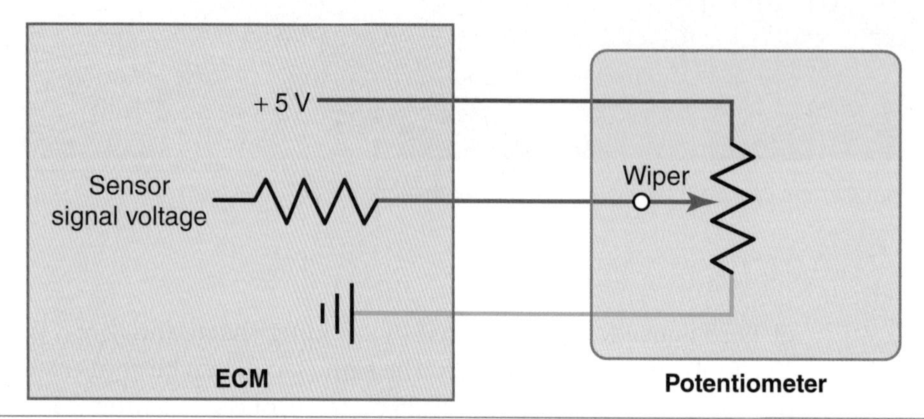

Figure 6-25. A potentiometer is a three-wire variable resistor. The sensor provides a variable voltage input to an ECM.

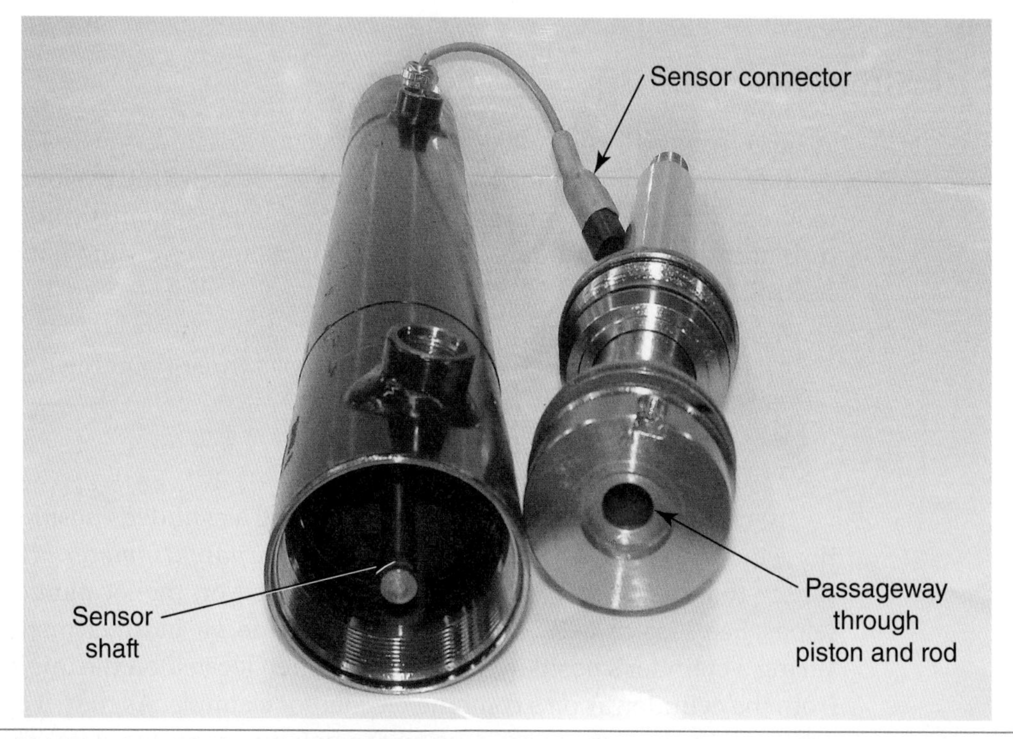

Figure 6-26. This steering cylinder uses magnetostrictive sensing technology that provides an input to an ECM. This technology is most common in machines with automatic guidance systems.

Cylinders can contain pressure sensors that measure a load placed on a cylinder. The feeder lift cylinders on a combine have a pressure sensor connected in parallel with the lift cylinders. The machine has a float setting, which measures the load on the cylinders using the pressure sensor. The combine's header float is designed to allow the header to float across the ground by means of sensing the feeder cylinders' pressure.

Agricultural tractors are frequently equipped with three-point hitches. The hitches often are configured with a three-point hitch rockshaft potentiometer and a hitch pin strain gauge sensor. The two sensors measure the position of the hydraulically controlled hitch and the draft load exerted on the hitch. The ECM will raise and lower the hitch's hydraulic cylinder based on input from those two sensors plus operator inputs, such as sensitivity and commanded position.

The ECM that monitors these various cylinder inputs will require a calibration process. The calibration is typically required after a sensor has been replaced, or if an error has occurred. Some calibrations take place every time the key has been cycled and the lift function is initiated.

Valving Designed for Cylinders

Machine manufacturers incorporate different valve technologies within mobile equipment to protect and control cylinders. Designers sometimes integrate valves directly into the cylinder housing. Some examples of pressure-control valves that can be mounted inside cylinder housings are pilot-operated check valves, counterbalance valves, thermal-relief valves, or pressure-relief valves. **Chapter 7** will detail the purpose and operation of pressure-control valves.

Regeneration

One valve technology that is designed specifically for a double-acting differential cylinder is regeneration. Cylinder ***regeneration*** takes place when an unloaded differential cylinder is extending and its return oil is rerouted to the cylinder's inlet. See **Figure 6-27**. The principle is similar to the way a positive-displacement pump generates a low pressure in a cavity that has an expanding volume and a higher pressure in a cavity with a decreasing volume.

When the two cylinder ports are connected together, the cylinder will extend due to the larger area on the cap side of the piston. Plus the cylinder will extend at a faster speed, because it is regenerating use of the oil that is normally routed to the reservoir.

Regeneration is sometimes used on loader circuits to increase the speed for dumping a bucket. A DCV in the regeneration position is illustrated in **Figure 6-28**. The regeneration portion of the spool is highlighted in yellow.

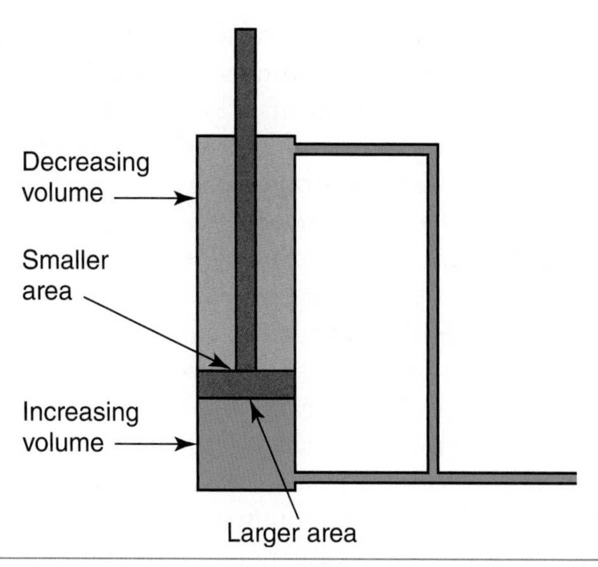

Figure 6-27. A differential cylinder can be extended faster when the DCV routes the return back into the cylinder's inlet.

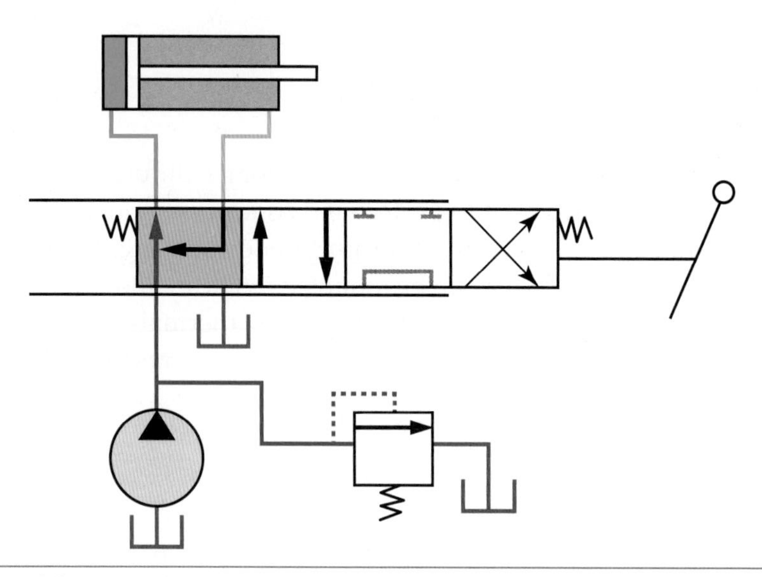

Figure 6-28. The portion of the control valve highlighted in yellow connects the cylinder's return oil back into the cylinder's inlet during regeneration.

Cylinder Drift

Cylinder drift occurs when a rod leaks down over an extended period of time. Several different types of valves can cause this problem if the valves are leaking or malfunctioning:

- Thermal-relief valves.
- Cylinder port relief valves.
- Pilot-operated check valves.
- Counterbalance valves.
- Lift check valves.
- Secondary poppet valves.

The operation of these valves will be detailed in **Chapter 7** and **Chapter 9**.

One misconception related to cylinder drift is that it is usually the result of a bad piston seal. However, a failed piston seal can cause a cylinder to drift only if the cylinder rod is pointing toward the ground. In Figure 6-29, the cylinder's rod is pointing upward. For this example, consider the cap side of the cylinder has an area of 10 in^2 and the cylinder's ring area is 9 in^2. If the piston's seal had a failure or was even removed, before the cylinder could drift back into the barrel one inch, it would require displacing 10 in^3 of oil from the cap side, and it would have to place that 10 in^3 in a 9 in^3 cavity in the rod side of the cylinder. As a result, a cylinder that has the rod facing upward cannot drift due to a piston seal failure.

However, if the double-acting cylinder has the rod facing toward the ground, a piston seal failure can cause the cylinder to drift. See Figure 6-30. Presuming the cylinder has the same areas, 10 in^2 on the cap side and 9 in^2 on the rod side, if the cylinder moved one inch, then 9 in^3 of oil from the rod side would easily fit in a 10 in^3 cavity on the cap side, and the cylinder would be allowed to drift. As the cylinder moved, the voided 1 in^3 cavity on the cap side would drop to a vacuum state.

The dozer in Figure 6-31 has two double-acting cylinders used to raise the blade. The blade has been removed. Notice the cylinder on the left has leaked down and the cylinder on the right remains secure in its barrel. **Chapters 21** and **22** will detail equipment and methods used for diagnosing cylinders.

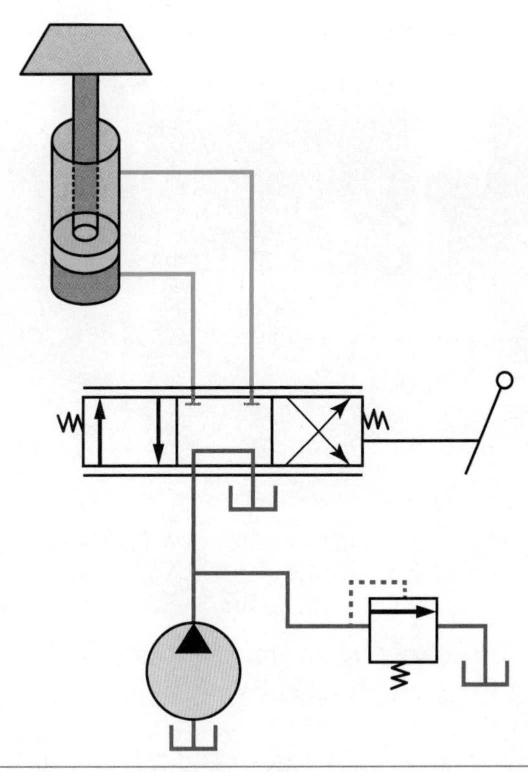

Figure 6-29. A double-acting cylinder cannot drift back into the barrel due to a bad piston seal if the rod is pointing upward. The amount of oil that is displaced from the cap side cannot fit into the smaller cavity in the rod side.

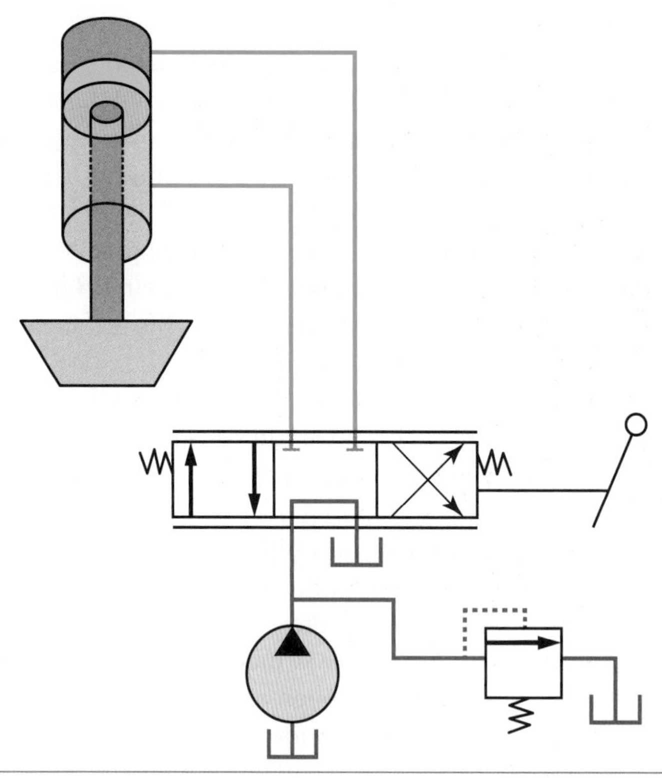

Figure 6-30. A double-acting cylinder that has the rod facing toward the ground can drift out of the barrel due to a faulty piston seal.

Figure 6-31. The dozer blade cylinder on the left has drifted and the cylinder on the right remains secure. Because the rods are pointing toward the ground, a faulty piston seal can cause the cylinder to drift.

Metering-In and Metering-Out Overrunning Loads

Manufacturers choose different methods for controlling the speed of hydraulic cylinders. Metering-in and metering-out are two methods used for controlling cylinder speeds. If a cylinder is actuating an overrunning load, then both metering-in and metering-out will cause problems.

A ***metering-in*** circuit will restrict the oil flowing into a cylinder. If the cylinder has an overrunning load, the weight of the cylinder can cause the rod to extend faster than the pump can supply oil, resulting in cavitation. See Figure 6-32. Make-up valves and counterbalance valves provide solutions to resolve this problem. Those valves will be explained in **Chapter 7** and **Chapter 8**.

If the needle valve was placed in the other working line of the cylinder to limit the cylinder's return oil, the cylinder's speed would be controlled in a ***metering-out*** fashion. See Figure 6-33. When a restriction is placed in the return line of an extending double-acting cylinder, the cap side of the cylinder will cause pressure on the rod side to intensify. The pressure can intensify enough to damage components. As mentioned, **Chapter 7** will explain counterbalance valves, which are used to resolve this problem.

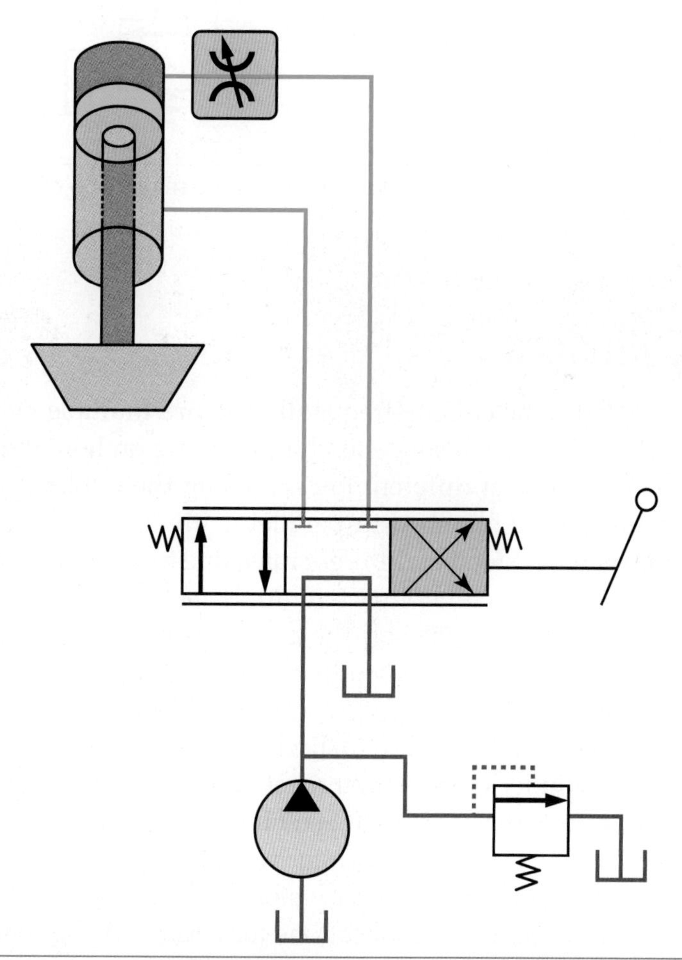

Figure 6-32. The needle valve is being used to control the speed of the cylinder. The cylinder has an overrunning load, which will cause cavitation in the cylinder.

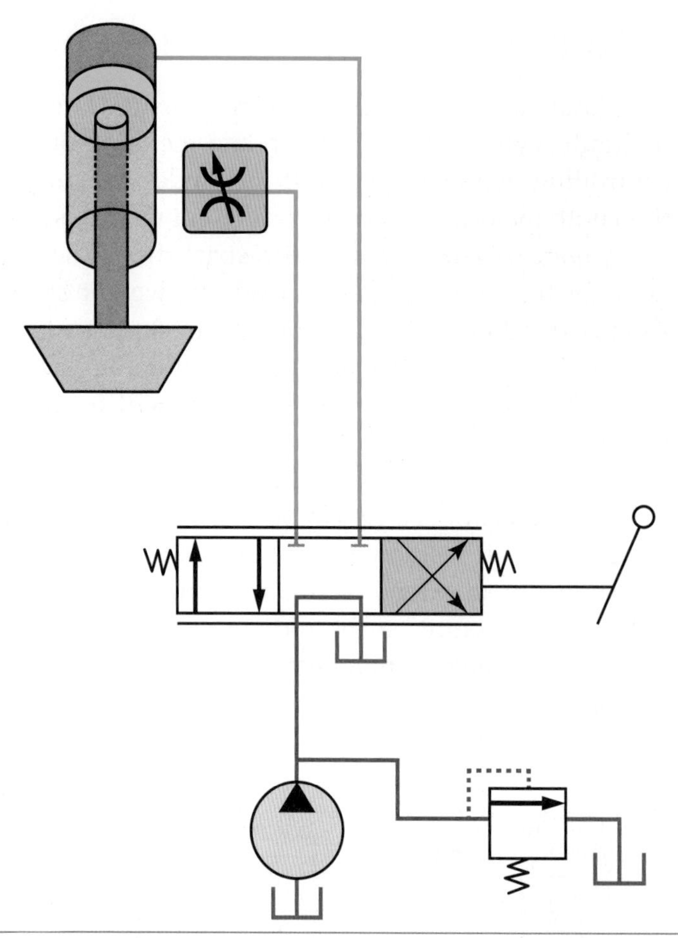

Figure 6-33. When the speed of a cylinder is controlled by restricting the return side of an extending cylinder, pressure intensification will occur.

Synchronizing Cylinders

Hydraulic circuits frequently require multiple cylinders to extend simultaneously to lift a single load. Depending on how the cylinders are plumbed, they can rise at different rates, causing the implement to lift unevenly. If the cylinders are plumbed in parallel, without any additional provision, the cylinders will extend at different rates due to variations in load. See Figure 6-34.

Mobile machinery uses multiple methods for keeping cylinders synchronized and for rephasing the cylinders back into synchronization. The most common method of extending two cylinders simultaneously is to plumb the cylinders in a master/slave series configuration. See Figure 6-35.

The master and slave cylinders must have a specific area relationship. The slave cylinder's piston area must equal the difference of the master cylinder's area minus the rod's area. See Figure 6-36. For example, a planter that uses a 3.25″ diameter master cylinder with a 1.25″ diameter rod, will require a slave cylinder with a 3″ diameter piston.

Three common applications for series cylinder configurations are reel lift on a combine header, reel fore and aft on a combine header, and lift cylinders on agricultural implements such as planters.

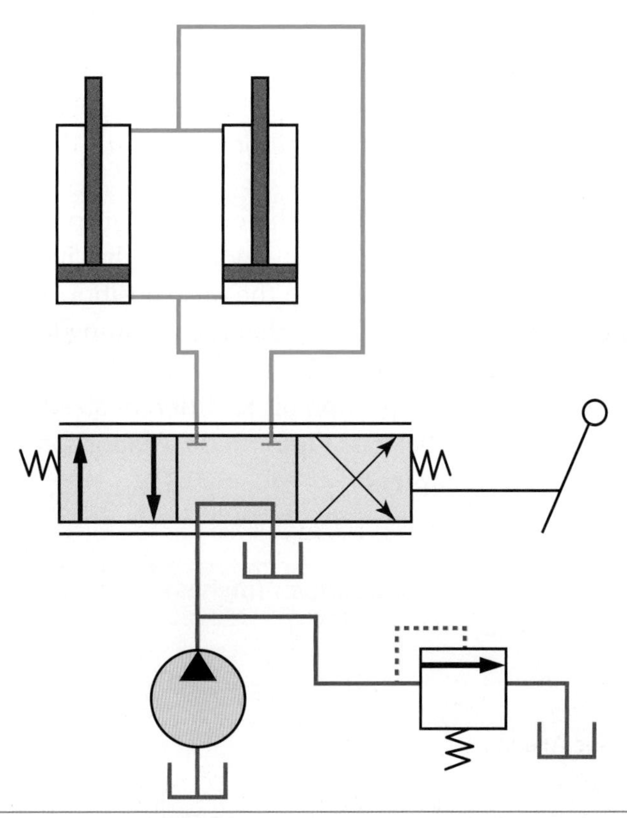

Figure 6-34. Two cylinders that are plumbed in parallel will not extend simultaneously, due to the differences in friction and loads on the cylinders.

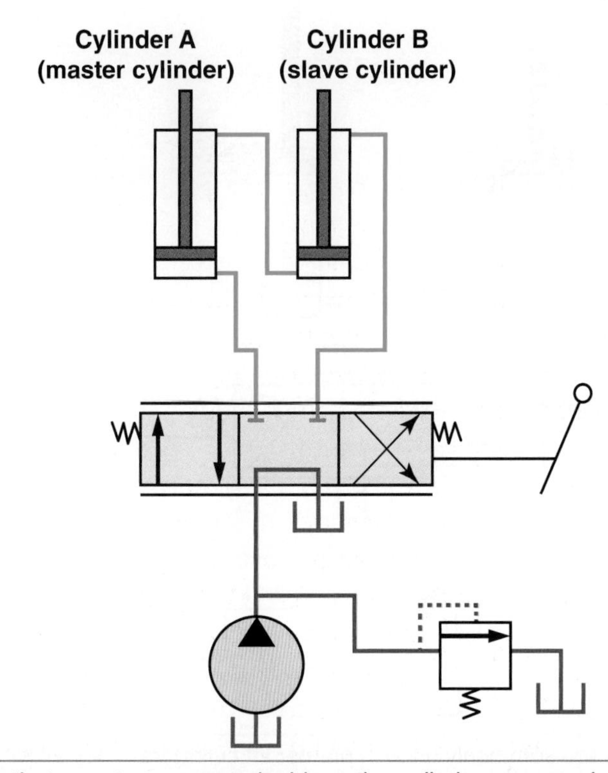

Figure 6-35. Mobile machinery that uses two or more double-acting cylinders commonly connect the cylinders in series.

If the series cylinders leak or get out of synchronization, it can be challenging to resynchronize the cylinders, also known as rephasing. Some planters have rephasing orifices in the cylinders. The operator rephases the cylinders by holding the DCV in the raise position until the cylinders are placed back in synch. See Figure 6-37.

Some implements have the cylinders in parallel and use rephasing relief valves. If one of the parallel cylinders reaches its end of travel and the operator continues to hold the DCV in the raise position, the rephasing relief valve supplies oil to the other cylinder that is continuing to extend. The rephasing valves are commonly used in conjunction with a flow divider valve. The flow divider valve is used to proportion oil to different sized implement cylinders that are plumbed in parallel. See Figure 6-38. Proportional flow divider valves will be explained in **Chapter 8**.

°	Diameter (inches)	Area (in^2)
Master Cylinder Piston	3.25	8.292
Master Cylinder Rod	1.25	1.227
Difference in Master Cylinder Areas	°	7.065
Slave Cylinder Piston	3	7.065

Figure 6-36. This table shows that the difference in areas between the master cylinder's piston and rod must equal the slave cylinder's piston area.

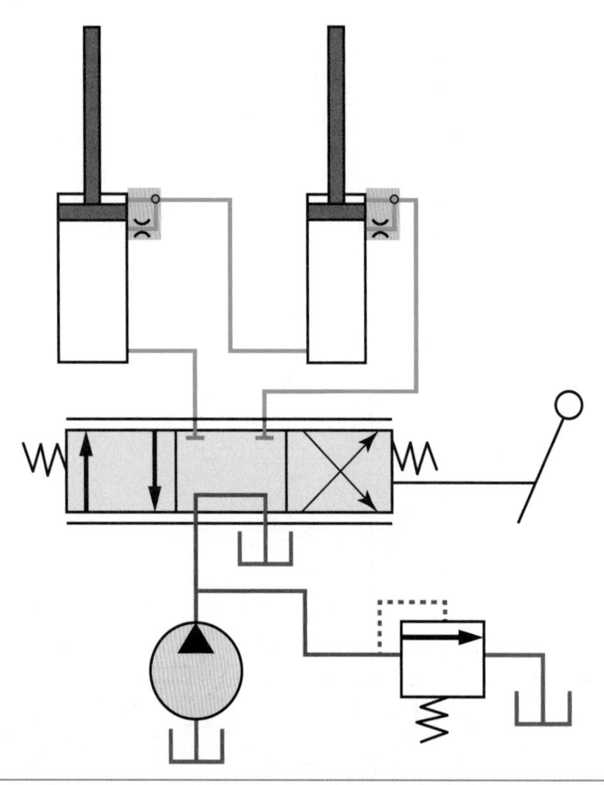

Figure 6-37. Rephasing orifices are commonly used in planter lift cylinders. They allow an operator to place the cylinders back into synchronization by holding the DCV in the raise position until the cylinders are synchronized.

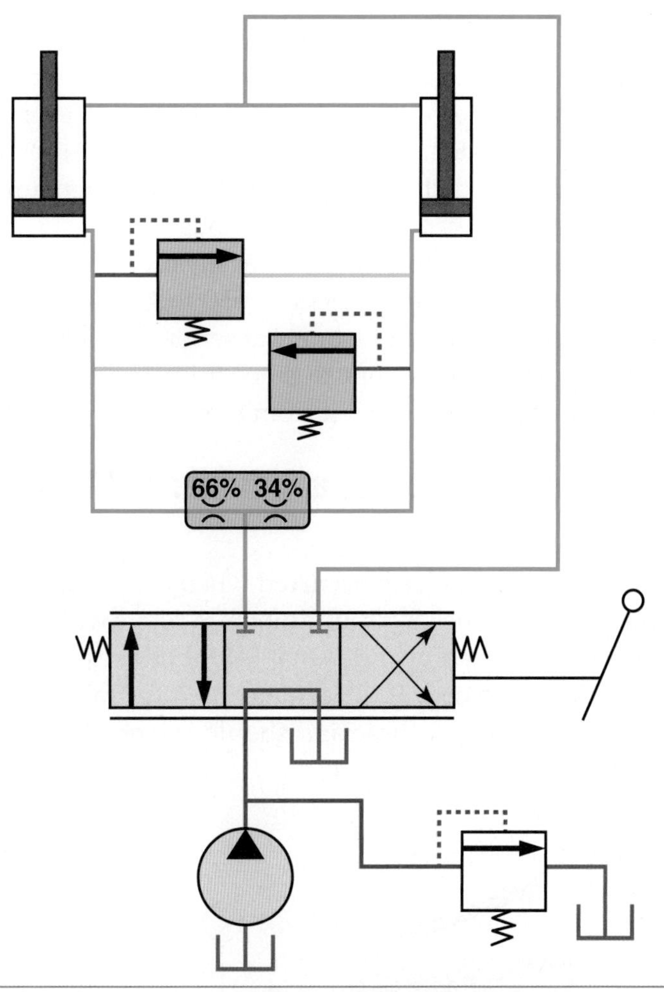

Figure 6-38. Some planters use rephasing relief valves when the cylinders are connected in parallel. When one cylinder reaches the end of travel, the operator can continue to hold the DCV control valve in the raise position, and the remaining oil will be routed to the other cylinder that is continuing to extend.

Cylinder Safety

Technicians are frequently tasked with repairing a hydraulic cylinder. It is common for the cylinder's piston or gland to be seized, making it difficult to remove, even after the snap ring has been removed and the gland has been unthreaded. Do not attempt to use compressed air or heat the cylinder's barrel with a torch.

A September 2001 *Hydraulics and Pneumatics Journal* article by Rory McClaren recommends placing the hydraulic cylinder with the rod downward inside a bucket. See Figure 6-39. A vise might be needed to hold the cylinder in place. At the opposite end, fill the cylinder with hydraulic oil. With the cylinder's snap ring removed, use a hydraulic hand pump to slowly push the piston rod and gland assembly out of the barrel. Observe the pressure gauge while applying pressure with the hand pump. If the pressure exceeds the cylinder's specification, the cylinder must be scrapped. The bucket is used to catch the oil when the rod is freed from the cylinder. Use only a hand pump; do *not* use a running hydraulic system.

Do Not Attempt to Stall a Hydraulic Cylinder or Motor under Pressure

While diagnosing and servicing hydraulic machines, technicians have at times been tempted to mechanically limit or prevent an actuator from moving while the actuator receives full hydraulic pressure and flow. This action must be strictly avoided. While hydraulic components might appear to be small, their output power can be tremendous.

A small hydraulic motor can easily produce 10 hp. Considering that a single horsepower is the equivalent to 550 foot pounds per second, any attempt to stall an actuator puts personnel at serious risk of injury. For this reason, technicians must never attempt to mechanically stall, bind, or hold a hydraulic cylinder or motor to prevent it from moving.

Do Not Attempt to Mechanically Actuate a Cylinder or Motor while It Is under Pressure

Other injuries have occurred when technicians have attempted to mechanically assist a stuck hydraulic motor to rotate while the motor is under pressure, or assist a seized cylinder that is under pressure. These actions should also be strictly avoided. If an actuator is inoperable, pressures and flows should be checked, but technicians should never attempt to mechanically assist a hydraulic actuator to move.

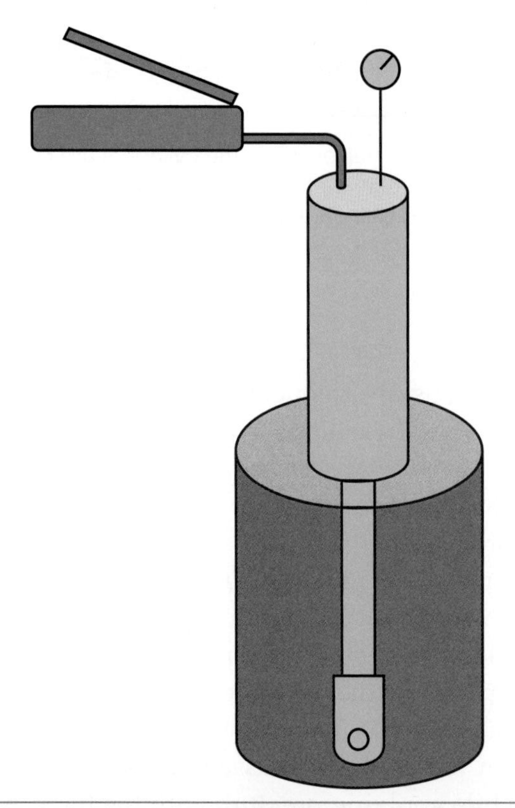

Figure 6-39. A seized cylinder piston or gland can be difficult to remove. Hold the cylinder in a vise. Fill the cylinder with oil. Install a pressure gauge and use a hydraulic hand pump to push the seized piston or gland out of the cylinder. If the pressure reaches the cylinder's limit, a new cylinder will need to be ordered and the old cylinder will need to be scrapped.

Summary

- ✓ Single-acting cylinders use fluid pressure in only one direction.
- ✓ A double-acting, single-rod cylinder exhibits different forces when extending and retracting under a given pressure and different speeds when extending and retracting using a given flow.
- ✓ Telescoping cylinders provide long strokes, and compact lengths when retracted.
- ✓ A ram's cylinder rod is the same diameter as the cylinder's piston.
- ✓ The gland guides the cylinder rod during extension and retraction.
- ✓ Wear rings prevent metal-to-metal contact during cylinder operation.
- ✓ Piston seals can be metal, Teflon, lip design, or O-ring.
- ✓ Rod seals retain the cylinder's fluid.
- ✓ Rod wipers prevent outside contaminants from entering the cylinder.
- ✓ Cylinder shock loads can be dampened with cushioned cylinders, orifices, accumulators, or programmable kick-outs.
- ✓ Computing a cylinder's speed requires knowing the cylinder's area and input flow.
- ✓ Potentiometers and magnetostrictive sensors measure the position of a cylinder.
- ✓ When the cap end and rod end of a double-acting cylinder are plumbed together, the cylinder will regenerate.
- ✓ Cylinders can drift due to a bad pressure control valve.
- ✓ A bad piston seal can cause cylinder drift only if the rod points toward the ground.
- ✓ Manufacturers use master/slave series designs to synchronize two cylinders.
- ✓ Rephasing orifices and rephasing relief valves are used to place cylinders back into synchronization.
- ✓ Compressed air and heat should never be used to remove a seized cylinder piston or gland.
- ✓ Do not mechanically restrict or stall an operating actuator.
- ✓ Do not mechanically assist an actuator that is under pressure.

Technical Terms

cap end
cushion
differential cylinder
double-acting cylinder
gland
head end
lip seal
magnetostrictive cylinder sensor
metering-in
metering-out
potentiometer
programmable kick-outs
ram
regeneration
rod seal
single-acting cylinder
spanner wrench
telescoping cylinder
tension load
thrust load
tie bolt cylinder
wear ring
wiper

Review Questions

1. A double-acting, single-rod cylinder has 500 psi pressure applied simultaneously to both the rod and cap side of the cylinder. What will the result be?
   A. The cylinder will retract.
   B. The cylinder will be hydraulically locked.
   C. The cylinder will extend.
   D. Answer varies, depending on the type of rod seal installed.
2. During the extension of an excavator's boom, stick, or bucket, which of the following will occur?
   A. The boom, stick, and bucket will exhibit maximum force.
   B. The boom, stick, and bucket are extended by extending the cylinder.
   C. The boom, stick, and bucket will exhibit the slowest speed.
   D. The boom, stick, and bucket will have reduced force while the arms are extending.
3. If a needle valve is used to actuate a hydraulic cylinder with an overrunning load in a "metering-out" mode, what will occur?
   A. Cavitation.
   B. Pressure intensification.
   C. Both A and B.
   D. Neither A nor B.
4. If a needle valve is used to meter oil into a cylinder with an overrunning load, which of the following could occur?
   A. Cavitation.
   B. Pressure intensification.
   C. Both A and B.
   D. Neither A nor B.
5. Which of the following cylinders would be the best choice for a steering application?
   A. Single-acting cylinder.
   B. Double-acting, single-rod cylinder.
   C. Telescoping cylinder.
   D. Double-acting, double-rod cylinder.
6. If a cylinder pulls a load, what type of load is it?
   A. Tension.
   B. Stress.
   C. Thrust.
   D. Inertia.
7. If a cylinder pushes a load, what type of load is it?
   A. Tension.
   B. Stress.
   C. Thrust.
   D. Inertia.
8. Which of the following should be used to remove a seized piston from a cylinder?
   A. Acetylene torch with rosebud tip.
   B. Compressed air.
   C. Hydraulic hand pump.
   D. Fluid from a running tractor.
9. Technician A states that it is okay to use a lever to extend a frozen cylinder while applying full system pressure to the cylinder. Technician B states that it is okay to check for cylinder leaks at mid-stroke by holding the cylinder using a stop. Who is correct?
   A. Technician A.
   B. Technician B.
   C. Both A and B.
   D. Neither A nor B.
10. Which of the following terms refers to a plunger blocking off a portion of a cylinder's flow at the end of its travel?
    A. Main relief valve.
    B. Rejuvenation.
    C. Cushion.
    D. Regeneration.
11. Which of the following terms refers to routing the oil from the rod side of a differential cylinder to the cap side?
    A. Main relief valve.
    B. Rejuvenation.
    C. Cushion.
    D. Regeneration.

12. As a cylinder rod moves back and forth, it is guided and supported by which of the following?
    A. Gland.
    B. Piston.
    C. Barrel.
    D. Cap.
13. What is the advantage of connecting two cylinders in series?
    A. Keeps the cylinders synchronized.
    B. Prevents cavitation.
    C. Generates higher lifting force with less input pressure.
    D. Lowers the system's pressure.
14. Which of the following cylinders has a rod with the same diameter as the piston?
    A. Single acting.
    B. Double acting.
    C. Telescoping.
    D. Ram.
15. Which of the following cylinder input devices sends a pulse-width modified signal through the center of the rod and uses a magnet inside the piston?
    A. Potentiometer.
    B. Rheostat.
    C. Thermistor.
    D. Magnetostrictive sensor.
16. The length of a fully collapsed telescopic cylinder can be as little as _____ of its extended stroke length?
    A. 10 to 20%
    B. 20 to 40%
    C. 40 to 60%
    D. 60 to 80%
17. All of the following are used to dampen a cylinder to prevent shock loads, *EXCEPT*:
    A. orifice.
    B. programmable kick-out.
    C. cushion.
    D. wear ring.
18. Technician A states that cylinder glands contain a wear ring. Technician B states that cylinder pistons contain a wear ring. Who is correct?
    A. Technician A.
    B. Technician B.
    C. Both A and B.
    D. Neither A nor B.
19. Technician A states that telescoping cylinders can be double acting. Technician B states that telescoping cylinders can only be single acting. Who is correct?
    A. Technician A.
    B. Technician B.
    C. Both A and B.
    D. Neither A nor B.
20. What type of speed will a telescoping cylinder exhibit during the initial extension?
    A. Slow.
    B. Medium.
    C. Fast.
21. What type of force will a telescoping cylinder exhibit during the final stage of extension?
    A. Small.
    B. Medium.
    C. Large.
22. Which of the following has the responsibility of preventing a cylinder from leaking oil?
    A. Piston seal.
    B. Rod seal.
    C. Rod wiper.
    D. Wear ring.
23. Which of the following has the responsibility of preventing contaminants from entering a cylinder?
    A. Piston seal.
    B. Rod seal.
    C. Rod wiper.
    D. Wear ring.

24. A piston seal is being installed; the service manual states to heat the seal prior to installation. What method is used to heat the seal?
    A. Butane torch.
    B. Acetylene torch.
    C. Heat gun.
    D. Hot water.
25. Technician A states that pipe wrenches are sometimes used to remove a cylinder's gland. Technician B states spanner wrenches should never be used to remove cylinder glands. Who is correct?
    A. Technician A.
    B. Technician B.
    C. Both A and B.
    D. Neither A nor B.
26. Prior to assembly of a rebuilt cylinder, what should be used to machine the cylinder's barrel?
    A. 1000-grit sandpaper.
    B. Emery cloth.
    C. Flex hone.
    D. Die grinder.
27. External aluminum clamp-on stops are commonly used in what application?
    A. Motor graders.
    B. Backhoes.
    C. Agricultural implements.
    D. Agricultural combines.
28. A cylinder with a five-centimeter diameter is receiving 30 lpm. What is the cylinder's extension speed?
    A. 0.255 m/sec
    B. 0.355 m/sec
    C. 0.455 m/sec
    D. 0.555 m/sec
29. Rephasing relief valves are used in what type of cylinder circuit?
    A. Cylinders in series.
    B. Cylinders in parallel.
    C. Both A and B.
    D. Neither A nor B.
30. Rephasing orifices are used in what type of cylinder circuit?
    A. Cylinders in series.
    B. Cylinders in parallel.
    C. Both A and B.
    D. Neither A nor B.

Chapter 7

Pressure-Control and Load-Holding Valves

Objectives

After studying this chapter, you will be able to:

- ✓ Explain the construction and uses of the different types of pressure-relief valves.
- ✓ Describe the operation of a pressure-reducing valve and list a common application.
- ✓ List the different types of unloading valves.
- ✓ Explain the operation of pressure-sequence valves.
- ✓ List the different types of load-holding valves and briefly describe their construction.

Introduction to Pressure-Control and Load-Holding Valves

Mobile machinery uses several different types of pressure-control and load-holding valves for a variety of purposes, including:

- Protecting circuits.
- Reducing pressures.
- Maintaining a fixed pressure.
- Unloading a pump.
- Holding loads.

Pressure-control valves and load-holding valves are used across a wide range of mobile machinery. From the simplest to the most advanced hydraulic systems, nearly all mobile hydraulic systems use some form of pressure control valves, and most employ some type of load-holding valves. Pressure-control valves, such as a pressure-relief valve, not only protect operating hydraulic systems, but also protect circuits that are not operating. The pressure-control valves and load-holding valves are placed inside pumps, directional control valves, actuators, and filters. They can also stand alone, inside their own valve block.

Pressure-Relief Valves

A ***pressure-relief valve*** is a normally closed valve that will open to allow fluid to pass once the pressure reaches the valve's setting. The schematic of a nonadjustable pressure-relief valve in Figure 7-1 shows that a spring holds the valve shut. Inlet pressure works against the closing pressure of the spring. Once the inlet pressure overcomes the spring's closing pressure, the pressure relief opens, allowing fluid to pass to the reservoir.

Note

In valve terminology, a normally closed valve may also be referred to as a normally nonpassing valve. A normally open valve may be called a normally passing valve.

Pressure-relief valves are primarily used to protect a hydraulic circuit in the event of excessive pressure buildup, preventing a ruptured hose or line, or damage to a component. For example, if hydraulic pressure rises within a circuit due to a clogged oil filter or cooler, the relief valve would open, allowing oil flow to bypass back to the tank. Pressure reliefs can also be used to feed a low-pressure circuit, such as a transmission's lubrication circuit. The pressure settings of relief valves can range as low as 15 psi (1 bar) all the way above 7500 psi (517 bar).

Direct-Acting Relief Valves

Relief valves can be designed as a simple relief, also known as a direct-acting relief, or a compound relief, also known as a pilot-operated relief. A ***direct-acting relief valve*** uses only one spring in its construction. See Figure 7-2. The valve can be designed as an inline valve or at a right angle, and can use a check ball or a poppet valve. Compared to pilot-operated relief valves, direct-acting relief valves provide faster reaction times, but have the disadvantage of a higher pressure override.

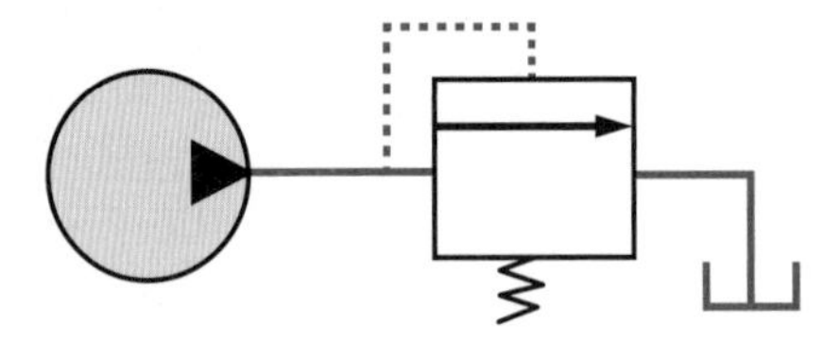

Figure 7-1. A pressure-relief valve is a normally closed valve that opens when pressure becomes excessive. When the valve opens, it diverts pump flow to the reservoir, decreasing pressure. When the pressure falls back to a normal level, the pressure-relief valve closes again.

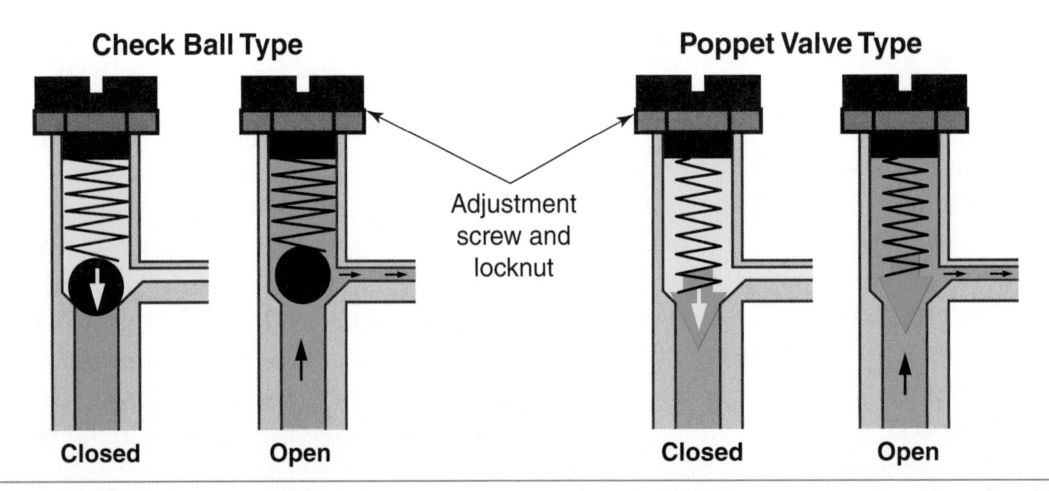

Figure 7-2. A simple direct-acting relief uses only one spring with a check ball or poppet valve. Note that this relief valve has a threaded adjustment for varying the relief valve's pressure setting.

Pressure override equals the full-flow pressure setting minus the valve's cracking pressure setting. ***Cracking pressure*** is the pressure setting at which the relief valve initially begins to separate from its seat, allowing oil to flow. ***Full-flow pressure*** is the pressure value at which the relief valve completely opens and is allowing the system's maximum flow. A graph of these pressure settings in relation to system flow for both direct-acting and pilot-operated reliefs is displayed in Figure 7-3. Relief valves that exhibit high pressure override are opening too soon. Any time fluid is dumping over a relief, the hydraulic energy—instead of performing useful work—is being converted to wasted heat energy.

Protecting Trapped Oil Circuits

Hydraulic systems contain trapped oil circuits. This type of circuit consists of hydraulic lines located between an actuator and the DCV spool. Outside forces acting on the actuator can cause the oil pressure within the trapped circuit to rapidly increase to an excessive level. Cylinder port relief valves, thermal relief valves, and accumulators provide protection for trapped oil circuits.

The quick response of a direct-acting relief valve makes it suitable for protecting trapped oil circuits. ***Cylinder port relief valves***, also known as line-relief valves or actuator-relief valves, are a type of direct-acting relief located between a DCV spool and the actuator, as shown in Figure 7-4.

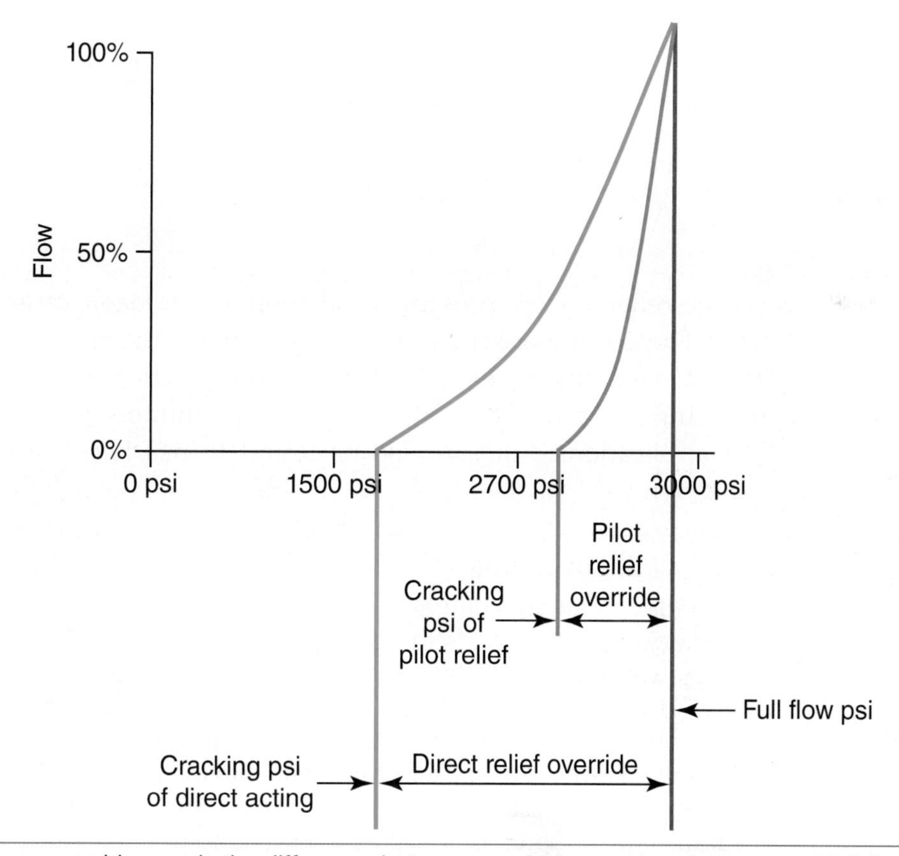

Figure 7-3. Pressure override equals the difference between a valve's cracking pressure and full-flow pressure. A direct-acting relief valve has more pressure override than a pilot-operated relief valve. The direct-acting relief valve can open with as little as 50% of the full-flow pressure. A pilot-operated relief valve will open closer to 90% of the full-flow pressure value.

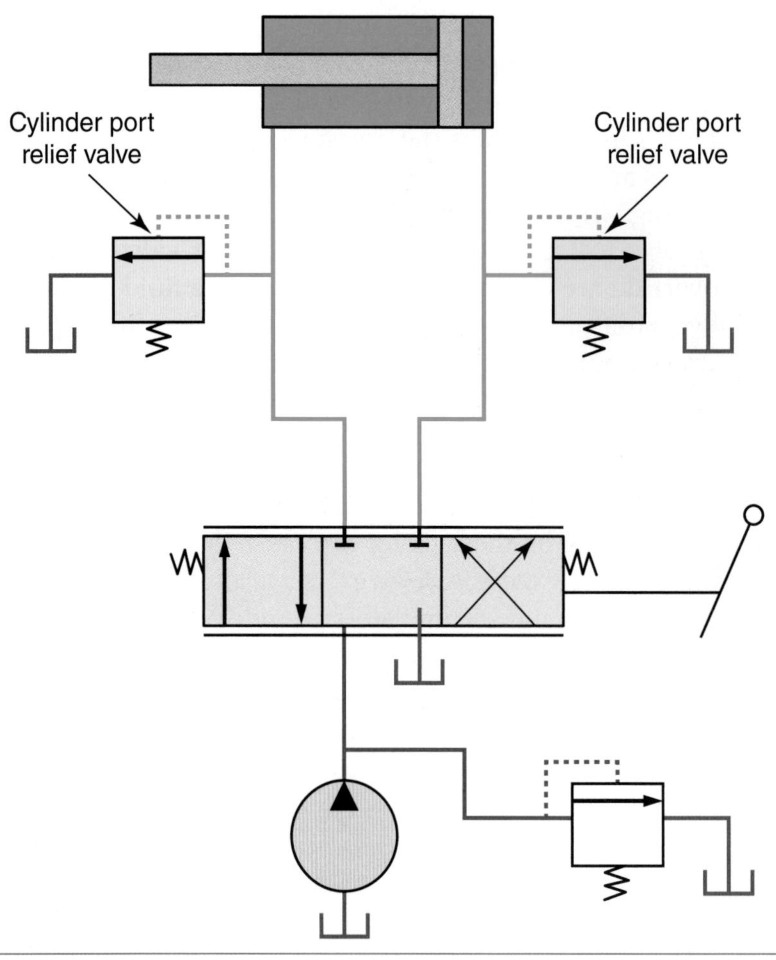

Figure 7-4. Cylinder port relief valves are located between the DCV spool and the actuator. They protect the cylinder from external forces and shock loads.

Line-relief valves provide protection from loads caused by external forces. These actuator reliefs are necessary when the operator's hand is off of the DCV, meaning the DCV is in the neutral position. Using relief valves for protecting trapped oil circuits is especially common in construction equipment applications, but can also be found in nearly all mobile machinery applications.

Case Study

Danger of Shock Loads

A loader on an agricultural compact utility tractor was configured with teeth on the bucket's cutting edge to help excavate dirt. The operator had shut off the machine with the bucket facing downward. Later, when the operator climbed back onto the tractor, he forgot that the bucket's teeth were pointing toward the ground. In a rush, the operator started the tractor and pressed the hydrostatic forward lever. The bucket's teeth dug into the ground, stopping the forward movement of the tractor. As a result of the external load placed on the circuit, a hydraulic hose ruptured, a hydraulic line ruptured, and multiple O-rings in the fittings were destroyed, as pictured in Figure 7-5. In this example, the manufacturer had not configured the loader circuit with any type of protection from external loads.

Figure 7-5. This hose and hydraulic tube both failed when the bucket teeth of the loader were driven into the ground. The trapped oil circuit did not contain any type of actuator circuit protection.

Several different types of external forces can cause a shock load on circuits. The following examples demonstrate such forces:

- Running a loader into a stump, boulder, or into the ground.
- Loading the bucket by driving into a stockpile of dirt.
- A heavy load in the bucket, bouncing up and down as a tractor is driving through a field.
- A heavy load falling into the bucket as it is being filled.
- Pushing over a tree with a loader.

Accumulators can also be used to absorb external shocks, and they will be explained in **Chapter 13**.

A thermal relief valve is another example of a pressure-relief valve that is located in a trapped oil circuit. On a hot, sunny day, the thermal effect of the sun will build high circuit pressures. A ***thermal relief valve*** will prevent the cylinder and hoses from rupturing due to this heated oil expansion by dumping the overpressurized oil to the tank. Thermal relief valves are set substantially higher than the system's relief. For example, if the system pressure is set at 2860 psi (197 bar), the thermal relief valve might be set as high as 4000 to 4500 psi (276 to 310 bar).

Like a line-relief valve, a thermal relief is located between the DCV spool and the actuator. If the circuit schematic fails to label the relief valve, it is difficult to know if the purpose is for thermal protection or for shock load protection. The thermal relief valve does not have to respond as fast as a cylinder port relief valve.

Both the thermal relief valve and the cylinder port relief valve can cause a cylinder to drift. Figure 7-6 shows a cylinder port relief that was replaced on a skid steer. The loader would leak down in less than three seconds.

Figure 7-6. This cylinder port relief valve was replaced on a skid steer because the loader was drifting to the ground.

Case Study

Cylinder Drift and Restoring Customer Satisfaction

One manufacturer produced a technical service bulletin (TSB) for diagnosing a cylinder drift problem. The TSB instructed the technician to remove and cap the thermal relief's drain line. Next it instructed the technician to measure the amount of oil drops exiting the thermal relief valve block. The specification was 4 to 6 drops of oil per minute. An oil leakage rate that exceeded specification indicated the need to replace the thermal relief valve. The TSB also provided an engineering specification for an allowable amount of cylinder drift of 1.0 inch drift (25.4 millimeters) per hour measured at the cylinder.

Although the TSB did not mention the justification for a 1″/hr cylinder drift rate, it is probable that the manufacturer justified this limit due to the stack-up of tolerances. For example, **Chapter 6** lists seven different items that can cause a cylinder to drift. If a machine was equipped with three or four of those components, and if each of those components were functioning at the minimal threshold, the stack-up of those tolerances could result in the cylinder drift.

A dealership might find itself in a dilemma when its unsatisfied customer purchases a $350,000 machine with a drifting cylinder. The manufacturer might state that the cylinder drift is within the engineering specification and therefore is unwilling to warranty the problem. The dilemma can continue to escalate when the neighbor's two machines, or the customer's other three machines, all operate with zero cylinder drift.

Normally this problem is resolved by the manufacturer's local territory representative, who has a budget, called *policy dollars*, for maintaining customer satisfaction. In this scenario, the territory representative can provide policy dollars to the dealership to cover the expense of the repair. The policy dollars are allocated to the territory representative to maintain the company's good faith or good will with the customer.

Pilot-Operated Relief Valves

A ***pilot-operated relief valve*** contains two different poppets that work in unison. The small poppet is the pilot valve. System pressure is often routed to the pilot valve through a drilled orifice. In Figure 7-7 the drilled orifice is through the large poppet.

When the system pressure is low, pressure acts on both sides of the large poppet resulting in a balanced poppet. However, the large poppet's spring will force the poppet closed.

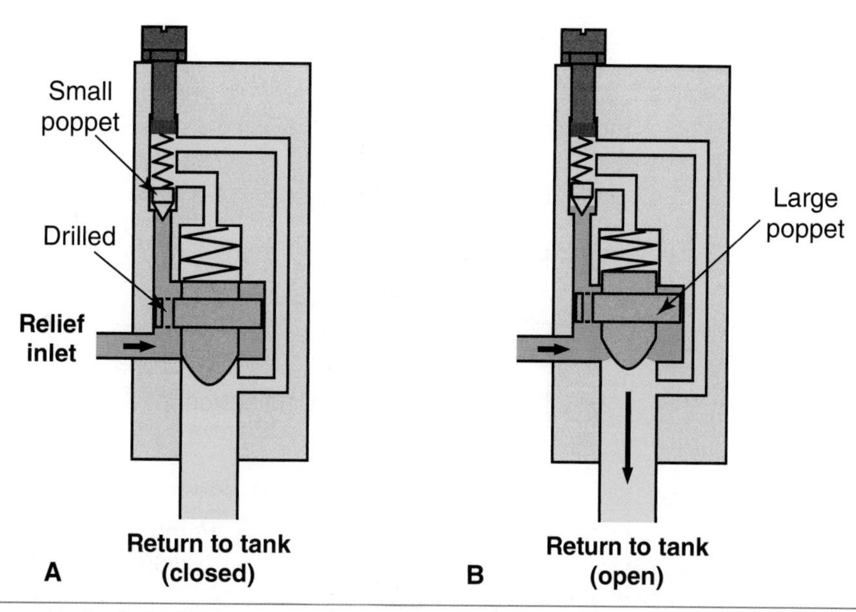

Figure 7-7. A pilot-operated relief valve uses a small pilot poppet along with a larger poppet. A—The large poppet remains closed as system pressure has not unseated the pilot poppet. B—When the pilot poppet opens, it creates an unbalanced large poppet that is opened by the high system pressure.

The pilot valve sets the cracking pressure of the relief valve. Once the system pressure can overcome the pilot valve's spring tension, it opens and dumps the pilot valve's oil to the reservoir. With oil flowing to the reservoir after passing through the drilled orifice, a pressure drop occurs across the large poppet. As the pressure behind the large poppet drops, it allows the high system pressure to open the large poppet, which dumps the system pressure to the reservoir. Refer to Figure 7-7.

A pilot-operated relief valve is also known as a compound relief valve or a balanced relief. It exhibits less pressure override than a direct-acting relief and is commonly used as a main system relief for controlling large amounts of oil flow. Figure 7-8 illustrates a DCV that is attempting to send all of the oil from a fixed-displacement pump to extend a cylinder. When the cylinder reaches the end of travel, the oil must have a path to flow or something will burst. The main system relief valve provides that protection for actuators when the actuators are said to be ***dead-headed***. This means that the oil has no other path to flow because the actuator has stalled and is not moving.

Main System Relief Valves

Main system relief valves are located between the pump and the DCV spools, and are frequently placed as close to the pump as possible. Sometimes the main relief is located inside the pump housing, as in Figure 7-8. However, in some variable-displacement pump applications, a main system relief can be located at the opposite end of the tractor, which is detailed in a case study in **Chapter 17**.

Many agricultural machines that are equipped with load-sensing, pressure-compensating (LSPC) systems do not use a main system relief valve. **Chapter 18** will explain the operation of fixed- and variable-displacement load-sensing, pressure-compensating systems.

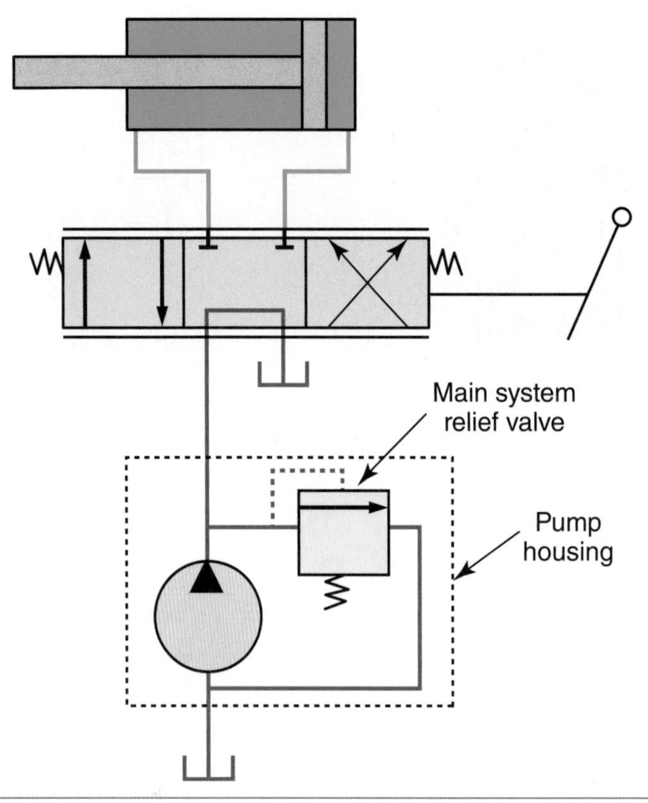

Figure 7-8. The main system relief is usually located as close to the pump as possible, sometimes inside the pump housing.

Adjusting Valves

Many relief valves are variable. The most common type of valve adjustment uses a jam nut to lock an internal adjustment screw, as pictured in Figure 7-9. The adjustment screw may require the use of an Allen head wrench or a flat tip screwdriver.

Multiple mistakes can be made while adjusting a relief valve:

- Adjusting the relief while the system is operating.
- Not holding the threaded adjustment screw while tightening the jam nut.
- Overtightening the jam nut.

Warning

As mentioned in **Chapter 1**, serious injury or even death can occur as a result of injected fluid. It does take a little more time to shut off the machine, make an adjustment, and then run the machine again to check the adjustment. However, an injury will cost even more time and money.

After the machine is shut off and the hydraulic pressure has been relieved, a wrench is required to loosen the jam nut. Most adjustment screws need to be threaded inward (clockwise) in order to increase the pressure, and threaded outward (counterclockwise) to lower the pressure.

Figure 7-9. This main system relief valve was internally leaking and required replacement. The adjustable valve was used in a John Deere loader backhoe. If only an adjustment was needed, a technician could loosen the jam nut with the open end of the combination wrench before threading the adjustment screw in or out with the Allen wrench.

The adjustment screw must be held stationary while the jam nut is torqued to specifications. Inexperienced technicians tend to forget this step, accidentally increasing instead of decreasing system pressure by inadvertently turning the adjustment screw as the jam nut is torqued.

Caution

Be sure to tighten the jam nut to the manufacturer's torque specification. The threaded adjustment screw can be pulled into two pieces if the jam nut is overtorqued.

Some valves are adjusted by adding or removing shims. Shims are offered in different thicknesses, and an adjustment might require adding or removing multiple shims. This style of valve adjustment requires physically removing a cap or plug, and sometimes a spring, before changing the number of shims. Be sure to recheck the pressure setting after making the final adjustment to confirm that the adjustment was performed correctly.

Pressure-Reducing Valves

A ***pressure-reducing valve (PRV)*** is a normally open valve that limits its outlet pressure by sensing downstream oil pressure. The schematic symbol in **Figure 7-10** looks similar to a pressure-relief valve, except the arrow depicts that the valve is open and the sensing line is located after the valve.

PRVs not only limit pressure, but they will maintain that fixed amount of pressure as long as the valve's inlet pressure does not drop below the spring value. PRVs are frequently used for pilot controls by supplying a low constant pilot pressure. See **Figure 7-11**.

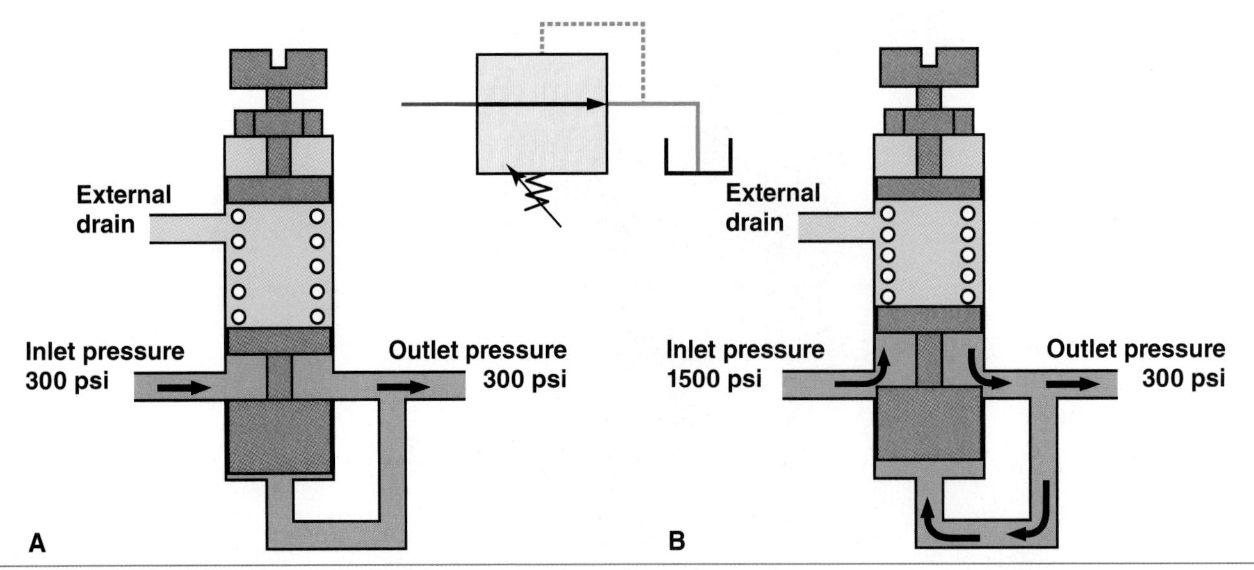

Figure 7-10. Pressure-reducing valves sense the outlet pressure and will reduce its outlet pressure to a set value. A—Valve is normally open. B—Valve partially closed to reduce pressure at outlet.

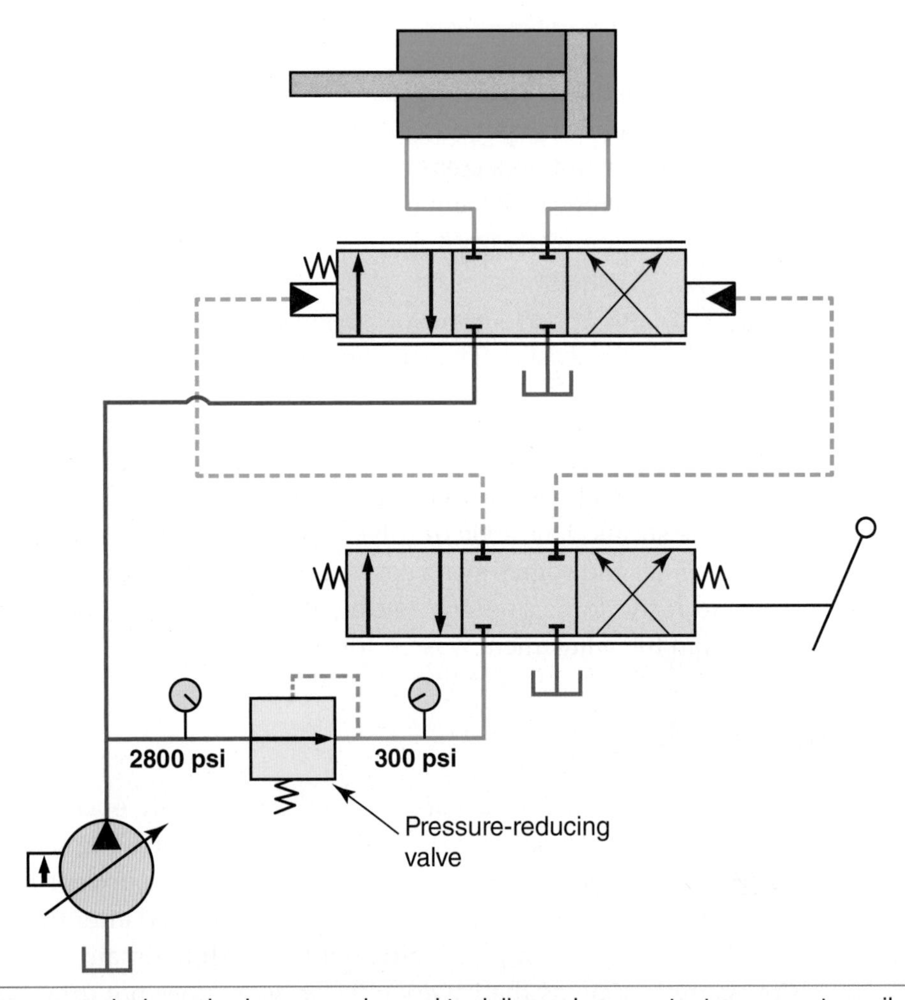

Figure 7-11. A pressure-reducing valve is commonly used to deliver a low constant pressure to a pilot control system. This is an example schematic using a variable-displacement pressure-compensating pump, which will be explained in **Chapter 17**.

Pilot-operated DCVs are used in numerous mobile machines. The pilot-control valve directs pilot oil to operate a DCV spool. **Chapter 9** will explain DCVs in more detail.

PRVs are also used as pressure compensators inside DCVs. This role of the PRV will be explained in **Chapter 8** and **Chapter 18**.

Unloading Valves

An ***unloading valve*** is a pressure-control valve that is used to unload a hydraulic pump, allowing the system to operate at a low pressure value. Most unloading valves are used in conjunction with fixed-displacement pumps in an open-center hydraulic system. However, at least one construction equipment manufacturer uses an unloading valve on a dozer with a variable-displacement pump.

Unloading valves can have several different configurations:

- Non-load-sensing unloading valves.
  - Fixed value (nonadjustable spring).
  - Variable valve (adjustable spring). See Figure 7-12.
- Load-sensing unloading valves.
  - Fixed value used in conjunction with a load-sensing signal.
  - Variable valve used in conjunction with a load-sensing signal.
- Solenoid operated.

A log splitter is an example of a hydraulic system that uses a non-load-sensing unloading valve. The pump assembly shown in Figure 7-13 contains two gear pumps, a small-displacement pump and a large-displacement pump. The unloading valve is designed to unload the large-displacement pump when the system pressure increases above a set limit, 500 psi (34 bar) for example. This two-stage open-center system will be explained in further detail in **Chapter 16**.

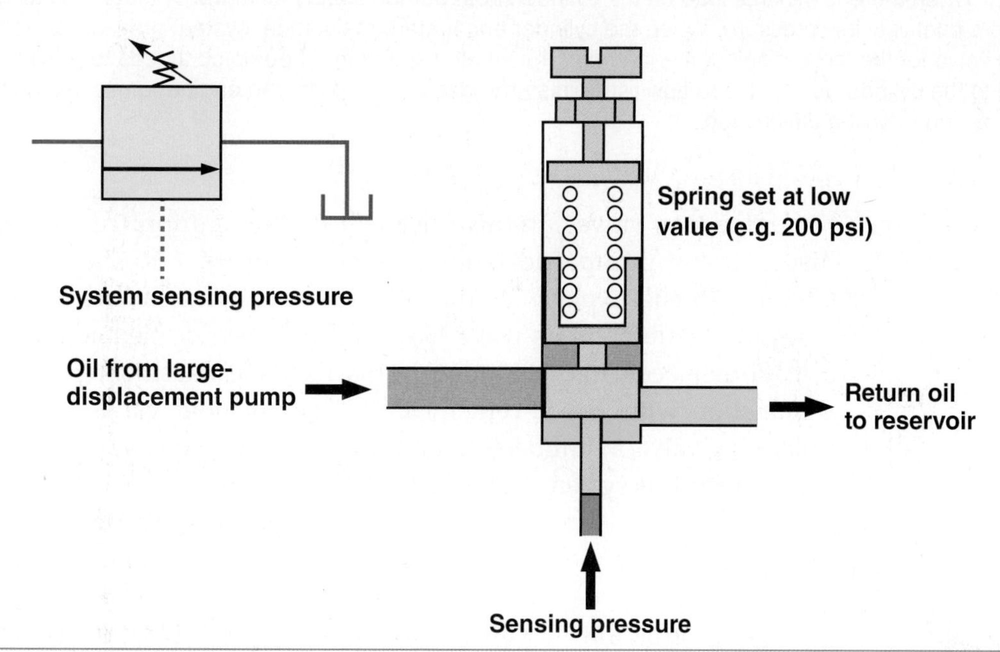

Figure 7-12. An unloading valve will dump system flow causing a drop in system pressure once the pressure reaches the unloading valve's spring value.

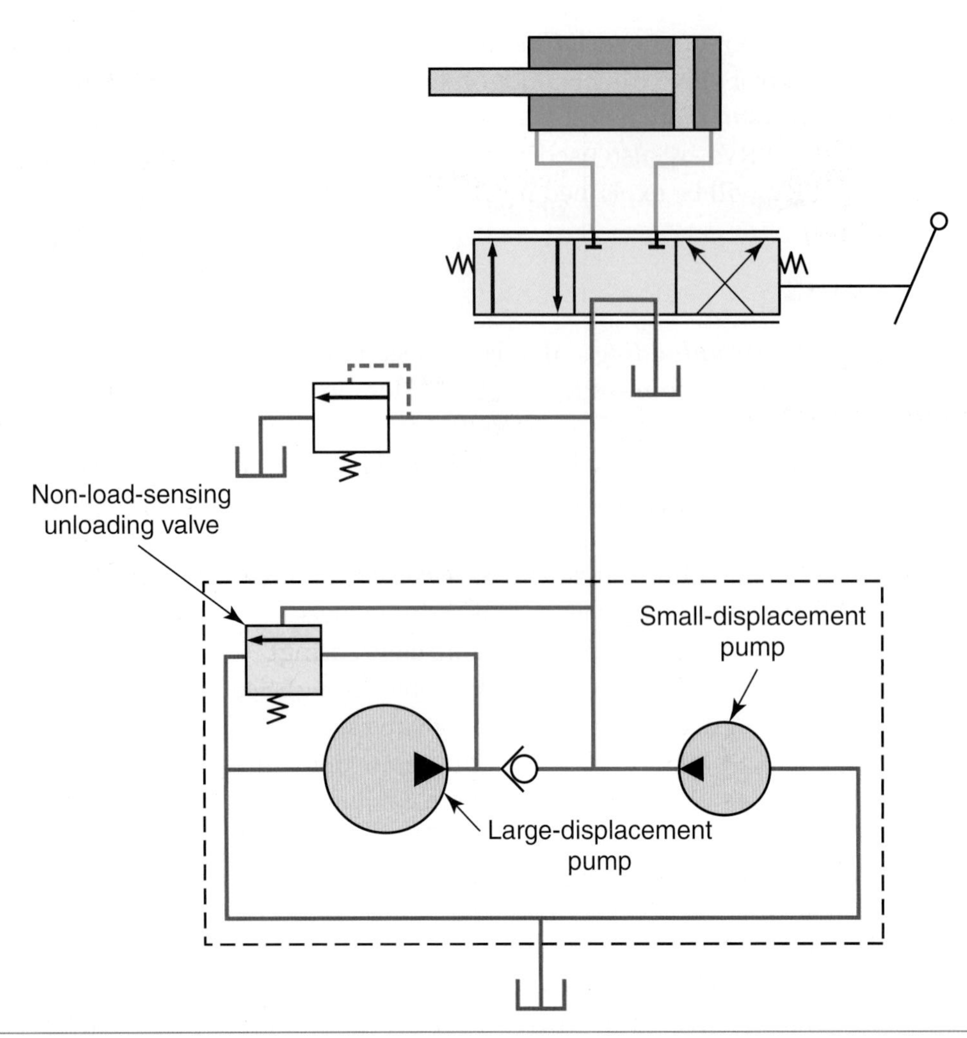

Figure 7-13. Log splitters use unloading valves to dump the large-displacement pump flow when the splitter reaches a stubborn log. When there is minimal load on the cylinder, both pumps supply oil to the cylinder, resulting in high system flow at a relatively low pressure. When the cylinder encounters resistance, system pressure builds, opening the unloading valve for the large-displacement pump. The small-displacement pump continues to provide higher-pressure fluid to the cylinder at a reduced flow rate. This arrangement prevents the engine from stalling when the splitter is attempting to split a difficult log.

Unloading valves are also used in mobile equipment that has a fixed-displacement pump and is load sensing. Figure 7-14 shows their design. **Chapter 18** will explain the use of load-sensing unloading valves.

Some manufacturers use a two-position electric solenoid valve to unload a fixed-displacement gear pump. This fixed-displacement pump is used in conjunction with closed-center DCVs. The solenoid valve can be called an unloading valve, a dump valve, or a jammer valve. See Figure 7-15. **Chapter 16** will explain this system and valve in further detail.

Pressure-Sequence Valves

A ***pressure-sequence valve*** is a pressure-control valve that will direct flow to a secondary circuit after system pressure has risen enough in the primary circuit to overcome the closing pressure of the spring within the valve, Figure 7-16.

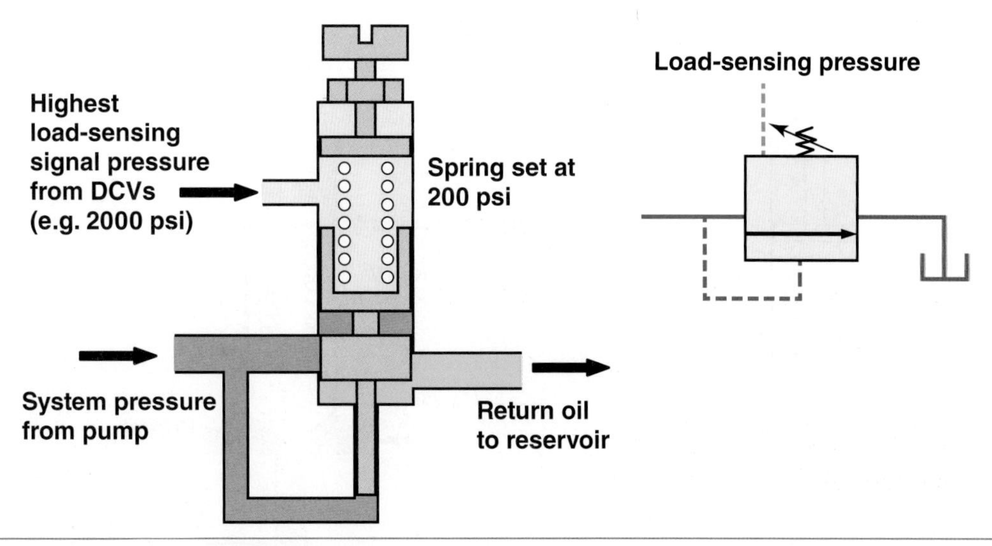

Figure 7-14. Manufacturers use unloading valves to enable a fixed-displacement pump to be used in a load-sensing application. **Chapter 18** details load-sensing hydraulic systems.

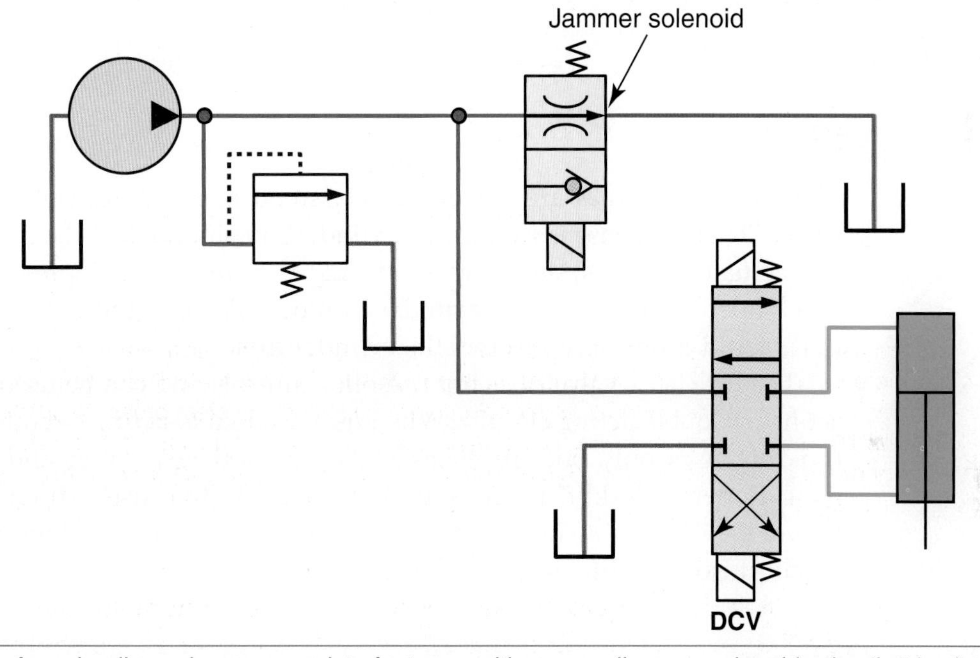

Figure 7-15. An unloading valve can consist of a two-position normally open solenoid valve that is electric actuated. This valve will be explained in further detail in **Chapter 16**.

Pressure-sequence valves are commonly used on agricultural planters. When the operator commands the planter to lift, all of the oil is sent to the planter lift cylinders. After the planter lift cylinders have reached their end of travel, the pressure-sequence valve opens, allowing oil to flow to the secondary circuit. The secondary oil is used to fold the planter's markers.

Sequence valves are also used in trash trucks. First, the truck's gate cylinders are actuated to close. After the cylinders reach the end of travel, the pressure-sequence valve directs fluid to the compactor to compress the load.

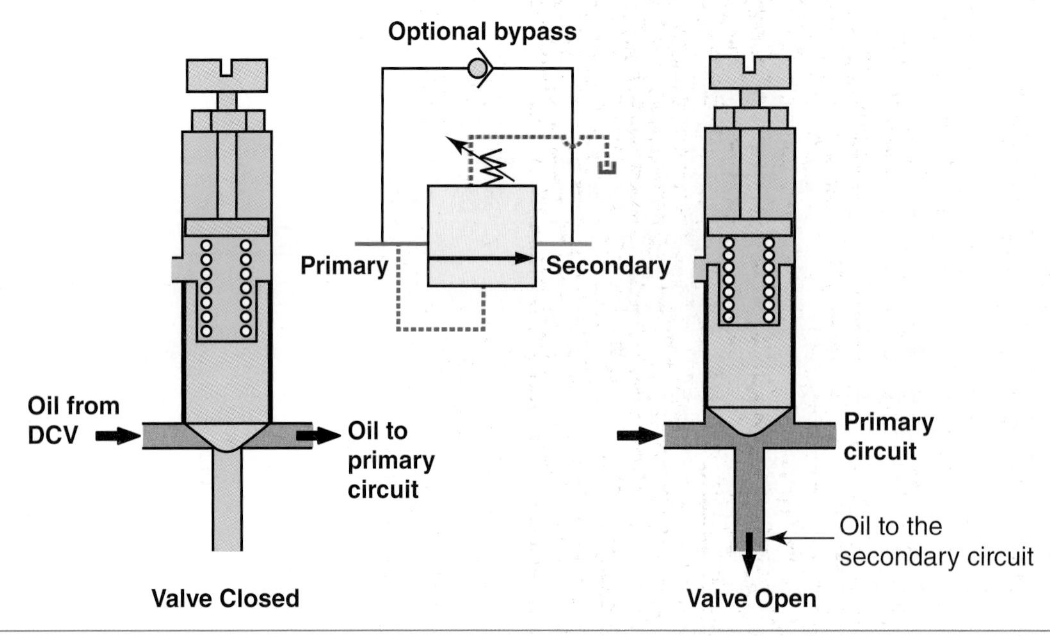

Figure 7-16. A pressure-sequence valve will direct fluid to a secondary circuit after the primary circuit pressure has risen—commonly when the cylinder in the primary circuit has reached its end of travel—to the valve's spring value.

Lift Check Valves

Lift check valves are sometimes classified as a type of pressure-control valves. Their purpose is to hold the cylinder's position when the DCV spool is first actuated. Once system pressure is higher than the load's pressure, the lift check opens and causes the cylinder to move. Without these valves, the cylinder could drop slightly before the cylinder lifts.

The lift check valve does not use pilot controls and can be used in single-acting or double-acting circuits. When used in double-acting circuits, the DCV typically uses only one lift check valve, for both extending and retracting the cylinder. A forklift is an example application that uses lift check valves. Figure 7-17 shows a simplified forklift schematic. Some combine header lift circuits also use a lift check valve.

Load-Holding Valves

Load-holding valves, sometimes also classified as pressure-control valves, prevent an actuator from moving or drifting when the DCV is in a neutral position. If a cylinder was allowed to drift, it could damage equipment, injure personnel, or even cause death. The two types of load-holding valves used in mobile equipment are pilot-operated check valves and counterbalance valves.

Pilot-Operated Check Valves

Pilot-operated check valves can have two different designs, pressure-to-open and pressure-to-close. Pressure-to-close valves are not used for load

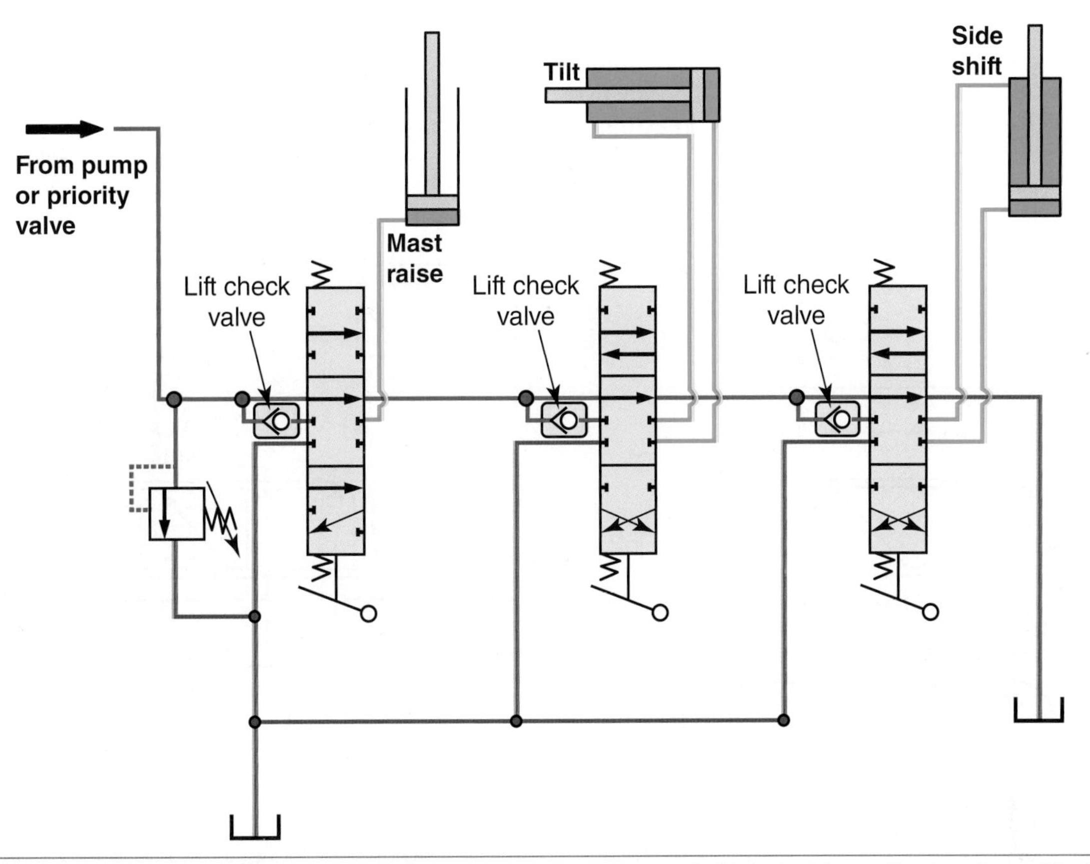

Figure 7-17. Forklifts commonly have three DCVs for lift, tilt, and side shift. Note in this application that a lift check is used in all three DCVs, regardless if the cylinder is single acting or double acting.

holding. Instead they can be used to pressurize a hydraulic reservoir during machine operation and will depressurize the reservoir when the machine is shut off.

The pressure-to-open pilot-operated check valve is the style commonly used in mobile machinery for holding a double-acting actuator stationary. See Figure 7-18. The pilot-operated check valves are located between the DCV spool and the actuator. It requires pilot oil pressure to open the valve before the cylinder's return oil can be routed to the reservoir.

A double pilot-operated check valve is pictured in Figure 7-19. It was designed to hold a combine's unloading auger swing cylinder in both directions. This valve contains a single pilot piston and two poppet assemblies that are threaded into the valve block. The pilot piston can be actuated from either direction—cylinder extend or cylinder retract. When the piston moves, it pushes open a poppet, compressing a spring, which allows the return oil to be routed back to the reservoir. Figure 7-20 shows the poppet and spring.

Engineers have to factor in several variables when choosing a pilot-operated check valve, including the following three:

- Load created by the cylinder.
- Ratio of the pilot piston area to the poppet piston area.
- Ratio of the cylinder's cap end area to the effective area of its rod end.

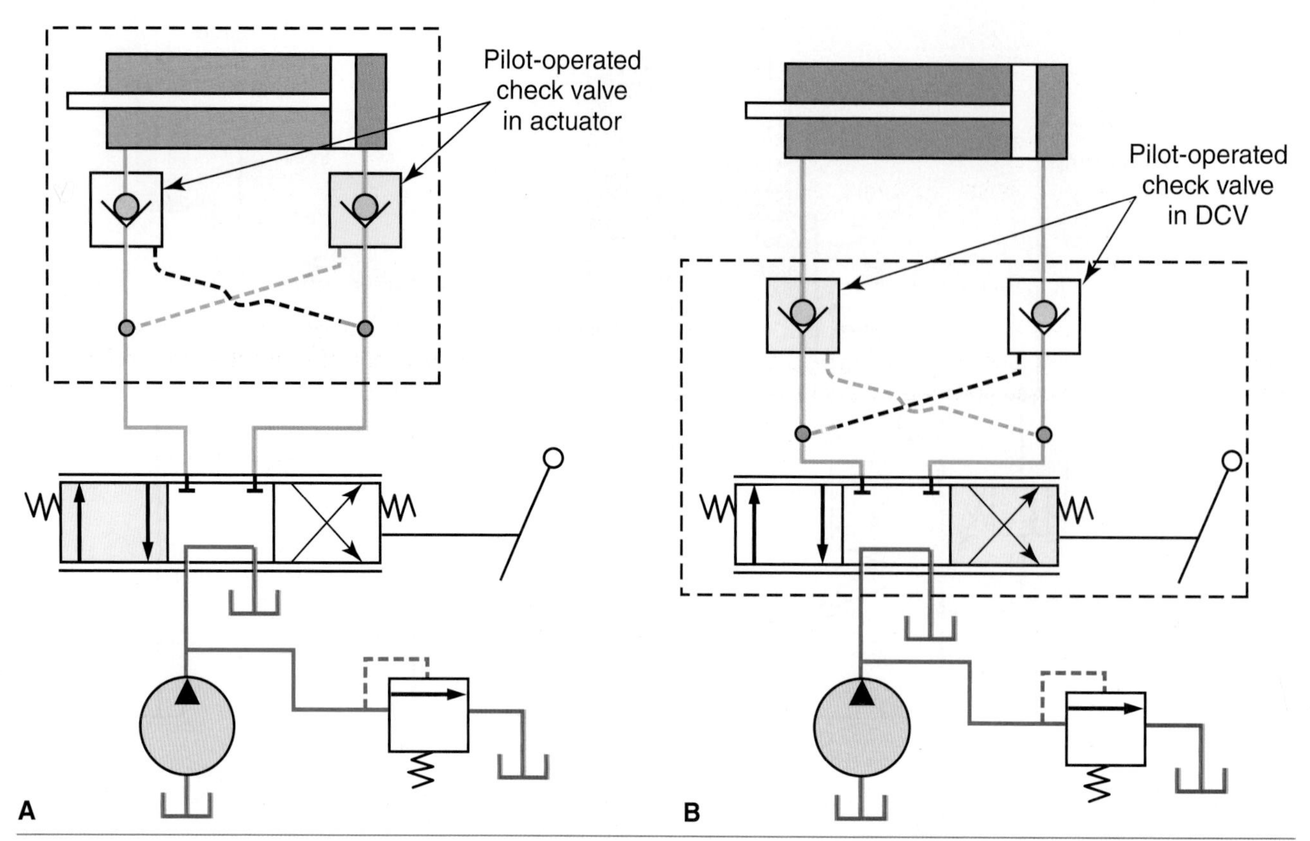

Figure 7-18. Pilot-operated check valves can be located inside the DCV block or in the actuator within a hydraulic circuit. A—Check valves are located at the cylinder for retraction mode. B—Check valves located inside the DCV block for the extend mode.

Figure 7-19. This double pilot-operated check valve assembly is used for holding a combine unloading auger swing cylinder stationary. A pilot piston is located in the center of the valve block. A load check poppet assembly is threaded into each end of the valve block.

At a minimum, the ratio of the pilot piston area to the poppet piston area must be greater than the ratio of the cylinder's cap end area to the effective area of the cylinder's rod end, or the valve will remain closed, Figure 7-21.

If the pilot-operated circuit is designed poorly, it can cause the cylinder's rod seal to fail. Engineers use software for computing the potential pressure intensification, and will choose the appropriate ratio for the pilot-operated check valve.

Figure 7-20. A cartridge that contains a load check poppet assembly is installed at each end of the valve block. A spring holds the poppet closed, preventing the cylinder from drifting. When the pilot piston is actuated, it pushes the poppet open allowing oil to flow freely back to the reservoir.

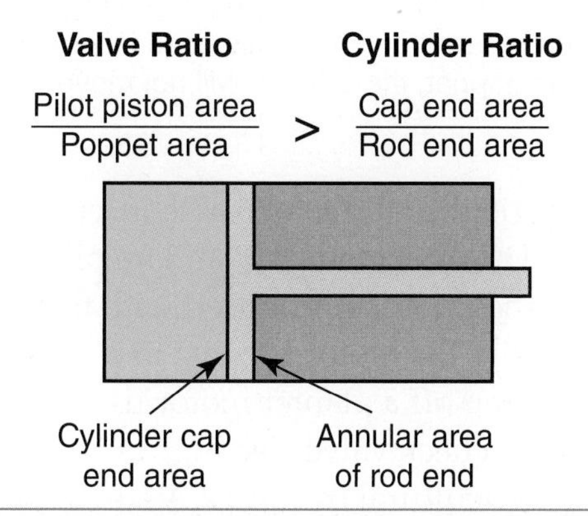

Figure 7-21. A pilot-operated check valve must have a larger ratio of the pilot piston area to the poppet area than the cylinder piston cap's end area to the effective area of the cylinder's rod end.

The pilot-operated check valves operate very similar to silicon-controlled rectifiers (SCRs) in electronic circuits, Figure 7-22. They require voltage (pressure) to open the device, allowing electrons to flow.

It is commonly stated that the pilot-operated check valves will hold an actuator stationary if a hydraulic hose fails. However, this depends on where the pilot-operated check valves are located in the circuit. If the pilot-operated check valves are located at the cylinder, as shown in Figure 7-23, and a hose fails, the cylinder would remain stationary. If the pilot-operated check valve is located inside the DCV, the cylinder could drift if a hose failure occurred between the cylinder and the pilot-operated check valve.

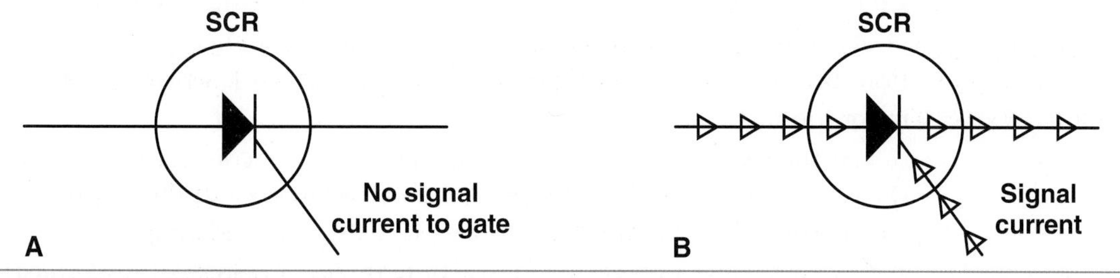

Figure 7-22. A silicon-controlled rectifier (SCR) in an electrical circuit operates in the same manner as a pilot-operated check valve in a hydraulic circuit. An SCR requires a certain signal current, applied to the gate, to open and will remain open as long as this current level is maintained. A—SCR is "off" as no signal current is being applied. B—SCR is "on" and electrons are allowed to pass through the diode.

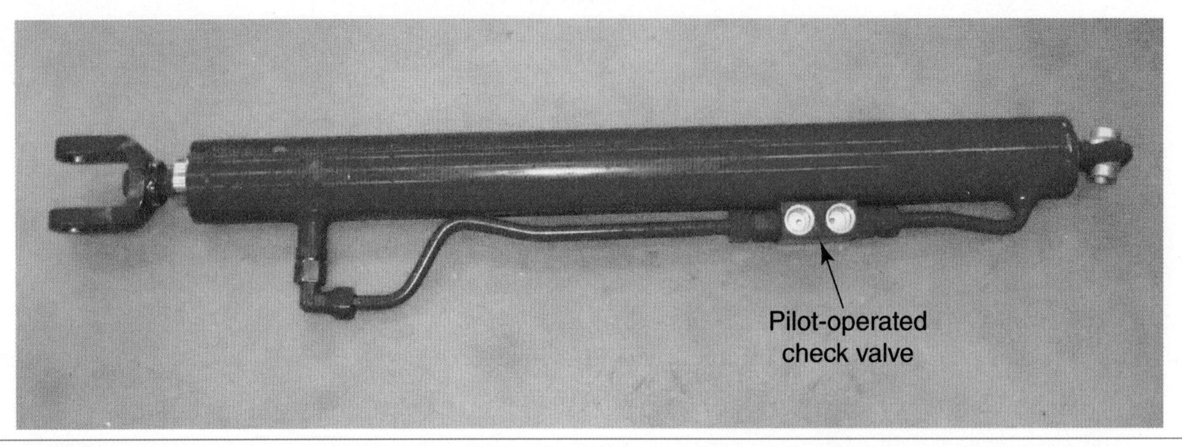

Figure 7-23. This unloading auger swing cylinder has the pilot-operated check valves located at the cylinder. It provides load holding of the cylinder in the event of a hydraulic hose failure. Unless the pilot-operated check valve receives fluid pressure to unseat the poppet, the cylinder will not move.

Many agricultural tractors are equipped with just one pilot-operated check valve in the DCV's extend port, and no pilot-operated check valve in the return port. When the DCV is configured with float, the operator can shut off a planter motor, allowing the motor to coast to a stop. **Chapter 9** will explain the strategies for shutting off a planter motor using a DCV equipped with float and one pilot-operated check valve. See Figure 7-24.

Not all agricultural tractor DCVs are equipped with a pilot-operated check valve. Some manufacturers require a customer to specify, at the time of purchase, that a DCV be equipped with one or two pilot-operated check valves. If the customer failed to purchase a machine with the check valves, they can typically add a pilot check valve later.

Many construction machines use two pilot-operated check valves with each actuator. The pilot check valves are frequently located inside the DCV but can be located at the actuator.

Counterbalance Valves

Pilot-operated check valves should only be used in stable applications for holding a load, and should not be used in overrunning applications. **Chapter 6** explained that metering-in an overrunning load causes cavitation and metering-out an overrunning load causes pressure intensification. ***Counterbalance valves*** should be used for controlling loads when the cylinder is operating an overrunning load. The counterbalance valve prevents cavitation, provides load-holding ability, and prevents a load from extending a cylinder too fast. See Figure 7-25.

As the DCV attempts to retract the cylinder, the counterbalance valve senses the pressure leaving the cap end of the cylinder. When pressure is high enough to overcome the counterbalance spring value, the valve opens, allowing fluid to return to the reservoir. This prevents the load from running ahead too fast. The counterbalance valve also requires the use of a reverse flow check that enables the cylinder to be extended. Counterbalance valves are used in forklift tilt spools, dozer blade lower controls, and hydraulic winch applications.

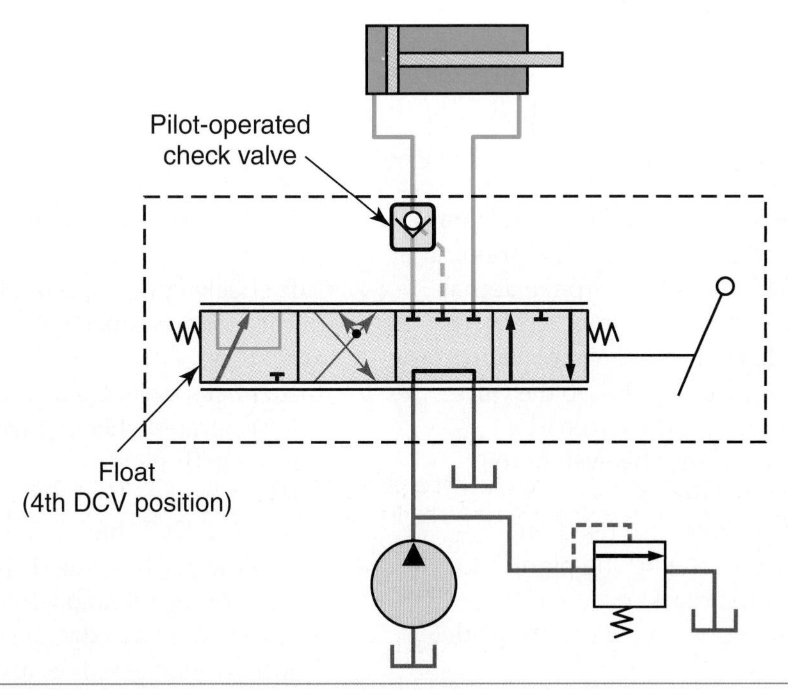

Figure 7-24. Many agricultural tractors have DCVs with only one pilot-operated check valve. The check valve is typically located in the DCV's cylinder extend port.

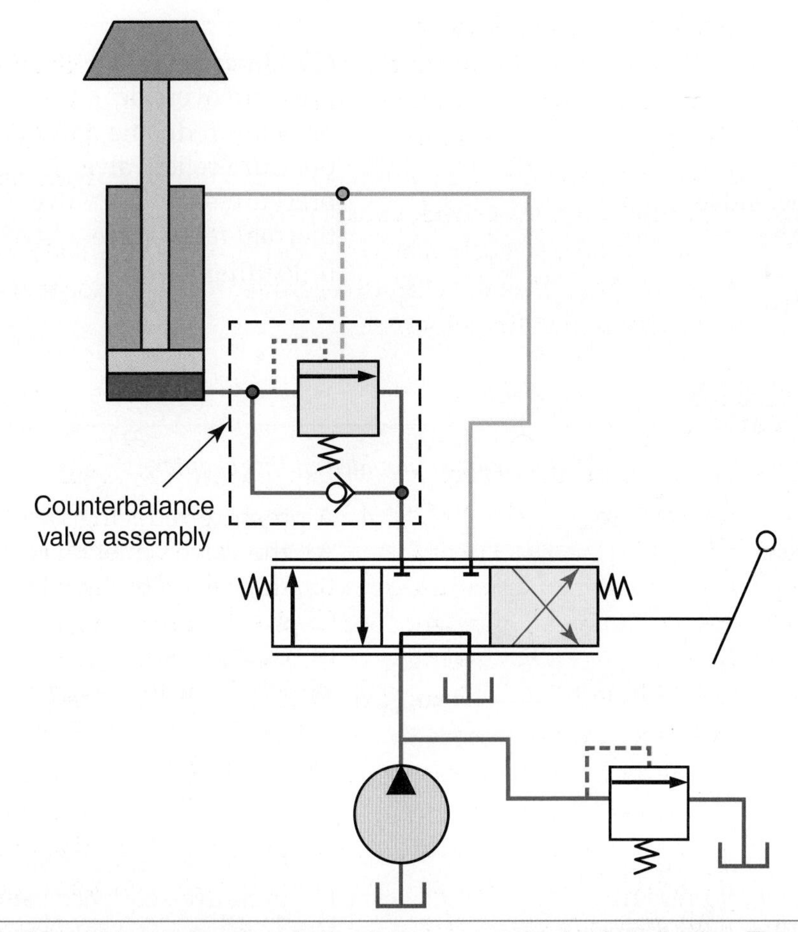

Figure 7-25. Counterbalance valves prevent cavitation in an overrunning load application. The check valve is used to enable reverse flow when the cylinder is extended.

Summary

- ✓ Pressure-relief valves have two designs: direct acting and pilot operated.
- ✓ Pressure-relief valves can be used for main system protection or for trapped oil protection.
- ✓ Relief valves located between the DCV spool and actuator provide protection from external loads but can cause the cylinder to drift.
- ✓ Most adjustable relief valves use a jam nut and an adjustable screw that threads into the valve.
- ✓ Unloading valves are used to unload a hydraulic pump, enabling the system to operate at a low pressure.
- ✓ Pressure-sequence valves are used on agricultural planters causing the planter to rise before it folds the marker.
- ✓ PRVs are normally open and sense the outlet pressure.
- ✓ PRVs will maintain a constant lower pressure, assuming the inlet pressure is higher than the PRV spring pressure.
- ✓ PRVs are used to set a machine's pilot pressure.
- ✓ Lift checks, pilot-operated check valves, and counterbalance valves are used to hold actuators.
- ✓ Lift checks do not use pilot oil.
- ✓ Pilot-operated check valves are used to hold a double-acting cylinder from drifting.
- ✓ Pilot-operated check valves can be located inside the DCV block or at the cylinder.
- ✓ When the pilot-operated check valve is located at the cylinder, it holds the actuator stationary in the event of a hydraulic hose failure.
- ✓ Counterbalance valves are used for load holding in overrunning applications.

Technical Terms

counterbalance valves
cracking pressure
cylinder port relief valves
dead-headed
direct-acting relief valve
full-flow pressure
lift check valves
load-holding valve
pilot-operated relief valve
pressure override
pressure-reducing valve (PRV)
pressure-relief valve
pressure-sequence valve
thermal relief valve
unloading valve

Review Questions

Answer the following questions using the information provided in this chapter.

1. Pressure-relief valve designs are _____.
   A. normally open
   B. normally closed
   C. partially obstructed
   D. None of the above.
2. Pressure-reducing valve designs are _____.
   A. normally open
   B. normally closed
   C. partially obstructed
   D. None of the above.
3. A pressure-relief valve senses _____.
   A. the valve's inlet oil pressure
   B. the valve's outlet oil pressure
   C. case pressure
   D. return pressure
4. A pressure-reducing valve senses _____.
   A. the valve's inlet oil pressure
   B. the valve's outlet oil pressure
   C. system pressure
   D. case pressure
5. Which of the following valves is designed to maintain a constant low pressure, such as pilot oil pressure?
   A. Unloading valve.
   B. Pressure-relief valve.
   C. Pressure-reducing valve.
   D. Pressure-sequence valve.

6. Which of the following valves will send oil to a secondary circuit after the primary cylinder reaches its end of travel?
   A. Unloading valve.
   B. Pressure-relief valve.
   C. Pressure-reducing valve.
   D. Pressure-sequence valve.
7. Which of the following valves is designed to dump the system oil flow, allowing a hydraulic system to operate at a low system pressure?
   A. Unloading valve.
   B. Pressure-relief valve.
   C. Pressure-reducing valve.
   D. Pressure-sequence valve.
8. Which of the following valves is designed to protect a system or circuit from high-pressure spikes?
   A. Unloading valve.
   B. Pressure-relief valve.
   C. Pressure-reducing valve.
   D. Pressure-sequence valve.
9. Which of the following valves is commonly used on agricultural planters and trash trucks for controlling the order of cylinder operation?
   A. Unloading valve.
   B. Pressure-relief valve.
   C. Pressure-reducing valve.
   D. Pressure-sequence valve.
10. Which of the following can be defined as the difference between opening pressure and full-flow system pressure?
    A. Cracking pressure.
    B. Holding pressure.
    C. Pressure override.
    D. Pilot pressure.
11. Which of the following is defined as the pressure at which a pressure-relief valve initially begins to open?
    A. Cracking pressure.
    B. Full-flow pressure.
    C. Pressure override.
    D. Pilot pressure.
12. Which of the following terms can be described as the pressure at which a relief valve is fully open?
    A. Cracking pressure.
    B. Full-flow pressure.
    C. Pressure override.
    D. Pilot pressure.
13. Which of the following valves will have a high pressure override?
    A. Direct-acting relief.
    B. Pilot-operated relief.
    C. Both A and B.
    D. Neither A nor B.
14. Which of the following valves is considered fast acting?
    A. Direct-acting relief.
    B. Pilot-operated relief.
    C. Both A and B.
    D. Neither A nor B.
15. All of the following components can be used to protect a circuit from shock loading, *EXCEPT*:
    A. main system relief valve.
    B. cylinder port relief valve.
    C. accumulator.
    D. actuator-relief valve.
16. All of the following can cause a double-acting cylinder, with a rod facing to the ground, to drift, *EXCEPT*:
    A. cylinder port relief valve.
    B. thermal relief valve.
    C. cylinder piston seal.
    D. main system relief valve.
17. A thermal relief valve is used to _____.
    A. prevent a cylinder fire
    B. relieve cylinder pressure when temperatures rise
    C. relieve severe shock pressure on cylinders
    D. relieve main system pressure
18. A DCV spool is in a neutral position. Technician A states that the main system relief will protect the cylinder from high-pressure spikes. Technician B states that the main system relief will only provide cylinder pressure protection when the DCV is actuated. Who is correct?
    A. Technician A.
    B. Technician B.
    C. Both A and B.
    D. Neither A nor B.
19. What is the purpose of a cylinder port relief valve?
    A. Relieve main system pressure.
    B. Protect an actuator from external forces and shocks.
    C. Holds an actuator stationary in the event of a hose failure.
    D. To unload a pump.

20. All of the following can cause a double-acting cylinder, with a rod facing upward, to drift *EXCEPT*:
    A. cylinder port relief valve.
    B. thermal relief valve.
    C. cylinder piston seal.
    D. a leaky pilot-operated check valve.
21. Which of the following can be used to adjust a pressure-relief valve's setting?
    A. Adjustable screw.
    B. Shims.
    C. Either A or B.
    D. Neither A nor B.
22. Technician A states that main system relief valves can be located inside the pump's housing. Technician B states that all mobile hydraulic machines use a main system relief valve. Who is correct?
    A. Technician A.
    B. Technician B.
    C. Both A and B.
    D. Neither A nor B.
23. Where is a cylinder port relief valve located in a circuit?
    A. Between the pump and the DCV spool.
    B. In the return line.
    C. Between the DCV spool and the actuator.
    D. In the suction line.
24. Where is a main system pressure-relief valve located in a circuit?
    A. Between the pump and the DCV spool.
    B. In the return line.
    C. Between the DCV spool and the actuator.
    D. In the suction line.
25. A technician is having trouble lowering a main system relief valve's pressure. Each time the pressure is checked, the value is the same. What is most likely the cause?
    A. A failed O-ring.
    B. Broken set screw.
    C. Not holding the set screw when torquing the jam nut.
    D. Adjusting the set screw too far.
26. Which of the following load-holding valves is used to hold a double-acting cylinder stationary in applications that are not overrunning?
    A. Lift check valve.
    B. Pilot-operated check valve.
    C. Counterbalance valve.
    D. None of the above.
27. Which of the following load-holding valves is used to hold a double-acting cylinder and will prevent cavitation in a double-acting cylinder with an overrunning load?
    A. Lift check valve.
    B. Pilot-operated check valve.
    C. Counterbalance valve.
    D. None of the above.
28. Which of the following load-holding valves is located before the DCV spool and is primarily used to prevent the load from lowering when the spool is initially actuated?
    A. Lift check valve.
    B. Pilot-operated check valve.
    C. Counterbalance valve.
    D. None of the above.
29. Where can pilot-operated check valves be located?
    A. Inside the DCV block.
    B.. At the actuator.
    C.. Either A or B.
    D.. Neither A nor B.
30. Technician A states that all pilot-operated check valves prevent a cylinder from moving in the event that a hose bursts. Technician B states that pilot-operated check valves must be located inside the DCV in order to hold a cylinder stationary when a hose bursts. Who is correct?
    A. Technician A.
    B. Technician B.
    C. Both A and B.
    D. Neither A nor B.

Chapter 8

Flow-Control Valves

Objectives

After studying this chapter, you will be able to:

- ✓ Explain the three functions that flow-control valves can perform.
- ✓ List the different types of non-pressure-compensated flow-control valves and their differences.
- ✓ Explain why pressure-compensated flow control is used in mobile hydraulic systems.
- ✓ Describe the most common method of pressure-compensated flow control in mobile hydraulic systems.
- ✓ Describe how a temperature-compensated flow-control valve operates.
- ✓ List application examples of priority circuits.
- ✓ Identify the three common ports found on a priority valve.
- ✓ Describe a mobile hydraulic application for using a proportional flow divider.
- ✓ Explain why velocity fuses and quantity fuses are used in hydraulic systems.

Functions of Flow-Control Valves

Flow-control valves are also known as volume-control valves. Manufacturers use flow-control valves to:

- Regulate speed.
- Regulate priority.
- Divide oil in proportions.

Regulating Speed

As explained in **Chapter 2**, the speed of a hydraulic actuator is directly proportional to the oil flow it is receiving. Manufacturers use several different types of flow-control valves for controlling the speed of actuators. Many flow-control valves, such as an orifice or a needle valve, are ***non-pressure-compensated***. If the oil feeding that non-pressure-compensated control valve has an increase in flow or a change in pressure, the flow control valve's output will change. In mobile machinery, this fluctuation of flow creates problems.

For example, if a farmer is planting along a crooked tree line or creek, the tractor must steer around the terrain. As the tractor steers, it causes a change in hydraulic system flow and pressure. The planter uses a fan that is driven by a hydraulic motor. The purpose of the fan is to maintain a constant vacuum or a constant positive air pressure. When the tractor is steering, or performing any other hydraulic function, the speed of the planter fan's hydraulic motor is subsequently increasing or decreasing. This change in the planter's fan speed could negatively affect the planter's ability to accurately plant the seed.

Construction machines are also negatively affected by non-pressure-compensated controls. For example, a motor grader can have more than 12 different hydraulic controls. Every time another hydraulic control is activated, it adversely affects the speed of the previously activated control.

Pressure compensation enables a DCV to maintain a constant actuator speed (cylinder or hydraulic motor) based on a fixed position (opening) of the DCV spool. Also note that pressure compensation is a term used to define a variable-displacement pump control system, which will be explained in **Chapter 17**.

Regulating Priority

Flow-control valves are used to regulate the priority of oil flow. ***Priority valves*** are found on most mobile machinery as a safety device. The valve ensures that the demand of the primary circuit is met first, before any remaining oil can be sent to secondary circuits. The two most common priority circuits in mobile machinery are steering and brakes.

Dividing Flow

The last function of flow-control valves is to divide the oil flow in proportions. ***Proportional flow divider valves*** are designed to divide a quantity of oil flow in a fixed ratio, supplying flow to two branches. Manufacturers design the valves to divide the oil flow in various different proportions. See **Figure 8-1**.

Flow dividers are frequently used on agricultural implements. A planter might be equipped with multiple hydraulic cylinders that are different sizes. The divider valve will proportion oil flow to different branches to meet the supply demands of the different cylinder sizes. Flow dividers are also commonly used in open-center hydraulic systems, which will be explained in **Chapter 16**.

Non-Pressure-Compensated Flow-Control Valves

An ***orifice*** is a non-pressure-compensated flow-control valve that consists of a small passageway that limits the flow of oil. See **Figure 8-2**. The orifice is the simplest flow-control valve found in mobile machinery. Three variables affect the amount of oil flow through an orifice:

- Oil viscosity.
- Pressure drop across the orifice.
- The size of the orifice.

Oil ***viscosity*** is a term used to describe the oil's resistance to flow based on its thickness. For example, honey and gear oil have high viscosities because their thickness prevents them from flowing freely. In contrast, water and vegetable oil have low viscosities and flow much easier. Temperature influences a fluid's viscosity. In general, as temperature increases, viscosity decreases. **Chapter 10** will explain fluids and their properties in further detail.

The pressure drop across an orifice quantifies how much oil is attempting to be pushed through the orifice. The greater the difference in pressures, the more resistance the inflowing oil is encountering. The size of an orifice also affects the flow of oil through it. A larger orifice will allow more oil flow than a smaller orifice.

The needle valve, globe valve, and gate valve are all adjustable flow-control valves that share the same hydraulic symbol, which is a variable orifice. See Figure 8-3. These non-pressure-compensated valves must be readjusted to maintain their prescribed flow if the inlet oil flow or pressure changes. The three valves are economical, simple, and provide more flexibility than a fixed orifice.

The ***gate valve*** is an adjustable flow-control valve that allows oil to flow straight through its passageway when the gate is opened. See Figure 8-4. Although the size of the passageway changes with the gate position, this valve is not capable of providing fine metering.

The ***globe valve*** is similar to the gate valve. However, oil is not allowed to flow straight through the block. The oil must make a right-angle turn through the block when the globe is opened. See Figure 8-5.

Input Oil Flow	Divider Valve Percentages	Branch A	Branch B
20 gpm	50% & 50%	10 gpm	10 gpm
20 gpm	30% & 70%	6 gpm	14 gpm
15 gpm	33.3% & 66.6%	5 gpm	10 gpm
1 gpm	40% & 60%	0.4 gpm	0.6 gpm

Figure 8-1. Flow divider valves receive a quantity of oil flow and divide that flow in proportions based on the divider's design.

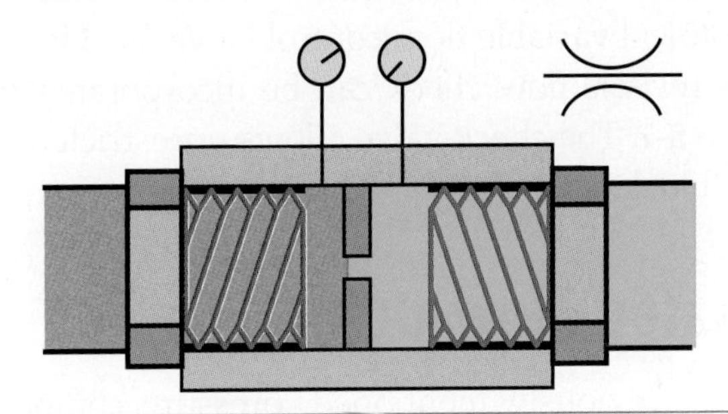

Figure 8-2. An orifice is a small opening in a component that is used to meter oil into a circuit.

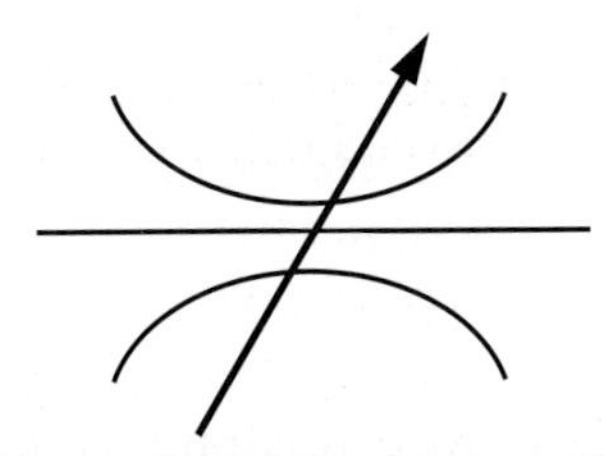

Figure 8-3. A variable orifice symbol is used to depict a needle valve, gate valve, or a globe valve.

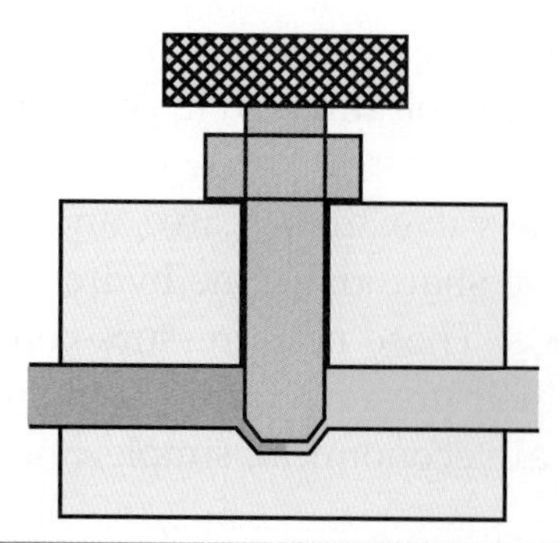

Figure 8-4. A gate valve provides a straight-through passageway when the gate is open.

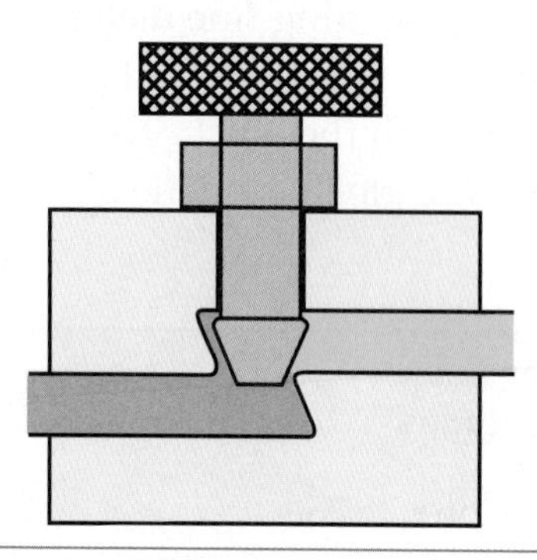

Figure 8-5. A globe valve is a variable flow-control valve that routes the oil through the block at a 90° angle.

The ***needle valve*** looks similar to the globe valve, with the oil making a right angle turn within the valve block. The fine threads on the valve stem allow small, precise adjustments to the valve's orifice size and the resulting pressure and flow. See Figure 8-6. The needle valve is the most popular and economical variable flow-control valve used in mobile machinery.

A reverse flow check can be incorporated into a needle valve block. See Figure 8-7. The check valve allows unrestricted reverse flow of oil through the valve block.

Pressure-Compensated Flow-Control Valves

As previously mentioned, pressure compensation prevents an operator from having to readjust the opening of a DCV spool every time the inlet flows and pressures change. Pressure-compensated valves can be configured in one of the three forms:

- Two-way (non-adjustable or adjustable).
- Three-way (non-adjustable or adjustable).
- Pressure-reducing valves.

The ***two-way pressure-compensated flow-control valve*** has only two hydraulic ports, an inlet port and an outlet port. The valve is sometimes called a restrictor. Figure 8-8 illustrates a simplified cutaway view of one style of restrictor flow-control valve.

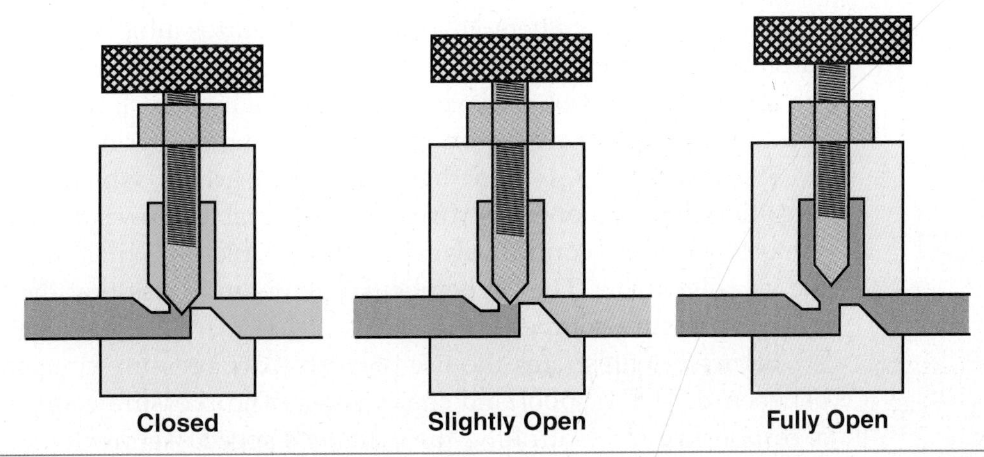

Figure 8-6. The needle valve is a popular variable-orifice flow-control valve used in mobile machinery. It allows for minor adjustment of flow, but it is not pressure compensated.

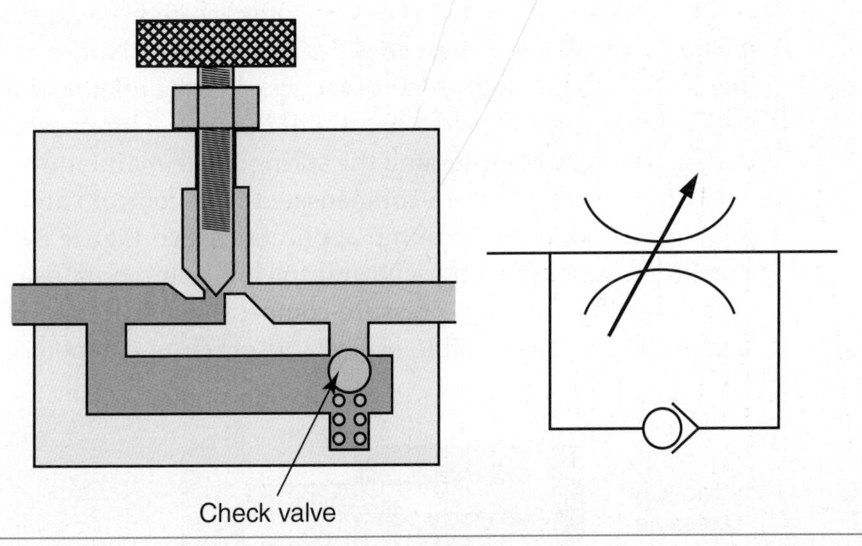

Figure 8-7. Reverse flow checks allow free flow of oil in reverse.

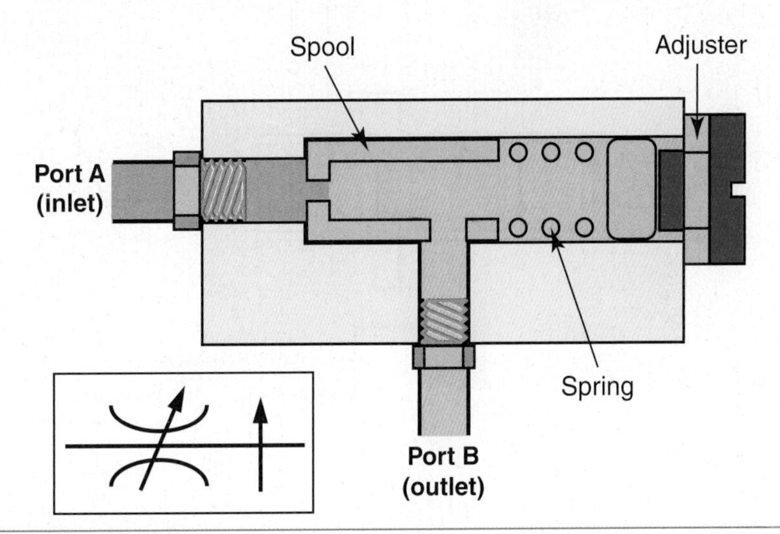

Figure 8-8. The two-way pressure-compensated valve has only two hydraulic ports, an inlet and an outlet. The adjuster changes the spring tension placed on the valve's spool.

As the inlet flow attempts to increase, it causes a higher differential pressure across the compensator's spool. This higher differential pressure causes the compensator spool to move away from the inlet port, restricting a portion of the oil exiting the outlet port and compressing the spring.

The schematic's symbol has an angled arrow, which normally indicates that the component is variable or adjustable. However, some pressure-compensated flow-control valves are not variable, but will still be drawn with the variable arrow. The perpendicular arrow indicates that the flow-control valve is pressure-compensated.

Figure 8-9 illustrates the use of a two-way pressure-compensated valve between the DCV spool and the cylinder. The pressure-compensated valve is responsible for controlling the cylinder's speed. As previously mentioned, metering oil into a cylinder can cause cavitation if the actuator has an overrunning load. Therefore, the configuration in Figure 8-9 is not commonly used in mobile equipment.

A two-way adjustable pressure-compensated flow-control valve can also include a variable needle valve. See Figure 8-10. Notice that the spool valve directs the inlet oil to the end of the spool. If the inlet flow increases, the pressure drop across the needle valve will increase. This causes the spool valve to shift to the right, compressing the spring while maintaining a constant flow.

The two-way pressure-compensated flow-control valve was used in older John Deere radial piston pump applications. See Figure 8-11. The older Deere pressure-compensated agricultural tractors commonly used a variable orifice in series with a two-way pressure-compensated flow-control valve that fed the tractor's DCV. The variable orifice and pressure-compensated valve were all

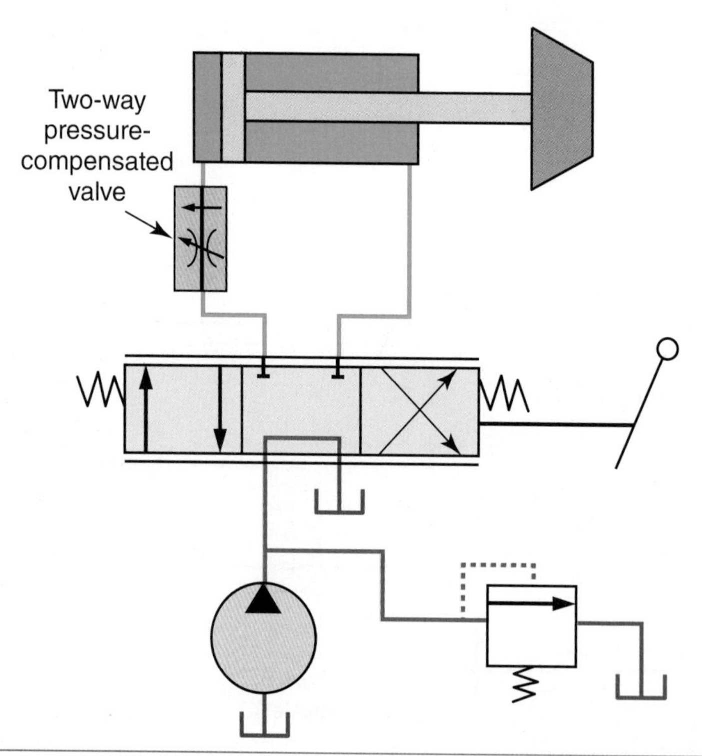

Figure 8-9. This schematic shows a two-way pressure-compensated flow-control valve located between the DCV spool and the actuator.

located in the valve block, which John Deere calls the ***Selective Control Valve (SCV)***. The Deere SCV term is another descriptor for DCV.

The variable orifice is used to adjust the rate of flow through the SCV, and the pressure-compensated valve is used to maintain a constant actuator speed. The John Deere variable-displacement pressure-compensated pumps will be explained in **Chapter 17**.

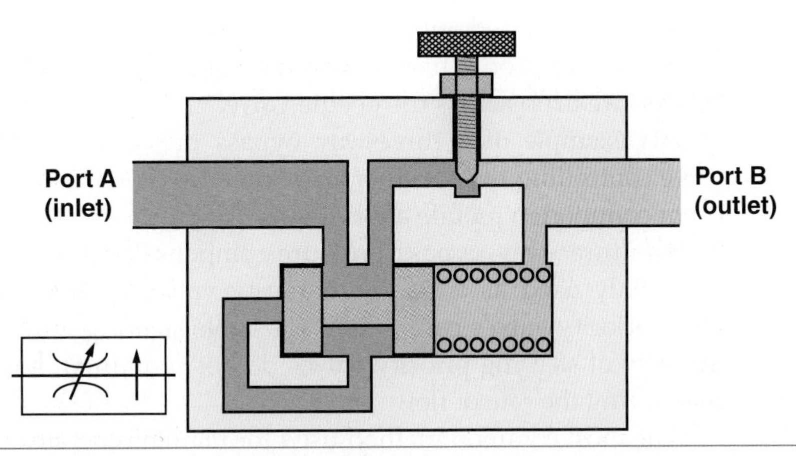

Figure 8-10. This two-way variable pressure-compensated flow-control valve uses a needle valve to vary the speed of the actuator. The farther the valve is unthreaded, the more flow will be sent to the actuator.

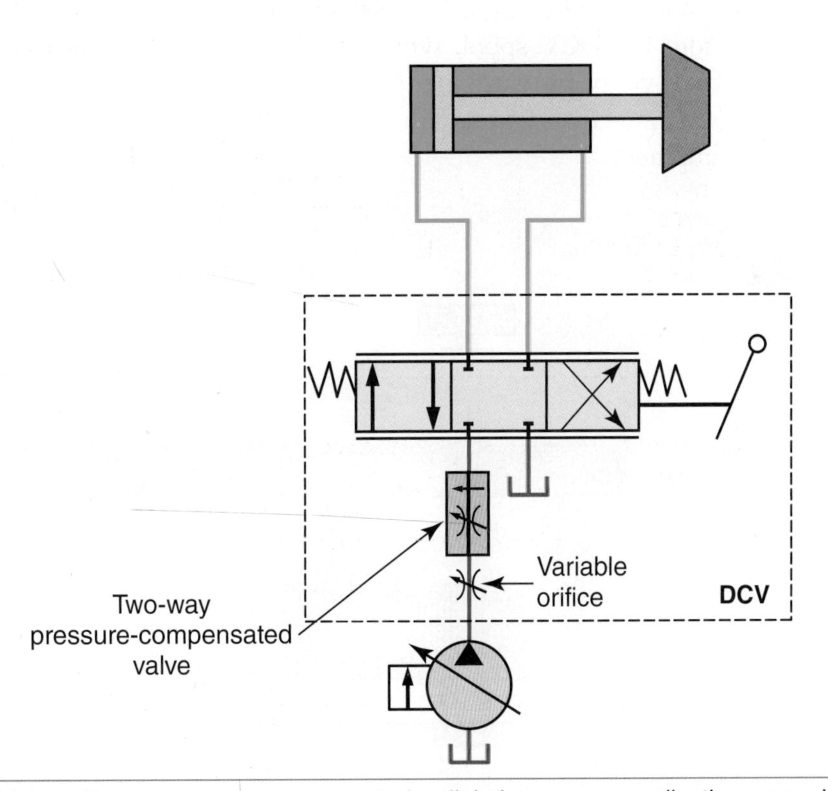

Figure 8-11. In the old John Deere pressure-compensated radial piston pump applications, a variable orifice and a two-way pressure-compensated valve were placed in series prior to the DCV poppet valves. Although the schematic resembles a DCV spool, the control valve actually used four poppet valves to control the oil in and out of the valve block. DCVs will be further explained in **Chapter 9**.

The ***three-way pressure-compensated flow-control valve*** has three hydraulic ports: an inlet port, an outlet port, and a bypass port. The valve is sometimes called a bypass flow-control valve. See **Figure 8-12**.

The spring biases the piston to the left, preventing oil from initially flowing out of the bypass port. The pressure drop across the needle valve will equal the value of the bias spring. As system flow increases, the pressure on the left side of the compensator piston will increase, causing the piston to shift to the right. When the compensator piston uncovers the bypass port opening, the ***excess flow (EF)*** is dumped out of the bypass port. The pressure on the left side of the compensator piston is higher than the right side of the compensator piston due to pressure drop across the needle valve.

An example of a three-way bypass pressure-compensated flow-control valve controlling the speed of an actuator is shown in **Figure 8-13**. The circuit is not common in mobile applications.

The three-way bypass pressure-compensated flow-control valve is more commonly used as a steering priority valve. **Chapter 25** will explain that some priority valves contain an adjustable steering relief valve. However, the majority of steering priority valves do not contain an adjustable needle valve for varying the rate of flow.

The most common method used for maintaining a constant cylinder speed in late-model mobile machinery is using a pressure-reducing valve (PRV) that senses working pressure. See **Figure 8-14**. A shuttle valve is used to send a signal of the actuator's working pressure to the PRV. The PRV will maintain a constant cylinder speed based on the DCV spool opening. Notice that the PRV is placed before the DCV spool, which is sometimes called ***pre-spool compensation***. This style of pressure compensation is the main focus of **Chapter 18**, and is commonly found in advanced agriculture equipment and some construction equipment.

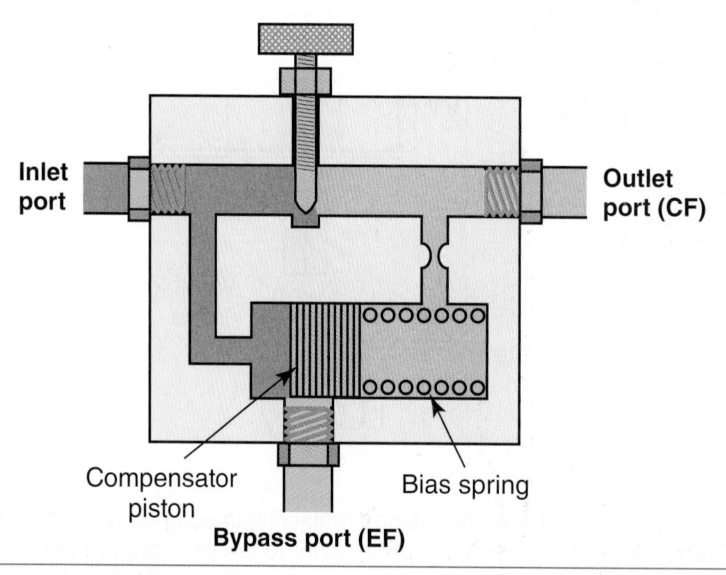

Figure 8-12. A three-way pressure-compensated valve has an inlet port, an outlet port, and a bypass port. This valve is shown with an adjustable needle valve.

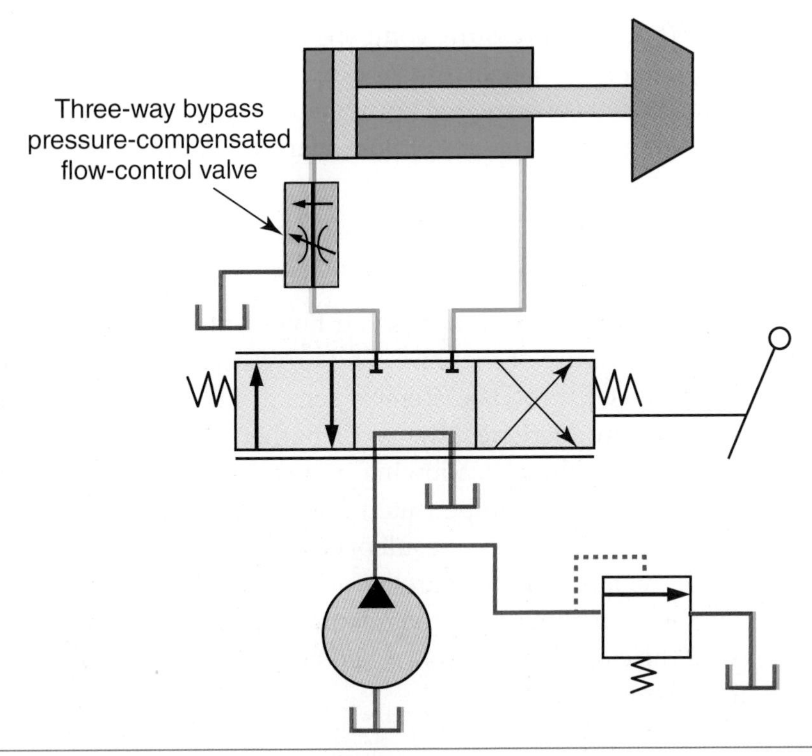

Figure 8-13. The three-way bypass pressure-compensated flow control valve is illustrated controlling the speed of an actuator. This application is not commonly used in mobile equipment.

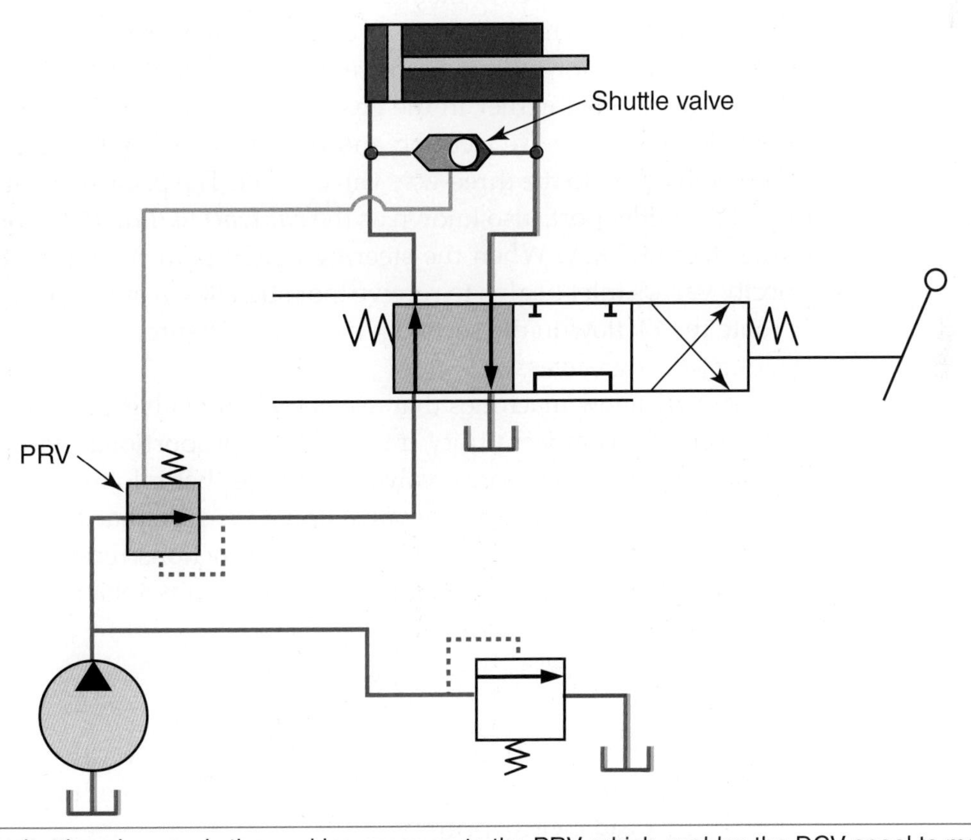

Figure 8-14. A shuttle valve sends the working pressure to the PRV, which enables the DCV spool to maintain a constant actuator speed based on the position of the DCV spool. **Chapter 18** further explains pre-spool compensation.

When the pressure compensator is placed after the DCV, it is called ***post-spool compensation*** or flow sharing. Flow sharing is commonly used in advanced construction equipment systems. **Chapter 19** explains the fundamentals and advantages of flow sharing.

Temperature-Compensated Valves

A temperature-compensated flow-control valve will adjust flow based on the temperature of the fluid. The valve uses a temperature-compensating rod. The bi-metallic rod is extremely sensitive to changes in oil temperature. As the temperature increases, the rod lengthens, reducing the flow. When the oil is cold, the rod shrinks, allowing more oil to flow.

Temperature-compensated flow-control valves are rarely found in mobile machinery, mainly as a result of electronic controls. Today, machines can determine the fluid temperature by using temperature sensors. The ECM uses this information along with the operator's request to actuate the cylinder, and can vary the electronically controlled DCV based on the operator's request and the fluid temperature.

Priority Valves

Many different types of mobile machinery use some type of priority valve to direct oil flow. The priority valve ensures that a primary circuit has its needs met first, before allowing the remaining oil to flow to the secondary circuits. As mentioned earlier in the chapter, the most common priority circuits in mobile machinery are steering and brakes. Refer back to **Figure 8-12**. The pump delivers oil to the three-way valve's inlet. The priority oil is routed out of the valve's outlet port, also known as the ***controlled flow (CF)*** port, to the inlet of the steering DCV. When the steering's DCV is in the neutral position, the priority valve's inlet begins to receive too much flow, causing the priority valve to route the EF flow to the secondary circuits. **Chapter 25** will further explain steering priority valves.

There are a few machines that use no priority valve. For example, an older John Deere 955 compact utility tractor uses a proportional flow divider valve in place of a steering priority valve. Also note that a John Deere 410D loader backhoe uses a steering priority valve only to ensure that steering receives oil before oil is sent to the loader. The backhoe functions receive non-prioritized pump flow, because there is no reason that an operator should be steering and operating the backhoe simultaneously.

Excavator Priorities

Some machines, such as an excavator, use priority valves for something other than steering or brakes. Three examples of excavator priorities are:

- Cutterhead control priority.
- Boom priority.
- Swing priority.

Cutterhead Priority

If an excavator is attempting to operate a cutterhead that has a high demand for hydraulic flow, every time the excavator arm is moved, it would cause the cutterhead to slow. Therefore, a priority valve can be used to ensure the cutterhead attachment receives priority oil flow. This style of priority valve is normally adjustable so that the valve can be adjusted for the precise amount of flow the cutterhead requires.

Boom Priority

When an excavator is operating in a mode that requires a short swing radius, some excavator manufacturers enable an operator to select a boom priority setting, which allows the machine to prioritize the oil for the boom, decreasing a boom's cycle time. ***Cycle time*** is the amount of time it takes an actuator to either fully extend, fully retract, or both fully extend and fully retract. Cycle times are a key performance indicator of construction machines and are used during diagnostics to determine if a machine is performing poorly.

Swing Priority

When an excavator is operating in a mode that requires a wide swing radius, some excavator manufacturers allow an operator to select a swing priority setting. The swing priority setting directs more oil for the swing function, which reduces the swing cycle time.

Excavator priorities are often controlled through the pilot control circuits and not the traditional main system flow circuit.

Proportional Flow Divider Valves

As stated earlier in the chapter, proportional flow dividers are used to divide a given amount of flow in a set proportion, such as 40/60, or 50/50, or 30/70. There are two types of flow divider designs: spool divider valves and gear motor dividers.

Divider valves are commonly used in agricultural implements. In Figure 8-15, the flow divider will deliver a larger proportion of oil flow to port B, for example 70%, than port A, for example 30%. Port B supplies oil to a cylinder with an area that requires 70% of the oil, and port A supplies oil to a cylinder with an area that only needs 30% of the oil.

Some flow dividers use gear motors for proportioning oil flow, as shown in Figure 8-16. The only purpose the motors serve is to proportion oil flow. The motors can be designed with different displacements to provide different proportions of oil flow. The motor-type flow divider is not commonly used in mobile applications. Compared to spool-type flow dividers, motor-type dividers are efficient and are used in circuits that have intensified oil pressures.

Some manufacturers use cartridge-style valves in priority circuits or flow divider circuits. A cartridge valve threads into the bore of a valve block. The valve block contains cross-drilled passageways. See Figure 8-17.

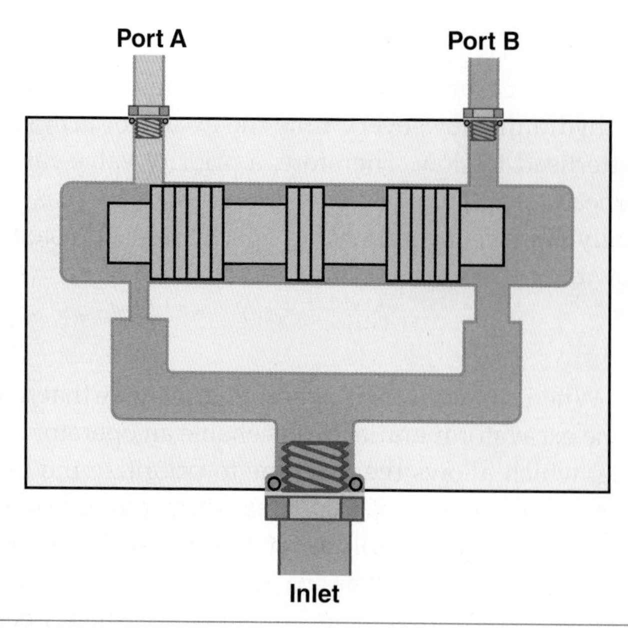

Figure 8-15. A flow divider proportions oil in a set ratio, sending the oil into two different circuits.

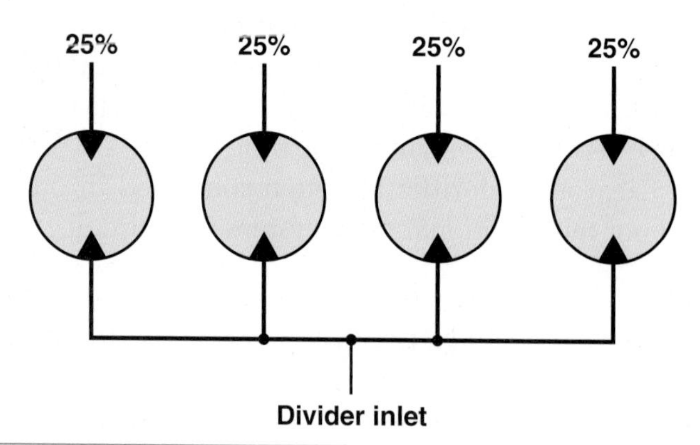

Figure 8-16. Gear motors can be used as proportional flow dividers. In this example, the inlet flow is divided into four equal portions.

Figure 8-17. Cartridge valves are threaded into a valve block. This cartridge valve is a steering priority valve. Planters commonly use cartridge valves, including flow divider cartridge valves.

Excessive Flow Protection

Electrical circuits use fuses and circuit breakers to prevent damage to the circuits when current flow increases. Hydraulic systems can also incorporate protection from excessive flow with the use of a ***velocity fuse***, also known as a flow fuse. The device works on the principle of a pressure drop across an orifice. See Figure 8-18. In the event that a hydraulic hose bursts, the circuit's velocity will dramatically increase. The increase in flow will cause an increase in differential pressure across the orifice, which will result in the fuse shifting the check valve to block off the circuit's flow. The fuse will prevent damage to machines and injuries to personnel.

Like electrical fuses, hydraulic velocity fuses must allow for the fluctuation of system flow. For example, a 20-amp electrical fuse might be used to protect a circuit that is normally flowing 15 or 16 amps. The hydraulic velocity fuse typically actuates when the circuit's flow increases 30% over the circuit's normal flow rate.

The velocity can be adjustable or fixed. The fuse in Figure 8-18 is adjustable, and will automatically reset if the flow rate returns to normal. Velocity fuses are much less common than electrical fuses, primarily due to their initial cost. However, fuses might become more popular in the future if manufacturers consider the potential savings associated with preventing machine damage and injuries to personnel.

One manufacturer uses the velocity fuse in a combine's header lift circuit. If a hose failed, the fuse would prevent the header from dropping abruptly.

A ***quantity fuse*** is similar to a velocity fuse, except that it is designed to be used in a low-flow circuit with a fixed flow rate, for example a 0.5 gpm circuit. The fuse contains two orifices that measure a fixed flow rate; if the flow increases, the fuse blocks off the circuit. The schematic symbol is illustrated in Figure 8-19.

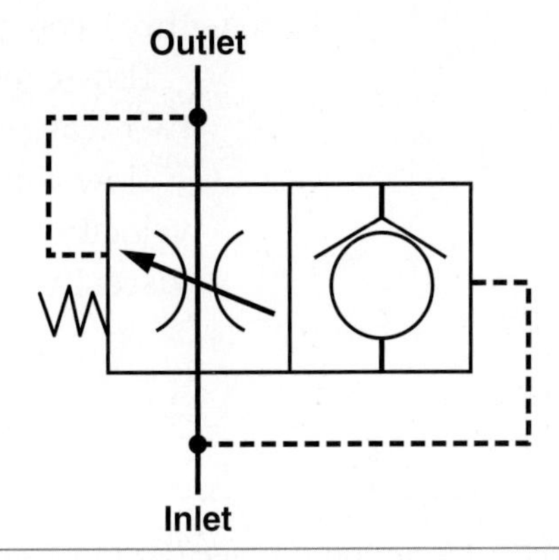

Figure 8-18. A velocity fuse contains an orifice that senses a differential pressure. If a circuit's velocity increases too much, the differential pressure causes the fuse to close, blocking off the circuit's flow.

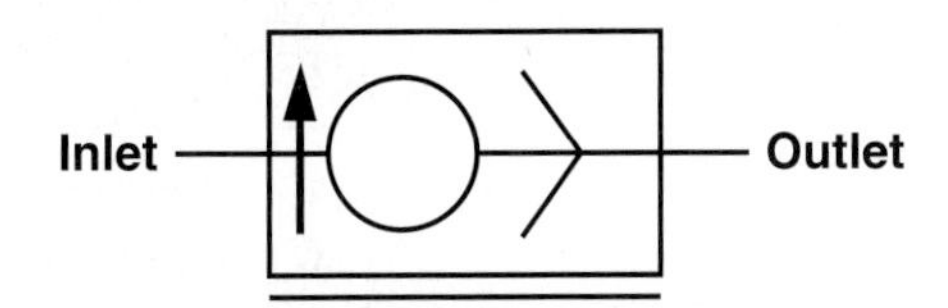

Figure 8-19. A quantity fuse is used in low-volume circuits with a fixed flow rate. The perpendicular arrow indicates the fuse is self-resetting.

Summary

- ✓ Flow-control valves are also known as volume-control valves.
- ✓ Flow-control valves are used to regulate speed, regulate priority, and divide oil flow proportionally.
- ✓ Non-pressure-compensated flow-control valves are economical, simple, but must be readjusted if the inlet pressure changes.
- ✓ Pressure-compensated flow-control valves are used to maintain a constant actuator speed.
- ✓ Temperature-compensated flow-control valves adjust flow based on changes in fluid temperature.
- ✓ Priority valves ensure the priority circuit needs are met before sending oil to the secondary circuits.
- ✓ Steering and brakes are the popular priority circuits.
- ✓ Priority valves are also used in mobile machinery for circuits other than steering and brakes, such as boom lift or excavator swing.
- ✓ Flow divider valves distribute oil into different branches in set proportions.
- ✓ Agricultural implements use flow divider valves to distribute oil to different size lift cylinders in the proper proportion.
- ✓ Velocity fuses and quantity fuses are used to block circuits when flow increases too much. Their function is similar to that of an electrical circuit breaker in an electrical circuit.

Technical Terms

controlled flow (CF)
cycle time
excess flow (EF)
gate valve
globe valve
needle valve
non-pressure-compensated
orifice
post-spool compensation
pre-spool compensation
pressure compensation
priority valves
proportional flow divider valves
quantity fuse
Selective Control Valve (SCV)
three-way pressure-compensated flow-control valve
two-way pressure-compensated flow-control valve
velocity fuse
viscosity

Review Questions

Answer the following questions using the information provided in this chapter.

1. Flow control valves are used to perform all of the following functions, *EXCEPT*:
   A. regulate speed.
   B. regulate priority.
   C. relieve pressure.
   D. divide oil in proportion.
2. Which one of the following flow-control valves would most likely be used to control a planter's fan speed?
   A. Orifice.
   B. Needle valve.
   C. Gate valve.
   D. Pressure-compensated flow-control valve.

3. Which of the following would be used on a planter to proportion oil to different size lift cylinders?
   A. Priority valve.
   B. Divider valve.
   C. Pressure-compensated flow-control valve.
   D. PRV.
4. Which of the following is most likely used to direct oil to a steering DCV?
   A. Priority valve.
   B. Divider valve.
   C. Pressure-compensated flow-control valve.
   D. PRV.
5. A 40/60 proportional flow divider is receiving 20 gpm in the valve's inlet. What will the valve's outlet flow be?
   A. 3 and 7 gpm.
   B. 4 and 6 gpm.
   C. 5 and 5 gpm.
   D. 8 and 12 gpm.
6. All of the following will affect the flow through an orifice, *EXCEPT*:
   A. oil viscosity.
   B. brand of oil.
   C. pressure drop across the orifice.
   D. size of the orifice.
7. Viscosity is a term used to describe oil's _____.
   A. thickness
   B. temperature
   C. color
   D. weight
8. Which of the following is *not* considered variable or adjustable?
   A. Needle valve.
   B. Gate valve.
   C. Globe valve.
   D. Orifice.
9. Which of the following provides a straight path through the valve?
   A. Gate valve.
   B. Globe valve.
   C. Needle valve.
   D. All of the above.
10. Which of the following provides the ability to make a gradual precise adjustment?
    A. Gate valve.
    B. Globe valve.
    C. Needle valve.
    D. All of the above.
11. Which of the following flow-control valves is most common in mobile machinery?
    A. Gate valve.
    B. Globe valve.
    C. Needle valve.
    D. Ball valve.
12. Which of the following is *not* used as a pressure-compensated flow-control valve?
    A. Two-way restrictor.
    B. Needle valve.
    C. Three-way bypass valve.
    D. PRV.
13. Which of the following is the most popular pressure-compensated flow-control valve in advanced agricultural equipment?
    A. Two-way restrictor.
    B. Needle valve.
    C. Three-way bypass valve.
    D. PRV.
14. Which of the following is used as a pressure-compensated flow-control valve on old John Deere PC hydraulic systems?
    A. Two-way restrictor.
    B. Needle valve.
    C. Three-way bypass valve.
    D. PRV.
15. Agricultural equipment uses what type of DCV pressure compensation?
    A. Pre-spool.
    B. Post-spool.
    C. Both A and B.
    D. Neither A nor B.
16. Construction equipment uses what type of DCV pressure compensation?
    A. Pre-spool.
    B. Post-spool.
    C. Both A and B.
    D. Neither A nor B.
17. Which of the following pressure-compensated flow-control valves is used as a priority valve?
    A. Two-way restrictor.
    B. Needle valve.
    C. Three-way bypass valve.
    D. PRV.

18. Which of the following materials is the key material for enabling a temperature-compensated flow-control valve to vary oil flow based on oil temperature?
    A. Bi-metallic.
    B. Carbon.
    C. Bronze.
    D. Phenolic.
19. Which port on a priority valve delivers oil to the secondary circuits?
    A. Inlet.
    B. EF.
    C. CF.
    D. None of the above.
20. Technician A states that flow dividers use spool valves. Technician B states that flow dividers use gear motors. Who is correct?
    A. Technician A.
    B. Technician B.
    C. Both A and B.
    D. Neither A nor B.
21. Which of the following valves will proportion oil flow to two circuits?
    A. Needle valve.
    B. Priority valve.
    C. Globe valve.
    D. Divider valve.
22. Which of the following is used to ensure that a circuit will receive the primary oil flow?
    A. Needle valve.
    B. Priority valve.
    C. Globe valve.
    D. Divider valve.
23. Speed control through a hydraulic system is achieved by controlling which of the following?
    A. Pressure.
    B. Flow.
    C. Load.
    D. Temperature.
24. Which of the following is sometimes placed in a flow-control valve to enable the free flow of oil in one direction?
    A. Reverse-flow check valve.
    B. Needle valve.
    C. Pressure-compensated flow-control valve.
    D. Orifice.
25. Technician A states that hydraulic systems have no device that serves a function similar to that of an electrical fuse or circuit breaker. Technician B states that hydraulic fuses are commonly used in most mobile hydraulic systems. Who is correct?
    A. Technician A.
    B. Technician B.
    C. Both A and B.
    D. Neither A nor B.
26. How does a velocity fuse work?
    A. It senses pressure drop across an orifice.
    B. It uses electrical flow-sensing technology.
    C. It uses electrical pressure-sensing technology.
    D. It uses the magnetostrictive principle.
27. A velocity fuse is being used to protect a circuit that is flowing 7 gpm. The appropriate velocity fuse should close at a flow rate of _____.
    A. 5 gpm
    B. 7 gpm
    C. 10 gpm
    D. 14 gpm
28. Which of the following is used to protect a circuit that has a low fixed flow rate?
    A. Quantity fuse.
    B. Velocity fuse.
    C. Pressure fuse.
    D. Relief valve.
29. A quantity fuse is being used to protect a circuit. Which of the following flow rates is an example of a circuit that would use a quantity fuse for protection?
    A. 0.5 gpm.
    B. 5 gpm.
    C. 10 gpm.
    D. 50 gpm.
30. Technician A states that priority valves can use a cartridge valve design. Technician B states that flow divider valves can use a cartridge valve design. Who is correct?
    A. Technician A.
    B. Technician B.
    C. Both A and B.
    D. Neither A nor B.

Chapter 9 Directional-Control Valves

Objectives

After studying this chapter, you will be able to:

- ✓ Explain the purposes of DCVs.
- ✓ List the different classifications of DCVs.
- ✓ Describe characteristics of the different types of hydraulic control systems.
- ✓ List the different methods used for actuating a DCV.
- ✓ Explain the common functions of DCVs.
- ✓ List the common number of spool positions used in mobile machinery.
- ✓ Describe the differences in DCV block design.
- ✓ List the different types of valve designs used in DCVs.
- ✓ Explain the purpose of power beyond.

Purpose of Directional-Control Valves

Directional-control valves (DCVs) are designed to control hydraulic actuators in the same fashion that electrical switches control electrical actuators. The DCVs receive pump oil flow, direct oil to the cylinder, then route the return oil back to the reservoir.

DCVs can be categorized by the following classifications:

- Type of control system.
- Type of actuation.
- Functions.
- Number of positions.
- Block design.
- Valve design.
- Flow and pressure capacities.

Types of Hydraulic Control Systems

A DCV must be designed to match the machine's type of control system. Mobile machines use the following types of hydraulic control systems:

- Open-center system.
- Pressure-compensating system.
- Load-sensing pressure-compensating system (pre-spool compensation).
- Load-sensing pressure-compensating system (post-spool compensation).
- Excavator controls.

Open-Center Systems

The traditional ***open-center hydraulic system*** uses a fixed-displacement pump that is matched with open-center DCVs as shown in Figure 9-1. When the DCVs are in a neutral position, an open passageway allows the pump's flow to be routed to the reservoir at a low pressure—for example 150 psi (10 bar). Open-center hydraulic systems are described in greater detail in **Chapter 16**.

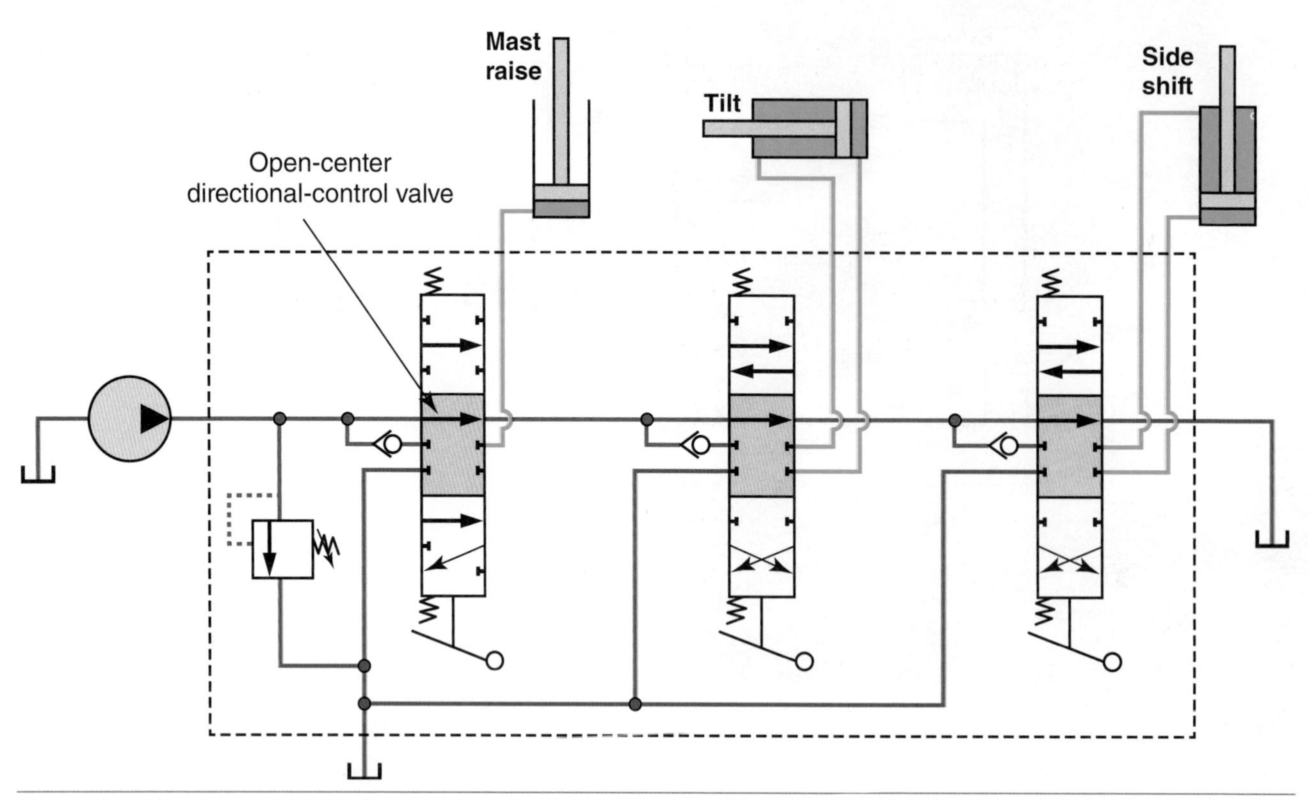

Figure 9-1. Open-center hydraulic systems provide an open passage for a fixed-displacement pump to flow back to the reservoir when the valves are in a neutral position. This forklift schematic is an example of an application commonly equipped with an open-center system.

Pressure-Compensating (PC) Systems

A *pressure-compensating (PC) hydraulic system* uses a variable-displacement pump that is matched with closed-center DCVs, shown in Figure 9-2. When the DCVs are in a neutral position, the pump's flow is blocked and not allowed to return to the reservoir. PC systems run at a high system pressure—for example 2800 psi (193 bar). **Chapter 17** explains PC systems in further detail.

Load-Sensing Pressure-Compensating (LSPC) Systems—Pre-Spool Compensation

A *load-sensing pressure-compensating (LSPC) hydraulic system* with pre-spool compensation can use either a fixed- or variable-displacement pump. The control valves are closed center. The pressure-reducing valves (PRVs) are placed before the DCV spools. The DCVs contain shuttle valves that are used for sensing the actuator's working pressure. See Figure 9-3.

A *shuttle valve* senses two different working pressures and sends the higher pressure to the next destination. Figure 9-4 shows how the valve operates. Within a load-sensing system, that next destination can be another shuttle valve, a pump's margin spool, or an unloading valve. A margin spool is used in variable-displacement pumps and has the responsibility of controlling the pump's flow based on the highest system working pressure.

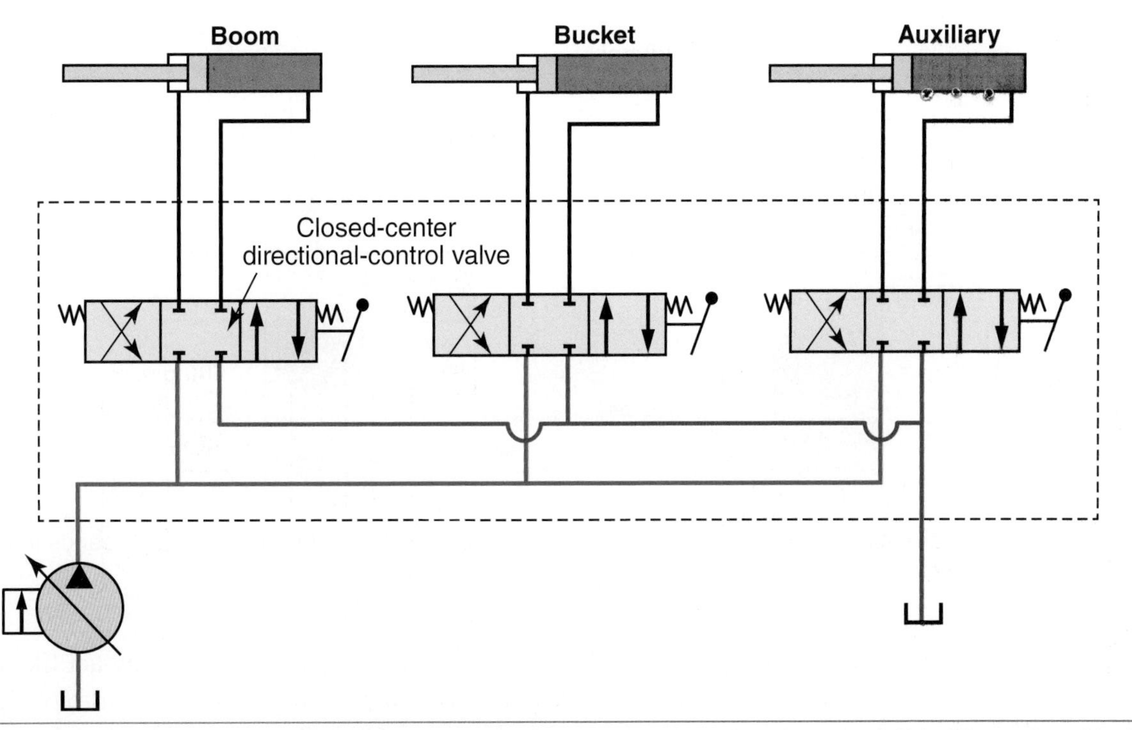

Figure 9-2. PC systems use a variable-displacement hydraulic pump and closed-center DCVs. The system blocks pump flow when the valves are in a neutral position and operates at a high system pressure. This simplified schematic is an example of the loader controls used on a loader backhoe.

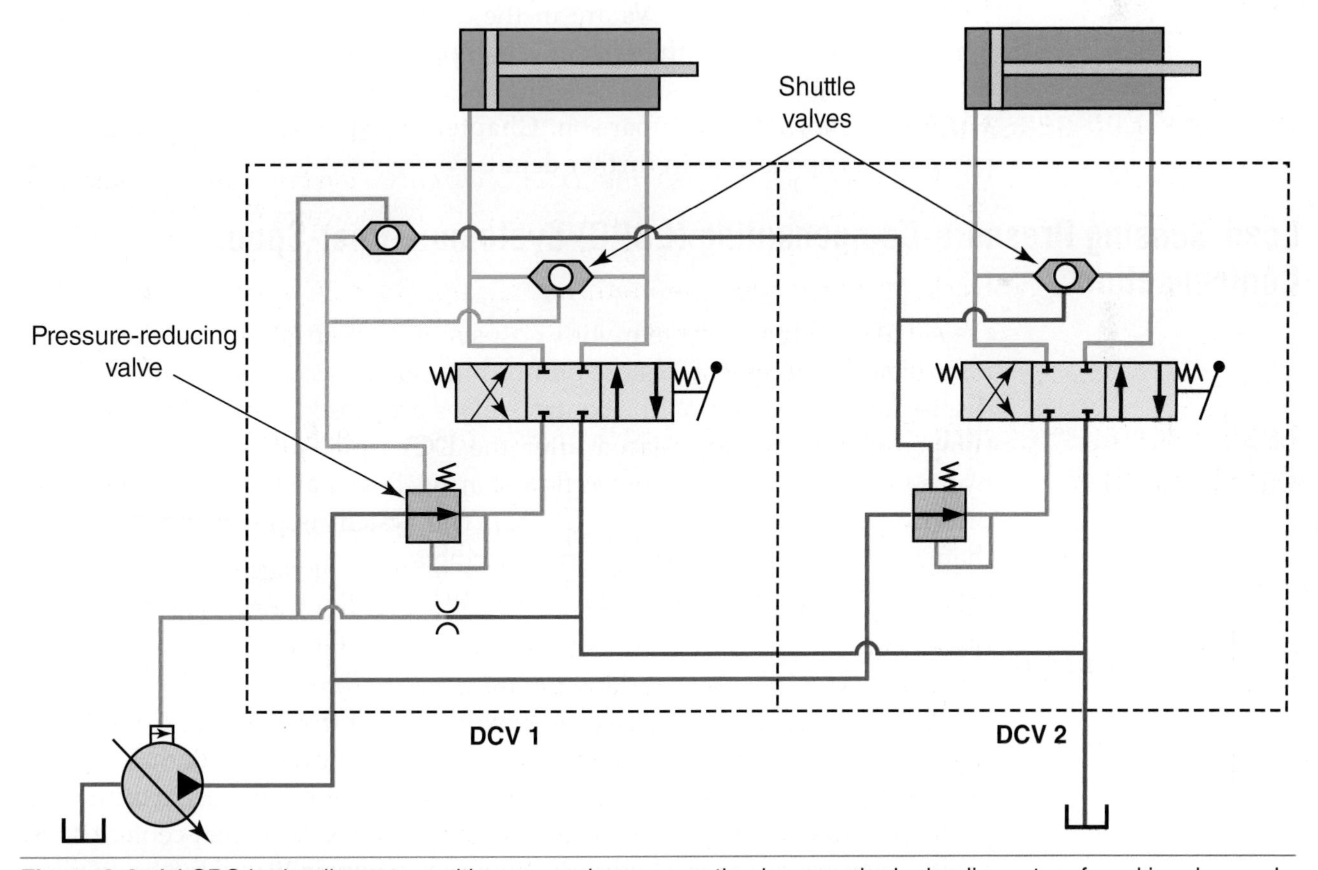

Figure 9-3. A LSPC hydraulic system with pre-spool compensation is a popular hydraulic system found in advanced agricultural equipment and some construction equipment. Both DCV 1 and 2 contain their own primary shuttle valve.

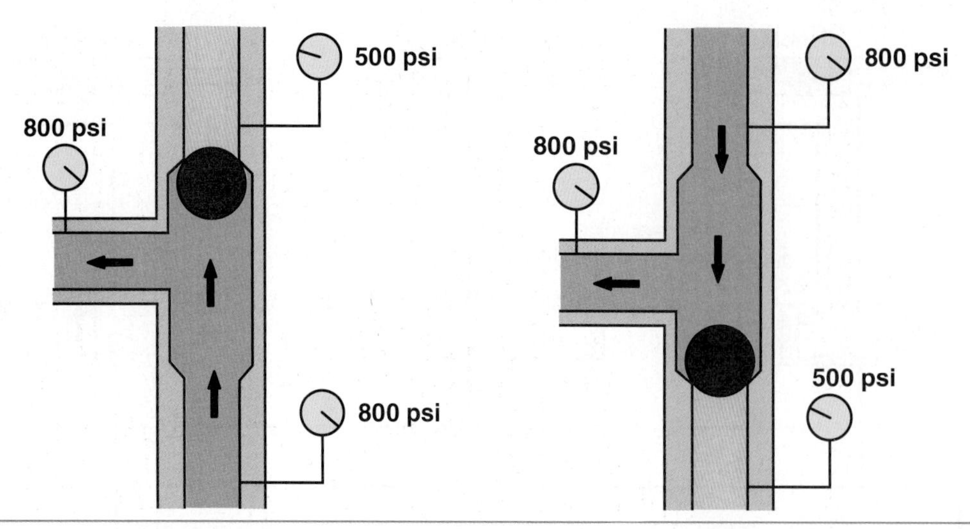

Figure 9-4. A shuttle valve chooses the highest working pressure among two different circuits and sends the highest working pressure downstream.

One-way check valves are similar to shuttle valves. They act like a diode in an electronic circuit, allowing flow in one only direction and blocking flow in the opposite direction. See Figure 9-5. Most one-way check valves contain a spring to return the poppet or ball to its seat, but some do not. The check valves can be used in load-sensing applications and are also found inside the hydraulic pilot computer on a positive flow control (PFC) excavator.

When load-sensing DCVs are in the neutral position, the hydraulic system's flow will be low if the system is equipped with a variable-displacement pump and high if the system is equipped with a fixed-displacement pump. Figure 9-6 shows this comparison. **Chapter 18** explains LSPC systems with pre-spool compensation in further detail.

Load-Sensing Pressure-Compensating (LSPC) Systems—Post-Spool Compensation

A LSPC hydraulic system with post-spool compensation commonly uses a variable-displacement piston pump, but can be equipped with a fixed-displacement gear pump. The control valves are closed center. The pressure-compensator valves are placed after the DCV spools. See Figure 9-7. The hydraulic system is also known as flow sharing, flow matching, or proportional priority pressure compensation (PPPC). The system is much more complex than a pre-spool LSPC system. Flow-sharing systems are found on advanced construction equipment. **Chapter 19** will further explain the system.

Excavator Controls

The two most common excavator hydraulic control systems are negative flow control (NFC) and positive flow control (PFC) systems. Both NFC and PFC systems use variable-displacement pumps that vary between a minimum pump flow and a maximum pump flow. NFC systems use open-center DCVs. PFC systems use a complex hydraulic computer. **Chapter 20** will further explain excavator pump controls.

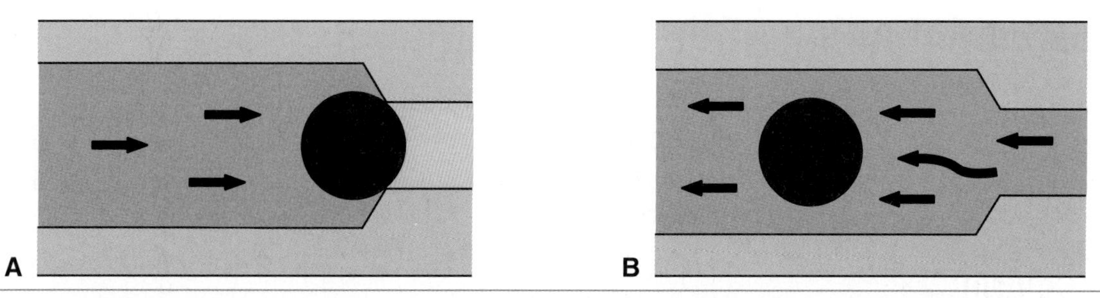

Figure 9-5. One-way check valves allow oil flow in one direction and block oil flow in the opposite direction, similar to an electronic diode. A—Ball in seat blocks flow in one direction. B—Ball lifts off seat to allow oil flow in opposite direction.

Control System	System Flow with DCVs in Neutral Position	PSI with DCVs in Neutral Position
LSPC pre-spool fixed-displacement pump	Maximum flow	300 to 450 psi
LSPC pre-spool variable-displacement pump	Practically no flow	300 to 450 psi

Figure 9-6. Pre-spool LSPC hydraulic systems will have different system flows when the DCVs are in the neutral position, based on the type of pump the system is using.

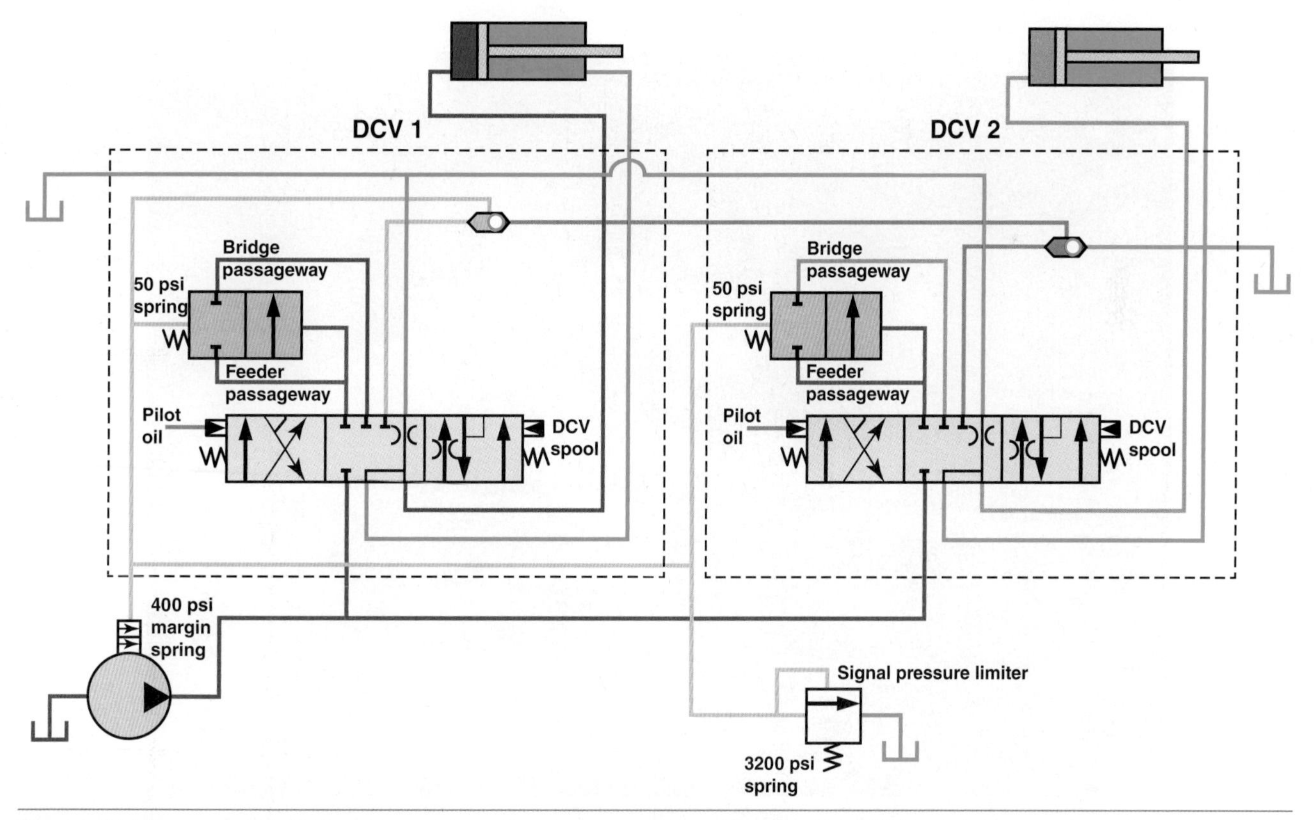

Figure 9-7. Post-spool LSPC systems place the compensator valve after the DCV spool. The system also uses a signal pressure limiter valve. **Chapter 19** is dedicated to explaining this complex hydraulic system.

Methods of Actuating DCVs

DCVs are actuated by one of three methods:

- Manual control lever.
- Pilot control.
- Electricity or electronic.

Manual Control

Manually controlled DCVs, Figure 9-8, have been the most common method for controlling DCVs for decades. Manual levers are the simplest means for controlling a DCV, eliminating the need for electrical circuits or pilot-control circuits. Examples of applications that use manual control DCVs include:

- Loader controls on compact utility tractors.
- Log splitters.
- DCVs on economy utility tractors.

DCV Hydraulic Detent Kickout

Some manually operated DCVs have detented levers that will mechanically hold the spool in a fixed position. Once a high pressure is reached, the lever will return to neutral, shutting off pump flow, Figure 9-9. The detent kickout pressure is adjustable. Prior to electronically controlled DCVs, many agricultural tractors used this type of DCV control. The mechanisms had advantages as well as disadvantages.

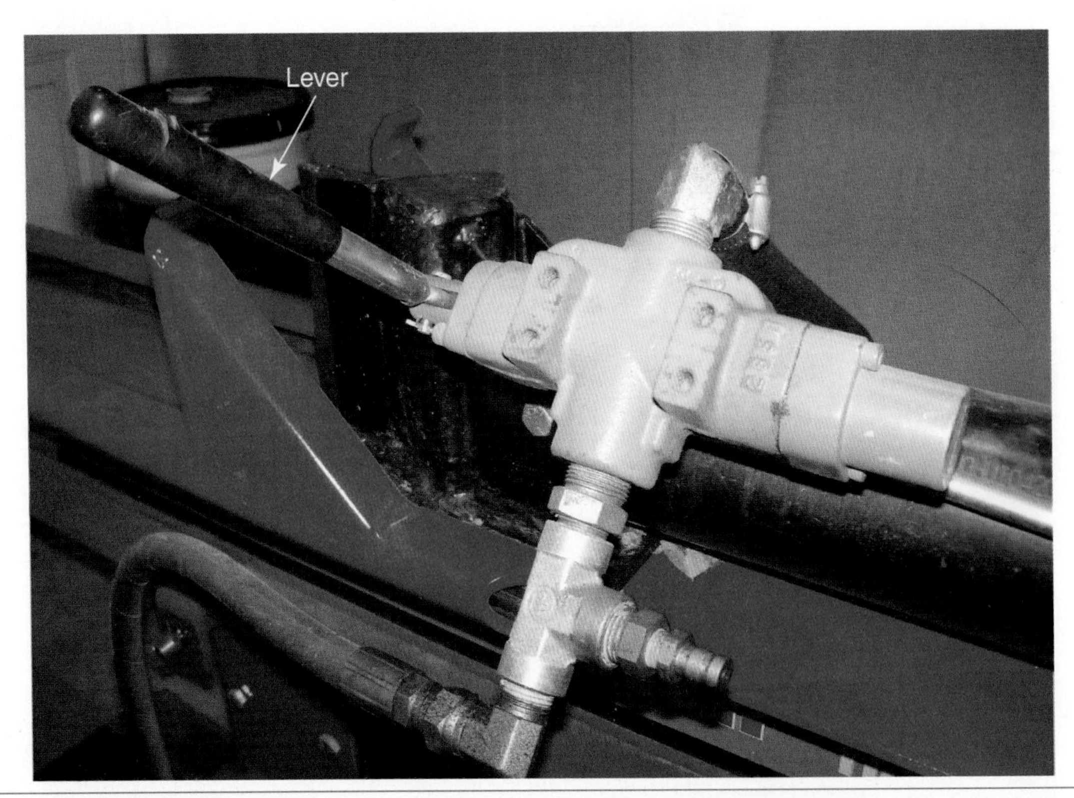

Figure 9-8. Log splitters use manual control levers for actuating the double-acting cylinder. The DCV is usually an open-center, three-position valve, and is sometimes detented for cylinder retraction.

The advantage of the detent is that, when adjusted correctly, it allows the operator to place the lever in either the extend or retract position, and once the cylinder reaches the end of travel, the lever is returned automatically to the neutral position. With this feature, the operator can focus on other immediate tasks at hand rather than only watching the cylinder extend or retract.

One negative is that customers might have the detent kickout relief valves adjusted too low. In this case, the DCV might not build a high-enough pressure to do the necessary work, such as raise a round baler's door or raise a planter. In scenarios where the pressure is adjusted too low and the customer needs continuous flow, the customer might tie a strap to the lever to hold it in the extend position. This poor solution causes system inefficiency because oil pressure is dropping across the detent kickout relief valve, and the oil should have been routed to the actuator. Many customers have called dealerships because their hydraulic system was malfunctioning, and the technician only had to properly adjust the detent kickout pressure.

The easiest way to set the detent is to install a hydraulic flowmeter in the DCV couplers, actuate the DCV spool, and begin to close the flowmeter's load valve. The technician watches the flowmeter's pressure gage. Once the detent pressure is reached, flow immediately stops and the DCV returns to the neutral position. If the pressure is set too low or too high, the technician adjusts the valve accordingly and retests the pressure using the flowmeter.

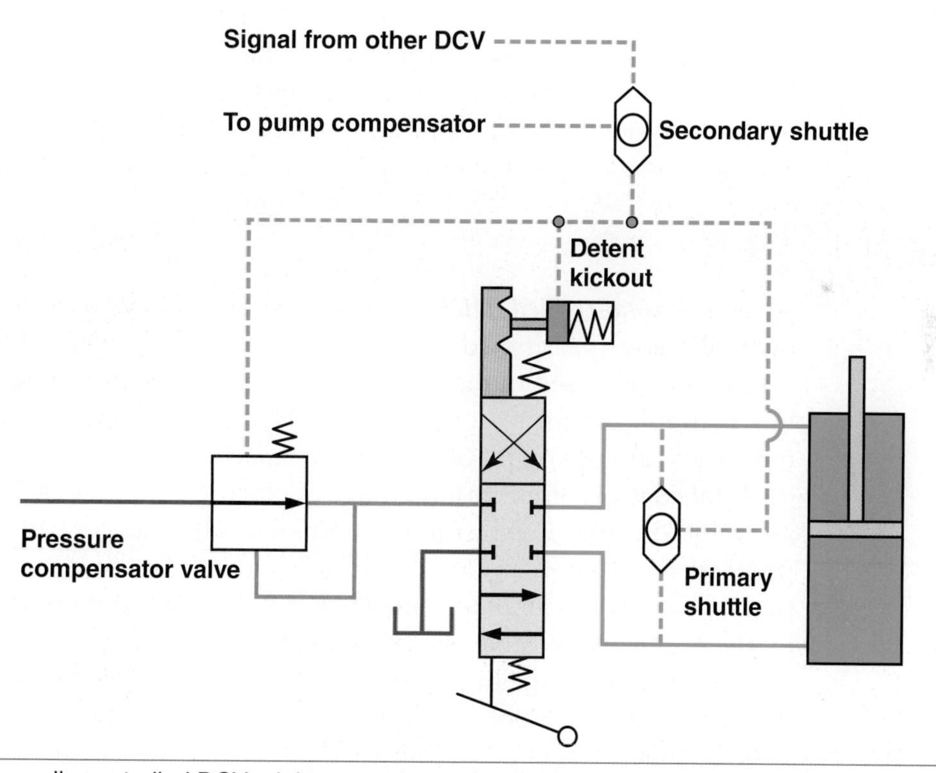

Figure 9-9. A manually controlled DCV might use a detent that holds the lever in position. Detents commonly use a relief kickout that will return the lever to neutral once the cylinder reaches its end of travel. The primary shuttle has the responsibility of choosing the highest working pressure from either the extend port or retract port, then sending it to three locations: (1) secondary shuttle valve, (2) detent kickout, and (3) the pressure-compensator valve. **Chapter 18** will explain LSPC systems.

John Deere economy 6030 series tractors, 7030 series tractors, and 6M series tractors with 300 series DCVs, have detent pressure kickout reliefs. However, these kickout reliefs all sense the same pump pressure instead of the individual work pressure. This can make it more difficult to set the kickouts when trying to run one DCV for continuous motor flow, one DCV for continuous planter down force, and when attempting to have the other DCVs kick out when the cylinders reach the end of travel. The two DCVs with the need for continuous flow would require the kickouts be set above the pump's relief pressure. The DCV controlling the cylinder would require the kickout be set just below the pump's relief pressure.

Many of today's agricultural tractors with electronically controlled DCVs use timers as a means for shutting off flow. The DCV timer is adjusted by means of a potentiometer sending a variable voltage to an ECM, or by means of a monitor's touch panel. This input is varied based on operator preference. When the operator actuates the DCV, the spool will remain in the extended position (or retracted position) based on the time established by the operator. If the time is set correctly, once the cylinder fully extends or fully retracts, the ECM will de-energize the DCV's solenoid.

On agricultural tractors, there are a few unique circumstances that require continuous flow of oil. For these situations, the timer is set for an infinite amount of time and the DCV remains actuated. A couple of examples are hydraulic down force on a drill or a planter, or a hydraulically driven fan on a row crop planter.

On late-model Caterpillar machines, for example a wheel loader, the kickouts are programmed by the operator so that the ECM de-energizes the DCV solenoid before the bucket cylinder reaches the end of travel. This prevents a hydraulic system pressure spike. As mentioned in **Chapter 6**, the operator can program the point at which the cylinder is de-energized.

Pilot Control

A ***pilot-controlled DCV***, or pilot-operated DCV, uses a pneumatic or hydraulic low pressure to remotely control a larger DCV. Pneumatic pilot control is rarely used in mobile machinery, but is frequently used for controlling hydraulic systems on semi-trucks. Hydraulic pilot controls are found in numerous different types of off-highway machines.

Pilot controls offer the advantage of placing a smaller DCV valve, known as a pilot controller, inside the operator's cab. The pilot controller sends a pilot pressure to actuate a larger DCV located outside of the machine, and potentially closer to the actuator. See **Figure 9-10**. The pilot pressure developed by the pilot-control valve is proportional to the lever's position. The further the lever is moved, the higher the pilot pressure placed on the control valve. By placing smaller pilot controllers in the cab and locating the larger pilot-operated DCVs outside of the cab, cab hydraulic noise and heat can be reduced.

Hydraulic pilot controls operate similar to electrical relays and electronic transistors, **Figure 9-11**, which use a small amount of pressure and flow to control a larger amount of pressure and flow.

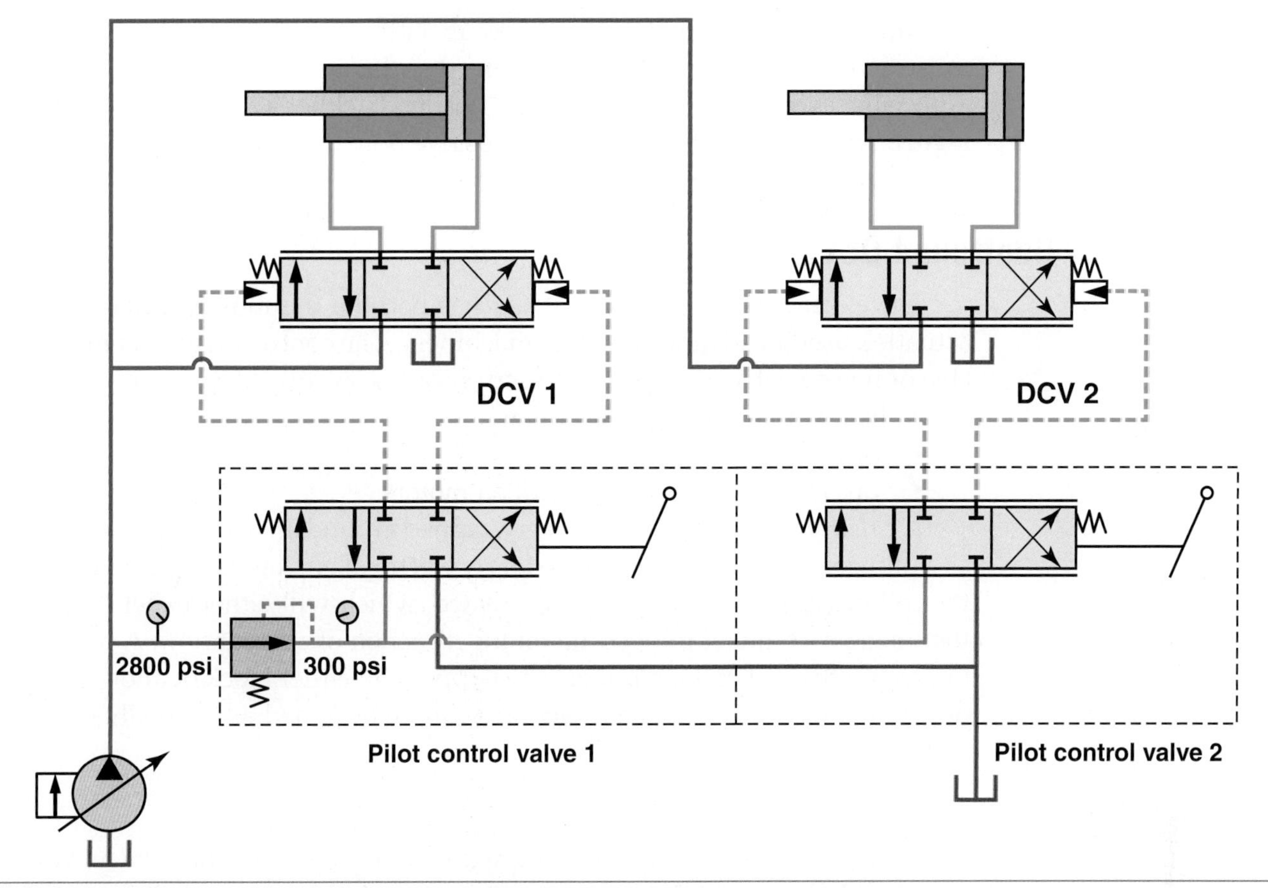

Figure 9-10. Pilot controls commonly place small pilot-control valves inside the cab for the purpose of actuating larger control valves located outside of the cab. Small diameter hoses connect the pilot-control valves to the larger control valves. Pilot control valve 1 actuates DCV 1 and pilot control valve 2 actuates DCV 2.

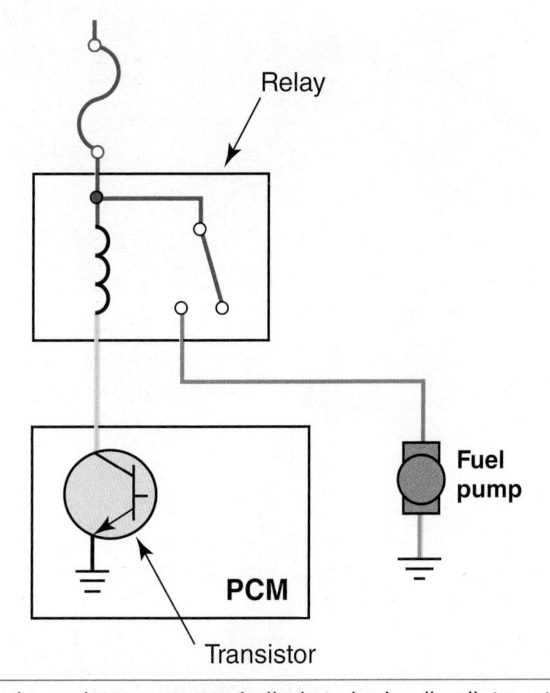

Figure 9-11. An electrical relay and transistor operate similarly to hydraulic pilot controls. In this schematic, a transistor controls a small amount of current passing to ground. The control of this small current causes the relay to actuate a larger amount of current for operating the fuel pump.

Although pilot controls are found in both agricultural equipment and construction equipment, they are more popular in construction machinery, especially excavators, motor graders, loader backhoes, and wheel loaders, Figure 9-12. **Chapter 19** will explain positive flow control (PFC) used on excavators which use a complex pilot system.

Electronically Controlled DCVs

The electronically controlled DCV has become the most popular type of actuation used in advanced mobile machinery. The controls will either be variable or non-variable. Examples of electronically controlled DCVs are:

- Non-variable (on/off) solenoids.
- Variable solenoids.
- Variable electronically controlled motors.

A ***solenoid*** is a coil of wire that is wrapped around an iron core, also known as a plunger or armature, Figure 9-13. As electrical current is directed through the coil of wire, a magnetic field is created, which will either expel or retract the solenoid's plunger depending on the direction of current flow. A poppet is located on the end of the plunger. The poppet will either block or allow oil flow depending on its design. The solenoids can be designed as normally open or normally closed.

On/Off Solenoid

Non-variable solenoids are designed to receive a constant voltage, which is normally battery source voltage, 12 or 24 volts. A simple on/off solenoid

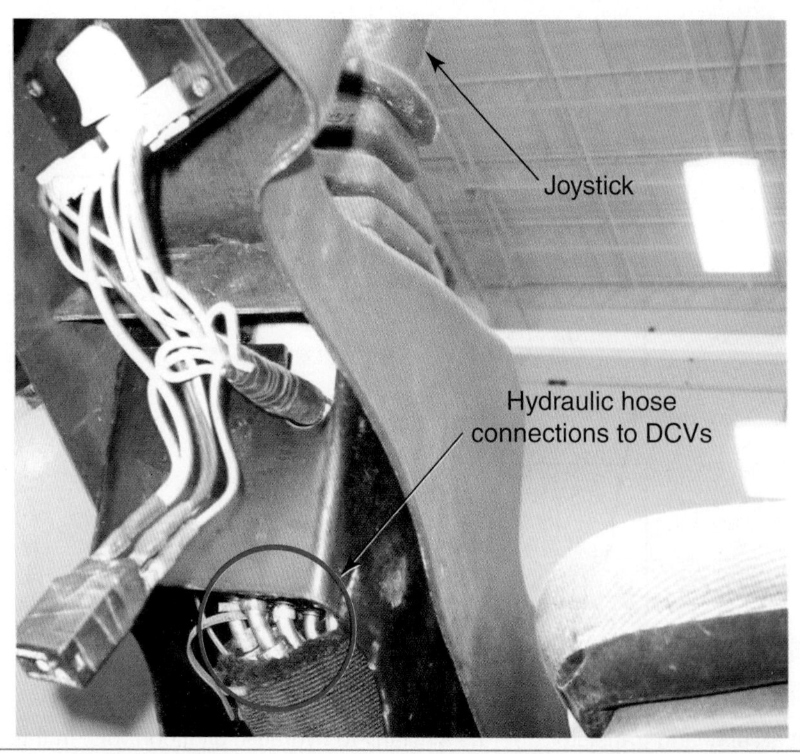

Figure 9-12. Excavators commonly use pilot-controlled joysticks also known as pilot controllers. The joysticks connect to the DCVs through hydraulic hoses for directing pilot pressure to operate the DCV spools.

that is used to directly control an actuator is sometimes called a "bang-bang solenoid." This term refers to the relatively fast speed at which the solenoid opens and closes and the fact that the actuator is set in motion abruptly. A two-position on/off solenoid controlling system pressure is commonly used as a jammer solenoid, unloading valve solenoid, or a dump valve solenoid, which will be further explained in **Chapter 16** and **Chapter 18**.

A simple on/off solenoid can be operated with simple switches and relays, or controlled with electronic control modules (ECMs). On/off solenoids usually receive full battery voltage. Some solenoids are labeled constant duty and use a coil with a high resistance value. Some solenoids are labeled intermittent duty and use a coil with a low resistance value. However, it is possible to find an intermittent duty cycle solenoid that receives full battery voltage and has a high coil resistance value. See **Figure 9-14** for some examples of solenoid coil resistance values.

Pulse-Width Modulated Solenoid

The most common solenoid is the variable ***pulse-width modulated (PWM) solenoid***, which is sometimes called a ***proportional solenoid valve*** or a rapid on-off solenoid valve. A PWM solenoid has the electricity pulsed on or off by an ECM at a variable rate based on the need for different pressures or flows. The rate of the PWM is measured in ***duty cycle***. See **Figure 9-15**.

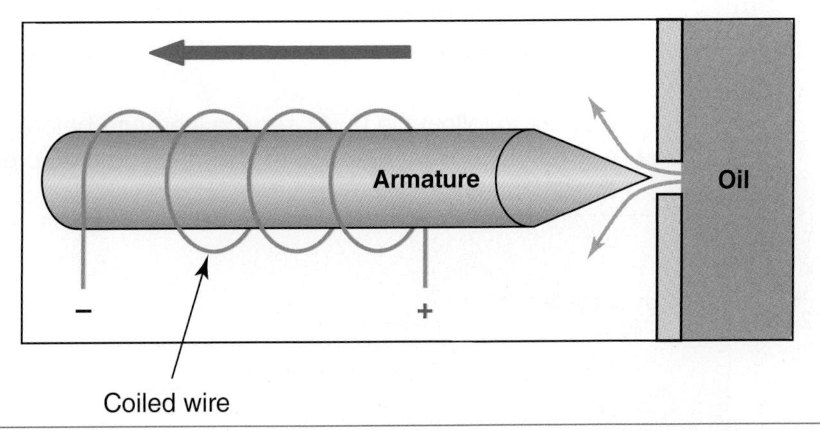

Figure 9-13. A solenoid consists of a coil of wire wrapped around an iron core, also called an armature. Depending on the direction that current flow is applied to the armature, a magnetic field causes the armature to extend or retract.

Combine Solenoid Function	Type of Solenoid	Resistance Value
Older header raise and lower	On/off	13.5Ω
Newer header raise and lower	PWM	6 to 7Ω
Older unloader swing	On/off	3.5Ω
Separator engage	On/off	7 to 8Ω

Figure 9-14. Solenoid coil resistance values will vary depending on the manufacturer. This figure illustrates an example range from a low 3.5Ω up to 13.5Ω. Some proportional solenoids use a solenoid coil with a low resistance value. Some on/off solenoids use a solenoid with a higher resistance value.

The larger the percent of duty cycle, the more the solenoid is actuated, similar to a manual lever being moved further. For example, a 100% duty cycle would equal battery source voltage and the solenoid being fully energized. A 0% duty cycle will result in the solenoid being de-energized.

A PWM solenoid is based on a fixed frequency. An example of a fixed frequency would be 50 hertz (or cycles per second). Figure 9-15 shows a fixed frequency, because the modulation periods are all the same length of time, while the percent of on and off time is different.

The ECM can pulse the solenoid's ground or the solenoid's power wire. An electronically controlled DCV that is used to operate a double-acting cylinder will usually employ a solenoid to extend the cylinder and a second solenoid to retract the cylinder. Figure 9-16 illustrates that PWM solenoids are commonly used in conjunction with pilot controls.

During repair, the solenoid's coil can usually be replaced while leaving the solenoid armature assembly inside the valve's manifold. If the coil is replaceable, it contains an outside jam nut that must be removed before the coil can be slid off the solenoid's armature, Figure 9-17.

Sometimes technicians swap coils for diagnostic purposes. For example, if the header raise and lower coils were swapped but wiring left unchanged, a technician could perform a simple test to check for proper coil operation. After the swap, pressing the header-lower switch should raise the header, and pressing the header-raise switch should lower the header. Swapping coils in this example—it should be noted—is likely to cause an error code.

Warning

Be sure to follow the OEM service literature before performing any test to prevent damage to the machine or to prevent injury to personnel.

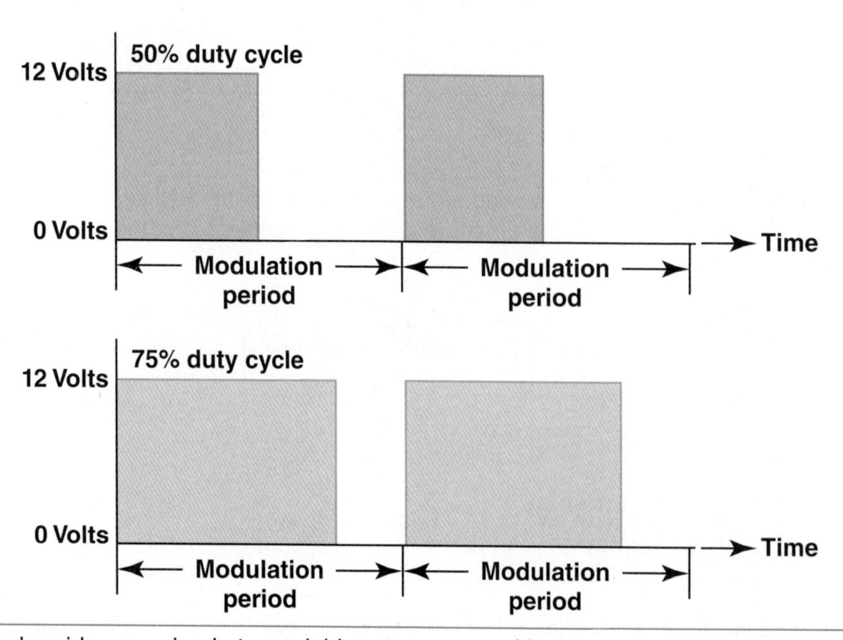

Figure 9-15. PWM solenoids are pulsed at a variable rate measured in duty cycle. The larger the percent of duty cycle the more the solenoid is actuated. This actuation period is shown by the width of the bars on the graph.

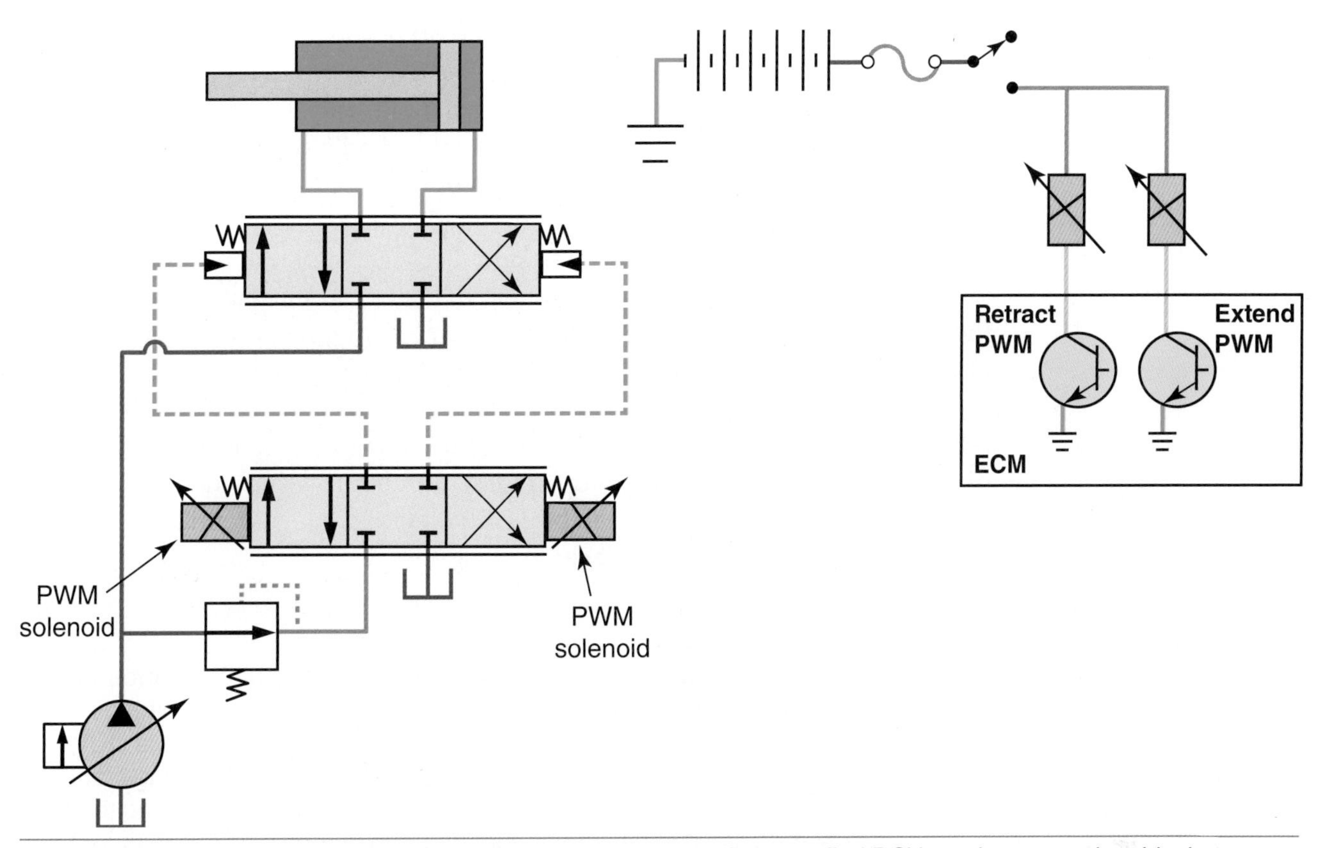

Figure 9-16. A double-acting cylinder operated by an electronically controlled DCV requires two solenoids that contain two wires on each solenoid, a power and a ground. The ECM in this illustration is responsible for pulsing the solenoids' ground wires to complete the appropriate electrical circuit. This actuates one of the solenoids' armatures for movement of the DCV valve.

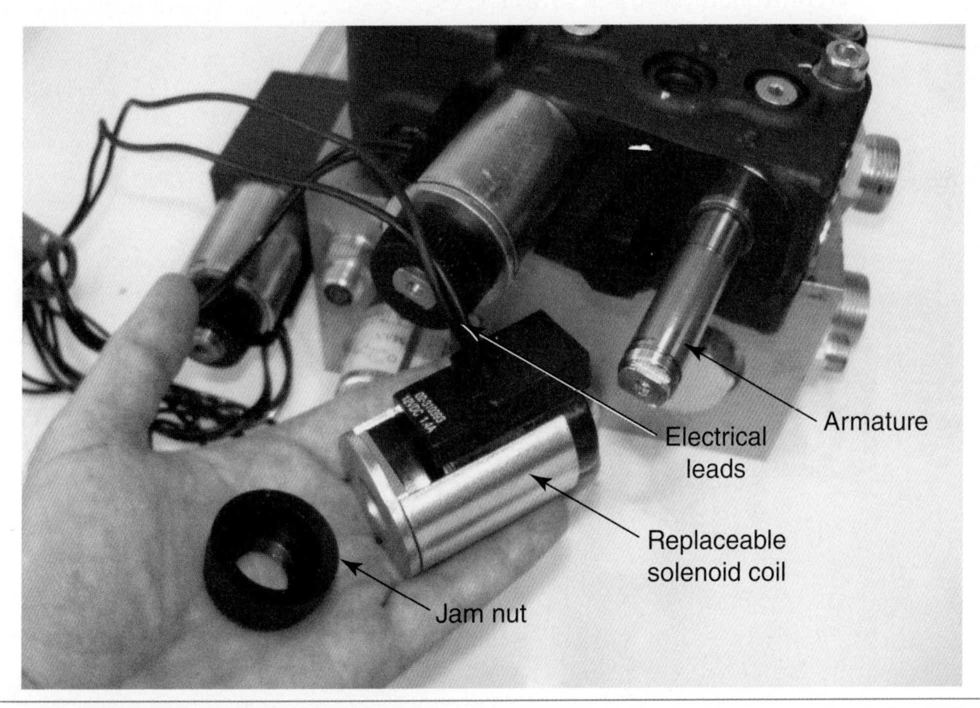

Figure 9-17. Most PWM solenoids allow the coil to be replaced without having to remove the solenoid's armature. After the outside jam nut is unscrewed, the coil can be slid off the armature.

It is also possible that the jam nut can work loose and fall off the solenoid, causing the coil to slide off the armature. The fix can be as simple as ordering a new jam nut, sliding the coil back on the solenoid, and tightening the jam nut. Follow OEM service procedures, especially if thread sealant is recommended.

Sometimes, a hydraulic control fails to operate, but a need exists to actuate the DCV. Some electronically controlled DCVs have an internal override for this purpose. A technician can manually activate the DCV by using a small screwdriver to press on the pintle inside the solenoid's armature. This allows the technician to act as the electrical coil and manually actuate the valve, as pictured in Figure 9-18. Be sure to follow the OEM service literature.

Manual override can enable an implement to be lowered if it is stuck in a raised position and the engine will not start. The override is also helpful for diagnostics. If the DCV will not operate, but functions when the manual override is pressed, the hydraulics can be eliminated from the potential list of faults and the electrical control circuit can be diagnosed.

Warning

Special caution must be taken when working on hydraulic valves. Personnel have been injured and machines have been damaged because a technician carelessly removed a component from a DCV. If an implement is suspended in the air, a 200-foot telescoping crane boom for example, it could be lethal to remove a spool, causing the boom to come crashing down to the earth.

Electronically Controlled Motors

Mobile hydraulic systems can use three different types of electronically controlled motors within DCVs:

- Force motor.
- Torque motor.
- Stepper motor.

Figure 9-18. A screwdriver is sometimes used to manually override a proportional solenoid to determine if the problem is electrical or hydraulic.

These three reversible motors do not spin like a common direct-current fan motor. Instead the motors are positioning devices. They move with a small range of limited motion. The motors are used in conjunction with other devices, such as a pilot spool valve, a flapper valve, or a jet pipe, which will control the oil pressure acting on a control piston or spool.

A ***force motor*** is a reversible linear electric motor that is used to actuate a pilot valve. The motor contains a core push-rod assembly. Permanent magnets hold the push rod in a neutral position. When the single coil is energized, it will shift to the push rod to the right or left based on the polarity of the coil's current. The push rod is used to control pilot oil. Notice in Figure 9-19 that the pilot oil operates a secondary spool valve. The secondary spool valve will direct oil to the actuator and direct the actuator's return oil to the reservoir.

A force motor has the advantage of needing only two electrical wire leads for controlling a double-acting actuator, as opposed to a proportional solenoid that commonly uses four leads (two per solenoid) for controlling a double-acting actuator. Although a force motor can be used in mobile machinery, a proportional solenoid is much more common.

A torque motor is commonly used within servo valves. A servo valve is more expensive to manufacture than a variable solenoid. Both the PWM solenoid and the servo valve can be considered an electronic proportionally controlled DCV. However, a ***servo valve*** is much more precise—having tighter internal tolerances requiring filtration down to 3 microns; has a faster response; and has less hysteresis and less valve overlap than a proportional solenoid valve.

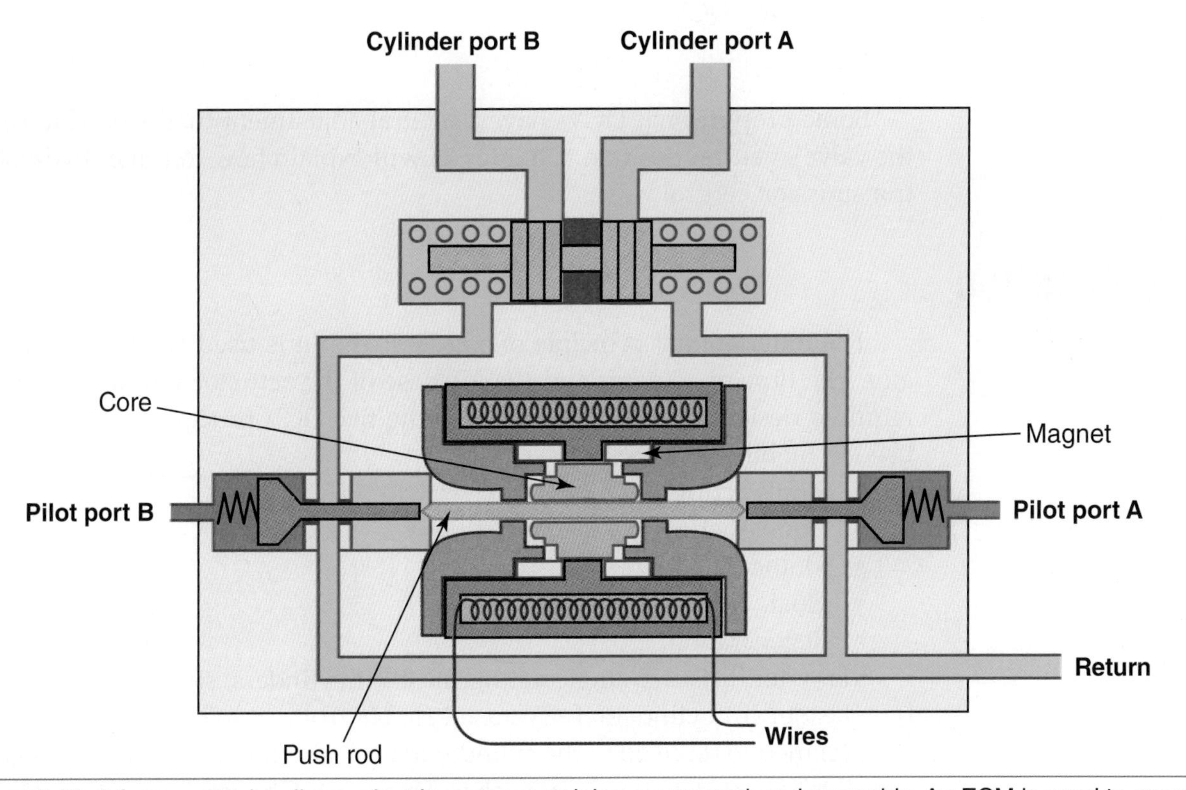

Figure 9-19. A force motor is a linear electric motor containing a core push-rod assembly. An ECM is used to operate a single coil. As with any electromagnet, the direction of the current flowing through the coil will determine which direction the push rod moves.

The term ***hysteresis*** refers to the lag that occurs when the ECM commands the solenoid or torque motor to actuate, but that command is delayed due to stiction forces placed on the valve. As a result, the electronically controlled valve does not respond in a smooth linear motion as commanded by the ECM. The hysteresis effect is also significantly noticeable when comparing how the valve responds to rising current versus how it responds to decreasing current.

To overcome hysteresis, engineers implement electronic ***dithering***, which causes the valve to be in a continuous motion or agitation. This constant motion prevents stiction and minimizes hysteresis.

Valve overlap describes how far a DCV spool must move from its centered position before it initially opens the valve's port(s). Servo valves also require the use of a feedback system. Feedback links will be covered in **Chapter 24**.

A ***torque motor*** is an electronically controlled electric motor that uses a "T-shaped" armature that pivots when activated. The motor will actuate either a flapper or a jet pipe inside the servo valve. See **Figure 9-20**. The motor uses only two wire leads. The electrical current is reversed to operate the actuator in the opposite direction. Proportional solenoids are more common than torque motors on mobile machinery.

A ***stepper motor*** commonly uses four wire leads to pulse electromagnets that surround the motor's armature. This on/off pulsing of the electromagnets divides the movement of the armature into controlled segments. An ECM is required to control the position of the stepper motor armature. Most stepper motors do not use feedback, because the ECM remembers the last commanded position. Proportional solenoid valves are used more often than stepper motors for controlling DCVs.

Centering DCVs

Some proportional DCVs have a neutral adjustment that is used to center the valve's neutral position. **Chapter 24** will explain how to center hydrostatic transmission control valves.

DCV Function

The foundational principle of a DCV is that it is used to operate an actuator. The type of actuator and the purpose of the actuator will determine the required design of the DCV. The following five DCV functions are commonly used in mobile machinery:

- Single acting.
- Double acting.
- Motor.
- Float.
- Regeneration.

Many forklifts use single-acting mast lift cylinders, similar to some combine header lift cylinders. DCVs used for controlling a single-acting cylinder are designed to route oil to the cylinder to lift it and the DCV only has to dump the oil to retract the ram. Refer back to **Figure 9-1**. **Chapter 16** will provide further details regarding this forklift DCV example, including what takes place when multiple functions are requested by the operator.

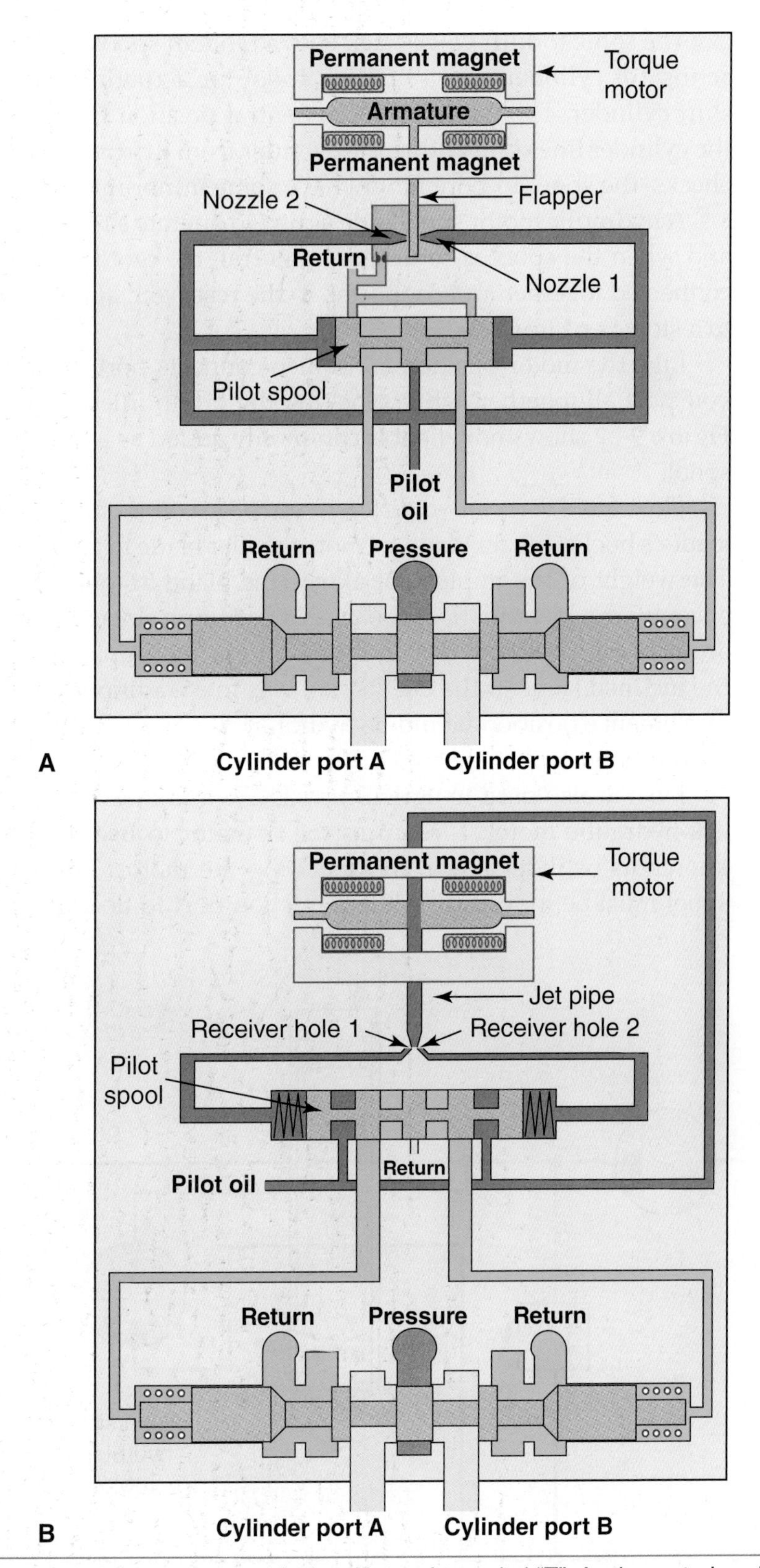

Figure 9-20. A torque motor has an armature in the shape of a capitol "T". As the motor's coils are energized, the armature pivots to the right or to the left. A—In a flapper nozzle system, as the torque motor moves the flapper moves closer to a nozzle, the pressure in that nozzle increases. This causes pressure on the corresponding side of the pilot spool to increase, moving the spool. B—In a jet pipe system, when the jet pipe is centered between the receiver holes, the pressure is equal on both ends of the pilot spool. When the nozzle is directed toward one of the receiver holes, pressure in the corresponding end of the pilot spool builds, shifting the spool.

The same forklift valve stack uses a tandem spool for actuating the double-acting tilt cylinder and a tandem spool for actuating the double-acting side shift cylinder. The tandem spool's neutral position traps the fluid pressure in the cylinder lines preventing the cylinder from drifting. However, without load checks, the spool-to-bore fit will have some minor internal leakage.

A hydraulic motor spool is designed to operate a reversible hydraulic motor, and when the spool is returned to neutral, the motor's port A and port B are connected together and connected to the reservoir, allowing the motor to coast to a stop. See **Figure 9-21**.

Like the motor position, ***float*** also connects port A, port B, and the reservoir port all together, while blocking the pump inlet from the actuator ports. **Figure 9-22** shows how float is commonly added as a fourth position to a DCV spool.

Float often serves two different purposes on mobile machinery. It allows a loader's bucket, dozer blade, or motor grader blade to glide along a hard surface. The weight of the implement allows the blade to lower when the implement encounters a low spot or approaches a declining slope. As the blade approaches an inclined slope, the float position allows the implement to rise and follow the inclined slope. If the DCV spool was in a traditional centered position, the implement's position would be hydraulically locked. The operator would have to continuously actuate the DCV spool.

Float is also used in agricultural DCVs as a means for shutting off a planter's hydraulic motor. It requires the operator to use the retraction port, also known as cylinder lower, for operating the motor. This is because the DCV spool must be moved from the retraction port to float and not through to the

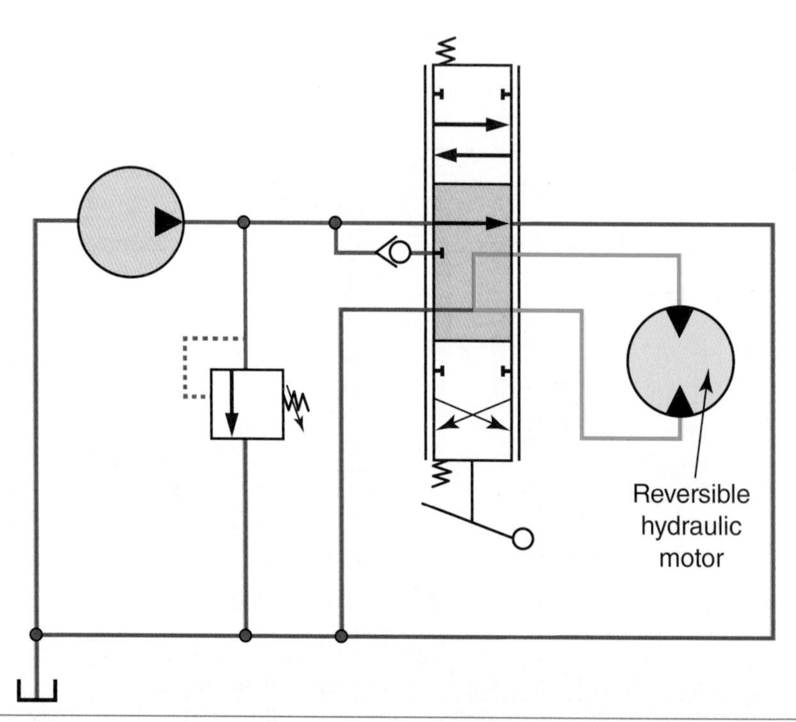

Figure 9-21. A motor spool allows the valve to operate a reversible hydraulic motor. When the valve is returned to neutral, port A, port B, and the reservoir port are all connected together, allowing the motor to coast to a stop without cavitating.

spool's neutral position. If the spool was operating the motor from the extend port, also known as cylinder raise, the DCV spool would have to be moved through a closed-center neutral position and the retracted position before it could be placed into the float position. The extend port also frequently contains a pilot-operated check valve that would lock the motor if the DCV spool was returned from the extend position to the neutral position, preventing the motor from coasting to a stop.

As explained in **Chapter 6**, regeneration connects the cap side of a hydraulic cylinder to the rod side while supplying pump flow to the cylinder. It allows the cylinder to reuse the oil exhausting from the rod side, causing the cylinder to extend faster by receiving a flow rate greater than what the pump alone is capable of supplying.

Number of DCV Positions

Directional-control valves can be configured with a different number of positions. An unloading valve, jammer valve, or dump valve, is a common two-position DCV. Remember that a jammer solenoid has one position that is normally open, allowing the pump to flow freely to the reservoir. The valve's second position blocks pump flow, which causes the system to build pressure to the value of the relief valve.

Double-acting cylinders are commonly controlled by a three-position valve. One position is the neutral position, and the other two positions are used to extend and retract the cylinder. A log splitter is an example of a system that uses a three-position DCV.

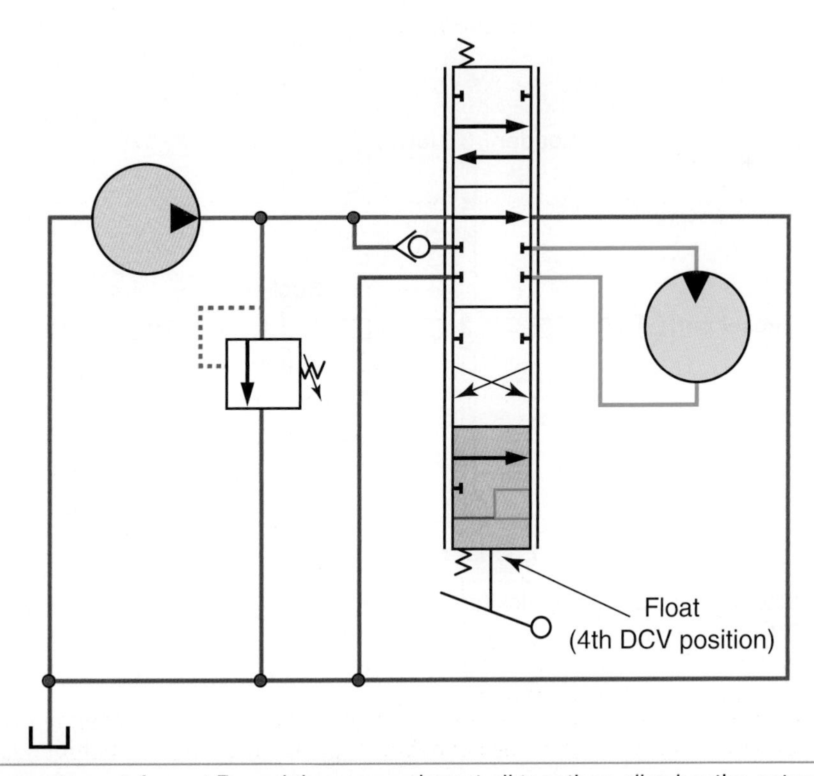

Figure 9-22. Float connects port A, port B, and the reservoir port all together, allowing the actuator to freely float.

Several mobile hydraulic systems use four-position DCVs. A tractor's loader commonly uses one joystick to control two different four-position DCVs. Figure 9-23 maps shift positions of the joystick control for a loader. One DCV is used for the loader lift, and the valve's position will be raise, neutral, lower, and float. The second DCV is used for controlling the bucket's curl and contains positions for dump, neutral, curl, and regeneration.

Block Design

Control valve blocks can consist of a single whole-block design, sometimes called monoblock, or a group of multiple valves stacked adjacent to each other, sometimes called sectional or valve stack. The valve stack design is popularly used in mobile machinery and provides the flexibility to add an extra valve section later or replace a bad valve block if it fails. In Figure 9-24, each rectangular valve block is stacked together in a row, with necessary hydraulic lines and electrical connections attached. The drawbacks of a valve stack are the potential for oil leaks and a low tolerance for improper bolt torque applied to the valve stack's bolts. An over-torqued valve stack can prevent the valves from moving within their bores.

The DCV in Figure 9-25 contains two block assemblies that are mated together. The black valve block is a self-contained combine header DCV. This DCV contains controls for header raise, header lower, a signal shuttle valve and a lift check valve. The larger rectangular block assembly contains multiple valve assemblies.

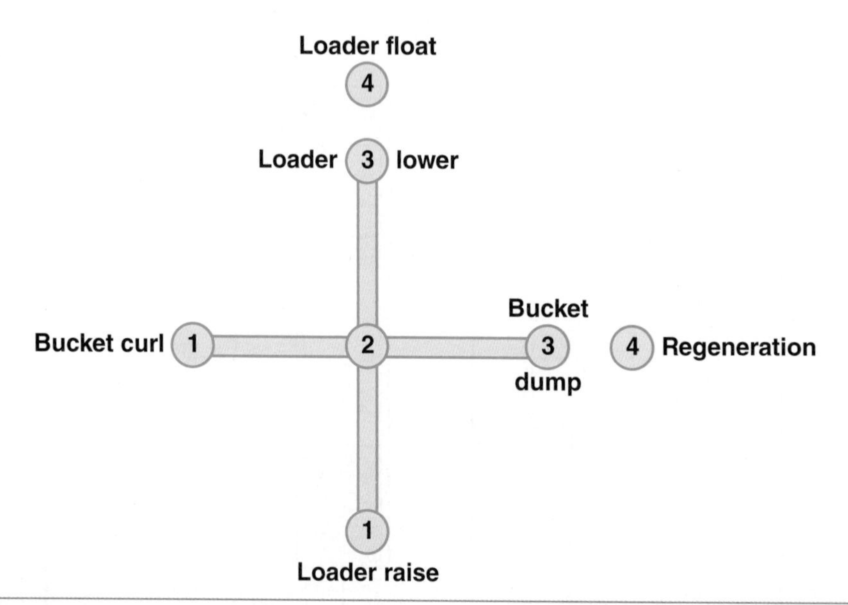

Figure 9-23. A tractor loader uses a single joystick for lifting and dumping the bucket. The joystick actuates two different DCVs. One DCV is responsible for (1) loader raise, (2) neutral, (3) loader lower, and (4) loader float. The float position is often detented. The second DCV controls the bucket: (1) bucket curl; (2) neutral; (3) bucket dump; and (4) fast bucket dump, also known as regeneration.

Figure 9-24. This valve stack is from a combine that was built into a simulator for students to troubleshoot. The stack of valves contains a secondary flow divider, a secondary relief, header lift, header lower, unloading auger swing, reel lift, and reel fore and aft control. If the customer's combine was not equipped with header tilt, then a whole goods kit could be ordered for a technician to install as an additional DCV.

Figure 9-25. This directional-control valve contains the controls for an agricultural combine. The smaller valve block controls the combine header. The gold, rectangular manifold contains multiple cartridge valves. Both are connected as one DCV.

Valve Design

The internal components of a DCV can have four different valve designs:

- Traditional spool valve.
- Seat and poppet valve.
- Rotary valve.
- Cartridge valve.

Spool Valve

The traditional ***spool valve*** contains spaces and lands. The spools are usually chrome plated. Figure 9-26 illustrates a simplified spool valve that has two lands and one interland space separating the two lands. Spool valves are located inside the bore of a valve block. The valve will shift based on a control force. Oil pressure, a manual lever, or an electrical armature is used to move the spool valve. An opposing force, such as a return spring or oil pressure, works against the control force. Whichever force is stronger will cause the spool to shift.

Spool valves often have several grooves and a few slots machined into the lands on the spool. Figure 9-27 magnifies these features on a spool valve illustration. The grooves help lubricate the spool and form an oil film on the outside surfaces of the spool to keep it centered in the bore, reducing the friction when the spool moves. These grooves are sometimes called balancing grooves, centering grooves, or lubricating grooves. The machined slots are sometimes called metering slots or throttling slots. As the spool changes position, a slot will reach the port in the bore before the interland space does. Since less fluid can flow through the slot than can flow through the interland space, this enables the operator to slowly meter the oil, rather than suddenly shifting maximum oil to the cylinder or motor.

Although the spool-to-bore valve fit clearance can be as tight as 0.0002″, the valve cannot provide leak-free operation when the valve is in a neutral position.

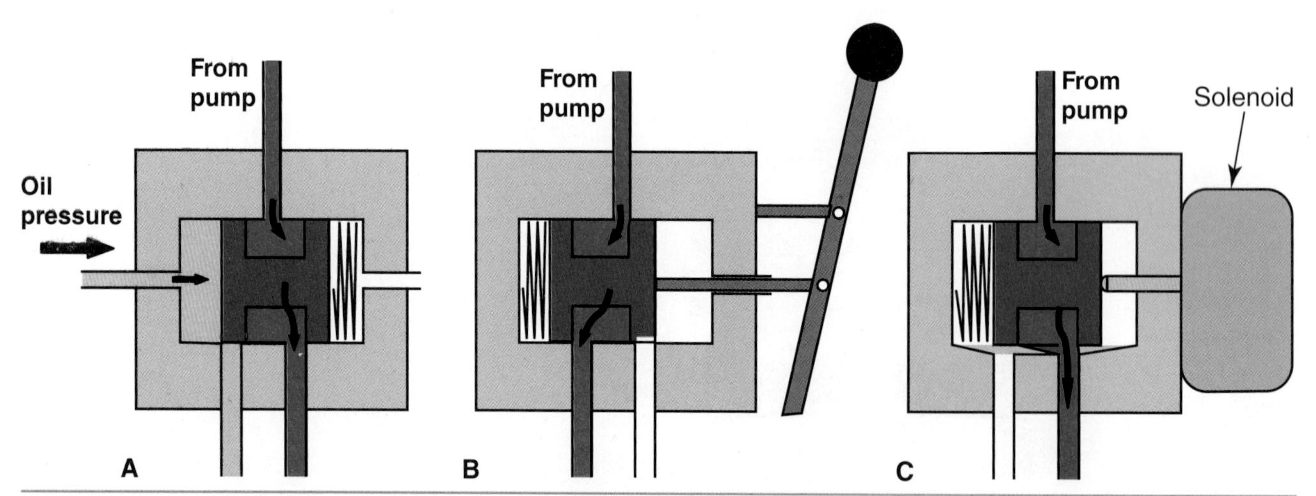

Figure 9-26. Spool valves have lands that are separated by spaces. The valves are placed inside the bore of a valve block. Spring force, oil pressure, a manual lever, or an electrical armature is used to force the spool valve to shift. A—Oil-pressure actuated. B—Manual-lever actuated. C—Solenoid actuated.

Seat and Poppet Valve

Compared to a spool valve, a ***seat and poppet valve*** provides a better seal with practically no internal leakage. Based on this fact, it is often used as the metering-out portion of the valve. The poppet valve design allows engineers to separate the metering of port A and the metering of port B, which are tied together on the same spool with spool valve designs. DCVs that contain poppet-style valves commonly use one poppet valve for port A's return and one poppet valve for port B's return.

Some DCVs contain a combination of both spool valves and poppet valves. One common configuration uses a spool valve for the metering-in portion of the DCV, and poppet valves for controlling the metering-out of the DCV. Both spool and poppet valves are commonly operated by pilot oil from a solenoid valve.

In Figure 9-28, the solenoid actuates a small spool valve. The small spool valve directs pilot oil to operate a larger spool valve which has the responsibility of metering oil into a single-acting cylinder. Another solenoid spool valve is used to operate a seat and poppet valve, which has the responsibility of metering oil out of the cylinder.

If the DCV is controlling a double-acting cylinder, one solenoid (extend solenoid) is responsible for actuating the single metering-in spool valve in one direction so that it sends oil to extend the cylinder as well as actuate the metering-out poppet valve so that it can direct oil returning from the cylinder to the reservoir. The retraction solenoid similarly actuates the same metering-in spool valve in the opposite direction so that it sends oil to retract the cylinder as well as actuating the second metering-out poppet valve so it can direct oil returning from the cylinder to the reservoir.

Rotary Valve

A steering DCV is a common mobile application that uses a rotary-style DCV valve. **Chapter 25** will cover the steering DCV and rotary valve in detail.

Cartridge Valve

Cartridge valves are frequently threaded into a manifold. Refer back to Figure 9-25 to view a manifold (the larger rectangular valve block). The cartridge valve design provides engineers the ability to compartmentalize several hydraulic controls into a compact area. A popular mobile application that uses cartridge valves is agricultural planter controls.

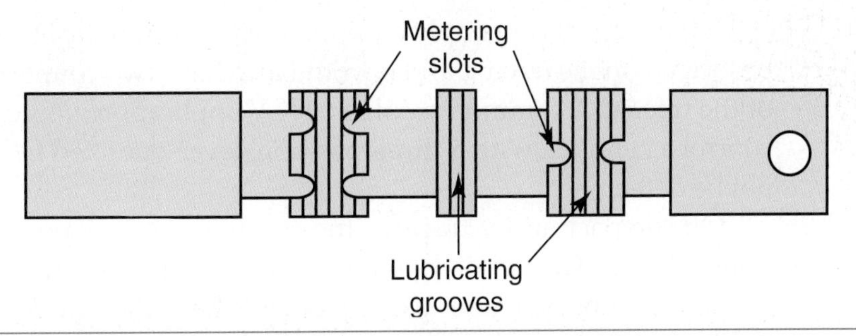

Figure 9-27. Spool valves are typically chrome plated. They have machined centering grooves for lubricating the spool and reducing friction. Machined metering slots slightly ease actuator movement as the spool shifts.

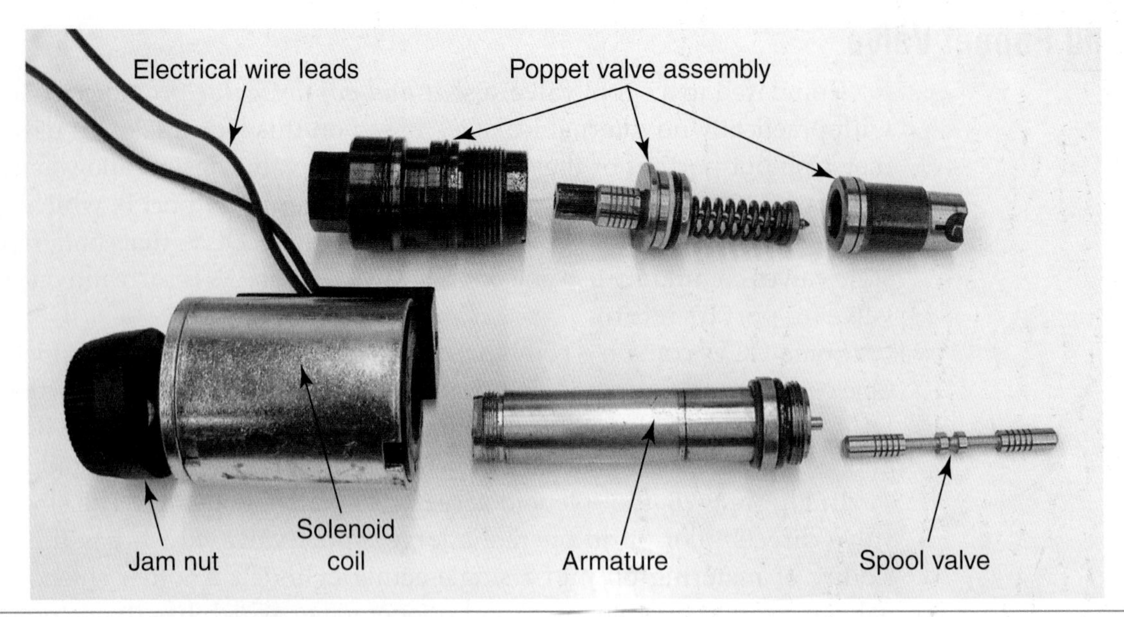

Figure 9-28. A disassembled combine header DCV. The header lower valve uses a solenoid to actuate a small spool valve. The spool valve directs pilot oil to operate the poppet valve.

Flow and Pressure Capacities

When attempting to add an extra DCV, there are other factors to consider in addition to the type of hydraulic control system, method of actuation, DCV function, and block design. The valve must also match the system's flow and pressure settings. For example, a DCV added to a compact utility tractor handles significantly less flow than a mining wheel loader's DCV. The DCV should also be rated to handle the appropriate pressure capacity.

Power Beyond

Power beyond is a common term used in the agricultural industry to describe a hydraulic feature that can be added to a tractor. This option enables additional hydraulic control valves to be added to the tractor's hydraulic system; however, the additional DCVs are usually located on the implement. On open-center agricultural tractors, a power beyond kit looks like a jumper hose at the back of the tractor, coupled together with a single coupling as pictured in **Figure 9-29**. This option is much different than an extra DCV, sometimes called an auxiliary DCV.

If the tractor was simply equipped with an auxiliary DCV, that DCV would be located on the tractor, and it would also have two couplers located at the rear of the tractor. A common auxiliary DCV application would be used to directly control a cylinder, with a three-position, lever-operated DCV, that may or may not be detented.

On compact utility tractors, the auxiliary DCV is not designed for continuous oil flow. The auxiliary DCV therefore is not the best option for log splitters and backhoes. These two implements have their own DCVs located on the implement and need live, continuous hydraulic power, which is where the term *power beyond* originates.

Figure 9-29. This John Deere 955 compact utility tractor has an open-center hydraulic system. The jumper hose is the machine's power beyond.

Installing power beyond in an open-center circuit can consist of simply installing a plug, two hydraulic lines, and one coupling assembly. In **Figure 9-30** the plug diverts oil to the power beyond coupler. In this example, the power beyond returns oil in series to the three-point hitch.

Some open-center power beyond systems stipulate using a different return line, **Figure 9-31**. The power beyond return line sends oil directly back to tank, eliminating the supply oil to the three-point hitch, which is a safety feature. For example, if an operator is located near the three-point hitch while operating a log splitter, the operator could be injured if the three-point hitch DCV was actuated and the hitch arms were raised.

Note

An open-center tractor equipped with power beyond provides a quick and easy method for flow rating the fixed-displacement pump. For that reason alone, it is nice to have this option installed on an open-center tractor.

Caution

Most open-center tractors equipped with power beyond have a caution label placed next to the coupling assembly. If the coupling is left disconnected and the tractor is operated for long periods of time, the system will overheat. The system is designed to operate at a constant flow. That constant flow cannot return to the reservoir at a low pressure value due to the disconnection, so it must go over the main system relief valve, generating excessive heat.

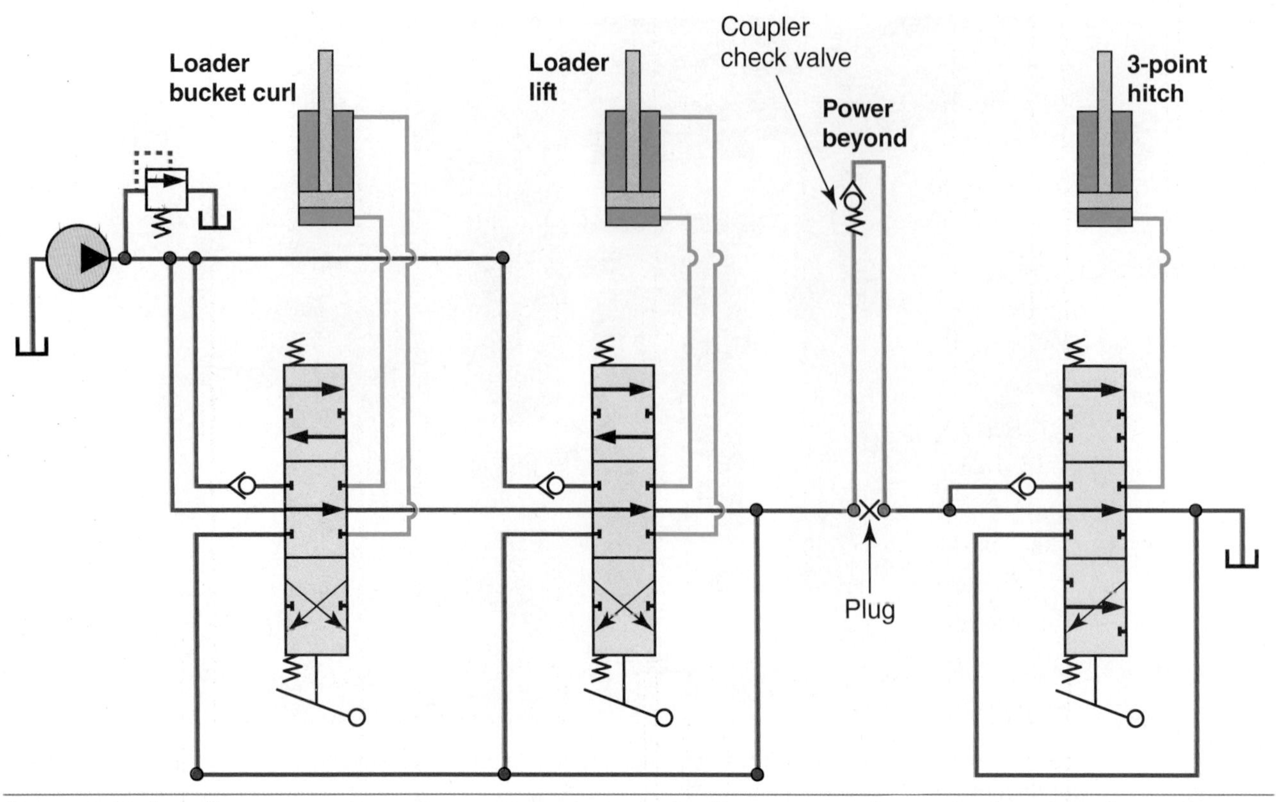

Figure 9-30. Installing a power beyond on an open-center hydraulic system can be as simple as installing a plug, two hoses, and one coupler. The coupler's check valve illustrates that the coupler's tip must be depressed within the coupler in order for oil to flow through the coupling.

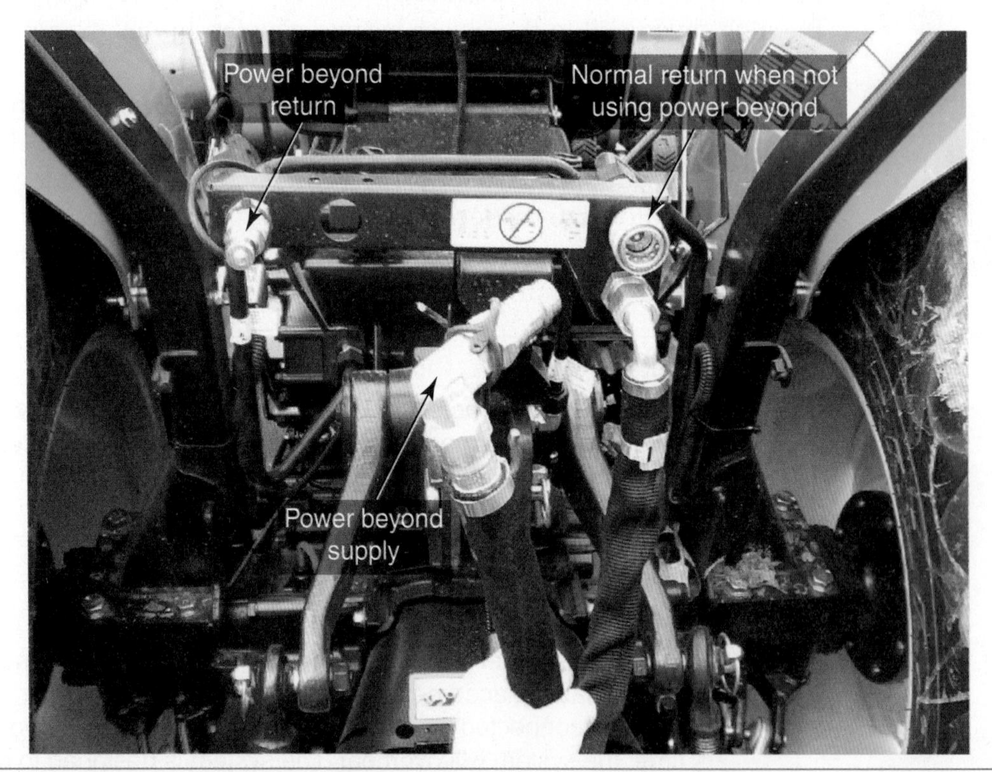

Figure 9-31. This John Deere 4105 compact utility tractor has power beyond. When using this tractor's power beyond, the implement's return line needs to be coupled to the power beyond return, rather than the traditional normal return. This eliminates the three-point hitch control.

Summary

- ✓ DCVs are used to operate hydraulic cylinders and motors.
- ✓ DCVs are designed to be used in open-center, PC, LSPC, flow-sharing, NFC, or PFC hydraulic systems.
- ✓ DCVs can be actuated by a manual lever, pilot oil, electronic solenoid, or an electronically controlled motor.
- ✓ Manually controlled DCVs frequently have detents that hold the spool in position and can return the spool back to neutral once the cylinder reaches its end of travel.
- ✓ DCV solenoid valves are designed as on/off valves or PWM solenoids.
- ✓ Electronically controlled DCVs can use force motors, torque motors, or stepper motors.
- ✓ DCVs are designed to operate single-acting cylinders, double-acting cylinders, and motors.
- ✓ DCVs can be designed with a float function or a regeneration function.
- ✓ Mobile DCVs commonly have two, three, or four positions.
- ✓ Valve stack DCVs provide the most flexibility, but have the risk of leaks or being overtorqued.
- ✓ Spool valves are commonly chrome plated, with centering grooves and throttling slots machined into the spools.
- ✓ Seat and poppet valves prevent internal leakage.
- ✓ Rotary valves are used in steering DCVs.
- ✓ Cartridge valves allow designers to place multiple DCVs in a small compartment.
- ✓ Power beyond allows an additional control valve on an implement to receive live pump flow.

Technical Terms

dithering
duty cycle
float
force motor
hysteresis
load-sensing pressure-compensating (LSPC) hydraulic system
manually controlled DCVs
one-way check valves
open-center hydraulic system
pilot-controlled DCV
power beyond
pressure-compensating (PC) hydraulic system
proportional solenoid valve
pulse-width modulated (PWM) solenoid
seat and poppet valve
servo valve
shuttle valve
solenoid
spool valve
stepper motor
torque motor
valve overlap

Review Questions

Answer the following questions using the information provided in this chapter.

1. A traditional open-center hydraulic system uses what type of pump?
   A. Fixed-displacement.
   B. Variable-displacement.
   C. Either A or B.
   D. Neither A nor B.

2. A traditional open-center hydraulic system operates at what type of relative pressure when the DCVs are in a neutral position?
   A. Low.
   B. High.
   C. Continuously fluctuating.
   D. None of the above.

3. A pressure-compensating hydraulic system operates at what type of relative pressure when the DCVs are in a neutral position?
   A. Low.
   B. High.
   C. Continuously fluctuating.
   D. None of the above.
4. A LSPC hydraulic system uses what type of pump?
   A. Fixed displacement.
   B. Variable displacement.
   C. Either A or B.
   D. Neither A nor B.
5. Which of the following hydraulic control systems requires the use of shuttle valves?
   A. Open center.
   B. Pressure compensated.
   C. LSPC.
   D. None of the above.
6. Which of the following components is designed to sense the higher of two different pressures and send that pressure to the next circuit destination?
   A. One-way check valve.
   B. Shuttle valve.
   C. DCV.
   D. Solenoid valve.
7. Detent kickout reliefs are commonly used with DCVs that are controlled with _____.
   A. manual levers
   B. hydraulic pilot actuation
   C. PWM solenoids
   D. servo valves
8. What happens when a DCV kickout relief valve opens?
   A. The DCV valve is shut off.
   B. The DCV lever is moved to the neutral position.
   C. Both A and B.
   D. Neither A nor B.
9. Late-model agricultural tractors with electronically controlled DCVs use what method to shut off DCV flow?
   A. Mechanical detent kick out relief valves.
   B. Electronic timers.
   C. Mechanical spring.
   D. Solenoid.
10. All of the following are examples that need continuous oil flow on an agricultural operated DCV, *EXCEPT*:
    A. hitch lift.
    B. planter down force.
    C. planter fan motor.
    D. drill down force.
11. Which of the following is *not* used to describe a type of electronically controlled DCV motor?
    A. Stepper.
    B. Solenoid.
    C. Force.
    D. Torque.
12. An electronically controlled DCV is using proportional solenoids to actuate a double-acting cylinder. How many wire leads does the DCV have?
    A. One.
    B. Two.
    C. Three.
    D. Four.
13. What is the unit of measurement for a PWM solenoid's operational rate?
    A. Amperes.
    B. Voltage.
    C. Hertz.
    D. Duty cycle.
14. Which of the following terms is defined as "the distance a spool must move from its centered position before the valve starts to open"?
    A. Hysteresis.
    B. Valve overlap.
    C. PWM.
    D. Dithering.
15. Which of the following terms is defined as "a valve's lagging response due to stiction forces"?
    A. Hysteresis.
    B. Valve overlap.
    C. PWM.
    D. Dithering

16. Which of the following terms is defined as "a constant agitation to help prevent a valve from lagging"?
    A. Hysteresis.
    B. Valve overlap.
    C. PWM.
    D. Dithering.
17. All of the following are true statements about electronically controlled DCV motors, *EXCEPT*:
    A. they are used as position devices.
    B. they activate a device to control pressure.
    C. they spin at a specific rpm.
    D. they are an electronic proportional-controlled DCV.
18. Which of the following devices uses a T-shaped armature?
    A. Stepper motor.
    B. Torque motor.
    C. Force motor.
    D. PWM solenoid.
19. Which of the following devices is the most popular method for electronically controlling a DCV on a mobile machine?
    A. Stepper motor.
    B. Torque motor.
    C. Force motor.
    D. PWM solenoid.
20. Which of the following statements is true about the motor spool position or float spool position on a DCV?
    A. Port A and Port B are blocked.
    B. Port A and Port B are tied together, but isolated from the reservoir.
    C. Port A, Port B, and the tank port are tied together.
    D. Port A, Port B, and the tank port are all blocked.
21. A tandem spool is designed to operate a _____.
    A. single-acting cylinder
    B. double-acting cylinder
    C. reversible hydraulic motor
    D. single-acting motor
22. A tractor has two DCVs for controlling the loader. Technician A states that the float position is used for bucket curl. Technician B states that regeneration is used for loader lower. Who is correct?
    A. Technician A.
    B. Technician B.
    C. Both A and B.
    D. Neither A nor B.
23. Which of the following valve designs provides the best chance of leak-free operation?
    A. Spool valve.
    B. Seat and poppet valve.
    C. Rotary valve.
    D. Cartridge valve.
24. The grooves machined on a spool valve land serve all of the following purposes, *EXCEPT*:
    A. lubrication.
    B. metering.
    C. balancing.
    D. centering.
25. The throttling slots machined on a spool valve serve what purpose?
    A. Lubrication.
    B. Centering.
    C. Balancing.
    D. Metering.
26. Spool valve-to-bore clearance specifications can be as tight as _____.
    A. 0.0002″
    B. 0.0200″
    C. 0.2000″
    D. 2.0″
27. A rotary valve is commonly used in which mobile machinery application?
    A. Three-point hitch control.
    B. Steering control.
    C. Transmission control.
    D. Power beyond.

28. A power beyond option has been added to a tractor. What else is typically required in order to operate the power beyond hydraulics?
    A. An extra pump.
    B. An extra DCV.
    C. A separate ECM.
    D. Nothing else is needed.
29. In an open-center hydraulic system, if the tractor is operated for a prolonged period with the power beyond coupler disconnected, what could be a possible result?
    A. Nothing, the system will run as normal.
    B. System will overheat.
    C. Pump flow rate will drop drastically.
    D. OEM actuators will not operate correctly.
30. A customer has 5 DCVs and is using each of those DCVs. The customer still needs more hydraulic controls; however, the machine has no more room for another DCV. What do some manufacturers offer to provide the customer more hydraulic controls?
    A. A secondary pump.
    B. Power beyond.
    C. Another DCV.
    D. All of the above.

Chapter 10

Fluids

Objectives

After studying this chapter, you will be able to:

✓ Explain the different functions of a hydraulic fluid.

✓ Explain the importance of each of the common hydraulic fluid properties.

✓ Describe the advantages and disadvantages of the different types of hydraulic fluid.

Function of Fluids

A machine's hydraulic fluid is considered the life blood of a hydraulic system. By most standards, the fluid itself is perhaps the most important component of the system. Costly damage can occur if the wrong fluid is placed in a machine.

The hydraulic fluid must be able to perform several important functions:

- Transmit power.
- Assist in removing contaminants and heat.
- Seal clearances.
- Lubricate mechanical parts.

Transmit Power

As mentioned in **Chapter 2**, liquids are practically incompressible. For this reason, fluids provide an effective means for transmitting power. Petroleum-based fluids will compress approximately 0.5% when the liquids are pressurized to 1000 psi (69 bar).

The measure of a fluid's ability to resist compression is the fluid property called *bulk modulus*, also known as the stiffness of a fluid. Bulk modulus is measured in units of pressure, for example psig, bar, or Pascals.

If bulk modulus is plotted on a graph with the fluid volume on the X axis and fluid pressure on the Y axis, the plot demonstrates a few points:

- Bulk modulus increases as the pressure increases.
- Bulk modulus is the lowest at a lower pressure level.
- The rate of bulk modulus increase is plotted on a curve. For example, if the volume compresses 0.5% at 1000 psi, the fluid volume compression at 6000 psi would be less than 3% because the rate is not a straight line. See **Figure 10-1.**
- In addition to fluid pressure, three other factors affect bulk modulus:
  - Air trapped in the fluid.
  - Fluid temperature.
  - Type of fluid.

The higher the quantity of air trapped in the fluid, the easier it is to compress the fluid. The hotter the fluid's temperature, the more the fluid will compress.

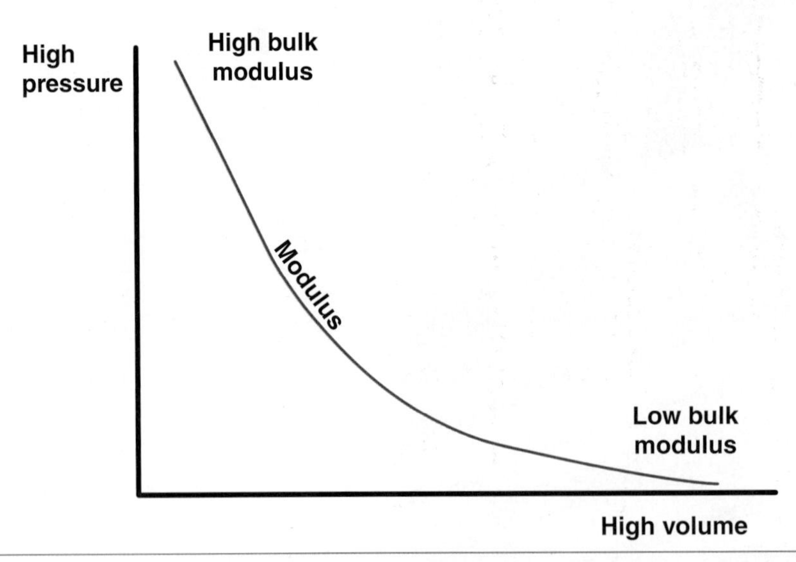

Figure 10-1. The plot of bulk modulus results in a curve. The fluid volume will be higher if the fluid compresses less. The fluid's volume will be less as the fluid is further compressed.

Different types of fluid have a different bulk modulus. For example, water is more compressible than a petroleum-based fluid. Some fire-resistant fluids, HFC and HFD, have a slightly higher bulk modulus than mineral-based petroleum oil. Fire-resistant fluids will be discussed later in this chapter.

Additives cannot be added to the oil to improve bulk modulus. However the base oil does affect the bulk modulus. Naphthenic petroleum-based hydraulic oil will have a higher bulk modulus than a paraffinic petroleum-based hydraulic oil. Petroleum-based fluids will be discussed later in the chapter.

Tangent modulus is a fluid's stiffness based on the fluid's rate of change in volume for a specific fluid pressure. This method of measuring resistance to compression is used in applications with a narrow range of pressures, like a power shift transmission for example. Secant modulus is stiffness of the fluid based on the overall change in the fluid's pressure compared to the overall change in the fluid's volume. Secant modulus is used for applications that operate with a wide range of system pressures, for example a hydrostatic transmission or excavator.

Bulk modulus is very important when working with high-pressure hydraulic systems that contain large volumes of fluid. Poorly designed systems can exhibit the following negative attributes when large quantities of fluid are compressed to high system pressures:

- Delay in the actuator's response.
- The actuator can continue to move after the DCV is closed.
- Loss of system efficiency.
- Hydrostatic transmission cavitation.

A hydrostatic transmission with long drive hoses is an example of an application that has the potential to operate at pressures above 6000 psi (414 bar) and have large fluid volumes between the pump and motor. As the pressure increases, the fluid can compress, and it will compress at a much more severe rate if air is trapped and if the fluid temperature is hot. If the system does not have a charge pump with an adequate displacement to overcome this fluid volume compression, the transmission will cavitate, leading to a pump failure.

Therefore, the following factors should be followed to avoid bulk modulus concerns:

- Use a quality hydraulic fluid that has a good bulk modulus.
- Eliminate air intrusion into the oil.
- Avoid overheating the oil.
- Minimize the fluid volume between the DCV and the actuator.

A fluid's bulk modulus can be computed by using the following equation:

$$\text{Secant Bulk Modulus } (B_S) = (\Delta \text{ pressure} \times \text{initial volume}) \div \Delta \text{ volume}$$

The opposite of a fluid's stiffness is a fluid's compressibility. Mathematically, ***compressibility*** is the reciprocal of bulk modulus and is typically measured in negative units. The two fluid properties have an inverse relationship. See **Figure 10-2**.

Remove Heat and Contaminants

Fluids must also carry away harmful contaminants and heat. Although hydraulic systems have components such as reservoirs, coolers, and filters designed to assist in these needs, it is the fluid itself that carries the contaminants to the filter and the fluid itself that transfers the heat from the components to the cooler and reservoir.

Seal Oil Clearances

Many components in hydraulic systems must have proper clearances between their parts in order to operate properly without leaking. The clearance between a DCV spool valve and its bore and the clearance between a pump's piston and its bore are examples of critical clearances. The thickness of the fluid assists in eliminating leakage between these parts. A fluid that is too thin will leak and perform poorly at sealing these clearances, which causes system inefficiency.

Lubricate Moving Components

Hydraulic components contain numerous moving parts. Fluids must function as a good lubricator between those moving parts. A fluid with good ***lubricity*** has the following characteristics:

- Provides a good oil film to separate moving parts.
- Maintains that film of oil even during high operating pressures.
- Sticks to individual components, especially during higher operating temperatures.

Figure 10-2. A fluid's bulk modulus has an inverse relationship to a fluid's ability to compress. The higher the bulk modulus, the less compressible the fluid is.

In a hydraulic system, a fluid with good lubricity will perform multiple functions:

- Reduce component wear.
- Lower the chances of ***galling***, which refers to material lifting off of one component and adhering to the other component due to friction and lack of lubrication.
- Prevent the moving components from seizing.

Fluid Properties

A quality hydraulic fluid must have good fluid properties. The fluid properties describe unique attributes of the fluid that, if properly applied, will reduce component wear and failure.

Excellent fluid properties are essential to maximizing the machine's life, improving the machine's performance, and improving fuel economy. The following are commonly considered important properties of hydraulic fluids:

- Viscosity.
- Viscosity index.
- Pour point.
- Stability.
- Anti-wear.
- Bulk modulus.

Viscosity

As explained in **Chapter 8**, viscosity is the term used to describe the resistance to fluid flow, which is based on the thickness of the fluid. The International System of Units (SI) unit for measuring viscosity is ***centistokes (cSt)***. Centistokes equals number of seconds it takes for a sample of oil to flow through a capillary tube of a given size times a constant. The constant is dependent on the size of the capillary tube. A kinematic viscometer is used to measure the oil's centistokes. See Figure 10-3.

The capillary tube is placed in a heated bath of oil to hold the sample to a constant temperature of 104°F (40°C). Note that engine oil viscosities are measured at a higher temperature, at 212°F (100°C).

The weight and the temperature of an oil will affect its viscosity. A thick oil will have a higher cSt because it takes more time to pass through the capillary tube. A thinner oil will have a smaller cSt because it takes less time to pass through the capillary tube.

Viscosity has a tremendous effect on hydraulic pumps. Most pump manufacturers will provide a minimum and maximum viscosity specification. Figure 10-4 provides examples of recommended cSt based on the type of pump.

Saybolt Universal Seconds (SUS), or Saybolt Seconds Universal (SSU), is the United States alternative measure to cSt. It is also considered an older measure of viscosity. George Saybolt developed his viscometer in 1919 to measure a fluid's resistance to flow.

Figure 10-3. The kinematic viscometer uses a capillary tube that contains the oil being measured. The oil's centistoke (cSt) value equals the time it takes for the oil to drain through the capillary tube (in seconds) times its kinematic constant. A—A kinematic viscometer. B—Closeup of a capillary tube.

Type of Pump	Example of Recommended cSt
Vane	14 to 160 cSt
Piston	10 to 160 cSt
External Gear Pump	12 to 300 cSt

Figure 10-4. A pump manufacturer will specify a fluid viscosity range that is recommended for the pump based on a fluid operating temperature, such as 104°F (40°C).

The meter contains a sample of oil that is preheated to 100°F (38°C). At the bottom of the meter is a 0.0695″ orifice, called a universal orifice, which is used to fill a 60 milliliter (60 cm^3) container. A cork stopper is removed from the contained oil and the number of seconds it takes for the oil to pass through the orifice and fill the 60 milliliter container is known as the SUS or SSU.

Note

Another unit of measure, called Saybolt Furol Seconds, is used to measure high-viscosity liquids. With this scale, an orifice with a diameter of 0.1240″ (called a Furol orifice) is installed in place of the universal orifice.

Figure 10-5 illustrates examples of how SUS relates to cSt. Note that the oil viscosities must be determined at a fixed temperature, such as 100°F (38°C).

SUS and cSt are common viscosity ratings, however, there are other measures of viscosity. For example, dynamic (or absolute) viscosity is measured in units called centipoise. However, pump manufacturers commonly specify a recommended fluid viscosity in cSt or SUS.

ISO Viscosity Grade

The International Standards Organization (ISO) has established an ***ISO viscosity grade (ISO VG)*** for fluids. The ISO VG is a fluid's viscosity range based on a temperature of 104°F (40°C). The actual ISO VG number is an averaged viscosity value for that ISO range, which is illustrated in the left column of **Figure 10-6**. Notice that the viscosity range is given in centistokes. Sometimes equipment manufacturers recommend fluids using the ISO VG.

Viscosity Index

Temperature greatly affects a fluid's viscosity. Unfortunately, cold oil is thicker than warm oil, which causes concern when a machine is started in a cold environment. ***Viscosity index (VI)*** is a rating assigned to an oil to indicate how much or how little the oil's viscosity changes across a wide temperature spectrum. A fluid with a viscosity that has little change across the spectrum is considered to have a high VI, for example 200 VI. An example of a fluid with a lower VI rating is 130 VI. Customers will pay a higher price for a hydraulic fluid that has a higher VI rating.

Saybolt Seconds Universal	Equals Approximately	Centistokes
80 SUS	~	15.6 cSt
90 SUS	~	18.3 cSt
100 SUS	~	20.6 cSt
110 SUS	~	22.9 cSt
120 SUS	~	25.2 cSt
130 SUS	~	27.5 cSt
140 SUS	~	29.7 cSt
150 SUS	~	31.9 cSt
160 SUS	~	34.2 cSt
170 SUS	~	36.5 cSt
180 SUS	~	38.6 cSt
190 SUS	~	40.8 cSt
200 SUS	~	43.0 cSt

Figure 10-5. For a given temperature, the SUS will be higher than the cSt due to the difference in the testing devices.

ISO Viscosity Grade	Minimum cSt	Maximum cSt
2	1.98	2.42
3	2.88	3.52
5	4.14	5.06
7	6.12	7.48
10	9.00	11.0
15	13.5	16.5
22	19.8	24.2
32	28.8	35.2
46	41.4	50.6
68	61.2	74.8
100	90	110
150	135	165
220	198	242
320	288	352
460	414	506
680	612	748
1000	900	1100
1500	1350	1650

Figure 10-6. The ISO viscosity grade represents a viscosity range in centistokes, based on a fluid temperature of 104°F (40°C). The specific ISO VG is listed in the left column and is an averaged value for the range.

Machines that are started in cold climates and operated at high operating temperatures must have oil with a good VI rating. Excavators are examples of machines that require oil with a high VI. Many machine manufacturers sell different grades of oil, one with an adequate VI based on average climate temperatures, and a fluid with a higher VI rating for operating in a wider temperature range.

Another example of a machine operating in extreme temperature conditions would be a forklift picking up a load outside in extreme cold temperatures, then bringing it back inside a foundry that is casting molten iron. **Figure 10-7** illustrates two fluids that have different VI ratings.

Hydraulic oils labeled with a high VI value are sometimes lower VI oil to which VI improvers have been added. Caution should be taken when selecting an oil with a high VI. Mobile machines typically use a smaller reservoir, causing a slower rate of air separation. Aeration degrades VI improvers over time, which can cause problems if the base fluid has too low of a minimum-permissible viscosity value. Be sure to consult the manufacturer to ensure that the oil meets the recommended viscosity specification.

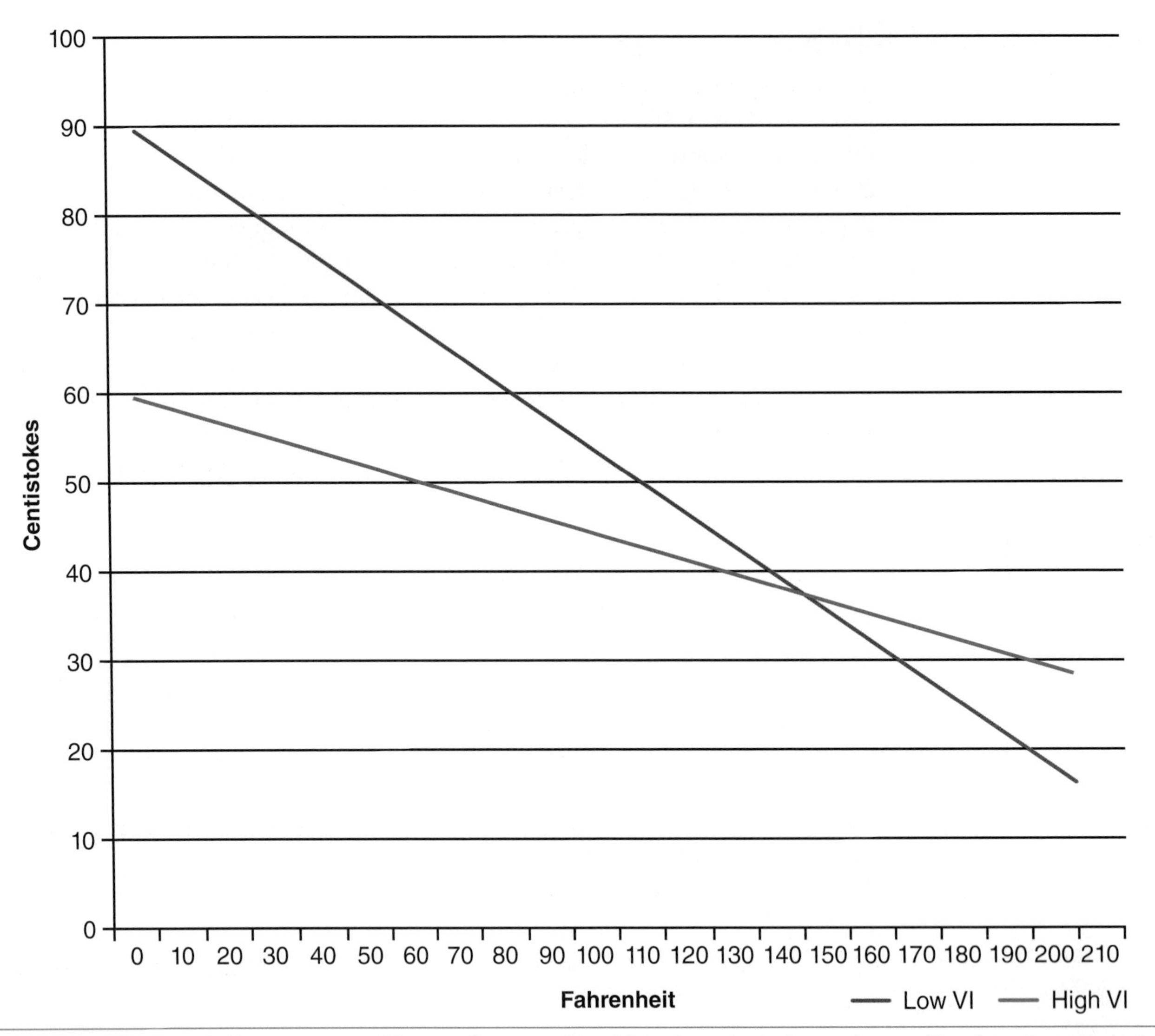

Figure 10-7. The graph illustrates two different oils, one with a high viscosity index and one with a low viscosity index. The best oil will have a flatter graph, meaning that its viscosity changes less over a wider temperature range. The more economical oil will have a steeper graph because its viscosity will change more across a wide temperature range.

SAE Viscosities

The Society of Automotive Engineers (SAE) provides fluid viscosity ranges for engine oils specific to summer and winter operation. SAE oil viscosities are based upon a 212°F (100°C) fluid temperature. Examples of summer weight oils are: SAE 50, SAE 40, SAE 30, or SAE 20. Winter grade oils have a W added to the number, for example SAE 10W.

If only one grade of oil is specified, the oil is considered a single grade or mono-grade. If the SAE oil has two weights listed, such as 5W30, the oil is considered a multi-grade or multi-weight oil.

Although it is rare, an occasional hydraulic oil specification may refer to SAE engine oil viscosity ranges. However, ISO VG ranges, straight cSt, or SUS viscosities values are much more common. See **Figure 10-8** for a comparison between ISO VG and SAE viscosities.

Pour Points

Manufacturers refer to another fluid property related to viscosity, called ***pour point***. A fluid's pour point is the lowest temperature at which the fluid will flow. A common rule of thumb is to use an oil with a pour point that is 20°F (11°C) below the coldest ambient temperature. Fluids can have pour point depressants added to improve their ability to flow at cold temperatures. The data sheet for an oil typically includes its pour point value.

A pump manufacturer is more likely to specify a start-up viscosity, and not pour points. See Figure 10-9. Following a pump manufacturer's maximum start-up viscosity recommendation will indirectly ensure that the pour point requirement is met. Care should be taken when starting a cold hydraulic system with a thick viscosity fluid. Even if the fluid meets the maximum start-up viscosity rating, the pump should not be operated at high rpm until the fluid is warmed.

Stability

In addition to viscosity, a hydraulic fluid should have good stability. A fluid with good stability will resist chemically reacting with other substances, such as oxygen and water. A fluid's resistance to reaction with oxygen is called ***oxidation stability***, and its resistance to reacting with water is known as ***hydrolytic stability***.

Oxidation is the chemical reaction that takes place when oxygen chemically combines with oil and its additives. Just as an apple turns brown after it is sliced and exposed to the atmosphere, hydraulic fluid will degrade when it chemically combines with oxygen.

Increases in operating temperature can greatly accelerate oxidation of the oil. For every 18°F (10°C) that the average fluid temperature rises above the recommended operating temperature, the fluid's oxidation rate will double. See Figure 10-10.

SUS 100°F	cSt 40°C	ISO VG	SAE Engine Oil	cSt 100°C	Sus 210°F
1150	220	220	50	19	98
780	150	150	40	15	80
500	100	100	30	11	65
220	46	46	20, 15W, or 20W	6.5	47

Figure 10-8. Most manufacturers use ISO viscosity grades for specifying fluid viscosities. This chart provides an approximation of the relationship between ISO VG and SAE oils.

Pump	Minimum Operating Viscosity	Optimum Operating Viscosity	Maximum Operating Viscosity	Maximum Start-up Viscosity
Pump XYZ	65 SUS 12 cSt	100–250 SUS 21–54 cSt	300 SUS 65 cSt	1000 SUS 216 cSt

Figure 10-9. An example of viscosity recommendations made by a pump manufacturer.

Operating Temperature	Recommended Fluid Replacement Interval
140°F (60°C)	1000 hours
158°F (70°C)	500 hours
176°F (80°C)	250 hours
194°F (90°C)	125 hours
212°F (100°C)	62 hours
230°F (110°C)	31 hours
248°F (120°C)	15 hours
266°F (130°C)	8 hours

Figure 10-10. Due to the oxidation of the fluid, the life of a fluid is cut in half for every 18°F (10°C) rise in operating temperature above the recommended threshold.

Preventing Oxidation

Steps can be taken to reduce the oil oxidation, such as running the hydraulic system at a lower operating temperature. Be sure that the cooler is not plugged and the cooler's fan is operating properly as well.

Equally important is to prevent fluid aeration. Petroleum-based hydraulic fluids at rest contain 8% to 12% dissolved air. When temperature rises or pressures fall, that air can form foam. Additional air can enter a hydraulic system through leaky fittings, leaking pump inlets, and leaky actuators. Air bubbles will form in the fluid if the hydraulic system is poorly designed and has a high amount of fluid turbulence. Using appropriately sized conductors with large-radius bends lowers turbulent oil flow. Ideally, designers prefer ***laminar oil flow,*** which occurs when fluid is flowing smoothly, but it is quite difficult to achieve. Designers compute the smoothness of fluid flow using the ***Reynolds number (Re)*** as the unit of measure.

Certain additives will help combat oxidation. For example, an additive may improve ***air separation ability (ASA),*** which is the ability of a hydraulic fluid to separate entrained air. Anti-foaming agents help speed the process of separating undissolved air within the hydraulic oil, which takes place inside the reservoir.

Rust and oxidation inhibitors are found in most hydraulic fluids and help prevent rust and oxidation. Some tractor supply stores sell hydraulic fluid labeled R&O, signifying the oil has rust and oxidation inhibitors. If oil becomes oxidized, it must be replaced by flushing the system and refilling it with clean, new fluid.

Note

Unfortunately mobile hydraulic systems use small reservoirs that do not help alleviate aeration problems. Companies have developed deaerators that can help improve the process of eliminating entrained air in hydraulic fluid. One type of deaerator uses centrifugal force to speed the separation of air from hydraulic fluid.

Anti-Wear

Today's mobile hydraulic systems that operate at pressures higher than 3000 psi (207 bar) require additives to help ensure a film of oil is maintained between moving components. A constant film is needed to reduce rubbing, wear, friction, heat, galling, and component seizure. The additives are called anti-wear (AW), wear resistant (WR), or extreme pressure (EP). It is possible to purchase a cheap hydraulic fluid that does not have the high-pressure additives. This type fluid, which offers minimal protection at high pressures, should be used only in a system that has a significantly less operating pressure than 3000 psi.

Anti-wear additives are especially important for ***boundary lubrication,*** a film that develops when two sliding components come into close contact with each other, to the point that tiny edges on their surfaces, known as asperities, can break-off and cause friction and heat. The desired condition is ***full film lubrication,*** which occurs when the two moving components have a complete film of oil separating their surfaces.

Demulsifiers

Due to poor maintenance and poor system design, moisture can find its way into hydraulic fluid. If the reservoir contains a significant amount of moisture, it can freeze, causing a pump to cavitate. Demulsifier additives help separate water from the oil in the reservoir so the water can be removed. Petroleum-based fluids are up to 14% lighter than water (a specific gravity of approximately 0.86), which causes separated water to fall to the bottom of the reservoir, where it can be drained.

Note that some detergents in oil will keep the water and other contaminants suspended within the fluid. In this scenario, it is the role of the filters to remove the contaminants.

Fluid Types

A hydraulic system can use one of several different types of fluids:

- Petroleum-based fluids.
- Synthetics fluids.
- Fire-resistant fluids.
- Environmentally acceptable fluids.

Petroleum Fluids

Petroleum fluids, also known as mineral fluids or hydrocarbon-based fluids, are the most common type of fluid used in mobile machinery. The fluid is obtained by refining crude oil. Oil additives, such as R&O inhibitors, AW agents, anti-foaming agents, and VI improvers are blended into the oil.

Petroleum-based fluids can be further classified as paraffin-based oil or naphthene-based oil. The major difference between the two is the hydrocarbon structure. See **Figure 10-11** for a comparison between paraffin- and naphthene-based oils. Both fluids can be made more stable by adding hydrogen to the composition, making them hydrogenated fluids.

Petroleum-Based Oil	Attributes
Paraffin	Long, pipe-like molecular structures.
	More stable (greater resistance to heat stress/ deterioration).
	Sensitive to cold temperatures, causing wax deposits when cold.
	Has solubility problems when mixed with other fluids, such as synthetics.
	More economical.
Naphthene	Ring or ball-like molecular structures.
	Slightly shorter life expectancy.
	Higher costs than paraffin.
	Less stable.
	Used in systems exposed to cold temperatures.
	Used when there is concern about solubility problems with mixed fluids, because it has a higher fluid solvency.

Figure 10-11. The table provides a comparison on paraffin-based oils and naphthene-based oils.

Petroleum fluids offer good lubricity, good stability, good bulk modulus, and are more economical than synthetic fluids. Two of the largest drawbacks to petroleum fluids are toxicity and the potential for environmental contamination. The ISO uses the following labels for petroleum-based hydraulic fluids: HL, HLP, HLPD, HVLP, and HLPD.

Synthetic Fluids

Synthetic fluids are designed by engineers, meaning that they are not simply refined from crude oil. Synthetics are stable and have good lubricity. They are desirable when the hydraulic system is operating at low or high temperatures and are also good for higher operating pressures. Synthetics cost more than petroleum-based fluids. They too can be toxic to the environment, but they can also be designed to be environmentally friendly. Some synthetic fluids are not compatible with certain types of seals.

Fire-Resistant Fluids

Hydraulic fluids have a flash point rating and a fire point rating. ***Flash point*** is the temperature at which the oil will temporarily ignite when exposed to an ignition source, such as a flame. Petroleum-based hydraulic fluid flash points can range from approximately 340°F (170°C) to 500°F (260°C) range. ***Fire point*** is the temperature at which the fluid can sustain a continued fire after the ignition source as been removed. Fire point is normally 30°F to 50°F higher than the flash point.

Some industries require ***fire-resistant hydraulic fluids***. These fluids are denoted by the ISO with a HF prefix, and have a third letter that represents, in general, the quantity of water contained in the fluid, such as HFA, HFB, HFC, and HFD. See Figure 10-12. The fire-resistant fluid classification can also contain a fourth letter, for example to distinguish between water emulsions (HFAE) and synthetics (HFAS).

The first HFD fluids used phosphate esters, but they are used less today, due to environmental concerns, price, and reduced compatibility. HFD fluids with phosphate esters are still used in aircraft and some power generation systems. HFD fluids today are more likely to contain polyol esters because they are more compatible with seals and are more environmentally acceptable.

Fire-resistant fluids are necessary in some industrial environments, especially in foundries with molten iron. Traditional petroleum-based fluids are more likely to burn than HF fluids in the event of a fire. HF fluids are less common in the mobile hydraulic industry.

Environmentally Acceptable Hydraulic Fluids

Some customers request ***biodegradable fluids***, which are less toxic and less harmful to the environment. A biodegradable fluid will degrade much faster than traditional hydraulic fluids. The ISO labels these fluids "environmentally acceptable hydraulic fluids" and uses the following letter classifications:

- HEPG.
- HEES—partially saturated.
- HEES—saturated.
- HEPR.
- HETG.

In order for a hydraulic fluid to be classified as an ISO environmentally acceptable hydraulic fluid, 60% of the oil must be biodegradable or the fluid must meet toxicity requirements related to fish toxicity, daphnia toxicity, or bacteria toxicity.

ISO Label	Subclassification	Structure	Water Content (%)
HFA	HFAE	Oil in water emulsions	80% or more Most are 95%
HFA	HFAS	Synthetic aqueous fluids	80% or more Most are 95%
HFB		Water-in-oil (also known as invert emulsion)	40% or more
HFC		Water polymer solutions (also known as water-glycol)	35% or more
HFD	HFDR	Phosphate esters	Zero
HFD	HFDU	Synthetic anhydrous fluids	Zero

Figure 10-12. Fire-resistant fluids are designated by the ISO with the letters HF. HFA fluids have the most water content, HFB and HFC fluids have less water content, and HFD fluids have no water content.

Vegetable-based oils and some synthetic fluids are biodegradable. Some biodegradable fluids are also fire resistant. Vegetable-based fluids can be made from rapeseed (also known as canola oil), sunflower oil, coconut oil, or soybean oil. Fluid manufacturers have used rapeseed and soybean oils more than the other vegetable oils.

Manufacturers have been making improvements in vegetable oils, which is critical considering the high costs of the alternative, synthetic biodegradable fluids. Vegetable fluids are more expensive than petroleum-based fluids, but are cheaper than synthetic biodegradable fluids. Petroleum-based fluids have better hydrolytic stability, oxidation stability, can operate at higher temperatures, and perform better in colder temperatures than vegetable-based fluids.

Specific Gravity

Hydraulic fluids have different specific gravities. As mentioned in **Chapter 2**, specific gravity is the ratio between the weight of the fluid and the weight of water. A number less than 1 means the fluid is lighter than water; a number greater than 1 means the fluid is heavier than water. See Figure 10-13. Notice that petroleum-based fluids are the lightest oils, meaning that they will float on top of water.

A heavier hydraulic fluid is more susceptible to pump cavitation if the reservoir is located below the pump's inlet. Some manufacturers might specify that the system be designed with a flooded pump inlet when using a fluid with a high specific gravity. A ***flooded pump inlet*** is a pump inlet that is located below the reservoir, which ensures that gravity is helping feed fluid into the inlet.

Type of Fluid	Specific Gravity
Petroleum oils	0.86 to 0.90
Synthetic ester	0.92 to 0.926
Rapeseed oil	0.92
Water	1.00
HFC	1.08
HFD	1.13

Figure 10-13. Hydraulic fluids have different weights, which are specified by their specific gravities. Specific gravity is the ratio of the weight of a substance compared to the weight of water.

Summary

- ✓ Hydraulic fluids transmit power, help remove heat and contaminants, seal oil clearances and lubricate components.
- ✓ Bulk modulus is a measure of a fluid's ability to resist compression.
- ✓ Viscosity is measured in centistokes (cSt) and Saybolt Universal Seconds (SUS). Both units indicate how many seconds it takes a specified volume of oil to pass through a capillary tube or orifice.
- ✓ A high cSt or SUS indicates a thick oil.
- ✓ Temperature greatly affects viscosity.
- ✓ A good hydraulic fluid will:
  - Have a high bulk modulus.
  - Prevent rust and resist oxidation.
  - Allow air to easily separate.
  - Allow water to easily separate.
  - Have a high VI.
  - Have a pour point 20°F below the ambient temperature.
- ✓ A high VI has less change in viscosity over a broad temperature range.
- ✓ Fluid with a high VI costs more than fluid with a lower VI.
- ✓ ISO VG is a fluid viscosity range, and the actual viscosity number is an averaged value within that range, in units of cSt.
- ✓ The ISO labels petroleum-based hydraulic fluids as HL, HLP, HLPD, HVLP, and HLPD.
- ✓ The ISO labels environmentally acceptable hydraulic fluids as HEPG, HEES, HEPR and HETG.
- ✓ An oil must meet toxicity specifications or be 60% biodegradable in order to be classified by the ISO as an environmentally acceptable hydraulic fluid.
- ✓ The ISO labels fire-resistant hydraulic fluids as HFAE, HFAS, HFB, HFC, and HFD. HF indicates that the fluid is fire resistant, and the third letter indicates the percentage of water content.
- ✓ Petroleum-based hydraulic fluids are the most economic, provide good lubricity, have good bulk modulus, are stable, but are more harmful to the environment.
- ✓ Some environmentally acceptable fluids are fire resistant.

Technical Terms

air separation ability (ASA)
biodegradable fluids
boundary lubrication
centistokes (cSt)
compressibility
fire point
fire-resistant hydraulic fluids
flash point
flooded pump inlet
full film lubrication
galling
hydrolytic stability
ISO viscosity grade (ISO VG)
laminar oil flow
lubricity
oxidation stability
petroleum fluids
pour point
Reynolds number (Re)
rust and oxidation inhibitors
Saybolt Universal Seconds (SUS)
synthetic fluids
viscosity index (VI)

Review Questions

Answer the following questions using the information provided in this chapter.

1. Which of the following fluid properties can be improved with the use of additives?
   A. Viscosity.
   B. Bulk modulus.
   C. Both A and B.
   D. Neither A nor B.
2. A petroleum-based hydraulic fluid will compress approximately _____ at 1000 psi (69 bar)?
   A. 0.5%
   B. 1.0%
   C. 1.5%
   D. 2.0%
3. Which of the following is considered the opposite of bulk modulus?
   A. Anti-wear.
   B. Oxidation stability.
   C. Compressibility.
   D. Viscosity index.
4. All of the following will affect bulk modulus, *EXCEPT*:
   A. air trapped in the fluid.
   B. fluid temperature.
   C. type of fluid.
   D. viscosity index.
5. When attempting to minimize bulk modulus, all of the following should be performed, *EXCEPT*:
   A. eliminate air intrusion into the oil.
   B. avoid overheating the oil.
   C. minimize the fluid volume between the DCV and the actuator.
   D. place the reservoir below the pump's elevation.
6. Technician A states that bulk modulus is measured in units of pressure. Technician B states that compressibility is measured in negative units of measurement. Who is correct?
   A. Technician A.
   B. Technician B.
   C. Both A and B.
   D. Neither A nor B.
7. All of the following are functions of a hydraulic fluid, *EXCEPT*:
   A. lubricate parts.
   B. reduce noise.
   C. transmit power.
   D. seal clearances.
8. A lack of lubricity can cause all of the following, *EXCEPT*:
   A. galling.
   B. component seizure.
   C. heat reduction.
   D. increased wear.
9. A fluid with good lubricity must be able to perform all of the following, *EXCEPT*:
   A. provide a film of oil separating moving parts.
   B. maintain oil film on components during high operating pressures.
   C. stick to individual components.
   D. increase fluid pressure.
10. At what temperature are ISO VG ranges determined?
    A. 40°C
    B. 40°F
    C. 80°F
    D. 210°F
11. ISO VG ranges are based on what unit of measurement?
    A. Centipoise.
    B. Centistoke.
    C. Saybolt Universal Second.
    D. Viscosity index.
12. Which of the following units of measurement uses a capillary tube for measuring viscosity?
    A. Centipoise.
    B. Centistoke.
    C. Saybolt Universal Second.
    D. Viscosity index.

13. Which of the following units of measurement uses a standard sized orifice and a 60 milliliter container of oil for measuring viscosity?
    A. Centipoise.
    B. Centistoke.
    C. Saybolt Universal Second.
    D. Viscosity index.
14. Which of the following oils is considered the thickest oil?
    A. 10 cSt.
    B. 10 SUS.
    C. 20 cSt.
    D. 20 SUS.
15. Hydraulic oil viscosities are commonly determined at what temperature?
    A. 38-40°C.
    B. 40°F.
    C. 140°F.
    D. 210°F.
16. An ISO VG value represents which one of the following?
    A. A specific viscosity.
    B. The minimum viscosity value of a viscosity range.
    C. An averaged value for the viscosity range.
    D. The maximum viscosity value of a viscosity range.
17. What is the minimum operating pressure that requires the inclusion of anti-wear additives in the hydraulic fluid?
    A. 1000 psi.
    B. 2000 psi.
    C. 3000 psi.
    D. 6000 psi.
18. All of the following affect oil oxidation, *EXCEPT*:
    A. bulk modulus.
    B. aeration.
    C. temperature.
    D. fitting leaks.
19. Which of the following oils provides the least change in viscosity over a wide temperature range?
    A. 100 VI.
    B. 140 VI.
    C. 160 VI.
    D. 200 VI.
20. Which of the following hydraulic fluids exhibits the most change in viscosity when exposed to a wide range of temperatures?
    A. 100 VI.
    B. 140 VI.
    C. 160 VI.
    D. 200 VI.
21. An excavator will be started at a temperature of 0°F. What is the recommended pour point of the fluid?
    A. 10°F.
    B. 20°F.
    C. –10°F.
    D. –20°F.
22. Which of the following types of oils can have an ISO classification of HL, HLP, HLPD, HVLP, and HLPD?
    A. Petroleum based.
    B. Fire resistant.
    C. Environmentally acceptable.
    D. All of the above.
23. Which of the following types of oils can have an ISO classification of HEPG, HEES, HEPR, and HETG?
    A. Petroleum based.
    B. Fire resistant.
    C. Environmentally acceptable.
    D. All of the above.
24. Which of the following types of oils can have an ISO classification of HFAS, HFB, HFC, and HFD?
    A. Petroleum based.
    B. Fire resistant.
    C. Environmentally acceptable.
    D. All of the above.

25. Which of the following hydraulic fluids has the most water content?
    A. HFAS.
    B. HFB.
    C. HFC.
    D. HFD.
26. A hydraulic fluid classified as an ISO HFAE, has a large amount of what fluid?
    A. Vegetable oil.
    B. Water.
    C. Synthetic fluid.
    D. Petroleum oil.
27. What is the minimum amount of biodegradability allowable in a hydraulic oil classified as an ISO environmentally acceptable fluid?
    A. 60%.
    B. 70%.
    C. 80%.
    D. 100%.
28. Which of the following fluids is the most sensitive to cold temperatures and has a limited upper temperature threshold?
    A. Vegetable-based hydraulic oil.
    B. Petroleum-based hydraulic oil.
    C. Synthetic-based hydraulic oil.
29. Which of the following liquids is more susceptible to pump cavitation if the system does not have a flooded pump inlet?
    A. Petroleum oil.
    B. Synthetic ester.
    C. Rapeseed oil.
    D. HFD fluid.
30. Which of the following is the lightest fluid?
    A. Petroleum oil.
    B. Rapeseed oil.
    C. Water.
    D. HFD fluid.

Chapter 11

Filtration

Objectives

After studying this chapter, you will be able to:

- ✓ Explain the uses and disadvantages of suction strainers.
- ✓ Describe the difference between surface filtration and depth filtration.
- ✓ Explain the difference between a filter and a strainer.
- ✓ List the advantages and disadvantages of breathers, suction filters, return filters, and high-pressure filters.
- ✓ Explain the difference between proportional flow and full flow filtration.
- ✓ Explain the operation of a filter bypass indicator.
- ✓ List factors to consider when choosing a hydraulic filtration method.

Introduction to Filtration

Mobile machinery can use many different types of devices to remove contaminants. The devices can be classified based on their construction, such as a strainer or a filter. ***Strainers,*** also called screens, are designed to remove large particles. ***Filters*** are designed to remove much smaller particles, so small that the human eye cannot see the contaminant.

The devices can also be labeled based on their location in the circuit, for example, a suction strainer or a return filter. Filtration design will differ between manufacturers, and may be different even among different machines from one manufacturer. The different styles of filtration designs have advantages and disadvantages. Figure 11-1 shows multiple locations where filters and strainers can be placed in a hydraulic circuit. As mentioned in **Chapter 3**, a filtration symbol is the same for a filter and a strainer.

Strainers

Strainers are constructed of wire mesh screens. See Figure 11-2. They are unable to remove fine particles. Strainers are classified as surface-type filtering devices. ***Surface-type filtering devices*** direct oil through a single thin surface layer. See Figure 11-3. The surface media of a strainer is comprised of a woven wire mesh.

Most strainers are placed inside the hydraulic reservoir. The strainers are also known as suction strainers or suction screens. They are used to prevent any type of large particles inside the reservoir from entering the pump's inlet.

Suction screens are configured as either an in-tank design or a side-mounted design. An in-tank design requires a reservoir to be fitted with

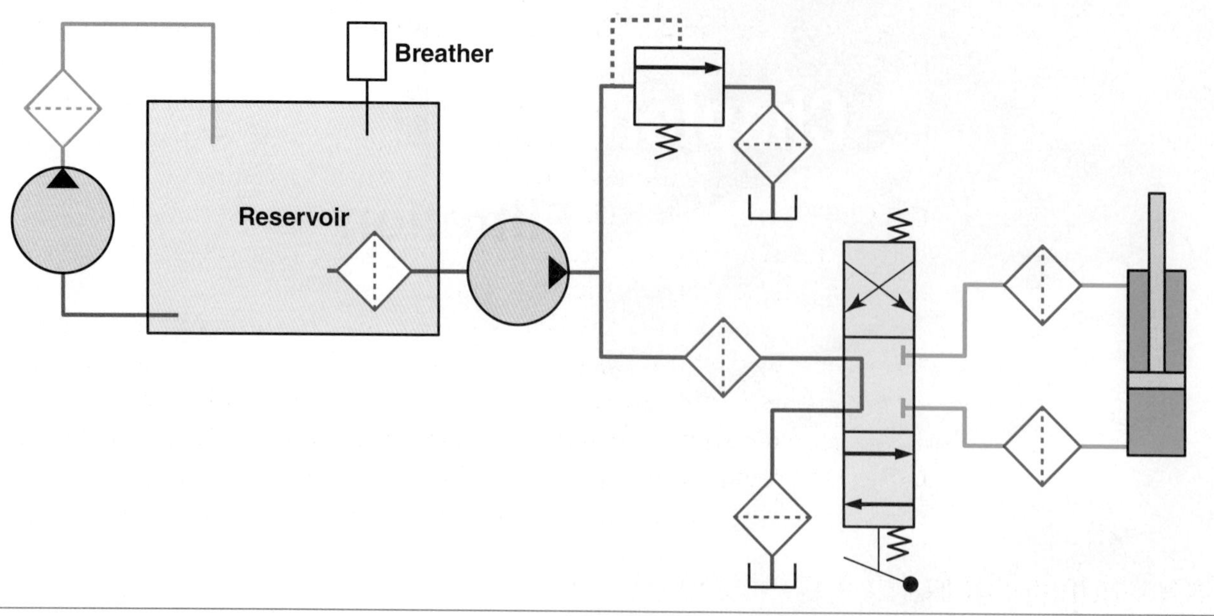

Figure 11-1. This schematic illustrates the numerous different locations where filters and strainers can be placed to remove contamination. It is unrealistic to see a machine with this amount of filtration.

Figure 11-2. A strainer consists of a wire mesh screen, which is designed to remove the large contaminants that are visible to the human eye.

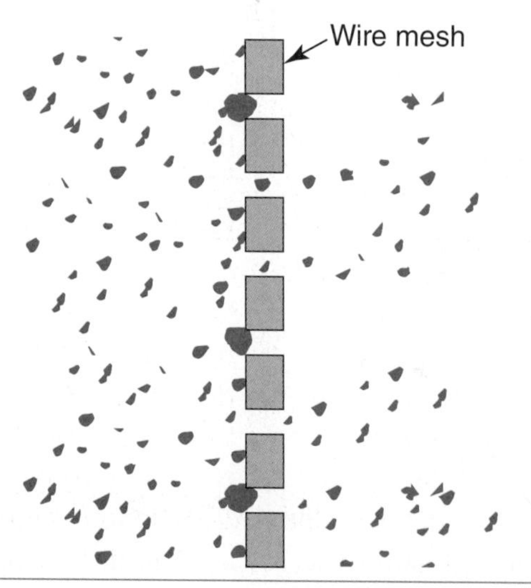

Figure 11-3. Surface filtering devices force oil through a single thin layer.

a removable side plate. The side plate provides access to remove and replace the in-tank suction screen.

An access panel is not needed to remove a side-mounted suction screen because the strainer threads directly into the side of the reservoir. See **Figure 11-4**. Side-mounted suction strainers commonly use pipe threads. **Chapter 15** will explain the disadvantages of using pipe threads. A threaded suction hose can be awkward to remove, potentially making the in-tank screen design a little less of a burden to replace.

Some suction screens can be located outside of the reservoir. See **Figure 11-5**. These strainers require their own housing. If the reservoir does not have an access panel or if the reservoir's suction port is too small for a side-mounted suction screen, such an external suction screen can be added. The housing is fabricated from common plumbing fittings purchased from a local hardware store. It is placed outside the reservoir, between the reservoir and the pump.

Disadvantages of Suction Strainers

Although suction strainers are relatively inexpensive, their disadvantages far outweigh their economy. Their disadvantages include:

- Usually no indication when it is plugged.
- Cumbersome to replace.
- Often contains no bypass, which causes pump cavitation when the screen is plugged.

Most suction screens have no provision for letting an operator know that the screen has become plugged and needs to be replaced. As a result, a plugged suction screen will cause the pump to cavitate and fail, costing the owner substantial repair costs.

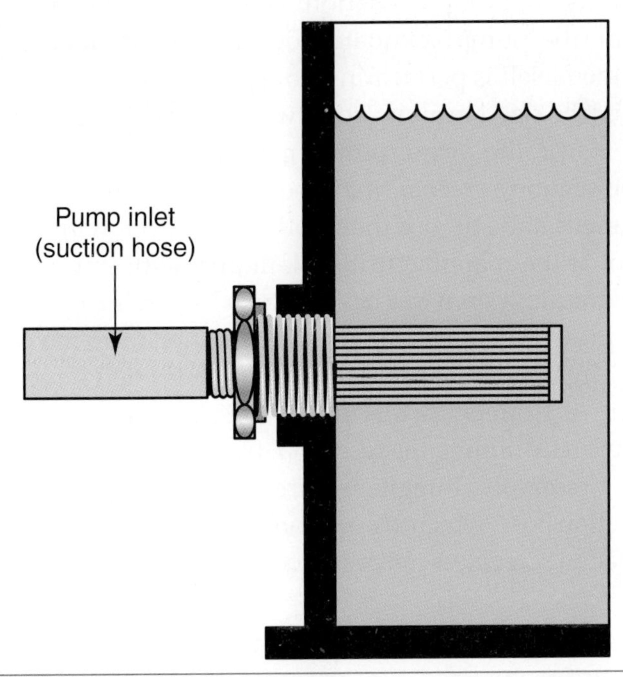

Figure 11-4. Side-mounted suction strainers thread into a port on the side of a hydraulic reservoir. The suction hose threads into the suction strainer.

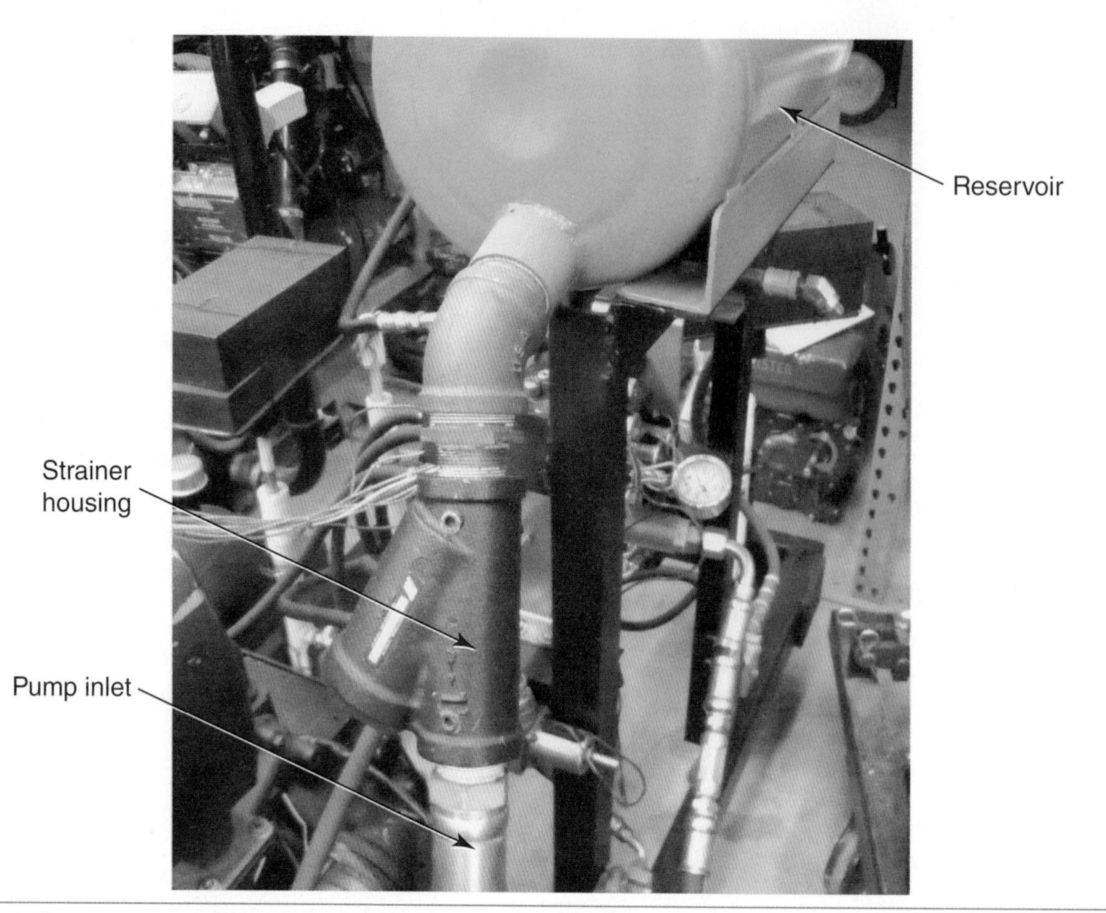

Figure 11-5. Some suction strainers can be placed outside of the reservoir, between the reservoir and the pump's inlet.

It is possible to incorporate a bypass valve with the suction screen to lower the risk of pump cavitation. However, the bypass valve must be rated to provide the pump adequate flow. When the bypass valve is open, the suction screen itself is performing no useful function. See Figure 11-6. Suction screens must be sized to allow flow two to four times greater than the pump's flow rate.

Over the years, pump manufacturers have provided recommendations for suction screens. Many equipment manufacturers have also used suction screens and still use them today. However, with improved contamination control at the manufacturing plant and with improved maintenance practices, a hydraulic system can have a much longer life by eliminating the suction screen in order to reduce the potential for cavitating the pump.

In addition to not knowing if the suction screen is plugged, another downside of strainers is that personnel sometimes avoid replacing them because it requires draining the reservoir. In addition, if the screen is threaded into the side of a reservoir, it might require unthreading a long and cumbersome hydraulic suction hose, which is one more reason personnel sometimes avoid replacing the suction screen. In contrast, a common oil filter is quick and easy to replace.

Strainer Ratings

Suction screens are commonly rated with a wire mesh number and a micron rating. A ***micron (µm)*** equals one millionth of a meter and is also known as a micrometer. A micron is equal to 0.0000394", or roughly 0.00004". The micron rating indicates the screen's largest opening.

A strainer's ***wire mesh number*** is equal to one-half of the total number of wires in each square inch of the strainer. For example, every square inch of a 60 wire mesh screen contains 60 lateral wires and 60 vertical wires, leaving openings between the wires that are 250μm by 250μm. See **Figure 11-7**.

Figure 11-8 lists examples of strainer wire mesh numbers and their corresponding openings, given in both microns and ten-thousandths of an inch. Notice that a larger wire mesh number equals a smaller opening within the strainer.

Suction strainers are also specified by their port size, their flow capacity, and whether or not they have a bypass valve. An example of a strainer bypass value is 3 psi, allowing the oil to bypass the screen at just 3 psi. Strainers are designed to flow oil with as little as 0.5 to 0.7 psi pressure drop. Flow capacities vary among strainers. For example, it is possible to find 100 mesh/149μm strainers with flow capacities that range from 5 gpm to 100 gpm.

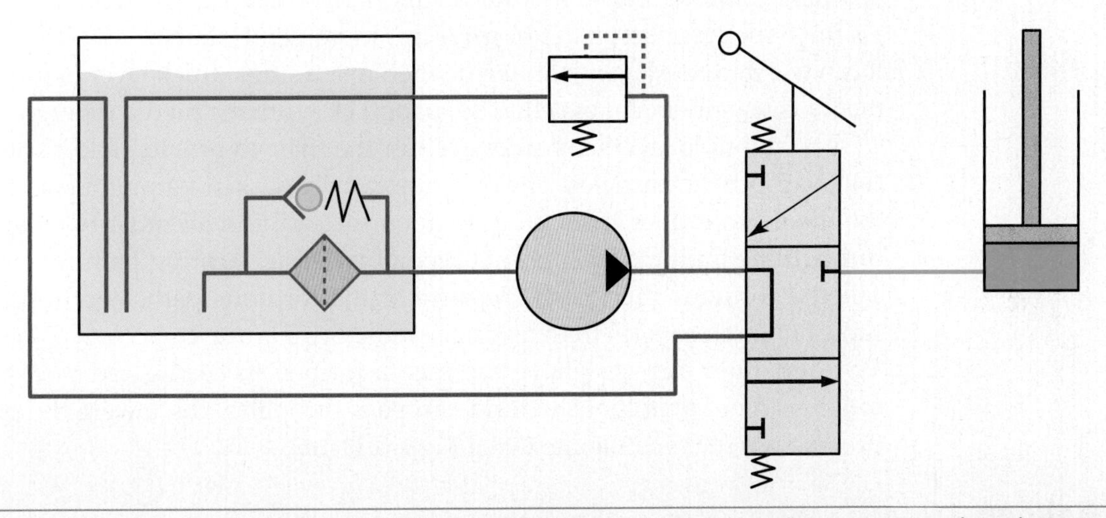

Figure 11-6. If an owner is adamant about using a suction screen, be sure that the screen contains a bypass valve that is large enough to provide a free flow of oil to the pump's inlet when the screen becomes plugged. The suction screen located inside this reservoir contains a bypass valve to prevent cavitation.

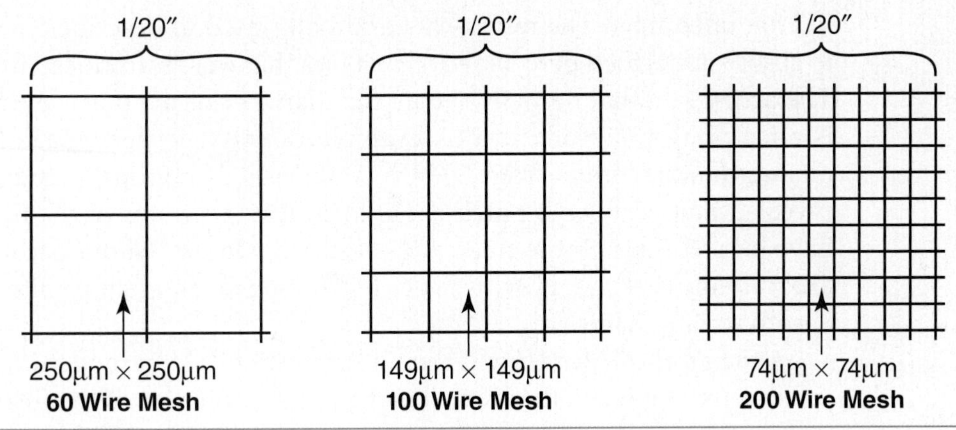

Figure 11-7. Strainers are specified by one-half of the total number of wires contained in a square inch. The number of wires determines the size of the openings between the wires. A 60 wire mesh screen has 250μm openings between the crossing wires. A 100 wire mesh screen has 149μm openings, and a 200 wire mesh screen has 74μm openings.

Wire Mesh Number	Opening (Micron "µm" rating)	Opening (Inches)
30	595µm	0.0232″
60	250µm	0.0098″
100	149µm	0.0059″
200	74µm	0.0029″
325	44µm	0.0017″

Figure 11-8. Common strainers are 100 and 200 mesh screens, with a corresponding 149µm and 74µm opening.

Filters

Filters are used in mobile machinery to remove smaller particles that strainers cannot remove. Hydraulic filters are often classified as depth filters. As the term describes, a ***depth filter*** has more depth than a surface-type filter. It can use multiple layers of filtering media or a single thick layer of media. The media is usually made of cellulose (paper) or synthetic media (fiberglass).

The multiple layers are interlaced in the filter to provide additional material to absorb the particles. The oil cannot easily pass in a straight path through the filter like it does through a strainer. Instead, the oil must twist and turn through the multiple layers. This type of twisting, turning path is sometimes called a tortuous fluid path. Because of the tortuous path, depth filters will capture more particles than a surface filter. See Figure 11-9.

Some filter elements have the media formed in corrugated pleats, which provides more area that the fluid can pass through. This lowers the pressure drop across the filter element. See Figure 11-10.

Filter Ratings

Filters have used several different rating systems over the years, such as nominal, absolute, and the ISO Beta rating. A filter's ***nominal rating*** is given by the filter manufacturer and is specified in microns, such as a 10 nominal micron rated filter. The rating was originally used to describe a filter's ability to absorb a specified percentage of particles that were introduced into the filter. The actual nominal number equals the diameter of the particle that the filter is attempting to absorb, in this example, 10µm. The percentage has ranged among manufacturers from 60% to 98%. Tests of nominal rated filters have revealed that very large particles, 10 to 20 times the size of the filter's rating, have passed through the filter. As a result, a nominal rating is considered an arbitrary number that provides very little value for assessing a filter's ability to absorb contaminants.

A filter's ***absolute rating*** is the size of the filter's largest opening, measured in microns. Stated another way, it is the largest pore size of the filter's element. Although this rating is an improvement over a nominal filter rating, it too is assigned by the filter manufacturer and does not require the manufacturer to complete a standardized filter efficiency test.

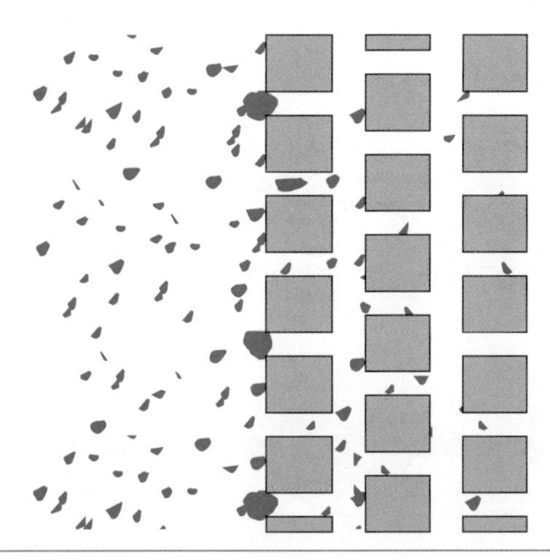

Figure 11-9. Depth filters sometimes use multiple layers of media to absorb contaminants.

Figure 11-10. A filter element can be formed in corrugated pleats to increase the surface area of the filtering media.

The ISO has developed a much better method for assessing a filter's ability to absorb contaminants. The test is known as the ISO 16889 multi-pass test and is essentially a filter efficiency test. The ISO test injects a known quantity of standard-sized contaminants upstream of the filter being tested. The test is completed when the filter becomes full, as noted on the differential pressure gauge. A particle counter is then used to measure the number of contaminants that were allowed to pass through the filter. See **Figure 11-11**.

Based on how the filter performs, it is given a Beta rating, also known as a ***Beta ratio*** and denoted with the symbol R or β. The β ratio equals the number of particles injected upstream compared to the number of particles that were able to pass through the filter and made it downstream.

A filter that has received a β ratio of $\beta_5 = 10$ had contaminants that were 5μm and larger injected upstream of the filter. The filter's ratio indicates that for every 10 particles that were injected upstream, the filter allowed one particle to pass downstream.

A filter's efficiency can be computed by using the following formula:

$$\text{Filter Efficiency} = [(\beta - 1) \div \beta] \times 100$$

A filter with a Beta ratio of 10 would equal the following efficiency:

$$\text{Filter Efficiency} = [(10 - 1) \div 10] \times 100 = 90\% \text{ efficient filter}$$

The correlation between Beta ratio and efficiency can be seen in **Figure 11-12**.

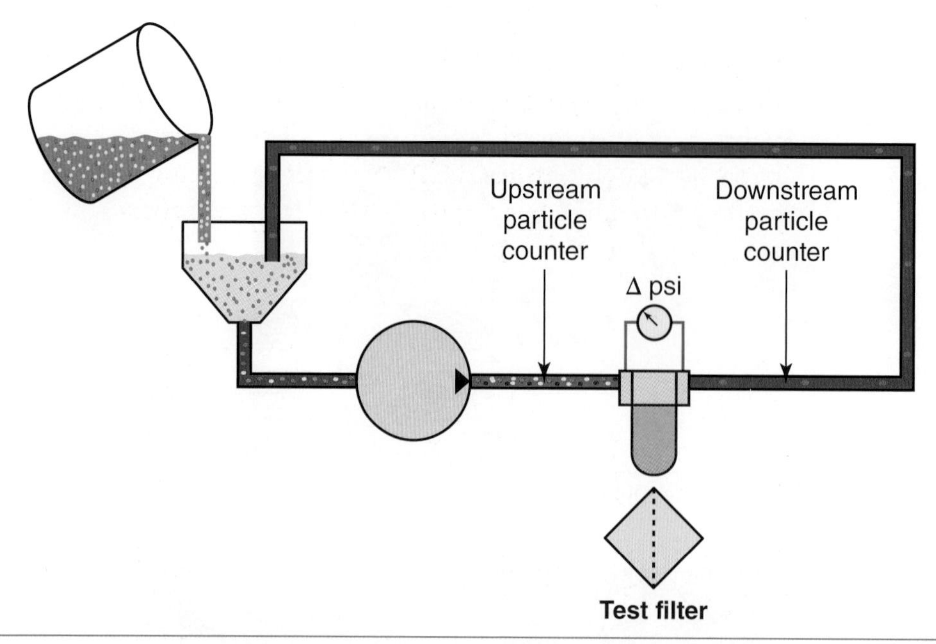

Figure 11-11. The ISO multi-pass filter test measures contaminants upstream and downstream from the filter. A Beta ratio is assigned to the filter, providing an indication of the filter's ability to absorb contaminants.

Beta Ratio	Filter Efficiency	Example Number of Particles Upstream	Example Number of Particles Downstream
1	00.0%	10,000	10,000
2	50.0%	10,000	5000
4	75.0%	10,000	2500
10	90.0%	10,000	1000
20	95.0%	10,000	500
40	97.5%	10,000	250
60	98.3%	10,000	166
75	98.6%	10,000	133
100	99.0%	10,000	100
200	99.5%	10,000	50
1000	99.9%	10,000	10

Figure 11-12. Beta ratios are established based on how well a filter captures the contaminants. The ratio equals the number of particles upstream of the filter, divided by the number of particles measured downstream from the filter.

Filters are also rated in pressures and flows. The filter's ***element collapse pressure*** is the pressure at which the internal media collapses. When the media collapses, the filter loses its ability to retain contaminants, and the collapsed media can also contaminate the system. Some manufacturers provide a working pressure, which is the equivalent of the system's operating pressure. Filter manufacturers may also list a burst pressure for spin-on filters. The burst pressure is the pressure at which the filter will deform to the point of failure. See **Figure 11-13**.

A ***filter bypass valve*** is often used to prevent the filter from collapsing or bursting. When a filter becomes full, clogged, or has difficulty flowing oil, the differential pressure increases across the filter assembly, causing the bypass valve to open. This allows oil to freely flow through the filter assembly. If a bypass valve is not installed or is malfunctioning, it is possible for the filter assembly to burst. See **Figure 11-14**. When the bypass valve is seated, the filter is able to filter the circuit's full flow.

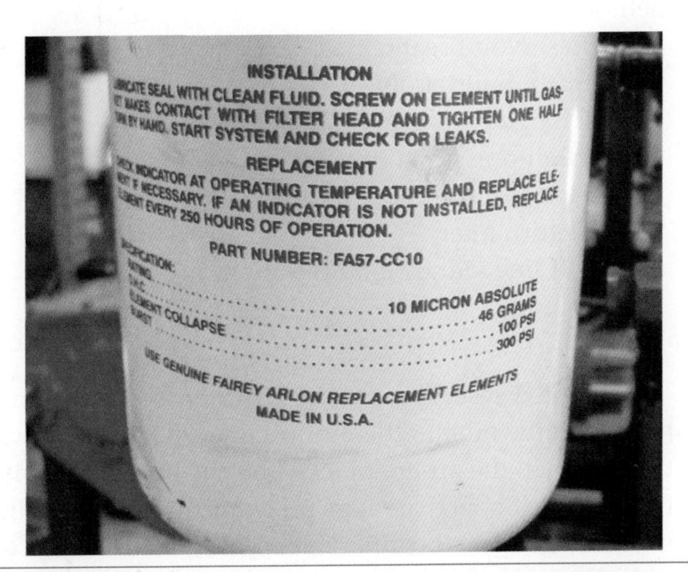

Figure 11-13. This filter has a 100 psi element collapse pressure rating and a 300 psi burst pressure rating.

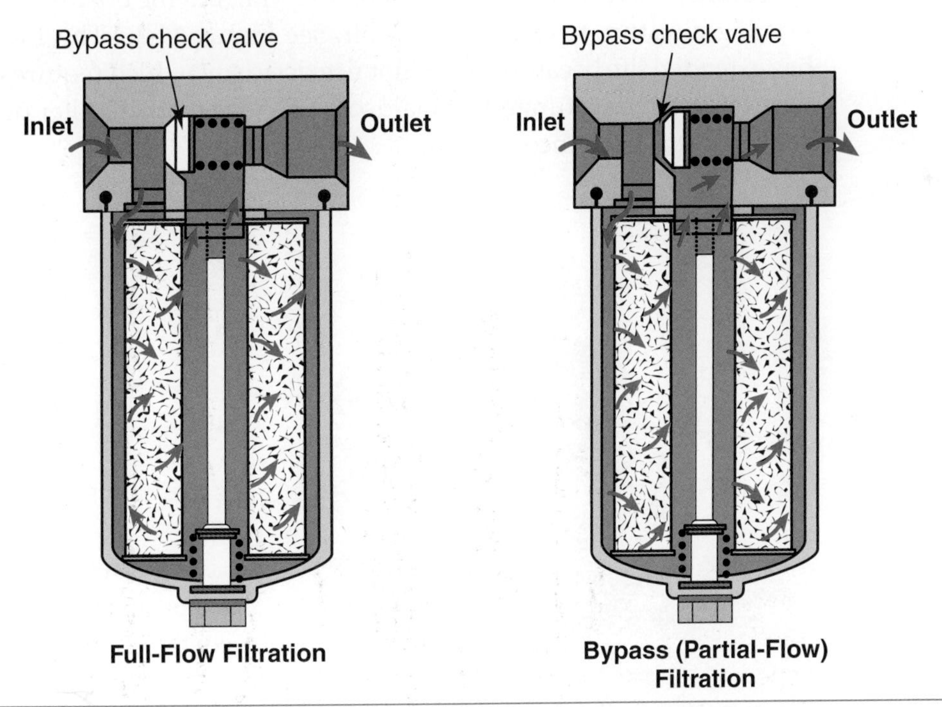

Figure 11-14. Filter elements are threaded onto a filter head assembly that usually contains a bypass valve. The bypass works on a pressure differential. If the filter becomes clogged or the fluid viscosity is too thick, the bypass valve opens. This allows the oil to bypass the filter element.

Filters are also sized by their flow capacity. Flow surges must be considered when adding a filter to a hydraulic system. The effect of flow surges on filter sizing will be discussed later in this chapter.

Filtration by Location

As illustrated in Figure 11-1, equipment manufacturers can choose to place filters in a variety of different locations in the hydraulic system. The filtration methods are classified as:

- Reservoir breather.
- Suction filtration.
- Charge pressure filtration.
- High-pressure filtration.
- Actuator filtration.
- Return filter.
- Off-line kidney loop filtration.

Breathers

A vented reservoir uses a ***breather*** to allow atmosphere into the tank to help push the oil into the pump's inlet. As air is drawn into the reservoir, contaminants in the air are blocked by the breather media. The breather is one of the most cost-effective filters on a machine, considering how small of particles it can filter. Some breathers prevent moisture from entering the reservoir as well. Donaldson manufactures a T.R.A.P.™ breather, which stands for thermally reactive advanced protection. The T.R.A.P. breather is not only designed to prevent moisture from entering the reservoir, but the breather will also allow existing moisture to exit the reservoir. See Figure 11-15. The T.R.A.P. breather has extended life because it does not use desiccant, which requires replacement when it becomes saturated. Donaldson T.R.A.P. breathers filter out particles as small as 3μm from incoming air.

Figure 11-15. A Donaldson T.R.A.P.™ breather prevents moisture and other contaminants from entering the reservoir.

Suction Filters

A *suction filter* protects the pump's inlet. It is similar to a suction screen, but protects the pump from smaller particles. Some hydrostatic transmissions use a suction filter to filter the oil supplied to the transmission's charge pump inlet. See **Figure 11-16**.

In addition to filtering smaller particles, the suction filter is easier to replace than a suction screen. The element is spun off and onto the filter head assembly. **Chapter 14** will explain methods for placing a vacuum on a reservoir, which might be required to prevent oil from leaking out the suction filter head during replacement.

Like a suction screen, a suction filter can cause a pump to cavitate if it becomes plugged and has no bypass valve. However, a suction filter is more likely than a suction screen to use a bypass valve. A common suction filter bypass valve setting is 2 to 3 psi.

Charge Pressure Filters

A gear pump is sometimes used to supercharge a piston pump in mobile equipment implement systems, power shift transmissions, and hydrostatic transmissions. Charge pumps were briefly mentioned in **Chapter 4** and will be further discussed in **Chapter 23**. Some manufacturers choose to use a ***charge pressure filter*** to protect components located downstream from the charge pump. An example operating range of charge pressure is 240 psi to 350 psi (16 bar to 24 bar). See **Figure 11-17** and **Figure 11-18**.

Figure 11-16. Suction filters are sometimes used in hydrostatic transmissions. They will cause a pump to cavitate if they become plugged and have no bypass valve.

The advantage of filtering the charge oil in a hydrostatic transmission is that it prevents contaminants from entering the piston pump, which is a very costly component to replace. The disadvantages of a charge filter are high initial, maintenance, and replacement costs. The filter assembly is more expensive than a suction filter or a return filter because it must be able to withstand a higher pressure. Notice in Figure 11-17 that the filter has a bypass valve to allow oil to freely bypass the filter if it becomes plugged.

High-Pressure Filtration

High-pressure filters are the most expensive type of filter. They are usually housed in a heavy housing that is designed to withstand the highest system pressure, sometimes up to 6000 psi (414 bar). Notice in Figure 11-19 that the high-pressure filter is located after the pump and before the DCV. It is used to protect sensitive components that have tight tolerances, like expensive servo valves for example.

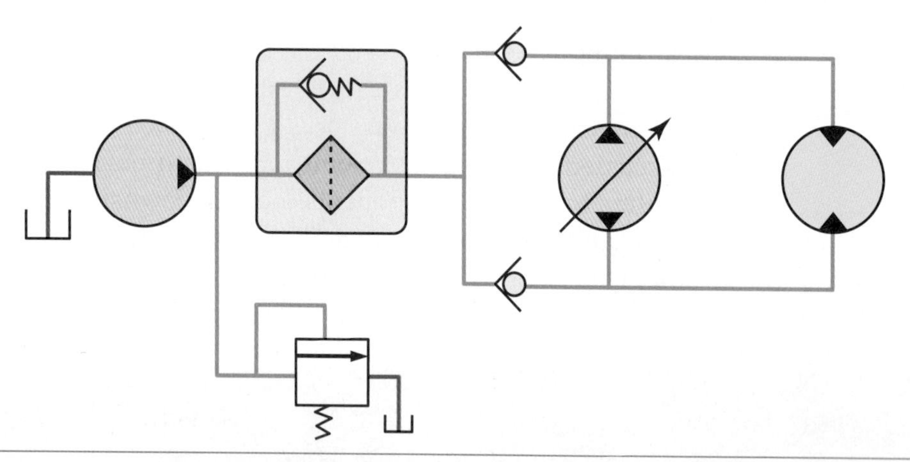

Figure 11-17. Charge oil is sometimes filtered to prevent contaminants from entering the main piston pump.

Figure 11-18. This power shift transmission uses a charge filter. The transmission operates at approximately 240 psi (16 bar). This filter has a burst rating approximately 2.5 times higher than the system operating pressure.

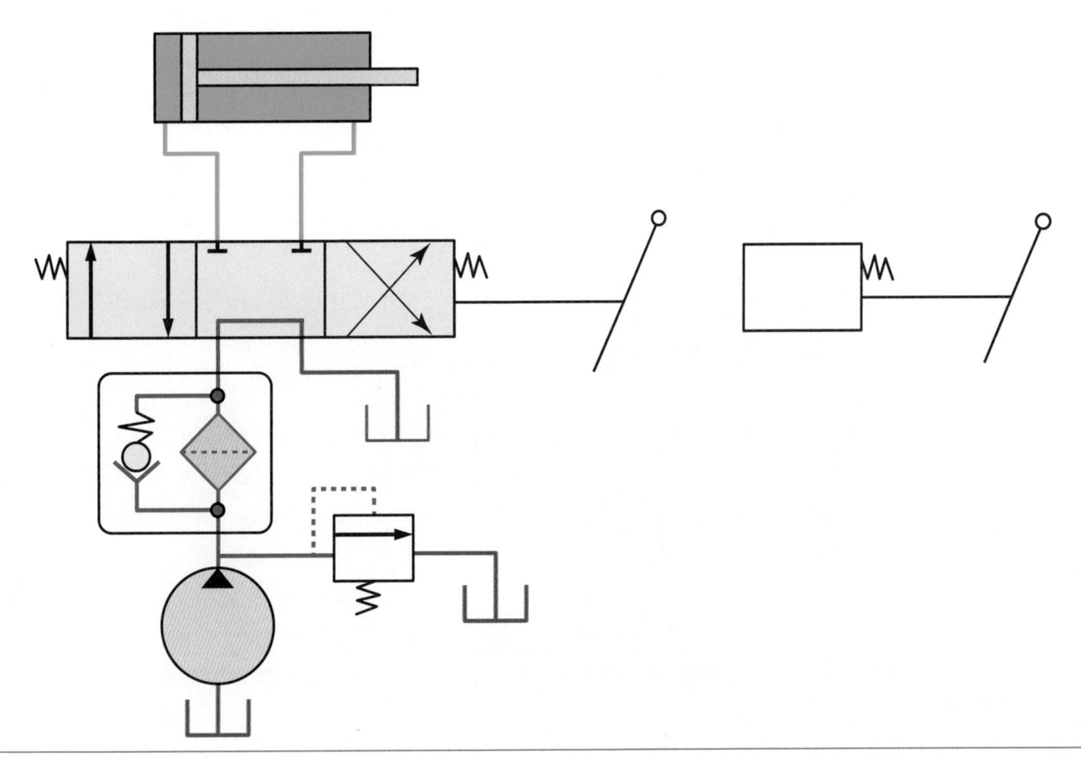

Figure 11-19. A high-pressure filter must be able to withstand the system's highest operating pressure and is located after the pump assembly.

High-pressure filters offer the advantage of filtering very small contaminants, as small as 2µm. However, the high pressure will reduce a filter's efficiency. As the pressure increases, it is more difficult for the media to hold small contaminants that are being injected. High-pressure filters are not commonly used in lower-priced mobile equipment.

Actuator Filtration

Actuator filters have the same characteristics as high-pressure filters. They are heavy, expensive, can filter contaminants down to very small particles, and operate at very high pressures. In addition, actuator filters must allow the oil to flow in both directions, reverse and forward flow. The filters are located between the DCV and the actuator. See Figure 11-20. In a hydrostatic transmission, which will be discussed in **Chapter 23**, the filter is located between the pump and motor.

Return Filters

One of the most popular types of filters used in mobile machinery is the ***return filter.*** It is designed to remove contaminants before the oil re-enters the reservoir. See Figure 11-21. These filters have the benefit of using the system's relatively low return oil pressure to push the contaminants into the filter media. Suction filters must rely on pump vacuum to pull the contaminants into the filter's media, and the suction filter can cause pump cavitation. High-pressure filters have the challenge of trapping contaminants that have media-piercing capability due to pressures that reach up to 6000 psi.

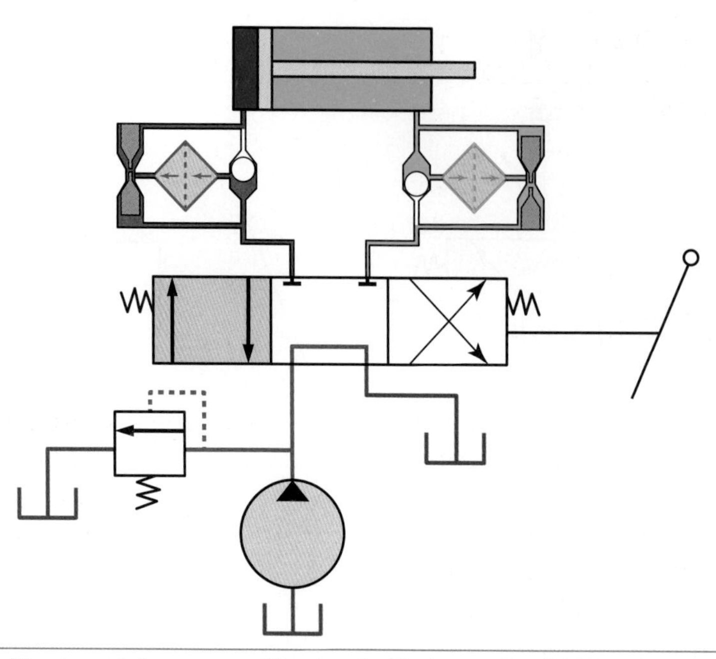

Figure 11-20. An actuator filter is a high-pressure filter located between the DCV and actuator. The filter must allow oil to flow in both directions.

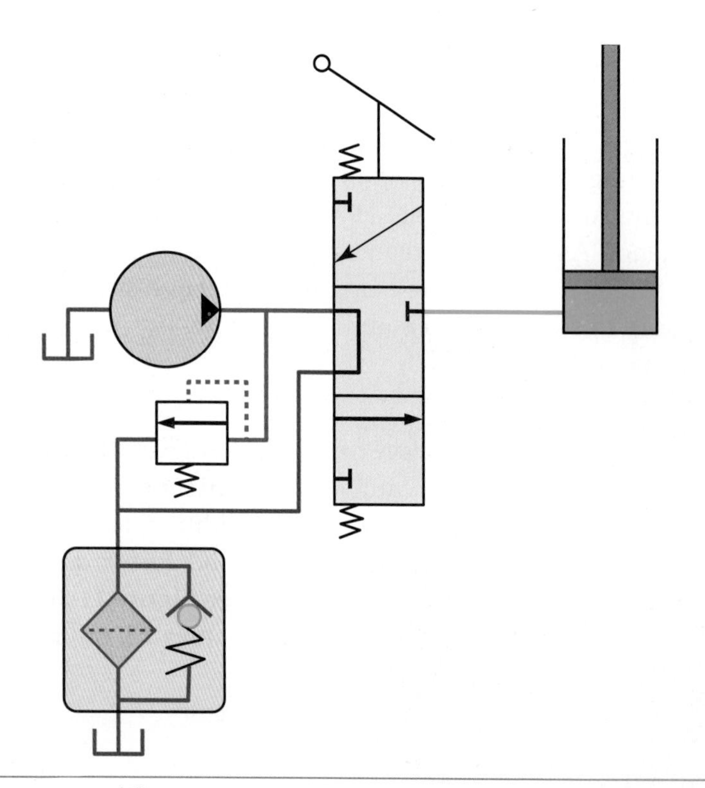

Figure 11-21. Return filters are one of the most common types of filters used in mobile equipment.

The return line generally operates at a lower system pressure, for example 30 to 60 psi (2 to 4 bar). Some return filters are designed to operate at pressures up to 150 psi (10 bar). Return filters are much more affordable than high-pressure filters.

One important factor when adding a return filter to a hydraulic system is flow surges, especially if the hydraulic system uses a fixed-displacement pump and large single-acting cylinders, such as a forklift's mast raise circuit. If the hydraulic system operates at 30 gpm (113 lpm), and if a person attempts to install a return filter with a 35 gpm (132 lpm) capacity, the filter might rupture when the mast lowers. The pump volume of 30 gpm (113 lpm) plus the mast return volume, which could easily be more than 5 gpm (19 lpm), would exceed the return filter's capacity. Flow surges also occur during the operation of a double-acting cylinder, due to the difference in cylinder areas. A typical return filter is shown in **Figure 11-22**.

Off-Line Kidney Loop Filtration

Most suction filters and return filters are unable to filter the smallest particles. If a smaller micron rated filter element is installed, the smaller pore size will increase the differential pressure across the filter unless the new element has more surface area, a thinner fluid is used, or the flow rate is reduced. If the differential pressure is increased, it is likely that the oil will bypass the element through the bypass valve.

An off-line ***kidney loop filter***, a form of bypass filtration, is a filtration device that allows a customer to use a filter with a very small micron rating and a high filter efficiency rating without having too large of a differential pressure. The system is designed for the sole purpose of slowly filtering the fluid inside the reservoir. See **Figure 11-23**.

Kidney loop carts often use a separate power source, such as a shop's 120 VAC wall outlet. See **Figure 11-24**. The cart can be plugged into a hydraulic reservoir when oil samples start showing high particle counts, or when convenient. For example, it might be used when the machine is in the shop for maintenance or repairs. Kidney loop filtration carts are already popular in the mobile equipment industry and will continue to grow in popularity.

Figure 11-22. A return filter is frequently used in mobile machinery. The filter removes contaminants before they enter the reservoir. The return hose on the left is an unfiltered case drain line.

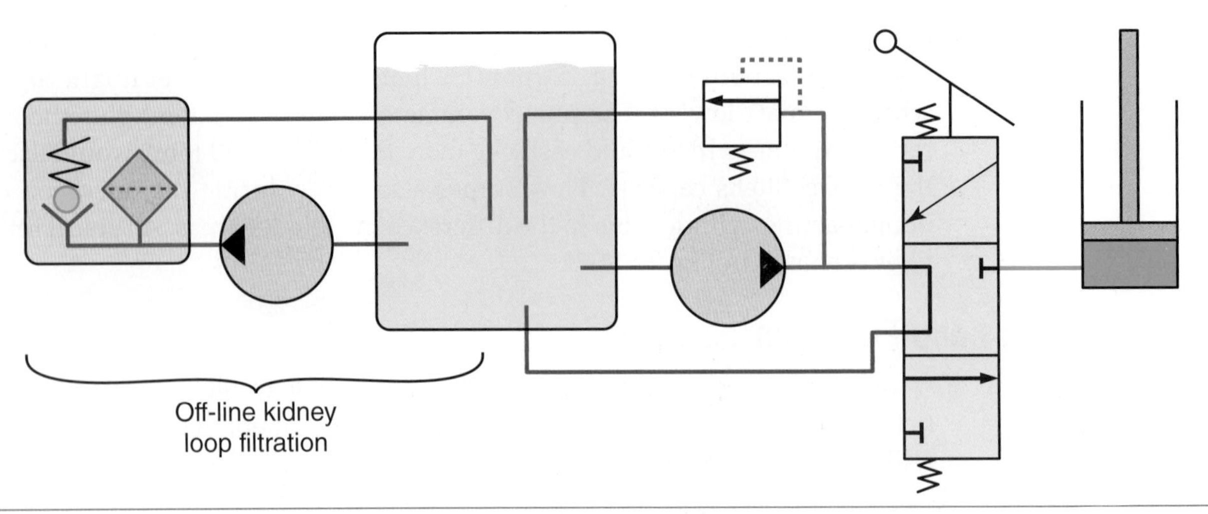

Figure 11-23. Off-line kidney loop filtration carts are built for the sole purpose of filtering the reservoir's oil to a lower micron rating using higher-efficiency filters.

Figure 11-24. This kidney loop filtration cart is powered by a 120 VAC wall outlet. It has been configured with quick disconnect couplers to quickly plug into a reservoir. It has two very efficient filters. The oil is first sent through a filter rated at $\beta_{11} = 1000$. Then, the oil is routed through a filter rated at $\beta_4 = 1000$.

Kidney loop filtration carts cost owners additional money for the initial purchase. It is also convenient to configure the reservoirs with quick couplers, which is an additional expense. One other potential negative is the cart is used only when the machine is shut down, meaning that the cart can filter only the oil inside the reservoir, not the oil in the rest of the hydraulic system. However, with repeated operations, the filtration cart will eventually clean all the fluid as it is cycled back into the reservoir.

Note

Some extremely high-priced equipment has a kidney loop filtration system installed onboard the machine.

Types of Filtration by Flow

Filtration can also be defined based on how much of the circuit is filtered. ***Proportional flow filtration,*** also known as bypass filtration, filters only a partial amount of the hydraulic system's oil flow. See Figure 11-25.

Full flow filtration filters the entire system's oil flow. Full flow filters can be placed before the pump or in the return circuit. See Figure 11-26.

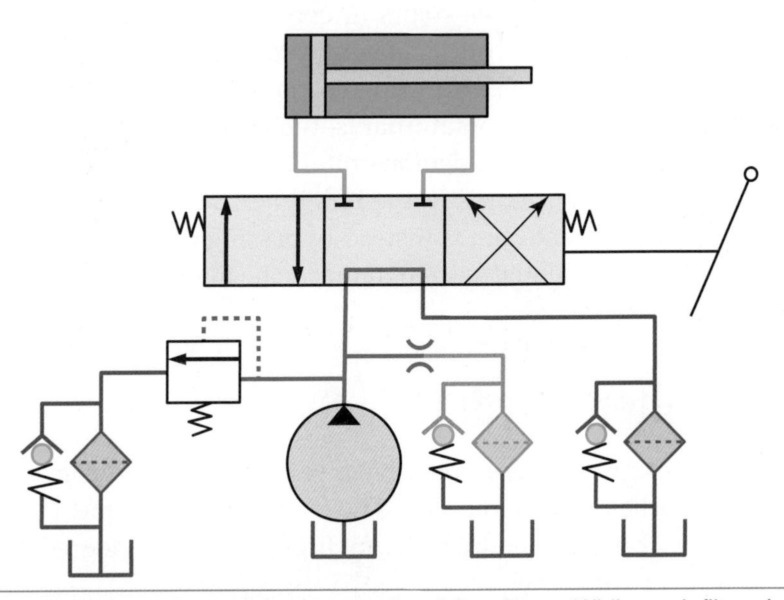

Figure 11-25. All three filters in the schematic are proportional flow filters. While each filter cleans only a portion of the system's oil flow, together they filter all of the oil flow.

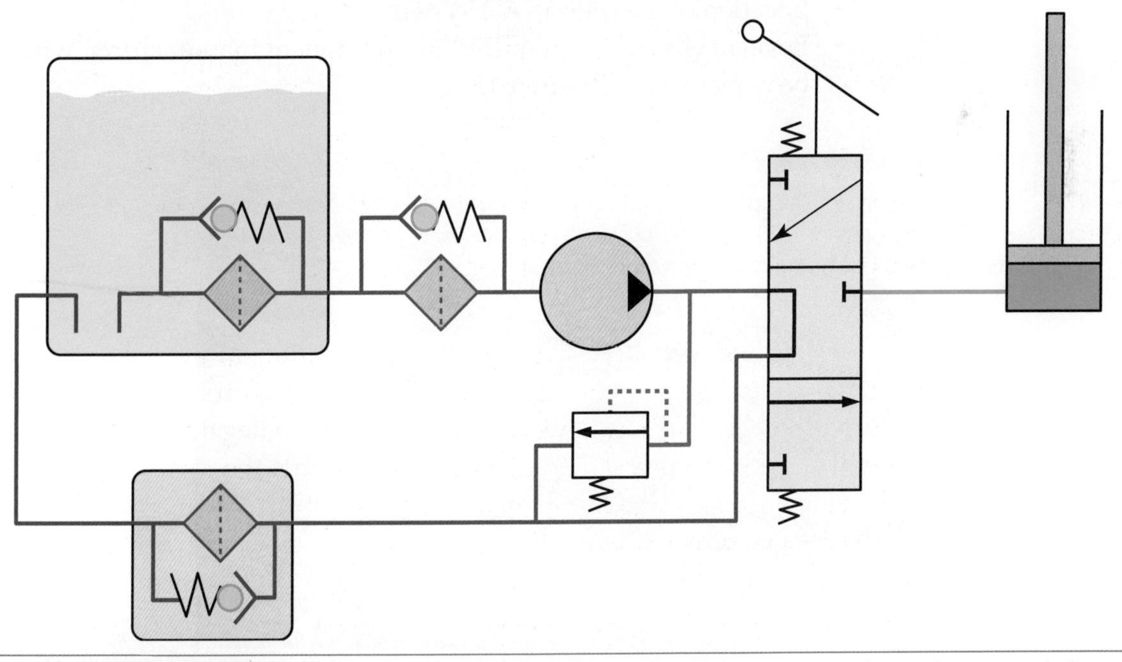

Figure 11-26. The three filtration devices in the schematic are full flow filters, because the entire system's flow passes through each of the filters. This schematic contains a suction screen in the reservoir, a suction filter located before the pump, and a return filter after the DCV and relief valve.

Filter Indicators

Many machine owners replace filters at a given machine-hour interval. However, replacing the filters during a machine-hour interval usually results in one of two scenarios, replacing a filter too early or waiting too long to replace a full filter.

Filter indicators are devices that measure the incoming pressure and the outgoing pressure. The indicators often have two colors (yellow and red), see Figure 11-27, or three colors (green, yellow, and red). When the indicator is in the green on a three-color indicator, the filter's differential pressure is low and the element is performing its job by capturing contaminants. As the filter begins to fill with contaminants, the indicator will move into the yellow portion on the scale. This demonstrates that the filter will soon need to be replaced. If the filter indicator is in the red, it indicates the filter is no longer capturing the contaminants, but is instead bypassing dirty oil into the hydraulic system. To prevent potential system damage, the filter must be replaced before the indicator is in the red.

Choosing How to Filter a Hydraulic System

Designers and owners have several factors to consider when choosing a hydraulic filtration method. The following factors must be considered:

- Type of fluid (petroleum, fire resistant, biodegradable).
- Fluid viscosity.
- Temperature.
- System flow, pressure, and duty cycle.
- Selection of components to protect.
- Location of the filter in the system.
- Level of cleanliness required by component manufacturer, which will be explained in **Chapter 12**.

Figure 11-27. Filter indicators sense the differential pressure across a filter's element. This indicator uses two colors, a yellow check valve and a red scale indicating when the oil is bypassing the filter. The green arrow depicts the direction of oil flow.

Summary

- ✓ Strainers remove only large particles.
- ✓ Strainers, also called screens, are commonly used to filter oil supplied to the pump.
- ✓ Using suction strainers increases the risk of pump cavitation.
- ✓ Filters remove smaller particles that are invisible to the human eye.
- ✓ Reservoir breathers are cost effective means of filtering incoming and outgoing air to the reservoir to remove contaminants, including moisture.
- ✓ Suction filters remove small contaminants from entering a pump's inlet.
- ✓ Charge pressure filters are sometimes used in power shift transmissions and hydrostatic transmissions.
- ✓ High-pressure filters are very expensive, heavy, and are rated to filter small particles.
- ✓ High-pressure filters are located after the pump.
- ✓ Actuator filters are high-pressure filters that must be able to flow oil in both directions.
- ✓ Return filters are low-pressure filters that are commonly used in mobile machinery.
- ✓ Kidney loop filter carts are the best means of filtering a reservoir's fluid to the smallest micron rating.
- ✓ Filter indicators signal when a filter is filtering, full, or bypassing fluid.

Technical Terms

absolute rating
actuator filters
Beta ratio
breather
charge pressure filter
depth filter
element collapse pressure
filter bypass valve
filter indicators
filters
full flow filtration
high-pressure filters
kidney loop filter
micron (µm)
nominal rating
proportional flow filtration
return filter
strainers
suction filter
surface-type filtering devices
wire mesh number

Review Questions

Answer the following questions using the information provided in this chapter.

1. One micron equals one millionths of a(n) _____.
   A. centimeter
   B. inch
   C. meter
   D. millimeter
2. A micron is also known as a _____.
   A. microcentimeter
   B. micrometer
   C. microinch
   D. micromillimeter

3. Which of the following filter devices will provide the finest particle filtration?
   A. 60 wire mesh.
   B. 100 wire mesh.
   C. 200 wire mesh.
   D. 325 wire mesh.
4. Which of the following filter devices will have the largest micron rating?
   A. 60 wire mesh.
   B. 100 wire mesh.
   C. 200 wire mesh.
   D. 325 wire mesh.
5. All of the following are disadvantages of a strainer/screen, *EXCEPT*:
   A. difficult to replace.
   B. difficult to determine if it is plugged.
   C. likely to cause pump cavitation if it is plugged.
   D. expensive.
6. Technician A states that all suction strainers are located inside the tank. Technician B states that many suction strainers are located inside the tank. Who is correct?
   A. Technician A.
   B. Technician B.
   C. Both A and B.
   D. Neither A nor B.
7. Which of following strainers will require a reservoir with an access panel?
   A. Side-mounted suction strainer.
   B. In-tank suction strainer.
   C. Both A and B.
   D. Neither A nor B.
8. An ISO multi-pass filter test is being performed on a filter. The number of upstream contaminants equals 30,102 and the number of downstream contaminants equals 2545. Compute the filter's Beta ratio.
   A. .084.
   B. 8.45.
   C. 10.8.
   D. 11.8.
9. An ISO multi-pass filter test is being performed on a filter. The number of upstream contaminants equals 30,102 and the number of downstream contaminants equals 2545. Compute the filter's efficiency.
   A. 84.5%.
   B. 90.8%.
   C. 91.5%.
   D. 95.1%.
10. Compute the efficiency of filter with a rating of $\beta_5 = 15$.
    A. 90.0%.
    B. 93.3%.
    C. 95.0%.
    D. 98.3%.
11. A filter has a $\beta_{10} = 20$ rating. What does the number 10 equal?
    A. The ratio of upstream particles to downstream particles.
    B. The size of the contaminants, measured in microns.
    C. The number of times the contaminant must pass through the filter.
    D. The surface area of the filter in square inches.
12. A filter has a $\beta_{10} = 20$ rating. What does the number 20 equal?
    A. The ratio of upstream particles to downstream particles.
    B. The size of the contaminants, measured in microns.
    C. The number of times the contaminant must pass through the filter.
    D. The surface area of the filter in square inches.
13. Which of the following is the best measure of a filter's proficiency in capturing contaminants?
    A. Nominal rating.
    B. Absolute rating.
    C. Beta rating.
    D. Collapse rating.

14. Which of the following devices removes only large particles?
    A. Suction screen.
    B. Suction filter.
    C. Return filter.
    D. Kidney loop filter.
15. Technician A states reservoir breathers can prevent small contaminants from entering the reservoir. Technician B states that reservoir breathers can prevent moisture from entering the reservoir. Who is correct?
    A. Technician A.
    B. Technician B.
    C. Both A and B.
    D. Neither A nor B.
16. Which of the following filtration devices is used to prevent small particles from entering a pump's inlet?
    A. Suction screen.
    B. Suction filter.
    C. High-pressure filter.
    D. Actuator filter.
17. Which of the following filtration devices operates at a lower pressure range?
    A. Actuator filter.
    B. High-pressure filter.
    C. Charge pressure filter.
    D. Return filter.
18. Which of the following filtration devices is most sensitive to flow surges?
    A. Actuator filter.
    B. High-pressure filter.
    C. Charge pressure filter.
    D. Return filter.
19. Which of the following filtration devices is designed to remove the smallest particles from a hydraulic reservoir?
    A. Return filter.
    B. Kidney loop filter.
    C. Suction screen.
    D. Pressure filter.
20. High-pressure filters can operate at pressures as high as _____.
    A. 3000 psi (207 bar)
    B. 4000 psi (276 bar)
    C. 5000 psi (345 bar)
    D. 6000 psi (414 bar)
21. What is a reasonable example of a maximum pressure for a return filter?
    A. 15 psi (1 bar).
    B. 150 psi (10 bar).
    C. 1500 psi (103 bar).
    D. 3000 psi (206 bar).
22. Which of the following filtration devices is designed to operate at high system pressures?
    A. Actuator filter.
    B. Kidney loop filter.
    C. Return filter.
    D. Suction filter.
23. Which of the following values is a reasonable example of a suction filter bypass valve rating?
    A. 2 to 3 psi (0.14 to 0.20 bar).
    B. 15 to 30 psi (1 to 2 bar).
    C. 30 to 60 psi (2 to 3 bar).
    D. 60 to 90 psi (3 to 4 bar).
24. When should filters be replaced?
    A. Every 500 hours.
    B. Every 1000 hours.
    C. Every 1500 hours.
    D. Just before the filter starts to bypass oil.
25. A filter indicator will display what color when the filter is bypassing oil?
    A. Green.
    B. Yellow.
    C. Red.
    D. Purple.
26. Which color is *not* commonly used on filter bypass indicators?
    A. Green.
    B. Yellow.
    C. Red.
    D. Purple.

27. Which of the following will filter all of a system's oil flow?
    A. Bypass filter.
    B. Proportional filter.
    C. Full flow filter.
    D. All of the above.

28. Technician A states that suction screens and suction filters can be considered full flow filtration. Technician B states that some return circuits can be considered full flow filters. Who is correct?
    A. Technician A.
    B. Technician B.
    C. Both A and B.
    D. Neither A nor B.

Chapter 12 Contamination Control

Objectives

After studying this chapter, you will be able to:

- ✓ List the possible sources of hydraulic system contamination.
- ✓ Explain the reasons for analyzing hydraulic fluids.
- ✓ Explain how to properly take an oil sample with a vacuum pump and an onboard sampling valve.
- ✓ List the types of data that are included on an oil sample label.
- ✓ Explain ISO range codes used to measure hydraulic oil cleanliness.
- ✓ Explain NAS codes used to measure hydraulic oil cleanliness.
- ✓ List the different types of analyses performed on an oil sample.
- ✓ Explain methods for improving contamination control.
- ✓ Describe the possible reconditioning procedures necessary following the major contamination of a hydraulic system.

Source of Contamination

It is critical that machines operate with a clean hydraulic system. The clearances within some hydraulic components are as small as 5μm or less. Contamination can cause hydraulic systems to become sluggish and may cause components to fail. 70% of all hydraulic system failures can be attributed to fluid contamination.

For safe and efficient operation of a hydraulic system, contaiminated hydraulic systems must be identified, the source of the contamination must be determined, and the problem must be remedied. Contamination occurs:

- During the manufacturing process of the components and the machine.
- From normal, everyday wear during operation.
- As a result of poor maintenance.

Contamination from Manufacturing

Touring any manufacturing facility will reveal the challenge that manufacturers face in producing a new machine with a clean hydraulic system. Airborne dust can easily migrate into components at each stage of the manufacturing process, such as foundry, machining, welding, metal forming, and assembly.

Many types of contamination can occur during the manufacturing process. Dust, dirt, sand, metal shavings, welding slag, paint chips, nuts, bolts, screws, paper, and plastic are common contaminants. Some of the dirtiest components are new components, such as a freshly cut hose, a fabricated reservoir, or even a new 55 gallon drum of hydraulic oil.

Most manufacturers have invested substantial funds in improving the quality of components. Machine manufacturers have high expectations for the component manufacturers. If anytime during the manufacturing process personnel become too relaxed toward preventing contamination, the lapse can easily cost the equipment owner tens of thousands of dollars in future service and repairs.

Contamination from Operation

As hydraulic systems operate, they produce contamination from the moving components. Due to internal moving components, it is virtually impossible to build a hydraulic system that will not wear. Cylinder pistons slide in and out of their barrel. Spool valves move within their valve block. Pumps have rotating members and/or reciprocating components. The dirtier the fluid is, the more the components wear from everyday operation.

Contaminants can also migrate into an operating hydraulic system. Faulty reservoir breathers allow contaminants and moisture into the reservoir as air is pulled through the breather. Cylinders can allow dirt to enter the system if the cylinder has a bent rod, chrome plating that is cracked or flaking, or if the rod seal is leaking. Nearly any component with a leaking seal can allow contamination to enter the system.

Poor Maintenance

Poor service and maintenance habits also contaminate hydraulic systems. Dirty shops, exposed components, and poor maintenance practices all contribute to contaminating hydraulic systems. Investing a small amount of extra time in preventing contamination can greatly improve a hydraulic system's cleanliness. The following are examples of poor maintenance practices:

- Failing to cap components.
- Storing 55 gallon drums of oil outside in a vertical position without covers.
- Failing to clean new hydraulic hoses.
- Failing to filter new oil.
- Choosing not to use kidney loop carts.
- Working in a dirty shop environment.
- Using filters with an inadequate Beta rating.

Reasons for Fluid Analysis

Farmers, large contractors, and even small firms have a variety of reasons for analyzing the hydraulic fluids within their machines. Perhaps the most important reason is to maximize the machine's availability. Most owners have a lot at stake when a machine breaks down. For example, farmers are battling Mother Nature to get their crop planted in the ground or harvested before the weather takes a turn for the worse. Large contractors or mines have deadlines to meet. A contractor might be tasked with removing an international airport's runway and repaving the runway all in 60 days.

When a machine fails unexpectedly, it may cause a chain reaction, forcing numerous other machines to remain idle. A mine can lose hundreds of thousands of dollars per hour if the shovel fails. Contractors can lose million-dollar bonuses if the essential crane or paver breaks down and the job is unable to be completed early. Knowing that a hydraulic pump or motor is on the brink of failure could prevent costly downtime. Many contractors are forced to rent other machines if the repair time is lengthened.

When an unexpected failure occurs and the company has much to lose from any downtime, owners must pay whatever it takes to get the machine operational. Unexpected failures always cost more money and time than planned component replacement. At times, it might require a special freight to get a component shipped across the country on short notice. The unexpected repair can require special fabrication or tools, which must be purchased or rented at the maximum retail price. The initial failure begins a domino effect, causing additional components to fail and accruing a seemingly never-ending repair bill. If an owner recognizes that a component is in distress and expects it to fail soon, the owner is able to better plan for the repair, reducing the total costs of the repair.

The gradual wear in a hydraulic system causes a loss of machine productivity. In relation to this, cycle times (the time it takes for actuators to extend or retract) increase. Unfortunately, tests have revealed that veteran operators may fail to realize that a machine is underperforming. Oil analysis can help owners maximize their machines' productivity by removing harmful contaminants from control valves and pumps with tight tolerances. The increase in machine productivity will increase company profits.

Fluid analysis can also help owners maximize component life. Knowing a hydraulic system is running with too much fluid contamination allows the owner to take action before it is too late.

Thinking Green

Today's hydraulic fluids are very costly. Analyzing the fluid allows owners to maximize the useful life of the hydraulic fluid. The simple conclusion to extend the oil change interval on a tested piece of equipment not only saves money but is also more environmentally friendly.

Oil Sampling

One of the first steps in contamination control is determining the cleanliness of the oil by taking an oil sample. A sample, however, is not an exact representation of the hydraulic fluid as a whole but rather just a localized segment. Using proper sampling techniques can help ensure that the sample is as representative of the entire system's fluid as possible.

Note

A popular term technicians sometimes use to describe oil sampling is *scheduled oil sampling*, or *SOS* as it is commonly abbreviated. The term and acronym were originally coined by the Caterpillar company.

There are two common methods used for taking a hydraulic sample. The oldest method is using a ***vacuum pump***, which is a manually actuated pump used to draw oil from the hydraulic reservoir. This method is the lesser preferred of the two methods because it is slow to perform and less reliable. See Figure 12-1. The other method uses an onboard sampling valve.

The following tips should be considered when taking an oil sample:

- Operate the machine and bring the fluid to operating temperature. The best case scenario is to take the sample immediately after the machine has finished for the day.
- In addition to operating temperature, be sure to cycle the hydraulic controls to circulate the oil.
- Use a clean sample bottle that has been stored in a clean, closed storage bag away from the elements. ISO 3722 provides details regarding the cleanliness of sample bottles.
- Always use new tubing that has been stored in a clean environment. Ensure that the tube ends have been protected or capped during storage.
- Minimize contaminants from the surrounding environment from entering the sample. If a kit is being used, leave the bottle inside the plastic bag while taking the sample. If a kit is not being used, place the bottle in a clean Ziploc bag before heading to the field.
- Most sampling kits recommend cleaning the sample bottle by taking an initial oil sample. Complete the following steps to minimize contamination in this manner:
  - Partially fill the bottle approximately two-thirds full using the oil from the hydraulic system which is being sampled.
  - Cap the bottle.
  - Shake the bottle vigorously to remove any contaminants in the bottle.
  - Dispose of the oil in the waste oil container so that it can be properly recycled as explained in **Chapter 1**.

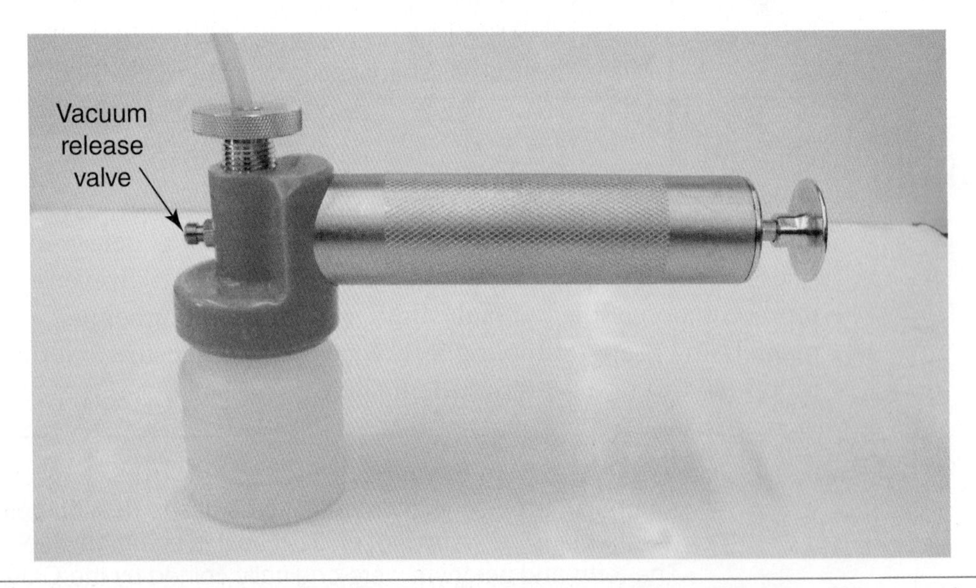

Figure 12-1. The EasyVac Vampire® brand vacuum pump is a popular oil sampling pump. This pump has a vacuum release valve that allows a technician to stop the oil being drawn into the sample bottle.

- After the sample bottle is cleaned, take the actual sample, but do not completely fill the bottle. Use the following International Standards Organization viscosity grade (ISO VG) ranges as a guide:
    - ISO VG of 32 cSt or less—fill bottle 3/4ths full.
    - ISO VG of 32 to 100 cSt—fill bottle 2/3rds full.
    - ISO VG higher than 100 cSt—fill the bottle half full.

Be sure to ship the sample to the lab that same day. Any delay in analysis can cost the owner money. If the analysis report arrives after a problem has already manifested, the test has done no good.

Using a Vacuum Pump

When using a vacuum pump, several mistakes can be made. The following steps will help ensure that the sample is representative of the system's fluid:

- After the machine has been warmed and the fluid has been cycled, shut off the machine.
- If the hydraulic reservoir has a dipstick, cut the tubing for the vacuum pump the same length as the dipstick.
- If the reservoir does not contain a dipstick, cut the tubing so that it reaches the halfway level of the oil in the reservoir.
- Insert the tubing into the reservoir.
- Do not let the tubing touch the bottom of the reservoir.
    - When inserting the sample tubing into the oil reservoir, be sure that the tubing does not touch the reservoir walls or bottom, which can be difficult because most tubing wants to curl back to its original coiled state. Some technicians tie a welding rod onto the tubing when working in small oil compartments to ensure that the tubing stays centered inside the oil compartment, for example when sampling a machine's final drive.
- Although it is possible to use a single vacuum pump for sampling different fluids such as coolant, fuel, engine oil and hydraulic oil, it is very easy to contaminate the pump by overfilling the bottle or allowing the pump to tip on its side. A contaminated pump will provide erroneous results. Therefore, it is best to use a vacuum pump that is dedicated for pulling hydraulic oil samples, not other fluid samples.
- Insert the tubing into the vacuum pump head assembly.
- The tube should protrude 1″ (2.5 centimeters) beyond the pump head assembly.
- Tighten the retainer to hold the tubing in position.
- Install the sample bottle.
- Hold the pump in a vertical position.
- Pull oil into the bottle by pumping the handle. Be careful not to overfill the sample bottle.
- Do not allow the sample of oil to touch the end of the tubing inside the sample bottle. Any oil on the outside of the tubing will contaminate the pump's head during disassembly.
    - If the tubing touches the sample, cut off the contaminated tubing before removing the tube from the vacuum pump.

Many manufacturers incorporate an oil sampling valve on their machine. The ***sampling valve*** is designed for the sole purpose of providing fast and reliable oil samples. Engineers strategically place the sampling valve so it will provide a representative sample using the system's turbulent oil flow. This live sampling valve is the preferred method for taking samples and provides a more accurate sample than one drawn from a reservoir.

The valve looks similar to a hydraulic quick-coupler test port, except it has a hole in the center of the sampling valve. A ***sampling probe*** is fitted to the end of the tubing as pictured in Figure 12-2. The probe is inserted in the sampling valve, which enables the oil from the hydraulic system to flow through the valve to begin filling the sample bottle. The bottle's lid is different than the type used on a vacuum pump. The bottle used with a sampling valve must have a small hole to vent the trapped air inside the bottle. Therefore, two lids are used when working with a sampling valve. The first lid has a hole for the tube and a hole for the vent, while the second lid contains no holes. This lid is used to securely cap the bottle after the sampling process is completed. Sampling valves are normally used at pressures below 500 psi (34 bar).

Using Onboard Sampling Valves

Note

If a sampling valve is being added to a hydraulic system, the valve needs to be located downstream from hydraulic components, but ahead of the filter to provide a representative sample.

In addition to the general tips previously listed for taking oil samples, the following should be performed when taking a sample using an onboard sampling valve:

- If a kit is being used, it will contain the tubing, sample bottle, two lids, a probe, and Ziploc bags.

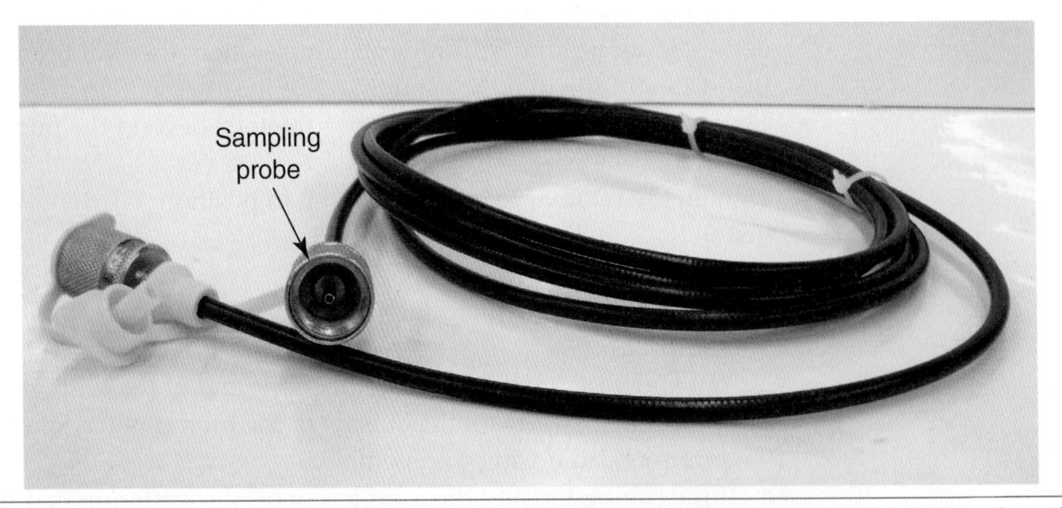

Figure 12-2. If the system is equipped with a sampling valve, a sampling probe must be inserted into the valve in order to take a sample. Sampling valves speed up the process of oil sampling.

- If a reusable, brass probe is being used, cut a new piece of tubing approximately 6″ long (15 centimeters).
- Remove the dust cap from the oil sampling valve and use a clean, lint-free towel to clean the oil sampling valve.
- Follow the manufacturer's specification for machine operation to ensure the hydraulic system pressure is low.
- Insert the probe into the sample valve.
- Fill the bottle approximately 2/3rds full. Follow the rest of the sample bottle cleaning steps outlined earlier. This first fill is used to flush the sampling valve and clean the interior of the sample bottle.
- After disposing of the oil from the initial fill into the proper waste container, use the bottle to take the actual sample.

Note

Only take samples using tried and proven sampling techniques, such as using a live sampling valve or a vacuum pump. A sample should never be taken from a filter or an oil drain plug because the sample will be contaminated.

Bottles

Sample bottles range in size from 1 to 6 ounces. Metrically sized bottles range from 50 to 200 milliliters. **Figure 12-3** shows one bottle size.

The bottles are made from plastic or glass. To prepare the sample for shipment, the oil sample and a label are both placed inside a round shipping container.

Keep in mind that the lab can only work with the data that it receives. If it receives bad data, it will generate bad data. Therefore, it is helpful to supply the lab with as much detailed information as possible. An example of an oil sample label can include the following information:

- Company name.
- Company address.
- Date.

Figure 12-3. This 2 oz (60 ml) sample bottle is manufactured by EasyVac, Inc.

- Contact person's name.
- Machine model number.
- Machine serial number.
- Name of fluid compartment (i.e. "hydraulic system").
- Sample location (i.e. "after the pump" or "before the return filter").
- Number of machine hours.
- Oil hours (since last oil change).
- Type of oil (brand, weight).
- Was the oil changed?
- Was oil added, how much?
- Job site name or number.

Frequency and Original Sample

Many machine owners that have never taken an oil sample will hope that taking just one oil sample will be the cure all for their machine problems. Oil sampling must be ***trended***, meaning that a sample needs to be taken and sent for analysis at regular scheduled intervals—for example, every 250 hours.

The sample should always be taken before the fluid is changed. A sample taken after the fluid is changed will not accurately indicate how much dirt was in the hydraulic system.

A sample must also be taken of the new oil. This is often called the ***reference oil***. As the machine is operated over long periods of time, it is important to know if the fluid properties or additives have been depleted. The only way of obtaining this information is to trend multiple oil samples and have an analysis of the original reference oil. Without the original reference oil sample, there is no way of establishing a baseline for future samples.

Cleanliness Standards

The hydraulic industry has two standards for specifying the cleanliness or dirtiness of a tested hydraulic system or the target cleanliness range for a particular hydraulic system. The most common standard followed in the mobile equipment industry is the ***ISO range code***. The ***National Aerospace Standard (NAS)*** is another standard that is sometimes referenced.

Many oil analysis laboratories will generate reports listing both ISO range codes and a NAS number. The labs use a ***particle counter***, which is a machine that counts the number of contaminants based on the size of the contaminants. The particle counter will report data using ISO range codes and NAS codes.

ISO Range Codes

The ISO range codes have been used since the 1970s. The first ISO range code used two numbers that represented the number of contaminant particles that were 5μm in diameter or larger and those that were 15μm in diameter or larger. Later the code was given a third range number that represented particles that were 2μm in diameter or larger.

Today's ISO range code (4406) continues to use three range numbers. However, the three numbers represent the number of contaminant particles that fit among the following size designations:

- 4μm in diameter or larger.
- 6μm in diameter or larger.
- 14μm in diameter or larger.

In addition to the size of the contaminant particles, the range numbers quantify how many of each size contaminant particles are in one milliliter of fluid. Refer to **Figure 12-4**.

For example, an oil sample was sent to an oil analysis lab and the lab sent back an analysis stating the oil's cleanliness equaled an ISO range code of 21/19/16. This code can be deciphered by tracing each number to its contaminant particle range outlined in the ISO cleanliness standard:

- ISO range number 21 = 10,000–20,000 contaminant particles that were 4μm or larger per one milliliter of oil.
- ISO range number 19 = 2500–5000 contaminant particles that were 6μm or larger per one milliliter of oil.
- ISO range number 16 = 320–640 contaminant particles that were 14μm or larger per one milliliter of oil.

Range of Contaminant Particles in One Milliliter of Fluid	ISO Range Code
5,000,000–10,000,000	30
2,500,000–5,000,000	29
1,300,000–2,500,000	28
640,000–1,300,000	27
320,000–640,000	26
160,000–320,000	25
80,000–160,000	24
40,000–80,000	23
20,000–40,000	22
10,000–20,000	21
5000–10,000	20
2500–5000	19
1300–2500	18
640–1300	17
320–640	16

Range of Contaminant Particles in One Milliliter of Fluid	ISO Range Code
160–320	15
80–160	14
40–80	13
20–40	12
10–20	11
5–10	10
2.5–5	9
1.3–2.5	8
0.64–1.3	7
0.32–0.64	6
0.16–0.32	5
0.08–0.16	4
0.04–0.08	3
0.02–0.04	2
0.01–0.02	1

Figure 12-4. Each ISO range code represents how many contaminant particles are in one milliliter of fluid.

The largest number in the three-digit range code represents particles that are 4µm and larger. The next largest number in the code includes all particles that are 6µm and larger. The smallest number in the code refers to only particles that are 14µm or larger.

Sometimes a range code is only given two range numbers, for example 18/15. In this case, the range code of 18 represents particles that are 6µm in diameter and larger. The range code of 15 represents particles that are 14µm in diameter and larger.

The goal is to achieve the smallest ISO range codes practical for a given application. Component manufacturers list target range codes based on the type of component and the operating pressure, **Figure 12-5**.

Industry contamination specialists and some machine manufacturers provide ISO range code recommendations. For example, the targets can be based on an existing operational hydraulic system or the oil used to fill that system. An ISO code of 18/15 is an example target for an operational hydraulic system. An ISO code of 14/11 is an example of a target for *fill oil*, which requires filtering brand new oil before filling the hydraulic system.

The life of a hydraulic system can be extended by bringing the system's overall cleanliness into cleaner ISO ranges. For example, if a hydraulic system that is currently operating at an ISO range code of 19/16 is cleaned to a target of 17/14, the life of the system can be extended by up to 50% according to some industry estimates. Contamination specialists have reference charts for extending the life of a hydraulic system.

National Aerospace Standard (NAS)

The NAS 1638 contamination classification code was first introduced in the 1960s as a way for aircraft manufacturers to classify contamination in aircraft systems. The NAS code was the very first contaminant classification method used in the hydraulic industry. The code is based on the contaminant particles found in 100 milliliters of oil.

Figure 12-6 shows that NAS has 14 codes that range from 00 to 12. The code is given based on a maximum allowable number of contaminants in five particle classifications:

- 5µm–15µm.

Component	1500 psi (103 bar)	3000 psi (207 bar)	4500 psi (310 bar)
Piston pump	18/16/13	17/15/12	16/14/12
Vane pump	19/17/14	17/15/13	16/14/12
Gear pump	20/18/15	18/16/14	17/15/13
Servo valve	17/15/13	16/14/12	15/13/11
Proportional DCV	18/16/13	17/15/13	15/13/11

Figure 12-5. This table provides an example of target cleanliness ISO range codes based on types of components and their operating pressure.

- 15µm–25µm.
- 25µm–50µm.
- 50µm–100µm.
- 100µm and larger.

Like the ISO range codes, the higher the NAS code, the more contamination exists in the system. For example, the highest NAS code, 12, would specify a maximum of 1,024,000 particles that are 5µm–15µm in size, 182,400 particles that are 15µm–25µm in size, 32,400 particles that are 25µm–50µm in size, 5760 particles that are 50µm–100µm in size, and 1024 particles that are 100µm and larger in size.

Types of Analysis

Fluids can be analyzed for a variety of purposes. Several measuring instruments are used to examine multiple, varying attributes. **Figure 12-7** lists common hydraulic fluid attributes that are measured in an oil analysis, the test(s) or instruments used to measure the attribute, and the purpose of the analysis.

	Maximum Number of Contaminant Particles in 100 Milliliters of Fluid				
NAS Code	**5–15µm**	**15–25µm**	**25–50µm**	**50–100µm**	**100µm and larger**
00	125	22	4	1	0
0	250	44	8	2	0
1	500	89	16	3	1
2	1000	178	32	6	1
3	2000	356	63	11	2
4	4000	712	126	22	4
5	8000	1425	263	45	8
6	16,000	2850	506	90	16
7	32,000	5700	1012	180	32
8	64,000	11,400	2025	360	64
9	128,000	22,800	4050	720	128
10	256,000	45,600	8100	1440	256
11	512,000	91,200	16,200	2880	512
12	1,024,000	182,400	32,400	5760	1024

Figure 12-6. NAS codes vary from 00 to 12. The code is based on the maximum allowable contaminant particles, grouped in five diameter ranges, in 100 milliliters of oil.

Attribute	Test Instrument	Purpose of Analysis
Number and size of particles	Particle counter	Wear metals, contamination
Particle origin identifier	Microscope	Wear metals, contamination
Fluid viscosity	Kinematic viscometer	Fluid properties
Elemental analysis	Atomic emission spectroscopy (AES) (optical emission spectrometry (OES)): rotating disc electrode (RDE) or inductively coupled plasma (ICP)	Wear metals, contamination, fluid properties
Presence of additives	Fourier transform infrared spectroscopy (FTIR)	Fluid properties
Presence of water	Hot plate crackle test, or Karl Fischer titration test, or Fourier transform infrared spectroscopy (FTIR)	Contamination
Acid number (AN)	Fourier transform infrared spectroscopy (FTIR) coupled with statistical prediction software	Fluid properties

Figure 12-7. Several different fluid attributes can be analyzed with oil sample testing. This table lists some examples of fluid attributes and the devices used to measure those attributes.

Particle Counting

As mentioned earlier in the chapter, particle counting is a measure used to estimate the size and quantity of particles found within a hydraulic system. Oil analysis reports list particles in ISO range codes and NAS codes. Some particle count reports list quantities of particles in a range of different sizes, for example 4μm, 6μm, 10μm, 14μm, 21μm, 25μm, 38μm, and 70μm.

Particle counters, **Figure 12-8**, are not only found in high-tech oil analysis labs, they are also available as a portable device. The portable version enables a technician to measure a sample in the field for an instant report.

Prior to use, a particle counter should be flushed with mineral spirits. The instructions for the particle counter in **Figure 12-8** specify flushing the particle counter before and after measuring a sample. Remember to follow manufacturer's instructions for cleaning these machines.

Particle Origin Identification

One drawback of particle counting is that the measure does not identify the type of material of the contaminants. Labs use powerful microscopes to identify the origin of the particle. The microscope is commonly employed when the sample contains a high quantity of large particles and the lab technician wants to identify the material of the particles. After using a microscope, the lab generates a statement based on their findings. Such a statement may read as follows:

"The lab scope found large-sized iron particles, 80μm in diameter. Flush the system. Check the filter for iron particles. If filter has iron particles, check pumps and motors for wear. Resample system in 100 hours."

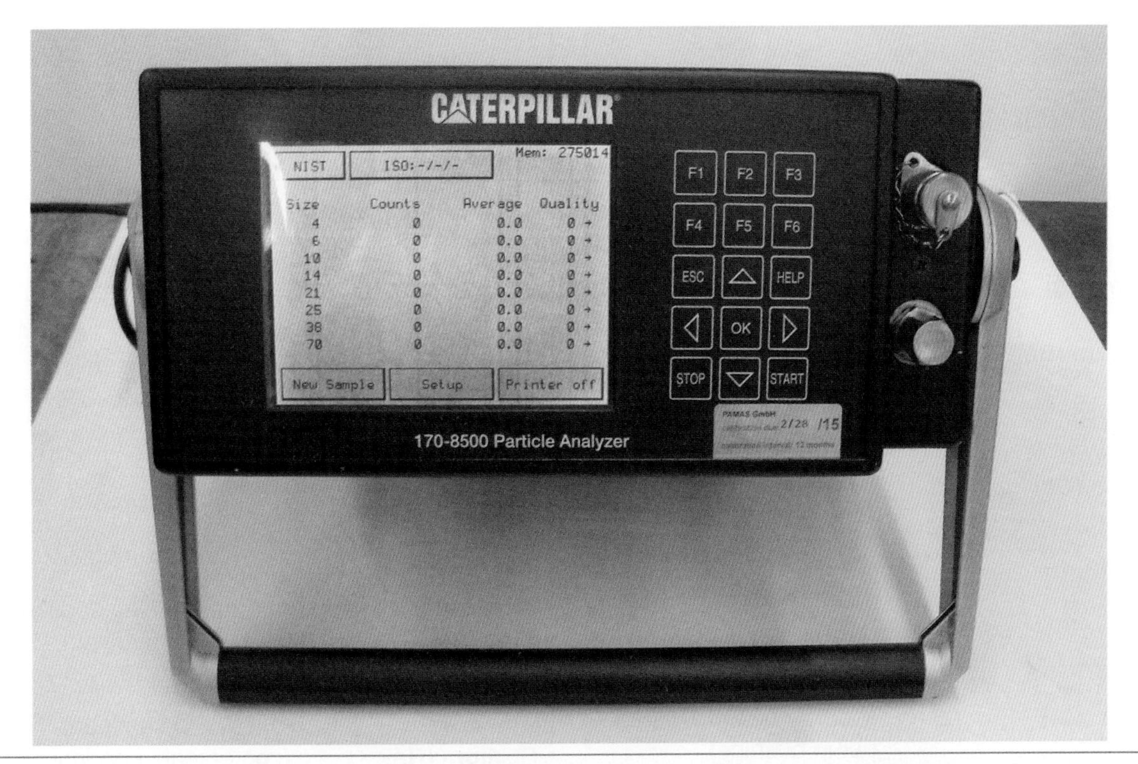

Figure 12-8. This particle counter generates an ISO range code and a NAS code. It can pull the oil from a sample bottle, or it can be directly connected to a machine's sampling valve.

Fluid Viscosity

Chapter 10 explained the procedure for measuring a fluid's viscosity. A fluid analysis report might list the fluid's viscosity in centistokes measured at 104°F (40°C) as well as provide a viscosity index (VI) number. The viscosity analysis is often labeled as an oil physical test. Three other tests labeled as fluid physical tests are moisture content, fuel content, and glycol content. Fuel and glycol content are more common for engine oil analysis.

Elemental Analysis

Labs perform ***elemental analysis***, also called elemental wear analysis, to identify traces of elements within the oil. The analysis has been carried out for decades by oil analysis laboratories. The test process is sometimes called atomic emission spectrocopy (AES) or optical emission spectrometry (OES).

The analysis consists of heating the oil to a high temperature, which causes the oil to atomize. Different waves of light energy are generated from the atomized oil. A ***spectrometer*** measures light waves for identifying the particles in an oil sample. Two different types of spectrometers can be used, a rotating disc electrode (RDE), or an inductively coupled plasma (ICP).

The RDE method can measure only particles that are less than approximately 8μm to 10μm in size. The ICP method can measure only particles smaller than 3μm. The lab report will list the elements found in the analysis in parts per million (ppm). Examples of elements included in oil analysis reports are listed in **Figure 12-9**. Some elements will be commonly found in oils other than hydraulic oils, such as gear box, transmission, or engine oils.

Element	Hydraulic System Source
Aluminum (Al)	Wear metals (cylinder gland, gear pump housing)
Antimony (Sb)	Wear metals (bearings) Contaminants (grease)
Barium (Ba)	R&O additive Contaminants (grease)
Beryllium (Be)	Wear metals
Boron (B)	Contaminants (dust, water) Oil additive
Calcium (Ca)	Oil additive
Chrome/ Chromium (Cr)	Wear metals (cylinder rod, roller bearing, spool valve)
Copper (Cu)	Wear metals (cooler leaching, gear pump side plate, piston pump slipper or bearing plate, pump bushing) Oil additive
Iron (Fe)	Wear metals (cylinders, bearings, pumps)
Lead (Pb)	Wear metals (bearings) AW additive

Element	Hydraulic System Source
Lithium (Li)	Contaminant (grease) Oil additive
Magnesium (Mg)	Oil additive
Molybdenum (Mo)	Oil additive
Nickel (Ni)	Wear metals
Phosphorous (P)	Oil additive
Potassium (K)	Contaminant
Silver (Ag)	Wear metals
Silicon (Si)	Anti-foaming additive Contaminant (dirt, grease, sealant)
Sodium (Na)	Contaminant (cooler leak, condensation) Oil additive
Tin (Sn)	Wear metals
Titanium (Ti)	Wear metals
Zinc (Zn)	Oil additives

Figure 12-9. Elements found in elemental wear analysis. The trace materials can be broadly categorized as contaminants, wear material, or oil additives, with each group encompassing a variety of sources.

When studying an elemental analysis report, having reference oil data is very important. Without having a good baseline of common additives used in the hydraulic oil, it will be difficult to determine if the trace element is a result of wear, contaminants, or an oil additive. The samples must also be trended over a period of time to see the rate of change in the elements. This can help determine if the additives are depleting over a period of time.

An additional test that can reveal oil additives is the Fourier transform infrared spectroscopy (FTIR). This test also uses a spectrometer. The oil sample is subjected to infrared radiation. Depending on the elements in the sample, a portion of the radiation is absorbed and the remaining radiation transfers through the oil. The spectrometer reads the wavelengths to determine the presence of elements.

Note

Some manufacturers produce analyzers that are compact and can be taken to the field. The Spectroil M/C-W is a compact RDE optical spectrometer. A2 Technologies manufactures a PAL series of FTIR portable analyzers.

Water Content

Three different tests can be performed by a lab to measure moisture in oils: hot-plate crackle test, Karl Fischer titration test, or Fourier transform infrared spectroscopy (FTIR). The hot-plate crackle test uses a plate similar to a kitchen frying pan. The plate is heated to 320°F (160°C) and a drop of oil is placed on the hot plate. If the oil crackles, or sputters, there is moisture in the oil. It is possible for the test to find water content as low as 500 ppm (0.05%) in the oil. The crackle test yields a yes (Y) or no (N) test result.

If the crackle test determines the presence of moisture, the Karl Fischer titration test is usually performed to obtain a more accurate reading of the exact water content. The Fischer test provides results of water content levels in units of parts per million (ppm). The minimum amount of moisture content for setting a lab alarm can be as little as 50 ppm. The Karl Fischer titration test is the most popular moisture test used by labs.

Acid Number

Labs sometimes report an oil sample's ***acid number (AN)***. The number is derived by inputting the results from a FTIR analysis into statistical software. The program generates an AN value. The value is used for non-engine oils, such as hydraulic fluids. It is a measure of the fluid's acidic contamination, additive depletion, and an indirect measure of the fluid's rate of oxidation. The AN value can be listed as TAN, which stands for total acid number.

Additional Tests

Oil analysis laboratories conduct additional tests on fluids. For example, labs measure the oil's base number (BN) and run a ferrous density test on the sample. The BN test is normally performed only on crankcase oils. The ferrous density test is normally performed only on non-filtered gearboxes.

Improving Contamination Control

Numerous actions can be made to improve contamination control within hydraulic systems. The actions should focus on preventive measures and reactive measures, both of which are useful and necessary. When contamination control actions are executed correctly, they can be some of the most cost-effective actions taken by equipment owners.

Storage of Components, Tools, and Test Equipment

Consider all of the dirt-generating activities that take place inside the average repair shop. Grinding, filing, welding, machining, cleaning, servicing, and sweeping all create or spread dirt and contaminant particles. The proper storage of components, tools, and test equipment is critical to prevent dust and dirt from entering components. New parts, disassembled components, and stored components must all be housed in a fashion to prevent contamination from entering the hydraulic ports.

Plastic and steel caps and plugs, as well as lids, sacks, and containers are all used to prevent dust and dirt from migrating into the internal passages of fittings, hoses, tubing, valves, pumps, motors, and cylinders. When a machine is being serviced, a steel plug or cap will outperform a plastic cap at preventing contamination. When components are not in use, for maximum protection they should be capped, placed inside a clean, sealed container, and stored inside a clean box in a clean and dry storage environment.

Some hydraulic personnel have filled hydraulic components such as cylinders full of hydraulic oil prior to storing the component. Their intent is to prevent corrosion, which can occur due to the trapped air left inside a cylinder. Keep in mind that if a cylinder is filled full of hydraulic oil and capped with steel caps during a cold winter day, then transported to a hot summer job site, the thermal effect from the sun can cause the pressure to increase tremendously within the cylinder. When the steel caps are removed, this extreme pressure buildup will be released, possibly causing personnel injury or equipment damage. Temperature must be considered if cylinders are going to be stored with hydraulic fluid inside of them.

Temperature fluctuations can also create issues with test equipment. For example, a technician plugged both ends of a flowmeter together in the cool atmosphere of a shop. While at the job site, the thermal effect of the sun increased the pressure in the flowmeter's leads, making it impossible to unfasten the flowmeter's couplings. The flowmeter had to be set in the shade to cool before it could be uncoupled and used on the tractor.

New Hoses

One of the dirtiest components found in a hydraulic system is a new hydraulic hose. Most hydraulic hoses use steel reinforcement in their construction, which requires the use of an abrasive saw to cut the hose. The saw blade generates a tremendous amount of residue and smoke as it cuts through the multiple layers of the hose. See **Figure 12-10**.

All new hoses must be cleaned prior to installation. The most successful industry practice is to use a hose cleaning gun, which uses air pressure to shoot a foam projectile directly through the inside of the hose. **Figure 12-11** shows a complete projectile hose cleaning kit

Caution

Although not advisable, some technicians use just a spray of compressed air through a hose or tube to clean it. Using compressed air does not remove the small iron and rubber particles that adhere to the walls of the hose or tube. This method is a shortcut to proper hydraulic hose cleaning. Lack of care taken at this step could lead to system damage and future service problems for the machine.

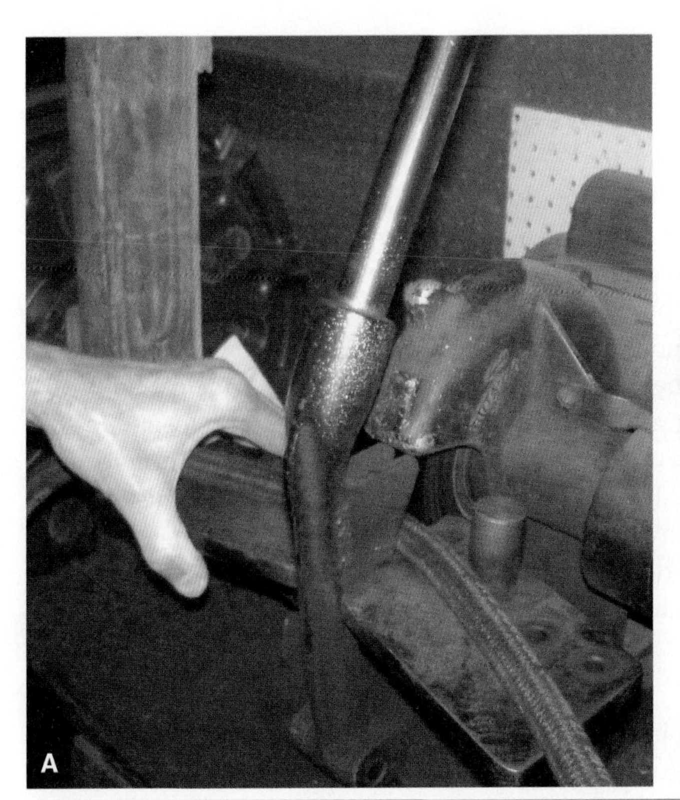

Figure 12-10. Cutting a hydraulic hose generates contaminants. A freshly cut hydraulic hose must be properly cleaned and then capped with plugs until installation. A—An abrasive disc cuts quickly, but produces a lot of residue. B—A steel waffle blade cuts more slowly, but produces less residue.

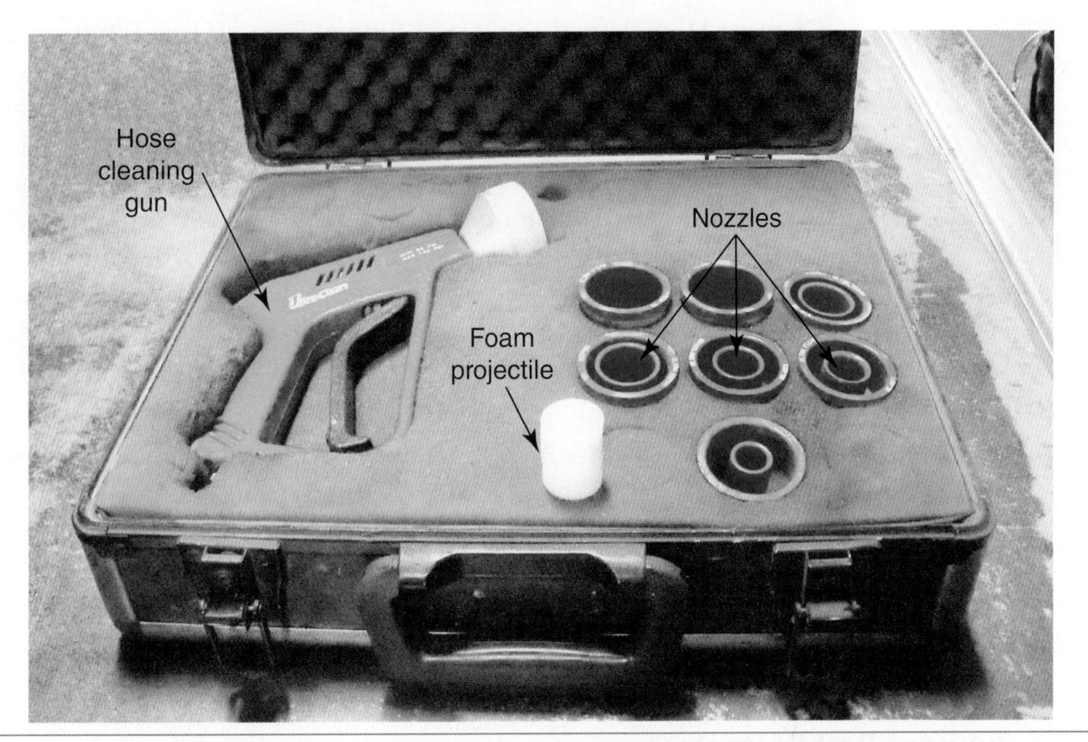

Figure 12-11. A hose cleaning gun, also called a launcher, is used to clean the contaminants from a freshly cut hydraulic hose. It uses air pressure to push a foam projectile completely through the inside of the hose. Each nozzle fits onto the barrel end of the gun and slides into the end of the dirty hose to create an airtight seal before the projectile is launched.

The projectiles are sized slightly larger than the inside diameter of the hose. It is a good practice to shoot the projectile through the hose after it is cut, but before the hose is crimped. A new foam projectile should be shot through the hose in both directions. Continue using new foam projectiles, alternating the direction the projectile is shot through the hose, until the projectile comes out of the hose clean. After the fittings are crimped onto the hose, the hose should be cleaned again using the cleaning gun. Immediately following this, the hose ends should be capped to prevent contamination.

Technicians that are most concerned with contamination control will also use a spray or liquid cleaner on a hose after the hose has been cleaned with a projectile from a cleaning gun. Before spray or liquid cleaner is used, be sure that it is approved and will not cause compatibility problems with the hydraulic fluid or component seals. Unfortunately, many technicians have used brake cleaner when cleaning a hydraulic hose. Brake cleaner will damage O-rings and seals. A common solution is to use a lower-viscosity hydraulic fluid that is compatible with the original oil.

Filtering New Oil

As mentioned earlier in this chapter, manufacturers and contamination specialists provide cleanliness targets for new hydraulic oil, such as an ISO 14/11. Brand new oil will not meet this stringent requirement. Therefore, new hydraulic oil should be cleaned before it is added to a hydraulic system.

Filter Carts

Chapter 11 explained the use of kidney loop off-line filtration carts. The carts can be used as a reactive measure anytime the system becomes contaminated, and needs to be cleaned. The carts should also be used as a preventative measure, and anytime the machine is in the shop for a repair or service. Fitting the reservoir with quick connectors increases the chances that a filter cart will be used. An ***ecology valve*** (or ***drain valve***) is another style of fitting used. The ecology valve has a built-in check valve that is opened when the mating fitting is threaded onto the drain valve and closes when the fitting is loosened. This allows for a leak-free connection, making it easier to drain or filter the reservoir.

Oil Sampling Ports

Earlier in the chapter it was mentioned that using a vacuum pump to pull oil samples from the reservoir is the least preferred oil sampling method. If a machine does not contain oil sampling valves, consider adding the valves to the machine to perform the following functions:

- Increase the speed of the sample taking process.
- Ensure the sample is representative of the hydraulic system.
- Increase the chances that samples will be taken.

Hydraulic Cylinder Rod Protectors

Some machine work environments have more airborne dust and potentially corrosive contaminants than other environments. Some manufacturers recommend the use of protective boots and bellows to be placed over the hydraulic cylinder rods to prevent dust from entering through the cylinder gland.

Pressure Taps

Most of today's late-model machines come already configured with several hydraulic test ports for measuring system pressures. See **Figure 12-12**. Although the ports do have a slight chance of leaking or allowing dirt to migrate into the system, if the dust covers are properly used, the ports are a helpful preventative measure in contamination control. The test ports enable a technician to quickly check a system pressure without having to open up the hydraulic system by removing a threaded plug.

Oil Drum Covers

Large drums of oil are bulky and cumbersome. Unfortunately, due to these facts, they are sometimes stored outside in a bad environment. Ideally, the drums should be stored on their side as described in **Chapter 1**. If drums must be stored in a vertical position, they should be kept in a clean and dry environment indoors and covered with drum covers.

Reconditioning a Contaminated Hydraulic System

Several conditions can warrant the serious action of reconditioning a hydraulic system back to a clean condition:

- The fluid has been overheated, degradation has taken place, the oil is badly oxidized, the oil has varnish, or the oil is full of contaminants.
- An unexpected, catastrophic pump or motor failure has taken place, sending large quantities and large-sized contaminants throughout the hydraulic system.

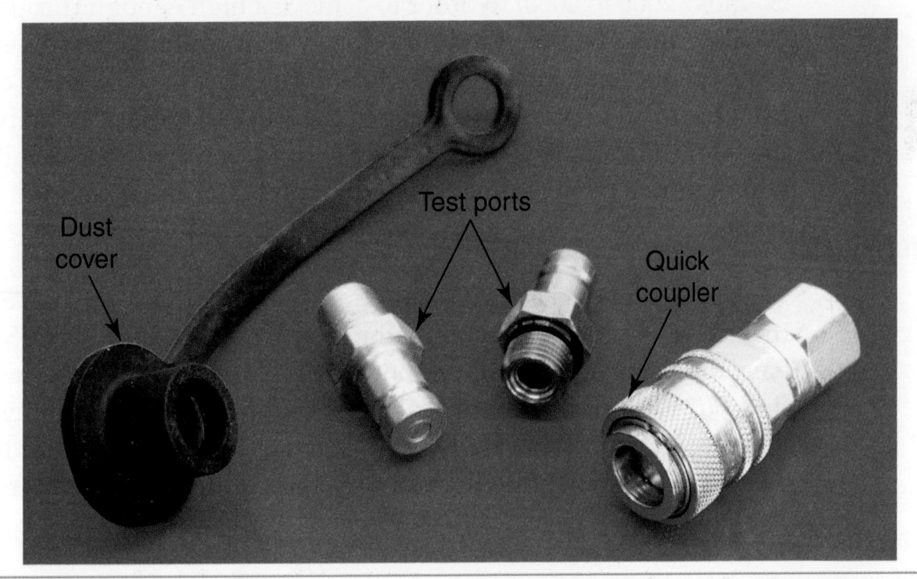

Figure 12-12. Hydraulic test ports are commonly used in most mobile machinery as a fast method for testing hydraulic pressures. The male tip is left on the machine, covered by a dust cover. A quick coupler remains on the pressure gauge so it can be quickly coupled to the machine. **Chapter 21** will provide more information about hydraulic test equipment.

Owners are left with the tough decision of how to recondition the hydraulic system. The following tactics are examples of the actions necessary to restore a hydraulic system back to a normal, clean condition:

- Double oil change.
- Power flush.
- Chemical power flush.
- Physically clean the system.

Caution

Always read equipment manufacturer recommendations prior to reconditioning any hydraulic system. Procedures and steps will vary by manufacturer. Some practices or products may damage certain hydraulic systems.

Double Oil Change

Owners that choose to perform a double oil change will first change the oil after it has been heated to operating temperature. The fluid and filters are changed and the system is cycled to ensure the old oil has had a chance to be returned to the reservoir. Then the fluid and filters are changed again. Depending on the volume of fluid in the hydraulic system, replacing the fluid and filters twice can be very costly and is less environmentally friendly.

Power Flush

A filter cart or a flushing machine can be connected to a hydraulic system for the purpose of flushing a fluid through the system. The goal is to use turbulent oil flow. Sometimes low-viscosity oil that is compatible with the original oil is used. Using a lower-viscosity oil with a higher-volume pump can increase the fluid turbulence. Flushing experts provide a target Reynolds number (for example, 2000 to 4000) in order to achieve a high enough turbulent flow.

Care must be taken anytime the system contains servo valves and other components with micron-rated tolerances. The servo valves can be isolated from the flush to prevent damage to the valves. Filters are also replaced. Check the filter indicator to see if the filter is full prior to the flush. If it is, the system will require a filter change both before and after the flush.

Chemical Power Flush

Using a chemical in conjunction with a power flush has serious risks regarding the chemical's compatibility with O-rings and seals. Some personnel have unwisely used brake cleaner or diesel fuel to clean a system, only to later find out it caused multiple seal failures due to incompatibility. If a chemical flushing method is desired, get expert advice to ensure that the chemical is compatible and that the chemical has been completely rinsed from the system.

Physically Clean the System

One common technique for reconditioning a system is to physically clean the system. All of the oil is drained from the reservoir, hoses, and other compartments. The reservoir is internally cleaned. Filters and strainers are replaced. Other components near the contamination site are also disassembled and cleaned. For example, if one of the machine's pumps had a catastrophic failure, and if the pump fed a steering priority valve, the steering priority valve must also be disassembled and cleaned. Foam projectiles are used to clean the interior of hoses.

Care must be taken to not cavitate the hydraulic system during start-up. Be sure to have the system pressure gauge in sight when restarting the system. Immediately shut the system off if the pressure is too low or too high. Be sure the system is started under no load. Pay close attention to the fluid level in the reservoir during the restart process, not allowing it to drop too low.

Changing Fluid Types

When an owner decides to make a complete change of hydraulic fluid—moving from petroleum fluids to environmentally acceptable fluids or fire-resistant fluids—a complete system flush is often required. Seek special assistance to avoid cross-contamination when changing the base type oil of a hydraulic system.

Case Study

Contamination Control in the Repair Shop

A closed-loop hydrostatic transmission suffered a catastrophic transmission failure. Due to his knowledge of the machine's working conditions, the owner demanded that the repair process be completed with a strong emphasis on proper contamination control techniques. Technicians began the repair with the following contamination control steps at the forefront of their minds.

Education: All dealership personnel, including parts, service, sales, and managers, must be properly educated on contamination control, including the causes of contamination, the consequences, and the correct measures for preventing contamination.

Cleaning: Prior to bringing the machine into the shop, the entire machine, including all of the components, needs to be meticulously cleaned to ensure that all mud, grime, and dirt are removed. Mud removal requires a high volume of water. Grease and challenging dirt requires pressurized hot, soapy water.

Shop Environment: Contamination control requires a new standard for shop cleanliness. Past shop practices, such as cluttered, dirty, and greasy working areas with poor lighting, must be eliminated. Today's shop must emulate a new standard in shop cleanliness.

- The shop's wash bay, apron leading into the wash bay, and path into the shop must be thoroughly cleaned after every machine or component has been cleaned.
- The shop doors should be fully operational.
- The shop floors should be clean and sealed.
- Machine stalls and work areas are clearly defined with stripes that are not flaking or peeling.
- Clean component stands and cribbing blocks are effectively used to enable staff to keep a clean shop floor.

- Workbenches have clean protective tops to provide a clean working environment.
- Shop practices that generate constant debris, such as grinding, welding, cutting, and deburring are performed in a different shop location. Activities that require intermittent debris are confined (i.e. using screens to contain contaminants within the immediate area).

Component Removal: Prior to removing components, technicians must have access to a clean stock of plugs, caps, and plastic wrap.

- Hose ends and tubing must be capped with the proper plug or cap.
- Components with machined surfaces must be wrapped in plastic that contains a rust inhibitor, especially if components are stored long term or in a climate with high moisture content. Flash rust can appear on a machined surface relatively quickly.
- During breaks such as lunches and shift changes, components must be covered with clean towels or cloths.
- Keep new components in their original, clean shipping packaging until they are ready to be installed on the machine.

Clean Rooms: Individual components with tight tolerances such as pumps, motors, and valves are disassembled, inspected, and rebuilt using clean rooms.

- Clean rooms are specially designed rooms that use a high-efficiency particulate air (HEPA) filter to remove contaminants from the air.
- They use specially designed heating, air conditioning, and ventilation systems to ensure that the room has a clean, dry atmosphere with a positive air pressure to prevent dirt and dust from entering when the door is opened.

Cleaning Solution: Components are cleaned only with approved cleaners, such as mineral spirits or clean solvent. Brake cleaner and other non-approved cleaners are harmful to seals, O-rings, and hoses.

- Solvent tanks are filtered to meet stringent ISO cleanliness levels, which were discussed in **Chapter 11**.
- Air guns must use air that has been filtered for moisture and contaminants. Power tools use clearly marked air lines that contain oilers.

Catastrophic Failures: In the event of a catastrophic failure, such as this closed-loop hydrostatic transmission failure, the entire system must be methodically disassembled and the components must be meticulously cleaned. Only lint-free towels should be used for cleaning hydraulic components.

- The reservoir must be thoroughly cleaned by removing the clean-out plate and thoroughly cleaning the reservoir's walls.
- Use a hose cleaning gun with a foam projectile to clean hydraulic hoses and tubing. Shoot a new foam projectile through the line in both directions until the projectile comes out clean. At a minimum, the ends of the lines should be disconnected even if the machine has dozens of hoses.

- If components are not systematically cleaned, the machine will most likely experience a repeat failure.
- Flush and back flush the hydraulic oil cooler.
- Replace the filters and strainers.

Installing Oil:

- If using new oil or reusing oil, filter the new oil so that it is two or more ISO range codes cleaner than the machine or component's cleanliness target.
- Assembly oil and lubricants must also be filtered to two or more ISO codes cleaner than the machine or component's cleanliness target.

Additional Contamination Control:

- New and rebuilt components are contained on their shipping stand and remain wrapped in plastic with all ports plugged and capped.
- Immediately clean all spilt fluid. Do *not* use granular floor dry.
- Clean the floors at the end of each work shift and additionally as needed.
- Measure the system's particle count. If the count is too high, use a kidney loop cart to bring the machine's hydraulic system back into the ISO cleanliness specification before releasing the machine to the customer.
  - Proper use of the kidney loop cart requires bringing the oil up to operating temperature and cycling the hydraulic implements. Thirty to forty percent of the system's oil is located in the plumbing and actuators.
  - A rule of thumb is that the oil must pass through the filter cart 35 times to account for the mixing of clean oil with the dirty oil and for the filter to achieve its efficiency rating.
    - For example, if the machine contained a total volume of 30 gallons of hydraulic oil, and if the filter cart pump had a flow rate of 4 gpm, the time the filter cart must run would be calculated as follows:

      30 Gallons × 35 = 1050 gallons

      1050 gallons/4 gpm = 263 minutes (assuming the implements were being cycled with the oil at operating temperature)
  - Oil that is seldom cycled back to the reservoir will take longer to be filtered because the kidney loop cart is cleaning only the reservoir's fluid.
  - Pumps and motors that are tested on test benches must also be measured to ensure they meet the ISO cleanliness target and flushed if needed.

After complying with all of the proper contamination control procedures, the technicians completed the repair on the hydrostatic transmission. The machine was returned to the owner who was satisfied with the repair. The technicians instructed the owner on how to develop an oil sampling schedule for the machine. Trended oil analysis can provide useful insight about the contamination conditions of the machine's hydraulic systems so preventive steps can be taken against possible future problems.

Summary

- ✓ Hydraulic system contamination can occur at the manufacturing plant, be caused by wear from everyday operation or by poor maintenance and bad service practices.
- ✓ Fluid analysis can help extend the useful life of a machine, maximize machine availability, reduce repair costs, and prevent a machine from losing hydraulic efficiency.
- ✓ ISO range codes are the gold standard for targeting the cleanliness of a hydraulic system.
- ✓ NAS codes and ISO codes are given by particle counters to describe the cleanliness of a hydraulic system.
- ✓ Microscopes are used to determine the origin of large particles found in fluid samples.
- ✓ A sample of the new oil is required for a future reference baseline to compare the depletion of additives.
- ✓ Oil samples must be trended, with samples taken and analyzed at scheduled intervals.
- ✓ Using a vacuum pump to draw oil samples out of the reservoir is the least preferred oil sampling method.
- ✓ Onboard oil sampling valves should be located downstream from the components and upstream from the filters.
- ✓ Emission spectrometers are used in elemental analysis to identify wear metals, contaminants, and additives.
- ✓ The crackle plate test and the Karl Fischer titration test are frequently used to test the moisture content of an oil.
- ✓ An acid number is generated using the FTIR, and statistical software provides an indication of a fluid's acidic contamination, additive depletion, and an indirect measure of the fluid's rate of oxidation.
- ✓ Equipment owners, shops, and service personnel can take multiple preventative and reactive actions for improving contamination control.
- ✓ If a hydraulic system is badly contaminated, owners have multiple choices for restoring the system back to a clean condition.

Technical Terms

acid number (AN)
drain valve
ecology valve
elemental analysis
ISO range code
National Aerospace Standard (NAS)
particle counter
reference oil
sampling probe
sampling valve
spectrometer
trended
vacuum pump

Review Questions

Answer the following questions using the information provided in this chapter.

1. What percentage of hydraulic failures can be attributed to fluid contamination?
   A. 40%.
   B. 50%.
   C. 70%.
   D. 90%.
2. Which of the following is the optimum method of taking a hydraulic oil sample?
   A. Using a vacuum pump to draw oil from the reservoir.
   B. Using a sampling valve before a return filter.
   C. Using a sampling valve in the pump suction line.
   D. Using a sampling valve after a pressure filter.
3. All of the following steps should be performed when taking a hydraulic oil sample, *EXCEPT*:
   A. wait one week before sending the sample to the laboratory to be sure that the machine has no unexpected failures.
   B. take the sample at the end of a machine's work shift.
   C. fill the bottle 2/3rds full for ISO VG 32 to 100 cSt.
   D. fill the bottle half full for ISO VG higher than 100 cSt.
4. All of the following steps should be performed when taking a hydraulic oil sample, *EXCEPT*:
   A. use a new, clean sample bottle.
   B. use new, clean tubing.
   C. collect the sample by removing the reservoir drain plug.
   D. warm the hydraulic fluid to operating temperature.
5. Which of the following steps should be performed when taking a hydraulic oil sample?
   A. Fill the sample bottle full.
   B. Keep the uncapped bottle in your shirt pocket until it is time to take the sample.
   C. Take the sample with the fluid at operating temperature.
   D. Take the sample before the machine starts its daily work shift.
6. When taking a hydraulic oil sample with a vacuum pump, Technician A states it is best to use a dedicated pump for sampling coolant and a dedicated pump for sampling oil. Technician B states that you should prevent the oil inside the sample bottle from touching the tube that is protruding out of the vacuum pump's head. Who is correct?
   A. Technician A.
   B. Technician B.
   C. Both A and B.
   D. Neither A nor B.
7. Most sampling valves are used in systems running at what pressure value?
   A. 60 psi (4 bar) and lower.
   B. 500 psi (34 bar) and lower.
   C. 1500 psi (103 bar) and lower.
   D. 2500 psi (172 bar) and lower.
8. Technician A states that an onboard sampling valve should be flushed before a sample is drawn from it. Technician B states that before a sample is drawn from a sampling valve, the dust cover must be removed and the valve wiped with a lint-free cloth. Who is correct?
   A. Technician A.
   B. Technician B.
   C. Both A and B.
   D. Neither A nor B.

9. Oil sample bottles range in what sizes?
   A. Quarter to a half ounce.
   B. 1 to 6 ounces.
   C. 20 to 30 ounces.
   D. None of the above.
10. Oil sample bottles are commonly made of what material?
   A. Plastic.
   B. Glass.
   C. Aluminum.
   D. Both A and B.
11. What does SOS stand for in regard to hydraulic oil contamination control?
   A. Scheduled oil sampling.
   B. Sampling oil system.
   C. Send oil sample.
   D. Saving oil sample.
12. Which of the following is considered a *reference oil*?
   A. A standard viscosity referenced by manufacturers.
   B. The oil fill level on a reservoir.
   C. A sample of new oil to be used in a hydraulic system.
   D. The type of oil recommended by an oil lab.
13. The latest ISO range codes classify what size of contaminant particles?
   A. 2µm, 5µm, 15µm
   B. 4µm, 6µm, 14µm
   C. 6µm, 8µm, 14µm
   D. 2µm, 4µm, 14µm
14. What does each number listed in an ISO range code indicate based on the location of the number within the code?
   A. The number of contaminant particles sized within a certain micron range.
   B. The ratio of contaminant size to particle number.
   C. The contaminant material's weight in grams.
   D. None of the above.
15. An oil sample received an ISO range code of 21/19/16. What does the 21 signify?
   A. Number of contaminant particles that are 2µm and larger.
   B. Number of contaminant particles that are 4µm and larger.
   C. Number of contaminant particles that are 6µm and larger.
   D. Number of contaminant particles that are 14µm and larger.
16. Which of the following ISO range codes is the cleanest?
   A. 20/18/16
   B. 19/17/15
   C. 18/16/14
   D. 17/16/13
17. ISO range codes are established based on the number of contaminant particles located in what quantity of oil?
   A. 1 milliliter.
   B. 10 milliliters.
   C. 100 milliliters.
   D. 1 liter.
18. NAS contamination code was initially developed for which industry?
   A. Agricultural industry.
   B. Aircraft industry.
   C. Construction equipment industry.
   D. Forestry industry.
19. Choose the NAS code that would contain the dirtiest hydraulic system.
   A. 11
   B. 10
   C. 4
   D. 3
20. A single NAS code provides contamination numbers for how many different contaminant particle classifications?
   A. Two.
   B. Three.
   C. Four.
   D. Five.

21. What type of fluid is used to flush a particle counter?
    A. Mineral spirits.
    B. Diesel fuel.
    C. Brake cleaner.
    D. Solvent.
22. Which of the following oil analysis test categories is responsible for measuring an oil sample's viscosity and water, fuel, or glycol content?
    A. Particle counting.
    B. Particle origin identification.
    C. Physical testing.
    D. Elemental wear analysis.
23. Which of the following oil analysis tests uses high-power microscopes?
    A. Particle counting.
    B. Particle origin identification.
    C. Physical testing.
    D. Elemental wear analysis.
24. Which of the following oil analysis tests lists the oil contaminants in micron sizes, for example 4μm, 6μm, 10μm, 14μm, 21μm, 25μm, 38μm, and 70μm?
    A. Particle counting.
    B. Particle origin identification.
    C. Physical testing.
    D. Elemental wear analysis.
25. Which of the following oil analysis tests is one of the oldest tests performed by oil labs and has three main categories: wear, contamination, and additives?
    A. Particle counting.
    B. Particle origin identification.
    C. Physical testing.
    D. Elemental wear analysis.
26. Elemental wear analysis uses what type of device?
    A. Particle counter.
    B. Microscope.
    C. Spectrometer.
    D. Viscometer.
27. Which of the following is the best procedure for providing accurate measurements of moisture content in a hydraulic oil sample?
    A. Particle origin identification.
    B. Physical testing.
    C. Elemental wear analysis.
    D. Karl Fischer titration test.
28. The rotating disc electrode (RDE) and inductive couple plasma (ICP) devices are used in what type of analysis?
    A. Particle counting.
    B. Particle origin identification.
    C. Physical testing.
    D. Elemental wear analysis.
29. What is recommended to clean a new hydraulic hose?
    A. Air-operated hose cleaning gun and a foam projectile.
    B. Can of brake cleaner.
    C. Can of diesel fuel.
    D. Solvent cleaning station.
30. A hydraulic system has a large amount of contamination. Which of the following actions has the most risk of harming O-rings and seals?
    A. Double oil change.
    B. Power flush.
    C. Chemical power flush.
    D. Physically cleaning the system.

Accumulators can be used for a number of reasons in a mobile hydraulic system. Wide sprayer booms use an accumulator to dampen the shock loads caused by the implement as it bounces up and down while traveling.

Chapter 13 Accumulators

Objectives

After studying this chapter, you will be able to:

- ✓ Describe the different functions of accumulators.
- ✓ Explain the advantages and disadvantages of the different types of accumulators.
- ✓ List safety precautions for working around and with accumulators.
- ✓ List the different methods that can be used for checking precharge pressure.
- ✓ Explain how to check the precharge pressure using a nitrogen service tool.
- ✓ Explain how to charge an accumulator with nitrogen gas.
- ✓ Describe the uses of additional accumulator tools and accessories (bottles, fuses, charging pumps, sensors).
- ✓ Explain the different types of hydraulic hybrid systems.

Functions and Applications

Accumulators are hydraulic energy storage devices, similar to capacitors that store electrical energy. The accumulator uses one of three mechanisms, gas pressure, springs, or weights, to resist hydraulic fluid entering the accumulator. When the oil pressure rises high enough to overcome the accumulator's resistance, hydraulic oil will enter the accumulator.

Accumulators can serve a variety of purposes:

- Storing energy for future use.
- Dampening pressure spikes.
- Easing the buildup of pressure.
- Maintaining a set pressure.
- Recovering energy.

Store Energy

One of the most popular functions of accumulators in the mobile industry is to store energy as a safety measure in case the machine loses the ability to supply hydraulic energy. Two common energy storage applications are for service brakes and steering. If the engine quits, or if the hydraulic pump fails, the accumulator is designed to provide enough energy to bring the machine to a safe stop. See **Figure 13-1**.

A check valve isolates the pump from the accumulator. Otherwise, when the system is shut off, the accumulator could force the pump to rotate backwards, causing damage to the pump.

Dampen Pressure Spikes

External loads can cause shock loading in actuator circuits. Accumulators are used to dampen the pressure spikes and absorb shock loads. Some examples of mobile applications include combine headers, sprayer booms, front suspensions, and tractor loaders.

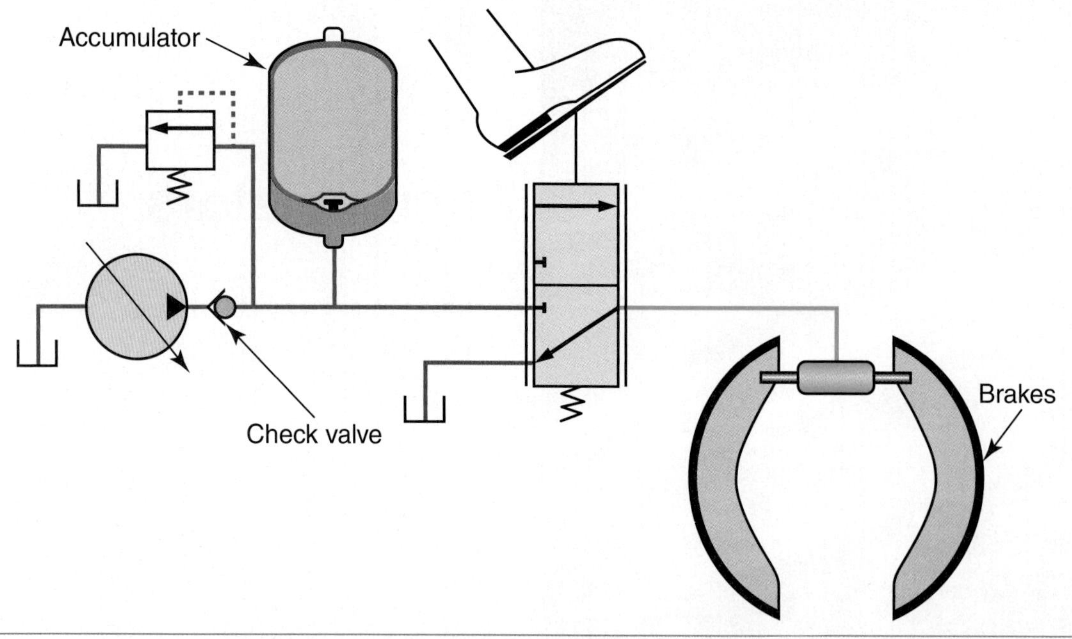

Figure 13-1. An accumulator is commonly used to store hydraulic energy for service brakes. In the event that the pump fails or the engine stops, the accumulator will provide energy to allow the operator to apply the brakes.

Combine headers and sprayer booms are wide implements that will cause shock loads while the machine is traveling. The accumulator is designed to dampen the shock as the implement bounces up and down. Figure 13-2 shows that a solenoid valve can be used to connect the accumulator to the circuit. In some applications, the machine designers enable the operator to de-energize the solenoid so that the hydraulic cylinder has a precision response with no delays, such as using automatic header height to harvest soybeans. However, during transport or when cutting standing crops, the solenoid can be energized to allow the accumulator to dampen the header lift circuit.

As mentioned in **Chapter 7**, actuator circuits might have no protection when the DCV is in the neutral position. An accumulator or a cylinder port relief valve will provide protection for the actuator circuit.

Ease the Buildup of Pressure

Accumulators are also used to soften the final extension of an actuator, such as a large press in an industrial plant or a clutch pack piston in a transmission. See Figure 13-3. Orifices are also sometimes used together with the accumulator or used in place of an accumulator to help ease the buildup of pressure.

Maintain a Set Pressure

Accumulators can be used to maintain a set system pressure. This application is more common for an industrial application using a weighted accumulator. However, Figure 13-4, provides an example of a combine separator clutch application. In this example, as in Figure 13-3, the accumulator will soften the engagement of the clutch, and will help maintain an apply pressure to the clutch. The clutch is controlled by a two-position solenoid control valve.

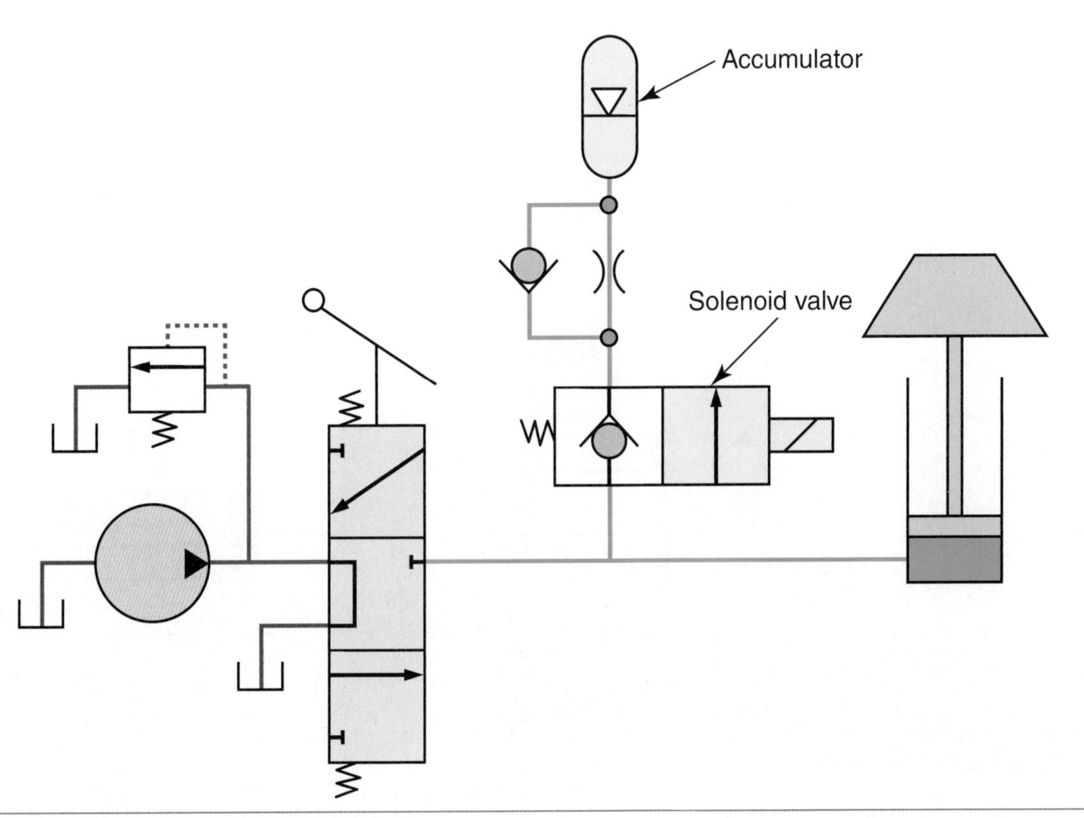

Figure 13-2. Many combines have an accumulator that is used to dampen the header. In this application, a two-position electrically controlled solenoid enables an operator to connect and disconnect the accumulator to the header lift circuit.

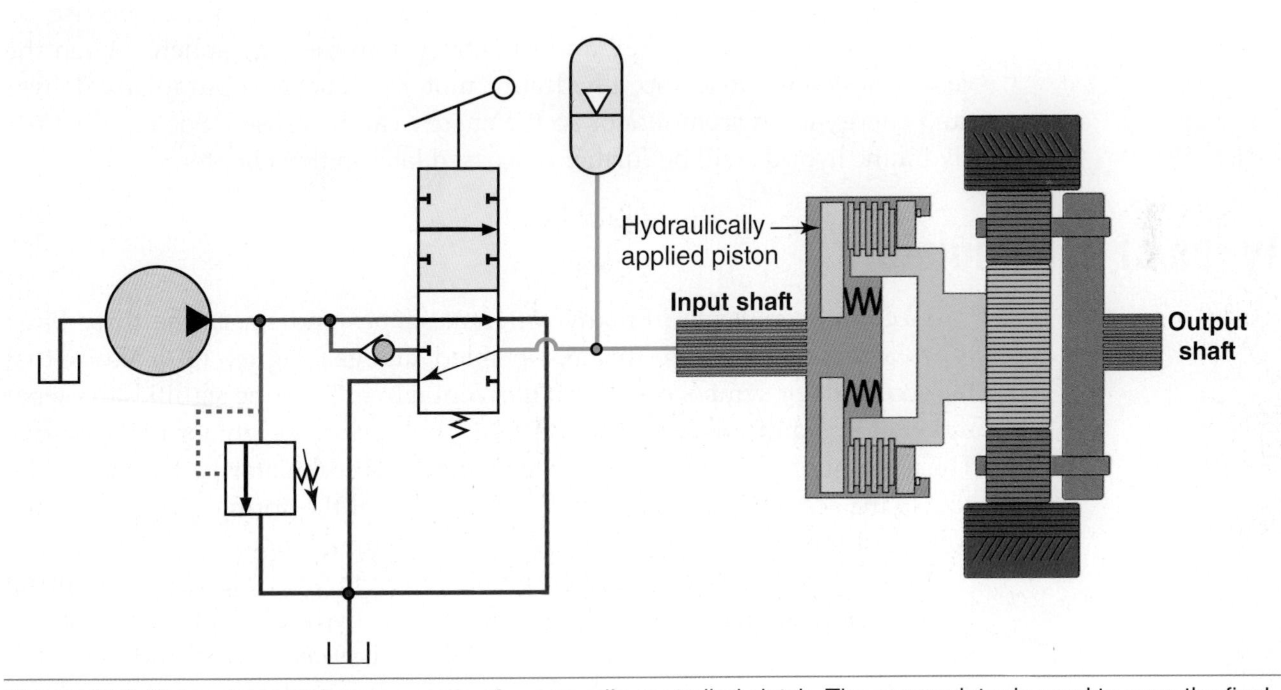

Figure 13-3. This schematic is an example of a manually controlled clutch. The accumulator is used to ease the final pressure buildup of the clutch engagement.

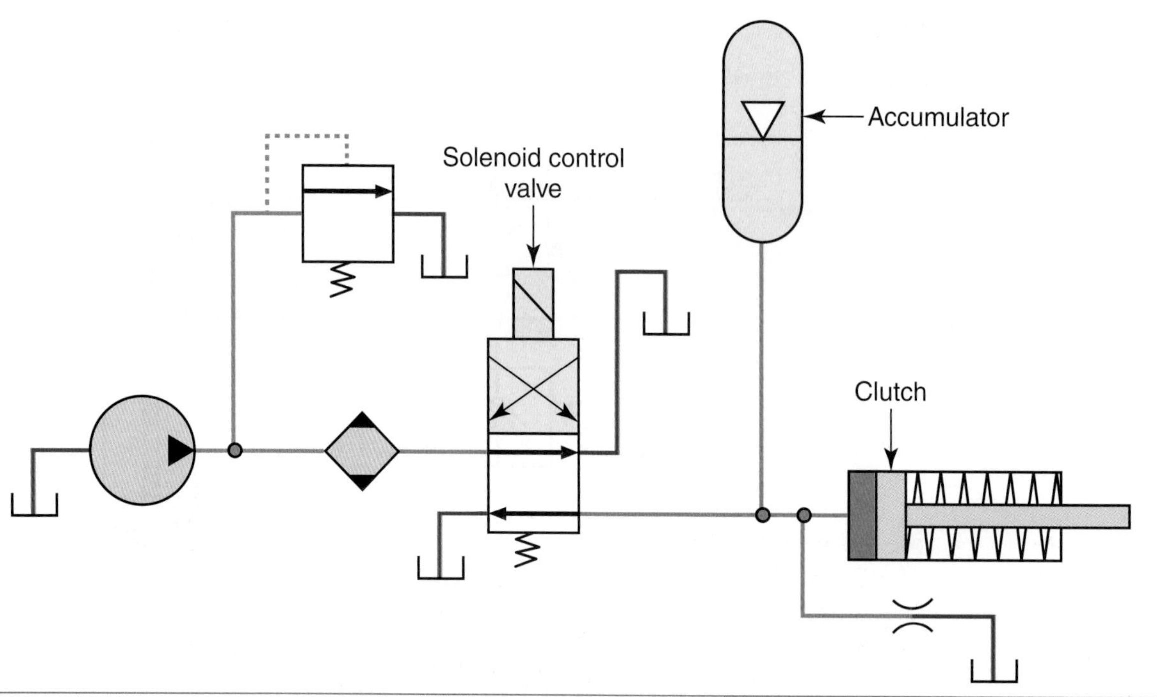

Figure 13-4. This is an example of a combine separator clutch that uses a two-position solenoid control valve. The accumulator helps maintain a clutch apply pressure on the separator clutch.

Recover Energy

Hydraulic hybrids use a pump and motor coupled with an accumulator for the purpose of recovering hydraulic energy that would otherwise be transformed into heat energy and dissipated into the atmosphere. When the vehicle is slowing to a stop, a hydraulic motor will act as a pump and deliver fluid energy to an accumulator so the energy can be reused. See **Figure 13-5**. Hydraulic hybrids will be further discussed later in this chapter.

Types of Accumulators

Accumulators come in many different shapes and sizes. The three basic types of accumulators are spring, weighted, and gas, **Figure 13-6**. Notice that the accumulator symbol contains a horizontal line. This line signifies the separation of hydraulic oil from the accumulator's spring, weight, or nitrogen gas. The amount of oil pressure required to fill the accumulator is determined by size of the accumulator piston and the strength of the spring, the mass of the weight, or the pressure of the nitrogen-filled accumulator.

Gas accumulators are used in mobile equipment. Weighted and spring accumulators are not commonly used in mobile hydraulic systems, although spring accumulators are used in some transmissions. Spring and weighted accumulators can be classified as mechanical accumulators because they do not use any type of gas. Each style of accumulator has its advantages and disadvantages.

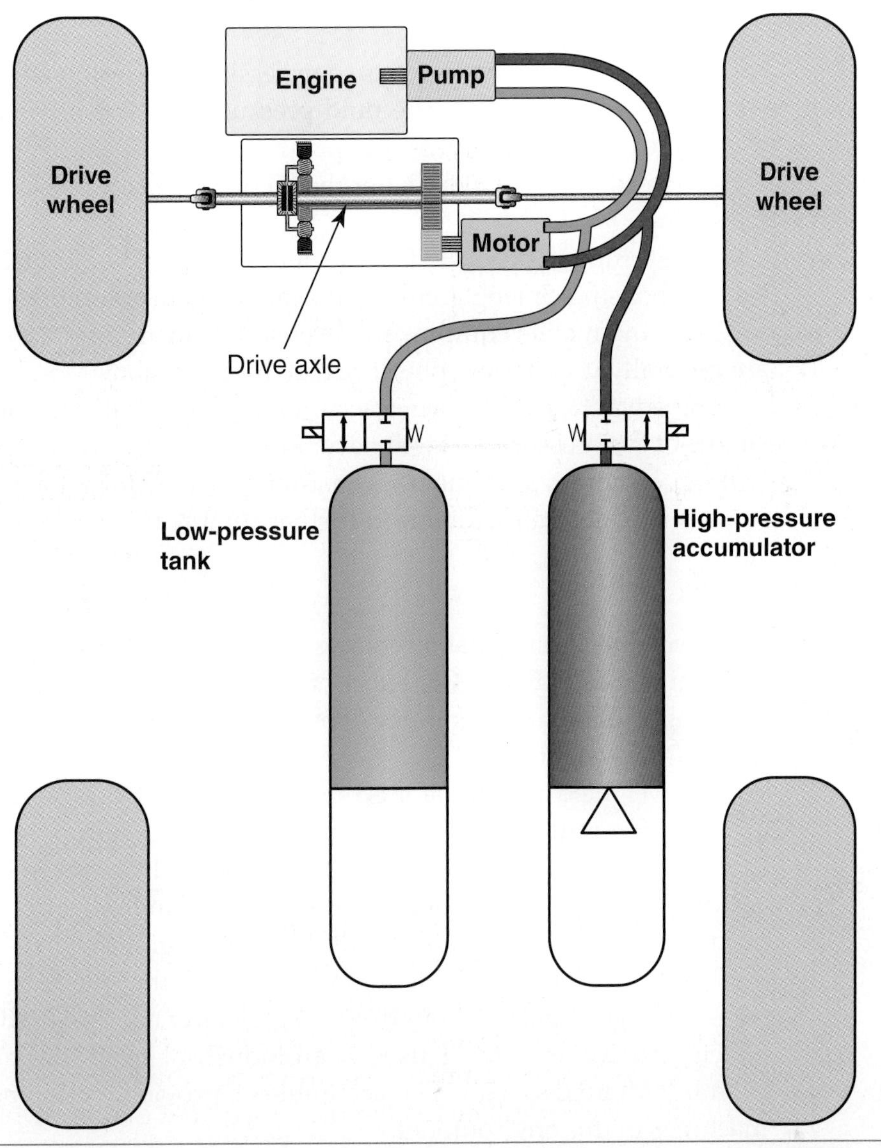

Figure 13-5. Hydraulic hybrids use a pump motor combination in conjunction with an accumulator. The accumulator recovers energy as the vehicle slows to a stop. The stored energy is then used to assist in propelling the vehicle from a stop. This example is a series hydraulic hybrid, which is explained later in this chapter. Note that hydraulic motor is connected to the drive axle, but the engine is not.

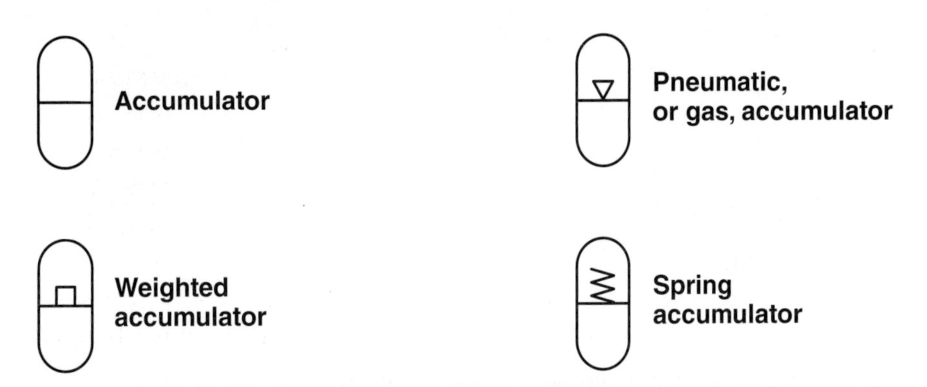

Figure 13-6. Three types of accumulators are spring, weighted, and gas.

Spring

Spring accumulators have a compressible spring on the opposite side of a piston. See Figure 13-7. The fluid pressure acting on the area of the piston causes the piston to compress the spring as the oil pressure increases. The more the spring compresses, the higher the oil pressure must be to further compress the spring.

The spring can be located inside of the accumulator housing or located outside the housing. Spring accumulators are used in automotive automatic transmissions and mobile equipment power shift transmissions. They are small and are generally used in low-pressure clutch applications.

The advantage of a spring accumulator is that it does not require a precharge of gas pressure and will never lose its gas. A drawback is that the spring will not compress at a smooth, consistent rate, meaning that the more it compresses, the more difficult it is to further compress.

Weighted

A weighted accumulator replaces the spring with a single weight or a series of weights. See Figure 13-8. Fluid pressure has to overcome gravity's pull on the weight. The oil pressure is dependent on the area of the piston and the mass of the weight.

The weighted accumulator is the oldest type of hydraulic accumulator. Like a spring accumulator, this accumulator does not require a gas pressure, and therefore will never leak any gas. The second advantage is that it is the only accumulator that can maintain a constant pressure. The accumulator must be used in a vertical position due to its reliance on gravity.

The drawback of weighted accumulators is that they are very large, with some having a capacity of hundreds of gallons of oil. One of the few places that a weighted accumulator is used is an industrial plant that has a single large central hydraulic system. The accumulator provides fluid energy to multiple machines in this environment.

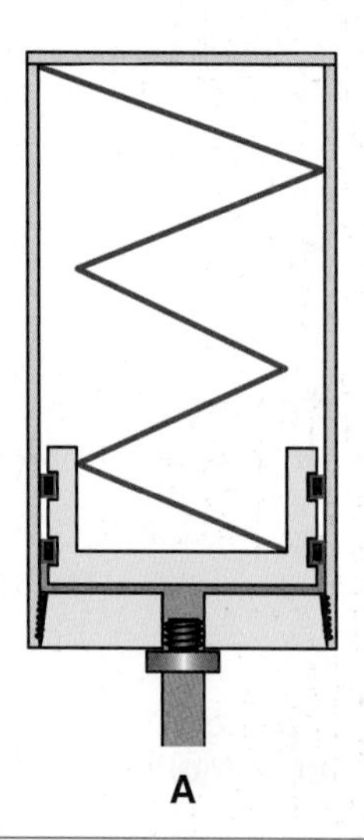

Figure 13-7. Spring accumulators have a compressible spring on the opposite side of the piston. A—Hydraulic pressure is not high enough to overcome spring pressure. B—Hydraulic pressure is high enough to compress the spring.

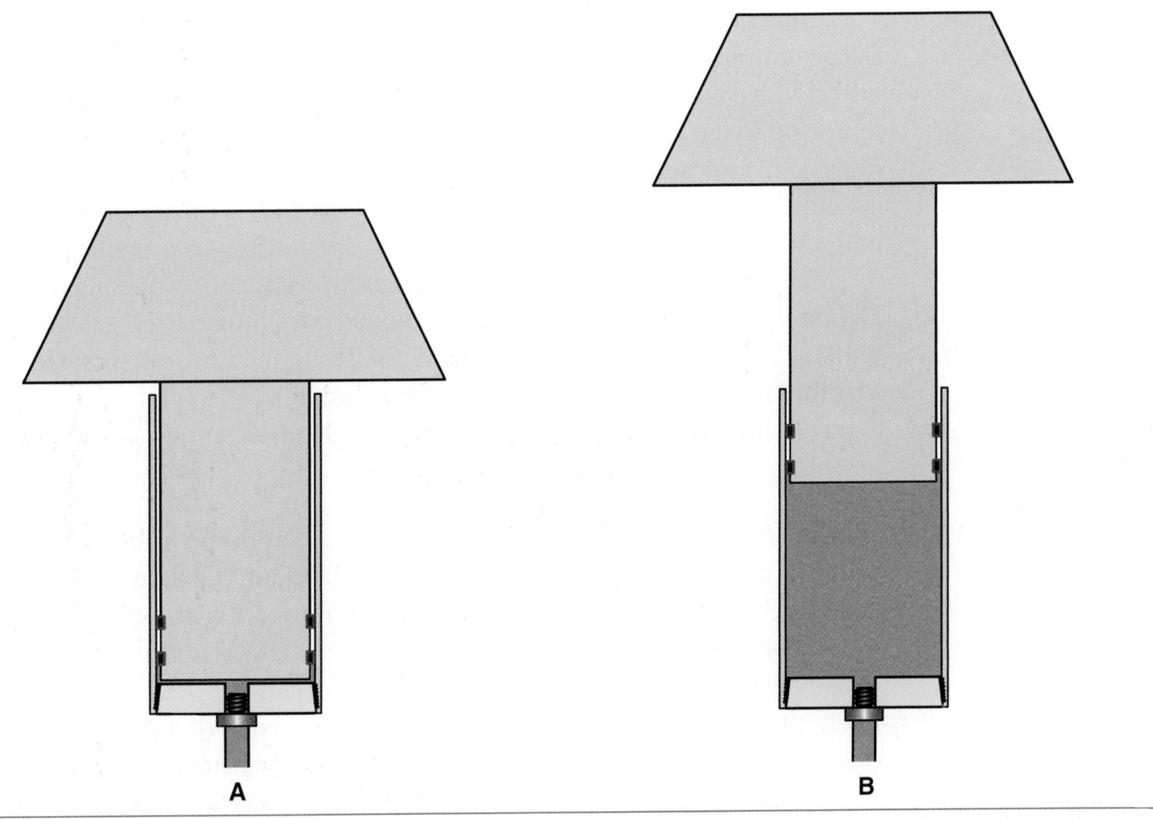

Figure 13-8. A weight accumulator uses a single weight or a series of weights. The mass of the weight can be used to provide a constant pressure to the hydraulic system. A—Hydraulic pressure is not high enough to overcome the mass of the weight. B—Hydraulic pressure is high enough to overcome the mass of the weight.

Gas

Gas accumulators can be called pneumatic accumulators or hydro-pneumatic accumulators. They are the type of accumulator most commonly used in mobile and industrial applications. The accumulator is filled with a gas to a ***precharge pressure,*** which is the level of gas pressure in the accumulator when no oil is present. The precharge pressure affects the volume of oil that is admitted into the accumulator.

Dry ***nitrogen gas*** is the preferred gas for filling accumulators, for multiple reasons:

- Nitrogen is an inert gas and will not react with the hydraulic fluid.
- Nitrogen is not a fuel source for a potential fire. Note that oxygen will react with hydraulic fluids and is a fuel source. Oxygen should *never* be used as a gas for precharging accumulators. Atmospheric air, which contains oxygen, must also *never* be used for charging accumulators.
- Nitrogen is economical. The atmosphere (by volume) is made up of approximately 78% nitrogen, 21% oxygen, and 1% argon and miscellaneous gases. Nitrogen, which is much more plentiful, economical, and compatible than other suitable gases, has proven to be the best solution. Therefore, use only nitrogen gas for charging accumulators.

A gas port is located on the accumulator housing for checking and filling the accumulator with nitrogen. Many accumulators use a Schrader-type valve with a valve core that is similar to those used on tires. And for this reason, it is tempting to the untrained technician to use shop air to charge an accumulator. However, atmosphere must never be used due to its oxygen content. Some accumulators use a different type of gas port that requires an adapter fitting and a wrench to check or fill the accumulator's nitrogen pressure.

Machine manufacturers use a host of different accumulator arrangements. Depending on the available space, sometimes one large gas accumulator is used, while at other times several smaller gas accumulators are plumbed together in parallel, called a bank of accumulators.

The three types of gas accumulators are bladder, piston and diaphragm.

Bladder

The ***bladder accumulator*** consists of a synthetic rubber bladder, also called a bag, inside a metal housing. The bladder is charged with nitrogen. See Figure 13-9. The accumulator can be equipped with one of two types of protection features, a safety poppet valve or a safety button. A safety poppet valve, also known as an anti-extrusion valve, can be located at the bottom of the housing. This spring-loaded poppet valve closes when the oil is exhausted out of the accumulator to prevent the rubber bladder from tearing.

A safety button is sometimes placed on the actual bladder. Whether the accumulator has a poppet valve or a safety button, both are used to protect the bladder as the bladder forces fluid out of the accumulator's housing. They also close off the accumulator, preventing oil from entering until the hydraulic pressure can overcome the bladder's precharge nitrogen pressure.

The bladder can be damaged if the precharge pressure is set too high or set too low. If precharge pressure is set too high, the bladder and poppet valve can be damaged. If precharge pressure is set too low, the bladder can be damaged by the gas charging valve when high-pressure oil fills the accumulator.

Bladder accumulators are the most popular style of gas accumulator. They are used in both mobile machinery and industrial applications. A bladder

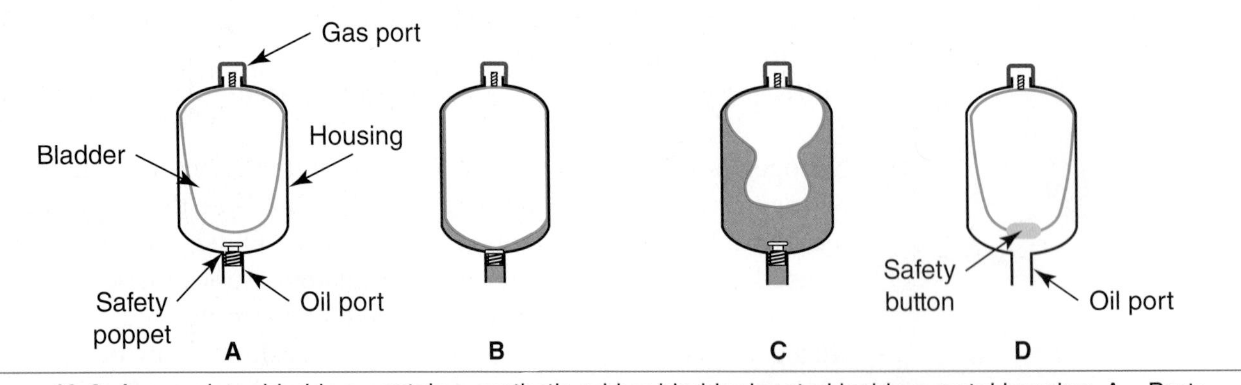

Figure 13-9. Accumulator bladders contain a synthetic rubber bladder located inside a metal housing. A—Parts of the accumulator. B—Bladder is fully charged with nitrogen, and hydraulic oil pressure is lower than nitrogen pressure. The bladder presses on the top of the safety poppet, closing it. C—Hydraulic oil pressure is high enough to reopen the safety poppet, allowing oil to fill the accumulator and compress the bladder. D—This bladder includes a safety button. The safety button is a reinforced area, larger than the oil port, that prevents the bladder from being ripped by the edges of the oil port.

accumulator is a quicker-responding gas accumulator than a piston gas accumulator. If the bladder accumulator is in a horizontal position, it is possible for a little fluid to become trapped between the bladder and the housing when oil pressure drops and oil is expelled from the accumulator. The bladder can wear unevenly when in the horizontal position, especially if the fluid is contaminated. The best orientation for bladder accumulators is the vertical position, with the gas port on top.

Bladder-type accumulators commonly range in sizes from 1/4 gallon up to 15 gallons. However, one manufacturer offers a 120 gallon low-pressure bladder. Some bladder accumulators are designed with a 10,000 psi (690 bar) pressure capacity.

Bladder accumulators can handle contamination better than piston accumulators. The process of charging a bladder with nitrogen should be completed slowly; otherwise the synthetic rubber bladder can become brittle due to the sudden cooling effect of the fast nitrogen charge.

Piston

A ***piston accumulator*** consists of a cylinder, with an internal piston and seals, and a gas fill valve. See **Figure 13-10**. The accumulator resembles a hydraulic cylinder without a rod. Piston seals are used to separate the nitrogen gas from the hydraulic oil.

Piston accumulators offer the largest volumes, up to 160 gallons in size. They can also be designed as small as 15 in^3. Piston accumulators have high pressure capabilities; some are capable of 15,000 psi (1034 bar).

A piston accumulator is most often mounted in a vertical position with the gas charging valve on top. However, the accumulator can be mounted in a horizontal position. Contamination control is the key to piston accumulator longevity.

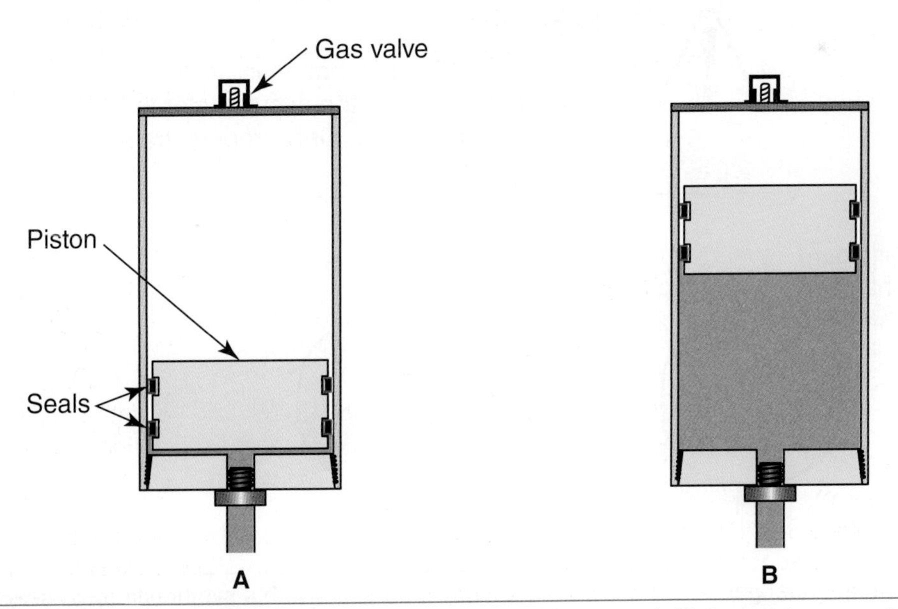

Figure 13-10. Piston accumulators resemble a hydraulic cylinder without a rod. Piston seals separate the nitrogen gas and hydraulic oil. A—Hydraulic pressure is not high enough to overcome nitrogen pressure. B—Hydraulic pressure is high enough to compress the nitrogen.

Piston accumulators can be twice as expensive as a bladder accumulator with the same capacity. Some designers prefer piston accumulators over bladder accumulators because they wear gradually over a period of time, whereas the bladder might have a sudden catastrophic failure. Both styles will wear due to contamination, but bladder accumulators are less sensitive to contamination.

Diaphragm

The ***diaphragm accumulator*** is a small vessel that contains a flexible metal diaphragm with a synthetic rubber diaphragm that separates the nitrogen gas and hydraulic oil. See Figure 13-11. The diaphragm flexes as the nitrogen pressure fills the vessel or as oil pressure compresses the nitrogen.

Diaphragm accumulators are more common in the aircraft industry due to their compact, lightweight nature. They range in volumes from 6 in^3 to 150 in^3. Some diaphragm accumulators have a pressure capability up to 5000 psi (345 bar).

Note

Because diaphragm accumulators are so small, they are mainly used for absorbing shock loads, and not for storing energy.

Accumulator Safety

Accumulators are considered one of the most dangerous hydraulic components on a machine. Any type of device that is capable of storing energy, such as a compressed spring, a charged capacitor, or a full accumulator, can cause serious injury or death. Accumulators, whether large or small, can contain a tremendous amount of stored energy.

Warning

An untrained technician that attempts to remove a charged accumulator can cause the accumulator to become a projectile, rocketing through the air. Only properly trained technicians should work on accumulators.

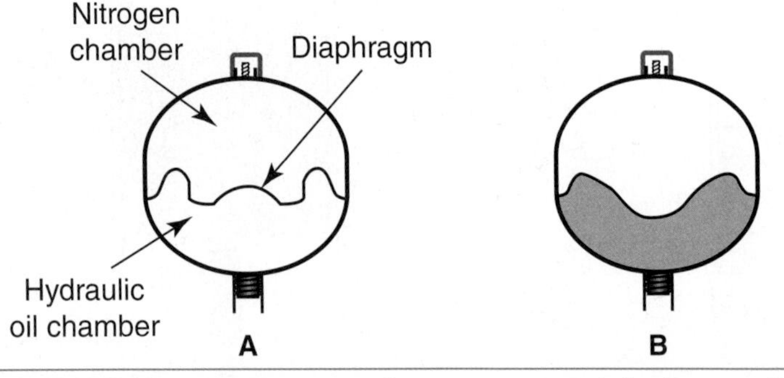

Figure 13-11. The diaphragm gas accumulator uses a metal diaphragm with a synthetic rubber seal to separate the nitrogen and hydraulic oil. A—When there is no hydraulic oil or nitrogen pressure, the diaphragm is relaxed. B—When nitrogen pressure exceeds hydraulic pressure, the diaphragm expands into the hydraulic oil chamber. C—As hydraulic pressure overcomes nitrogen pressure, the diaphragm deflects into the nitrogen chamber.

If a machine is shut off, but if the hydraulic system contains an accumulator that is charged full of oil, the hydraulic system must be treated as a live operating hydraulic system. The charged accumulator can supply oil pressure, causing actuators to move or fluid to be injected into the skin of a technician. Hydraulic systems must have the oil drained from the accumulator before the hydraulic system is worked on. Follow the manufacturer's literature to properly discharge the accumulator.

Accumulators are drained using a ***dump valve***, which routes the charged oil to the reservoir. One style of dump valve is the manual control dump valve, shown in **Figure 13-12**. Notice that an ***isolation valve*** is used to isolate the accumulator from the hydraulic system, and the dump valve drains the hydraulic oil to the reservoir.

Safety block is a term that is sometimes used to describe a dump valve and an isolation valve that are placed inside one valve block assembly. The safety block is used to safely drain hydraulic oil from an accumulator. The safety block might also include a relief valve and a pressure gauge.

Automatic dump valves can be solenoid controlled or pilot operated, see **Figure 13-13** and **Figure 13-14**. Both types of automatic dump valves are normally open, meaning that they take either electricity or pilot oil to close the valve. The valves are designed to drain the accumulator when the machine is shut off.

Do not presume the automatic dump valve is flawless. Spool valves can stick, either stuck open or closed. If dump valve is stuck closed, the accumulator will still be charged with oil pressure. That oil pressure can cause actuator movement or a technician to be injected with fluid.

After the oil has been discharged, the accumulator still requires the precharge of nitrogen to be drained before it can be rebuilt. Always follow the manufacturer's service literature for depleting the precharge nitrogen pressure.

Precharge an accumulator *only* if it is installed on the machine. Just as capacitors should not be handled or transported while charged with electricity, neither should accumulators be handled or transported when charged with nitrogen. The high-pressure gas can cause an injury if the bladder or seals burst. Many freight companies will not ship a charged accumulator due to the inherent risk of the pressurized vessel.

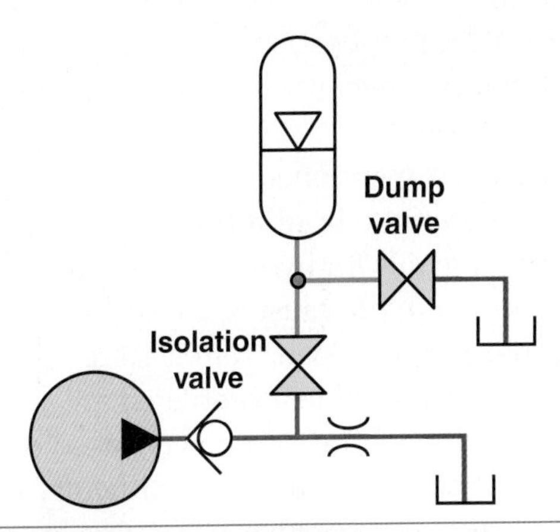

Figure 13-12. Two valves are used to drain an accumulator. One valve isolates the accumulator from the hydraulic system. The other valve is used to dump the hydraulic oil to the reservoir. The manual control valves can be a ball valve, gate valve or a needle valve.

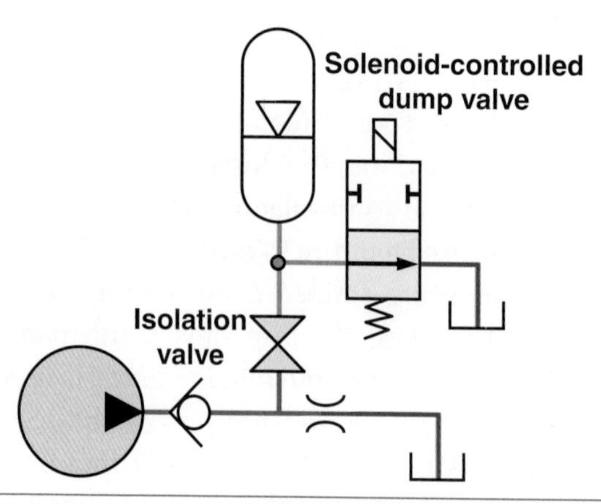

Figure 13-13. A solenoid-controlled dump valve is designed to drain the accumulator when the machine is shut down. The solenoid de-energizes, which causes the dump valve spring to drain the accumulator.

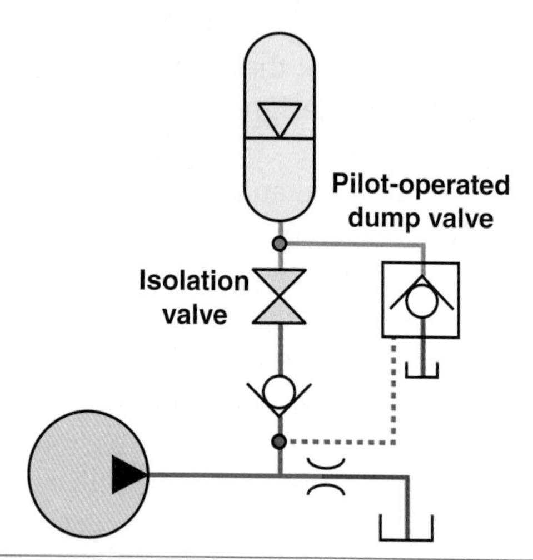

Figure 13-14. A pilot-operated dump valve will automatically drain the accumulator when the hydraulic pump is shut down. Notice the oil for the pilot circuit is located before the check valve. This is considered a 'pilot to close' check valve. Without pilot pressure, the valve is normally open, causing the accumulator oil to drain to the reservoir.

Accumulators should never be exposed to heat. A technician who disregards this critical precaution can cause the pressure vessel to explode, easily causing a fatality. Never use a welder, torch, or any other heat-generating device near accumulators.

As previously mentioned, never use oxygen or compressed air for charging an accumulator. Keep in mind that most mobile hydraulic systems use traditional petroleum oils for hydraulic fluids. The compressed oxygen would react with the petroleum oil, causing an explosion.

If discharging nitrogen from a large-volume accumulator, make sure the procedure is performed in a ventilated area. Do not discharge the nitrogen in a small, confined area. The nitrogen released could displace air in the space, and the absence of oxygen could cause difficulty in breathing.

Never use an automotive valve core in an accumulator. The automotive valve core is a lower pressure valve and is not designed to withstand the high pressures that are used in hydro-pneumatic accumulators.

Service and Repair

Gas accumulators will lose their nitrogen precharge over the course of time. Precharge pressure must be checked periodically. During installation, or if a leaking accumulator is suspected, the precharge pressure should be checked daily. If the accumulator is operating correctly, the precharge pressure should be checked once every 3 to 6 months or during a 500-hour preventative maintenance (PM) inspection.

If the precharge pressure is low, the accumulator is likely leaking. Soapy water is often used to look for nitrogen leaks. The source of the leak should be identified, and the accumulator should be repaired or replaced before it is precharged again. If the accumulator is empty, there is a chance that the bladder has burst and cannot hold any nitrogen pressure. Always follow the manufacturer's service literature for checking precharge pressure and precharging the accumulator.

Checking Precharge Pressure with a Nitrogen Service Tool

A couple methods can be used to check the pressure of the accumulator's precharge of nitrogen. The first method is to use a dedicated service tool that is designed to charge the accumulator with nitrogen gas. See **Figure 13-15**.

This method physically taps into the accumulator's valve core to accurately measure the nitrogen pressure. Although it is the most precise way to measure precharge pressure, there is a slight risk that the accumulator's valve core will stick in an open position, causing a loss of nitrogen. **Figure 13-16** shows the service tool connected to a nitrogen supply tank.

Warning

Nitrogen supply tanks contain high volumes of high-pressure nitrogen gas, with pressures exceeding 3000 psi (207 bar).

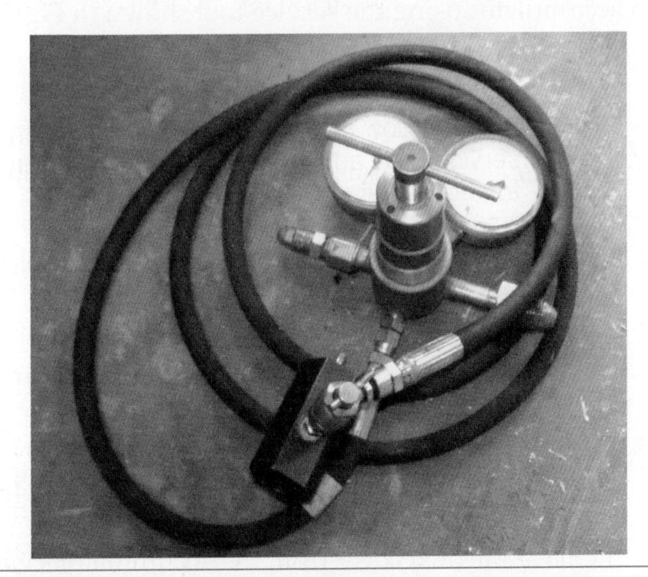

Figure 13-15. This accumulator service tool contains a high-pressure regulator valve with two gauges and a fitting that threads into the accumulator.

Figure 13-16. The accumulator service tool is connected to a nitrogen supply tank. The tool contains two pressure gauges, a high-pressure nitrogen regulator valve, a hose, and a valve chuck with a fitting to thread into the accumulator's Schrader valve.

When using the tool shown in Figure 13-16 to measure an accumulator's precharge pressure, it must be connected to a nitrogen supply tank. The tool does not have a check valve, which means if it was only connected to an accumulator, the tool would vent the accumulator's nitrogen gas to the atmosphere.

The following steps can be followed for checking the precharge pressure of an accumulator using the service tool shown in Figure 13-16.

Caution

Always follow the manufacturer's service instruction. Accumulator service tools and their proper operation vary from manufacturer to manufacturer.

1. Be sure that all of the oil pressure has been depleted from the accumulator.
2. Be sure the cylinder valve on the nitrogen supply tank is closed. You will know it is closed if the cylinder pressure gauge reads zero pressure. See Figure 13-16.
3. As a good habit, be sure the regulator handle is backed out (turned counterclockwise). The reason for this step will be explained in more detail later in the chapter.
4. Remove the dust cover from the accumulator's gas port.

5. On the service tool, prior to attaching the valve chuck to the accumulator, back out the T-handle fully by turning it counterclockwise. See **Figure 13-17**.
6. Thread the service tool's valve chuck fitting onto the accumulator's gas port. Many of these ports resemble a tire's valve stem core. See **Figure 13-18**.
7. Some accumulators will require extensions if the valve core is situated inside a deep valley, preventing the attachment of the service tool. See **Figure 13-19**.
8. Some accumulators will require special fittings to adapt to the accumulator's gas port. See **Figure 13-20**.
9. With the service tool attached to the nitrogen supply tank and the accumulator, both gauges should still read zero pressure.
10. Slowly turn the T-handle on the valve chuck clockwise. When the valve is fully open, the working pressure gauge (on the left in **Figure 13-16**) will read the accumulator's precharge pressure.

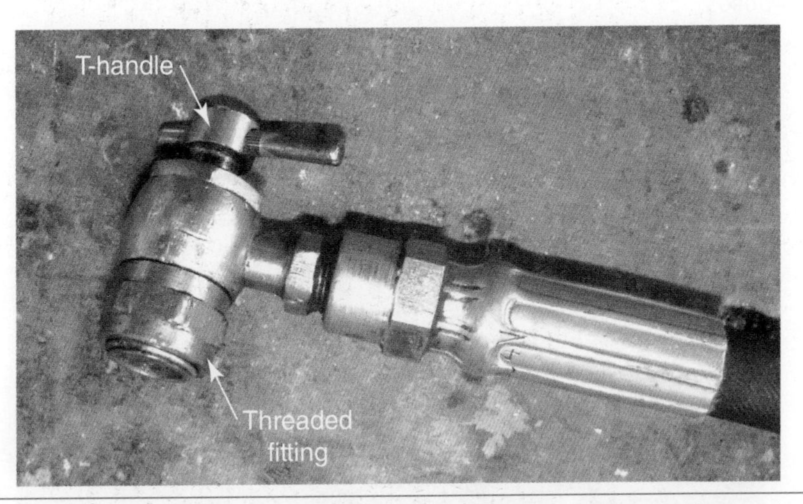

Figure 13-17. The service tool contains a threaded fitting on the valve chuck that attaches to the accumulator. Prior to attaching the fitting to the accumulator, turn the T-handle fully counterclockwise.

Figure 13-18. Some accumulators have a gas port that resembles the valve core of a car tire.

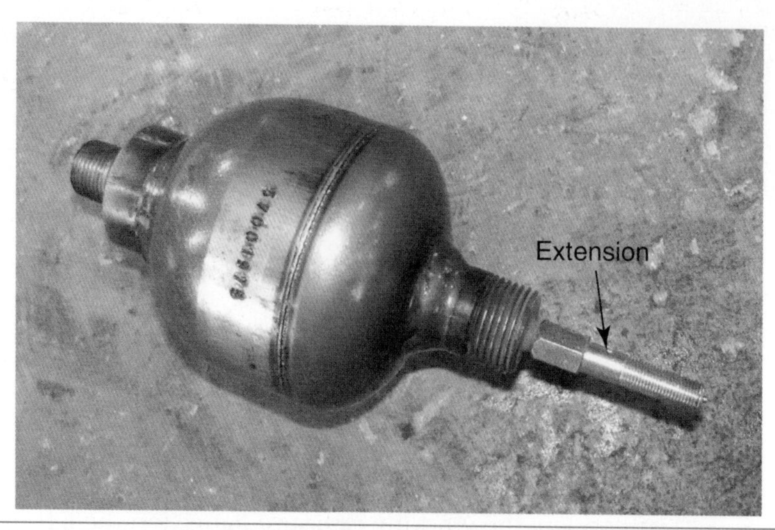

Figure 13-19. An extension was threaded on to this accumulator. If the accumulator's valve core is located inside a valley, it might require the use of an extension. This accumulator would not require the extension, but it has been added for illustration purposes.

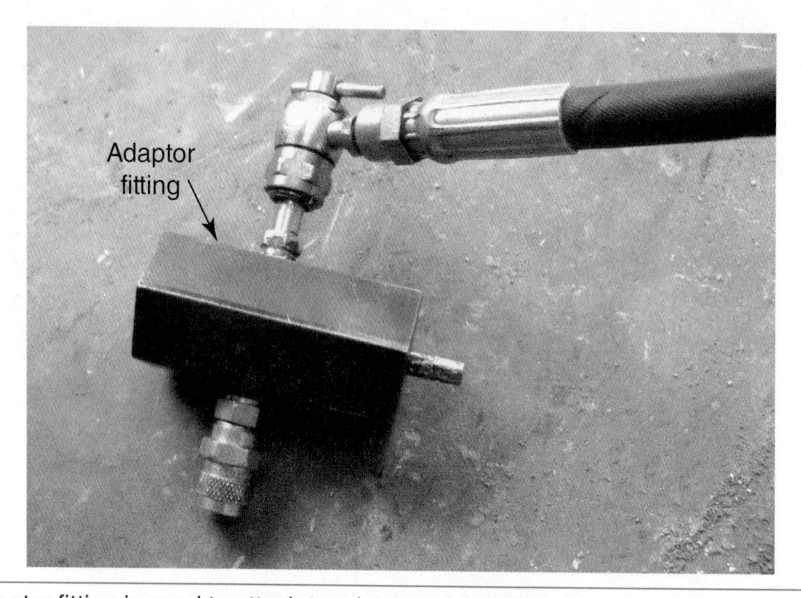

Figure 13-20. This adaptor fitting is used to attach to a late-model combine header accumulator.

Charging an Accumulator with Nitrogen

Continuing from step 10, the following steps can be followed to charge the accumulator with nitrogen gas.

Caution

As previously mentioned, precharge an accumulator *only* if it is installed on a machine. Some customers will bring in an empty accumulator, asking for it to be recharged. Not only is it a hazard to charge the accumulator with nitrogen while it is off the machine, if the bladder has previously failed, the precharge will spew residual oil out of the hydraulic oil port. See **Figure 13-21**.

Figure 13-21. Do not attempt to precharge an accumulator off of a machine. Leave the accumulator installed on the machine. If the bladder bursts, it can cause an injury. If the bladder has already failed, it can cause oil to spew out of the accumulator as shown.

1. *Prior* to charging the accumulator, the nitrogen tank's cylinder valve must be closed (turned fully clockwise), the regulator valve must be backed out (turned fully counterclockwise), and the cylinder pressure gauge (on the right) should read zero pressure. The working pressure gauge, on the left, will read accumulator precharge pressure. If the accumulator is discharged, it will read zero.
2. Double-check the precharge pressure specification.

Warning

If the regulator valve is mistakenly left fully open (turned clockwise until seated), high-pressure nitrogen will be immediately sent to the accumulator as soon as the cylinder valve is opened. For example, if the tank supply pressure is 3000 psi (207 bar) and if the accumulator precharge setting is 90 psi (6 bar), the accumulator will receive 3000 psi of nitrogen, which could rupture the bladder and accumulator housing, and potentially injure the technician.

3. Open the nitrogen supply tank's cylinder valve. The cylinder pressure gauge (on the right) will read the nitrogen tank pressure. If the tank pressure is lower than the required accumulator precharge pressure, the nitrogen supply tank will need to be replaced with one that is fully charged.
4. Slowly turn the service tool's regulator valve clockwise while closely monitoring the working pressure gauge (on the left).

Caution

Charge the accumulator at a slow rate to avoid damaging it.

5. When the precharge pressure has been met, immediately close the valve on the valve chuck by turning it counterclockwise. Then, close the cylinder valve.
6. Remove the valve chuck from the gas port on the accumulator. Keep in mind that a small amount of nitrogen pressure will be lost while removing the service tool.
7. If the charging tool has a bleeder valve, open it to deplete the nitrogen in the service tool.
8. Back out the regulator valve by turning it counterclockwise until the valve handle is loose.

Measuring Precharge Pressure with a Hydraulic Oil Pressure Gauge

A second method can be used to measure precharge pressure by monitoring the oil pressure in the accumulator circuit. Note that this method is not commonly used for shock-load accumulators because the accumulator precharge pressure could be higher than the operating pressure.

The hydraulic pressure method is less accurate, but does not require tapping into the accumulator. The oil pressure can be monitored as the accumulator is charged with oil or as the oil is depleted from the accumulator.

When oil pressure is measured while the accumulator is charging, the nitrogen precharge pressure equals the point at which the oil pressure gauge noticeably first lurches and then continues to climb. The first jolt in oil pressure equals the accumulator's precharge pressure and the subsequent climb in pressure is the amount of oil pressure required to compress the bladder.

Another way of using an oil pressure gauge is to monitor the pressure while draining oil from the accumulator. The hydraulic oil pressure gauge will drop slowly. The pressure reading right before the gauge suddenly drops to zero is the accumulator's nitrogen precharge pressure.

Repair

A common life expectancy of a gas accumulator is 12 years. If an accumulator develops a leak, it will require repair or replacement. Bladder, piston, and diaphragm accumulators are designed as repairable or non-repairable.

Repairable bladder accumulators can be configured as a bottom-repairable or top-repairable design. The bottom-repairable design is the older style, which requires removing the accumulator in order to replace the bladder. The top-repairable design allows a technician to replace the bladder while the accumulator is attached to the system. As previously mentioned, the oil must be drained from the accumulator and the nitrogen pressure must be drained from the accumulator.

Tools and Accessories

Hydraulic designers also have a host of other tools, accessories, and equipment that can be used in conjunction with accumulators. These include gas bottles, safety fuses, nitrogen charging pumps, and sensors to measure the amount of stored energy in an accumulator.

Gas Bottles

Industrial plants sometimes need additional accumulator capacity. Some industrial designers choose to add a gas bottle in parallel with a gas accumulator, which is normally a high-capacity piston accumulator. The purpose of the gas bottle is to provide a larger volume of compressible gas, which enables a smaller accumulator to store more energy. The bottle can be placed in a remote location away from the hydraulic system. This application is not used in mobile equipment.

Safety Fuse

Accumulator circuits can contain a ***safety fuse***, either a safety pressure fuse or a safety temperature fuse. The safety pressure fuse is designed to vent the nitrogen gas to the atmosphere in the event of a sudden over pressurization within accumulator or gas bottle. The fuse is a burst disc that ruptures at the designated pressure rating. Some safety pressure fuses are set at approximately 140% of the highest nitrogen operating pressure.

A safety temperature fuse is designed to vent the nitrogen to atmosphere if the temperature rises too high. High temperatures will cause an increase in gas pressures. High temperatures or a fire can cause the accumulator to explode. The temperature fuse is designed to melt at a designated temperature rating. An example range of melting points is 320°F (160°C) to 356°F (180°C).

Nitrogen Charging Pumps

Nitrogen bottles are not the only source of nitrogen for charging accumulators; ***nitrogen charging pumps*** are also available. HYDAC is one manufacturer who offers this technology. The device draws nitrogen from the atmosphere and filters out the oxygen and other gases. The pump can then supply nitrogen at pressures up to 5000 psi (350 bar) to charge accumulators. One advantage of this system is that it eliminates the need to periodically exchange gas bottles, because it harvests nitrogen directly from the atmosphere. For more information, visit the HYDAC website.

Accumulator Energy Sensors

One key to hydraulic hybrid technology is knowing how much energy an accumulator is storing. It is not as simple as using a pressure sensor to measure the oil pressure or nitrogen pressure, because nitrogen leaks, and temperature has a large effect on pressures. It is also a challenge to attempt to simply measure the position of an accumulator's piston.

One technology that works for small accumulators is the linear variable displacement transducer (LVDT). The sensor measures the piston position through an inductive signal. The challenge is that it does not work well for large accumulators that have long-traveling pistons. One company has mated the LVDT technology with a spool and lead screw. It has the capacity to measure up to 6 feet of piston travel to provide an accurate measure of the accumulator's stored energy.

Hydraulic Hybrid Applications

One of the most common hydraulic hybrid applications is the hydraulic launch assist used in large trucks. This technology has been tested in half-ton trucks all the way up to large trash trucks. The system can use series hydraulic hybrid technology or parallel hydraulic hybrid technology.

Series Hydraulic Hybrid

Referring back to **Figure 13-5**, this truck application is a ***series hydraulic hybrid*** application. The engine is designed to drive the pump, and the pump is designed to drive a motor. The motor is responsible for driving the wheels through a final drive and differential.

A series hydraulic hybrid operates similar to a train engine that uses an internal combustion engine to drive an electrical generator, which directly drives electric motors. In other words, the drive motor(s) are the only means of propelling the vehicle.

During normal operation, the engine is responsible for powering the vehicle by driving the hydraulic pump. During deceleration, the hybrid can recover energy that is normally dissipated to the atmosphere through hot brakes. It does this by storing the fluid energy that is generated by the motor while the vehicle is coming to a stop.

When the vehicle is ready to move again, the hybrid harnesses the high pressure stored in the accumulator to drive the hydraulic motor. After the accumulator's pressure has been depleted, the engine drives the pump to continue powering the vehicle. The system reduces wear on the service brakes and saves fuel by harnessing the energy that was recovered while the machine was slowed to a stop.

Thinking Green

In an EPA study, a prototype series hydraulic hybrid delivery truck was found to have 60–70% greater fuel economy, 40% fewer CO_2 emission, and 60% fewer particle emissions than an equivalent conventional delivery truck.

Parallel Hydraulic Hybrid

A ***parallel hydraulic hybrid*** also uses a pump/motor assembly in conjunction with an accumulator. However, the hydraulic motor is not the sole source responsible for driving the vehicle. A parallel hydraulic hybrid can be propelled solely by the engine and power train with no aid from the hydraulic motor, which would be uncoupled in this mode. See **Figure 13-22**. A parallel hydraulic hybrid operates similar to most automotive electric hybrid cars. The parallel hydraulic hybrid is not as common as the series hydraulic hybrid.

Hydraulic Hybrid Excavator

Caterpillar manufactures a hydraulic hybrid excavator, the 336EH excavator. The goal of the hydraulic hybrid excavator is to recover the wasted energy that is normally lost during the slowing of the upper structure during the swing mode. That energy is stored in the two hydraulic accumulators and is immediately available to be reused to accelerate the swinging of the upper structure back to the previous position. When the job site requires wider swing angles, such as 180° swing angles, the hybrid excavator will recover more energy.

Note

Komatsu and Hitachi hybrid excavators are electric hybrids that use capacitors, rather than hydraulic hybrids that use accumulators.

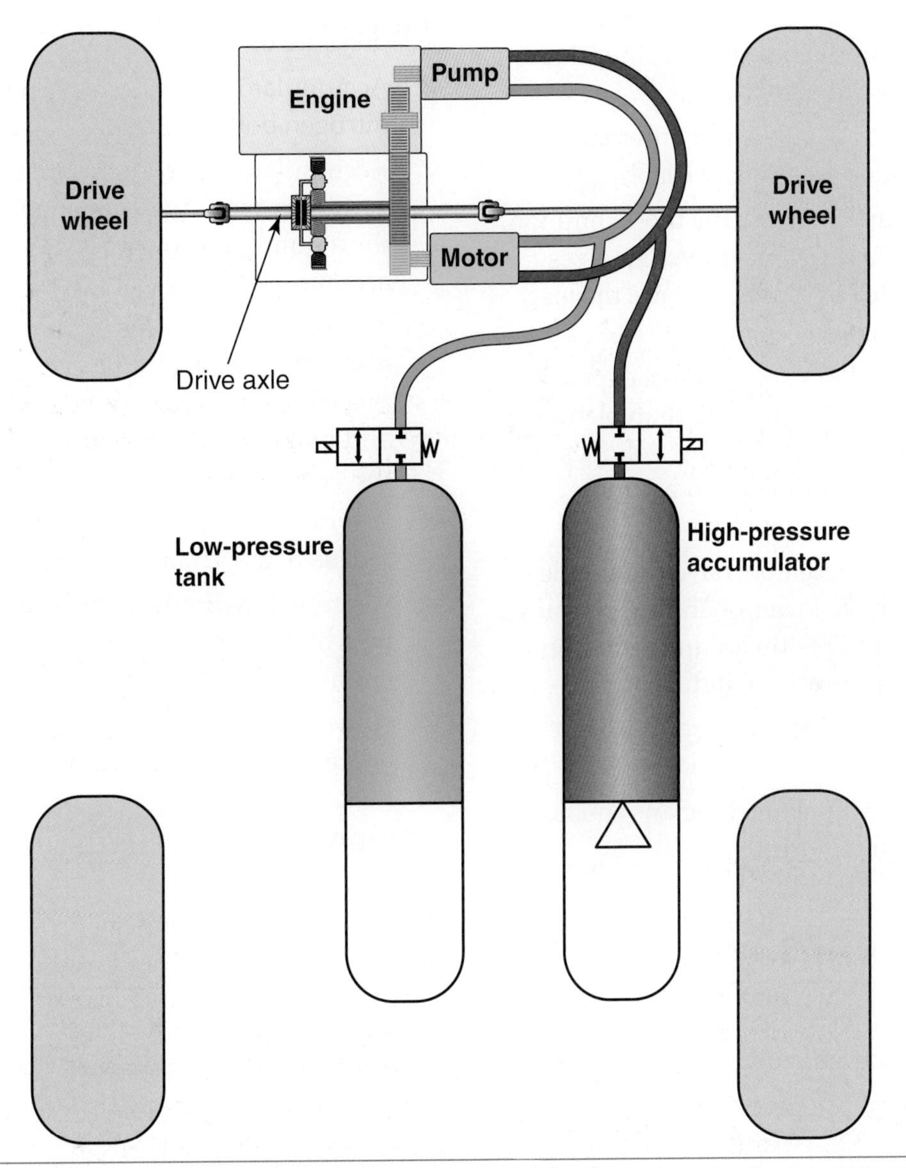

Figure 13-22. A parallel hydraulic hybrid can be propelled by a hydraulic motor or by the internal combustion engine or a combination of both. Note the gear train connecting both the engine and hydraulic motor to the drive axle. Compare this to the series hydraulic hybrid shown in **Figure 13-5**, in which only the hydraulic motor was mechanically connected to the drive axle.

Summary

- ✓ Accumulators are used to store energy, dampen pressure spikes, ease the buildup of pressure, maintain a set pressure, and recover energy.
- ✓ Accumulators have three designs: spring, weighted, and gas.
- ✓ Spring and weighted accumulators are mechanical accumulators that do not use gas.
- ✓ Spring accumulators are used in transmissions.
- ✓ Weighted accumulators are rare, found only in large industrial plants.
- ✓ Gas accumulators have three different designs: bladder, piston, and diaphragm.
- ✓ Bladder accumulators are the most common, are faster responding, have average sizes in volumes and pressures, and are the least susceptible to contamination.
- ✓ Piston accumulators are the most expensive, are a little slower responding than bladder accumulators, offer the highest volumes and pressure capabilities, and are more susceptible to contaminants.
- ✓ Diaphragm accumulators have the smallest gas volumes, have lower operating pressures than bladder and piston accumulators, and are used in the aerospace industry.
- ✓ Heat should never be used around accumulators.
- ✓ Only properly trained technicians should work on accumulators.
- ✓ Prior to repair, an accumulator must have the oil depleted and the nitrogen precharge pressure depleted.
- ✓ Accumulator precharge pressure can be checked with a nitrogen service tool or measured with an oil pressure gauge while the accumulator is charging or draining.
- ✓ Safety pressure fuses and temperature fuses vent nitrogen to prevent over pressurization of the accumulator due to increases in pressure or temperature.
- ✓ Accumulators can be charged using dry nitrogen bottles or a nitrogen charging pump.
- ✓ Accumulator energy sensors are used in hydraulic hybrid technology for measuring the amount of energy stored in the accumulator.
- ✓ A hydraulic hybrid uses a hydraulic pump, hydraulic motor, and an accumulator. The accumulator recovers energy during braking and makes it available to assist in launching the vehicle from a stop.
- ✓ A series hydraulic hybrid is propelled by means of a hydraulic motor.
- ✓ A parallel hydraulic hybrid can be propelled by a hydraulic motor or by the vehicle's engine and power train.
- ✓ A hydraulic hybrid excavator recovers lost energy when the excavator swing is slowing to a stop. The energy is reused to swing the excavator back to its original position.

Technical Terms

bladder accumulator
diaphragm accumulator
dump valve
gas accumulators
hydraulic hybrids
isolation valve
nitrogen charging pumps
nitrogen gas
parallel hydraulic hybrid
piston accumulator
precharge pressure
safety block
safety fuse
series hydraulic hybrid

Review Questions

Answer the following questions using the information provided in this chapter.

1. All of the following are functions of an accumulator, *EXCEPT*:
   A. absorb shock.
   B. create energy.
   C. maintain pressure.
   D. ease the buildup of pressure.
2. An accumulator serves a similar purpose as what electrical device?
   A. Capacitor.
   B. Lamp.
   C. Motor.
   D. Switch.
3. Which of the following accumulators would be the best solution for attempting to maintain a fixed or constant pressure?
   A. Bladder.
   B. Piston.
   C. Spring.
   D. Weighted.
4. All of the following are common accumulator applications in mobile machinery, *EXCEPT*:
   A. store energy for service brakes.
   B. store energy for steering.
   C. maintain a constant system pressure in a centralized large hydraulic system.
   D. absorb implement shock.
5. Which of the following accumulators can be classified as a mechanical accumulator?
   A. Bladder.
   B. Spring.
   C. Piston.
   D. Diaphragm.
6. Which accumulator is the most expensive to purchase?
   A. Piston.
   B. Bladder.
   C. Diaphragm.
   D. Spring.
7. Technician A states that a bottom-repairable bladder-type accumulator requires removing the accumulator before the bladder can be replaced. Technician B states that the top-repairable bladder-type accumulator has been around longer than the bottom design. Who is correct?
   A. Technician A.
   B. Technician B.
   C. Both A and B.
   D. Neither A nor B.
8. All of the following actions must be followed in order to perform safe accumulator service, *EXCEPT*:
   A. discharge oil pressure before service.
   B. discharge gas pressure before repair.
   C. apply heat to dislodge an accumulator with seized threads.
   D. never weld on an accumulator.
9. Which style of accumulator uses a safety button?
   A. Piston.
   B. Bladder.
   C. Weighted.
   D. Spring.
10. Which of the following accumulators uses a synthetic rubber bag?
    A. Piston.
    B. Bladder.
    C. Diaphragm.
    D. Spring.
11. Which style of accumulator is lightweight and operates at lower pressures?
    A. Weighted.
    B. Piston.
    C. Bladder.
    D. Spring.
12. Which of the following types of accumulators resembles a hydraulic cylinder without a rod?
    A. Piston.
    B. Bladder.
    C. Diaphragm.
    D. Spring.

13. Which of the following accumulators is used in automotive automatic transmissions and power shift transmissions?
    A. Piston.
    B. Bladder.
    C. Diaphragm.
    D. Spring.
14. If a gas accumulator is suspected of having a nitrogen leak, what can be used to detect the leak?
    A. Electronic leak detector.
    B. Butane torch leak detector.
    C. Soap and water.
    D. Dye and a black light.
15. Which of the following accumulators is the smallest in volume and has the lowest pressure capability?
    A. Bladder.
    B. Diaphragm.
    C. Piston.
16. Which of the following accumulators offers moderate volume sizes with moderate pressure capability?
    A. Bladder.
    B. Diaphragm.
    C. Piston.
17. Which of the following accumulators has the highest volume capacity and the highest pressure capability?
    A. Bladder.
    B. Diaphragm.
    C. Piston.
18. Which of the following gases is used to charge an accumulator?
    A. Oxygen.
    B. Atmosphere.
    C. Nitrogen.
    D. Acetylene.
19. Which of the following is the most popular type of accumulator used in mobile machinery?
    A. Diaphragm.
    B. Spring.
    C. Bladder.
    D. Piston.
20. Technician A states that piston accumulators are faster responding than bladder accumulators. Technician B states that bladder accumulators are more expensive than piston accumulators. Who is correct?
    A. Technician A.
    B. Technician B.
    C. Both A and B.
    D. Neither A nor B.
21. When setting the precharge pressure on a bladder accumulator, Technician A states a precharge pressure that is too low can cause damage to the bladder. Technician B states that a precharge pressure that is too high can cause damage to the bladder. Who is correct?
    A. Technician A.
    B. Technician B.
    C. Both A and B.
    D. Neither A nor B.
22. Which of the following is the most sensitive to fluid contamination?
    A. Bladder.
    B. Piston.
    C. Diaphragm.
23. When charging a bladder accumulator, what can occur if the accumulator is charged too fast with gas pressure?
    A. The bladder could overheat.
    B. The bladder could become brittle.
    C. The valve core could stick open.
    D. The valve core may not open.
24. When charging an accumulator, if the regulator is accidentally left open (with the handle turned fully clockwise), what will occur when the nitrogen tank is opened?
    A. The accumulator will instantly be charged at the full tank pressure value.
    B. The bladder might burst.
    C. Both A and B.
    D. Neither A nor B.

25. Accumulator safety pressure fuses are often set at what percentage of the highest operating nitrogen pressure?
    A. 90%
    B. 110%
    C. 140%
    D. 200%
26. An accumulator safety temperature fuse is set at what temperature range?
    A. 220°F (104°C) to 256°F (124°C)
    B. 320°F (160°C) to 356°F (180°C)
    C. 420°F (216°C) to 456°F (236°C)
    D. 520°F (271°C) to 556°F (291°C)
27. What is an estimated life expectancy for a gas accumulator?
    A. 5 years.
    B. 12 years.
    C. 22 years.
    D. 30 years.
28. Technician A states that accumulator dump valves can be manually actuated. Technician B states that accumulator dump valves can be automatically controlled, either through a solenoid or pilot operated. Who is correct?
    A. Technician A.
    B. Technician B.
    C. Both A and B.
    D. Neither A nor B.
29. Accumulators use two valves for draining the hydraulic oil. What are they called?
    A. Dump valves and isolation valves.
    B. Relief valves and pressure-reducing valve.
    C. Make-up valves and check valves.
    D. Shuttle valves and counterbalance valves.
30. Where should the accumulator be located while it is being charged with nitrogen?
    A. In a vise.
    B. In a bucket.
    C. Installed on the machine.
    D. Held in your hand.

A variety of factors influence the design, location, and size of the reservoir used in a hydraulic system on a mobile machine. The reservoir on this excavator is located behind the boom and close to the engine. Engineers design reservoirs to fit and function in the limited space available on mobile machines.

Chapter 14 Reservoirs and Coolers

Objectives

After studying this chapter, you will be able to:

- ✓ List the functions of a hydraulic reservoir.
- ✓ Explain the purposes of the components used in hydraulic reservoirs.
- ✓ Describe the different types of reservoir designs and the effects of the designs.
- ✓ Explain how heat is generated in a hydraulic system.
- ✓ Calculate the heat generated due to inefficiency.
- ✓ Describe the different types of oil coolers.
- ✓ Explain why OEMs use heaters.
- ✓ Identify the location of a heater on a machine.

Reservoirs

A hydraulic ***reservoir***, also called a hydraulic tank or sump, is a containment device that houses a machine's hydraulic fluid. The reservoir serves multiple purposes:

- Houses hydraulic fluid.
- Receives and dissipates heat.
- Supplies oil to the pump's inlet.
- Provides a place for entrained air to separate from the fluid.
- Provides a place for contaminants to settle from the fluid.

Housing

Mobile machinery use hydraulic reservoirs that are much different than those found in industrial plants. The reservoir is the hydraulic component that is affected the most by the machine's limited amount of space. As a result, mobile equipment manufacturers often use reservoirs that are small and sometimes oddly shaped. The container shape can range from a traditional rectangular cube, to a round tank, to an abstract shape. Many machines use a single reservoir, others use multiple separate reservoirs, and it is possible to find two reservoirs that are plumbed together in parallel. See the schematic in Figure 14-1.

Some machines, such as agriculture tractors, use the transmission or axle housing as the reservoir, which is referred to as an ***integral reservoir***. See Figure 14-2. These reservoirs have been used since the 1920s and hold more oil than most mobile reservoirs. However, integral reservoirs can have more contamination, heat generation from the gears, and aeration from the powertrain.

In a perfect world, a hydraulic reservoir would have a volume that equals at least three times the pump's maximum flow rate. For example, a system that flows 20 gpm (76 lpm) would ideally incorporate a 60 gallon (227 liter) reservoir. A large volume reservoir offers multiple advantages:

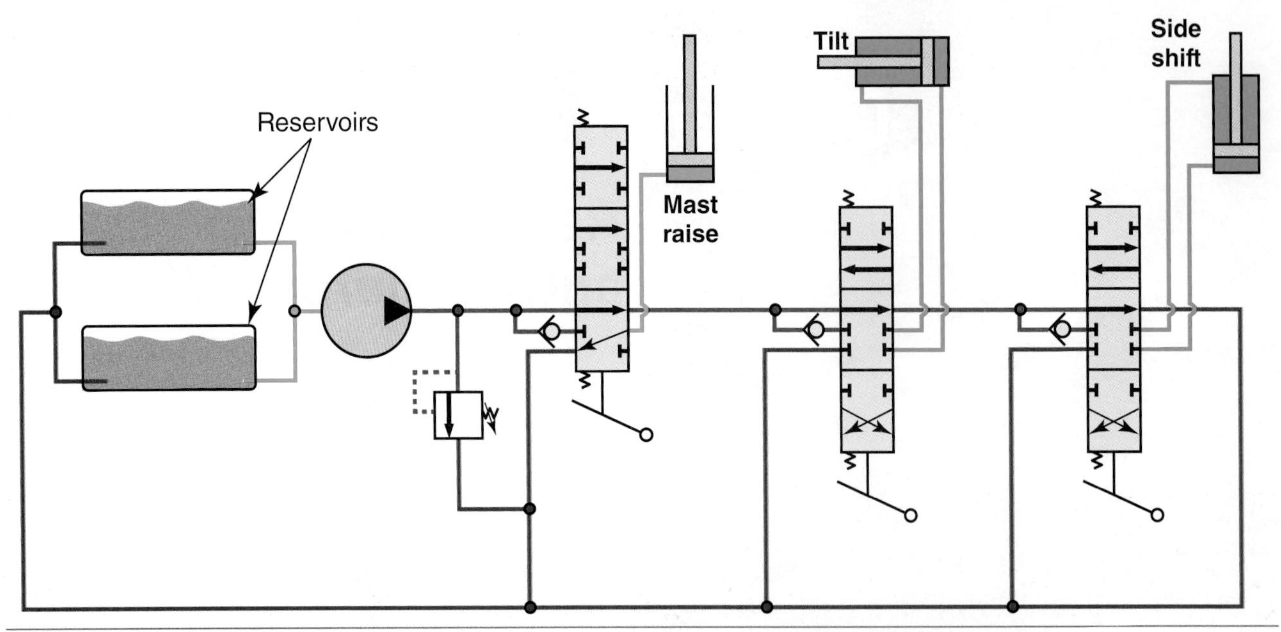

Figure 14-1. An example of a Case 586e forklift that uses two reservoirs that are plumbed together in parallel. Steering has been eliminated from the schematic to simplify the drawing.

Figure 14-2. This New Holland compact utility tractor uses an integral reservoir to supply hydraulic oil to the hydrostatic transmission and to the implement pump. The reservoir's fill cap and dipstick are located behind the seat near the three-point hitch.

- Improves oil cooling, by providing more time for the oil to cool between cycles and more tank surface area.
- Provides a larger supply of oil for a short period of time, regardless of returning oil.
- Due to the large volume of oil, reduces the risk of cavitation at the pump following a repair.
- Improves the separation of entrained air, which reduces the chance of pump failure due to aeration.

However, many mobile machines use small reservoirs. For example, a skid steer or a swather might use a reservoir that is two-thirds smaller than the implement pump's capacity. For example, the hydraulic system might flow 30 gpm and have a reservoir that contains 10 gallons of oil. Smaller reservoirs provide the following advantages:

- Cheaper to build.
- Less weight on the machine.
- Take less space.
- Have less oil to leak, causing less harm on the environment.

Later in this chapter, an example will be provided of a cylindrical reservoir that has a volume that is only a fraction of the system's flow rate.

Receives and Dissipates Heat

The reservoir receives the system's return oil, which is often heated. The reservoir is designed to dissipate heat by allowing the heat to transmit through the reservoir's thin walls. Tanks should be mounted in a manner that allows good air circulation to occur across all six of the reservoir's walls. When the reservoir's volume is less than twice the pump capacity, a hydraulic oil cooler will be required to prevent the oil from overheating.

Supplies Oil to the Pump's Inlet

The reservoir must deliver a good supply of oil to the pump's inlet. If the reservoir contains a suction screen, the pump will pull a vacuum if the screen becomes plugged, resulting in a premature pump failure. **Chapter 21** explains factors relating to measuring pump inlet vacuum based on the location of the vacuum gauge.

Provides a Place for Entrained Air to Separate from the Fluid

One of the most important functions of a hydraulic reservoir is to provide a good location for allowing entrained air to separate. Otherwise, the aerated oil will cause the pump to fail. A larger reservoir helps the air separation process. Later in this chapter, diffusers will be explained as a means to help air separation.

Provides a Place for Contaminants to Settle from the Fluid

Reservoirs also provide a place for contaminants to settle. Contaminants include wear particles, external contaminants, water, and air. The tank design has a large effect on the settling of contaminants, and will be discussed later in the chapter.

Reservoir Design and Components

Although a hydraulic reservoir appears to be a simple container to collect return oil and supply pump intake oil, the container is more complex than that. Numerous different components are strategically used in reservoirs to optimize the life of the hydraulic system. The use and placement of the following reservoir designs and components vary, based on the manufacturer and machine:

- Vented versus sealed.
- Variable-volume reservoir.
- Reservoir walls.
- Baffles.
- Diffuser.
- Visual indicator.
- Placement of hydraulic lines.
- Filler port.
- Breather.
- Drain plug.
- Clean-out plate.
- Reservoir shape.

Vented or Sealed

The reservoir housing can be vented or sealed. If the reservoir is vented, a breather is used to filter contaminants, including moisture, from the incoming air. This was described in **Chapter 11**. Sealed reservoirs can build pressure using one of two methods:

- Cap the reservoir, which enables the heated oil to compress the air within the reservoir.
- Use an external source of air pressure.

Pressurized reservoirs help provide a good supply of oil to the pump's inlet, especially if the pump is placed above the reservoir oil level or if the pump is located farther away from the reservoir. Pressurized reservoirs are also sometimes used on machines that have open-loop pumps.

The filler cap used on a sealed reservoir sometimes resembles a boat plug. See **Figure 14-3**. As the top of the cap is twisted clockwise, it compresses the rubber plug, which makes a tight seal in the reservoir's filler neck.

Sealed reservoirs can have a depressurization ***bleeder valve***, which is used to deplete the reservoir's air pressure. The bleeder valve can be manually operated or automatically controlled. If manually operated, it usually consists of a simple push button valve located on the reservoir. If the sealed reservoir does not have a bleeder valve, and if the reservoir was able to build pressure, the system will burp air as the reservoir's cap is removed.

Service manuals and operator manuals often reference pressing the manual bleeder valve prior to servicing the machine. If this is not done, the reservoir pressure can force oil to spray out of a fitting as it is loosened. If the bleeder valve is automatic, it acts similarly to the automatic accumulator dump valve that was explained in **Chapter 13**. The automatic bleeder valve will vent the air pressure from the reservoir after shutdown.

Only a small amount of pressure is needed for a pressurized reservoir, for example 3 to 4 psi. Based on Pascal's law, the reservoir will have to withstand a high amount force with only a small amount of air pressure. See **Figure 14-4**. If the reservoir's size is 36″ long and 18″ tall, one side will have an area of 648 in^2. A supply of 4 psi of air pressure will generate 2592 pounds of force on the side wall. The combination of that load along with the dynamic loads of the oil and G-forces that occur while operating in the field can cause work hardening of the reservoir wall, which will cause the wall to become brittle and potentially crack.

Note

Boot strap reservoirs are cylindrical pressurized reservoirs. The reservoir uses hydraulic fluid pressure to act on a small piston, which pressurizes the reservoir to 50 psi. They are used in aircraft to provide a good oil supply regardless of the plane's elevation, pitch, and speed.

Figure 14-3. This reservoir cap is for a sealed hydraulic reservoir. As the knob is turned clockwise, it compresses the rubber plug, making a tight seal.

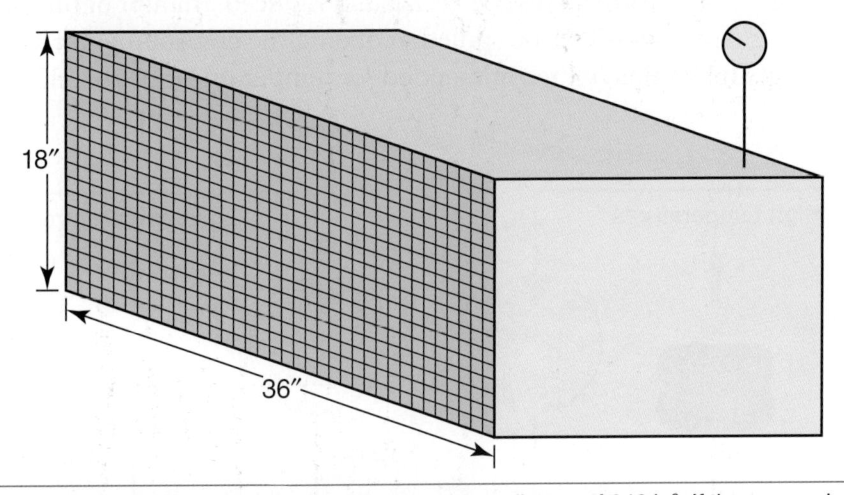

Figure 14-4. A reservoir that is 36″ long and 18″ tall will have a side wall area of 648 in^2. If the reservoir is pressurized at 4 psi, the side wall must be able to handle 2592 pounds of force.

Variable-Volume Reservoir

A ***variable-volume reservoir (VVR)*** is a sealed vessel that does not allow outside air into the reservoir. The VVR consists of a pressurized rubber bladder, resembling the double-convoluted air spring used on semi-trucks. The reservoir uses a coiled spring, but no air or nitrogen gas.

This unique reservoir operates similarly to a closed-loop hydrostatic drive. A closed-loop hydrostatic transmission returns the motor's oil to the pump's inlet instead of directing the oil to the reservoir. **Chapter 23** explains closed-loop hydrostatic transmissions in further detail. During normal operation, the return flow is directed immediately back to the pump's inlet. Oil enters or is extracted from the reservoir under two conditions:

- Changes in temperatures. See **Figure 14-5**.
- Changes in flow, such as those due to the difference in volumes in a differential cylinder. See **Figure 14-6**.

Charging a VVR

Charging a VVR is more complex than filling a traditional reservoir. A separate hydraulic pump, either a manual pump or an electric pump, is used to charge the closed hydraulic system and reservoir. The cylinders must be retracted while the system is being charged with oil. The reservoir is charged approximately 75% full, which allows for additional expansion due to an increase in temperature.

During the oil charging process, air is introduced into the hydraulic system. The system is equipped with a bleed valve. The bleed valve is used to bleed air from the closed system, which is necessary in the following situations:

- The system is filled with oil.
- A repair has been made.
- A faulty seal has allowed air to be drawn into the closed system.

The advantages of a VVR are:

- An absence of air inside the reservoir, which extends the life of the oil.
- Reduced oxidation.
- Reduced atmospheric contaminants.
- Compact reservoir size and reduced amount of oil.
- Less oil to be spilled, reducing the environmental risks.

The VVR is not recommended for temperatures below –13°F (–25°C).

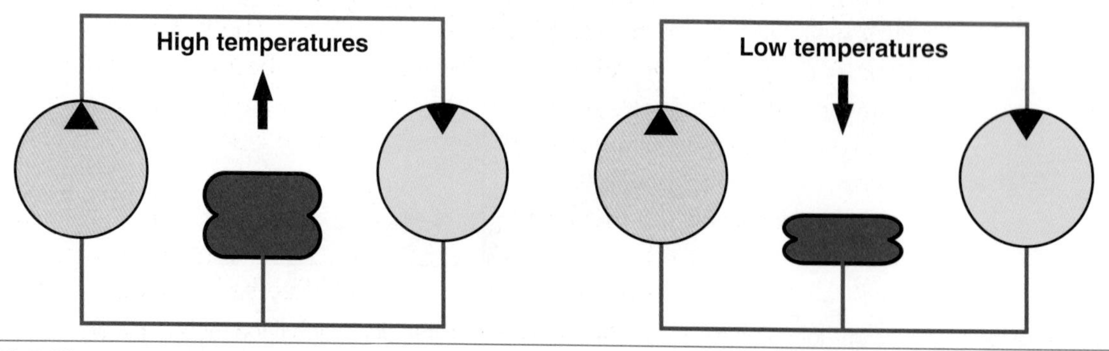

Figure 14-5. The reservoir's volume will expand when the system temperature is increased.

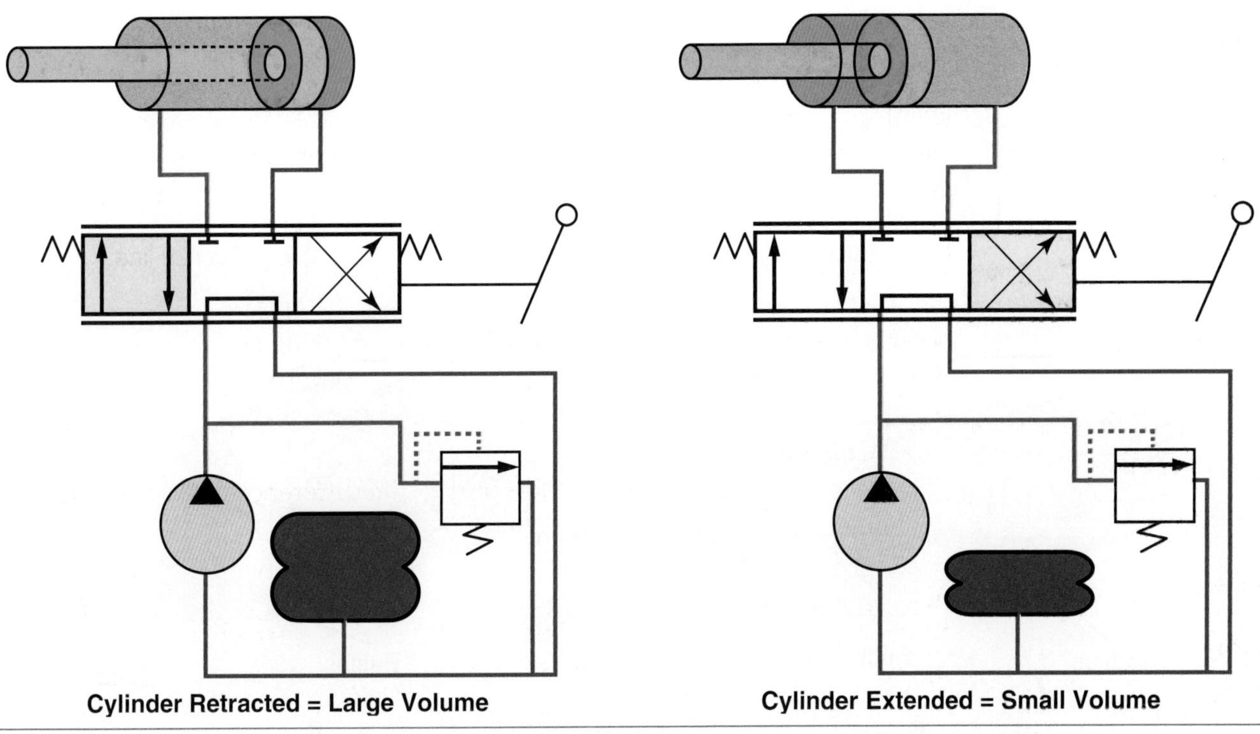

Figure 14-6. The reservoir will increase in volume during cylinder retraction and will decrease in volume during cylinder extension.

Reservoir Walls

Traditional reservoir walls are designed thin, which helps them dissipate the heat from the hot oil. Many tanks are made of common steel, but some are made of plastic. The average wall thicknesses are:

- 1/16″ thick for reservoir volumes of 25 gallons (94 liters) and smaller.
- 1/8″ thick for reservoirs that range from 25 to 100 gallons (94 to 378 liters).
- 1/4″ thick for reservoirs that hold more than 100 gallons (378 liters) of oil.

Baffles

A ***baffle***, also known as a dam, is a divider plate placed inside the reservoir. The baffle can be mounted on the bottom or on the top of the reservoir, perpendicular to the return and suction ports. The bottom baffle needs to be submerged 30% below the oil level. The top baffle must be elevated 30% above the floor of the reservoir. The baffle can also be placed parallel to the return and suction ports, causing the oil to make a U-turn inside the reservoir. See **Figure 14-7**.

Baffles serve the primary purpose of preventing the return oil from immediately cycling back into the pump's inlet. They also reduce the sloshing of the oil as the machine moves through the field. The baffles can help stiffen the rigidity of the tank walls, which is beneficial to strengthening a pressurized reservoir. The baffles also help trap contaminants in the reservoir.

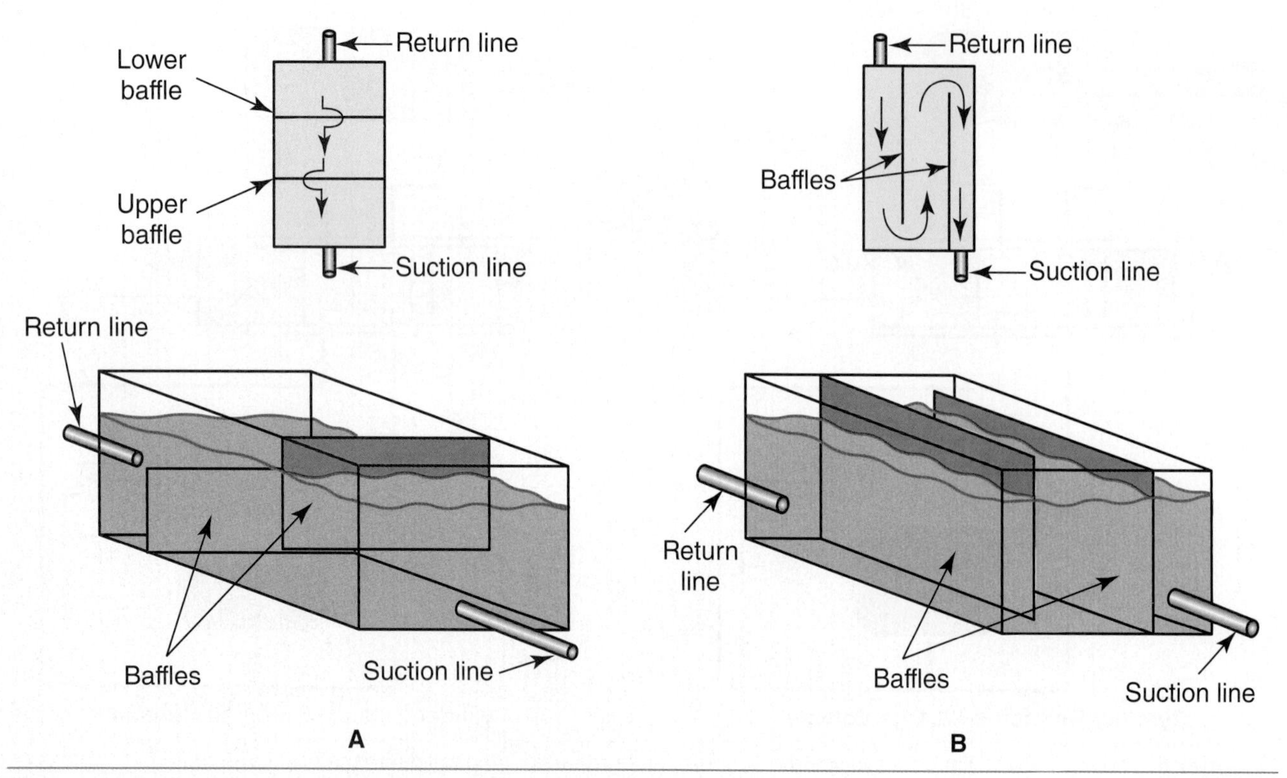

Figure 14-7. Baffles are plates placed inside the reservoir to prevent the oil from directly cycling back into the pump's inlet. A—Baffles perpendicular to ports. B—Baffles parallel to ports.

Diffuser

A ***diffuser*** is a perforated screen or a perforated housing that is placed inside a reservoir. The diffuser reduces the fluid velocity to as little as 1 ft/sec, which reduces fluid aeration. Diffusers are especially helpful if the hydraulic system has a small reservoir. See Figure 14-8.

Sight Glass or Dipstick

Reservoirs will either have a dipstick or a sight glass for measuring the reservoir's oil level. Some sight glasses include a thermometer. See Figure 14-9.

Note

Relying solely on the visual reading of a sight glass can result in false readings. The sight glass, over a period of time, will become foggy or discolored, especially if it is exposed to the sun. When encountering multiple hydraulic faults on a system that contains one reservoir, it is a good idea to obtain a positive reading by using a tape measure through the fill tube, in order to avoid wasting time with a fluid level misdiagnosis.

Some reservoirs use a dipstick, which is a more accurate method for measuring the reservoir's fluid level. See Figure 14-10. One drawback of a dipstick is that the hydraulic system is opened to contamination every time the dipstick is removed.

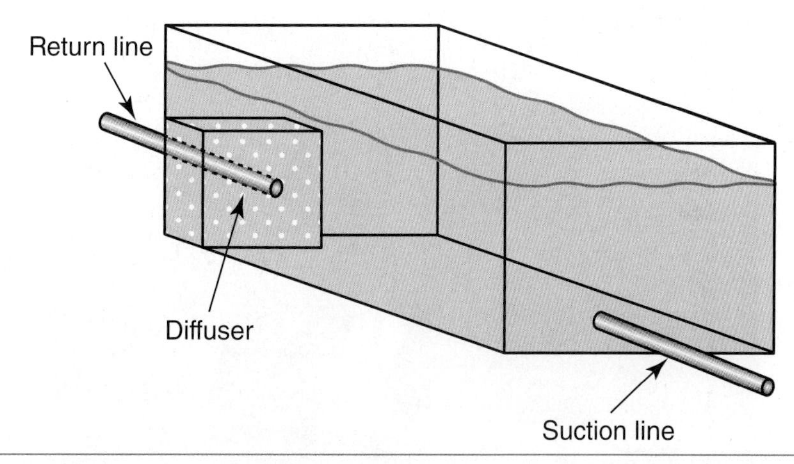

Figure 14-8. Diffusers are perforated plates or perforated housings placed inside the reservoir. They slow the oil velocity reducing fluid aeration.

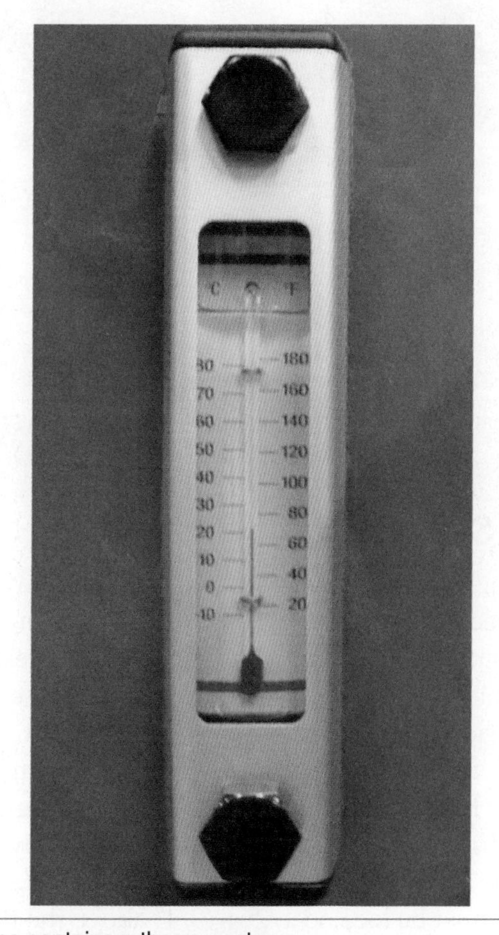

Figure 14-9. This reservoir's sight glass contains a thermometer.

Regardless of whether the reservoir is configured with a sight glass or a dipstick, be sure to follow the operator's manual instructions for checking the fluid level. The operator's manual will specify to place the machine on level surface, the proper position of the implements, and the operating temperature. All of those factors will affect the reservoir's oil level.

Figure 14-10. This reservoir uses a dipstick that is attached to the filler cap. The cap seals the reservoir.

Suction Line and Suction Strainer

The reservoir's outlet port, also known as the suction line, might be equipped with an internal suction screen. **Chapter 11** detailed the advantages and disadvantages of suction screens. If the reservoir is too small to accommodate an internal suction screen, a suction screen assembly might be attached to the outside of the reservoir on the outlet port. The reservoir's suction line may have a vacuum gauge to identify if the screen is plugged. **Chapters 21** and **22** provide further information about measuring suction pressures.

The suction lines need to be approximately 1/2″ to 3/4″ above the bottom of the reservoir. If the tank contains an internal suction screen, the bottom of the screen must be located 1/2″ to 3/4″ above the floor of the reservoir.

Return Lines

Reservoirs receive oil from different types of return lines:

- Pressurized return lines.
- Unpressurized return lines, also known as leakage lines or case drain lines.
- Scavenge lines.

Pressure Return Lines

Pressurized return lines are the main return lines that send low-pressure oil back to the reservoir. These lines must be located below the oil level, preferably 1″ from the bottom of the reservoir, and have a baffle isolating the oil from the suction line. These lines commonly have a small amount of back pressure, for example 10 psi (0.7 bar).

Leakage Lines

Some pumps and actuators use an unpressurized return line, also known as a leakage line. In a pump and motor circuit, the return is also called a ***case drain line***. In a cylinder circuit, the line can be called a gland drain line.

Caution

If the case drain line is filtered, the circuit will build back pressure, which can cause the shaft seal to fail and induce forces on the internal components. Filtering case drain lines is not recommended.

Multiple sources provide different recommendations for plumbing case drain lines into a reservoir. Leakage lines can contain entrained air, and for that reason, at least one source recommends returning the leakage line above the oil level. However, other sources recommend placing the case drain line below the reservoir's oil level to help keep the pump and motor case full of oil. The case drain line must be separate from the pressurized return line to keep the drain line at zero pressure.

The case drain line at the reservoir must be higher than the pump case or motor case to ensure that the reservoir always provides a positive supply of oil back to the component. Case drain lines must also be installed in a manner to prevent siphoning the oil from the pump's case back to the reservoir. Otherwise, the pump could be deprived of oil, causing it to fail.

Scavenge Lines

A ***scavenge pump*** is a pump that is designed to draw oil from a sump and return the oil back to a reservoir. Scavenge pumps pump oil that has a large portion of entrained air, for example 80% oil and 20% air.

Scavenge pumps are more common in engine applications, but are used in some off-highway powertrains, making it possible to be used in conjunction with an integral reservoir. Due to the large amount of entrained air in scavenge circuits, scavenge oil should be returned above the reservoir's oil level.

Filler

The reservoir needs some type of opening to fill the reservoir with oil. The filler is located on the top of the tank. It often contains a screen to capture large contaminants. Some personnel make the mistake of removing the screen because it slows the refilling process. This is a bad practice because it makes the reservoir more susceptible to contamination.

Quick couplers were mentioned in **Chapter 11**. They can be used with filtration tanks and carts for filling the reservoir. The actual filler cap can also be the reservoir's breather. See **Figure 14-11**.

Figure 14-11. The filler cap on this reservoir also serves as a breather.

Breather

As mentioned in **Chapter 11**, breathers are used in vented reservoirs for the purpose of removing atmospheric contaminants. The breathers are also designed to remove moisture. While some filler caps also function as a breather, it is possible to have a separate filler cap and breather.

Drain Plugs

Reservoirs can use different types of drain plugs. As mentioned in **Chapter 12**, some reservoirs use drain valves (also known as ecology valves) that make it easy to connect a filter cart. The most common type of drain plug is the traditional type that uses an O-ring to seal the plug. The plug can be magnetic to attract iron particles from the fluid.

Many mobile machines place the reservoir drain plug in an awkward location. One combine manufacturer plumbs multiple fluid compartments into a clear tube manifold that leads to the bottom of the combine. On that machine, a container is placed at the bottom of the clear tube to capture the oil when the reservoir's drain valve is opened. One skid steer manufacturer threaded a capped hose to the reservoir's drain port, because it was easier to remove the cap from the hose than to gain access to the reservoir's drain port. Industrial hydraulic systems use ball valves to drain the reservoir. The valve also has a plug installed, just in case the ball valve's handle is inadvertently bumped.

Clean-out Plate

As mentioned earlier in this chapter, reservoirs trap contaminants. The tank must have some type of ***clean-out plate*** to provide access for cleaning the reservoir. See Figure 14-12. As mentioned in **Chapter 11**, the reservoir must have a clean-out plate if the reservoir contains a suction screen inside the tank. However, some reservoirs do not have an access plate, making it difficult to thoroughly clean the contaminants, especially if the reservoir contains baffles.

Reservoir Shape

Some hydraulic specialists argue that the traditional rectangular reservoir can actually cause contamination problems by allowing the particles to settle on the flat bottom of the reservoir. The arguments are:

- Reservoirs are rarely cleaned and difficult to clean.
- When contaminants settle to the bottom of a rectangular reservoir, the contaminants are less likely to be routed to a filter for removal.
- The contaminants will continuously contaminate the oil, affecting the chemistry of the oil and its properties.

A cylindrical reservoir is now produced by manufacturers. The placement of the suction line and the return line causes the oil to rotate and contaminants to fall to the bottom. Two different styles promote the removal of the falling contaminants. One uses a suction filter that immediately removes the contaminants before they enter the pump. See **Figure 14-13**.

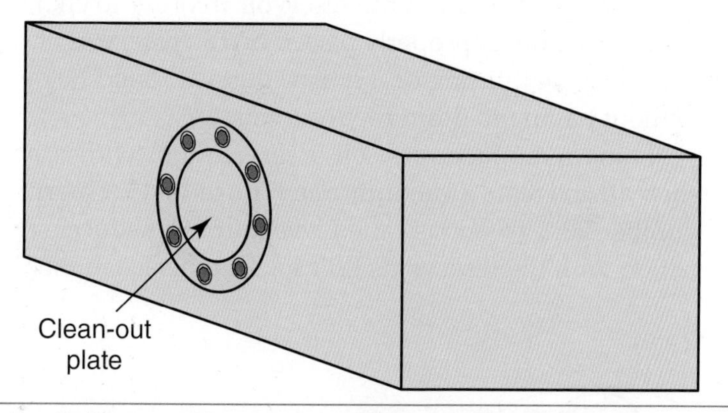

Figure 14-12. Reservoir clean-out plates provide access to the reservoir's interior for cleaning contaminants and replacing suction screens.

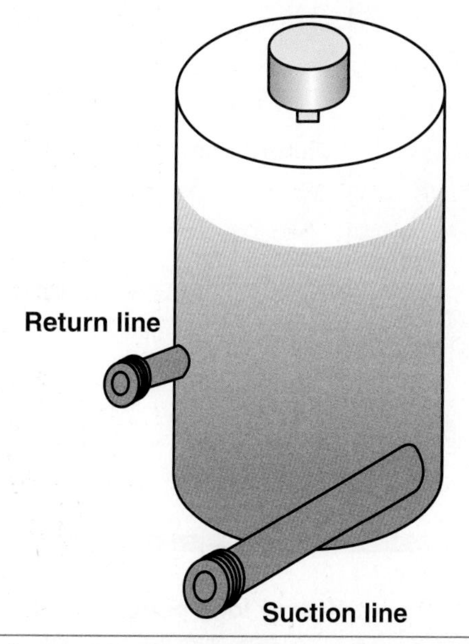

Figure 14-13. Cylindrical reservoirs were first designed for hydraulically driven fans in mobile equipment, requiring less fluid and lowering the machine's cost.

Note that the cylindrical reservoir is much smaller than the traditional rectangular reservoir. For example, one manufacturer has designed a cylindrical reservoir that holds only 3.75 gallons (14 liters) for use with hydraulic systems with flow rates of 20 to 40 gpm (75 to 151 lpm).

Another manufacturer has designed a cylindrical reservoir with a cone on the bottom. It uses a kidney loop filter to draw the contaminants from the bottom of the reservoir. See Figure 14-14. Using an onboard kidney loop filter to drawn oil from the bottom of the cone maximizes removal of contaminants from the reservoir.

Reservoirs and Maintenance

Many mobile machines tend to have smaller reservoirs than industrial equipment. However, some mobile systems use large hydraulic reservoirs. When the reservoirs contain 50 to 150 gallons of hydraulic oil, it becomes a significant task to drain the reservoir in order to change a hose or a fitting.

If a vacuum is properly placed on the reservoir, it can allow a technician to replace a small fitting or a sensor that is located below the reservoir's oil level without having to drain the reservoir.

Some manufacturers sell a service tool with a built-in venturi that uses shop air to create a vacuum. See Figure 14-15. Caterpillar's special tool is part number 5P-0306 and is called a vacuum transducer. The tool is used in combination with the reservoir's fill cap. The service manual will specify how much

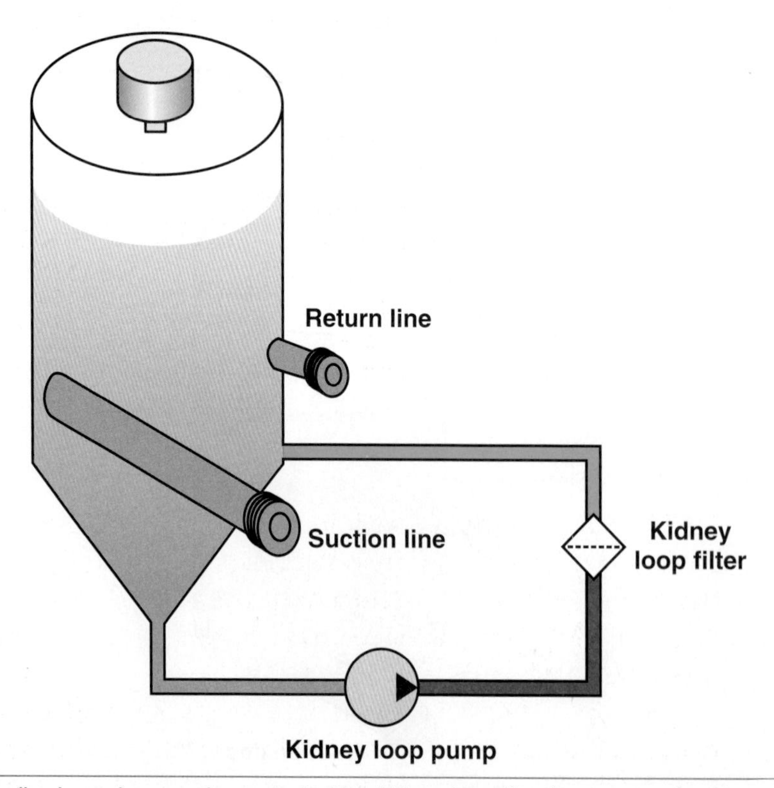

Figure 14-14. A cylindrically shaped reservoir coupled with a kidney loop filtration system that draws oil from the bottom of a cone will optimize reservoir cleanliness.

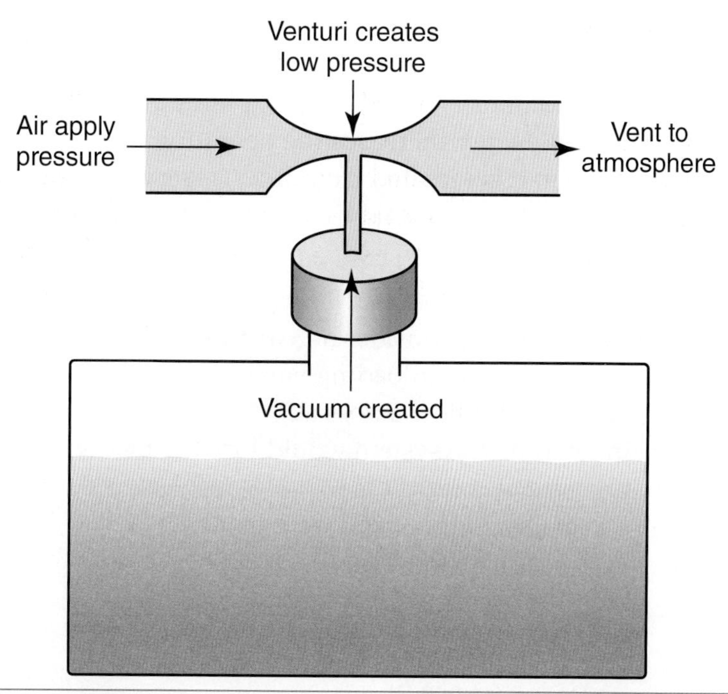

Figure 14-15. A special service tool is used to pull a vacuum on a reservoir. The venturi generates a vacuum in the reservoir when air pressure is applied to the tool.

air pressure must be applied to the vacuum transducer. A small oil tank may only require 20 psi, other tanks might require 40–60 psi, and the largest reservoirs will require more air pressure. If the tank has a breather that is separate from the fill port, a vacuum cannot be pulled on the reservoir.

Thinking Green

At least one forestry machine has a factory-installed vacuum pump that allows an operator to stop or slow the depletion of the hydraulic reservoir in the event of a fluid leak. The operator has to shut off the engine, depressurize the reservoir, and then press the vacuum apply switch. The vacuum pump will turn off and on while monitoring the vacuum inside the reservoir. This option is designed to be used only as a temporary solution while the leak is being repaired. Its use minimizes contamination to the environment in the event of a leak.

Heat

Heat is harmful to hydraulic systems. As mentioned in **Chapter 10**, increased fluid temperature causes the fluid to oxidize, shortening the life of the fluid. High operating temperatures will cause seals to degrade. The rate of degradation of a hydraulic seal will increase when the component's temperature rises above 180°F (82°C).

Heat in a hydraulic system is caused by inefficiencies. Any time the hydraulic energy is not being used to perform useful work, it is being transformed into heat energy, which is transmitted into the atmosphere.

$$\text{Lost Energy} = \text{Pump Input Energy} - \text{Actuator Output Energy}$$

Examples where fluid energy is transformed into heat energy are:

- Valves—relief valves, regulator valves, control valves.
- Restrictions—orifices, filters, plumbing.

The heat energy increases the operating temperature of the system, and is wasted when it is radiated from lines, reservoirs, and coolers. The quantity of energy wasted as heat can be computed using two different formulas:

$$\text{Horsepower (HP) Loss} = 0.000582 \times \text{Flow (Q)} \times \text{Pressure Drop } (\Delta P)$$

$$\text{Heat (Btu/hr)} = 1.48 \times \text{Flow (Q)} \times \text{Pressure Drop } (\Delta P)$$

As an example, consider a gear pump that is flowing 25 gpm (95 lpm). If the jammer valve or unloading valve malfunctioned and became stuck closed, the 25 gpm would flow across a 3000 psi (207 bar) relief valve, while the DCVs were in neutral. The system would lose the following amount of energy in the form of waste heat:

$$\text{HP Loss} = 0.000582 \times 25 \times 3000 \text{ psi}$$

$$\text{HP Loss} = 43.65 \text{ hp}$$

or

$$\text{Heat} = 1.48 \times 25 \times 3000 \text{ psi}$$

$$\text{Heat} = 111{,}000 \text{ Btu/hr}$$

Reducing system inefficiencies, such as losses of flow or pressure losses, will reduce heat in the system.

There are a few rules of thumb regarding system temperatures. At the exterior wall of the reservoir, the fluid temperature should not exceed 140°F (60°C). The temperatures will be higher at other locations, such as the pump's outlet. Therefore, the temperature of the system will depend on the location of the temperature reading. Some temperature sensors are placed at the pump's outlet. Others are located before the cooler.

Note

The term *heat loss* is commonly used to refer to the losses in efficiency that result in heat. In the context of hydraulic systems, this term must not be confused with the rate at which the system can dissipate heat.

Coolers

As mentioned earlier in the chapter, when the reservoir's volume is less than twice the pump capacity, a hydraulic oil cooler is recommended. Keep in mind that there are differences in the amount of heat generated in fixed-displacement systems and variable-displacement systems. **Chapters 16** through **19** will detail the differences in inefficiencies for common types of mobile hydraulic systems.

An ***oil cooler,*** also known as ***heat exchanger,*** is a device used to cool the hydraulic oil. The cooler is usually incorporated in the system's return line, which has a low operating pressure. Heat exchangers cool using either air or water.

Regardless of the type of heat exchanger, both perform better when the oil flow is turbulent or the fluid velocities are high. If a cooler consisted of a single large-diameter pipe, the oil flowing through the center would not be cooled, and the oil travelling next to the wall of the tubing would be cold. As a result, exchangers are often designed to disturb the oil, making it more turbulent so more of the oil comes into contact with cooling surfaces, improving cooling.

Water-Type Oil Coolers

Water-type oil coolers (or simply water coolers) use water or engine coolant to cool the oil. They are used for cooling automatic transmission fluid, diesel engine oil, or hydraulic oil. When engine coolant is used, the exchanger can also be called a ***temperature controller*** because the engine coolant can help warm the oil to operating temperature. Although water coolers are seldom used in mobile machinery, they are used in industrial hydraulic systems and marine hydraulic systems.

There are different types of water cooler designs. Figure 14-16 illustrates a water cooler that contains a bundle or group of small tubes placed inside a canister. Coolant is routed through the small tubes and oil is directed to flow across the tubes.

Figure 14-17 illustrates a plate-type water cooler. The oil is directed through a stack of plates and water is routed around the plates to continuously cool the oil. Oil travels through the plates from one side to the other while water cools the plates.

Water coolers have the risk of mixing the two different fluids. Some manufacturers use pressure sensors to determine if there is a loss in pressure in the exchanger. The electronic control module (ECM) can alert the operator when a pressure loss has been detected.

Figure 14-16. A bundled-tube water cooler routes water through the small tubes, and oil is directed to flow across the tubes.

Figure 14-17. A plate-type water cooler routes oil through a stack of flat plates, and coolant is directed to flow across the plates to cool the oil.

A water cooler is smaller and more efficient than an air cooler because water can transfer heat at a much greater rate than air can. A water cooler is quiet and is less affected by the ambient air temperature. However, most mobile hydraulic systems use air coolers rather than water coolers because coolant temperatures are too high. Also, land-based mobile machines do not have access to the large body of cold water available to marine applications. As a result, hydraulic water coolers are more common in marine applications.

Air-Coolers

An ***air cooler*** contains tubes with fins to dissipate heat. Hydraulic oil is routed through the tubes. A fan is used to blow air across the cooler assembly, similar to an engine's radiator. The fan can be driven mechanically by a belt, an electric motor, or a hydraulic motor. See **Figure 14-18**.

The advantage of using an air cooler is that if a leak occurs, it will not cross-contaminate the water or coolant. The disadvantage of an air cooler is that it is affected by the ambient air temperature. It is more difficult to cool the hydraulic oil when the system is operating in a hot climate. Also, the hydraulic air cooler can be located next to other air coolers, which are also dispersing heat, reducing its efficiency.

Warmers and Heaters

Equally important to cooling hot oil is that the oil be warmed to operating temperature. If a hydraulic system is operated with cold oil, it will cause the system to be sluggish (slow responding) and can damage the system. Oil heaters or warmers are required in some colder climates. ***Hydraulic warmers*** and ***hydraulic heaters*** use heat from engine coolant or an electrical heating element to warm the hydraulic oil. The heating element is commonly placed in the reservoir. See **Figure 14-19**.

Figure 14-18. An air cooler is the most popular type of hydraulic oil cooler used in mobile machinery. Oil is routed through the cooler's tubes. A fan blows air across the cooler to dissipate heat through the fins on the tubes.

Figure 14-19. Hydraulic warmers and hydraulic heaters are placed in the reservoir and are used to warm the oil to operating temperature.

Summary

- ✓ Reservoirs house the oil, cool the oil, provide a good supply to the pump inlet, and provide a place for air and contaminants to separate.
- ✓ Reservoirs are either sealed or vented.
- ✓ Baffles prevent return fluid from immediately cycling into the pump's inlet.
- ✓ Diffusers use a perforated screen to slow the velocity of return oil and prevent aerated fluid.
- ✓ Sight glasses can become discolored, causing a misreading of the reservoir's fluid level.
- ✓ Suction lines and suction screens are placed 1/2″ to 3/4″ off the bottom of the reservoir's floor.
- ✓ Filler caps are sometimes sealed or might be a breather cap.
- ✓ Breathers remove contaminants from the air as it is drawn into the reservoir including moisture.
- ✓ Clean-out plates enable a person to clean the interior of a reservoir and replace a tank-mounted suction screen.
- ✓ Reservoir shape influences contamination in the reservoir.
- ✓ A vacuum is sometimes placed on a reservoir to allow the fast change of a small fitting without draining the reservoir.
- ✓ Variable-volume reservoirs use a rubber bladder that changes in volume to meet demand.
- ✓ Heat is the result of hydraulic system inefficiencies.
- ✓ Oil coolers are required in mobile machinery because hydraulic reservoirs are too small to cool the oil.
- ✓ Air coolers are the most popular type of oil cooler used in mobile machinery.
- ✓ Heaters are placed in reservoirs when machines are operating in severe cold climates.

Technical Terms

air cooler
baffle
bleeder valve
case drain line
clean-out plate
diffuser
heat exchanger
hydraulic heaters
hydraulic warmers
integral reservoir
oil cooler
pressurized return lines
reservoir
scavenge pump
temperature controller
variable-volume reservoir (VVR)
water-type oil coolers

Review Questions

Answer the following questions using the information provided in this chapter.

1. A hydraulic reservoir performs all of the following functions, *EXCEPT*:
   A. receives and dissipates heat.
   B. relieves pump pressure.
   C. provides a place for entrained air to separate from the fluid.
   D. provide a place for contaminants to settle from the fluid.
2. Large reservoirs provide all of the following advantages, *EXCEPT*:
   A. improve oil cooling.
   B. provide excess oil for a short period of time regardless of any return oil.
   C. reduce the risk of pump cavitation after a component has been repaired or replaced.
   D. are lightweight.

3. Smaller reservoirs provide all of the following advantages, *EXCEPT*:
   A. are cheaper to build.
   B. take less space.
   C. improve separation of entrained air.
   D. have less oil to leak, causing less harm to the environment.
4. Manufacturers might pressurize a hydraulic reservoir for all of the following reasons, *EXCEPT*:
   A. the machine is using a charge pump.
   B. pump is located above the reservoir's oil level.
   C. pump is located a long distance from the reservoir.
   D. the machine is using an open loop pump.
5. All of the following are true statements about sealed hydraulic reservoirs, *EXCEPT*:
   A. if reservoir does not contain a bleeder, the system will burp air when the filler cap is removed.
   B. the reservoir must be located higher than the pump inlet.
   C. the bleeder valve should be depressed before the hydraulic system is serviced or repaired.
   D. the sealed reservoir filler cap is sometimes a rubber plug similar to a boat plug.
6. All of the following are true statements regarding sealed reservoirs, *EXCEPT*:
   A. some use a breather.
   B. some contain no bleeder valve.
   C. some use a manual bleeder valve.
   D. some use an automatic bleeder valve.
7. Hydraulic reservoir walls can be all of the following materials, *EXCEPT*:
   A. rubber bladder.
   B. fiberglass.
   C. steel.
   D. plastic.
8. All of the following names are used to describe the container for housing hydraulic oil, *EXCEPT*:
   A. reservoir.
   B. diffuser.
   C. tank.
   D. sump.
9. Which of the following terms can be described as a perforated plate or perforated housing?
   A. Baffle.
   B. Diffuser.
   C. Breather.
   D. Bleeder valve.
10. Which of the following is designed to reduce the velocity of the return oil for the purpose of reducing fluid aeration?
    A. Baffle.
    B. Diffuser.
    C. Breather.
    D. Bleeder valve.
11. Technician A states that baffles can be attached to the bottom of the reservoir. Technician B states that baffles can be attached to the top of the reservoir. Who is correct?
    A. Technician A.
    B. Technician B.
    C. Both A and B.
    D. Neither A nor B.
12. Which one of the following prevents the return oil from cycling straight through the reservoir to the pump's suction hose?
    A. Baffle.
    B. Diffuser.
    C. Breather.
    D. Bleeder valve.
13. A technician is checking the reservoir's oil level. All of the following can result in an erroneous reading, *EXCEPT*:
    A. using a tape measure through the fill port.
    B. machine not level.
    C. glancing at the sight glass.
    D. the implements are raised.
14. All of the following will affect the reservoir's oil level, *EXCEPT*:
    A. level of the machine.
    B. position of the implements.
    C. fluid temperature.
    D. color of oil.

15. What is the minimal amount of air pressure needed to pressurize a reservoir?
    A. 3 to 4 psi (0.2 to 0.27 bar).
    B. 30 to 45 psi (2 to 3 bar).
    C. 60 to 90 psi (4 to 6 bar).
    D. 90 to 120 psi (6 to 8 bar).
16. The reservoir's main return line should be placed where in the reservoir?
    A. 1/2" to 3/4" above the oil level.
    B. 1/2" to 3/4" below the oil level.
    C. 1/2" to 3/4" above the floor of the reservoir.
    D. 1/2" to 3/4" below the top of the reservoir.
17. The reservoir's suction screen or suction line should be placed where in the reservoir?
    A. 1" above the oil level.
    B. 1" below the oil level.
    C. 1" above the floor of the reservoir.
    D. 1" below the top of the reservoir.
18. All of the following are true statements about case drain lines, *EXCEPT*:
    A. they should be filtered.
    B. they are low- or zero-pressure lines.
    C. they should not siphon oil back to the reservoir.
    D. they deliver leakage oil back to the reservoir.
19. Which of the following can be described as an integral reservoir?
    A. Rubber bladder.
    B. Plastic rectangular housing.
    C. Thin wall steel housing.
    D. Axle housing.
20. What is the most popular type of heat exchanger used in mobile hydraulic systems?
    A. Air cooler.
    B. Bundled-tube water cooler.
    C. Stacked-plate water cooler.
    D. Mobile machines do not use heat exchangers.
21. All of the following are true regarding a variable-volume reservoir, *EXCEPT*:
    A. it uses a rubber bladder.
    B. it uses a coiled spring.
    C. it is pressurized to 1 to 9 psi.
    D. it uses nitrogen gas.
22. Technician A states that heaters are sometimes placed inside a reservoir. Technician B states that hydraulic heater/warmers can be warmed with coolant or an electric heating element. Who is correct?
    A. Technician A.
    B. Technician B.
    C. Both A and B.
    D. Neither A nor B.
23. All of the following are advantages of water cooler–type heat exchangers, *EXCEPT*:
    A. they are compact
    B. they are efficient.
    C. they eliminate the risk of cross-contamination of fluids.
    D. they are quiet.
24. A hydraulic system is flowing 50 gpm of oil at 500 psi. The hydraulic energy is not being used for any useful work. How much horsepower is being lost?
    A. 10.5 hp.
    B. 14.6 hp.
    C. 20.5 hp.
    D. 24.6 hp.
25. A hydraulic system is flowing 40 gpm of oil at 600 psi. The hydraulic energy is not being used for any useful work. Compute the amount of heat generated.
    A. 25,500 Btu/hr.
    B. 35,500 Btu/hr.
    C. 45,500 Btu/hr.
    D. 55,500 Btu/hr.

Chapter 15

Plumbing

Objectives

After studying this chapter, you will be able to:

- ✓ Explain the importance of understanding plumbing in mobile hydraulic systems.
- ✓ List examples of different types of plumbing used in mobile hydraulic systems.
- ✓ Explain the advantages and disadvantages of using hose, tubing, and pipe in a hydraulic system.
- ✓ List the guidelines to be followed when choosing and attaching a hose end fitting.
- ✓ Briefly summarize the steps for using a crimping machine to attach a fitting to the end of a hose.
- ✓ Explain the advantages and disadvantages of the different types of fittings.
- ✓ Describe the construction of the different types of fittings.
- ✓ List examples for using quick couplers.
- ✓ Describe actions for reducing noise and vibration in hydraulic systems.

The Importance of Understanding Plumbing in Mobile Hydraulic Systems

Although working fewer hours per year on average, mobile hydraulic systems typically operate at higher system pressures than industrial machinery. As a result of the high system pressures, mobile hydraulic systems typically need more service and repairs.

It is easy for technicians to fall into the practice of simply replacing components rather than performing proper diagnostic test procedures. Some think that installing a new component takes less time than connecting a pressure gauge or a flowmeter into the circuit and diagnosing the problem. Often, however, performing repairs using this method results in a longer machine downtime. Unfortunately, many hydraulic personnel find themselves replacing components not because it appears faster, but because they lack the knowledge required to tap into the system, are unsure of what fittings to order, and can be overwhelmed with where to start the testing process. A technician's ability to correctly identify the different types of fittings, hoses, and tubing will expedite the process of being able to test, diagnose, service, and repair hydraulic systems.

Technically speaking, if a hydraulic system could be operated in a perfectly efficient world, it would never need to be serviced or repaired, and could be rigidly fabricated with no need for disassembly. However, the reality is that these systems must be accessible. As a result, manufacturers around the world use a host of different types of plumbing to connect a hydraulic system's circuitry. This creates a complex puzzle for hydraulic personnel.

For example, some personnel in North America have had the challenge of working on machines that were custom built by a European or Asian manufacturer. When the machine arrived in North America, the technicians rarely had a service manual or parts manual because the machine was the only one of its kind. In this case, it is critical to know the difference between common design engineering metrics used to manufacture hydraulic system plumbing. This includes, but is not limited to, North American threads, metric threads, British Standard Pipe threads, and Japanese Industrial Standards threads.

Types of Plumbing

Plumbing can be classified as either one of the following types:

- Lines, also known as conductors.
- Port connections, also known as fittings.

Hydraulic conductors can be one of three different types:

- Hose.
- Steel tubing.
- Steel pipe.

All three conductors can be found on mobile machinery. However, pipe is the least common.

Different styles of fittings can seal with metal-to-metal sealing surfaces or can be sealed using an O-ring. Metal-to-metal fittings are categorized into three groups:

- Flare.
- Pipe.
- Flareless.

O-ring fittings are also classified into three groups:

- O-ring boss.
- O-ring face seal.
- Split-flange O-ring.

Each fitting has advantages and disadvantages.

Conductors

Manufacturers can choose to route oil through hose, tubing, or pipe. Pipe is difficult to prevent from leaking and is less common in mobile machinery. Hose and tubing are both used throughout mobile hydraulic systems. Both have advantages over pipe.

Hose

Hydraulic hose is the only type of conductor that can flex as an implement rotates, swivels, or pivots, such as during boom, stick, or bucket operation on a backhoe. However, some hydraulic hose is stiff due to the number of steel wires incorporated into it.

Hose is the preferred conductor when trying to eliminate noise or vibration problems. (These issues are detailed at the end of this chapter.) Hose also offers the advantage of fast fabrication.

Hose has its disadvantages as well, including the following characteristics:

- It is more expensive than pipe or tube.
- It must be replaced periodically, adding cost and contamination.
- During operation, it can undergo a length change of –4% to +2%.
- It will swell in volume, acting like an accumulator, when used in high-pressure applications.

All hydraulic hoses have a ***working pressure*** rating, which is the pressure value at which the hose can safely operate or work. A conductor must be rated at a working pressure that is higher than the system's highest operating pressure, including the effects of pressure intensification.

Manufacturers also specify a ***burst pressure*** for hydraulic hoses, which is the minimum pressure value that will rupture the hose. Manufacturers pressurize the same type of hose multiple times to establish its burst pressure value. The working pressure and burst pressure have a relationship called a ***safety factor***. The safety factor for hydraulic hoses equals four, which can be expressed in the following ways:

- Working pressure is one quarter of the hose burst pressure.
- Burst pressure is four times the value of the hose working pressure.

Hydraulic hose sizes refer to the conductor's inside diameter (ID). This diameter is normally given in a ***dash size***, which equals the diameter in 1/16ths of an inch. For example, a –12 hose equals 12/16ths of an inch, which equals a 3/4″ hose with an inside diameter of 0.750″.

The inside diameter of many hydraulic hoses has an inverse relationship with the hose's pressure capability. For a given type of hose construction, hoses with a smaller inside diameter will normally have a higher pressure rating. See **Figure 15-1**.

Hose Model and Size	ID	Working Pressure	Burst Pressure
Weatherhead H245-04	1/4″	5000 psi (345 bar)	20,000 psi (1379 bar)
Weatherhead H245-06	3/8″	4000 psi (276 bar)	16,000 psi (1241 bar)
Weatherhead H245-08	1/2″	3500 psi (241 bar)	14,000 psi (965 bar)
Weatherhead H245-10	5/8″	2750 psi (190 bar)	11,000 psi (758 bar)
Weatherhead H245-12	3/4″	2250 psi (150 bar)	9000 psi (620 bar)
Weatherhead H245-16	1″	2000 psi (138 bar)	8000 psi (552 bar)

Figure 15-1. Many hydraulic hoses with a small inside diameter will have a higher pressure rating than the same hose with a larger inside diameter. This Weatherhead® hose is an SAE 100R16 hose.

However, some hydraulic hoses will have the same pressure ratings across different hose sizes, such as the Weatherhead® H545 Rhino hydraulic hose. This hose is offered in sizes of –04, –06, –08, –10, –12, –16. All of the hose sizes are rated at 3000 psi (207 bar) working pressure and 12,000 psi (827 bar) burst pressure.

SAE classifies five categories of hose that all have a constant working pressure regardless of the hose's inside diameter:

- SAE 100R13 5000 psi.
- SAE 100R15 6000 psi.
- SAE 100R17 3000 psi.
- SAE 100R18 3000 psi.
- SAE 100R19 4000 psi.

SAE hose classification is detailed later in this chapter.

Hose Composition

Hydraulic hose is commonly made of synthetic rubber. This material is the most popular among manufacturers. Some hose is made of thermoplastic. Thermoplastic is commonly used in personnel lifts that are made to operate near high-voltage power lines. These types of lifts require ***nonconductive*** hoses to prevent the hoses from channeling electricity if they come into contact with a power line.

The hose is constructed of three structures:

- Inner tube.
- Reinforcement.
- Outside cover.

Inner Tube

The ***inner tube*** is in direct contact with the oil and must be compatible with the hydraulic oil. The inner tube can be made of the following oil-resistant materials:

- Thermoplastic.
  - Nylon.
  - Urethane.
  - Hytrel.
- Synthetic rubber.
  - Neoprene.
  - Nitrile rubber.

Reinforcement

The ***reinforcement*** of the hose is what gives the conductor its strength. This center structure uses a braided, spiraled, or helical design. **Figure 15-2** shows these three different reinforcement styles. A braided layer will use some type of fiber, in the form of yarn or cord, or use a metal steel composition. The braided layer will be interlaced together and sheathed around the inner tube to form a reinforcement-type sleeve.

A spiral-coiled reinforcement layer wraps around the inner tube in multiple layers. If the reinforcement contains multiple spiral layers, the layers will be wrapped in alternating directions for additional strength. A hose with braided reinforcement will be more flexible than a hose with spiral-coiled reinforcement.

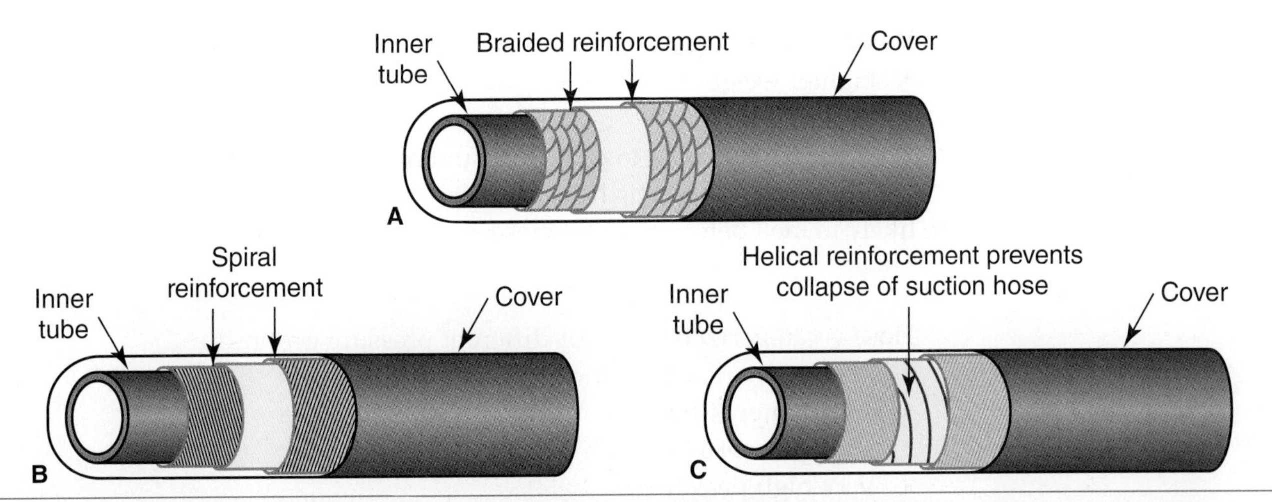

Figure 15-2. Hydraulic hoses are made of three structures: an inner tube, reinforcement, and the outside cover. Reinforcement of a hydraulic hose is typically designed in one of three patterns. A—Braided. B—Spiral. C—Helical.

Helical reinforcement uses a large, coiled wire that forms a helix between the tube and cover. Helical reinforcement is used in suction hose to prevent it from collapsing.

A separation layer, also called cushion stock, is sometimes placed in between multiple reinforcement layers.

Materials used for reinforcement include the following examples:

- Fiber braids.
- Steel braids.
- Spiral steel plies.
- Steel helical coils.
- Both steel ply and steel braid.
- Fiber braids and steel braids.

Cover

The outer hose layer is called the ***cover***. Its responsibility is to protect the hose from the elements, such as chemicals, weather, and abrasion. The cover can be made of synthetic rubber, thermoplastic, synthetic fabric, or nylon.

Manufacturers offer multiple types of products, such as the following, to further protect hoses:

- Abrasion sleeves.
- Poly hose guards.
- Round-wire spring guards.
- Flat-wire spring guards.
- Fire sleeve tape.
- Hose bend restrictors.

Nonconductive Hose

Personnel lifts that are designed to work around high-voltage power lines use hydraulic hose that is nonconductive. The outside cover is a non-perforated polyurethane cover used to keep moisture out of the hose. A non-conductive hose cover is orange and labeled *nonconductive*. Keep in mind, *not* all orange hoses are nonconductive hydraulic hose. The inner tube is made of a thermoplastic elastomer and the reinforcement uses a single braid of Kevlar™.

When compared to rubber hoses, thermoplastic hoses have some advantages:

- Lighter weight.
- Smaller diameter.
- React less severely to chemicals and oils.

However, thermoplastic hoses are less durable and, at high temperatures, are likely to melt before a rubber hose.

Hose Categories

Hose is categorized into four different pressure groups:

- Low pressure, which includes suction hose.
- Medium pressure.
- High pressure.
- Very high pressure.

Low-pressure hose and suction hose are similar because they are not engineered to contain high pressures. However, ***suction hose*** is tasked with supplying a hydraulic pump with a good supply of inlet oil, and may be subjected to vacuum. In such cases, atmospheric pressure outside the hose could cause it to collapse. The suction hose must be reinforced to prevent it from collapsing, **Figure 15-3**. Therefore, even though a suction hose is not subjected to high pressures, it is still a stiff, inflexible hose.

SAE Hose Classification

Hose manufacturers build hose to meet SAE J517 hose classifications. SAE has 19 different hydraulic hose categories. Each hose category has a specific pressure rating, hose construction, temperature rating, and fluid compatibilities. The temperature range is dependent on the fluid type, such as synthetic, petroleum, or water-based hydraulic fluids. **Figure 15-4** is a table housing this information for all of the 19 different SAE hydraulic hose categories.

Within one SAE category, a hose can be further classified. For example, a SAE 100R2 hose has Type A, Type AT, Type B, and Type BT. Type AT and Type BT are the same as Type A and Type B respectively, except that a portion of the cover does not have to be removed for installing the hose end on either hydraulic hose.

Figure 15-3. This hose is a –20 suction hose. A steel, helical reinforcement is used to prevent the suction hose from collapsing.

SAE Hose No.	Inner Tube	Reinforcement	Cover	Temperature Range	Working Pressure Ranges Based on Hose ID	Hydraulic Fluid Type
SAE 100R1 Medium Pressure	Oil-resistant type (Nitrile)	1 wire braid	Synthetic rubber	–40° to 212°F (–40° to 100°C)	3/16″=3000 psi 2″= 375 psi	Petroleum oil, water based
SAE 100R2 High Pressure	Oil-resistant synthetic rubber	Type A— 2 braids of steel wire Type B— 2 spiral wires and 1 braid Type AT— same as A, but has cover not requiring skiving Type BT— same as B, but has cover not requiring skiving	Synthetic rubber	–40° to 212°F (–40° to 100°C)	3/16″=5000 psi 2 1/2″=1000 psi	Petroleum oil, water based
SAE 100R3 Low Pressure	Oil-resistant synthetic rubber	2 braids of textile yarn	Synthetic rubber	–40° to 212°F (–40° to 100°C)	3/16″=1500 psi 1 1/4″=37 psi	Petroleum oil, water based
SAE 100R4 Low Pressure Return or Suction Line	Oil-resistant synthetic rubber	1 or more braided or woven textile fibers with a spiral wire, and may have helical reinforcement to prevent collapse	Synthetic rubber	–40° to 212°F (–40° to 100°C)	3/4″=300 psi 4″=35 psi (Note the large diameters)	Petroleum oil, water based
SAE 100R5 Medium Pressure	Oil-resistant synthetic rubber	2 textile braids separated by single wire braid	Oil- and mildew-resistant, polyester braid	–40° to 212°F (–40° to 100°C)	3/16″=3000 psi 3″=200 psi	Petroleum oil, water based
SAE 100R6 Low Pressure	Oil-resistant synthetic rubber	1 braid of textile yarn	Synthetic rubber	–40° to 212°F (–40° to 100°C)	3/16″=500 psi 3/4″=300 psi	Petroleum oil, water based
SAE 100R7 Medium Pressure Available in Non-Conductive	Thermoplastic	Synthetic fiber	Thermoplastic	–40° to 200°F (–40° to 93°C)	3/16″=3000 psi 1″=1000 psi	Petroleum oil, water based, synthetic oil
SAE 100R8 High Pressure Available in Non-Conductive	Thermoplastic	Synthetic fiber	Thermoplastic	–40° to 200°F (–40° to 93°C)	3/16″=5000 psi 1″=2000 psi	Petroleum oil, water based, synthetic oil

Figure 15-4. SAE hydraulic hose classification specifies the type of hose construction, pressure ratings, temperature range, and fluid compatibility for each of their 19 different categories. *(Continued)*

SAE Hose No.	Inner Tube	Reinforcement	Cover	Temperature Range	Working Pressure Ranges Based on Hose ID	Hydraulic Fluid Type
SAE 100R9 High Pressure	Oil-resistant synthetic rubber	4 wire spiral layers	Synthetic rubber	−40° to 212°F (−40° to 100°C)	3/8″=4500 psi 2″=2000 psi	Petroleum oil, water based
SAE 100R10 Very High Pressure	Oil-resistant synthetic rubber	4 wire spiral layers	Synthetic rubber	−40° to 212°F (−40° to 100°C)	3/16″=10,000 psi 2″=2500 psi	Petroleum oil, water based
SAE 100R11 Very High Pressure	Oil-resistant synthetic rubber	6 wire spiral layers	Synthetic rubber	−40° to 212°F (−40° to 100°C)	3/16″=10,000 psi 2 1/2″=2500 psi	Petroleum oil, water based
SAE 100R12 Very High Pressure	Oil-resistant synthetic rubber	4 wire spiral layers	Synthetic rubber	−40° to 250°F (−40° to 121°C)	3/8″=4000 psi 2″=2500 psi	Petroleum oil, water based
SAE 100R13 Very High Pressure	Oil-resistant synthetic rubber	Multiple spiral layers	Synthetic rubber	−40° to 250°F (−40° to 121°C)	All 5000 psi 3/4″to 2″	Petroleum oil, water based
SAE 100R14 Low Pressure Available in Non-Conductive	Type A—polytetrafluorethylene (PTFE) Type B—electrically conductive, prevents electrostatic charge	1 steel wire braid	Single braided stainless steel wire	−65° to 400°F (−54° to 204°C)	1/8″=1500 psi 1 1/8″=600 psi	Petroleum oil, water based, synthetic oil
SAE 100R15 Very High Pressure	Oil-resistant synthetic rubber	Multiple spiral layers	Rubber	−40° to 250°F (−40° to 121°C)	All 6000 psi 3/8″to 1 1/2″	Petroleum oil *only*
SAE 100R16 High Pressure	Oil-resistant synthetic rubber	1 or 2 steel wire braids	Synthetic rubber	−40° to 212°F (−40° to 100°C)	1/4″=5000 psi 1 1/4″=1625 psi	Petroleum oil, water based
SAE 100R17 Medium Pressure	Oil-resistant synthetic rubber	1 or 2 steel wire braids	Synthetic rubber	−40° to 212°F (−40° to 100°C)	All 3000 psi 3/16″to 1″	Petroleum oil, water based
SAE 100R18 Medium Pressure Available in Non-Conductive	Thermoplastic	Synthetic fiber	Thermoplastic	−40° to 200°F (−40° to 930°C)	All 3000 psi 3/16″ to 5/8″	Petroleum oil, water based, synthetic oil
SAE 100R19 High Pressure	Oil-resistant synthetic rubber	1 or 2 steel braids	Synthetic rubber	−40° to 212°F (−40° to 100°C)	All 4000 psi 1/4″ to 1″	Petroleum oil, water based

Figure 15-4. *(Continued).*

Hose manufacturers must also adhere to other hose standards such as European Norm (EN), Deutsches Institut fur Normung (DIN), Mine Safety and Health Administration (MSHA), Department of Transportation's Federal Motor Vehicle Safety Standards (DOT/FMVSS), and International Standards Organization (ISO).

Selecting a Hose End

Before a hose assembly can be fabricated, the correct hose end must be selected. The hose ends are chosen based on two categories: the attachment method (permanent or field attachable) and the style of fitting (pipe, flare, or O-ring).

An explanation of the different styles of fittings is explained later in the chapter, including the differences in thread selection, such as National Pipe Tapered (NPT), SAE, British, Japanese Industrial, or metric.

Hose Fabrication

A failed hydraulic hose has the risk of injuring personnel if it whips loose from its component. If the hose end is improperly attached, the hose end can also fly off the hose. Adhering to the following rules when fabricating a hydraulic hose can help prevent such problems from occurring:

- Do not mix-or-match a hose end from one manufacturer with a hose from another manufacturer.
- Always use the hose ends specified by the hose manufacturer.
- Do not install a new hose end on a used hose.
- Do not use a damaged field-attachable hose end.
- Do not over crimp or under crimp a fitting.
- Do not use a hacksaw to cut the hydraulic hose.
- Do not lubricate the hose cover in an attempt to insert the hose into the hose end, although some hose makers recommend lubricating the hose end nipple with 20W oil.

Assembling a hydraulic hose requires multiple steps. A technician must perform these actions:

- Choose the type of fitting (O-ring, pipe, and flare fittings are detailed later in the chapter).
- Choose the attachment method (permanent or field attachable).
- Determine the overall length and cutoff factor of the hose.
- Cut the hose.
- Clean the hose.
- Attach the hose ends.

The process of attaching the hose end not only imposes great risks if performed incorrectly, it can be a common source of fluid leakage and can reduce the service life of the hose. Hose end connections are a common source of hydraulic hose leaks.

Choosing the right type of hose end fitting will determine if an adapter fitting has to be added to the hose assembly. The adapter fitting will shorten the length of the hose, reducing its overall flexibility.

Field Attachable

Hose ends can be permanently attached by crimping or swaging. They can also be ***field attachable***, meaning that a technician can make a hose assembly in the field without the need of a crimping machine or a swaging machine. Not all field-attachable hose ends are reusable fittings. ***Reusable hose ends*** allow the fitting to be removed from an old hose and properly installed on a new hose. Field-attachable hose ends have several different classifications, but most can be classified as the following:

- Screw-on (threaded) type.
- Push-on or pull-on type (socketless).
- Clamp type.

The most common type of field-attachable fitting is the threaded type. The hose end contains an outside collar sometimes called a ferrule or a socket. The collar has internal left-hand threads. The collar is screwed onto the hose in a counterclockwise direction.

Warning

Prior to installing any hose end onto a hose, place the fitting next to the end of the hose, with the end tip of the hose flush with the shoulder of the fitting's defined collar. Then, be sure to make a reference mark on the hose as an indicator during installation to know when the hose has been fully seated into the fitting, Figure 15-5. If partially or improperly seated, the hose end could detach from the fitting while the system is operating, causing injury to personnel and damage to machines.

When the collar has been threaded onto the hose at the correct depth, the end of the hose will bottom out on the shoulder of the collar. At this point, carefully follow the manufacturer's assembly instructions. One hose manufacturer requires backing off the collar one turn after it has been fully inserted. The other end of the field-attachable fitting is threaded into the face of the collar in a clockwise direction until it contacts the collar. A bench-top vise is often used for assembling the field-attachable fitting.

One type of low-pressure field-attachable fitting is the push-on fitting, which Weatherhead calls Barb-Tite™ and Aeroquip® calls socketless. See Figure 15-6. The fitting has ribbed barbs. No clamp is used for this hose connection. The hose is simply pushed or pulled over the fitting's barbs. The hose cannot be pulled off the fitting once assembled. The hose end must be cut in order to remove it from the fitting.

Figure 15-5. Marking the hose prior to installing a two-part, threaded-type of field-attachable hose end. The outside collar is threaded onto the hose in a counterclockwise direction. Then, the stem end of the fitting is threaded completely into the collar in a clockwise direction to complete the assembly.

The other type of field-attachable hose end is a clamp style, which can use a hose clamp or a specially designed clamp to secure the fitting to the hose. These fittings are also used in low-pressure applications.

Permanent Attachment

Permanent fittings are attached to the hose using a crimping machine or a swaging machine. A swaging machine and a crimping machine can be either manually operated or electro-hydraulically operated.

Many personnel use the terms swaging and crimping incorrectly, believing that they are the same method of permanent attachment. ***Swaging*** is the process of forcing a hose end through a set of smooth-tapered dies. See **Figure 15-7**. Before the fitting is swaged, the dies are lubricated. The swaging machine forces the fitting through the dies, shrinking the fitting, which reduces the fitting's outside diameter to the specified dimension.

Crimping machines also use a pair of dies, sometimes called a collet. Unlike a swaging die with a two-piece smooth taper, crimping dies contain six to eight fingers, **Figure 15-8**. As the crimper is actuated, the collet's multiple fingers form pleats in the connection, shrinking the diameter of the fitting. The crimping dies are also designed to specifically fit the hose's diameter.

Note

Before attempting to use any swaging or crimping machine, remove all jewelry, secure all loose clothing, and wear the appropriate eye protection to avoid injury.

The following steps are performed to crimp a fitting using a Weatherhead Coll-O-Crimp® T-400 crimper:

1. Choose the hose based on pressure, temperature, and fluid compatibilities.
2. Choose the two hose ends with the correct fittings. Be sure that at least one of the two hose ends has a swivel fitting to ease the hose installation. Be sure that the hose ends are compatible with the hose crimper and the hose.

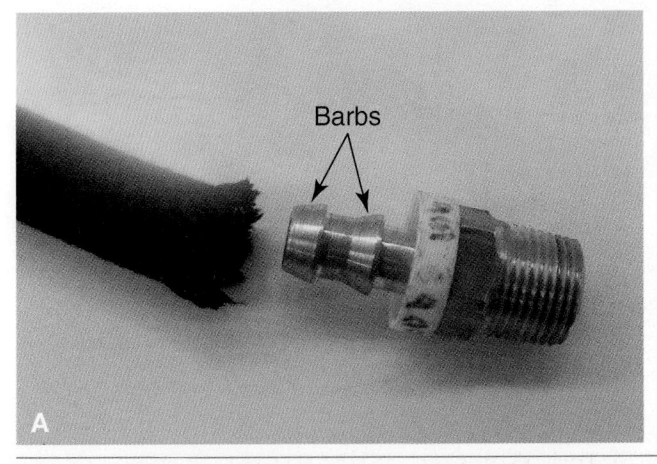

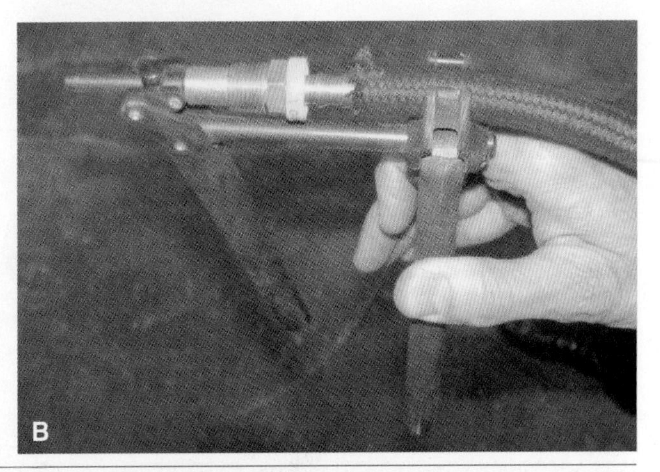

Figure 15-6. A threadless, push-on, field-attachable hose fitting. A—The fitting has barbs that grip the interior of the hose. B—A special tool is available to aid in installing the fitting into the end of a hose.

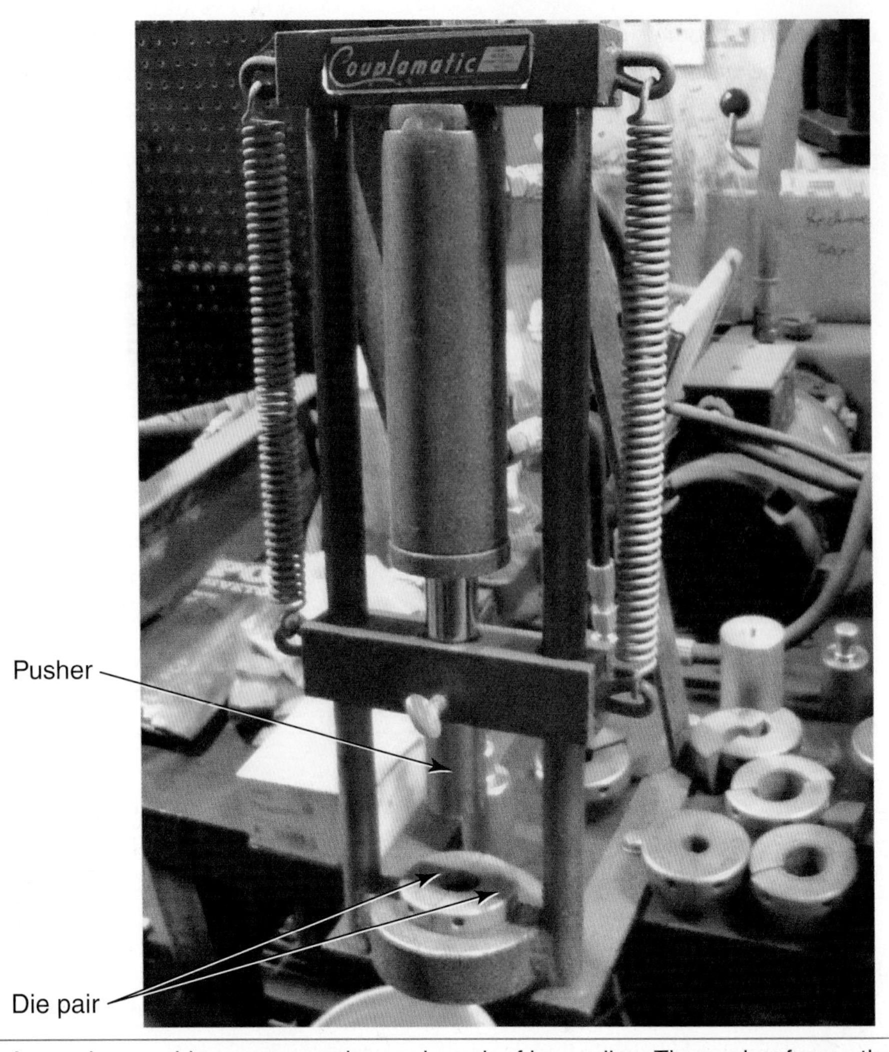

Figure 15-7. A swaging machine uses a pusher and a pair of lower dies. The pusher forces the fitting through the two dies until the fitting's diameter has been swaged to the specification.

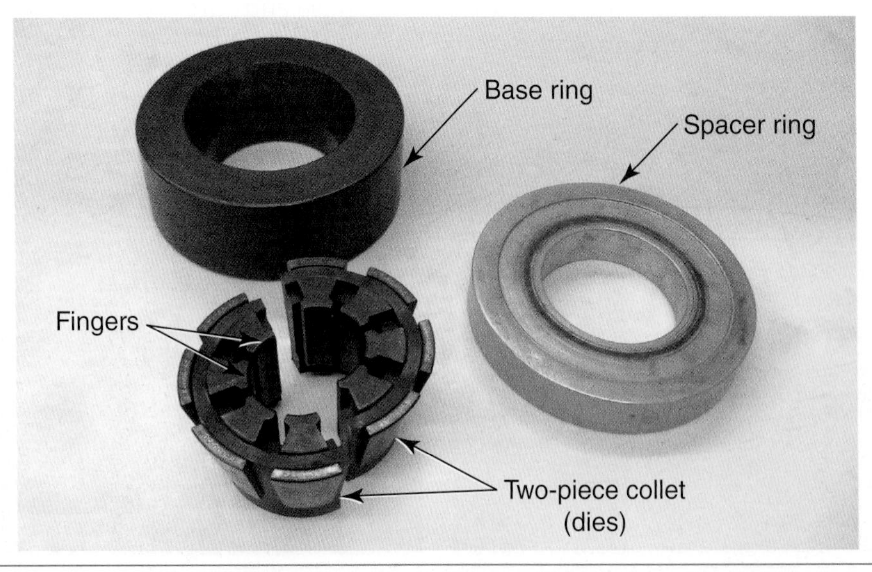

Figure 15-8. This hose crimping die is designed to crimp 3/4″ hose. The two dies fit inside the tapered, base ring and the spacer ring is placed on top of the collet.

3. Prior to cutting the length of the hose, check the service literature to determine the fittings' ***hose cutoff factor***, which is the amount of hose that must be removed to achieve the correct overall length for the hose assembly.

   For example, if a technician needs to build a hose assembly that is 18″ long, and if the two hose end fittings have a 1″ hose cutoff factor, subtract those two factors from the overall hose length. See Figure 15-9.

   18″ – (1″ cutoff factor for fitting A) – (1″ cutoff factor for fitting B) = Overall hose length of 16″

4. Cut the hose squarely using a hose cutter.

   Hydraulic hose contains some type of steel reinforcement: braided, spiral, or helical. Many technicians use a metal chop saw to cut the hose. For the best cut, use a hose cutter, as shown in Figure 15-10. This machine has two dowels and a center handle that is designed to place an equal force on both sides of the hose through the entire cut. The final cut must be straight. Most manufacturers and industry standards state the cut must be a straight, perpendicular cut, with a tolerance of ±5°.

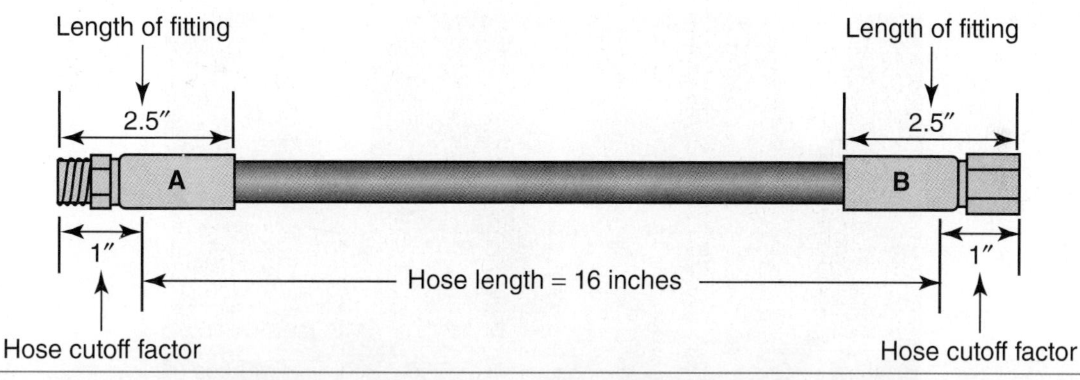

Figure 15-9. The hose service literature will specify the overall length of the fittings and their hose cutoff factor. Both are used to determine the length of hose to cut.

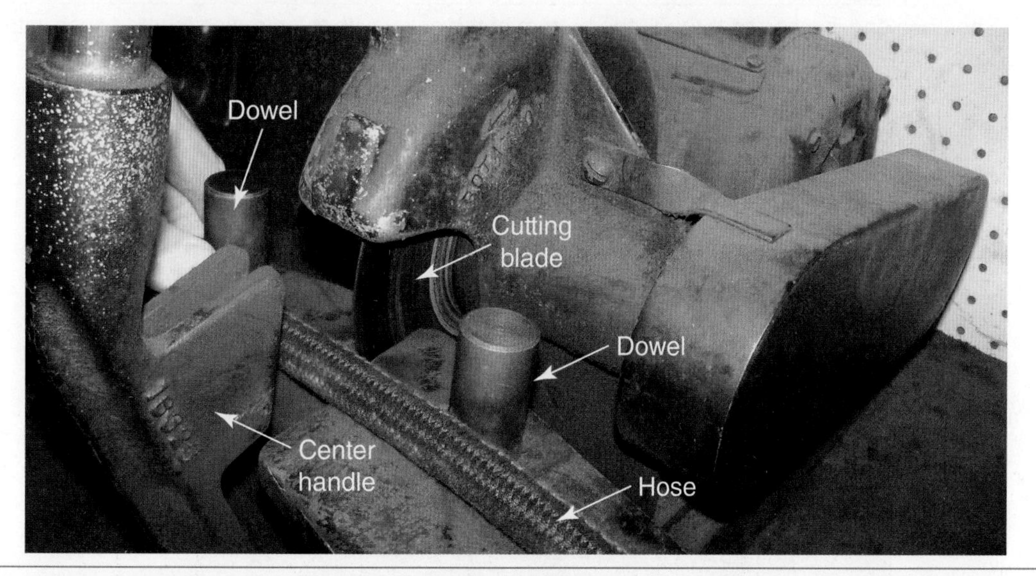

Figure 15-10. A hose cutter uses two dowels and a center handle that place an equal amount of force on both sides of the hose during the cut. This type of machine provides a straight, clean cut.

5. Clean the hose with a hose cleaning gun and foam projectile.
6. Mark a reference point on the hose that will align with the end of the fitting when the hose has been fully inserted into the hose end. Refer back to **Figure 15-5**.
7. Insert the hose into the hose end all the way to the reference mark. This will ensure the fitting is properly seated.
8. Insert the hose end through the bottom of the crimper's dies. Be sure to use the correct collet, specified with a part number and hose size labeled on the collet. The Weatherhead U series fitting contains dimples on the side of the fitting. The dimples must be located flush with the top of the collet, **Figure 15-11**.
9. Reference the service literature to determine the correct color of spacer ring and its orientation. For this example, when using H245-04 hose, the yellow spacer ring should be used, with the flat side positioned downward.
10. Slide the entire die, die holder, and spacer ring back against the two dowels at the back of the crimping machine. See **Figure 15-12**.

Figure 15-11. The hose end is inserted through the bottom of the collet with the dimples located flush with the top of the collet's surface.

Figure 15-12. Slide the collet assembly back against the two dowels.

11. With one hand, hold the portion of the hose that extends out of the bottom of the machine and actuate the machine's electric switch with the other hand. Remember to keep all hands and fingers free of the crimper's moving parts. The machine's switch is a momentary switch and must be held down until the crimper stops moving, **Figure 15-13**. Once the switch is released, a spring will retract the crimper's press.
12. Remove the crimped hose assembly from the collet.
13. Check the final crimp.
    a. The dimples must be within ±1/16″ from the edge of the crimp.
    b. Measure the outside diameter of the crimp as in **Figure 15-14**. Check the measurement with the specification.

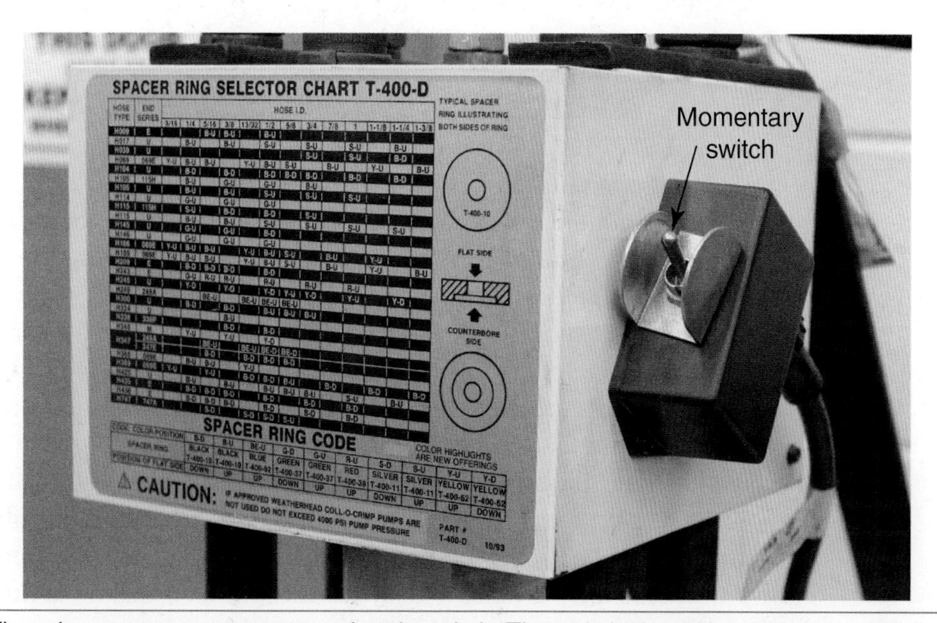

Figure 15-13. The crimper uses a momentary electric switch. The switch should be held down until the crimper stops moving. As soon as the technician releases the switch, the crimper will automatically retract.

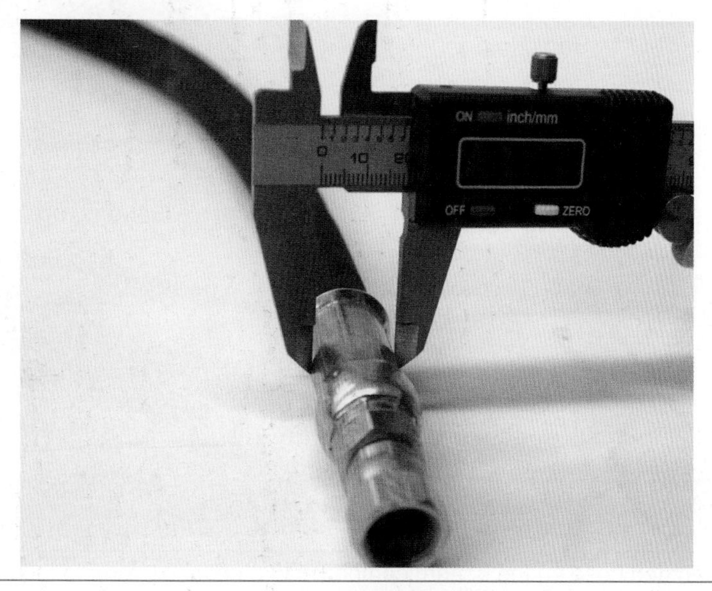

Figure 15-14. The outside diameter of a crimp can be measured with a dial caliper to determine if it is within specifications.

c. Inspect the crimp to see if it is uniform around the entire surface. Figure 15-15 shows a crimp that was made on two pieces of 3/8″ hose, one done correctly and one done using a collet with too small of a diameter.

14. Re-clean the hose after it has been assembled to eliminate contamination that could have occurred during the assembly process.
15. Cap or plug the hose ends after the assembly has been re-cleaned.

When choosing a hose crimper to purchase, consider the following factors:

- Capacities.
  - Number of hose wires.
  - Diameter of hose.
- Method of actuation.
  - Manual.
  - Electro-hydraulic.
    - 12 VDC.
    - 120 VAC.
    - 220 VAC (amperage).
  - Air-actuated hydraulic (commonly called air-over-hydraulics).

Some hoses require ***skiving***, which consists of removing the outer covering from the end of a hose that will be inserted into fitting. The reinforcement layer of the hose should not be marred if the process is performed correctly. A hose skiving tool resembles a speed handle and is inserted into the end of a hose. Skiving tools are made for each diameter of hose, Figure 15-16A. The tool has a sharp knife that will cut off the outside cover of the hose as the knife is revolved around the exterior diameter of the hose, Figure 15-16B.

As an aid for technicians, some hoses have their part number on an aluminum tag that encircles the hose. Figure 15-17 shows this type of identification band. This tag provides a fast reference for technicians to quickly find and give to the parts store, speeding the hose replacement process.

Figure 15-15. A correctly crimped hose on the bottom compared to a crimp performed incorrectly on the top. If a smaller-sized collet is accidentally used, it will ruin the fitting.

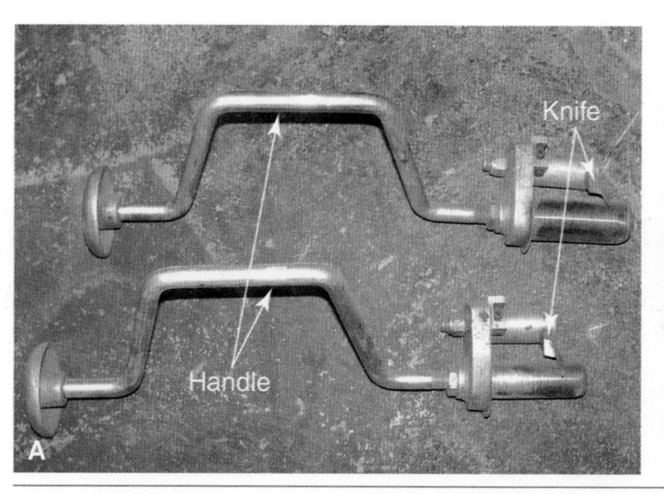

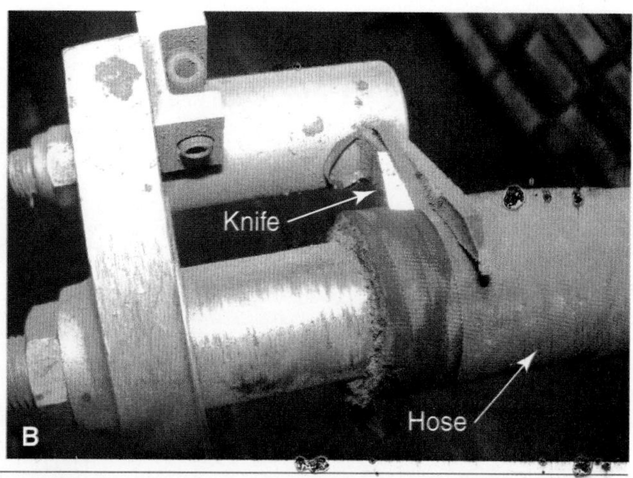

Figure 15-16. Skiving tools. A—Skiving tools are available for different size hoses. B—A skiving tool being used to remove the outside cover of a hose prior to installation of the fitting.

Figure 15-17. This hose has an aluminum band, or tag, that provides quick identification. A technician simply has to read the tag's part number and provide it to a parts supplier to order the correct replacement hose.

Hose Installation

Installing a hydraulic hose is not as straight forward as it appears. A few minor mistakes can dramatically reduce the life expectancy of the hose.

Lay Line Twisting

Hoses have a ***lay line*** drawn across its side. The line indicates if the hose has been twisted during its installation. See **Figure 15-18**. A hose should have no twist after its final installation. A small twist, as little as 10%, can reduce the life of the hose by 90%.

Minimum Bend Radius

Hose manufacturers specify a ***minimum bend radius*** in units of inches or centimeters. The bend radius is specified for hoses that are bent 180° as in **Figure 15-19**. After making the 180° bend, to determine if the hose has the minimum amount of bend radius, measure the *inside distance between the two hose ends,* which equals the diameter. Then, divide the diameter by 2. The answer equals the hose's bend radius.

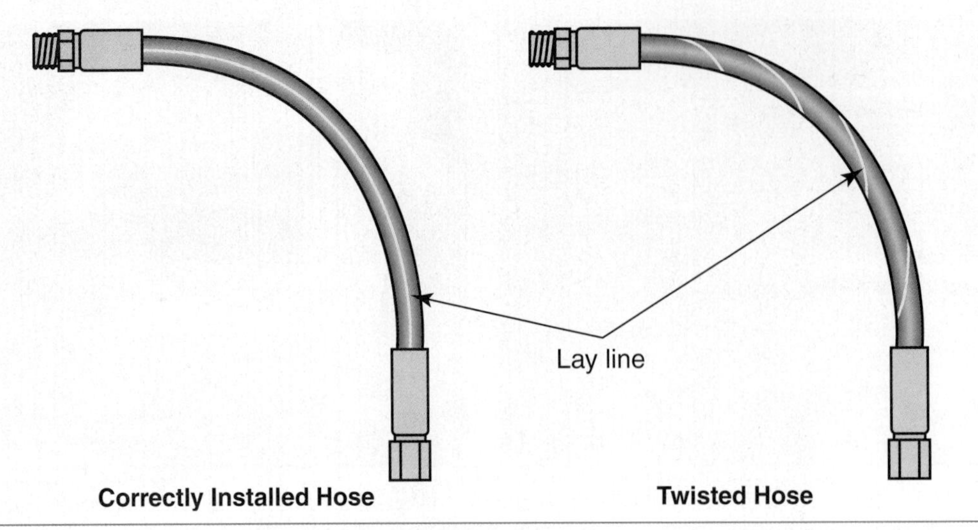

Figure 15-18. A hose must not have any twist after its installation. As little as 10% twist can reduce the hose life by 90%. Note how the lay line on the twisted hose easily identifies this problem.

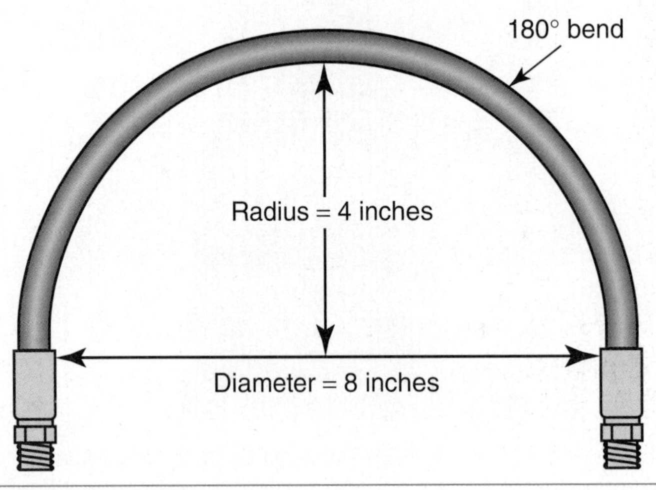

Figure 15-19. A hose's minimum bend radius is listed in the hose catalog. For a 180° bend, measure the distance between the inside of both hose ends and divide that number by 2, which will equal the bend's radius.

Suction hose and high pressure hose are very stiff. If a high pressure hose has been bent past its minimum bending radius, it is likely to break the hose's reinforcement, causing the hose to burst when subjected to high pressure. If a suction hose is bent past its minimum bend radius, it is likely to kink, Figure 15-20, leading to pump cavitation.

Length Change

As mentioned earlier in the chapter, a hose can change in length, ranging from +2% to –4% during operation. When installing the hose, be sure to leave some slack in the hose, Figure 15-21. A hose with no slack is subjected to fatigue as the hose lengthens and shortens during normal operation.

Tubing

Tubing is a thin-walled, non-flexible steel conductor designed to be bent and shaped as required by its application within the system. It is commonly

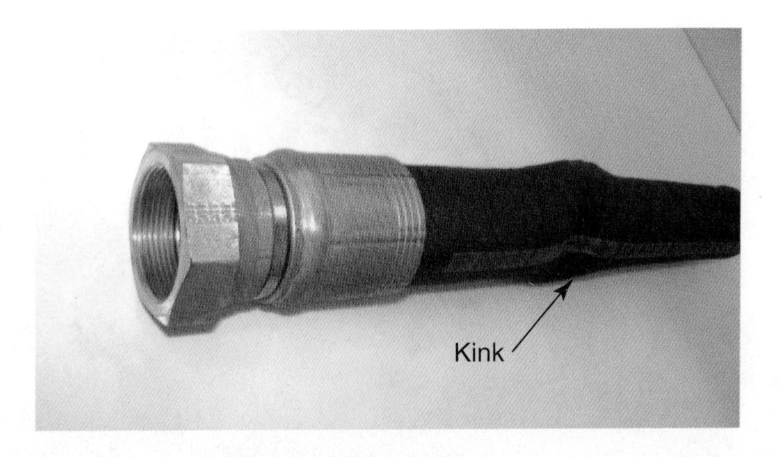

Figure 15-20. This suction hose was bent past its minimum bend radius and kinked.

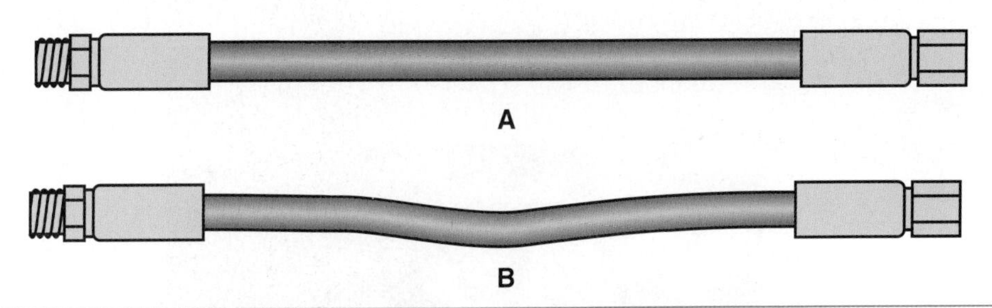

Figure 15-21. Hoses must have some slack to allow for changes in length during operation. A—An improperly installed hose, with no slack. B—A properly installed hose should have a slight amount of slack.

used in mobile machinery. One advantage of tubing is that, unlike hose, it will not swell in volume when subjected to high pressure. Tubing also provides manufacturers the ability to bend the conductor at sharper angles than a hose, but with a more gradual curve than a traditional elbow fitting.

Consider how difficult it would be to replace the tube in **Figure 15-22** with a high pressure hose containing multiple wires of reinforcement. It would be nearly impossible to make a hose snake its way around components in the shape of this tube. If a group of fittings were used in place of the tube, the sharp elbows would cause a significant pressure drop.

Like hose, tubing diameter is sized in a dash number equaling the number of sixteenths in one inch. However, the dash size for tubing indicates the tube's outside diameter, whereas the hose dash size indicates the inside diameter of the hose. A –16 tube has a nominal outside diameter of 1". **Figure 15-23** shows that a –16 tube can be inserted inside a –16 hose.

Tubing is also specified in wall thicknesses. For example, a –16 tube is offered in twelve different tube wall thicknesses, ranging from 1/32" (0.89 mm) to 1/4" (6.35 mm). Charts are available, based on SAE standards, that list working pressures dependent on the tubing dash size and wall thickness. A –16 tube has working pressures that range from 1087 psi (75 bar) to 10,428 psi (718 bar).

Tubing has a safety factor of 4, similar to hydraulic hoses. The tubing burst pressure value is four times the amount of the working pressure value.

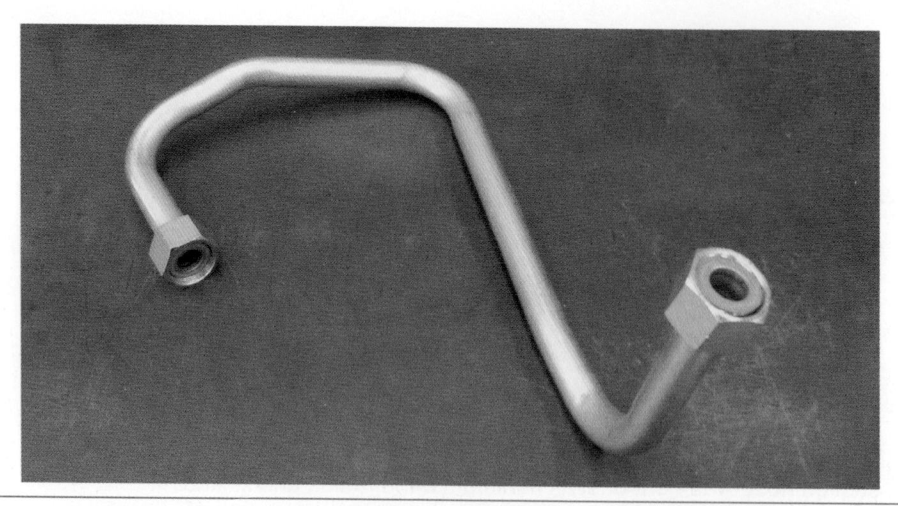

Figure 15-22. Tubing is a rigid conductor that is designed to be bent and shaped as needed.

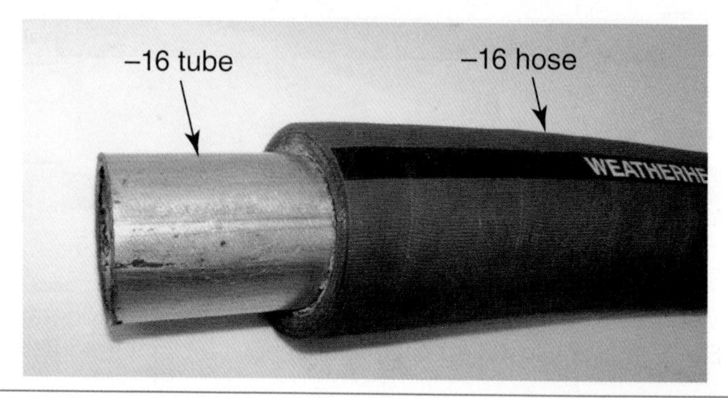

Figure 15-23. The nominal diameter of tubing is the outside diameter. The nominal diameter of hose is the inside diameter. This means that a tube of a particular dash size is capable of being inserted inside a hose of the same dash size.

Hydraulic shops commonly fabricate new tubing as needed. The tube can be flared, have a flat face formed, or have a steel sleeve fitting silver soldered to the tube. Thick wall tubing cannot be flared. Flare-type fittings are explained later in this chapter.

When a new tube is being fabricated, it must be cut square to a tolerance of ±1°. The tube must be deburred without chamfering the end of the tube, and it must be free of oil and chemicals.

Tubing also has a minimum bending radius. The radius is measured from the center point of the bend to the centerline of the tubing. See **Figure 15-24**. A rule of thumb for standard tubing bending radius is:

Minimum Tube Bending Radius (inches) = 3 × Tube OD.

For example a –16 tube (1″ OD tube), would have a minimum bend radius equal to 3″, measured from the tube's center line.

Pipe

Pipe is a thick-wall, rigid conductor that is *not* designed to be bent to conform to an ideal position like steel tubing. Pipe is sized by diameter in units of inches or millimeters. Note that the diameter in **Figure 15-25** is labeled nominal. For example, a 1″ diameter pipe is a relative number, meaning that the outside diameter is not exactly 1″, nor is the inside diameter exactly 1″. The inside diameter is dependent on the tube's wall thickness, which is specified with a ***schedule number***. A lower schedule number specifies a thinner pipe wall, which equates to a lower working pressure. Pipe is offered in numerous, different schedules. Figure 15-25 lists the various working pressure for pipes with a 1″ nominal diameter but different schedule numbers.

Note

Many personnel incorrectly gauge the size of a pipe. For pipe sized 1 1/4″ in diameter and smaller, a rule of thumb is to measure the outside diameter of the pipe's threads, then subtract 1/4″ from that measurement to obtain the correct size of the pipe. For example, if a pipe thread outside diameter measures roughly 1″, the pipe's actual specified diameter would be 3/4″.

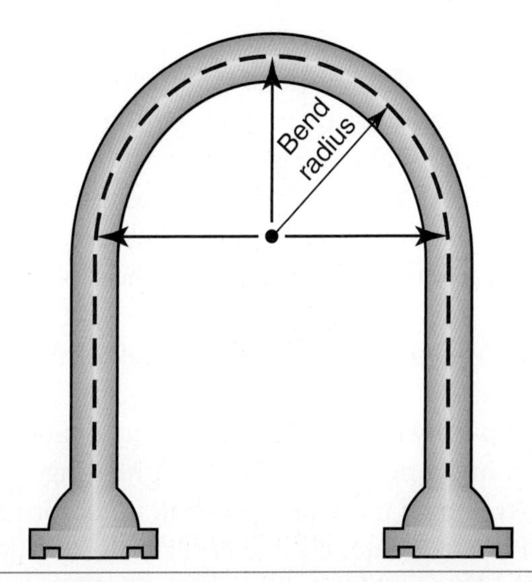

Figure 15-24. Tubing minimum bending radius is measured using the centerline of the tubing.

Nominal Diameter	Schedule	Outside Diameter	Inside Diameter	Working Pressure	Wall Thickness
1″	SCH 40 and Standard	1.315″	1.049″	622 psi (43 bar)	0.133″
1″	SCH 80 and Extra heavy (EH)	1.315″	0.957″	1595 psi (110 bar)	0.179″
1″	SCH 160	1.315″	0.815″	3187 psi (220 bar)	0.250″
1″	Double extra heavy (XXH)	1.315″	0.599″	5103 psi (352 bar)	0.358″

Figure 15-25. This table shows how different pipe schedule numbers correlate to variations in pipe working pressure for matching nominal diameter pipe. The variation depends on the pipe's wall thickness, which is specified by its schedule number. This example is based on ANSI B31.3 for Grade B seamless steel pipe that is using a threaded connection. A welded connection has higher working pressures.

Pipe conductors form a metal-to-metal seal using tapered threads. The threads easily distort when they are tightened, making it more difficult to disassemble and reassemble without leaking or breaking. Heat and vibration can cause the pipe threads to eventually leak. For this reason, pipe threads are not recommended for use in hydraulic systems, especially in higher-pressure applications.

In some industrial applications that use high pressures and high flows—such as conductors with 2″ diameters—a steel pipe with a weld-on fitting and an O-ring seal might be used.

Pipe diameter can also be sized by DIN standards or in millimeters. For low-pressure applications, pipe is economical and convenient. Pipe is also more readily available than hydraulic hose or steel tubing, because most hardware stores and lumber yards stock it.

Metal-to-Metal Fittings

Regardless of the type of conductor, some type of fitting must be used to attach one conductor to another. Fittings can be classified as forming either a metal-to-metal seal or an O-ring seal.

Most ***metal-to-metal fittings*** seal through mating the two metal surfaces and do not use any other sealing substance. However, manufacturers do offer a variety of products to reduce the chances of leakage at a threaded pipe or flare fitting.

Compared to O-ring fittings, metal-to-metal fittings are defined with these qualities:

- Typically have a narrower torque value range, causing a greater chance of improper torque during tightening.
- Are much more likely to leak if not torqued correctly.
- Have a greater chance of distortion while being torqued.

Metal-to-metal fittings can be made of steel or brass. Brass fittings are used in lower-pressure systems such as automotive and household applications. For this reason, , the clerk at an auto-parts store might ask the purchaser of a pipe fitting, "Automotive or hydraulic?" with the understanding that the question is, "Brass or steel?"

Flare Fittings

A ***flare fitting***, also known as a flared tube fitting, mates a cone-shaped flare on the male fitting to a tube that has a matching angled seat. Figure 15-26 shows some flared fittings used in hydraulic systems.

Some different flare angles are listed in Figure 15-27. The JIC 37° flare fitting, shown in Figure 15-28, is the most common flare fitting used in North American mobile hydraulic systems. This metal-to-metal fitting is used in applications that have pressures up to 3000 psi (207 bar). Some Asian machines use the 30° Japanese flare fitting that is also known as a Komatsu fitting.

The 45° flare fitting is frequently used in on-highway trucks and automotive applications. It is sometimes called a SAE fitting, even though both SAE and ISO also have standards for the 37° flare fitting.

Figure 15-26. Various flared tube fittings. The 37° JIC flare fitting is the most popular flare fitting used in North American mobile hydraulic systems.

Flare Angle	Application	Common Descriptor
30°	Japanese hydraulic systems	Komatsu fitting, JIS fitting
37°	Mobile hydraulic systems	JIC fitting
45°	Pneumatic, refrigeration, automotive, and truck systems	SAE fitting

Figure 15-27. Flare fittings can be designed with different angles. Note the applications in which each angled flare fitting is commonly used.

Figure 15-28. A 37° flare fitting's dash size equals the size of the tube's outside diameter, not the threads. The threads will have a larger diameter dimension. A—Dash size. B—Thread size.

Flare fittings are sized the same as tubing. A flare fitting dash size is equal to the number of sixteenths of an inch of the outside diameter. The dimension equals the outside diameter of the tubing, not the outside diameter of the threads, which is the larger diameter. An example chart of JIC flare fitting dimensions is shown in **Figure 15-29**.

Dash Size	Tube OD	Thread OD	# of Threads/Inch
–4	.25″	7/16″	20
–6	.31″	9/16″	18
–8	.50″	3/4″	16
–10	.63″	7/8″	14
–12	.75″	1 1/16″	12

Figure 15-29. A 37° flare fitting is sized in dashes, which equals the outside diameter of the tube. Fitting manufacturers, like Aeroquip®, Gates®, Parker®, and Weatherhead/Eaton®, list the dimensions for common fittings in their catalogs.

Following the service literature for the correct toque specification is critical to achieving a leak-free connection. Torque specifications for 37° fittings are commonly provided in foot-pounds or Newton-meters.

The JIC 37° flare fittings can also have a torque specification that is referenced as ***flats from finger tight*** (FFFT). This sequence consists of tightening the fitting hand tight, then marking on the fitting and bordering nut's flat surface this initial seating position. Following the manufacturer's FFFT specification, turn the nut the appropriate number of flat faces to secure the connection. An example of a FFFT specification is turning a hex nut two more flats after it has been tightened finger tight.

The 37° flare fitting is a reusable fitting. It provides a better seal than threaded pipe fittings. However, due to manufacturing imperfections and poor service practices, it is still possible for the fitting to leak. A seal can be purchased for flare fittings to prevent leakage. A ***flaretite seal*** is a metal, cone-shaped seal with Loctite™ sealant baked onto it. It is designed specifically for flare fittings. The seal is placed on the tip of the male flare and mates against the tube's seat as the fitting's nut is tightened around it. It is designed to provide a leak-free, permanent seal.

Threaded Pipe Fittings

The most common types of threaded pipe fitting are ***National Pipe Tapered (NPT)*** and ***National Pipe Tapered Fuel (NPTF)***. Some types of pipe fittings are shown in Figure 15-30A. Both NPT and NPTF fittings seal on the principle of compressing the male component's tapered exterior threads into the mated port's or pipe's tapered interior threads during tightening. For this reason, the fittings are not well suited for repeated removal and installation. Interestingly, many pressure measuring devices of the past used tapered pipe fittings to connect to a hydraulic system. The threads of a tapered pipe fitting can become deformed after the fitting is installed and removed from a system multiple times. Manufacturers of late-model machines frequently install the SAE J1502 diagnostic pressure taps described in **Chapter 21** as the means for a technician to attach diagnostic equipment to a system.

Caution

Diagnostic pressure taps installed in modern hydraulic systems use quick couplers that are specifically designed for repeated installation and removal. However, the same safety precautions must be taken before attaching or detaching a piece of equipment to a hydraulic system regardless if the connection is made via a threaded pipe fitting or quick coupler.

NPT fittings use a coarse type of thread that requires the application of a thread sealant. An NPTF fitting is sometimes labeled as fine thread and can be assembled without applying thread sealant, although some personnel still apply it.

Many manufacturers specify using a specially designed pipe sealant that is less likely to harm the hydraulic system than the common Polytetrafluoroethylene (PTFE) tape, also called Teflon® tape. The tape is used for household water plumbing and should *not* be used on hydraulic systems. Hydraulic warranties can be voided based on the use of Teflon tape. Technicians have also been terminated from their jobs for using Teflon tape in hydraulic systems. Loctite and Permatex® are two suppliers that sell specially designed thread sealant for NPT fittings, **Figure 15-30B**.

Threaded pipe fittings are much more prone to leaking. Like other metal-to-metal fittings, they have a much narrower torque range. It can be difficult to judge when a fitting is tight, especially if it is a tee or elbow that requires a specific alignment with a neighboring fitting. Swivel pipe fittings are available to alleviate the common pipe alignment problems.

Figure 15-30. A—Steel or brass pipe fittings come in a variety of designs. If called to service a machine equipped with threaded pipe fittings, use only replacement steel pipe fittings that are designed for hydraulic systems. B—Loctite is one manufacturer that offers a pipe sealant for NPT fittings.

Other styles of threaded pipe fittings used on machinery are listed below:

- Metric parallel.
- Metric tapered.
- British Standard Pipe Parallel (BSPP).
- British Standard Pipe Tapered (BSPT).
- Japanese Industrial Standards (JIS).

The three international styles of threads are not as convenient to purchase as NPT and NPTF pipe fittings in the United States. NPT and NPTF can be purchased at many tractor supply stores and automotive part stores. Metric, BSPP, BSPT, and JIS have to be purchased from a hydraulic shop, and some of the fittings might have to be special ordered.

British pipe threads are often confused with NPTF threads. British pipe threads have a 55° thread angle and NPTF threads use a 60° thread angle. Use a thread gauge to identify the difference. It is also necessary to determine if the British threads are parallel or tapered.

Metric threads are easily confused with SAE threads. After determining if the threads are parallel or tapered, measure the outside diameter of the threads and use a thread gauge to identify the fitting.

As mentioned earlier in the chapter, pipe is used in manufacturing plants. Long runs of pipe can be difficult to disassemble unless the system contains a ***pipe union***. The fitting, Figure 15-31, is a three-piece component designed to speed the process of connecting and disconnecting two pieces of pipe. The union has a center nut and two end pieces that have a flare connection. One of the end pieces threads onto each of the pipe ends needed for the joint. The nut slides over one of the union's end pieces and threads onto external threads on the other end piece. Tightening the nut clamps the two end pieces together to seal the union. Loosening the nut allows the long run of pipe to be easily disassembled.

Flareless Fittings

Flareless fittings are the least popular steel fitting used in the off-highway industry. The fittings use a ferrule, also known as a sleeve, which is compressed into the outside diameter of a tube as the nut is tightened. Even though these fittings are able to withstand vibration better than other metal-to-metal

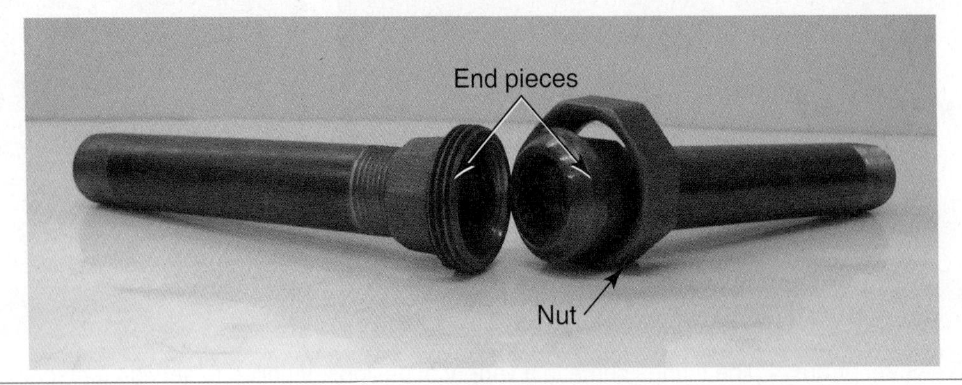

Figure 15-31. Pipe unions are used for connecting pipe in industrial plants. Tightening the nut seals the fitting. Loosening the nut allows the long run of pipe to be disassembled.

fittings, require very little tube preparation, and can handle higher pressures, closer to 3000 psi (207 bar), they are used far less often than 37° JIC flare fittings and O-ring fittings. The flareless fitting requires thicker wall tubing. See Figure 15-32.

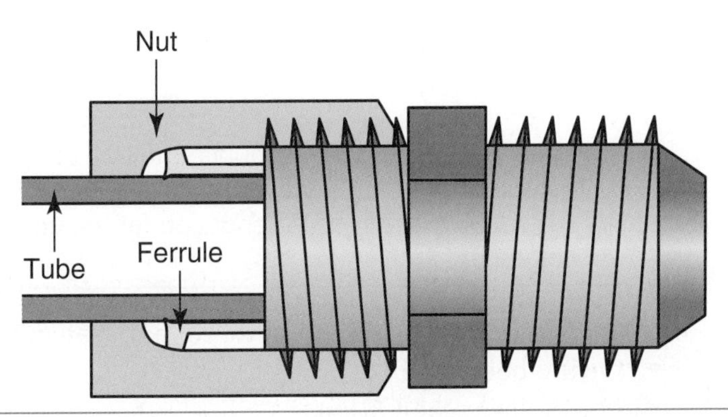

Figure 15-32. Flareless fittings are less common in mobile machinery. Tightening the nut causes the ferrule to bite into the tubing, forming a seal.

O-Ring Fittings

O-ring fittings are used throughout the mobile hydraulics industry. The fittings provide excellent sealing ability in systems that operate as high as 6000 psi (414 bar). The fitting is also less likely to be distorted during installation, compared to a metal-to-metal fitting. O-ring fittings have straight threads.

O-rings come in SAE sizes and metric sizes. Both SAE and metric use different thicknesses of O-rings as well. See Figure 15-33. O-rings should be kept in a contaminant-free environment. Prior to assembly, the O-ring should be lubricated with hydraulic oil. Most service literature recommends discarding the old O-ring and using a new O-ring whenever service is performed on a fitting.

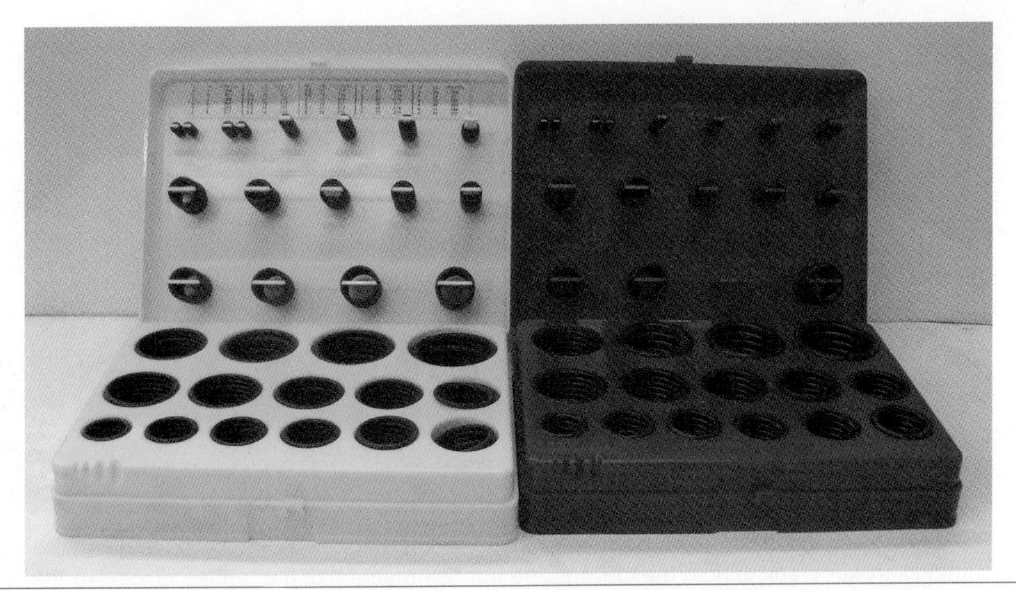

Figure 15-33. These two containers house bulk O-rings. The container on the right houses SAE-sized O-rings and the container on the left houses metric O-rings.

O-ring fittings cost more than most metal-to-metal fittings. During installation, the O-ring can cause multiple problems such as the following:

- Excessive ambient heat can harden the seal.
- The O-ring can fall out of the fitting.
- It can be pinched, cut, damaged, or contaminated.
- The wrong size O-ring might be used, causing a possible leak.

Note

Some manufacturers stipulate the use of specific O-rings that are stronger than a standard black O-ring. Be sure to follow the manufacturer's recommendation for the correct O-ring size and material.

O-ring fittings are classified in three main categories:

- O-ring boss (ORB).
- O-ring face seal (ORFS).
- Split-flange O-ring.

All three designs are commonly used in mobile hydraulic systems.

O-Ring Boss

An ***O-ring boss (ORB) fitting*** has a groove machined into the sealing surface of the male fitting to hold an O-ring. See **Figure 15-34**. The male fitting compresses the O-ring into a machined seat located in the female port of the housing. ORB fittings are used in medium- and high-pressure hydraulic systems. The O-ring creates the sealed connection.

The fitting is also called straight-thread ORB. They are sized with a dash number in the same manner as hydraulic hose and tubeing.

O-ring boss fittings can be adjustable. A washer and jam nut are used, as shown in **Figure 15-34B**, to adjust the fitting. The adjustable-type ORB fittings are normally elbows and T-fittings. Prior to installation, the jam nut is loosened.

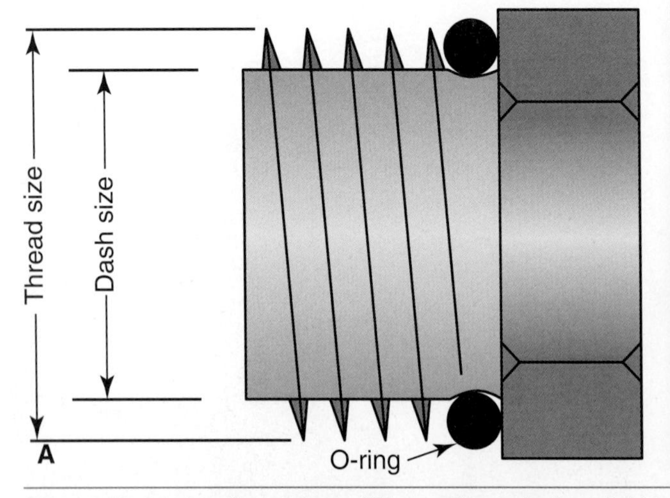

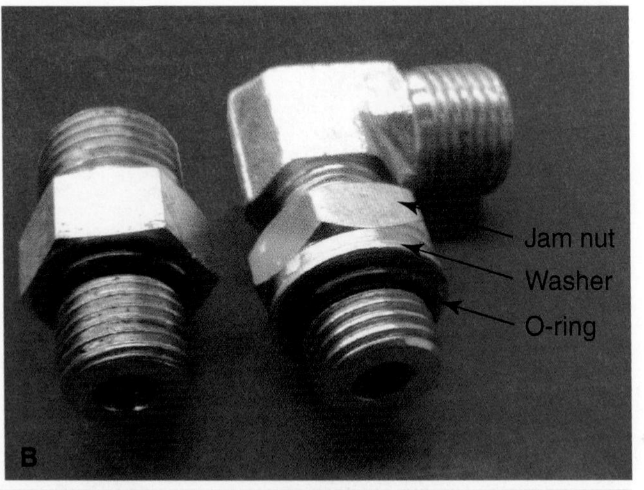

Figure 15-34. A—O-ring boss fittings, also called straight thread O-ring boss, have a groove cut in the outside diameter of the fitting's thread body to hold the O-ring. The opposing seat has a chamfer to receive the O-ring seal. B—A nonadjustable O-ring boss on the left and an adjustable O-ring boss elbow. Note the washer used between the jam nut and O-ring on the elbow fitting.

The fitting is threaded into the port until the back-up washer comes in contact with the port. The fitting can then be properly oriented by turning it up to one more full turn. Next, the jam nut is tightened to the specified torque.

O-Ring Face Seal

An ***O-ring face seal (ORFS) fitting***, also known as flat-face O-ring (FFOR), has a groove machined into the face of the thread body of the male fitting to hold the O-ring, Figure 15-35. The female fitting has a flat face. The male fitting compresses the O-ring into the female fitting's flat face when assembled. Figure 15-36 shows each half of the fitting before assembly. This style of fitting provides a very strong seal. It is similar to the split-flange fitting that also uses a flat face. However, ORFS fittings contain their own threads and do not use external clamps like the split-flange fitting. During the installation process, if the ORFS fitting is pointing toward the ground, special care must be taken so that the O-ring does not fall out of its groove before mating with the female half.

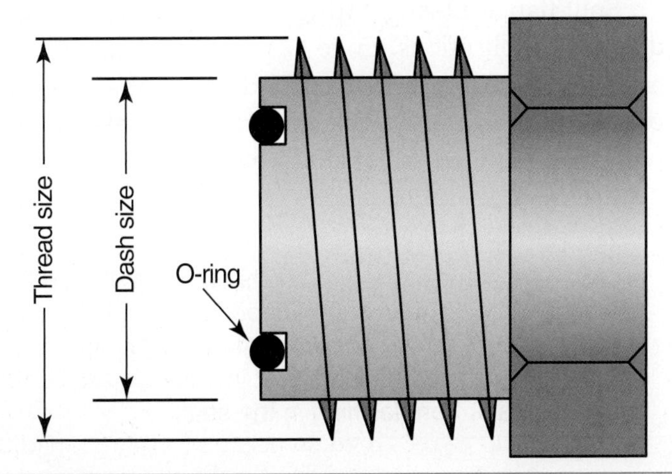

Figure 15-35. O-ring face seal fittings have a groove machined into the male fitting for holding the O-ring. The fitting is mated against a flat face to form a tight seal.

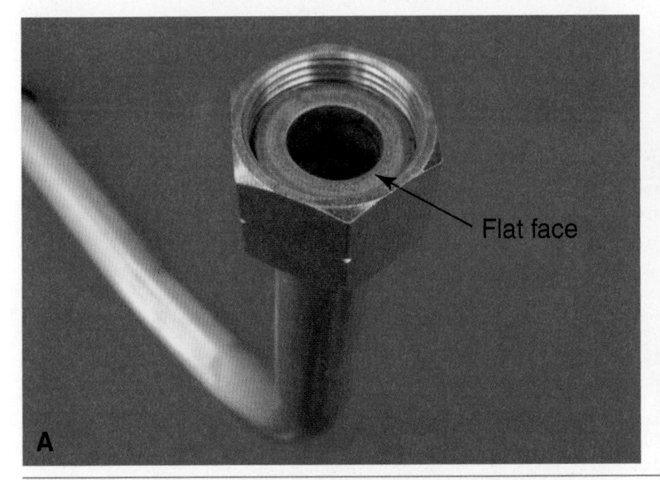

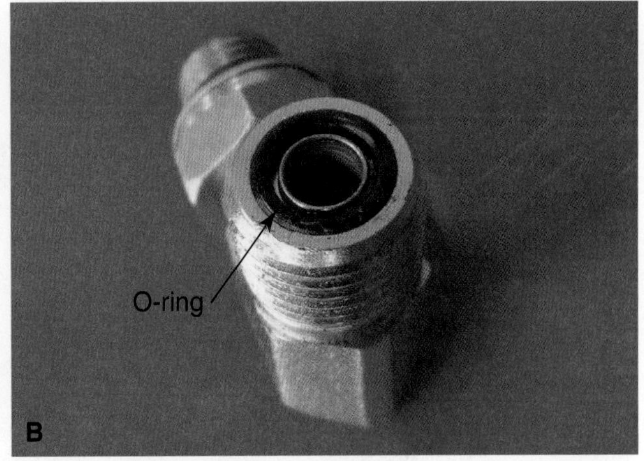

Figure 15-36. O-ring face seals use a female fitting with a flat face and a male fitting with a face groove that holds the O-ring in place. The fitting is secured with a threaded connection. A—The female half of the fitting. B—The male half of the fitting.

A machine can be used for forming a flat face onto a piece of tubing. The machine presses the end of the tubing to create the flat face, also called a flange. See Figure 15-37.

Split-Flange O-Ring

The ***split-flange O-ring fitting*** is similar to the ORFS fitting. It mates an O-ring to a flat face as a ORFS fitting does, but the split flange uses four bolts and two clamps to connect the fitting. See Figure 15-38 and Figure 15-39.

The split-flange fitting's biggest advantage is the ease of installation where large-diameter conductor is required. For example, if a hydraulic system used –24 hose and a traditional O-ring fitting, the size of the wrench required for tightening the fitting would be nearly two feet long. If the hose was located deep inside the machine, using a traditional O-ring fitting would limit a technician's ability to properly torque the fitting. A split-flange fitting uses two clamps that are secured with four substantially smaller bolts, eliminating the need for such large tools to tighten the fitting.

Split-flange O-ring fittings are commonly used in high-pressure, large-diameter applications. Code 62 pressure rating is designed for 6000 psi. Code 61 pressure rating is the standard series of split-flange fittings used in lower pressure applications. Some manufacturers also use split-flange O-ring fittings with square-edged seals rather than the standard, round O-ring.

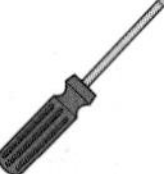

Note

When securing the bolts on a split-flange fitting, tighten them in an alternating star-shaped pattern, following the order of numbers listed on the bolts in Figure 15-39. Tightening the bolts in this sequence ensures that the clamps are tightened evenly, preventing gaps from forming in the seal.

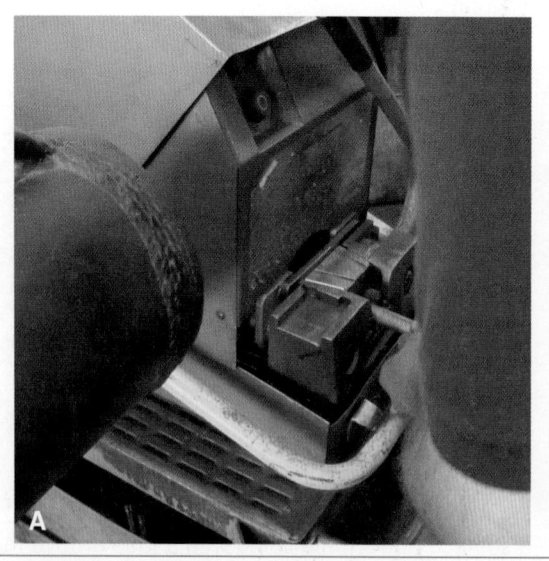

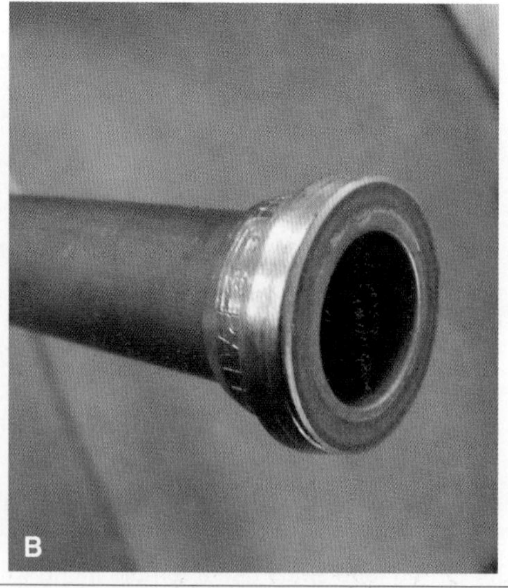

Figure 15-37. Forming a flat face on tubing. A—The machine used to form a flat face on the tubing. B—Closeup of the completed face.

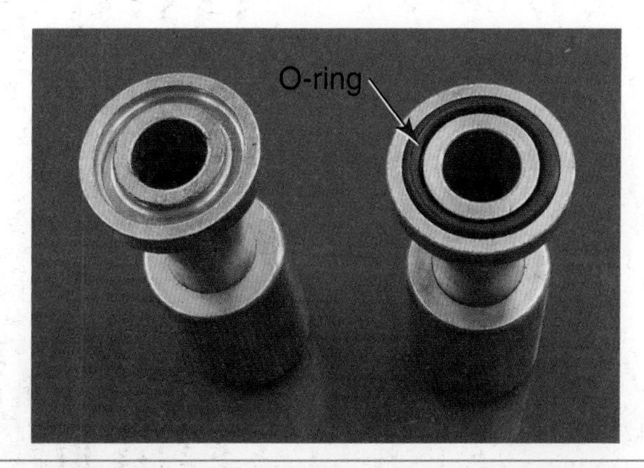

Figure 15-38. A split-flange O-ring fitting has an O-ring inside a fitting with a machined groove that is mated against a flat face, much like an ORFS fitting. Not pictured are the two clamps and four bolts that make the tight seal.

Figure 15-39. A split-flange O-ring fitting can hold some of the highest operating pressures and its design makes it much easier to install when connecting large-diameter hoses. Note the clamps and bolt tightening sequence.

Quick Couplers

Many mobile hydraulic systems use quick couplers. Some examples are shown in **Figure 15-40**. The fitting's design allows an operator to quickly attach or disconnect the fitting from a hydraulic system. The female end of the fitting contains a sliding collar of ball bearings that locks into a machined groove in the male end when the two ends are pushed or threaded together. This connection opens internal valves within each end to allow the flow of fluid. There are numerous different types and sizes. Most use a spring-loaded check to prevent leakage while the couplers are disconnected.

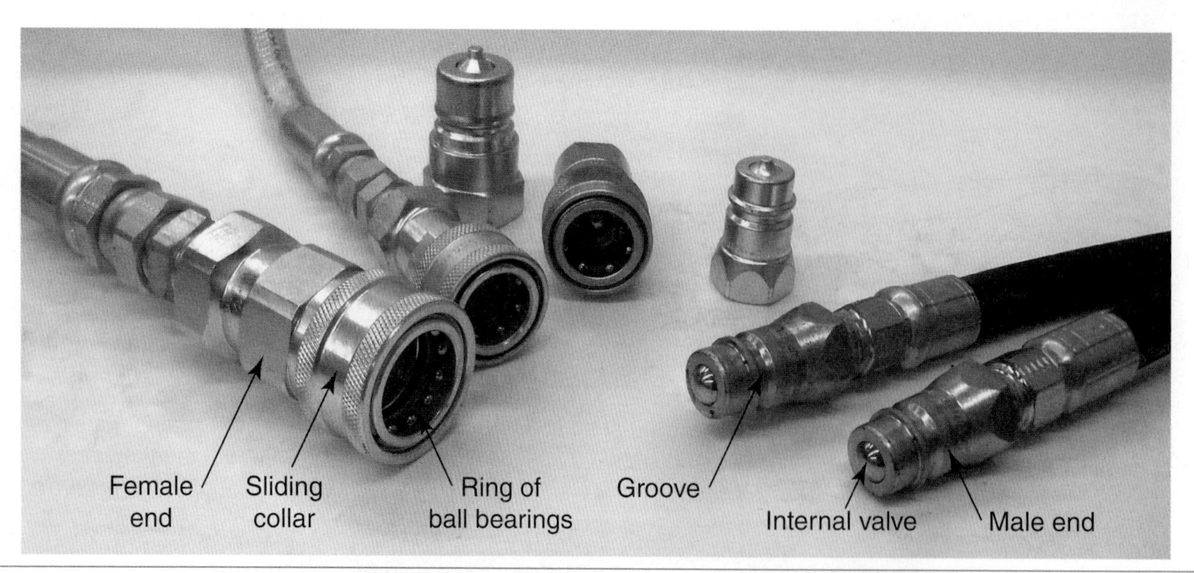

Figure 15-40. Quick couplers are used on mobile machinery as a fast method for attaching and detaching hydraulic equipment.

The couplers are commonly used on agricultural tractors for connecting to implements and on construction equipment for connecting to hydraulic attachments.

Warning

Be sure to shut off the machine and depressurize the hydraulic system before attempting to connect the quick couplers. Even after depressurization of the system, prolonged exposure to the heat of a hot summer day can build enough pressure in the hydraulic lines to cause a fluid injection injury. In such situations, the system may need to be depressurized again before fittings are connected or disconnected.

Noise and Vibration

Noise is generated in mobile hydraulic systems by hydraulic pump operation, fluid pulsations, fluid pressures, fluid velocities, and machine vibrations. Prolonged exposure to high-decibel noise can cause hearing loss in personnel. Unusual levels of noise from a hydraulic system can also signal impending equipment failure, leaks, or unsafe operating conditions. Hydraulic personnel have an arsenal of actions to reduce noise and vibrations. One of the most popular, but more expensive options is to add an attenuator or noise suppressor to a system.

Rigid Conductors and Noise

The hydraulic system's plumbing has a large influence on noise and vibration. Pipe and tubing generate much more noise than hydraulic hose because the rigid conductors tend to resonate and magnify vibration, shock, and noise. Rigid conductors are also unable to swell like a hose, which is described as a conductor's rate of ***volumetric expansion (VE).*** A hose with a high VE rating will swell more and generate less noise than a hose with a low VE rating. The effect is also known as the accumulator effect, although a hose will not be able to absorb as large a quantity of oil as an accumulator.

Manufacturers offer numerous styles of tubing mounts to reduce shock, vibration, and noise. The mounts can hold a single tube or multiple tubes and use split grommets made of rubber or thermoplastic. A metal clamp should never be used due to the pulsations that occur in rigid conductors. Clamp spacing recommendations are found in **Figure 15-41**.

In addition to mounting the tubing in rubber mounts, using a larger tube can reduce the fluid velocity, which reduces fluid-generated noise.

Hose and Noise

Installing a traditional synthetic rubber hose in place of a rigid conductor will reduce noise and vibration. If the traditional hose cannot solve the problem, some manufacturers use a thermoplastic fiber-reinforced hydraulic hose, which has a higher VE rating, to further reduce the noise and vibration.

Caution

Using hose with a high VE in closed-loop hydrostatic drives will cause cavitation. A large-volume charge pump must be used to compensate for the accumulator effect of the hose. **Chapter 23** and **Chapter 24** will explain closed-loop hydrostatic drive transmissions.

Case Study

A Hose Can Serve a Different Purpose

One excavator manufacturer had plumbed a hose in parallel with the hydraulic pump's outlet line as the excavator was being built. The hose was simply capped and did not lead to another circuit. The hose was used to dampen fluid-borne noise originating from the pump outlet. During a repair, a technician found that the hose was bad and realized that it did not feed another circuit. He chose to eliminate the hose. Shortly after receiving the machine back, the customer returned to the dealership with a noise complaint regarding the excavator. After closer inspection, the technician learned that the hose really did serve a purpose and now fully understands that the hose dampened fluid noise.

In this application the conductor is called a resonator hose. Some resonator hoses are deadheaded against the frame of the machine instead of being capped. Engineers try different lengths of resonator hoses while designing and testing the machines in order to achieve maximum noise reduction.

Diameter of Tubing	Space between Clamps
3/16″ to 3/8″	3′
1/2″ to 7/8″	5′
1″ to 1 1/4″	7′

Figure 15-41. Tubing mounts need to be evenly spaced by the listed distances that are based on the diameter of the tubing.

Summary

- ✓ A thorough understanding of hydraulic plumbing is essential for a technician to correctly test, diagnose, service, and repair hydraulic systems.
- ✓ Working pressure equals the pressure value at which a hose can safely operate.
- ✓ Burst pressure is the estimated value at which the hose will rupture.
- ✓ Safety factor is the ratio of working pressure to burst pressure.
- ✓ Hose and tubing have a safety factor of four.
- ✓ In the dash size measurement system, the number following the dash indicates diameter in 1/16ths of an inch.
- ✓ Hose diameter is measured by the hose's inside diameter.
- ✓ Hose is composed of three structures: an inner tube, reinforcement, and an outer cover.
- ✓ The reinforcement layer gives the hose its strength.
- ✓ The inner tube must be compatible with the hydraulic fluid.
- ✓ The outer cover protects the hose from the elements, weather, and abrasion.
- ✓ Nonconductive hose uses an orange cover, but not all orange hose is nonconductive.
- ✓ SAE has 19 different hose classifications.
- ✓ Hose ends can be field attachable, but not all are reusable.
- ✓ Crimping reduces the hose end's diameter by pleating the fitting.
- ✓ Swaging reduces the hose end's diameter by stretching.
- ✓ Some hoses may require skiving before a hose end can be attached.
- ✓ Tubing diameter is measured by the tube's outside diameter.
- ✓ Tubing is sized in diameter and wall thickness.
- ✓ Pipe wall thickness is specified with a schedule number.
- ✓ Pipe diameter is relative to the pipe's wall thickness and does not equal the inside diameter nor the outside diameter.
- ✓ To determine pipe diameter, measure the pipe thread's OD and subtract 1/4″ from that measurement.
- ✓ JIC 37° flare fittings are the most popular flare-type fitting used in mobile machinery.
- ✓ A 30° flare fitting is sometimes called a Komatsu flare fitting.
- ✓ NPT fittings require thread sealant, but Teflon tape should *never* be used in hydraulic systems.
- ✓ NPTF fittings use fine pipe threads and are considered a dry seal fitting.
- ✓ Metric parallel, metric tapered, British Standard Pipe Parallel (BSPP), British Standard Pipe Tapered (BSPT), and Japanese Industrial Standards (JIS) are examples of different threaded pipe fittings.
- ✓ O-ring fittings are labeled as either O-ring boss (ORB), O-ring face seal (ORFS), or split-flange fittings.
- ✓ ORB fittings seal with a seated O-ring on one end of the fitting compressing into a matching machined seat on the other end of the fitting.
- ✓ ORB elbows and T-fittings are adjustable fittings.
- ✓ ORFS and split flange use a flat face on one end of the fitting as the sealing surface.
- ✓ Split-flange fittings are used for large conductors and high-pressure applications.
- ✓ Quick couplers are used on numerous mobile hydraulic systems to enable operators to quickly attach or detach implements and attachments to the machine.
- ✓ Be sure the system is depressurized before connecting or disconnecting quick couplers.
- ✓ Pipe and tubing are more prone to noise and vibration than hose.
- ✓ The swelling of a hose is rated in volumetric expansion (VE).

Technical Terms

burst pressure
cover
crimping
dash size
field attachable
flare fitting
flaretite seal
flats from finger tight (FFFT)
hose cutoff factor
hydraulic hose
inner tube
lay line
metal-to-metal fittings
minimum bend radius
National Pipe Tapered (NPT)
National Pipe Tapered Fuel (NPTF)
nonconductive
O-ring boss (ORB) fitting
O-ring face seal (ORFS) fitting
pipe
pipe union
reinforcement
reusable hose ends
safety factor
schedule number
skiving
split-flange O-ring fitting
suction hose
swaging
tubing
volumetric expansion (VE)
working pressure

Review Questions

Answer the following questions using the information provided in this chapter.

1. A hydraulic hose has a 6000 psi (414 bar) burst pressure rating. What is the working pressure for the hose?
   A. 500 psi (34 bar).
   B. 1000 psi (69 bar).
   C. 1500 psi (103 bar).
   D. 2000 psi (138 bar).
2. Hydraulic hose has what safety factor rating?
   A. Two.
   B. Three.
   C. Four.
   D. Five.
3. Which of the following terms is the value at which a hose can be safely operated?
   A. Safety factor.
   B. Working pressure.
   C. Burst pressure.
   D. Line pressure.
4. Which of the following terms describes the relationship between normal hose operating pressure and a rupture pressure value?
   A. Safety factor.
   B. Working pressure.
   C. Burst pressure.
   D. Line pressure.
5. Hydraulic hose is sized by which of the following?
   A. Outside diameter.
   B. Wall thickness.
   C. Inside diameter.
   D. Both A and B.
6. Tubing is sized by which of the following?
   A. Outside diameter.
   B. Wall thickness.
   C. Inside diameter.
   D. Both A and B.
7. All of the following are advantages of hose compared to pipe or tube, *EXCEPT*:
   A. less expensive.
   B. faster to fabricate.
   C. less noise.
   D. less vibration.
8. Which conductor is the correct choice when the conductor needs to make multiple sharp turns?
   A. Hose.
   B. Pipe.
   C. Tubing.
   D. Multiple elbow fittings.

9. Many hydraulic hoses will have what type of pressure rating when the diameter of the hose is small?
   A. Low.
   B. Medium.
   C. High.
   D. None of the above.
10. What size is a –6 hose?
    A. 1/4″
    B. 3/8″
    C. 1/2″
    D. 3/4″
11. Which of the following hose structures has the most influence on a hose's working pressure?
    A. Inner tube.
    B. Reinforcement.
    C. Cover.
    D. None of the above.
12. Which of the following hose structures has the most influence on a hose's compatibility with the oil?
    A. Inner tube.
    B. Reinforcement.
    C. Cover.
    D. None of the above.
13. Which of the following hose structures is responsible for protecting the hose from the environment?
    A. Inner tube.
    B. Reinforcement.
    C. Cover.
    D. None of the above.
14. A technician is installing a threaded field-attachable hose end. How is the outside collar installed on the hose?
    A. Threaded counterclockwise.
    B. Threaded clockwise.
    C. Pushed on.
    D. None of the above.
15. Which term is used to describe removing the outside cover of the hose to allow the hose to fit inside the hose end?
    A. Skiving.
    B. Swaging.
    C. Flaring.
    D. Crimping.
16. Which of the following terms is used to describe attaching a permanent hose end by using a die that reduces the fitting's outside diameter by stretching the fitting?
    A. Skiving.
    B. Swaging.
    C. Flaring.
    D. Crimping.
17. Which of the following terms is used to describe attaching a permanent hose end by using a die that reduces the fitting's outside diameter by forming pleats in the fitting?
    A. Skiving.
    B. Swaging.
    C. Flaring.
    D. Crimping.
18. A personnel lift truck is used for working on high-voltage power lines. What is the color of the hydraulic hose?
    A. Black.
    B. Blue.
    C. Red.
    D. Orange.
19. A rule of thumb to obtain the minimum bend radius of tubing is to multiply the tube outside diameter by what factor?
    A. Two.
    B. Three.
    C. Four.
    D. Five.
20. A hose that is twisted 10% during installation can reduce the life of a hose by what percent?
    A. 60%
    B. 70%
    C. 80%
    D. 90%

21. Technician A states that ORFS seals use a flat face. Technician B states that split-flange fittings use a flat face. Who is correct?
    A. Technician A.
    B. Technician B.
    C. Both A and B.
    D. Neither A nor B.
22. Flaretite makes a cone-shaped steel seal for which fitting?
    A. JIC 37°.
    B. O-ring face seal.
    C. O-ring boss seal.
    D. Split-flange O-ring.
23. Which of the following fittings is used for high system pressures with large-diameter conductors?
    A. JIC 37°.
    B. O-ring face seal.
    C. O-ring boss seal.
    D. Split-flange O-ring.
24. Which of the following fittings uses four small bolts and a clamp to secure the fitting to the component?
    A. JIC 37°.
    B. O-ring face seal.
    C. O-ring boss seal.
    D. Split-flange O-ring.
25. Which of the following conductors is the least likely to generate noise or vibration?
    A. Pipe.
    B. Tubing.
    C. Hose.
    D. None of the above.
26. Which of the following terms is used to describe how much a hose will swell under high pressure?
    A. Volumetric expansion.
    B. Pressure expansion.
    C. Linear expansion.
    D. Velocity expansion.
27. Which of the following conductors will have the most swell at higher pressures?
    A. Pipe.
    B. Tubing.
    C. Synthetic rubber hose.
    D. Thermoplastic hose.
28. All of the following describe thermoplastic hydraulic hose compared to traditional synthetic rubber hose, *EXCEPT*:
    A. has a higher tolerance to heat.
    B. lighter weight.
    C. smaller diameter.
    D. react less severely to chemicals and oils.
29. Which of the following flare fitting angles is considered a Komatsu type fitting?
    A. 30°
    B. 37°
    C. 45°
    D. 58°
30. Which of the following flare fitting angles is considered a JIC flare fitting?
    A. 30°
    B. 37°
    C. 45°
    D. 58°

Open-center hydraulic systems have been used for decades by many agricultural machine manufacturers in the construction of their combine harvesters.

Chapter 16

Open-Center Hydraulic Systems

Objectives

After studying this chapter, you will be able to:

✓ Describe an open-center hydraulic system's flow and pressures when the DCVs are in a neutral position.
✓ Describe the operating differences of open-center DCVs in series and parallel.
✓ List methods used for varying actuator speeds in open-center systems.
✓ Explain how a jammer solenoid works in an open-center hydraulic system.
✓ Describe the system operation of a two-stage open-center log splitter hydraulic system.
✓ List examples of machines that have used open-center hydraulic systems.

Overview and Introduction to Open-Center Systems

The terms *open* and *closed* can have three different meanings in the context of hydraulic systems. In **Chapter 4**, it was explained that open and closed loop describes the inlet of a hydraulic pump. *Closed-loop* means the pump uses some type of assistance for supercharging the pump's inlet with oil. In an *open-loop* system, the pump does *not* receive assistance. **Chapter 23** will also discuss closed-loop hydrostatic transmissions in further detail.

Closed control is a term sometimes used by design engineers to describe an electrical or hydraulic system that uses some type of feedback, such as electrical feedback, mechanical feedback, or hydraulic feedback. The feedback is used by the system to vary or shut off either hydraulic flow or electrical current flow in the system. The term *open control* refers to a system that does *not* use feedback.

Open and closed center are two very popular terms that are used to describe a specific style of hydraulic DCV. The focus of this chapter is open-center DCVs, which are designed to route pump flow back to the reservoir when the valve is in a neutral position.

Although hydraulic systems are frequently classified as either open center or closed center, the next several chapters will illustrate that it is archaic to rely solely on those two traditional classifications, due to the variety of open- and closed-center systems found in today's equipment. Technicians or technical specialists interviewing for a job are often asked to explain the differences between open-center and closed-center hydraulics.

Open-center hydraulic systems are the oldest systems used in the off-highway equipment industry. In a traditional open-center hydraulic system, when the DCV is in the neutral (centered) position, the valve is open and free, enabling the valve to route the pump's flow of oil freely to its next destination. The next destination is generally back to the reservoir or to the next DCV. Said another way, when the DCV is in an off position, the DCV provides a pathway (preferably with little resistance and pressure drop) for the oil to travel back to the reservoir with as little energy consumption as possible. See **Figure 16-1**.

It does not matter if the DCV is manually operated, pilot-operated, or actuated with solenoids; all of those actuation methods can still use open- or closed-center valves. It also does not matter if the valve has spools or seats and poppets, the DCV can still be either open or closed center.

Traditionally, an open-center hydraulic system uses a fixed-displacement pump. The pump is normally a gear pump or a fixed-displacement vane pump. These types of systems produce a specific amount of flow for a given rpm. What does that statement mean? First, you must understand that most (but not all) hydraulic pumps in the off-highway industry are driven directly or indirectly by the engine. This means that the pump will be rotating in proportion to the engine speed. Many off-highway machines are designed to be operated in the field with the engine at wide-open throttle (WOT). A common application would be a combine harvester. If the harvester uses a fixed-displacement pump driven by the engine, the pump will be delivering the most flow it possibly can deliver for the speed at which the engine is rotating. Engine speed is commonly greatest at WOT.

Note

It is worth noting that some hydraulic pumps can be driven by an electric motor as well.

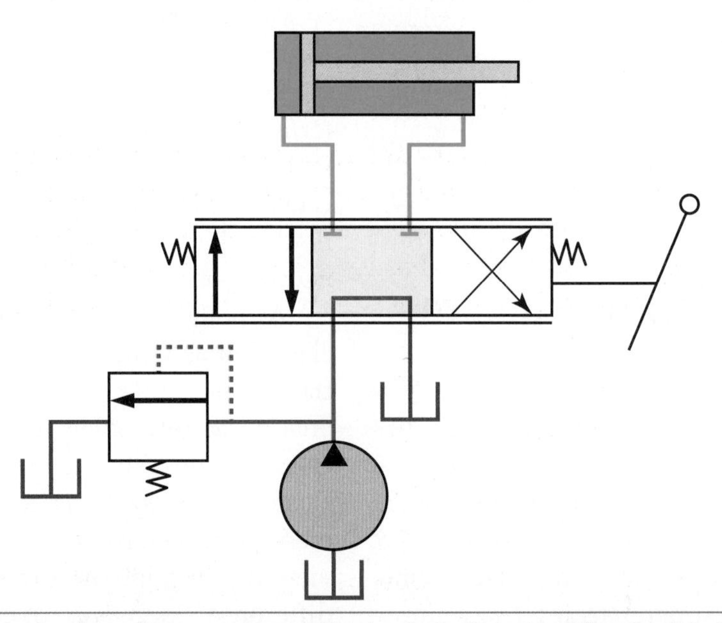

Figure 16-1. Open-center hydraulic systems use a control valve that will route oil flow back to the sump when the valve is in the neutral position. Open-center systems rely on a main system relief valve for pump protection. **Chapter 9** details relief valves. Later chapters will explain that some variable-displacement pumps do not use a main system relief valve.

When a DCV valve is in the neutral position, the open-center system exhibits low system pressures and high system flow compared to an equivalent closed-center system. See Figure 16-2.

Disadvantages of Open-Center Systems

An open-center hydraulic system has some inherent inefficiency. That inefficiency is due to the high amounts of constant flow. Keep in mind that two main factors are used for determining hydraulic horsepower. Those two factors are pressure and flow. Anytime either of those factors can be reduced, fuel efficiency will be improved.

What is a typical open-center hydraulic system pressure when all of the DCVs are in the neutral position? It depends on many factors. The following are some of questions that must be answered:

- Where is the pressure being measured (close to the pump outlet or closer to the reservoir)?
- Does the system have large unrestricting conductors?
- How many fittings and valves are in the system?
- What is the temperature and viscosity of the oil?
- How much oil is flowing through the system?

If a technician was asked the same question in a technical interview, a good average system pressure could range anywhere from 50 psi (3.4 bar) up to 300 psi (21 bar). However, keep in mind that the pressure could easily be more or less depending on the system design and the variety of factors previously listed.

Two different equations can be used for calculating hydraulic horsepower.

$$\text{Hydraulic Horsepower (hp)} = \text{Flow (gpm)} \times \text{Pressure (psi)} \div 1714$$

or

$$\text{Hydraulic Horsepower (hp)} = \text{Flow (gpm)} \times \text{Pressure (psi)} \times .000583$$

How much hydraulic horsepower is developed if an open-center hydraulic system delivers 30 gpm at 300 psi? For now, to keep things simple, do not calculate for any other inefficiency.

$$(30 \times 300) \div 1714 = 5.25 \text{ hp}$$

When the DCV is actuating a cylinder, the hydraulic power is used to perform useful work. Once the cylinder has reached the end of its travel, or if the DCV is in the neutral position, essentially all of the hydraulic horsepower is wasted in an open-center system. These inefficiencies can be minimized by using advanced types of hydraulic systems, which will be discussed in later chapters.

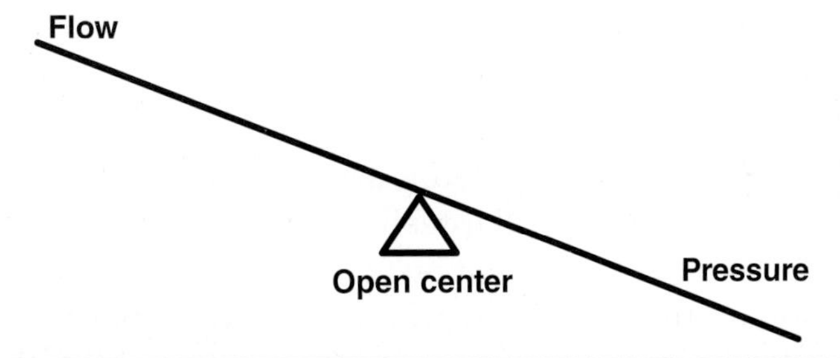

Figure 16-2. In an open-center hydraulic system, pressure and flow have an inverse relationship when the DCVs are in a neutral position. The system produces maximum flow and minimal pressure when DCVs are in a neutral position.

Advantages of Open-Center Systems

Open-center systems have been around for decades and are commonly the most easily understood system. These systems are easily interchangeable and, unlike advanced load-sensing hydraulic systems, need no complex signal network. Adding a hydraulic attachment or DCV is often easier to accomplish in an open-center hydraulic system than the complex systems discussed in later chapters.

Whenever the DCVs are in the neutral position, the low system pressure provides two additional advantages: reduced hydraulic noise and less heat generation.

Open-Center System Applications

Open-center systems can be found in practically any type of machine and application. These were the first hydraulic systems used in off-highway equipment. For example, in the agricultural industry, the first hydraulically operated combine harvesters were open-center systems. Case IH used open-center hydraulic systems as the main header control on their 1400 series combines in 1977, and used this type of system all the way up to 1994 on the late 1600 series combines. John Deere has used open-center hydraulic systems since the inception of hydraulically operated combines in the 1940s and still uses them in S-series combines. In 2011 (2012 model year), John Deere introduced the S-series combines. On the larger models (S680, S685, and S690) a LSPC closed-center system controls most of the machine's functions. However, the machines use open-center systems for a few functions, such as spreader drives and reel drives. The smaller late model John Deere S-series combines (S550, S660, and S670) use only open-center systems.

The more expensive a tractor is, the less likely it is that the machine uses open-center hydraulics. For example, most compact tractors (up to approximately 60 hp) use open-center systems. However, utility and small row crop tractors are less likely to use open-center systems. Large mechanical front-wheel drive (MFWD) and four-wheel drive (4WD) articulated tractors commonly use variable-displacement LSPC hydraulic systems, which will be discussed in **Chapter 18**. In the construction industry, it is even more rare to find an open-center hydraulic system used in a late-model machine.

Valves and Designs

Open-center hydraulic systems can be designed with series DCVs as well as parallel DCVs. Both have their own unique characteristics.

Series

In open-center hydraulic systems with DCVs connected in series, the location of the spools will establish an order of hydraulic priorities. When a DCV spool is moved into position to perform certain functions, the spool blocks flow to downstream DCVs to prevent other operations from being performed at the same time. For example, on a Case 586e forklift, the mast control valve has three separate functions: raise and lower, tilt, and side shift. See **Figure 16-3**.

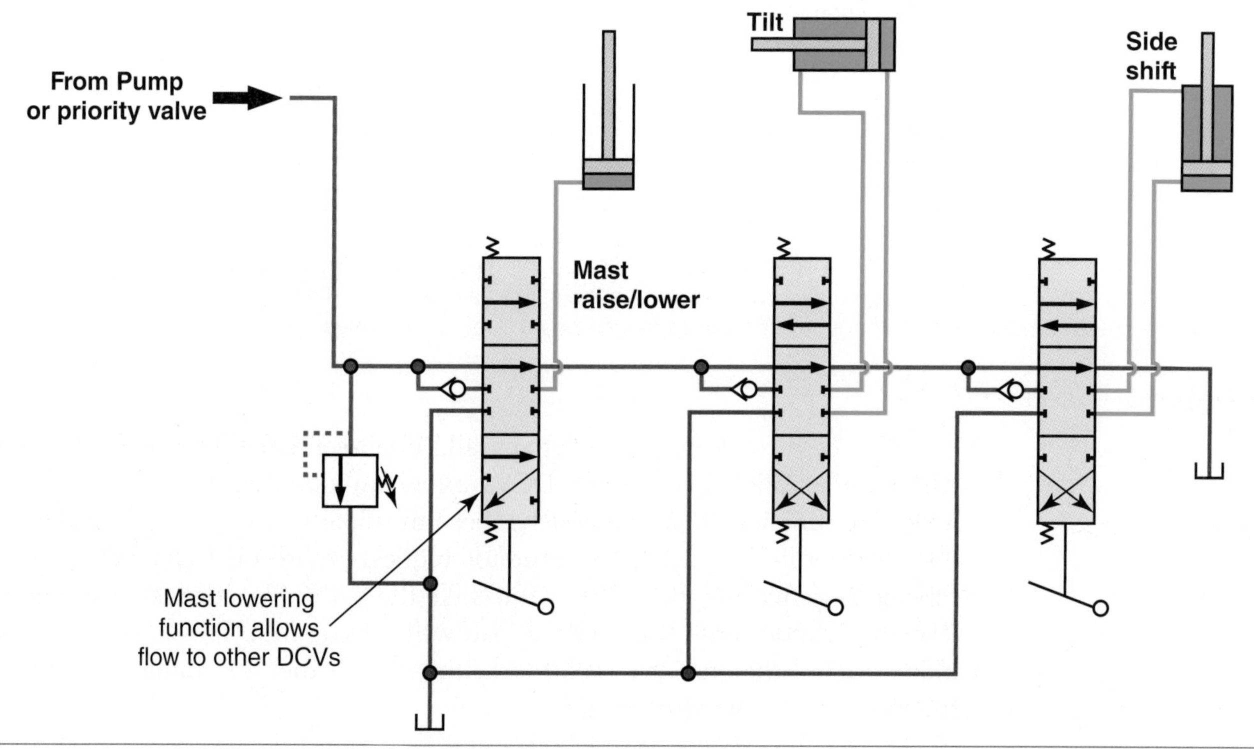

Figure 16-3. An example of a forklift using an open-center hydraulic system with three DCVs connected in series.

On this particular machine, you can only perform one function at one time, with the exception of lowering the mast. For example, the DCV spools will not allow the operator to raise the mast and side shift the mast simultaneously. Nor will the DCV spools allow the operator to tilt the mast and side shift the mast simultaneously. The spools will, however, allow the operator to lower the mast and either tilt or side shift. This is allowable primarily because pump flow is not used for lowering the mast. The mast has single-acting cylinders that are retracted by the weight of the mast.

In this particular set of DCVs, the first priority is raise, the second priority is tilt, and the last priority is side shift. Therefore, if the operator was tilting the mast, and attempted to raise the mast at the same time, the mast would quit tilting and begin to rise.

What is the advantage of eliminating the ability to use two hydraulic functions at the same time in a series circuit? In an open-center series circuit, the upstream valve (in this case, the mast raise DCV), feeds the downstream valves (in this case, tilt and side shift DCVs). If two hydraulic functions were selected at the same time, the pump would have to build enough pressure to overcome the sum of both loads.

To simplify this concept, consider three simple check valves that are connected together in series. See **Figure 16-4**. Notice that the pump must build enough pressure to overcome the load of all three check valve springs.

600 psi + 400 psi + 200 psi = 1200 psi

Therefore, in this case, the pump must be capable of building 1200 psi in order to push oil through all three check valves.

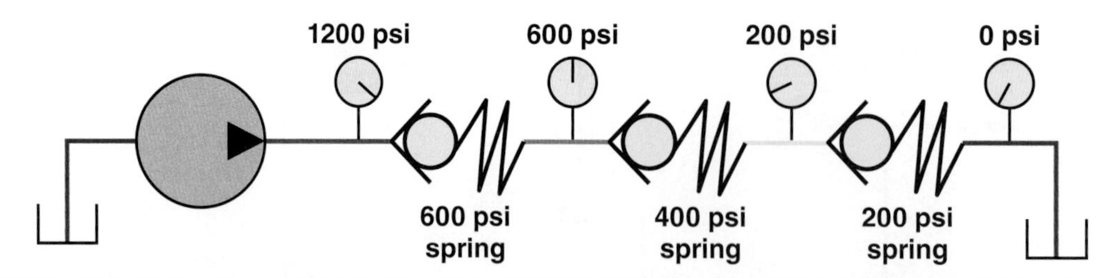

Figure 16-4. In a series circuit, if multiple operations were allowed to be performed simultaneously, the pump would need to be capable of building enough pressure to overcome the sum of all the loads. This is similar to the way the pump in the drawing must generate enough pressure to overcome all three check valve springs.

Parallel

DCVs can also be designed with parallel open-center DCVs. See **Figure 16-5**. Although parallel open-center DCVs allow an operator to select multiple hydraulic functions simultaneously, those multiple functions will not always operate simultaneously. If the operator requests more oil than the pump is capable of delivering, the oil flow will take the path of least resistance. Therefore, the cylinder with the smallest load will actuate first. Then, after that cylinder reaches the end of travel, the cylinder with the next lightest load will function, until that cylinder stops as well.

The one advantage of the parallel DCVs is that it is easy to convert from open-center to closed-center. All that is needed to convert an open-center valve block into a closed center valve block is the insertion of a plug. See **Figure 16-6**.

Flow Dividers and Priority Valves

One way to alleviate the problems that occur in series and parallel designs is to incorporate the use of priority valves or flow divider valves. As explained in **Chapter 8**, priority valves are commonly used with steering and service brake applications. They ensure that the demand of the steering and service brakes have been met before oil can be made available to the secondary functions. **Chapter 8** also explained that a proportional flow divider valve can be used to divide a given amount of flow into different proportions for different branches of a hydraulic system.

Jammer Solenoids

Jammer solenoid valves, also known as unloading valves or ***dump valves***, are used in both open-center hydraulics systems and closed-center LSPC hydraulic systems, as a means to build maximum system pressure. When a jammer solenoid is used in an open-center system, technically the DCVs are closed center. Some might think that this statement is a contradiction. How can closed-center valves be used in an open-center application? The jammer solenoid provides the open-center function for the hydraulic system. The jammer solenoid is normally open when the DCVs are not being used.

The DCVs continue to have the responsibility of directing oil to the actuators. However, the jammer solenoid has the responsibility of building pump pressure, which is also known as putting the system on demand. The jammer valve provides the path for the oil to flow back to the reservoir when the valve is de-energized, allowing oil to flow freely at a low pressure. See **Figure 16-7**.

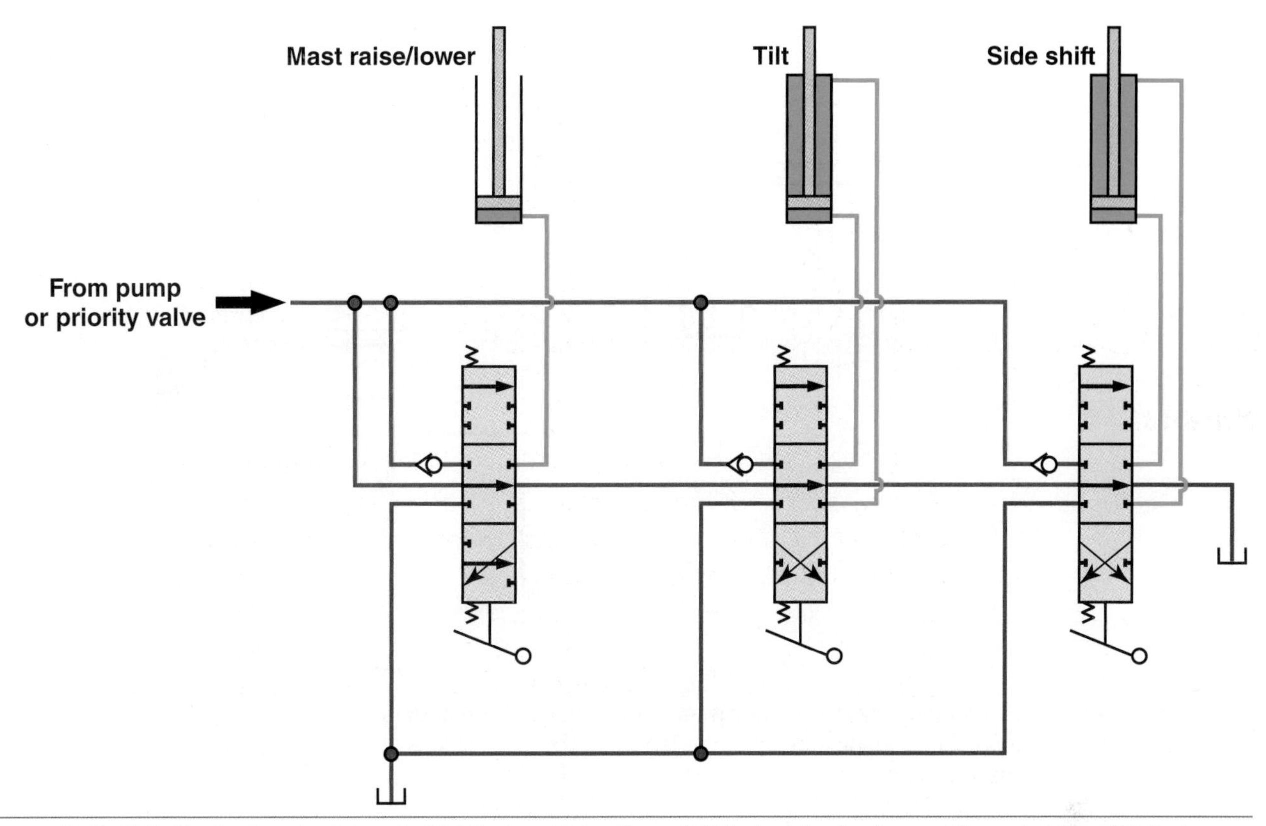

Figure 16-5. This illustration depicts a forklift that contains three DCVs (mast raise/lower, tilt, and side shift) that are connected in parallel.

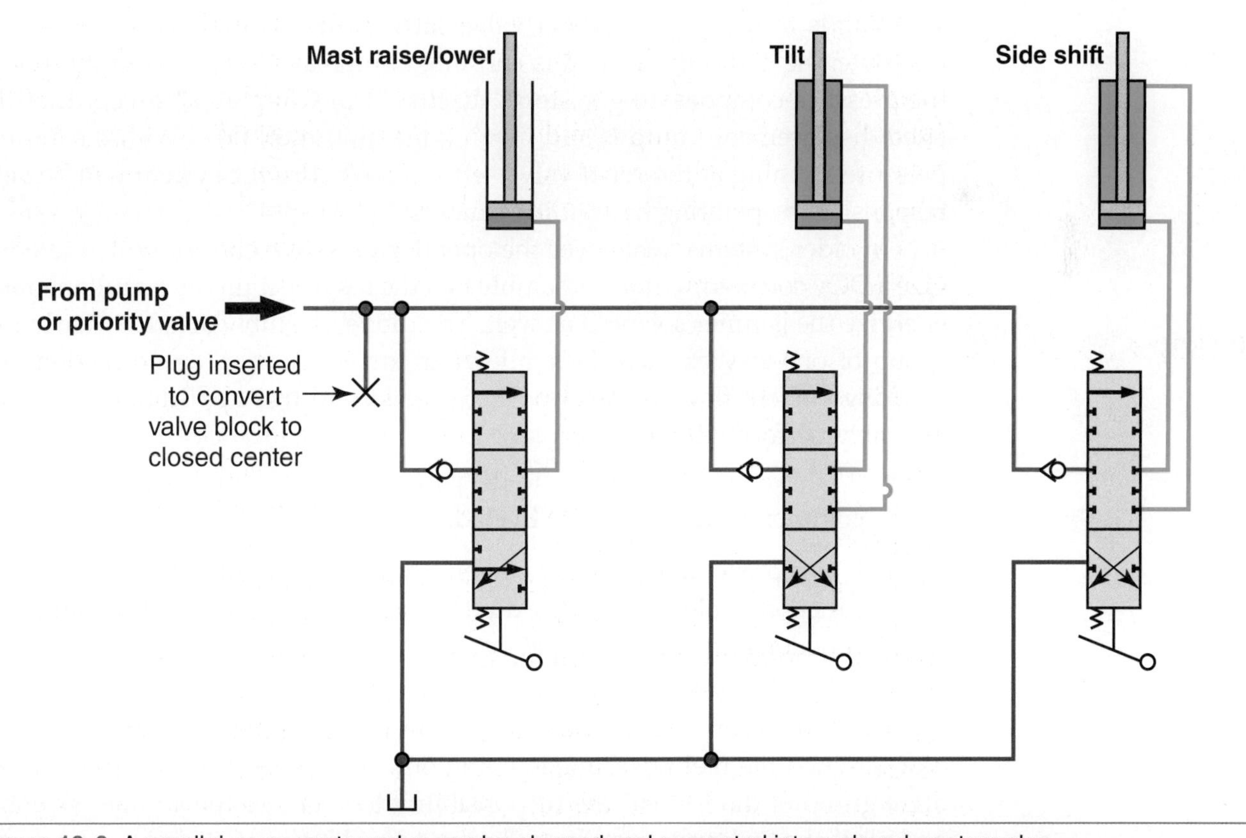

Figure 16-6. A parallel open-center valve can be plugged and converted into a closed-center valve.

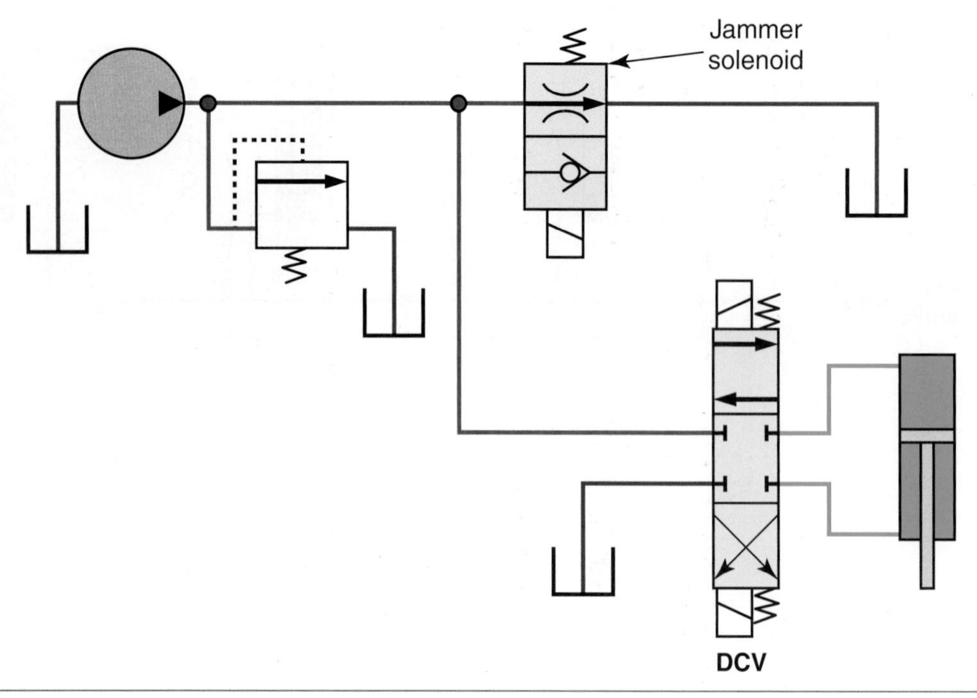

Figure 16-7. A jammer solenoid valve, also known as a dump valve or unloading solenoid valve, is used in hydraulic systems to build pump pressure. In this example, the jammer solenoid valve is acting as the open-center valve. Notice the DCV is actually a closed-center valve.

When the jammer solenoid is energized, the valve will block off oil flow, and because the oil can no longer freely flow to the reservoir, the oil pressure builds to the system's relief valve setting, thus jamming the system. In this state, the hydraulic system is running at high system pressure, similar to the pressure-compensating systems discussed in **Chapter 17**, except that the fixed-displacement pump is still flowing maximum oil flow. With the system pressure running at the relief valve setting, the hydraulic system will be very responsive to operating hydraulic cylinders.

On older systems, whenever the operator presses a control switch to energize a DCV solenoid, a diode assembly had the responsibility to simultaneously energize the jammer solenoid as well. The diode assembly acted similarly to a group of one-way check valves, allowing current flow in one direction and blocking current flow in the opposite direction. In this application the diode assembly performed two functions:

- It allowed current to flow to the jammer solenoid anytime one of the common functions was activated.
- It also prevented electrical current from back-feeding to the other four functions (solenoids). If the diodes were not used, all five hydraulic functions (solenoids) would operate anytime one hydraulic function (solenoid) was activated.

For example, on Case IH combines, from 1977 until 1994, anytime the operator actuated the reel raise, auger swing out, auger swing in, reel fore, or reel aft, a group of diodes had the responsibility to energize the jammer solenoid and place the hydraulic system on demand. See **Figure 16-8**.

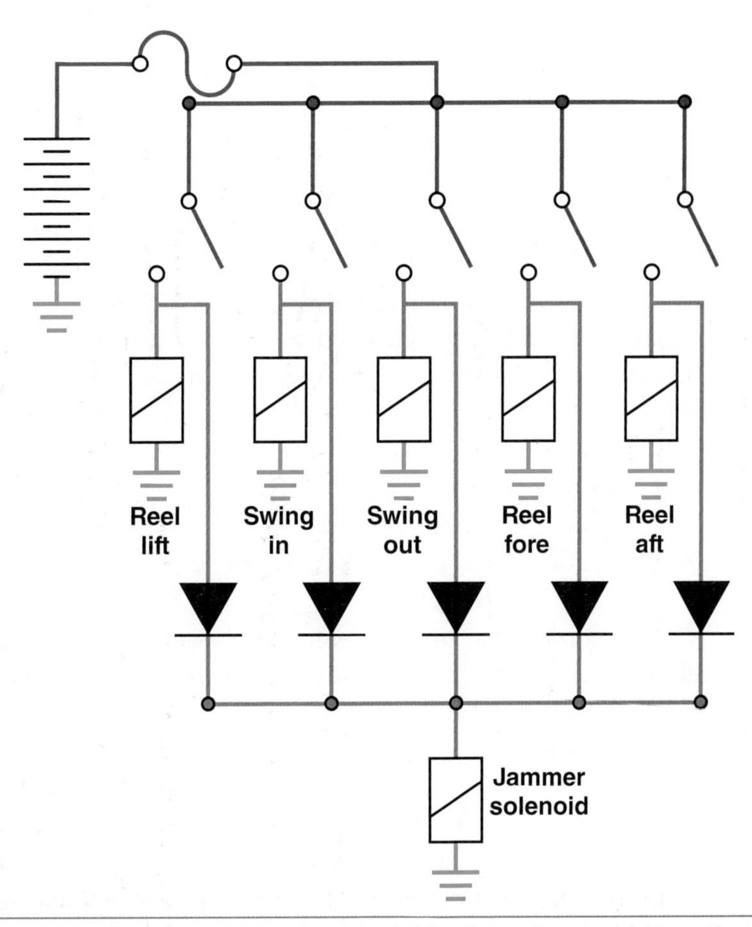

Figure 16-8.1 A group of diodes can be used in conjunction with a jammer solenoid to activate the jammer solenoid valve anytime one of the implement functions is activated. The diodes also prevent other hydraulic functions from activating by blocking the current flow. Late-model machines incorporate the diode function inside the ECM.

Case IH used the diode assembly to perform a similar function on 2100 and 2300 series combines. However, the jammer solenoid was being used in variable-displacement LSPC systems. **Chapter 18** will explain that jammer solenoids in LSPC systems act similarly to jammer solenoids in open-center systems. When one hydraulic function is energized, it will cause the hydraulic system to jump to high system pressure. The 2100 and 2300 series machines similarly used the group of diodes to simultaneously energize the jammer solenoid anytime one of the common hydraulic functions was activated, and also prevented the other functions from activating by preventing the electrical current from back-feeding and energizing the other solenoids. **Chapter 18** will explain that the significant difference in LSPC systems is that the jammer solenoid is normally closed, instead of normally open.

Note

Most late-model machines, whether equipped with fixed-displacement pumps or variable-displacement LSPC pumps, eliminate the diode assembly and directly control the jammer solenoid with an ECM.

The system in Figure 16-7 was highly simplified to help explain the principle behind a normally-open jammer solenoid. When a jammer solenoid was used on the Case IH open-center combines, a flow divider proportioned oil to the circuits, dedicating 1.5 gpm (5.7 lpm) to the jammer circuit. See Figure 16-9. If a problem existed in the 1.5 gpm circuit (reel raise, reel fore/aft, unloader swing in/out), a technician could not isolate all potential problems by simply dead-heading the header lift cylinder because the header lift cylinder and the 1.5 gpm circuit received their own dedicated oil from a flow divider valve. In this system, the jammer solenoid contained a primary poppet and a secondary poppet. If a problem existed in the jammer circuit, the failure could be caused by the following:

- Flow divider.
- Jammer primary poppet valve.
- Jammer secondary poppet valve.
- The dedicated relief valve for those functions (also known as the secondary relief).

AGCO manufactures late-model Massey Ferguson and Challenger brand self-propelled windrowers, for example WR9700, that utilize open-center hydraulic systems with a jammer solenoid. The jammer is used for reel fore and aft, header tilt left and right, and header raise. These machines do not employ a diode block, but a controller directly energizes the normally open jammer solenoid. The machines use a variable-displacement LSPC system for the knife drive, also known as header drive.

John Deere 9000 series, 10 series, and 50 series combines also used a hydraulic system similar to the jammer solenoid system in Figure 16-7. John Deere called the jammer valve a "dump valve". It had the same responsibility to block off the constant flow, and then a DCV would open to operate a hydraulic function. The dump valve was closed when the following operations were performed: header lift, header tilt, reel fore/aft, reel lift, and unloading auger swing in and out.

Varying Cylinder Speeds in Open-Center Hydraulic Systems

Earlier in this chapter it was mentioned that open-center hydraulic systems have been used for decades on combine harvesters. One characteristic of open-center systems is that the pump is always delivering a maximum amount of flow. At times, the operator will desire to slow an extending cylinder. Often, manufacturers use a variable orifice to slow an actuator. See Figure 16-10. The drawback of using a variable orifice is that the remaining oil flow will go over a relief valve, which causes high pressure, heat, noise, and wasted power.

For the Deere 60 and 70 series combines, the header raise and lower were given their own gear pump and proportional valve to allow for variable speed raise and lower. The same system is also used on the S660 and S670 combines.

With the use of LSPC systems this problem is solved. **Chapter 18** will explain the use of LSPC systems.

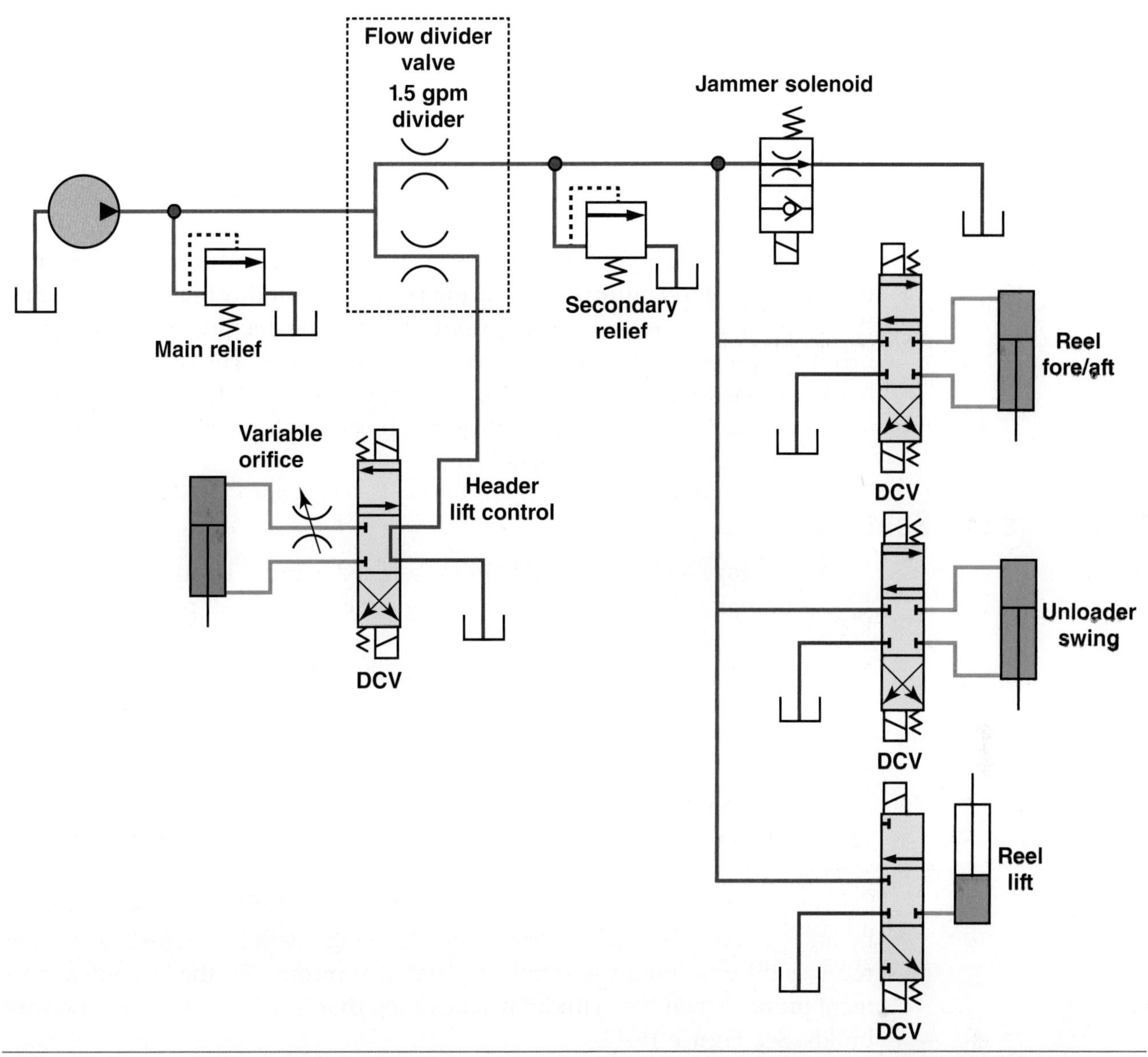

Figure 16-9. A simplified illustration of an open-center hydraulic system that contained a 1.5 gpm (5.7 lpm) flow divider for the reel lift, unloader swing, and reel fore and aft. Notice that the 1.5 gpm circuit contains its own secondary relief valve. This jammer solenoid valve has been drawn as a single solenoid to simplify the circuit.

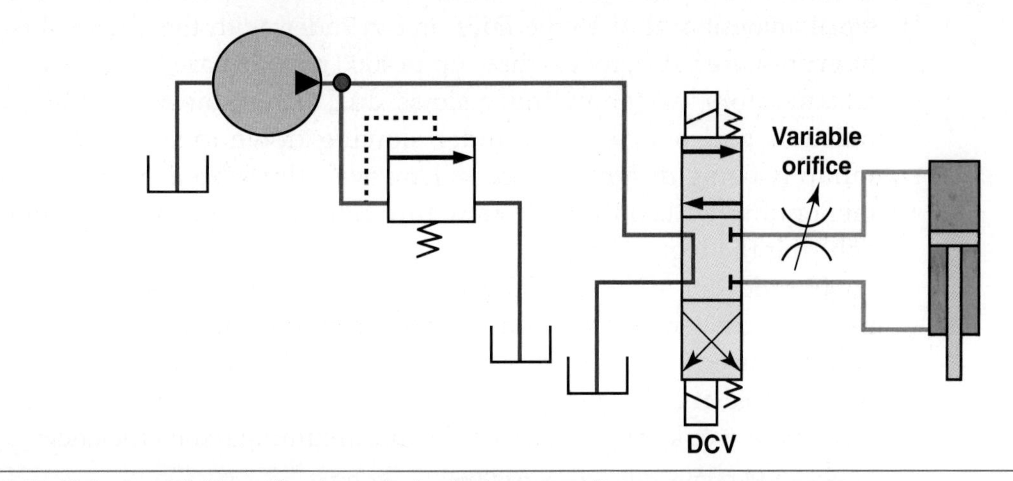

Figure 16-10. Using a variable orifice to control cylinder speed in an open-center system.

Tandem Two-Stage Systems

Log splitters and a few construction machines use a unique open-center hydraulic system consisting of a tandem two-stage hydraulic pump. The two-stage pump normally consists of two separate fixed-displacement gear pumps driven by the same input shaft. The pumps are always producing a maximum flow at high idle.

The system incorporates the use on an unloading valve that is designed to dump one of the pump's flow during certain conditions that require higher system pressure. This action prevents the engine from lugging or stalling, since high pressure and high flow requires more horsepower. The two-stage system harnesses the benefit of two pumps' flow anytime the system pressure is low and prevents engine lugging when system pressures are high.

Two-Stage Log Splitters

A two-stage hydraulic log splitter is designed to provide either fast cylinder speeds or high splitting force, while using a relatively small power source, for example 6 hp engine. See **Figure 16-11**.

One of the pumps provides large amounts of flow at relatively low pressure, for example 300 psi (21 bar). The other pump has a much smaller displacement and provides a small amount of flow, but with the potential for maximum system pressure, for example up to 4000 psi (276 bar). The unloading valve operates similarly to a relief valve and is normally closed. This unloading valve will open at a relatively low pressure value, for example 300 psi (21 bar).

During operation, as long as the cylinder's resistance can be overcome by an operating pressure of less than 300 psi (21 bar), the two-stage system receives oil from both the small-displacement pump and the large-displacement pump. When the cylinder reaches a log that is difficult to split, pressure builds. See **Figure 16-12**.

When pressure is high enough to open the unloading valve, the unloading valve will dump the large-displacement pump's flow to the reservoir. The check valve enables the small-displacement pump to continue to supply a small amount of flow to the DCV and cylinder, with the potential to reach a high pressure value, for example up to 4000 psi (276 bar).

Therefore, as the cylinder slows due to resistance from the log, the operator will notice the cylinder slowing down, because the high flow pump is being dumped to tank. However, fluid from the small-displacement pump will continue to flow through the DCV to the cylinder, slowly splitting the log.

In order for a log splitter with a single-stage pump to achieve the same maximum flow rate and maximum pressure provided by the two-stage pump, it would need to be driven by an engine with a much higher horsepower. For example, if a single-stage pump needs to flow 14 gpm (53 lpm) at 4000 psi (276 bar), it would require 32.6 engine hp, assuming 0% inefficiency:

$$14 \text{ gpm} \times 4000 \text{ psi} \times 0.000583 = 32 \text{ hp}$$

Although the two-stage pump does not simultaneously provide maximum flow and maximum pressure, it will provide either high flow at low pressures, or high pressures at low flow, whichever is needed at any given time. This example of a two-stage pump can be driven by a 6.76 hp engine.

For example if the high-flow pump was producing 12 gpm (45 lpm) at 300 psi (21 bar), it would require only 2.1 hp:

$$12 \text{ gpm} \times 300 \text{ psi} \times 0.000583 = 2.1 \text{ hp}$$

If the low-flow pump was producing 2 gpm at 4000 psi, it would require only 4.67 hp:

$$2 \text{ gpm} \times 4000 \text{ psi} \times 0.000583 = 4.66 \text{ hp}$$

The total engine horsepower required to drive both pumps would be 2.1 + 4.66, equaling 6.76 hp, assuming no losses for inefficiency.

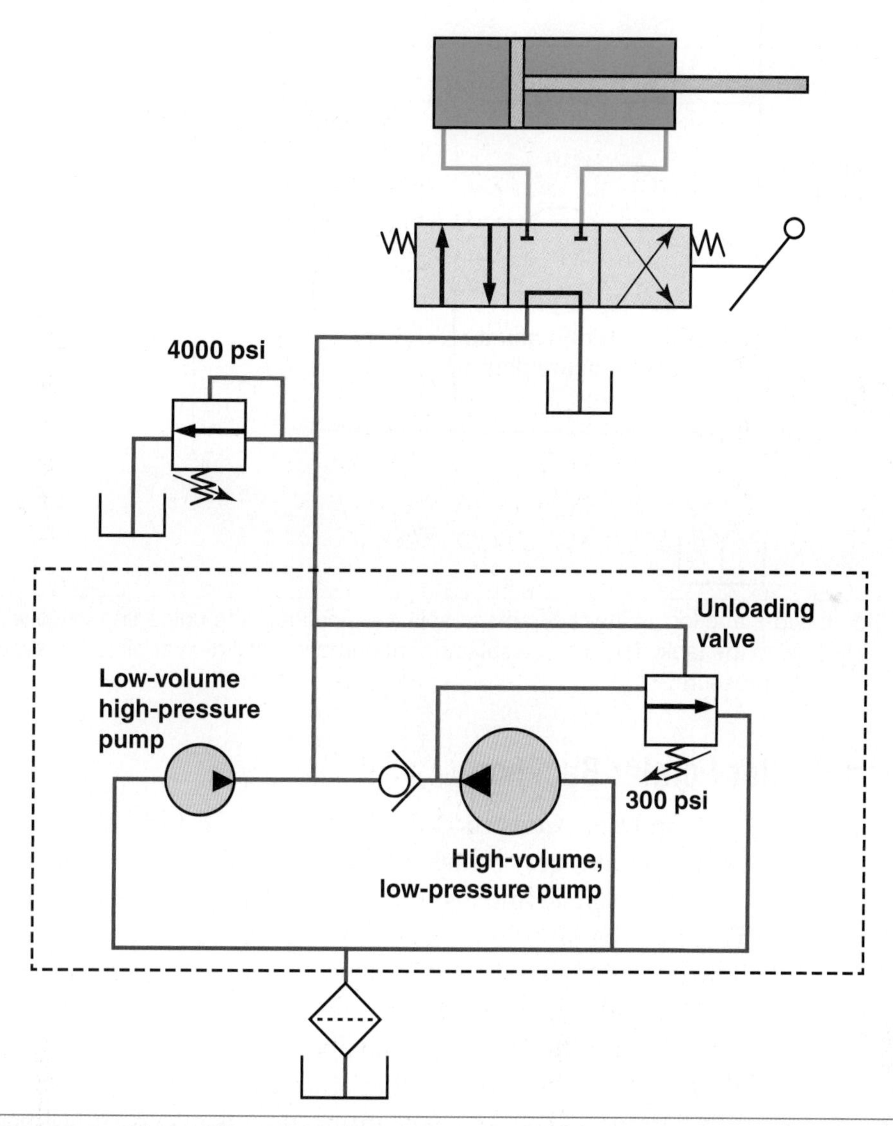

Figure 16-11. Most gas-powered log splitters use a two-stage pump, which contains a large-displacement pump that is unloaded at a low pressure value and a small-displacement pump that can be operated at high pressure. This log splitter uses a two-stage open-center hydraulic system with two gear pumps, one unloading valve, one relief valve, one DCV, one double-acting cylinder, and one check valve.

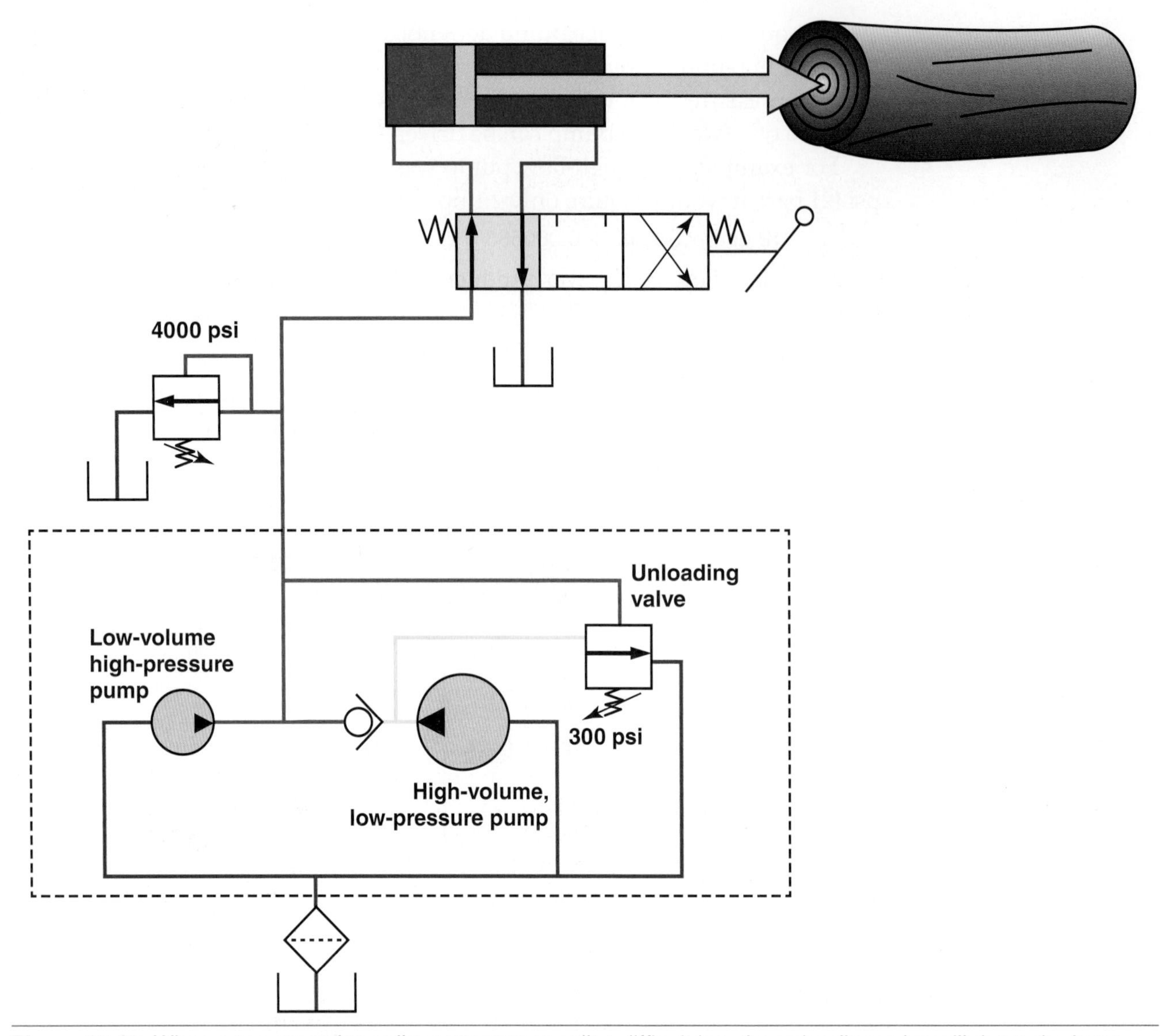

Figure 16-12. When a two-stage log splitter attempts to split a difficult log, the unloading valve will dump the large-displacement pump's flow to the tank. The small-displacement pump will still deliver oil to the system and it is capable of operating at a very high pressure.

Tandem Open-Center Loader Backhoe

John Deere Construction offers a 310K EP and 310SK loader backhoe, which uses a tandem gear pump with an unloading valve. The unloading valve consists of a solenoid and a spool. The tandem pumps consist of a main pump and a secondary pump. During certain applications, the secondary pump has the oil flow unloaded, meaning that the oil is routed back to the reservoir. In this scenario, only the main pump flow is available to do the work. During most backhoe operations, both the main pump and the secondary pump are supplying system oil flow, except for when using attachments that use one-way functions, such as a hammer. The secondary pump flow is also diverted back to the reservoir when the machine is using the loader, for example stock pile applications. In this scenario, the machine is using engine horsepower for propulsion and the system hydraulic flow is receiving oil only from the main pump.

Case Study

A Blocked Directional Control Valve in an Open-Center System

A technician received a call from a local veterinarian who needed help with a hydraulic system. No service literature was available. The hydraulic system was not mobile; it was permanently installed in the building. The customer stated that the hydraulic system sounded like it was going to blow up, but they continued to operate the system.

The hydraulic system was used to power several hydraulic functions, such as a squeeze chute, a calf-lifting table, and one other function that will be detailed soon. A different technician had previously replaced several components: a relief valve, the electric drive motor, and a hydraulic pump. The owner was considering replacing the oil with thinner viscosity oil.

The technician spent 30 to 45 minutes tracing the hydraulic lines throughout the building. The pump was located in one room and the hydraulic lines were routed back to the barn. The veterinarian was asked to actuate the hydraulic control valves, and he operated each of the visible hydraulic control valves. No change occurred. It was later determined that another control valve was hiding in a corner.

Each time the electrically driven pump was turned on, the hydraulic system sounded like it was on high pressure, similar to a pressure-compensated system (**Chapter 17**). The hydraulic system had two hydraulic lines that were routed from the power unit to the rear barn. One line was pump outlet and the other was return. The hydraulic system was a traditional fixed-displacement, constant flow, open-center hydraulic system. When the DCVs were in a neutral position, the system should exhibit constant flow at a low pressure.

The hydraulic system used household black pipe for transmitting fluid from the power unit to the DCVs. Union fittings enabled the disassembly of the lines and the installation of a hydraulic flowmeter. A flowmeter was installed in place of the first set of DCVs, which allowed the pump's flow to bypass back to the reservoir. The machine was turned on and the noise went away. Therefore, with no DCVs in the circuit, the hydraulic system exhibited normal open-center characteristics.

The hydraulic system consisted of several open-center DCVs that were installed in series, with each DCV supplying the next DCV, and finally returning the oil back to the power unit. The flowmeter was next installed after the first set of DCVs and retested, again the noise was silenced, and it was determined that the power unit and the first group of DCVs were okay. The process of installing the flowmeter after the next set of DCVs was repeated, until the last DCV was reached. It was then determined that another DCV existed, hiding in the back of the barn. This portion of the hydraulic system was designed and built by an engineer to dehorn cattle. The controls consisted of a 120VAC-operated DCV that actuated a small cylinder. The veterinarian was asked to actuate the newly discovered DCV, and at that time the hydraulic system noise subsided and the system returned to a low operating pressure. The DCV was designed to block off pump flow (jam the circuit) and enable high pressure oil to be directed to the actuator. Internally, the controls were sticking in the "blocked-off" mode, similar to the jammer solenoid. The immediate solution was to toggle the DCV anytime the hydraulic system became stuck in the blocked mode. The long-term solution was to return the controls to the engineer for further repair.

The important points to keep in mind are that open-center systems should be quiet and exhibit low system pressure when the DCVs are in the neutral position. When DCVs are actuated, the constant flow is no longer routed back to the reservoir at low system pressures, but is routed to the actuator. When a cylinder reaches the end of travel, the oil is sent over the main system relief valve at a high pressure. If a constant-flow, fixed-displacement system sounds like it is operating under high pressure, it means the low-pressure return has been blocked at some point in the circuit, preventing the oil from returning to tank, and forcing the oil to dump over a high-pressure relief.

Summary

- ✓ Open-center systems were some of the first hydraulic systems installed on machines.
- ✓ They have been used for many decades and are still used on some late-model machines.
- ✓ Some examples where open-center hydraulic systems are still used today include Toyota forklifts, most compact and utility tractors, John Deere S660 and S670 combines, and Case New Holland skid steers.
- ✓ Traditional open-center hydraulic systems have the following characteristics:
  - Simple.
  - Constant, maximum flow.
  - Lower pressure (when DCVs are in neutral).
  - Fixed-displacement pump(s).
- ✓ When a jammer solenoid valve is used with fixed-displacement pumps, the jammer is designed to be normally open when it is de-energized. When the jammer solenoid is energized, it will block pump flow, causing the system to build high pressure.
- ✓ Two-stage open-center pumps use two fixed-displacement pumps and an unloading valve. During high load conditions, the unloading valve is designed to dump flow from one of the pumps, which prevents the engine from stalling.

Technical Terms

dump valves
jammer solenoid valves

Review Questions

Answer the following questions using the information provided in this chapter.

1. Which of the following systems has low pressure and sends pump outlet oil to the reservoir when the valves are in the neutral position?
   A. Open center.
   B. Closed center.
   C. Both A and B.
   D. Neither A nor B.

2. Which type of system is shown in the following drawing?
   A. Open center.
   B. Closed center.

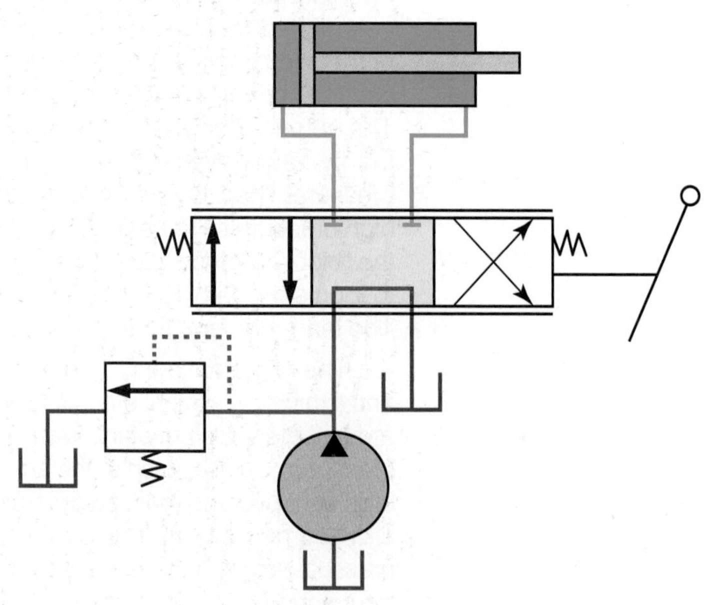

3. Which of the following groups of hydraulic terms is used to describe the directional control valve located directly after the hydraulic pump?
   A. Open/closed center.
   B. Open/closed loop.
   C. Open/closed control.
4. Which of the following terminologies describes the pump's inlet?
   A. Open/closed center.
   B. Open/closed loop.
   C. Open/closed control.
5. Which style of DCV routes oil back to the tank when the valve is in a neutral position?
   A. Open.
   B. Closed.
   C. Load sensing.
6. Traditional open-center hydraulic systems use what type of pump?
   A. Fixed displacement.
   B. Variable displacement.
   C. Either A or B.
   D. Neither A nor B.
7. When compared to an equivalent closed-center system, a traditional open-center hydraulic system generates what type of flow when the valves are in a neutral position?
   A. Lower flow.
   B. Higher flow.
8. Traditional open-center hydraulic systems exhibit what type of pressures when the valves are in a neutral position?
   A. Low pressure.
   B. High pressure.
9. Technician A states that open-center hydraulic systems must use manually operated DCVs. Technician B states that some open-center hydraulic systems use electronically controlled DCVs. Who is correct?
   A. Technician A.
   B. Technician B.
   C. Both A and B.
   D. Neither A nor B.
10. How much horsepower is produced by a hydraulic system delivering 20 gpm at 400 psi?
    A. 2.67 hp.
    B. 3.67 hp.
    C. 4.67 hp.
    D. 5.67 hp.
11. Technician A states that open-center hydraulic systems were the first type of hydraulic system used in the off-highway industry. Technician B states that manufacturers no longer use open-center hydraulic systems. Who is correct?
    A. Technician A.
    B. Technician B.
    C. Both A and B.
    D. Neither A nor B.
12. What is the drawback of using a needle valve for slowing cylinder speed in an open-center hydraulic system?
    A. Costly.
    B. Complex.
    C. Balance of the oil goes over the relief valve causing heat and noise.
    D. All of the above.
13. A traditional open-center valve block uses a parallel design. What will happen when an operator tries to raise a 2000-pound load and a 4000-pound load simultaneously?
    A. Both will be raised simultaneously.
    B. The 2000-pound load will be raised first.
    C. The 4000-pound load will be raised first.
    D. Neither will be raised.

14. Based on the following drawing, what is the minimum amount of pump pressure that must be developed in order to flow oil through the three check valves?
    A. 400 psi.
    B. 600 psi.
    C. 1200 psi.
    D. 2400 psi.

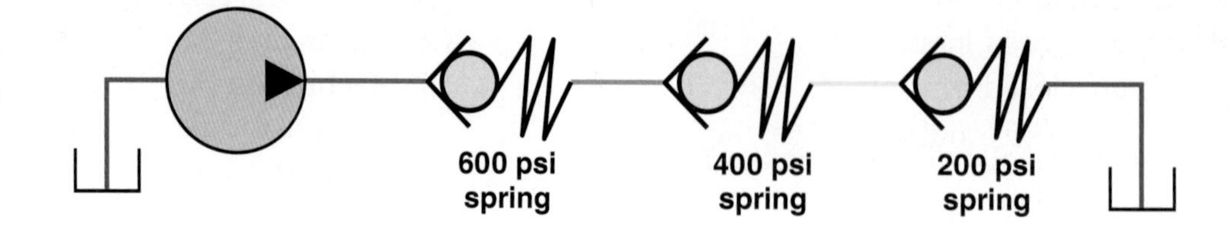

15. All of the items listed below can be commonly found in open-center hydraulic systems, *EXCEPT*:
    A. jammer solenoid.
    B. priority valve.
    C. load-sensing pump.
    D. divider valve.
16. An off-highway machine is using a gear pump and a jammer solenoid in conjunction with three DCVs. The sales brochure labels the machine as an open-center hydraulic system. Which of the components below is truly the open-centered component?
    A. Jammer solenoid.
    B. DCVs.
17. What happens in terms of flow when a two-stage log splitter's system reaches high pressure?
    A. Flow from the small gear pump is bypassed internally.
    B. Flow from the large gear pump is bypassed internally.
18. What happens in terms of pressure when a two-stage log splitter's system reaches high pressure?
    A. High pressure from the small-displacement gear pump builds and is made available to the cylinder.
    B. High-pressure from the large-displacement gear pump builds and is made available to the cylinder.
19. A technician is dispatched to troubleshoot a malfunctioning open-center hydraulic system. Prior to leaving the shop, the supervisor states that the open-center hydraulic system sounds like it is going to blow up. What should the technician be looking for when he or she arrives at the machine?
    A. A stuck open DCV.
    B. A blocked passage preventing oil from returning to the tank.
    C. Too high of viscosity fluid.
    D. None of the above.
20. All of the following are common characteristics of open-center hydraulic systems, *EXCEPT*:
    A. Simple.
    B. Constant maximum flow.
    C. Lower pressure (when DCVs are in neutral).
    D. Variable-displacement pump.
21. Which of the following is most likely to be the approximate system pressure when an open-center system's DCVs are in the neutral position?
    A. 150 psi (10 bar).
    B. 600 psi (41 bar).
    C. 1000 psi (69 bar).
    D. 2500 psi (172 bar).

22. What type of electrical component is sometimes used in conjunction with a jammer solenoid in order to operate multiple DCV functions?
    A. Relay.
    B. Diode assembly.
    C. Circuit breaker.
    D. Double pole double throw switch.

23. All of the following are true statements about traditional open-center hydraulic systems when the DCVs are in a neutral position, *EXCEPT*:
    A. The system pressure is low.
    B. Noise is reduced.
    C. Heat generation is reduced.
    D. It is one of the more complex systems.

With the 1960 introduction of a variable-displacement pressure-compensating hydraulic system, John Deere became a pioneer in tractor hydraulics. Their 50 series agricultural tractors came equipped with the system decades after its initial release.

Chapter 17

Pressure-Compensating (PC) Hydraulic Systems

Objectives

After studying this chapter, you will be able to:

- ✓ Describe the first closed-center hydraulic system used in the agricultural industry.
- ✓ Describe the system flow and pressure when a PC hydraulic system has the DCVs in a neutral position.
- ✓ Explain how a John Deere PC radial piston pump destrokes and upstrokes.
- ✓ Explain how an inline axial PC pump upstrokes and destrokes.
- ✓ List the advantages and disadvantages of a pressure-compensating hydraulic system.
- ✓ Explain why open-center hydraulic systems are not operated at a constant high system pressure.

Introduction to Closed-Center Systems

John Deere was one of the first mobile equipment manufacturers to use a closed-center hydraulic system. In 1960, with the introduction of the 10 series agricultural tractors, John Deere revolutionized tractor hydraulics by introducing a variable-displacement *pressure-compensating (PC) hydraulic system.* It was designed to operate at a constant high pressure with practically no flow when the closed-center DCVs were in a neutral position. See **Figure 17-1**.

John Deere continued using this type of hydraulic system for nearly thirty-five years, in both agricultural equipment and construction equipment. From 1960 until 1994, John Deere produced 10, 20, 30, 40, 50, 55, and 60 series agricultural tractors that used PC hydraulics. They also produced numerous other agricultural and construction machines with PC systems.

The PC systems were the first type of closed-center hydraulic system used in the agricultural industry. They are unique, and today, practically no agricultural or construction manufacturer use a traditional PC system, although some John Deere sprayer hydraulic systems have some close similarities. Eaton calls the PC hydraulic system a ***pressure-limiting hydraulic system.***

The PC system uses a variable-displacement pump that is designed to shut off flow once the pump reaches the pressure compensator's spring value. In reality, the pump will be generating a small amount of flow that is required to maintain a constant high pressure, while overcoming small system inefficiencies, such as leaks in the pump controls and leaks at the DCV spools. The pump will reach the high pressure value once the DCV is returned to neutral or after a hydraulic cylinder reaches its end of travel.

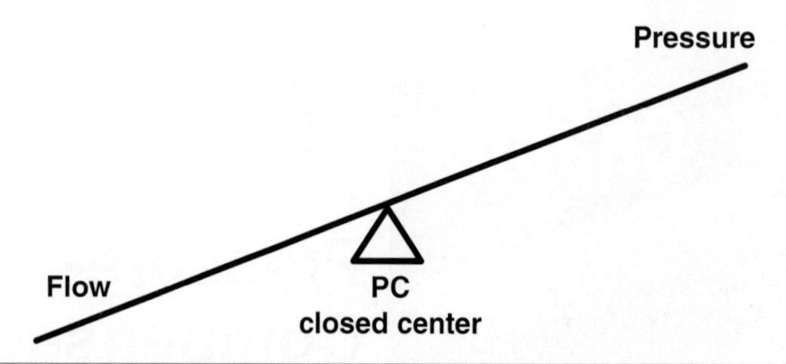

Figure 17-1. A PC closed-center hydraulic system has practically no flow and maximum pressure when the DCVs are in a neutral position.

PC systems and open-center systems exhibit opposite characteristics when their DCVs are in a neutral position. Open-center systems have high flow and low pressures, while PC systems have high pressure and low flow.

Pressure-Compensating Pump Symbols

A pressure-compensating pump symbol can vary depending on the choice of the manufacturer. See Figure 17-2. Pressure-compensating pumps sense the pump's output pressure. When the pump's output pressure reaches the value of the pressure-compensated valve, the pump will destroke. These three PC pump symbols are examples that can be used to depict a pressure-compensating pump. In the first two symbols, the arrow within the rectangle represents the pressure-compensated valve. In the last symbol, the pilot line is operating the pressure-compensated valve.

Pump Frame Designs

Although a true PC hydraulic system is not commonly found in mobile machinery today, many industrial manufacturing facilities still use a true PC hydraulic system. There are also more than three decades of John Deere agricultural and construction machines that are still in operation today using these systems.

The PC hydraulic systems can use any type of variable-pump design, such as:

- Variable vane.
- Bent-axis piston.
- Inline axial piston.
- Radial piston.

John Deere Radial Displacement PC Systems

The John Deere PC systems used radial displacement pumps. See Figure 17-3. John Deere named their pumps 1000, 2000, and 3000 series pumps. They were used in thousands of agricultural and construction machines. The Deere PC pumps are normally mounted at the front of the machine and driven off the front of the engine's crankshaft, rotating in a counterclockwise direction. The pumps can be designed with a single bank of eight pistons, or with a dual bank containing two sets of eight pistons. See Figure 17-4 for pump displacements.

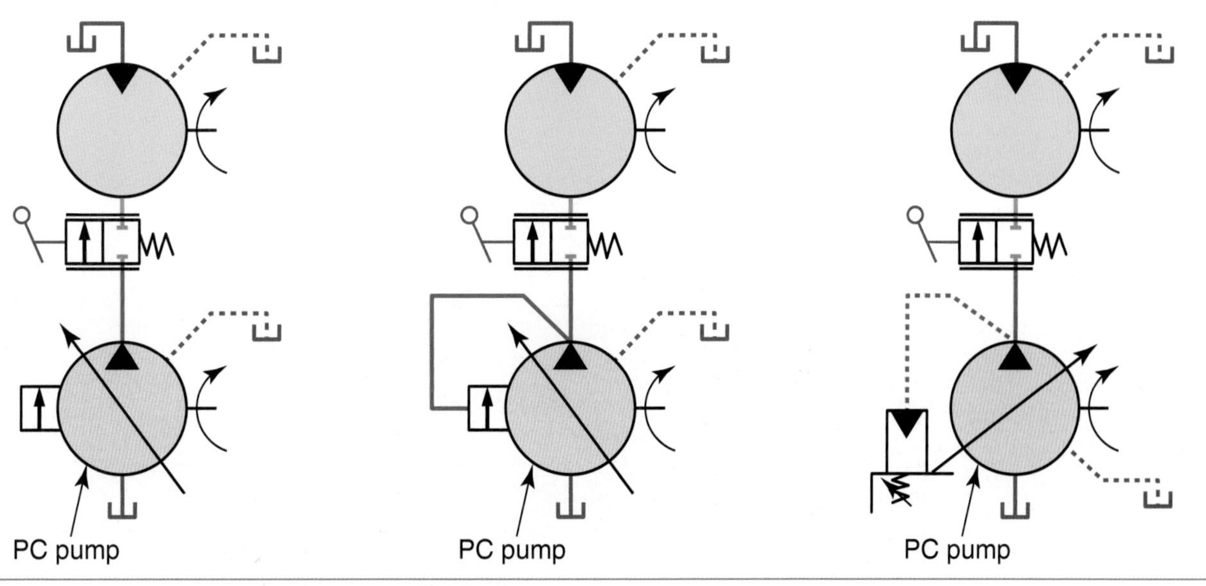

Figure 17-2. Pressure-compensating pump symbols.

Figure 17-3. From 1960 until the mid 1990s, John Deere used radial displacement PC piston pumps. The pump on the right is a John Deere 2000 series pump, and the pump on the left is a 3000 series pump.

As explained in **Chapter 4**, radial displacement pumps can be classified as rotating cam or rotating piston. The John Deere PC pumps were a rotating cam pump. In a rotating cam pump, an eccentric cam is mounted on the pump's input shaft. See Figure 17-5A. As the input shaft rotates, the cam pushes the pistons into their bore. This is the pumping stroke of the piston. As the cam rotates away from the piston, a spring pushes the piston out of its bore. This is the piston's intake stroke.

Each of the pump pistons has its own inlet and outlet check valves. The inlet check valves are mounted inside a round ***oil gallery***, which receives oil from a charge pump and distributes oil to each of the inlet valves. The inlet check valves allow the charge oil into the piston's bore during the piston's intake stroke.

Pump Model	Displacement
1000 Series	
Single Bank	11 cm^3
Single Bank	23 cm^3
2000 Series	
Single Bank	40 cm^3
Single Bank	50 cm^3
Single Bank	66 cm^3
Double Bank	100 cm^3
Double Bank	130 cm^3
3000 Series	
Single Bank	40 cm^3
Single Bank	52 cm^3
Single Bank	66 cm^3
Double Bank	104 cm^3
Double Bank	115 cm^3
Double Bank	130 cm^3

Figure 17-4. John Deere PC radial displacement piston pumps range in displacement depending on the pump series and whether the pump uses a single or double bank of pistons.

The outlet check valves are also mounted inside an oil gallery. The outlet check valves are smaller than the inlet valves. The outlet check valves allow oil to leave the piston's bore during the pumping stroke. The outlet gallery receives the oil from the outlet valves and directs the oil to the pump's outlet port.

Pump Compensator

The pump uses a ***pump compensator*** to shut off the pump flow when the system reaches high pressure. The pump compensator is the brain of the variable-displacement pump. Pump compensators can use one or more valves for controlling the pump's flow.

The John Deere radial compensator performs two functions:

- Directs fluid into the pump case to shut the pump flow off.
- Directs fluid out of the pump case to turn on the pump flow.

The 1000 series and 2000 series pumps use two valves for controlling the pump, a stroke control valve and a crankcase outlet valve. The ***stroke control valve*** has the responsibility to destroke the pump by delivering pressurized oil into the pump's case. The ***crankcase outlet valve*** has the responsibility to upstroke the pump by draining oil out of the pump's case.

When the pump controls direct pressurized oil into the pump's case, the oil pressure holds the pistons in a retracted position, the input shaft/cam assembly will continue to spin at engine rpm, but because the pressure is holding the pistons retracted, no flow will occur. When the pump controls drain the case pressure, the pistons are allowed to reciprocate inside their bores, causing the pump to flow oil. The extent to which the pistons are held retracted in their bores will dictate how much flow is produced by the pump. For example, if the pistons are held 50% retracted, the pump produces 50% pump flow.

Figure 17-5 is an image of a John Deere pump that has been cutaway to further illustrate pump operation. In Figure 17-5A, the top piston is pressed into the pump housing by the cam. That piston is at the end of its pumping stroke. In Figure 17-5B, the top piston has been pushed out of its bore by the piston's return spring. The piston is at the end of the intake stroke.

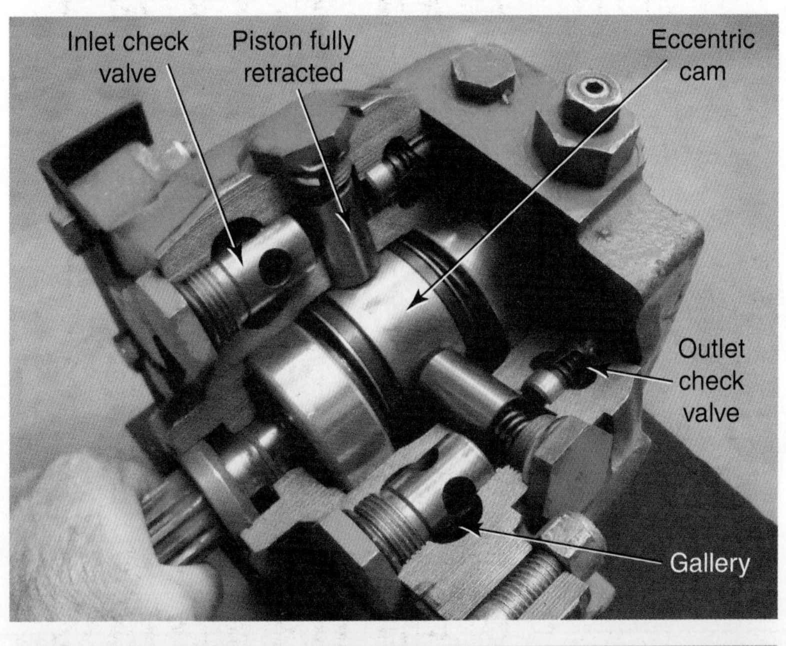

Figure 17-5. Rotating cam pump. A—The top piston is retracted by the cam shaft. B—The cam has been rotated. The top piston is now pushed out of its bore by the piston's return spring.

When the closed-center DCVs are in a neutral position, pump flow is blocked, causing the system pressure to build. As a result, the stroke control valve will open, which will send oil pressure into the pump's case to hold the pistons retracted. See Figure 17-6A.

When the operator actuates a DCV, requesting oil to flow to an actuator, pump outlet pressure will drop. As a result, the stroke control valve will close while the crankcase outlet valve will open. Because of the movement of these two valves, the crankcase oil pressure will drop, which causes the pump's piston to begin reciprocating again, causing the pump to flow oil. See Figure 17-6B.

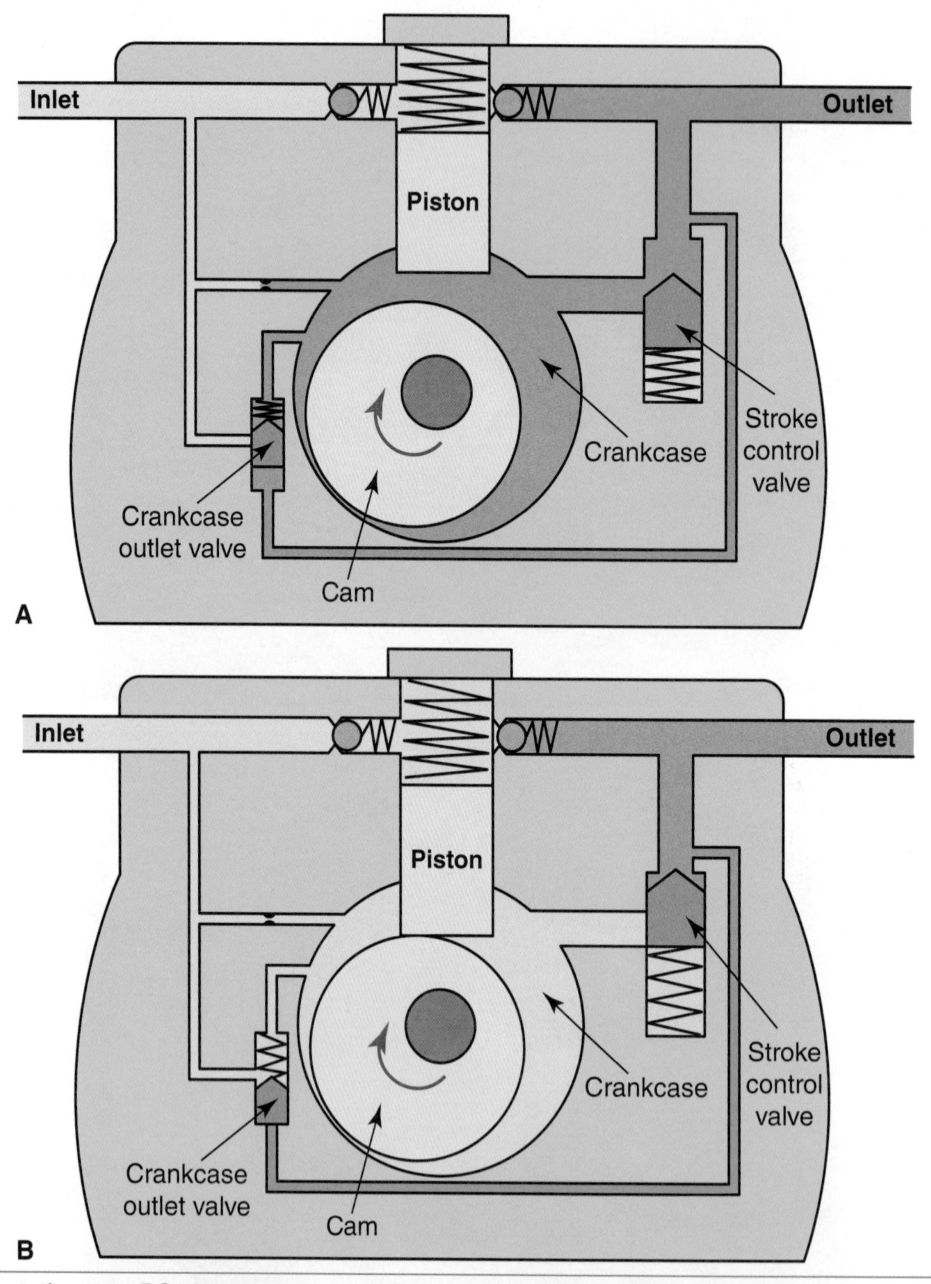

Figure 17-6. A rotating cam PC piston pump. A—Pressurized oil is directed into the pump's case to hold the pistons retracted, which causes the pump to destroke. B—As a DCV opens, the system pressure drops resulting in the stroke control valve closing and the crankcase outlet valve opening. The pistons are now allowed to reciprocate, which causes the pump to deliver oil flow.

Notice on this style of hydraulic system that the variable pump flow is controlled entirely by the pump's compensator control and that the system does not require any type of external sensing line, also known as a signal line. Late model variable-displacement pump controls typically require some type of feedback, which can be hydraulic or electronic. **Chapter 18** will explain that LSPC systems require the use of a hydraulic signal network consisting of a series of primary and secondary shuttle valves that form a signal network. This network is responsible for signaling the highest system working pressure to a margin spool, which controls the system's flow. Electronic controls consist of pressure sensors and swash plate angle sensors, which send feedback to a control module so it can accurately control the pump's electronic actuator.

Case Study

Flow and Pressure in a PC System

A technician assisted a contractor on a 1994 John Deere 410D loader backhoe that contained a 3000 series radial-displacement piston pump. The $5000 pump had already been replaced at 8000 hours. However, approximately 3000 hours later, the hydraulic system began having intermittent problems. The pump control schematic is shown in **Figure 17-7**. Notice that this 3000 series pump uses two valves, a crankcase relief valve, and a pressure-compensated valve, for controlling the pump's variable flow. The PC valve is used to control the pump flow. The crankcase relief valve is used to reduce pressure spikes.

The pump also utilized a ***destroking solenoid valve***. Some Deere PC pumps incorporated methods like the destroking solenoid to allow the tractor's diesel engine to start without a hydraulic load. Previously, an electrical short was found in the destroking solenoid circuit. The intermittent hydraulic problem had also started around the time the shorted destroking solenoid was discovered. See **Figure 17-8**.

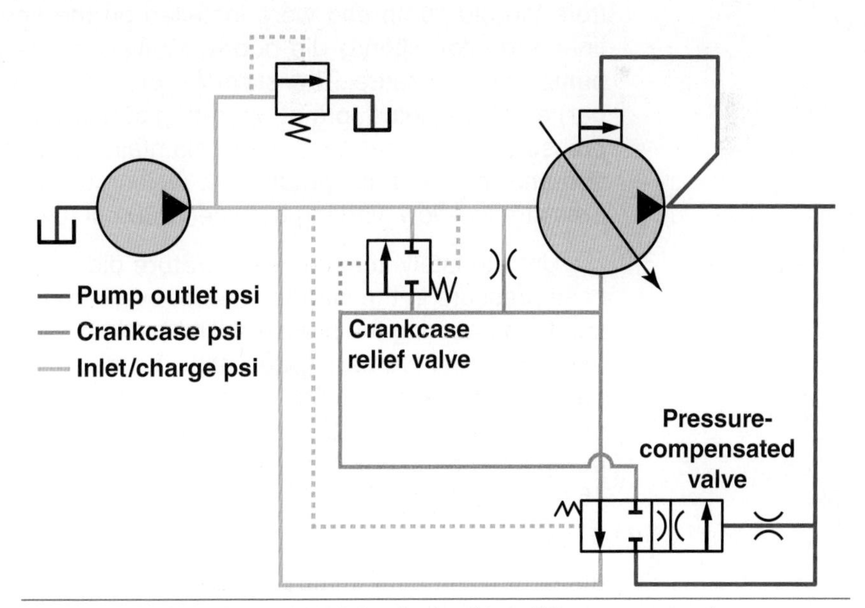

Figure 17-7. John Deere 410D loader backhoe PC pump controls use a pressure compensator valve and a crankcase relief valve.

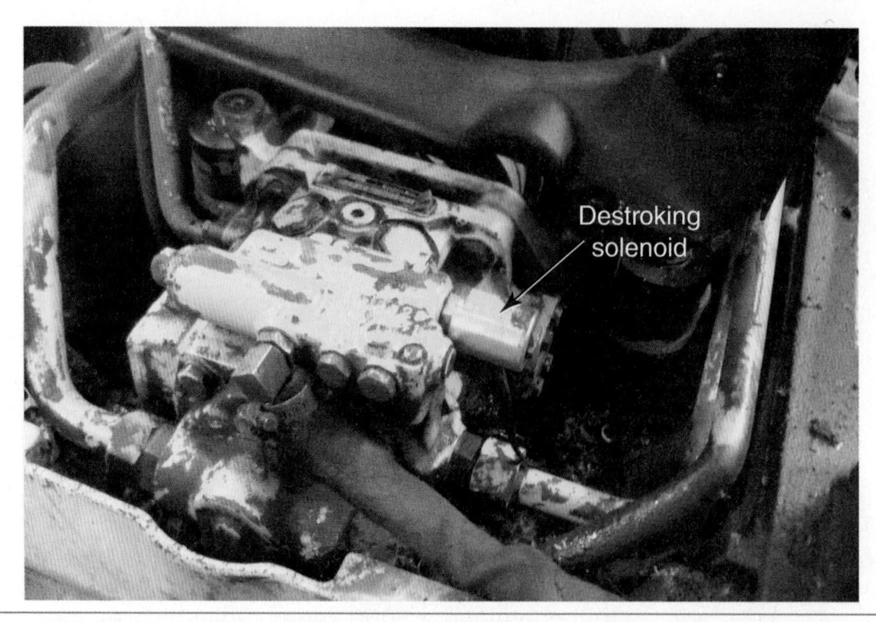

Figure 17-8. This John Deere 3000 series PC pump uses a destroking solenoid valve that holds the pump destroked during engine starting to ease the load on the starter.

Much time, care, and diagnosis was given to this portion of the pump control. The solenoid coil was physically removed from the solenoid's armature to eliminate the possibility of an electrical short destroking the pump. The tractor performed well for two days after the coil was removed from the solenoid's armature, and it was suspected that an electrical problem had been intermittently destroking the pump.

However, a few days later, the hydraulic system began to perform poorly once again. More time and energy was spent diagnosing the variable pump controls. The old pump was still accessible. Therefore, the pump controls were removed from the old pump and were installed on the new pump, but the backhoe continued to intermittently dig poorly. Multiple pressures were measured, including pump case pressure. The ***standby pressure***, specification was 2750 psi (190 bar) at 2000 engine rpm at operating temperature. Standby pressure is the high pressure setting that a PC pump maintains when the DCVs are in a neutral position and the pump is standing-by, waiting to perform work. The pump's standby pressure was low, varying between 900 psi (62 bar) and 1500 psi (103 bar).

Unfortunately, the service literature did not provide a specification for pump case pressure in the standby mode, even though the manual specified the location for measuring the pressure. The backhoe's case pressure measured approximately 150 psi (10 bar). Later, after the tractor was successfully repaired, it was determined that the pump case pressure was closer to 200 psi (14 bar) in the standby mode. Also note that this system was designed to have an unusually high pump case pressure. Many pump seals will leak if case pressure climbs above 15 to 30 psi (1 to 2 bar).

After multiple pressure measurements were taken, the technician thought about removing the pump and taking it to the dealership to be rebuilt. As it turns out, that would not have fixed the problem. The service manual, like most manuals, was not 100% comprehensive. It did not provide the critical step-by-step process needed to isolate the problem.

The customer mentioned a noise sometimes occurred at the back of the machine, but the pump was located at the front of the machine. After listening intently to the customer's description of the symptoms and analyzing the pump controls and hydraulic schematics, the technician dug further into the service literature and schematics. He began to suspect the system's main relief valve was dumping oil, acting like an open-center valve.

The main system relief valve was located at the back of the machine in the backhoe's valve block. Therefore, a flowmeter was installed at the pump's outlet, so that the pump could be isolated from all of the other valves. See Figure 17-9. **Chapter 22** will detail precautions about connecting a flowmeter to a hydraulic system and the risks associated with using flowmeters.

Note that service manuals do not always provide the easiest steps for installing a flowmeter, or provide multiple locations for installing a flowmeter. Also, note that installing a flowmeter can take a long time and might require the use of specially fabricated hoses or fittings. It is for that reason that technicians sometimes replace parts without a clear idea of the cause of the problem. Novice technicians often believe it is faster to simply replace parts than to install diagnostic test equipment that can properly diagnose a malfunctioning system. Unfortunately, this approach can result in unnecessary part replacements, delays in the machine's availability, and a drop in customer satisfaction.

Figure 17-9. A flowmeter was connected immediately after the PC pump. Two hoses were specially fabricated to enable the installation of the flowmeter.

Properly understanding the operation of PC systems enabled the customer and technician to diagnose and repair the problem. With the flowmeter installed and the DCVs in a neutral position, the flowmeter was measuring 20 gpm (76 lpm). Remember that standby pressure fluctuated, sometimes measuring as little as 900 psi (62 bar) and sometimes as high as 1500 psi (103 bar), even though the specification was 2750 psi (190 bar). The key to this problem is being able to answer the questions:

- What is a normal measurement of pump flow and pump pressure for a PC pump when the DCVs are in a neutral position?
- Should a PC pump be flowing 20 gpm (76 lpm)?
- Should standby pressure be 900 to 1500 psi (62 to 103 bar)?

PC systems should have practically no flow and maximum system pressure when the DCVs are in a neutral position. The flowmeter's load valve was slowly closed and, because there were no components between the pump and flowmeter, the pump's flow slowly dropped and standby pump pressure climbed, eventually reaching 2750 psi (190 bar) at 0 gpm (0 lpm).

As a result of installing the flowmeter, it was clearly determined that the pump compensator was performing as it should, and therefore the pump was not removed from the tractor. At this point, it was clear that, somewhere downstream of the pump, a valve was causing a leak in the hydraulic system. The main system relief valve was removed, inspected, and found to be faulty. Interestingly, a flashlight was placed at the end of the relief valve, and it was easy to see that the valve was not blocking off the light and that it was partially open, causing a leak. After a new main system relief valve was installed, the loader backhoe was put back into service and had excellent hydraulic performance. See **Figure 17-10** for a schematic of the tractor.

Note that many agricultural variable-displacement pumps commonly do not use a main system relief valve, but rely solely on the pump's compensator as the only method for relieving maximum system pressure. However, many construction machines that have variable-displacement pumps do have a main system relief valve. Such a valve should not open during normal operation, but is in the system as a backup in case the pump compensator fails.

Two other key points can be drawn from this case study. First, any time a technician is going to condemn a hydraulic pump, especially an expensive one, a flowmeter should first be used to isolate the pump and measure the pump's flow and pressure. The second important point is that it is imperative to know the fundamentals regarding what type of pressure and flow a PC pump should exhibit when the DCVs are in a neutral position. In this case, the technician found 20 gpm (76 lpm) and a standby pressure of 900 psi (62 bar). At first glance, the technician might think that the system was behaving like an open-center hydraulic system. This would have lead to an improper diagnosis of the problem. A technician should never jump to conclusions that a system is an open-center or a PC system by simply measuring pump flow and pressure. Correctly determine what style of hydraulic system the equipment uses before measuring pump flow and pressure.

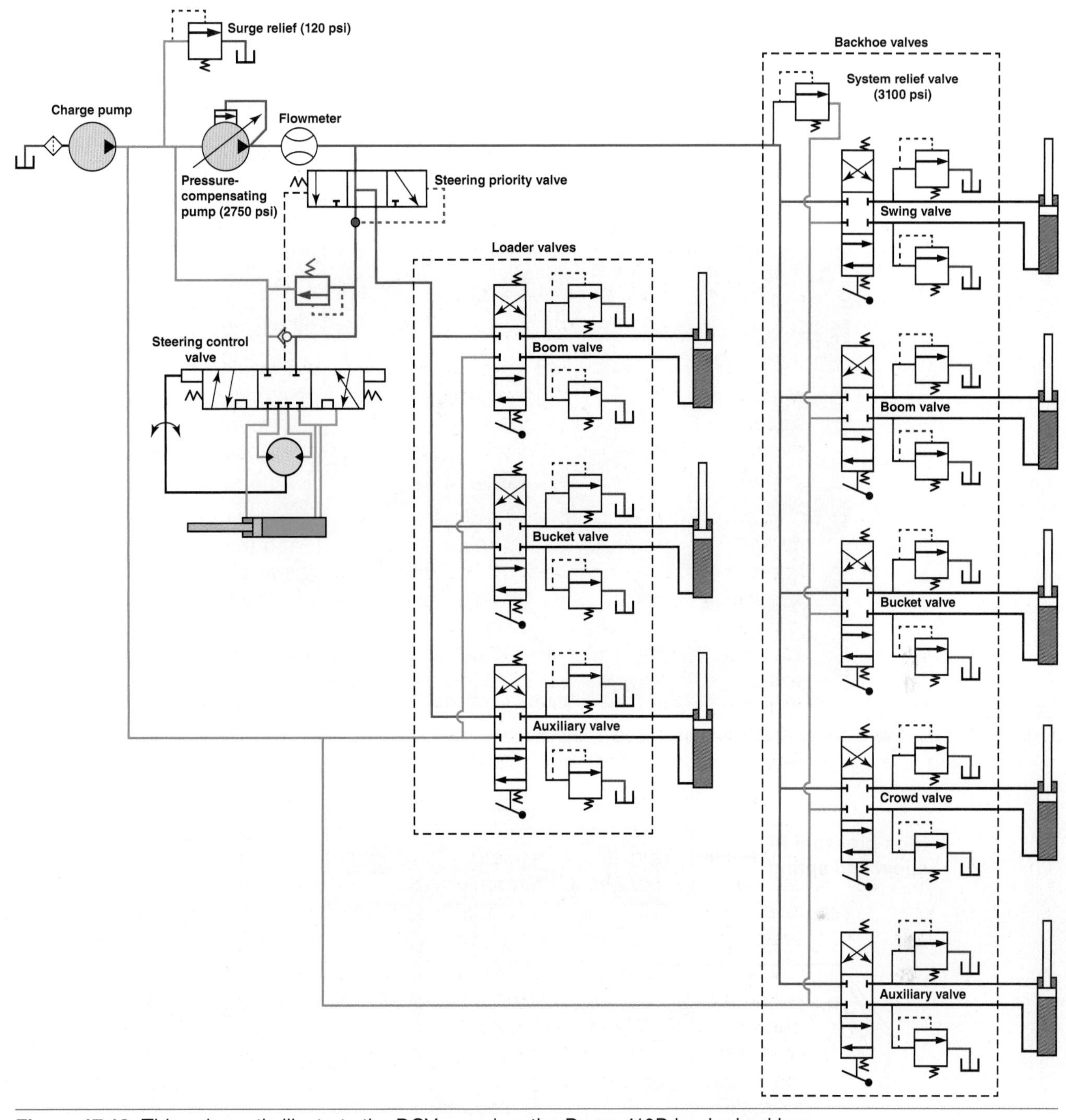

Figure 17-10. This schematic illustrate the DCVs used on the Deere 410D loader backhoe.

Inline Axial Piston Pump PC Systems

The pump compensator valve, as previously mentioned, can also be called a pressure limiter valve or ***pump cutoff valve***. These three different terms are used to describe the pressure compensator valve in inline axial piston pump applications. **Chapter 18** will explain that inline axial piston pumps used in LSPC systems continue to use the pressure compensating valve and have one or two additional control valves added to the pump: (1) flow control valve (also known as the margin spool) and, potentially, (2) a torque limiting spool.

When PC systems have an inline axial piston pump, the pump controls are simpler, using a single PC spool valve. See Figure 17-11. The compensator is set to the system's high pressure, for example 2700 psi (186 bar). See Figure 17-12. Once the system pressure reaches the value of the compensator spring value, the compensator spool shifts and directs oil to the control piston. **Chapter 4** explained that the control piston has the responsibility of shutting off the pump's flow by destroking the pump's swash plate to a neutral angle.

Figure 17-11. An inline axial PC piston pump uses a single spool valve to control the pump's flow.

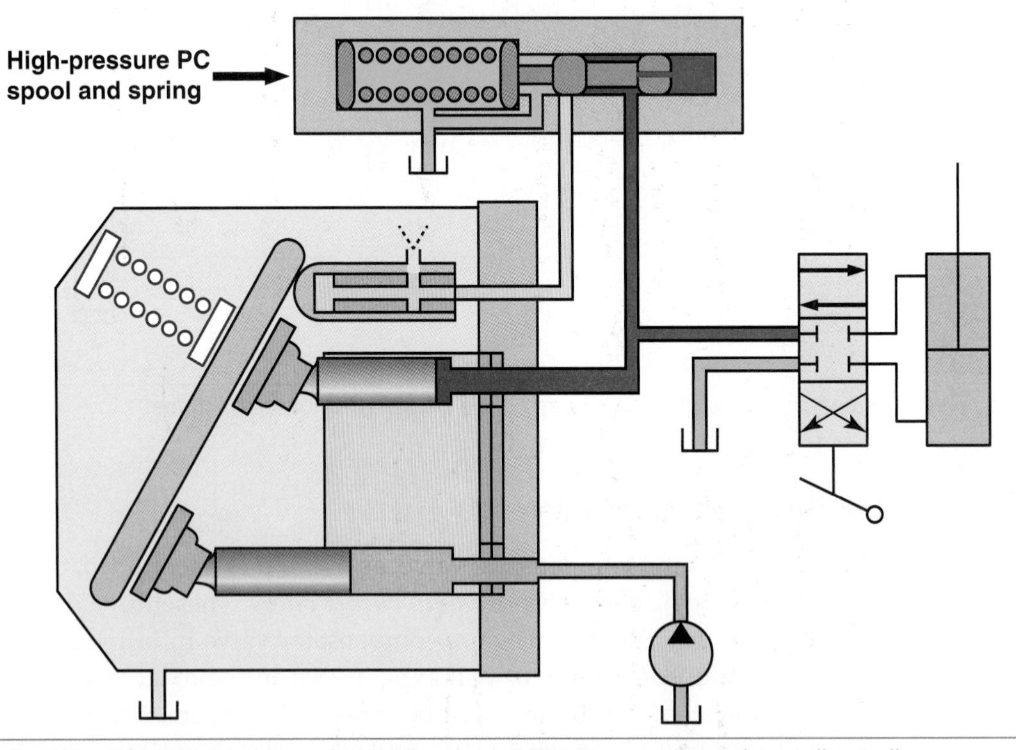

Figure 17-12. An inline PC piston pump used a single spool valve compensator valve to direct oil to a control piston to shut off the pump flow.

Older style inline axial pumps could be easily identified as PC or LSPC pumps based on whether they used one compensator spool or two. One spool indicated a PC pump, and two spools indicated a LSPC pump. However, in the next two chapters, you will learn that it is possible to have a LSPC pump with a single spool that uses an external signal limiter.

Many variable-displacement pumps are inline axial piston pumps. See Figure 17-13. Regardless if the inline axial piston pump uses PC, LSPC, or flow-sharing pump controls, the operation of the pump has some similarities. As explained in **Chapter 4**, when the pump's swash plate is at a maximum angle, known as ***upstroked***, the pump provides maximum oil flow. When the swash plate is close to a neutral angle, known as ***destroked***, the pump's flow is nearly shut off.

Most inline axial piston pumps use either a bias piston or a bias spring to upstroke the pump. Most pumps use a control piston to destroke the pump. The control piston receives the oil from the pump compensator. If the pump uses a bias piston and a control piston, the bias piston will have a smaller surface area than the control piston.

Note

It is possible for an external pressure compensator valve to be located in a remote location, away from the pump. Some manufacturers call this application ***remote pressure compensation***. Remote pressure compensators are rarely used in mobile machinery. They are used in industrial applications that require frequent adjustments to the pressure compensator or for applications that must have the ability to adjust the compensator away from the pump.

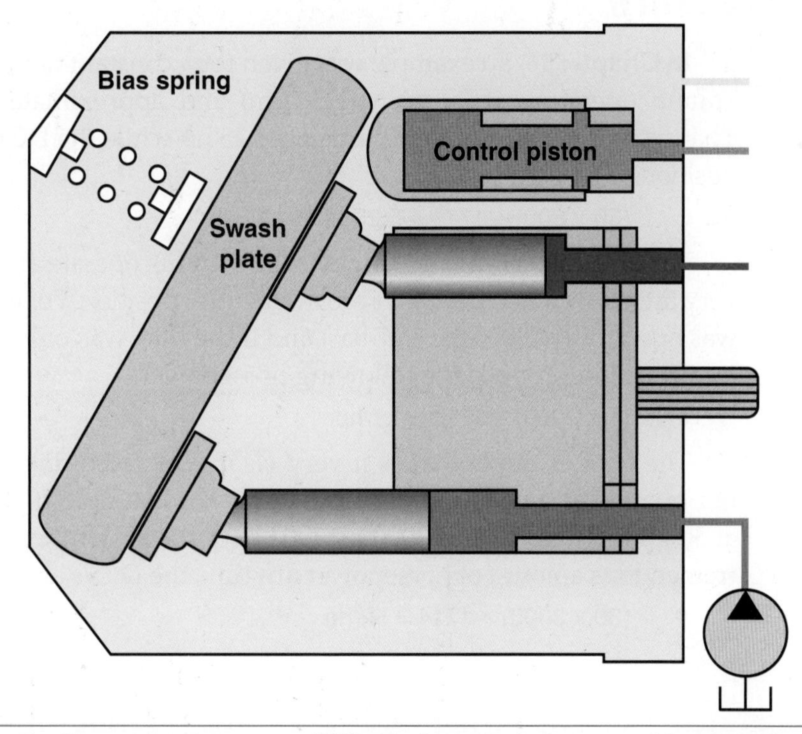

Figure 17-13. The swash plate angle of a pump will dictate the pump's flow rate. A large angle equals more flow. The closer the angle is to neutral, the less flow the pump will be producing.

Advantages and Disadvantages of PC Systems

Pressure-compensating systems are the simplest closed-center hydraulic system, because the variable-displacement pump does not require any type of feedback, such as a signal line or a network of shuttle valves. It does not need to sense the specific load at the cylinders. The pump can automatically upstroke or destroke by simply sensing its own outlet pressure.

The main system hydraulic pressure will vary based on system's load and how much oil is requested by the operator, which equals the amount of opening across the DCV.

Since PC systems are designed to operate at a constant high pressure, the hydraulic systems are very responsive. In 1960, when John Deere released the 10 series tractors, farmers were impressed with the hydraulic performance. It would be hard to design a hydraulic system that is more responsive than a system that is constantly running close to 2700 psi (186 bar). Technically, John Deere sprayers operate at high system pressures, but they use a LSPC pump and a jammer solenoid to maintain the high system pressure, which will be explained in the next chapter.

The constant high pressure within the hydraulic system also has drawbacks. It produces constant disturbing noise, generates unnecessary heat, always has a higher risk of fluid leakage, and wastes horsepower. Late-model variable-displacement controls solve these problems and will be discussed in the following chapters.

Why Open-Center Hydraulic Systems Do Not Operate at Maximum Pressure

In **Chapter 16,** an example was given for a constant flow, fixed-displacement pump operating at 30 gpm (113 lpm) and approximately 300 psi (21 bar). That system consumed approximately 5.25 hp while the DCVs were in a neutral position.

$$(30 \times 300) \div 1714 = 5.25 \text{ hp}$$

Variable-displacement PC hydraulic systems operate at a high pressure and very little flow when the DCVs are in a neutral position. For example, if the pump was operating at 3000 psi (207 bar) and if the flow was only 1 gpm (3.7 lpm), the pump would produce the following horsepower.

$$(1 \times 3000) \div 1714 = 1.7 \text{ hp}$$

The next example makes it very clear why fixed-displacement pumps do not operate at a constant high pressure. For example if the system operated at 30 gpm (113 lpm) and 3000 psi (207 bar), the system would be consuming a tremendous amount of horsepower anytime the DCVs are in a neutral position.

$$(30 \times 3000) \div 1714 = 52 \text{ hp}$$

Summary

- ✓ John Deere is well known for producing PC hydraulic systems from 1960 to 1994.
- ✓ PC systems operate at a constant high pressure and practically no flow when the DCVs are in a neutral position.
- ✓ Deere PC systems used a rotating cam radial-displacement piston pump that was driven off the front of the engine.
- ✓ Deere used three different styles of PC pumps: 1000, 2000, 3000 series pumps.
- ✓ In a rotating cam PC pump, when oil pressure is directed to the pump's case, the pump will be destroked. When the oil is drained from the pump's case, the pump will be upstroked.
- ✓ An inline axial PC piston pump uses a single spool valve for its pump compensator.
- ✓ PC systems can also be called pressure-limiting systems.
- ✓ The compensator valve can also be called the pump cutoff valve.
- ✓ Not all machines that use a variable-displacement pump will have a main system relief. Some systems rely solely on the pump controls, such as the pressure compensator valve.

Technical Terms

crankcase outlet valve
destroked
destroking solenoid valve
oil gallery
pressure-limiting hydraulic system
pump compensator
pump cutoff valve
remote pressure compensation
standby pressure
stroke control valve
upstroked

Review Questions

Answer the following questions using the information provided in this chapter.

1. Pressure-compensated hydraulic systems use what style of DCV?
   A. Open center.
   B. Closed center.
2. A pressure-compensated hydraulic system will exhibit _____ when the DCVs are in a neutral position.
   A. practically no flow
   B. moderate amount of flow
   C. high flow
   D. fluctuating flow
3. A pressure-compensated hydraulic system will exhibit _____ when the DCVs are in a neutral position?
   A. low pressure
   B. a moderate amount of pressure
   C. high pressure
   D. vacuum
4. John Deere used _____ variable-displacement pump control from 1960–1994 on their 10–60 Series Ag tractors.
   A. open center
   B. pressure-compensated
   C. LSPC
   D. None of the above.

5. John Deere tractors from 1960–1994 used a specific style of hydraulic system. What series of pumps were used in these systems?
   A. 500 and 1000 series pumps.
   B. 1000, 2000, and 3000 series pumps.
   C. 2000 and 4000 series pumps.
   D. 5000 and 10,000 series pumps.
6. Where did John Deere commonly place the PC pump?
   A. At the front of the machine.
   B. At the back of the machine.
   C. In the middle of the machine.
   D. At the highest point on the machine.
7. What type of variable-displacement hydraulic pump did John Deere use on their 10–60 series Ag tractors?
   A. Gear.
   B. Vane.
   C. Radial piston.
   D. Axial piston.
8. On older John Deere pressure-compensated pumps, what is used to shut off pump flow?
   A. Case pressure.
   B. Control piston.
   C. Bias piston.
   D. Servo piston.
9. In an inline axial piston pump, the *primary* responsibility of the _____ is to shut off pump flow.
   A. bias spring
   B. control piston
   C. swash plate
   D. relief valve
10. In an inline axial piston pump, the *primary* responsibility of the _____ is to turn the pump flow on.
    A. bias spring
    B. control piston
    C. swash plate
    D. relief valve
11. If a pump is equipped with a destroke solenoid, when is the solenoid actuated?
    A. When a cylinder reaches end of travel.
    B. When the engine is starting.
    C. Anytime the engine is at idle.
    D. When the operator presses the destroke switch.
12. The pressure compensator within an inline axial piston pump can also be called a _____.
    A. pump cutoff
    B. pressure limiter
    C. Both A and B.
    D. Neither A nor B.
13. Which of the following hydraulic systems can be described as having the following disadvantages: noisy, more prone to have an external leak, and wastes horsepower?
    A. Fixed-displacement pump.
    B. PC.
    C. LSPC.
    D. LSPC with torque limiting.
14. An inline axial variable-displacement pump that is used in a pressure-compensating hydraulic system has how many control spools (used for controlling the pump flow and pressure)?
    A. Zero.
    B. One.
    C. Two.
    D. Three.
15. Which of the following types of hydraulic systems does not require a signal network?
    A. Pressure compensating.
    B. LSPC.
    C. LSPC and torque limiting.
    D Flow sharing.
16. Hydraulic pumps that are used in closed-center systems will have what?
    A. Constant flow.
    B. Variable flow.
    C. Variable pressure.
    D. None of the above.

17. Which of the following hydraulic systems provides the advantage of being the most responsive?
    A. Open-center.
    B. PC.
    C. LSPC.
18. Technician A states that all PC hydraulic systems have a main system relief valve. Technician B states that many construction machines have a main system relief in conjunction with a variable-displacement pump. Who is correct?
    A. Technician A.
    B. Technician B.
    C. Both A and B.
    D. Neither A nor B.
19. Which of the following is a reasonable example of standby pressure in a Deere PC hydraulic system?
    A. 150 psi (10 bar).
    B. 450 psi (31 bar).
    C. 1500 psi (103 bar).
    D. 3000 psi (207 bar).
20. When a rotating cam PC piston pump has its case pressure drained, what is the status of pump flow?
    A. Upstroked.
    B. Destroked.
21. Regarding the 410D 3000 series pump, which one of the following was responsible for reducing pressure spikes?
    A. Crankcase relief valve.
    B. Pressure compensator valve.

This John Deere 6R series agricultural tractor uses a variable-displacement load-sensing pressure-compensating hydraulic system. Although more expensive to manufacture than other hydraulic system designs, variable-displacement LSPC systems are generally more efficient.

Chapter 18

Load-Sensing Pressure-Compensating (LSPC) Hydraulic Systems

Objectives

After studying this chapter, you will be able to:

- ✓ List the two different types of pre-spool-compensated LSPC systems.
- ✓ Explain the operation of a fixed-displacement LSPC hydraulic system.
- ✓ Describe the operation of a signal network.
- ✓ Define load sensing.
- ✓ Explain the two different meanings of pressure compensation in an LSPC system.
- ✓ List examples of fixed-displacement LSPC systems in mobile machinery.
- ✓ Explain the similarities between unloading valves and flow control valves; and pressure relief valves and pressure compensator valves within LSPC systems.
- ✓ Explain the operation of a variable-displacement LSPC hydraulic system.
- ✓ Describe examples of relief valves and the absence of relief valves in LSPC systems.
- ✓ Explain the benefit and operation of torque-limiting controls.
- ✓ Describe the application of jammer solenoids in variable-displacement LSPC systems.
- ✓ Explain what equipment is needed to add power beyond in an LSPC system.
- ✓ List all three items that must receive oil from the primary shuttle valve when a DCV uses a detent kick-out.
- ✓ List the advantages and disadvantages of a variable-displacement LSPC system.

Introduction to Load-Sensing Pressure-Compensating (LSPC) Hydraulic Systems

The previous chapter focused on pressure-compensating hydraulic systems, which use a variable-displacement hydraulic pump that maintains high pressure and practically no flow anytime the DCVs are in a neutral position. LSPC systems use closed-center DCVs and one of two hydraulic pump designs:

- Variable-displacement with variable flow.
- Fixed-displacement with constant flow.

In order to fully understand LSPC systems, it is helpful to break the system into two separate operations: load sensing and pressure compensation. This chapter will begin by explaining load sensing and pressure compensation separately in order to build a foundation to better understand the principles of load-sensing pressure-compensating (LSPC) systems.

Load-Sensing System

In Figure 18-1, a closed-center DCV valve is used with a fixed-displacement pump. In this scenario, anytime the DCV is in a neutral position, maximum hydraulic horsepower will be generated by constantly flowing oil over the main system relief valve at a high system pressure. As a result, the system will generate unnecessary heat, pressure, and noise. This is one reason this style of hydraulic system is not used in machinery today.

A ***load-sensing (LS) hydraulic system*** is designed to operate at a prescribed pressure value above the highest working pressure. See Figure 18-2. Two items have been inserted into the previous closed-center system schematic in order to convert the hydraulic system into a LS hydraulic system: an unloading valve and a shuttle valve. The term *unloading valve* means that the valve dumps the hydraulic pump's flow to the reservoir at a set low-pressure value anytime the closed-center DCV is in the neutral position. In this case, the unloading valve's spring value is set at 300 psi (21 bar).

When the DCV is actuated, the shuttle valve senses the load's working pressure and sends that signal pressure to the unloading valve. If the cylinder requires 1000 psi (69 bar) to extend, the shuttle valve sends that 1000 psi (69 bar) signal pressure to work in conjunction with the unloading valve's spring pressure. The combination of the spring pressure and signal pressure causes the system to operate at 1300 psi (90 bar). The difference between pump outlet pressure and signal pressure is known as ***differential pressure*** or ***margin pressure***. Margin pressure as it relates to variable-displacement LSPC systems is explained later in this chapter.

Note

The unloading valve's spring can be called the differential spring, margin spring, standby spring, low-pressure standby spring, or the dump valve spring.

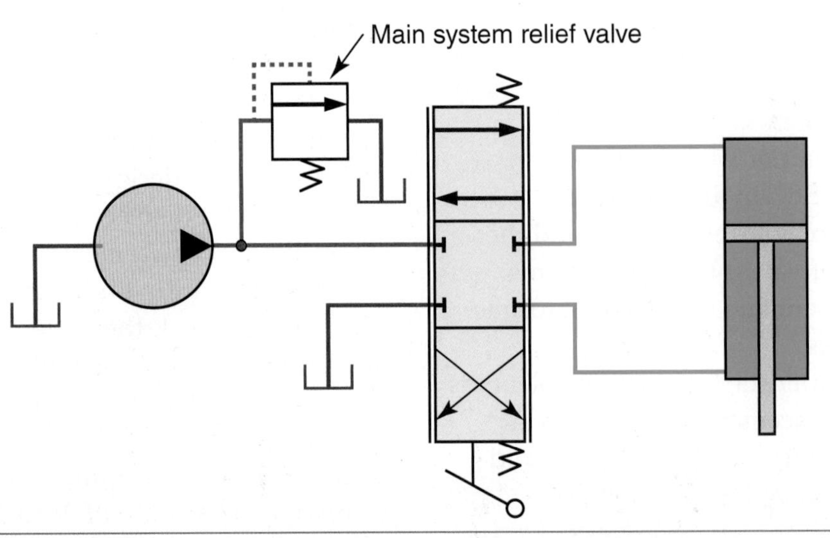

Figure 18-1. The fixed-displacement pump in this schematic operates at a constant high pressure. It wastes too much energy and generates excess heat when the DCV is moved to a neutral position.

At least one gear pump supplier, Concentric (formerly Haldex), manufactures a LS gear pump that places the unloading valve inside the pump housing. They advertise that this design has the pump directly controlling the flows and pressures, which will reduce valving losses and circuit inefficiencies.

Some individuals might be tempted to call the hydraulic system in Figure 18-2 an open-center hydraulic system. When the DCV is in the neutral position, the fixed-displacement pump delivers a maximum amount of flow at a relatively low pressure of 300 psi (21 bar). Technically, the DCV is closed center and the DCV sends a signal to an unloading valve. The hydraulic system is therefore more accurately described as a constant flow closed-center LS hydraulic system. This style of hydraulic system still wastes hydraulic horsepower when the DCV is in the neutral position. In this case, if the gear pump is delivering a constant 30 gpm (113.5 lpm) of flow at 300 psi (21 bar), the system will consume a certain amount of engine horsepower:

$$30 \text{ gpm } (113.5 \text{ lpm}) \times 300 \text{ psi } (21 \text{ bar}) \times 0.000583 = 5.2 \text{ hp}$$

This style of hydraulic system is still found in mobile machinery and examples of machines using this design are listed later in this chapter.

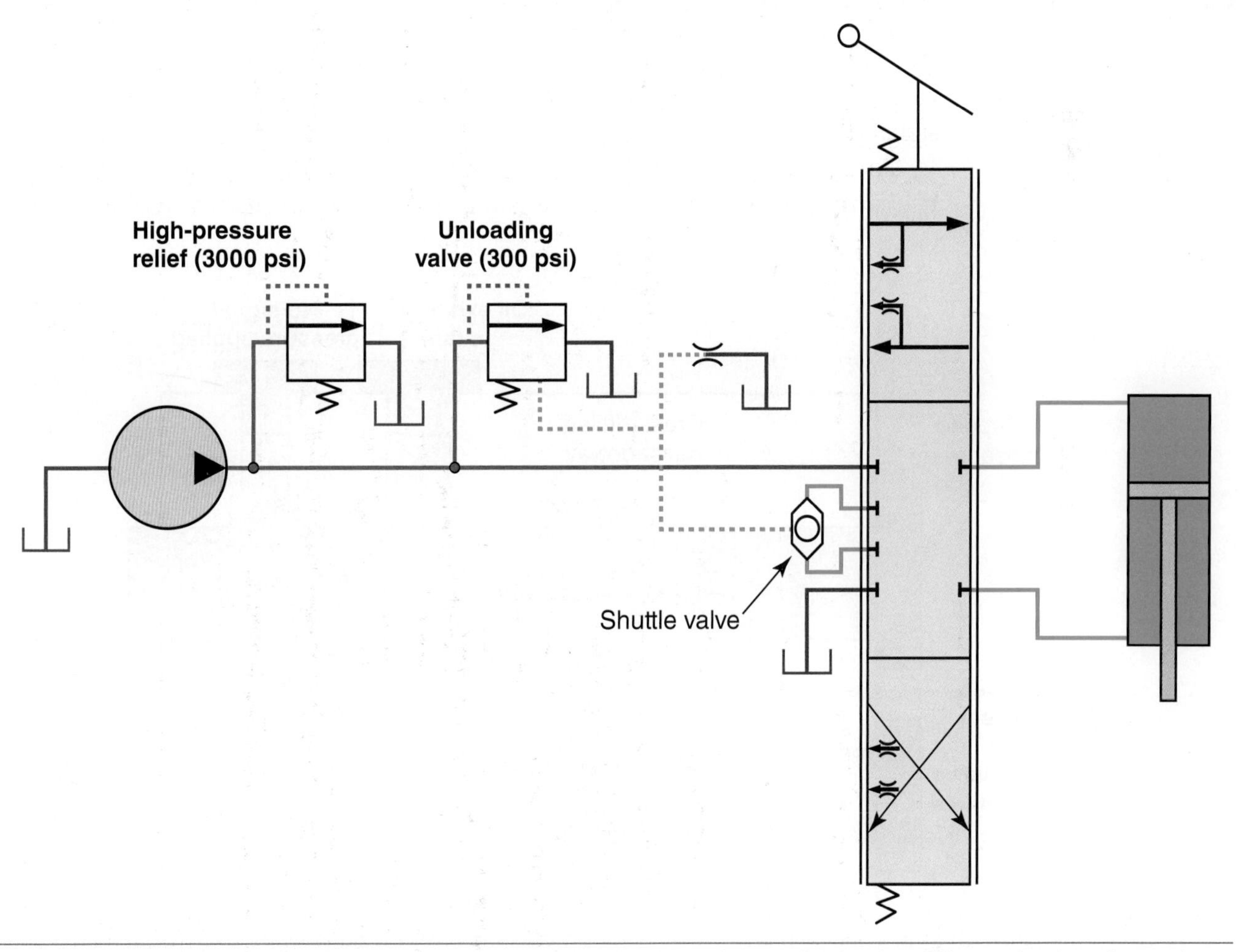

Figure 18-2. A fixed-displacement pump can be used in a load-sensing application by adding a shuttle valve and an unloading valve. The unloading valve keeps the pump operating at a set low system pressure when the DCV is in a neutral position.

Fixed-Displacement Pump Standby Mode

Figure 18-3 illustrates a cross-sectional drawing of an unloading valve that is used in a fixed-displacement LS hydraulic system. The spool valve has a cross-drilled passageway, which enables pump pressure to act on the left-hand side of the spool. Pump pressure opposes the margin spring. When the DCV is in the neutral position, signal pressure will equal zero. **Chapter 25** will explain that some steering circuits will have a residual steering signal pressure that will cause the standby pressure to be a little higher than the margin spring value.

Pump pressure will build to achieve the margin spring value—300 psi (21 bar) in this example—causing the spool to shift to the right, compressing the margin spring. As the spool shifts to the right, it opens the passageway allowing the pump's constant flow of oil to be dumped to the reservoir at the margin spring value of 300 psi (21 bar). While the DCV is in neutral, the system is in the ***standby mode***, also called low-pressure standby. LS systems normally operate at a pressure between 300 to 500 psi (21 to 34 bar) while in the standby mode. Standby mode as it relates to variable-displacement LSPC systems will be explained later in this chapter.

Fixed-Displacement Pump Working Mode

When a DCV is actuated, its shuttle valve directs the hydraulic cylinder's working pressure, also known as ***signal pressure***, to the unloading valve so that the signal pressure can work in conjunction with the margin spring. See Figure 18-4.

Anytime an actuator is moving, the hydraulic system is in the ***working mode***. While in the working mode, hydraulic pump outlet pressure will equal the

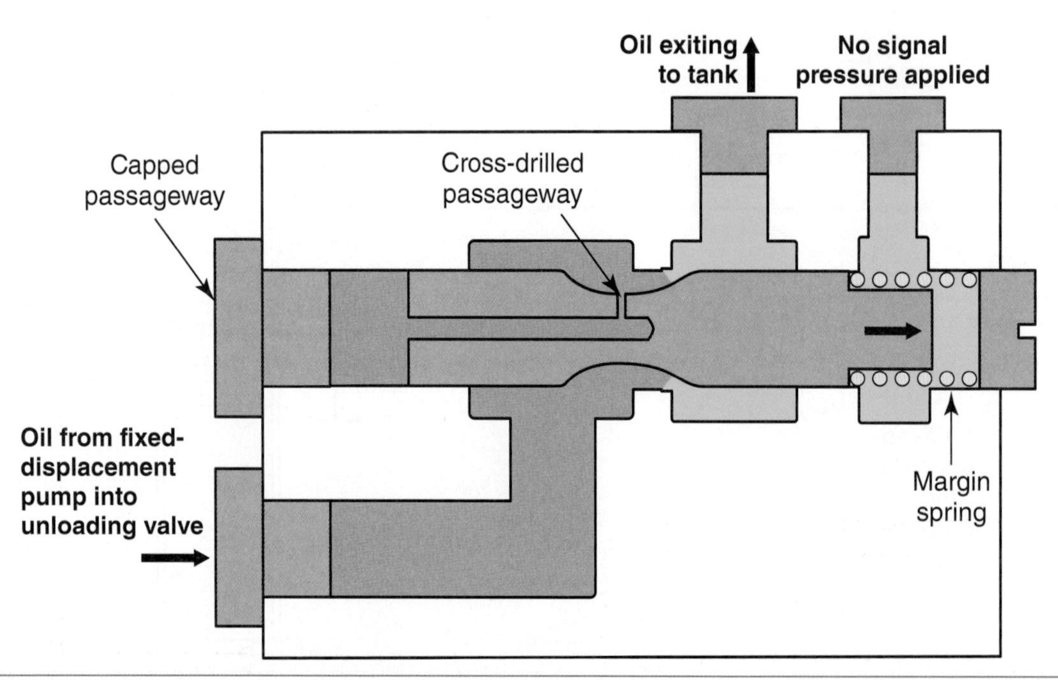

Figure 18-3. When the DCV is in a neutral position, the unloading valve receives no signal pressure. The pump will build enough pressure, such as 300 psi (21 bar), to overcome the unloading valve's margin spring pressure. This causes the unloading spool valve to shift to the right, allowing the constant flow of oil to be dumped to the reservoir at a relative low pressure value.

signal pressure plus the margin pressure. For example, if the cylinder requires 1000 psi (69 bar) to extend, and the margin pressure equals 250 psi (17 bar), the pump outlet pressure will equal 1250 psi (86 bar).

Differential Spring Value

A margin spring has a static spring pressure value, known as standby pressure, and a dynamic spring pressure value, known as margin pressure. The standby pressure is usually a higher value than the margin pressure. Some service literature will only specify adjusting the standby pressure, while other literature will specify only adjusting the margin pressure. Some service manuals will provide specifications for adjusting both the standby and the margin pressure. This topic will be discussed in more detail later in this chapter.

Referring to **Figure 18-4**, the function of the LS system's working mode is the heart of LS hydraulics. The system builds only the necessary pressure to do the work, plus the set pressure value of the margin spring.

Stall Pressure

When the cylinder reaches its end of travel, the LS system will be in its final mode, called the ***stall mode***. It is also called the high-pressure cutoff mode or high-pressure standby mode. In this mode, the hydraulic pump cannot build enough pressure to overcome both the signal pressure and the margin spring. As a result, the margin spool remains closed, blocking off pump flow, which causes the system pressure to climb to the highest system pressure value. In a fixed-displacement system, a high-pressure relief valve is used to set the pump's stall pressure.

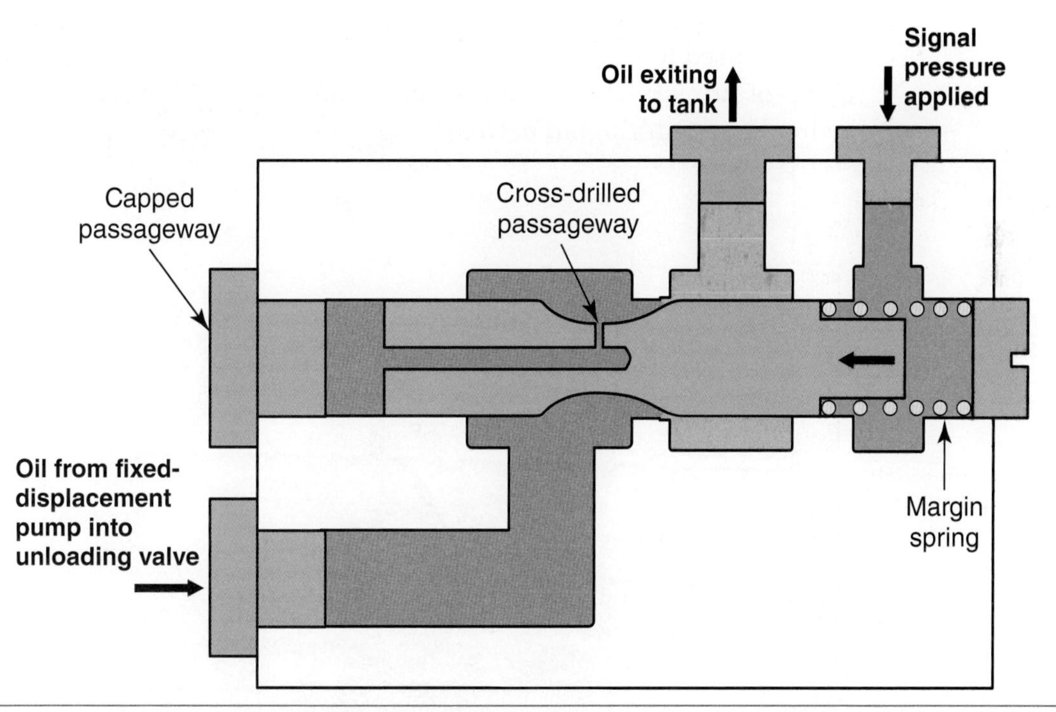

Figure 18-4. In the working mode, the DCV's shuttle valve sends the working signal pressure to the unloading valve. The system pressure builds only to the necessary pressure value, which equals signal pressure plus margin pressure. The remaining pump flow that is not used by the DCV is dumped over the unloading valve back to the reservoir.

LS systems can also use variable-displacement pumps. Systems using variable-displacement LS pumps are even more efficient than fixed-displacement LS systems. In variable-displacement LS systems, the unloading valve becomes the flow control valve or the margin spool. These systems are described later in the chapter.

Signal Network

LS hydraulic systems require a group of shuttle valves to sense actuator working pressure. As mentioned in **Chapter 9**, a shuttle valve is a T-shaped valve that determines the higher pressure of two inlet pressures and sends the higher pressure to a new destination within the system. See **Figure 18-5**. That destination can be another shuttle valve, a pump compensator assembly, or an unloading valve.

Primary Shuttle Valve

Shuttle valves can be designated as primary or secondary shuttle valves. A ***primary shuttle valve*** chooses an actuator's higher working pressure. This means the primary shuttle determines if the double-acting cylinder is extending or retracting based on the higher working pressure and sends the appropriate signal pressure to a secondary shuttle valve, as shown in **Figure 18-6**.

Secondary Shuttle Valve

A ***secondary shuttle valve*** is used to choose the higher working pressure between two different DCVs' working pressures. A series of secondary shuttle valves is used to send the highest working pressure to the margin spool. The group of primary shuttle valves and secondary shuttle valves makes up the hydraulic system's ***signal network***.

Figure 18-5. Most LS DCVs have the shuttle valves incorporated inside the DCV's valve block. Individual shuttle valves are also available for building a LS system when using simple closed-center DCVs.

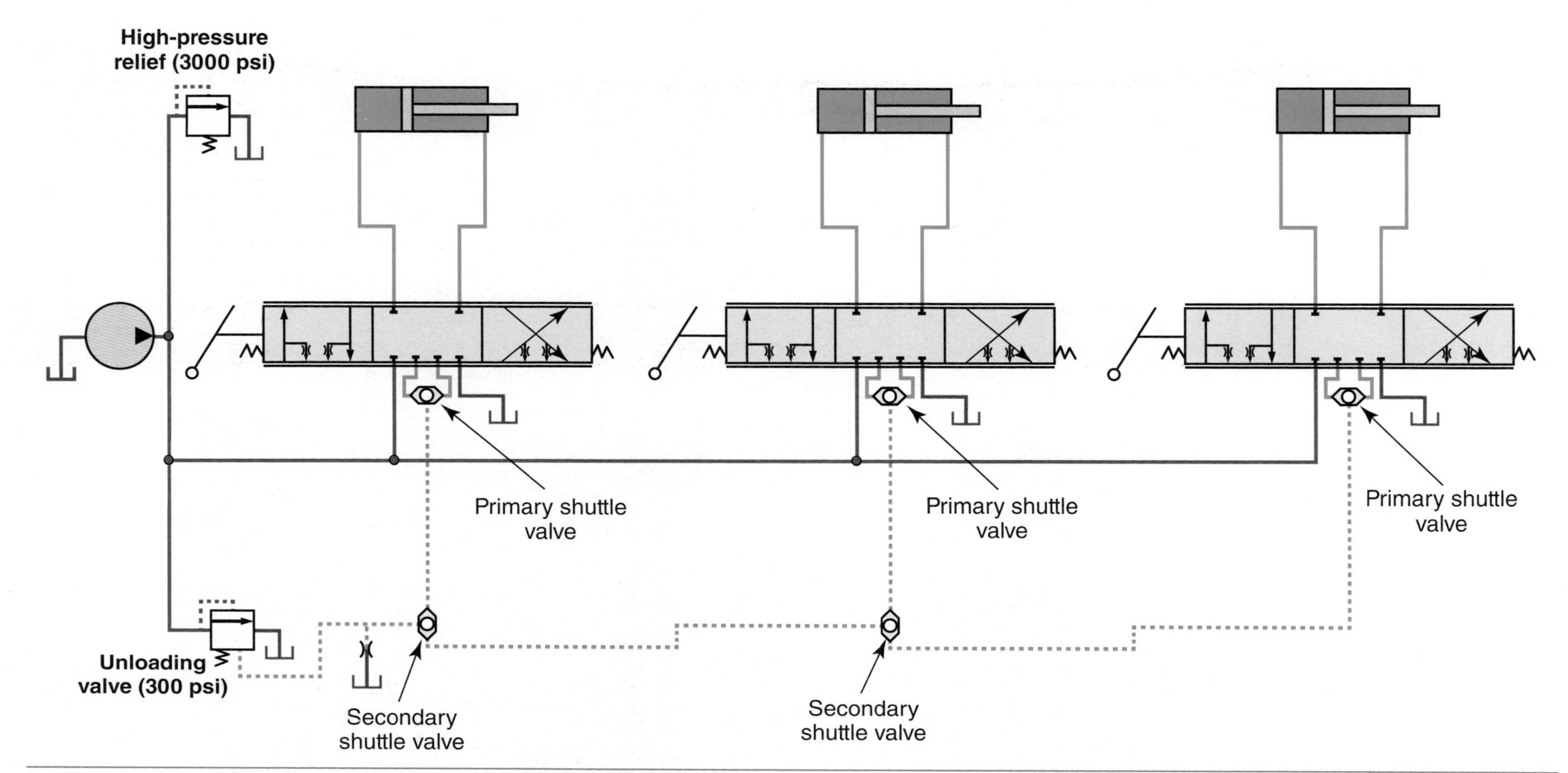

Figure 18-6. The primary shuttle valves in this schematic determine if the cylinder it monitors is extending or retracting and send the higher signal pressure to a secondary shuttle valve. The secondary shuttle valve chooses which DCV is operating at the higher pressure and sends that signal pressure to another secondary shuttle valve or to the unloading valve, depending on the design of the signal network.

For example, refer to Figure 18-7 and consider the following actuator requirements:

- DCV 3 requires 1500 psi (103 bar) to retract its cylinder.
- DCV 2 requires 1000 psi (69 bar) to extend its cylinder.
- DCV 1 requires 500 psi (34 bar) to extend its cylinder.

The following sequence of events would occur if the operator requested the previous actions from the cylinders:

1. The primary shuttle inside DCV 3 would direct 1500 psi (103 bar) to DCV 2 secondary shuttle.
2. The primary shuttle in DCV 2 would direct 1000 psi (69 bar) of signal pressure to the DCV 2 secondary shuttle.
3. The secondary shuttle in DCV 2 would choose the higher working pressure of 1500 psi (103 bar) and send it to the secondary shuttle located in DCV 1.
4. The primary shuttle in DCV 1 would direct 500 psi (34 bar) to the secondary shuttle inside DCV 1.
5. The secondary shuttle in DCV 1 would choose 1500 psi (103 bar) and send the 1500 psi (103 bar) to the unloading valve.
6. The unloading valve would maintain a system pressure of 1500 psi (103 bar), plus the value of margin pressure, 300 psi (21 bar).

Note that DCV 3 in Figure 18-7 does not contain a secondary shuttle valve. The last DCV does not have to distinguish the difference in signal pressure of a downstream valve that does not exist. However, if a customer wanted to add another DCV (such as DCV 4, power beyond, or a three-point hitch), a secondary shuttle valve must be installed inside DCV 3.

The signal network must have an orifice that allows the network to have a controlled drain back to the reservoir. If the signal network has no controlled drain back to tank it can cause one of two problems, depending on whether the system uses a variable-displacement pump or a fixed-displacement pump. In a fixed-displacement LS hydraulic system, a plugged LS network orifice will cause the signal network to hydrostatically lock, resulting in the unloading valve blocking system oil flow. When the unloading valve blocks off the fixed-displacement pump, the constant flow of oil must dump across the main system relief at high pressure. In a variable-displacement LS hydraulic system, a plugged orifice in the LS network will cause the variable-displacement pump to achieve high-pressure standby. This mode will be discussed later in the chapter.

Note

In John Deere agricultural machines, the shuttle valves are called ***dime valves*** because they resemble the shape of a dime, as shown in Figure 18-8. The shuttle valves are called ***isolators*** in John Deere construction machines. Caterpillar calls their shuttle valves ***resolvers*** because the shuttle valves resolve which pressure is the higher pressure.

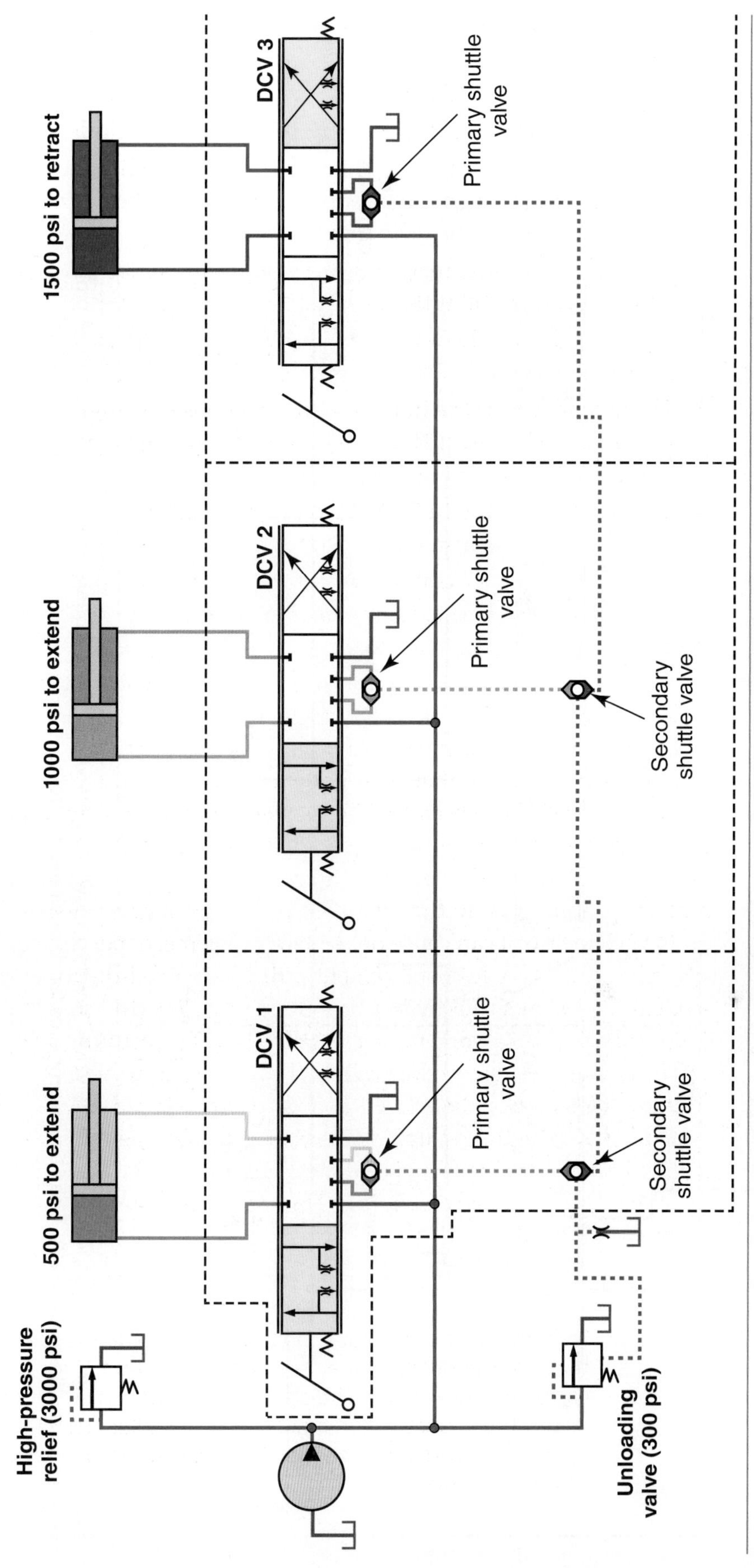

Figure 18-7. The initial secondary shuttle valve in a signal network sequence chooses the higher working pressure between its DCV and the working pressure it receives from the downstream DCV(s). It sends the higher signal pressure to the unloading valve, which maintains system pressure at 1800 psi (124 bar).

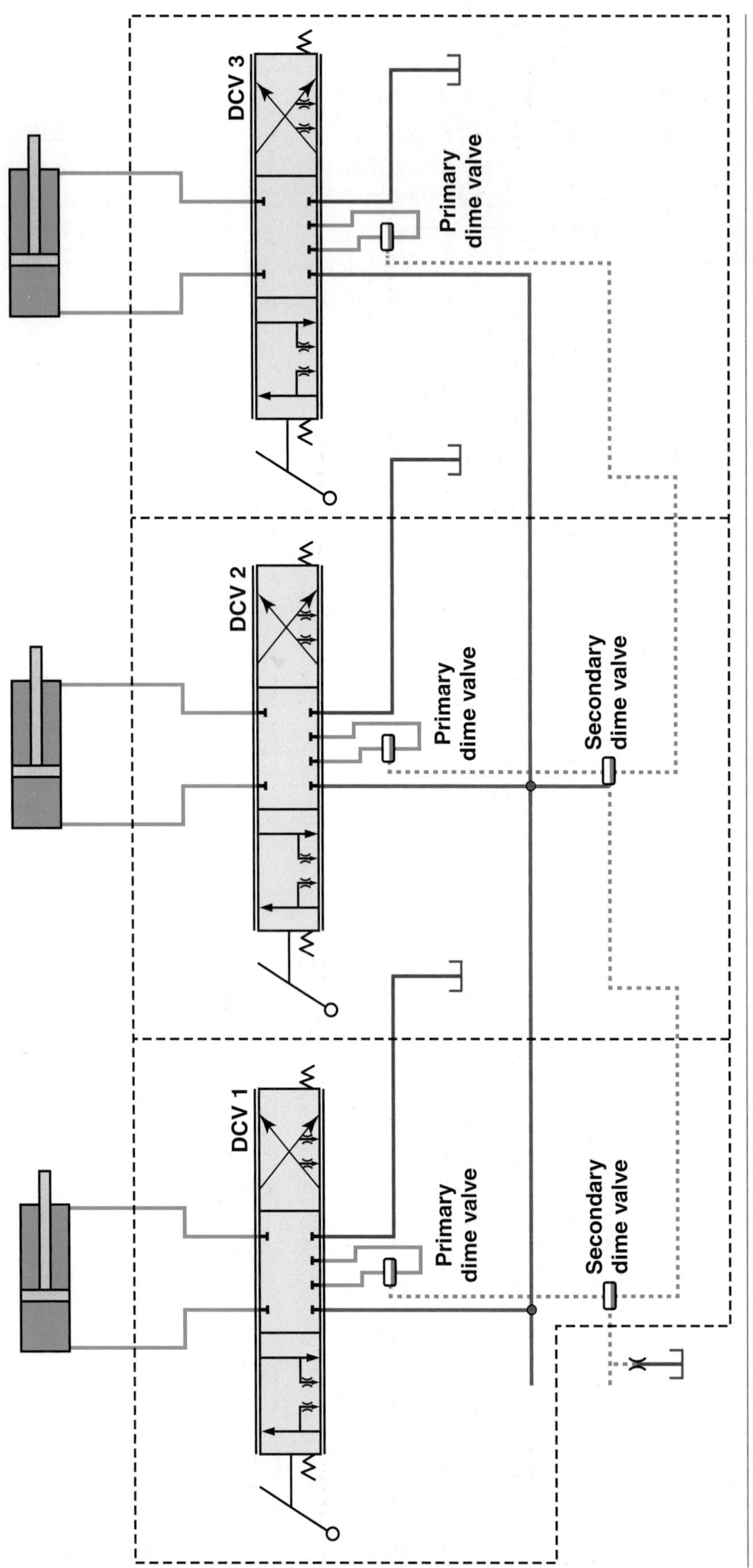

Figure 18-8. Due to their appearance, shuttle valves are also known as dime valves in John Deere agricultural equipment. A problem with an upstream secondary shuttle valve in DCV 1 can cause problems for downstream DCVs.

Diagnosing Secondary Shuttle Valves

Shuttle valves can cause malfunctions in hydraulic systems if they become stuck in one position. If a technician is faced with a downstream DCV that will not flow oil, it is possible that an upstream secondary shuttle valve is at fault. For example, in **Figure 18-8**, if the secondary shuttle valve in DCV 1 is stuck in the lifted position, it would cause problems downstream for DCV 2 and DCV 3. Initially, a technician might think that DCV 2 and DCV 3 are suspect because those are the valves having trouble flowing oil. However, the problematic shuttle valve is actually located upstream in DCV 1.

Load-Sensing System with Single-Acting Actuator

Figure 18-9 illustrates an LS system with one single-acting actuator, a unidirectional hydraulic motor. Note that neither a primary shuttle valve nor a secondary shuttle valve is necessary. Primary shuttles are required for choosing the higher working pressure of a double-acting actuator, for example choosing between the working pressures at the rod end and cap end of a double-acting cylinder. A single-acting actuator only has one possible working pressure. In this example, it is the hydraulic motor's forward rotation. If the hydraulic system contained multiple single-acting actuators, secondary shuttles would be required to choose the highest system working pressure among the multiple actuators so it could be directed to the unloading valve.

Pressure-Compensating System

Most LS hydraulic systems are also pressure compensating (PC). The combination of LS and PC can be called any one of the following names, depending on the manufacturer:

- LSPC—Load-sensing pressure compensation system (Caterpillar).
- PCLS—Pressure compensation load-sensing system (John Deere's construction division).

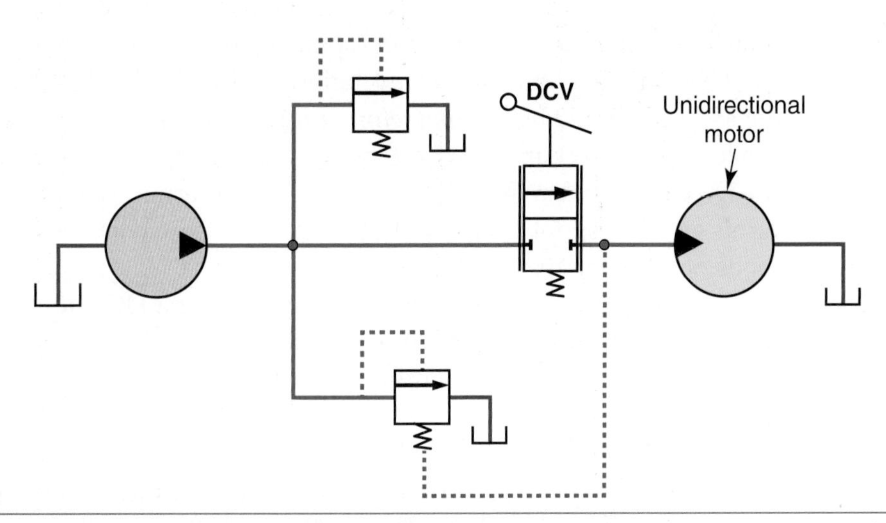

Figure 18-9. An LS system with one single-acting actuator does not require the use of a primary shuttle valve or a secondary shuttle valve.

- PFC—Pressure and flow compensated system (Case IH, Eaton, and John Deere's agricultural division).
- CCLS—Closed-center load-sensing system (New Holland Agriculture and AGCO).

The most problematic of the five descriptors is PFC. Although Case IH, John Deere agricultural equipment, and Eaton have all used PFC to describe a load-sensing pressure-compensating hydraulic system, there is another type of PFC hydraulic system known as positive flow control which is completely different than an LSPC system. Positive flow control hydraulic systems will be explained in **Chapter 20**. LSPC is the term adopted for this textbook. The term *pressure compensation*, in the context of an LSPC system, can have two different meanings.

Pressure-Compensating Hydraulic Pump

The first PC description is based on the pressure compensator valve located inside a variable-displacement pump, which was the focus of **Chapter 17**. The pump's PC spool operates in the same manner as described in **Chapter 17**. Within an LSPC variable-displacement pump, the PC spool has the responsibility to destroke the pump when a hydraulic cylinder reaches its end of travel. The PC spool maintains that constant high pressure until the DCV is returned to the neutral position.

Pressure-Compensated DCV

The second description for pressure compensation in an LSPC system focuses on a pressure compensator valve inside the DCV, which was briefly mentioned in **Chapter 8**. The valve is also called a flow compensator valve. It is essentially a pressure-reducing valve that senses working pressure. The valve uses this signal pressure to compensate the DCV's flow.

Without the compensator valve, anytime the system's flow or pressure changes, the operator would have to readjust the position of the DCV spool to attempt to maintain the same implement speed. In this context, *pressure compensation* is used to maintain a constant actuator speed (cylinder or hydraulic motor) based on a fixed position (opening) of the DCV spool.

The PC valve can be placed before the DCV spool, which is known as *pre-spool compensation* or ***upstream compensation***. Pre-spool compensation is the focus of this chapter. When the pressure compensator valve is placed after the DCV spool, it can be called many different terms. Two of the most common terms are post-spool compensation or downstream compensation, which is the focus of **Chapter 19**.

Consider an agricultural tractor that is using a DCV for operating a planter's fan motor. The hydraulic motor is used to develop either a precise positive air pressure or a precise vacuum pressure. The farmer must dial-in the speed of the hydraulic motor to achieve a specific air pressure or vacuum, otherwise the seeding mechanism will not plant accurately.

In this example, assume that planter motor operation needs precisely 4 gpm (15 lpm) at 1000 psi (69 bar). The challenge is that, as the operator uses other hydraulic functions, such as raising or lowering the three-point hitch for draft control or steering the tractor along a line of trees, the hydraulic system flow and pressures will vary. As the flow and pressures vary, the planter's hydraulic motor speed will be negatively affected. As a result, the speed of the planter's hydraulic motor will fluctuate, causing seed population problems. Inserting a compensator valve in-line prior to the DCV spool can alleviate this concern.

Most planter motors are unidirectional. However, the schematic in **Figure 18-10** shows a bidirectional hydraulic motor to illustrate how the primary shuttle is used to direct signal pressure of a double-acting actuator to the compensator valve.

Pre-spool compensation has the following attributes:

- The compensator valve must sense the actuator's working pressure. Therefore, a primary shuttle valve is used to send the actuator's signal pressure to the compensator valve.
- The compensator's spring value establishes the pressure drop across the DCV spool.
- The DCV spool must be closed center.

The pressures in **Figure 18-10** will equal the following values:

- Working pressure/signal pressure.
    - 1000 psi (determined by the load on the actuator)
- Pump outlet pressure.
    - 300 psi (unloading spring) + 1000 psi (working pressure) = 1300 psi
- Pressure drop across the compensator valve.
    - 1300 psi – (1000 psi + 40 psi spring) = 260 psi

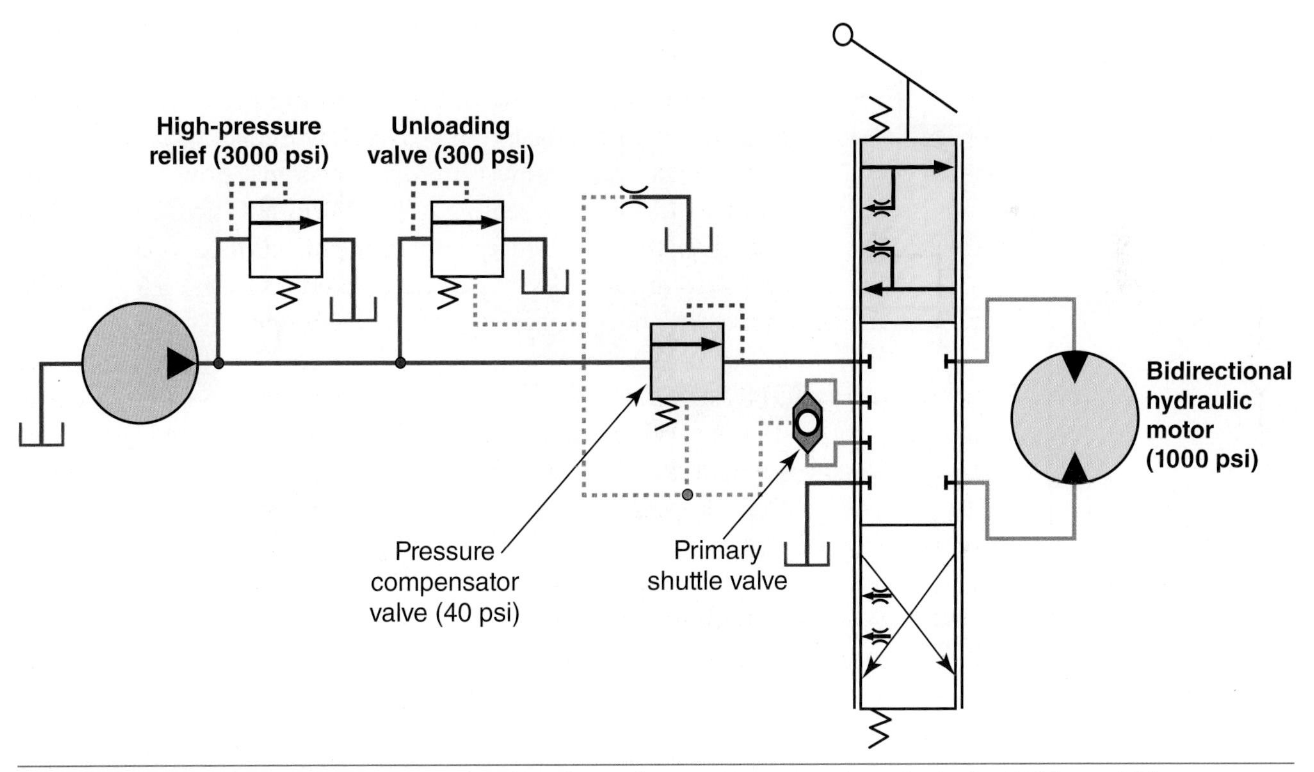

Figure 18-10. Pre-spool pressure compensation places the compensator valve prior to the DCV spool. The compensator valve senses the actuator's working pressure through the use of a shuttle valve to maintain a fixed implement speed.

- Pressure supplied to the DCV spool.
  - 1000 psi + 40 psi = 1040 psi
- Pressure drop across the DCV spool (equals the value of the PC spring).
  - 1040 psi – 1000 psi = 40 psi

The addition of the shuttle valve and the pressure compensator valve provides precise flow to the actuator. As the hydraulic system's flow and pressure changes, the pressure compensator valve will automatically compensate and adjust the flow so that the actuator remains at the same speed.

There is a limit to pressure compensation. If the engine-driven pump is slowed and is no longer delivering minimum flow as requested by the operator, the pressure compensator valve cannot magically magnify oil flow that is not there.

The shuttle valve and the pressure compensator valve also help alleviate an additional problem. As oil flows through a spool valve, oil velocities create forces that affect the position of the spool valve. See **Figure 18-11**.

The addition of the compensator valve results in a smaller pressure drop across the DCV spool, which minimizes the negative effect of flow forces. If the DCV spool is manually operated, the PC valve will help by lowering the amount of effort the operator must use to actuate the spool valve and the effort required to maintain the position of the spool valve.

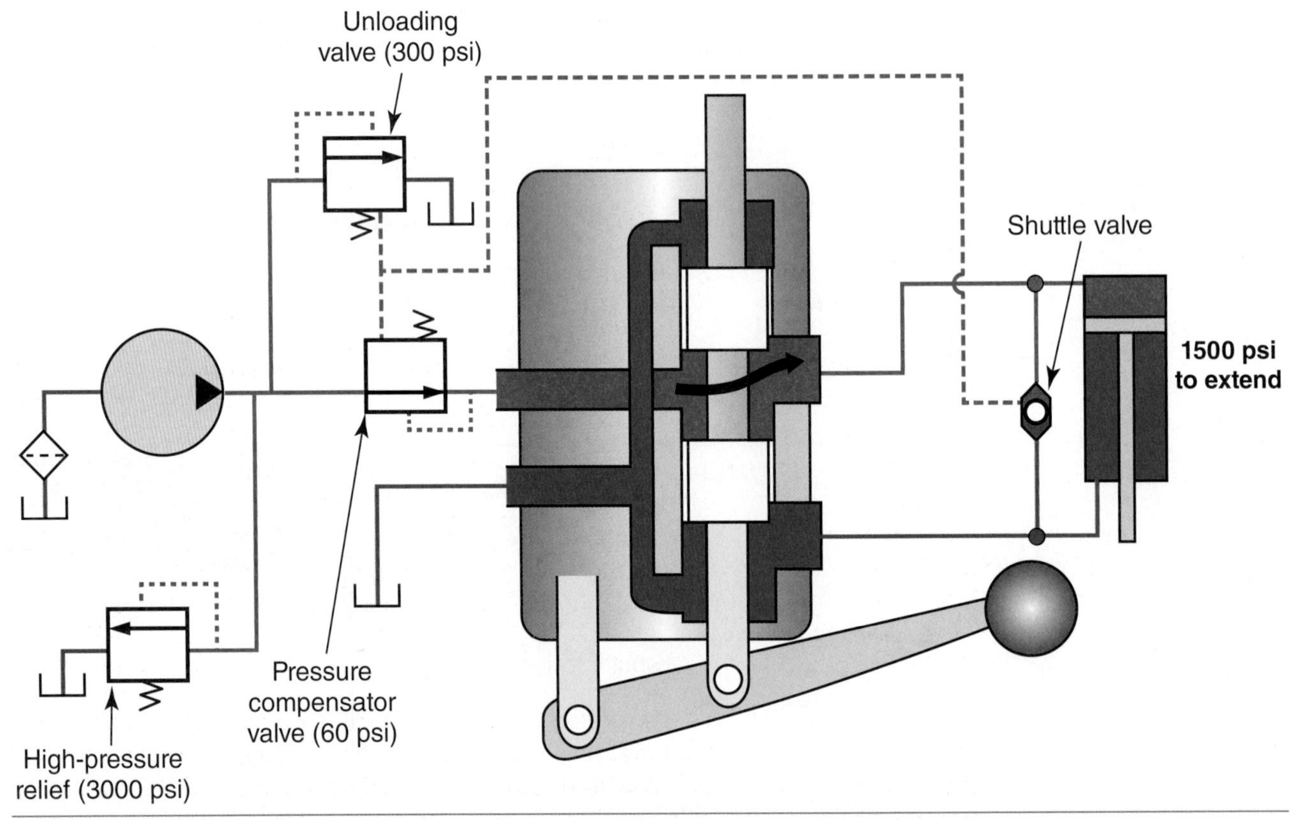

Figure 18-11. Oil flowing through a DCV spool creates forces, causing a resistance when attempting to open and close the spool valve. High-pressure pump flow is directed to extend the cylinder as the spool is opened. The shuttle valve and PC valve work in unison to keep the pressure drop across the DCV spool relatively low, 60 psi. Decreasing the size of the orifice between the pump input oil and cylinder extend oil ports by closing the spool increases these flow forces.

Note

Some personnel will label constant-flow LSPC systems (for example, **Figure 18-6**) as an open-center hydraulic system. They would point out that the system uses a fixed-displacement pump that delivers a constant flow at a relatively low pressure anytime the DCVs are in a neutral position. However, this system uses closed-center DCVs; has pressure compensator valves installed in series prior to the DCV spools; and uses an unloading valve that enables the hydraulic system to operate at a preset value above the actuator working pressure. Technically, the DCVs in this system are identical to the DCVs used in a variable flow LSPC system.

One drawback to a fixed-displacement LSPC system is that the pump produces a constant flow when the DCVs are in a neutral position. Even though the flow is at a relatively low pressure, this still results in wasted power. Variable-displacement LSPC hydraulic systems can reduce the hydraulic pump's horsepower consumption to only a fraction, sometimes as little as 0.15 hp.

Case Study

Examples of LSPC with Fixed-Displacement Hydraulic Pumps

LSPC systems that use a fixed-displacement pump are not as popular as LSPC variable-displacement pump systems. Some manufacturers use fixed-displacement LSPC systems in their economy model tractors and variable-displacement LSPC systems in their premium model tractors. A fixed-displacement LSPC system was used on older machines such as the Case IH 9390 Steiger 4WD tractor equipped with the high-flow hydraulic system, Caterpillar D6H and D7H track-type tractors, and Caterpillar 926A, 936A, 936E wheel loaders. Some examples of newer machines using this system design are Caterpillar D3K–D5K Series I and II track-type tractors and John Deere 6030, 7030, and 6M economy model tractors.

Case IH 9390 High Flow Steiger

The Case IH 9390 High Flow tractor contained two separate LSPC systems, one with a fixed-displacement pump and one with a variable-displacement pump. The tractor was designed so that DCVs 3 and 4 used the fixed-displacement 30 gpm steering gear pump, while DCVs 1 and 2 used the variable-displacement 30 gpm piston pump. The fixed-displacement gear pump contained an unloading valve that received a load-sensing signal pressure from DCV 3 and DCV 4. DCV 1 and DCV 2 sent a LS signal to the variable-displacement LSPC pump. Variable-displacement LSPC systems are discussed later in this chapter.

Caterpillar D3K–D5K Series I and II Dozers

The Caterpillar D3K–D5K dozers used a fixed-displacement pump and LSPC DCVs to control the implement hydraulic system. The unloading valve dumped the gear pump's constant flow at a low pressure value when the DCVs were in a neutral position. The DCVs sent a signal pressure to enable the unloading valve to build the pressure required for the DCVs to operate.

John Deere Base Model 6030 and 7030 Tractors

John Deere manufactured late-model tractors that used a fixed-displacement LSPC system. These were the 6030 and 7030 base models, produced from 2008 to 2011. The premium models used a variable-displacement LSPC system.

The base model 6030 and 7030 tractors used two pumps. The first was a low-pressure transmission pump (254 psi) and the second was a high-pressure hydraulic pump (2973 psi) for steering, brakes, and DCVs. The high-pressure circuit consisted of a gear pump that used a priority valve. Deere labeled the priority circuits as one and three. The priority-one circuit included steering and brakes. After priority-one hydraulic requirements had been met, oil was routed to the priority-three circuit, which included the three-point hitch, selective control valves (SCVs), independent control valves (ICVs), and power beyond. The service literature omitted reference to a priority-two circuit. The SCVs were essentially DCVs located at the rear of the machine, and ICVs were mid-mounted DCVs used for loader control functions. When all of the DCVs were in a neutral position, a pressure-regulating valve dumped oil to the reservoir at approximately 217 psi (15 bar). When a DCV was actuated, a load-sensing signal pressure was sent to the pressure-regulating valve, which enabled the system to build the necessary pressure to operate the actuator.

The John Deere 6M series tractors replaced the 6030 series. The 6M series tractor also uses a fixed-displacement LSPC hydraulic system. Note that the 6R uses a variable-displacement LSPC system.

LSPC Systems with Variable-Displacement Hydraulic Pumps

LSPC variable-displacement systems gained popularity in the off-highway industry in the late 1970s and early 1980s.

- Massey Ferguson introduced LSPC in 1978 on the MF60 loader backhoe.
- In 1979, Case introduced the 90 series Case tractors with LSPC.
- In 1981, International Harvester introduced the 88 series with a system called Power Priority Hydraulics, which was an LSPC system.
- In the mid 1970s, Caterpillar introduced the G Series motor graders that used load-sensing hydraulics.
- John Deere first introduced LSPC on 6000 and 7000 series tractors in 1992, and 8000 series tractors in 1994.

Many late-model machines use LSPC variable-displacement systems, including many of the agricultural tractors that retail for $150,000 or more. Case IH started using LSPC in their 2100 series Axial Flow® combines in 1995 and still use LSPC variable-displacement systems in their late-model combines. Caterpillar has used these systems in numerous machines in the past, but today many of their machines use proportional priority pressure compensated (PPPC), negative flow control (NFC), or positive flow control (PFC) systems.

Relating Constant-Flow LSPC to Variable-Flow LSPC

Referring back to Figure 18-6, if the high-pressure relief and the unloading valve were relocated to the compensator assembly inside a variable-displacement pump and relabeled as the ***pressure cutoff spool*** and the ***flow control spool*** respectively, the fixed-displacement LSPC system could be converted to an LSPC system that uses a variable-displacement pump. This LSPC system, Figure 18-12, is commonly used in thousands of old and new machines today.

A compensator assembly used on an inline axial variable-displacement LSPC piston pump is shown in Figure 18-13. The compensator assembly has two control spools. The smaller of the two spools is the flow control spool. The larger one is the pressure compensator spool. The flow control spool must receive the working pressure from the signal line, which is connected via the flat-face O-ring elbow fitting.

LSPC Variable-Displacement Hydraulic Pump Symbols

LSPC variable-displacement pump symbols can be depicted in different styles (or formats). Two simplified symbols used to represent the pump in a basic hydraulic system are shown in Figure 18-14. These two symbols are commonly found in older schematics.

Newer schematic symbols used to depict LSPC variable-displacement pumps sometimes show the pump's individual pressure compensator spool and load-sensing (flow control) spool in greater detail. See Figure 18-15.

LSPC Variable-Displacement Pump Modes of Operation

An LSPC variable-displacement piston pump will operate in one of three modes depending on the status of the DCV and the actuator:

- Low-pressure standby.
- Working mode.
- Stall mode.

Low-Pressure Standby

Low-pressure standby is the mode of operation when the DCVs are in the neutral position. For example, when the tractor is started and no DCVs have been actuated, the pump flow is destroked in the low-pressure standby mode. Refer to Figure 18-16.

The oil flow from the hydraulic piston pump is low, 0.5 to 1 gpm (1.9 to 3.8 lpm), which is the minimum amount of oil needed to maintain the minimal leakage inside the pump. The pump operates at a relatively low pressure, 300–500 psi (21–34 bar). With practically no flow of oil and a low system pressure, the pump uses very little horsepower while in the standby mode.

$$500 \text{ psi} \times 0.5 \text{ gpm} \div 1714 = .15 \text{ hp}$$

The flow control spool establishes the low-pressure standby pressure setting. On older machines, the flow control spool is usually the smaller of the two spools. This spool must sense the actuator's working pressure from the signal network.

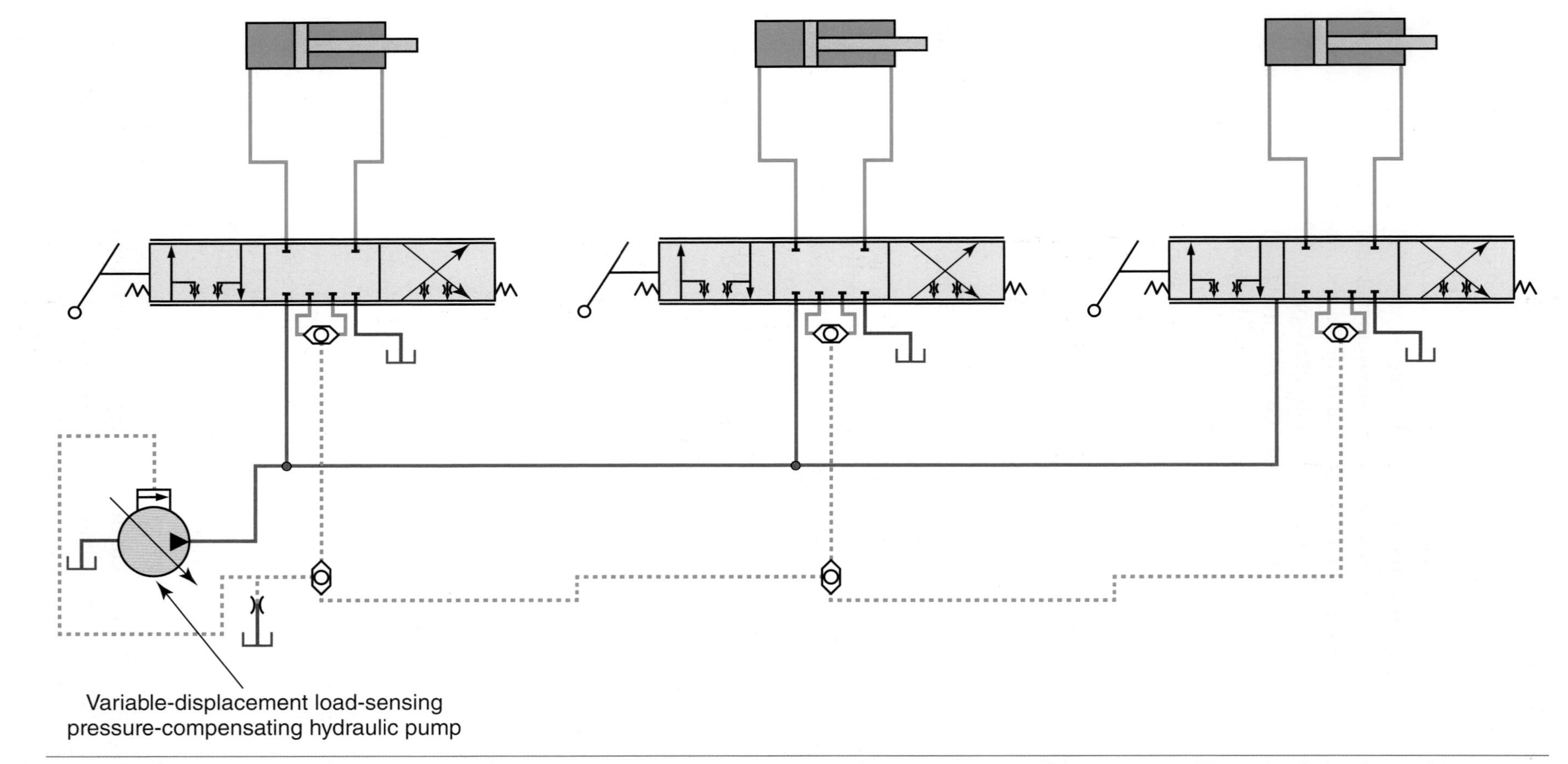

Figure 18-12. An LSPC system using a variable-displacement hydraulic pump simplifies the system and the schematic. The unloading valve and the main system relief are both incorporated into the pump's compensator assembly, which is represented on the schematic by the arrow inside the envelope attached to the pump symbol.

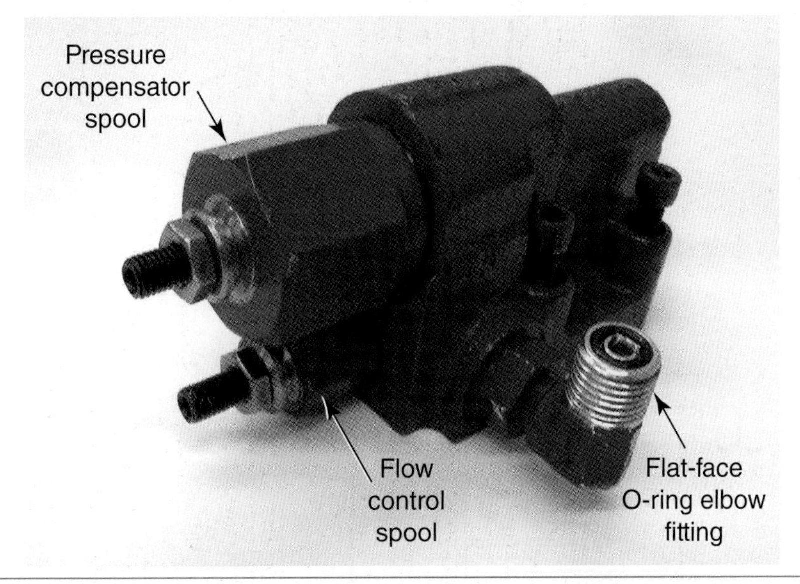

Figure 18-13. A compensator assembly from an LSPC variable-displacement piston pump uses two spools: a flow control spool and a pressure compensator spool. The compensator also receives a signal pressure through the O-ring elbow fitting.

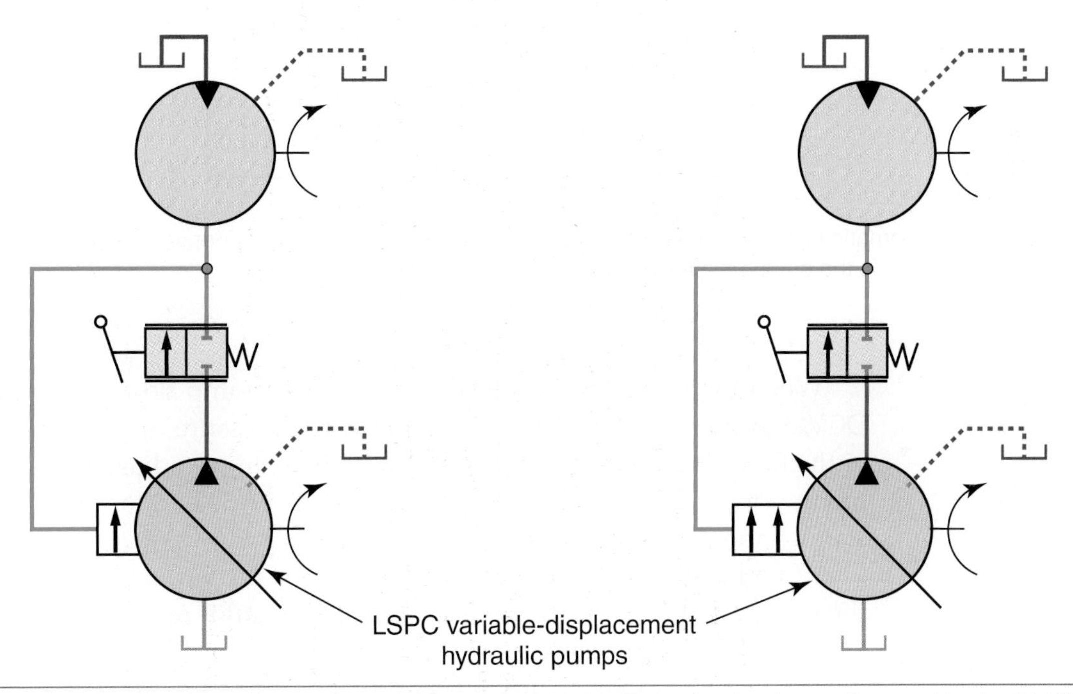

Figure 18-14. Both pump symbols shown within the simple hydraulic motor circuits can be used to depict LSPC variable-displacement hydraulic pumps. The pump senses the working pressure of the motor. Older Caterpillar schematics might contain two arrows within the envelope to directly represent the pump compensator spool and the load-sensing spool.

Note

The flow control spool can also be called the low-pressure standby spool, flow compensator spool, or the margin spool. In a fixed displacement LSPC system, the unloading valve takes the place of the flow control spool.

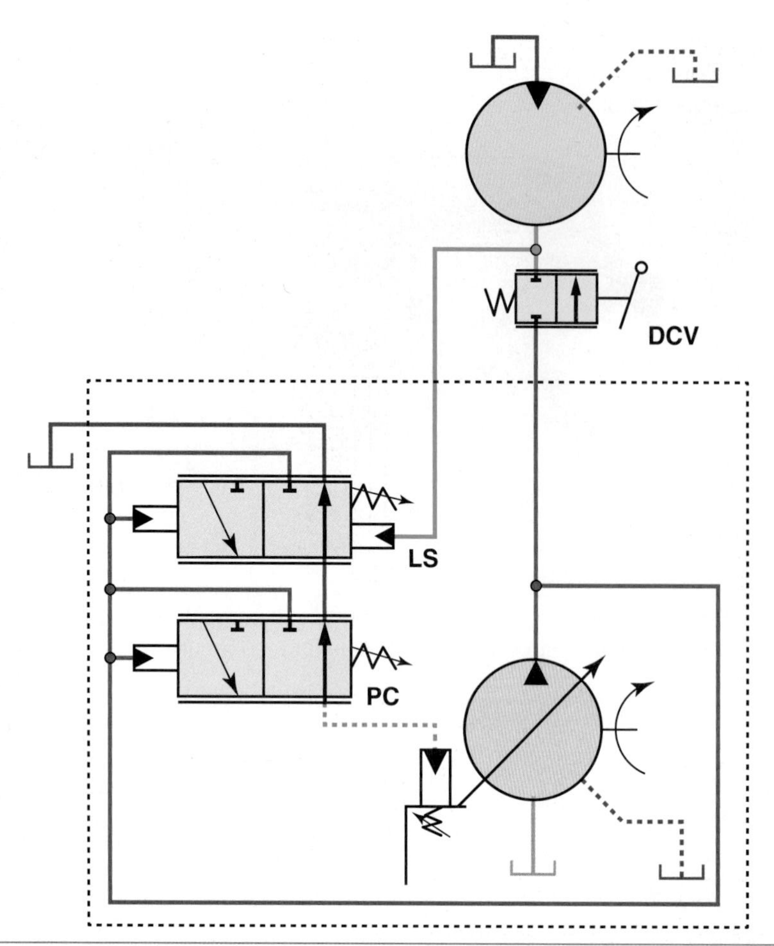

Figure 18-15. Newer schematic symbols sometimes detail the spool valves (pump compensator spool and load-sensing spool) housed within the pump assembly.

When the DCVs are in the neutral position, pump flow deadheads at the DCV, and pump pressure begins to build. The pressure increases to the value of the flow control spool's spring value. This spring is adjustable, and is commonly adjusted to a value between 300–500 psi (21–34 bar) for the low-pressure standby mode.

If the low-pressure standby is set at 300 psi (21 bar) and the DCVs are in the neutral position, pump pressure increases until it reaches 300 psi (21 bar). At this point, pump pressure overcomes the flow control spring value, causing the flow control spool to shift. Once the spool shifts, oil pressure is sent to the control piston, which destrokes the pump.

As the control piston pushes the pump's swash plate back to a neutral position, the control piston meters a small amount of oil into the pump's case through an orifice that becomes exposed as the control piston is extended. While in the standby mode, the swash plate will never reach a complete neutral angle, but will maintain a minimal amount of flow to overcome leakage and maintain the standby pressure. The orifice effectively stops the swash plate from returning to a completely neutral angle. As the oil dumps through the orifice, the oil pressure can push no further on the swash plate. The oil is commonly routed through a case drain back to the reservoir.

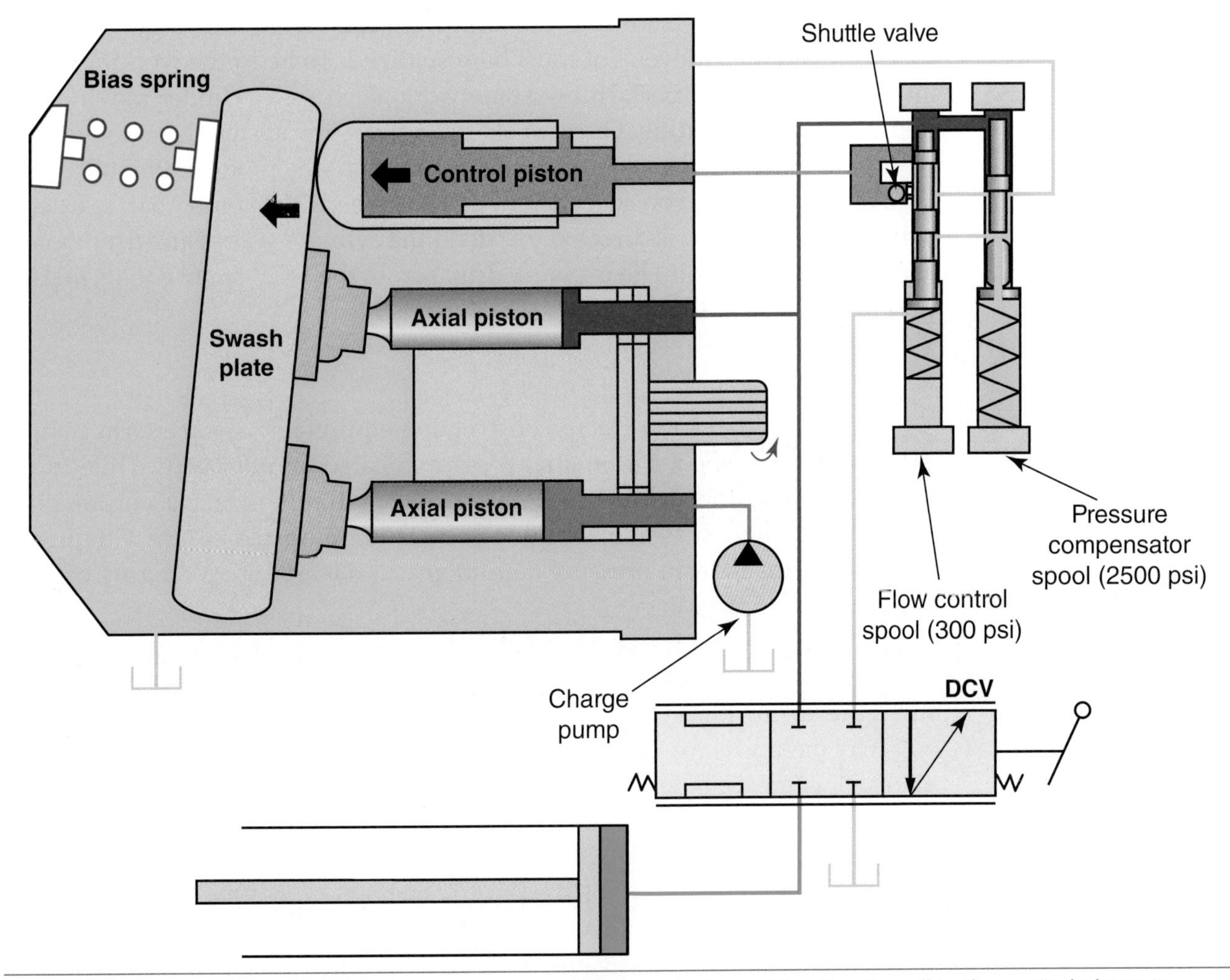

Figure 18-16. An inline axial LSPC piston pump uses a flow control spool for directing oil to the control piston to destroke the variable-displacement pump while in the low-pressure standby mode. The control piston extends against the swash plate, pushing the swash plate slightly against the resistance of the bias spring. Note the smaller charge pump used to provide low-pressure fluid to the axial piston pump.

If the variable-displacement pump uses a bias piston instead of a bias spring, the control piston will have a larger surface area, and it might be called the large piston. The control piston must be larger than the bias piston in this style pump because pump output pressure acts on both piston surfaces anytime the pump compensator is trying to destroke the pump. In this situation, the control piston's larger surface area must produce a stronger force to overcome the bias piston's force.

Working Mode

Whenever the operator actuates a DCV to request oil flow, the pump generates oil flow to perform some work. The DCV must be a load-sensing valve, meaning that as the oil is routed to the cylinder, the load-sensing DCV must also send a signal pressure to the pump's compensator so the pump can meet the system's demands. See the schematic in **Figure 18-17**.

The key to understanding the working mode shown in **Figure 18-17** is that the cylinder's movement must be visualized. To be in the working mode, an actuator must be performing some work, such as a cylinder moving or a hydraulic motor rotating. Once an actuator stops, the system is no longer in a working mode, and the pump enters either the standby mode or the stall mode.

For this example, consider that the cylinder needs 1000 psi (69 bar) to extend. While the pump flow is directed to extend the cylinder, a pressure drop occurs across the DCV spool. The pressure drop across the DCV spool equals margin pressure, 300 psi (21 bar).

Margin Pressure

Caterpillar and John Deere construction equipment use the term margin pressure to describe a differential pressure. Stated another way, while in the working mode, the difference between the hydraulic pump's output pressure and the highest system working pressure is margin pressure. **Chapter 19** expands on the margin pressure definition as it relates to flow sharing or post-spool compensation.

Both the low-pressure standby value and margin pressure value are changed by adjusting *one* flow control spring value, which is the smaller spool in **Figure 18-16**. It has the responsibility of setting the pump's reaction.

Two methods are used for adjusting the spool. The first method is a static adjustment made with the DCVs in the neutral position. The second, or dynamic, method is to operate a hydraulic actuator slowly while measuring the margin pressure. Both adjustments alter the pump's reaction. As more spring tension is placed on the spool, the pump's reaction time is shortened.

The flow control spool's spring setting is usually one adjustment that can make a substantial improvement to an LSPC hydraulic system. However, there are negative effects to adjusting the flow compensator spring value too high. As the spool's spring pressure increases, system inefficiency normally increases minimally. As an example, consider an LSPC system that is operating at 300 psi (21 bar) in low-pressure standby and presume the system is flowing approximately 1 gpm. The hydraulic horsepower could be determined with the following equation:

$$300 \text{ psi} \times 1 \text{ gpm} \times 0.000583 = .17 \text{ hp}$$

If the standby pressure was adjusted up to 500 psi, it would increase the hydraulic horsepower to the following value:

$$500 \text{ psi} \times 1 \text{ gpm} \times 0.000583 = .29 \text{ hp}$$

In addition, if the spool spring is adjusted too high, the machine can become too sensitive and overly reactive. For example, a loader backhoe operator complains that the tractor's hydraulic system is sluggish. The technician servicing the machine is in a hurry and adjusts the low-pressure standby value from 200 psi (14 bar) to 650 psi (45 bar) before going home for the day. The next day, the operator tries to slowly operate the boom swing while gingerly lowering a pipe into a trench. Now the operator finds that the machine is too sensitive and too hard to control. The operator calls the technician back to fix the new problem—a hydraulic system that is overly reactive. In addition to the downside of a callback repair, the technician's lack of ability to follow the manufacturer's service instructions may have voided the manufacturer's warranty, reduced the machine's normal life cycle, and placed personnel in danger due to the overly responsive hydraulic system.

Note

If a technician could substantially increase the low-pressure standby value to a high value such as 2000 psi (138 bar), the LSPC system would operate more as a traditional PC system than the traditional LSPC system. The system would be noisy, operate at a high pressure constantly, and its actuators would respond very quickly to DCV spool movements.

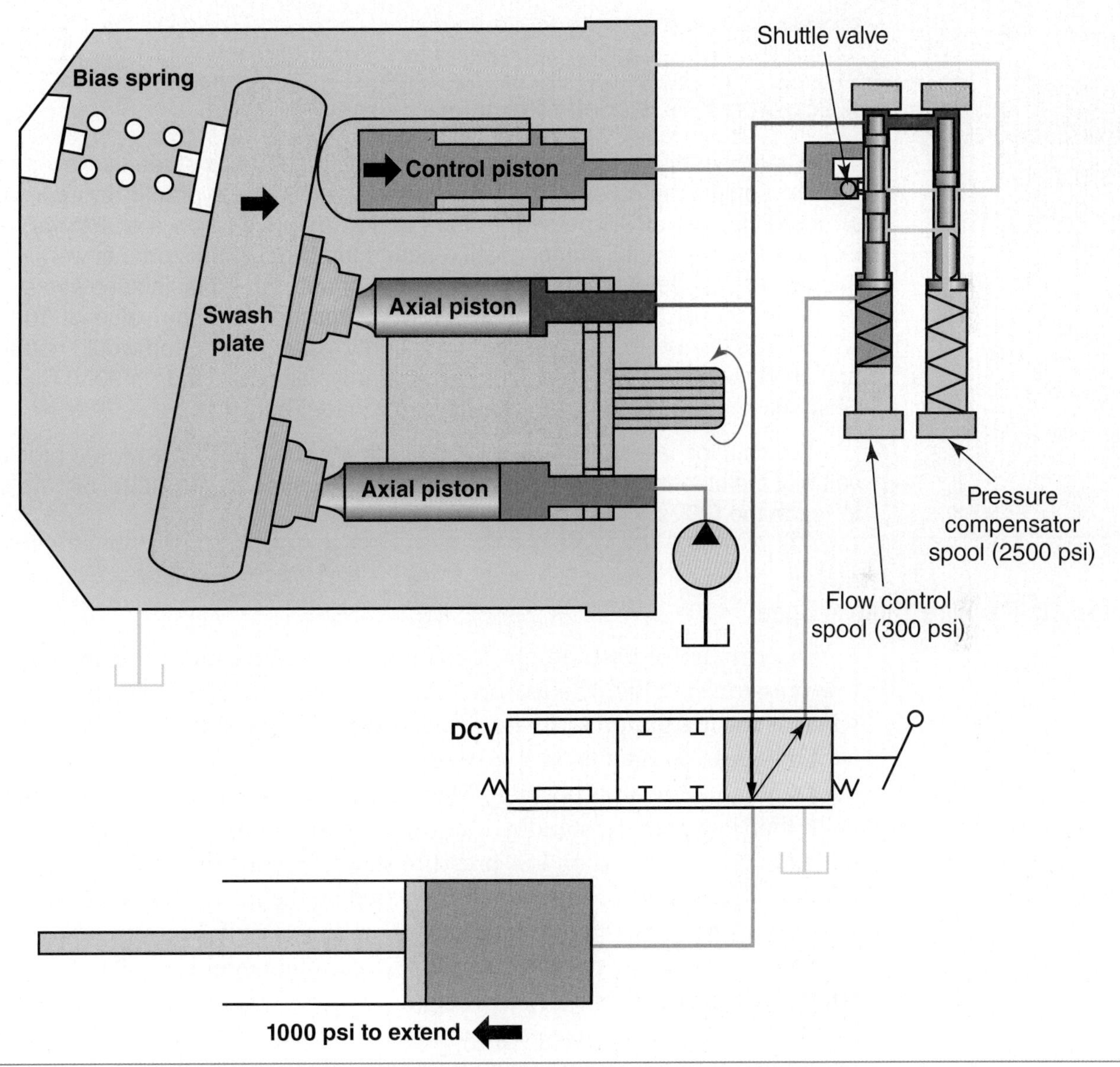

Figure 18-17. In the working mode, the axial piston pump output pressure will equal the working pressure plus the flow control spring value (also known as margin pressure). Bias spring pressure overcomes the resistance of the control piston, pushing the swash plate at a greater angle to increase the displacement of the rotating axial pistons.

PQ Curve Differences

Some manufacturers have their engineers graph a ***PQ curve***, which consists of graphing the hydraulic pump's flow rate (Q) at different operating pressures (P). To understand the effects of low-pressure standby on LSPC system performance, it is helpful to compare a PQ curve with a low value for low-pressure standby and a PQ curve with a higher value for low-pressure standby. **Figure 18-18** is a graph that students charted on an LSPC hydraulic system in a Pittsburg State University laboratory setting. The flow rate (Q) is listed on the left side of the graph, and the operating pressure (P) is on the bottom of the graph. Notice that the performance improved when the low-pressure standby (LPSB) was set at 550 psi (38 bar) versus 200 psi (14 bar).

When charting a PQ curve, be sure to follow all of the test procedures, as performance can drastically change based on many factors, especially pump speed. Manufacturers might provide correction factors to aid the diagnostic process.

Case Study

A Low-Pressure Standby Problem

In a laboratory setting, instructors can demonstrate to students how a mistake in setting the low-pressure standby can affect a hydraulic system. After an instructor set LPSB pressure close to 100 psi (7 bar) on a LSPC variable-displacement hydraulic pump, no oil reached the DCVs. The students working on the project later determined that the steering priority valve was biased at a spring value of 150 psi (10 bar). Therefore, a low-pressure standby value of 100 psi (7 bar) was too low to shift the steering priority's spool valve and would not allow any oil to flow to the DCVs. This type of priority valve is found on 9300 Case IH Steigers 4WD tractors and Caterpillar 420D loader backhoes.

The solution involved setting the LPSB to a value slightly above the pressure value of the steering priority spring. Once this adjustment was made, oil was able to reach the DCVs.

Adjusting the Flow Control Spool

Some manufacturers might focus more on setting margin pressure than low-pressure standby. The reason is that low-pressure standby is a static test, whereas the margin pressure test is a more dynamic test.

Low-pressure standby is a measure of the pressure the spool is set at when the DCV is in a neutral position. Margin pressure tends to have a little lower value than low-pressure standby. Margin pressure values have a tighter tolerance (for example, ± 15 psi) than low-pressure standby pressure values (for example, ± 100 psi). The margin pressure value is based on the system flowing oil and the spool balancing between pump output pressure and signal pressure.

For example, a low-pressure standby specification might be 410 psi ± 105 psi, while the margin pressure specification might be 305 psi ± 15 psi. If a technician accurately adjusts margin pressure, low-pressure standby is typically also within its specification. Some manufacturers might only provide a specification for low-pressure standby or margin pressure and not both. Other machines will have both specified.

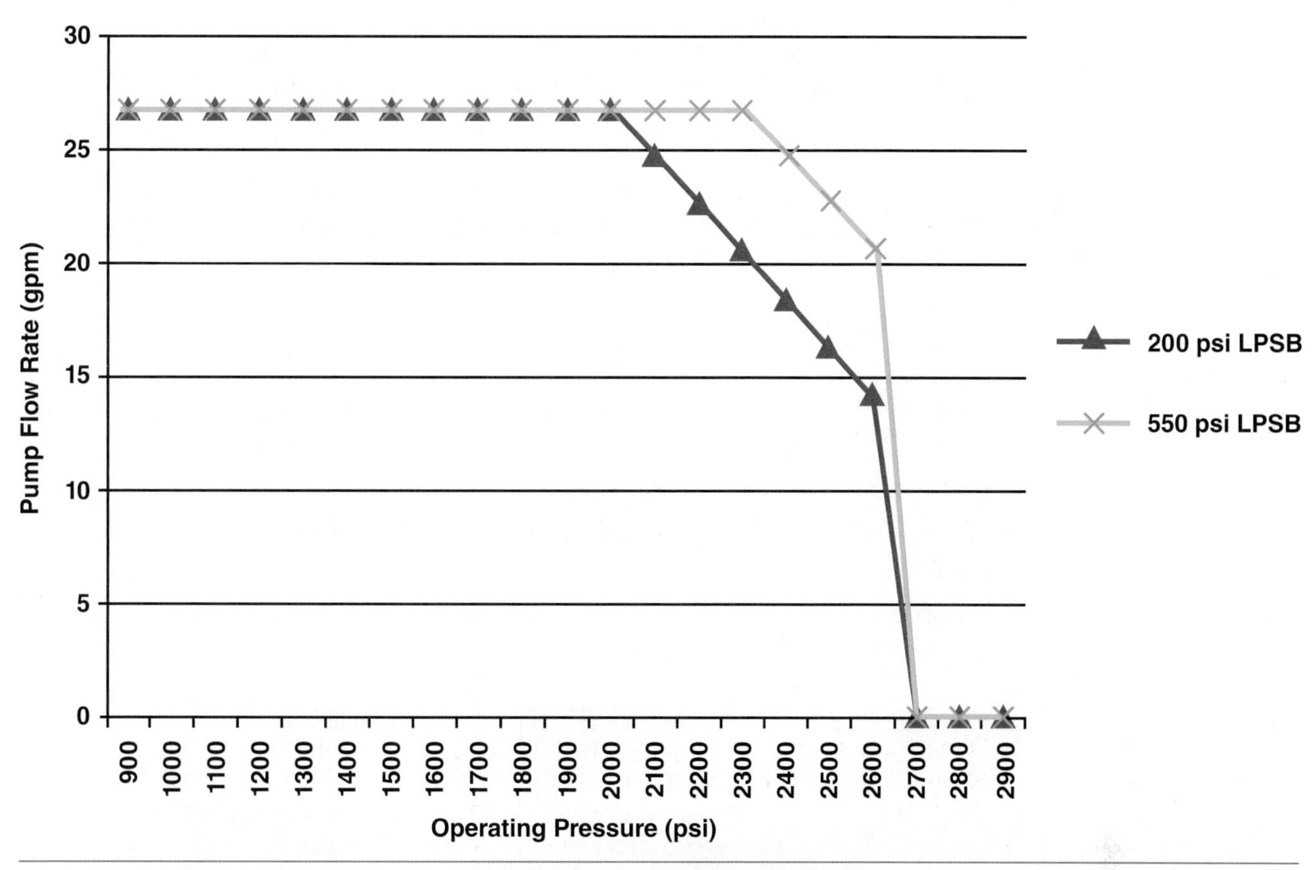

Figure 18-18. A PQ curve is a graphic representation of the relationship between a hydraulic system's flow rate and operating pressure. This graph has two different curves, one with a low value for low-pressure standby (LPSB) and one with a higher value. The higher value LPSB allows a greater pump flow rate over a longer operating pressure range compared to the lower value LPSB.

Chapter 21 will discuss hydraulic test equipment, including a differential pressure gauge. This type of gauge is essential for measuring margin pressure. Margin pressure requires reading two different pressures simultaneously while oil flows through a circuit and subtracting one pressure reading from the other pressure reading.

An example is measuring margin pressure in a bulldozer hydraulic system while actuating the dozer blade to lift. The problem is that the load on the blade will cause the pressure value to constantly change as the blade is moving. Operating the DCV at a relatively slow speed rather than a faster speed helps minimize this issue. As the pump outlet pressure varies so will the signal pressure change during the test. It is difficult to simultaneously watch two pressure gauges with fluctuating pressures and attempt to subtract the difference.

A differential pressure gauge solves this problem. The gauge displays the difference in pressure between two separate input pressure values, which, in this case, are the pump outlet pressure and signal pressure. One pressure line is connected to the left side of the gauge and the other is connected to the right.

Note

It is important to realize that a low pressure standby value might be influenced by more than just a pump's margin spring. Some LSPC systems have a warm-up signal pressure created in the steering circuit. Some manufacturers take that steering signal pressure into consideration, eliminating it or isolating it to achieve a true low-pressure standby value. **Chapter 25** explains warm-up signal pressures used in steering circuits.

Stall Mode

When a cylinder reaches its end of travel, the pump begins to stall, also known as pump cut off. The stall mode pressure is set by the PC spool spring. This is the same PC spool that was explained in **Chapter 17**. The PC acts similar to a main system relief valve. Anytime the system pressure deadheads, the PC spool shifts, allowing oil to be sent to the control piston to destroke the pump. See **Figure 18-19**. The PC spool does not need to sense signal pressure.

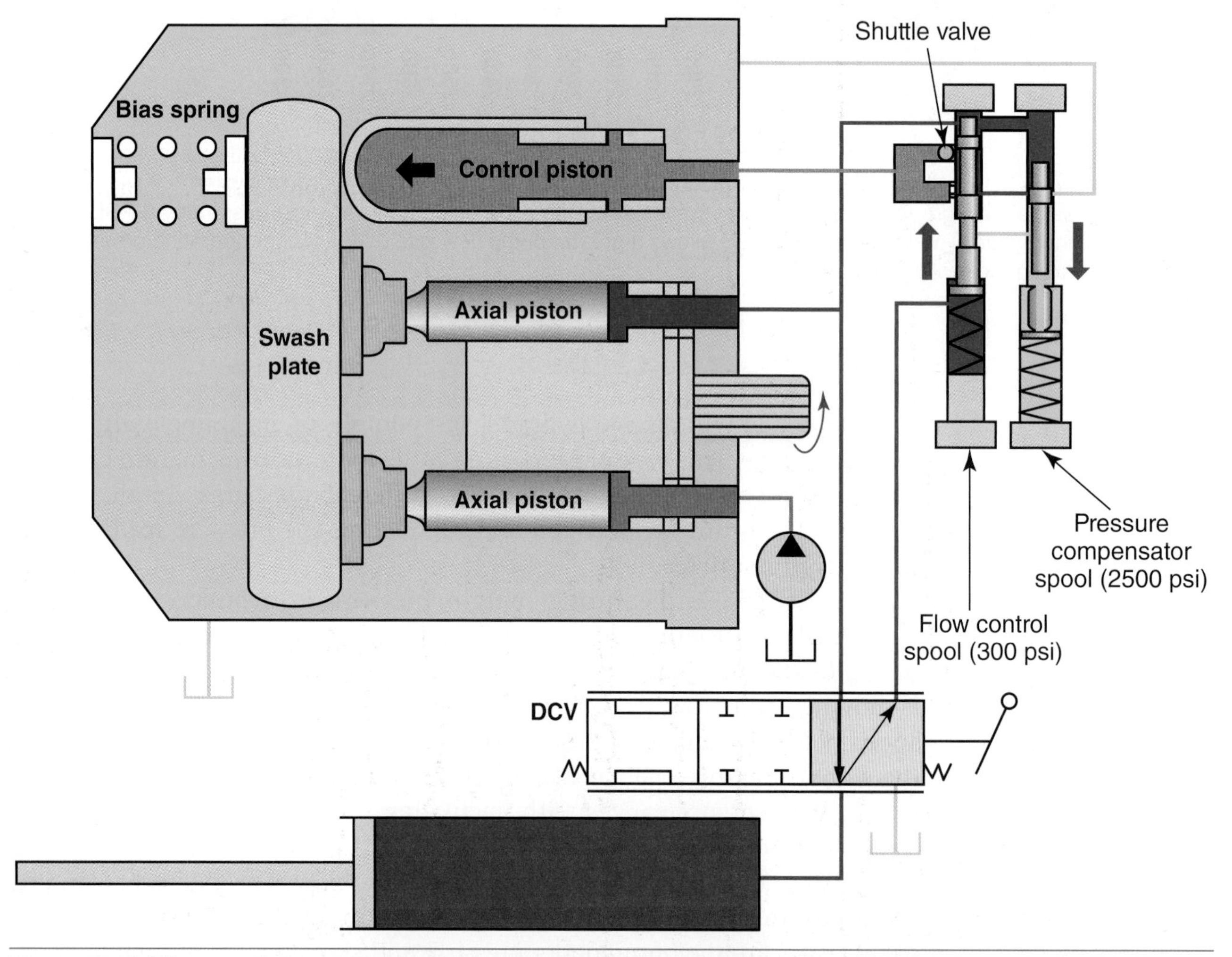

Figure 18-19. When a cylinder reaches its end of travel (stall mode), the axial piston pump will be destroked by the pressure compensator spool. Once the high pressure is directed to the control piston, the control piston's force is enough to move the swash plate parallel to the displacement pistons' axis of rotation. This reduces pump flow to nearly zero.

If the PC spool is set at 2500 psi (172 bar), once the pump reaches that pressure, the PC spool shifts down, directing oil to the pump's control piston to destroke the pump. The pump flow will be reduced to practically no flow but will remain at the high stall pressure value until the DCV is returned to a neutral position.

Another interesting point about stall mode on a variable-displacement LSPC pump is that the flow control spool is technically shifted up, while the pressure compensator spool is shifted down. How is this possible? The bottom of the flow control spool has a 300 psi (21 bar) spring setting and 2500 psi (172 bar) of pump outlet pressure acting on it, while the top of the spool only has pump outlet pressure, 2500 psi (172 bar), acting on it. As a result, the flow control spool is shifted upward due to the higher combined pressure value at the bottom of the spool. However, the pump outlet pressure can still make its way around the top of the flow control spool so that it can still act on the top of the PC spool, allowing the PC spool to shift down and destroke the pump.

Main System Relief in Variable-Displacement LSPC Systems

Many agricultural variable-displacement LSPC systems do not use any type of a main system relief valve, and rely solely on the pump's PC valve to protect the system during high-pressure conditions. Conversely, many construction equipment machines contain a main system relief valve in addition to the pump's PC valve. If the machine has both a main system relief and a pump PC, traditionally one of those components is used as a backup in case the primary fails. This means that one of the components serves primarily as a high-pressure relief while the other is not normally used but is set higher than the primary component. Normally the main system relief is set higher than the pump's PC. However, in a few rare cases, a machine can have the pump compensator set higher (for example, 4000 psi [276 bar]) than the main system relief (for example, 3600 psi [248 bar]).

Most systems set the main system relief valve as the secondary safety feature in the event that the pump PC is slow or is ineffective at destroking the pump. A Caterpillar D8R dozer has a main system relief set at 3900 psi (269 bar) and the pump pressure cutoff set at 3500 psi. In this example, if the main system relief valve is inaccurately set below the pressure compensator value, the LSPC system will be operating inefficiently. The inefficiency will be evident during stall mode, when the pump is destroked. If the main system relief valve takes command during high system pressures, the pump will not only be running at high pressure, but will also be flowing a lot of oil over the main relief valve, consuming unnecessary energy.

Industry experts recommend that when using a main system relief in conjunction with a variable-displacement pump, the main system relief value should be set 10 to 15% above the pressure compensator value.

Torque-Limiting Control

Some variable-displacement LSPC pump systems are designed to provide high amounts of flow or high amounts of pressure, but use an engine that does not have the capacity to deliver enough horsepower to provide both maximum system flow and maximum system pressure simultaneously. In these applications, a torque-limiting spool is incorporated into the pump compensator assembly. It destrokes the pump when the operator requests too much hydraulic horsepower.

Corner Horsepower

Corner horsepower is the point on a PQ curve that shows the total amount of horsepower required to deliver both maximum system hydraulic pressure and maximum system hydraulic flow simultaneously. See Figure 18-20. The corner horsepower is located in the upper right-hand corner of the PQ curve.

A pump with ***torque-limiting control*** shaves off the corner horsepower requirement by preventing the operator from requesting both maximum flow and maximum pressure simultaneously, Figure 18-21. Manufacturers save on production costs by designing a machine equipped with torque-limiting controls. The system allows a smaller engine with reduced fuel consumption compared to unequipped machines.

In addition to housing the pressure compensator and flow control spools, the variable-displacement LSPC pump in Figure 18-22 uses two components as torque-limiting controls: a piston that senses the angles of the pump's swash plate and a torque-limiting spool valve. The piston uses an orifice that drops oil pressure based on the angle of the pump's swash plate. When the pump swash plate is at a large angle, delivering high amounts of hydraulic flow, the swash plate sensing piston is pulled out of its bore resulting in little pressure drop. As a result, high pump outlet pressure causes the torque-limiting spool valve

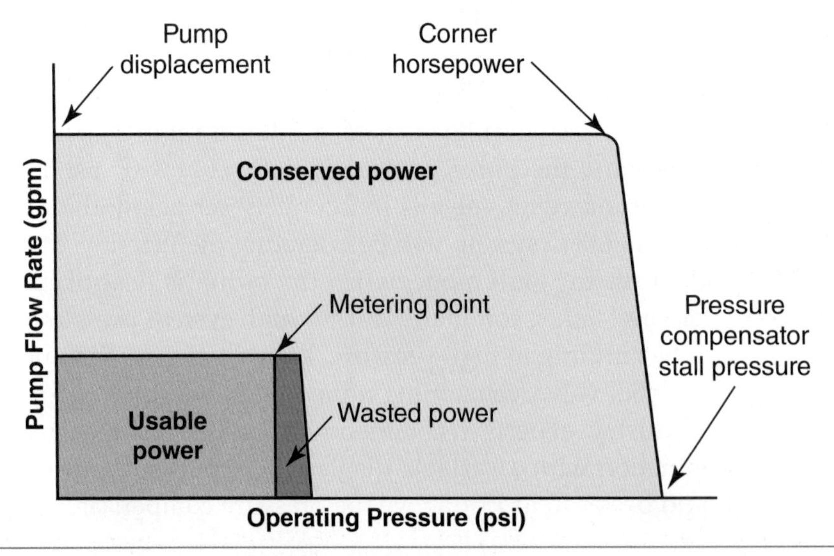

Figure 18-20. A hydraulic system that is producing maximum flow and maximum pressure simultaneously is depicted in the upper right-hand corner of a PQ curve.

to shift against its spring, opening a passage to allow oil to destroke the pump. A torque limiter acts like an infinitely variable relief valve that destrokes the pump anytime too much hydraulic horsepower is requested.

Torque limiting is a common descriptor used by many manufacturers. However, horsepower limiting is a more accurate descriptor because the control is limiting the total amount of hydraulic horsepower in an effort to prevent the engine from lugging, stalling, or even dying.

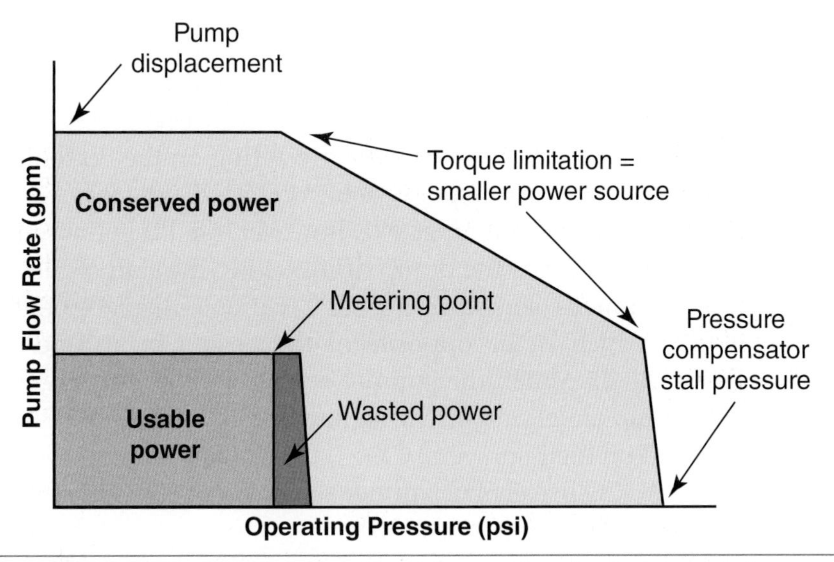

Figure 18-21. Torque-limiting control reduces the required amount of engine horsepower by preventing an operator from requesting maximum flow and maximum pressure simultaneously. Note the movement of the corner horsepower point on the PQ curve for a hydraulic system equipped with torque-limiting controls.

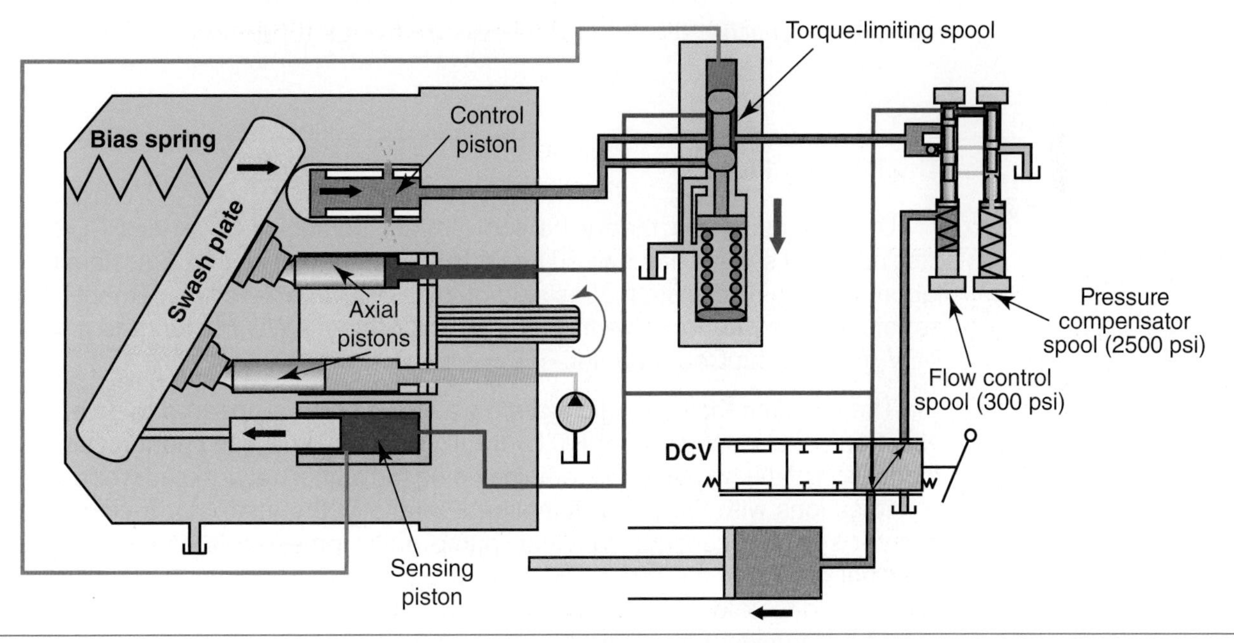

Figure 18-22. A hydraulic pump with torque-limiting control allows the operator to request maximum flow and maximum pressure individually but not simultaneously. A piston senses the angle of the swash plate, and when too much hydraulic horsepower is requested from the system, it causes a torque-limiting spool to destroke the pump.

Jammer Solenoids and the Implications of Connecting the Pump Outlet to a Signal Network

In **Chapter 16**, jammer solenoids were described in open-center hydraulic systems. Jammer solenoids are also used in LSPC variable-displacement systems. Even though the solenoids provide the same effect—placing the system on demand (maximum system pressure)—the jammer solenoid has an opposite design when used in LSPC variable-displacement pump systems.

In an LSPC variable-displacement hydraulic system, anytime the pump outlet pressure is directly connected to the signal pressure, the pump will suddenly enter the stall mode. Why is this? The flow compensator has pump outlet oil pressure acting on the top of the spool while pump outlet oil pressure and the flow control spring pressure are acting on the bottom of the spool. Therefore, the flow control spool remains shifted upward, asking for more oil. As the pump attempts to upstroke, however, the PC spool senses the high system pressure and holds the pump destroked in the stall mode.

A jammer solenoid in an LSPC variable-displacement pump is normally closed, which is the opposite of the design in an open-center system. See **Figure 18-23**. When the jammer solenoid valve is energized, it opens to connect pump outlet pressure to signal pressure. John Deere uses an LSPC jammer solenoid on their sprayers. Case IH uses LSPC jammer solenoids on their combines. AGCO uses LSPC jammer solenoids on their Massey Ferguson, Gleaner, Challenger, and Fendt combines.

Case IH used the 12-volt diode assembly that was discussed in **Chapter 16** on their 2100 and 2300 series combines. However, like most manufacturers, late-model machines have the jammer solenoid directly controlled by the ECM, allowing the manufacturer to eliminate an external diode module. The jammer solenoid is sometimes called a signal valve solenoid because it creates a false signal by connecting pump outlet pressure to signal pressure.

Case Study

The Missing Piece

Understanding the theory behind jammer solenoids in an LSPC variable-displacement system has helped an instructor diagnose an LSPC system built for laboratory use. The system consisted of an LSPC variable-displacement pump, a set of DCVs, a priority valve from a Case IH Steiger 4WD tractor, and a steering DCV from a John Deere tractor.

The first time the hydraulic system was operated, it appeared to be operating as a traditional closed-center PC hydraulic system. Time was spent reviewing the fabrication of the hydraulic system, including the signal network, as well as having conversations with the pump supplier. Eventually, the instructor focused on the jammer solenoid concept, "anytime pump outlet pressure is connected directly to signal pressure, the pump will enter the stall mode." He used cross-sectional drawings to look for a place where pump outlet pressure could be hydraulically connected to signal pressure. The instructor determined that if the steering priority valve compensator spool was missing, the pump would immediately enter the stall mode.

The donated priority valve had previously been used for corporate service training. After further inspection, he discovered that the spool had not been placed back into the priority valve. A new priority valve was ordered and the problem was solved.

Reasons to Use a Jammer Solenoid in an LSPC System

Manufacturers typically use jammer solenoids in LSPC systems for two reasons. The first reason is to force the hydraulic system to perform the same as a PC hydraulic system described in **Chapter 17**. Any time the jammer solenoid is activated, the LSPC pump operates at high system pressure and little flow.

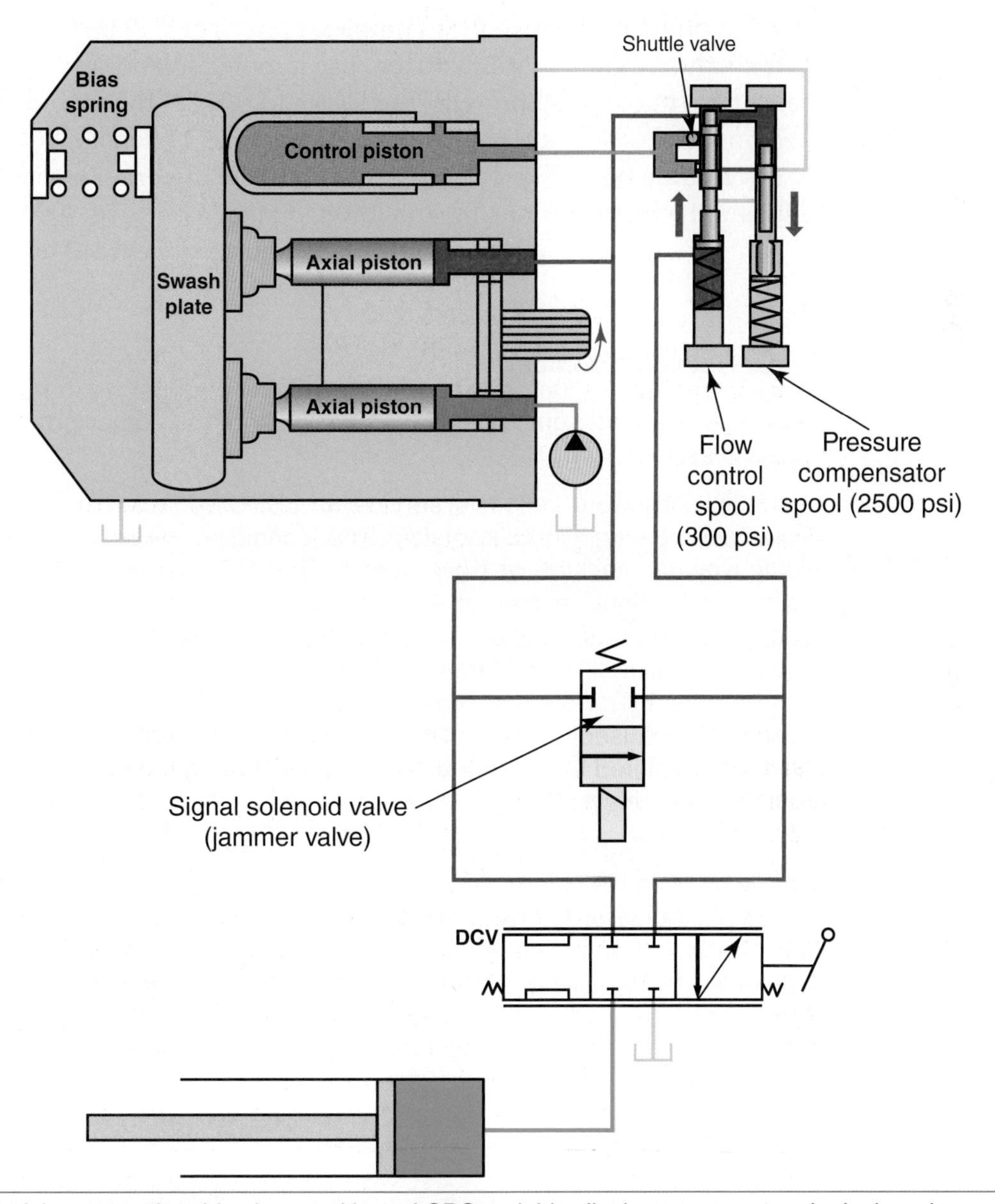

Figure 18-23. A jammer solenoid valve used in an LSPC variable-displacement system is designed to connect pump outlet pressure to signal pressure when energized, resulting in the LSPC system operating in the stall mode.

A John Deere 4630 sprayer uses a jammer solenoid that pressurizes the hydraulic system to a constant 3000 to 3100 psi (207 to 214 bar). In this application, the jammer solenoid is constantly energized. The sprayer takes advantage of the instantaneous response of the hydraulic system. One benefit of this system over a true PC system is that the jammer solenoid can be disabled during engine startup.

Jammer solenoids have the same effect whether a hydraulic system is open center or an LSPC variable-displacement system. Any time the jammer solenoid is energized, the hydraulic system exhibits the characteristics of a PC system: high system pressure, high noise, and responsive hydraulic controls.

The second reason for using a jammer solenoid is to eliminate the need of a signal network with numerous shuttle valves (refer back to **Figures 18-6**). A simple closed-center DCV can easily be incorporated into a variable-displacement LSPC system with the use of a jammer solenoid. This system design eliminates the use of a complex system of primary and secondary shuttle valves. When manufacturers use jammer solenoids, they sometimes use orifices in conjunction with the DCVs to control the cylinder speed.

John Deere 4940 and 4730 sprayers use a jammer solenoid, but not for all of the system functions. The 4730 sprayer uses the jammer for everything except adjusting the tread and raising and lowering the boom. The 4940 sprayer uses the jammer solenoid for everything except adjusting the tread and tilt.

Case Study

Signal Network Problems

A leak in a load-sensing signal network can cause other problems within a hydraulic system.

Technicians were called to service an LSPC variable-displacement tractor with an overheating hydraulic system. The technicians were having trouble determining why the machine was overheating. The OEM territory representative visited the technicians, examined the tractor, and requested that they follow the specified service information. After carefully following the specified diagnostic procedures outlined in the literature, the technicians were surprised at the source of the problem. The service brakes had developed a leak in the load-sensing circuit, which caused the pump to constantly run at higher flows and higher pressures than required by the actuators. This was the underlying cause of the overheating hydraulic system.

The technician found that the closed-center (service) brake control valve was not fully blocking the pump's flow when the foot pedal was released. The valve was allowing fluid pressure to leak into the signal line leading back to the pump's flow control valve, causing the pump to upstroke. The brake valve was not serviceable. A new service brake valve was ordered and installed, which remedied the overheating hydraulic system.

Power Beyond in LSPC Systems

As mentioned in **Chapter 9**, power beyond in open-center systems simply consists of a single coupler assembly. An LSPC power beyond option consists of three coupler ports as shown in Figure 18-24. One port taps directly into pump outlet pressure, one port allows for a return, and the third port requires sending a signal pressure to the signal network. Figure 18-25 shows two similar LSPC systems with one main difference in their designs. The signal network in Figure 18-25A does not have power beyond. The signal network in Figure 18-25B is equipped with power beyond and contains the necessary shuttle valve and couplers for proper operation.

DCV Hydraulic Detent Kickout in LSPC Systems

As explained in **Chapter 9**, some manually operated DCVs have detented levers that mechanically hold the spool in a fixed position during operation. Once a high pressure is reached, the lever returns to neutral. The detent kickout pressure is adjustable. The primary shuttle in this type of system has the responsibility of choosing the higher working pressure from either the extend or retract pressure and sending it to three locations: the secondary shuttle valve, the detent kickout valve, and the pressure compensator valve.

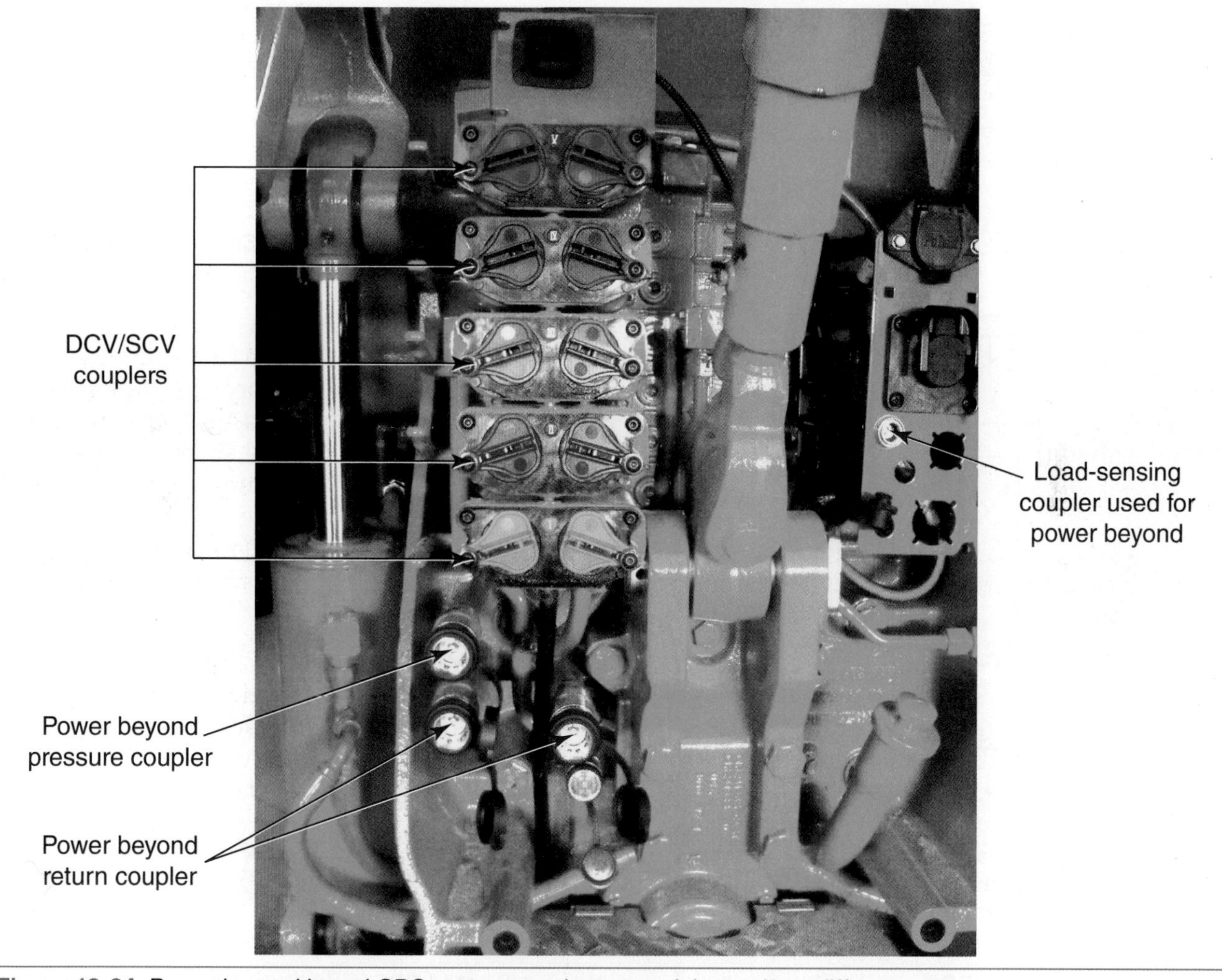

Figure 18-24. Power beyond in an LSPC system requires at a minimum three different ports: a pump pressure coupler, a return coupler, and a load-sensing coupler. Note that the power beyond return couplers are also known as hydraulic motor return couplers.

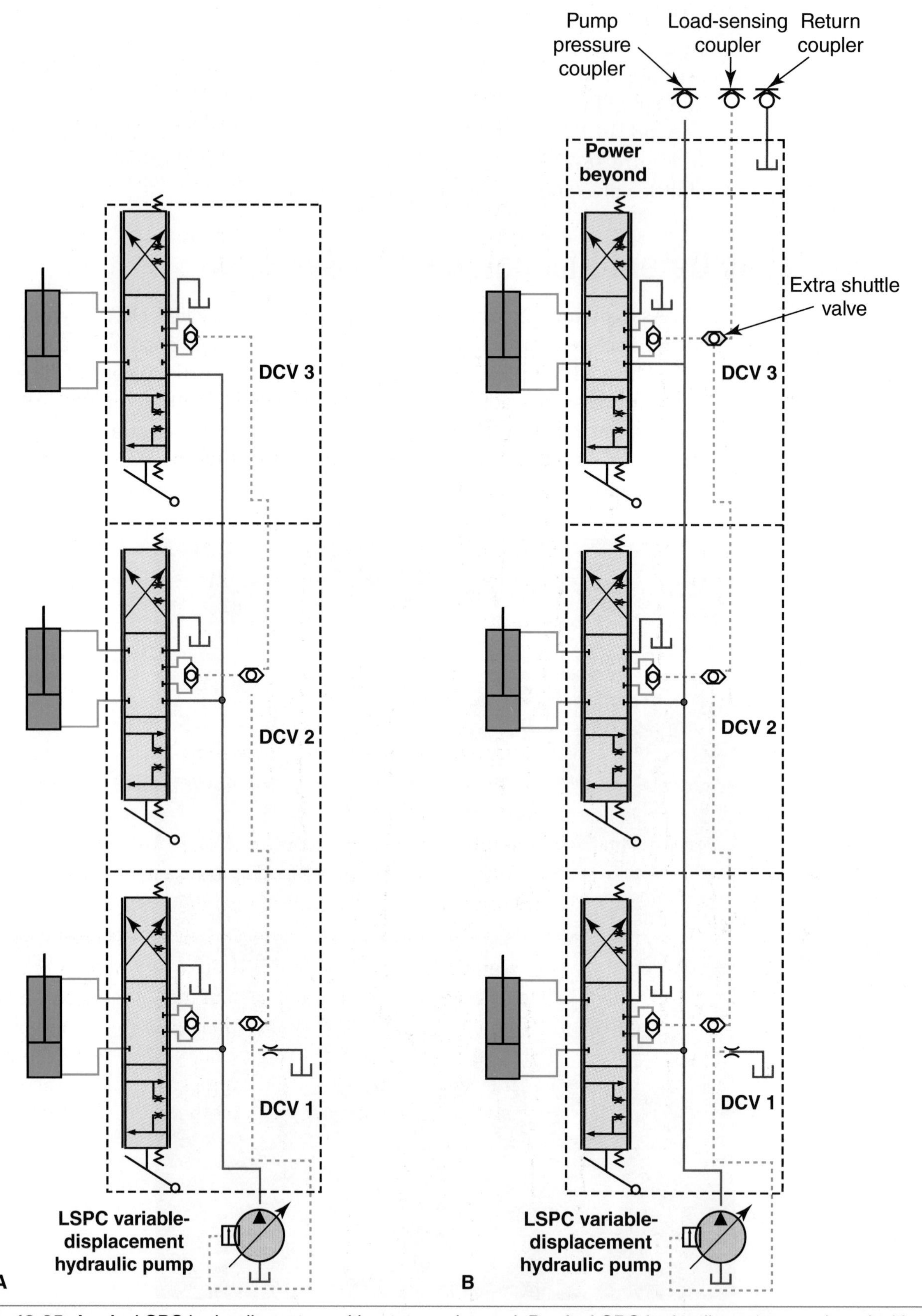

Figure 18-25. A—An LSPC hydraulic system without power beyond. B—An LSPC hydraulic system equipped with power beyond. Note the difference in signal networks between the two systems. An extra shuttle valve is required for the addition of power beyond.

Variable-Displacement LSPC System Advantages

Several decades ago, when hydraulic engineers began to design variable-displacement LSPC systems, they envisioned a hydraulic system that would deliver only the "pressure and flow" required to do the work and nothing more. Today, nearly four decades later, manufacturers are still using these systems on much more complex and expensive machines. LSPC variable-displacement hydraulic systems have the following advantages:

- The system runs at low pressures (300 to 500 psi [21 to 34 bar]) when the DCVs are in a neutral position and produces very little flow (perhaps less than 0.5 gpm).
- The hydraulic system uses only a fraction of a horsepower when the system requires no flow.
- With the DCVs are in a neutral position, the system generates very little heat and very little noise.
- The system can quickly ramp up to deliver hydraulic flow and pressure on demand.

The agricultural industry serves as one of the largest customers for machines equipped with variable-displacement LSPC hydraulic systems. The systems are commonly used on large agricultural tractors and some combine harvesters.

Variable-Displacement LSPC System Disadvantages

LSPC variable-displacement systems do have some disadvantages:

- The system costs more to manufacture than open-center hydraulic systems and PC systems.
- The system requires a complex signal network of primary and secondary shuttle valves that sense the highest system working pressure and send that signal back to the pump compensator.
- Perhaps the largest drawback to LSPC pre-spool compensation systems is that if an operator requests multiple hydraulic functions at one time and if that request was for more oil than the pump is capable of delivering, the actuators with the smallest load will receive the oil first, while the highest system loads will receive no oil. The next chapter will explain the benefit of using flow-sharing post-spool compensation, an advanced style of pressure compensation that will solve this problem.

Summary

- ✓ A load-sensing hydraulic system is designed to operate at a prescribed pressure value above the highest working pressure.
- ✓ Pressure compensation enables a hydraulic actuator to maintain a specific speed based on how far the operator has positioned the DCV spool.
- ✓ Pressure compensation is accomplished by using a primary shuttle valve to sense actuator working pressure and directing the actuator working pressure to a pressure compensator valve.
- ✓ An LSPC system has a pressure compensator in the variable-displacement pump.
- ✓ An LSPC system also has a pressure compensator within the DCV prior to its spool, an arrangement known as pre-spool compensation.
- ✓ The pressure compensator valve counteracts changes to system pressures and flow to help maintain the prescribed actuator speed as dictated by the position of the DCV spool that is set by the operator.
- ✓ When the DCVs in an LSPC fixed-displacement hydraulic system are in a neutral position, the system operates in the standby mode, and the pump delivers maximum flow at a low pressure value, for example 300 to 500 psi.
- ✓ When the DCVs, in an LSPC variable-displacement hydraulic system are in a neutral position, the system operates in the standby mode. The system generates very little flow and low pressure, wasting very little hydraulic horsepower.
- ✓ When an LSPC hydraulic system is in the working mode, an actuator will be moving and the system pressure will equal the highest signal pressure plus margin pressure.
- ✓ When a LSPC fixed-displacement system is in the stall mode, the unloading valve blocks the pump's flow, forcing the oil flow to dump over the main system relief valve. In this mode, the pump generates maximum pressure and flow, equaling maximum hydraulic horsepower.
- ✓ When a LSPC variable-displacement pump is in the stall mode, the flow control spool will be closed and the PC spool will open, directing oil to the control piston to destroke the pump, resulting in high system pressure with little flow.
- ✓ PQ graphs consist of mapping a hydraulic pump's flow rate (Q) at different operating pressures (P).
- ✓ Torque limiting is a pump control that destrokes the pump to prevent the engine from stalling any time the operator simultaneously requests high flow and high pressure.
- ✓ Jammer solenoid valves in variable-displacement LSPC systems will connect pump outlet pressure to signal pressure when activated. This puts the system in stall mode, so that high pressure is available immediately when the operator activates a control.

Technical Terms

corner horsepower
differential pressure
dime valves
flow control spool
isolators
load-sensing (LS) hydraulic system
margin pressure
PQ curve
pressure cutoff spool
primary shuttle valve
resolvers
secondary shuttle valve
signal network
signal pressure
stall mode
standby mode
torque-limiting control
upstream compensation
working mode

Review Questions

Answer the following questions using the information provided in this chapter.

1. Load-sensing pressure-compensating hydraulic systems can be called all of the following, *EXCEPT*:
   A. Closed-center load-sensing system.
   B. Negative flow control system.
   C. Pressure compensation load-sensing system.
   D. Pressure and flow compensated system.
2. Which of the following terms can be defined as a system that is designed to operate at a fixed value above working pressure?
   A. Load sensing.
   B. Pressure compensating.
   C. Stall mode.
   D. None of the above.
3. Which of the following terms can be defined as a system that provides a constant cylinder speed for a fixed position of a DCV spool?
   A. Load sensing.
   B. Pressure compensating.
   C. Stall mode.
   D. None of the above.
4. What is the name of the valve that chooses that higher working port pressure between two different control valves and sends the higher pressure to the pump compensator?
   A. Shuttle valve.
   B. Pressure-reducing valve.
   C. Pressure relief valve.
   D. Drop check valve.
5. What does John Deere's agricultural division call the valve that chooses the higher working port pressure between two different control valves and sends the higher pressure to the pump compensator?
   A. Nickel valve.
   B. Dime valve.
   C. Quarter valve.
   D. Isolator.
6. What does Caterpillar call the valve that chooses the higher working port pressure between two different control valves and sends the higher pressure to the pump compensator?
   A. Restitution valve.
   B. Reiterator valve.
   C. Resolver valve.
   D. Isolator.

7. In what time era did variable-displacement LSPC pumps begin appearing in agriculture equipment?
   A. Late 1950s-early 1960s.
   B. Late 1960s-early 1970s.
   C. Late 1970s-early 1980s.
   D. Late 1980s-early 1990s.
8. What is the name of the valve that has the responsibility of choosing the higher cylinder working pressure, either from the rod end or the cap end of a double-acting cylinder?
   A. Primary shuttle.
   B. Secondary shuttle.
   C. Intermediate shuttle.
   D. None of the above.
9. What is the name of the valve that has the responsibility of choosing the higher working pressure between two different DCVs?
   A. Primary shuttle.
   B. Secondary shuttle.
   C. Intermediate shuttle.
   D. None of the above.
10. A tractor with an LSPC hydraulic system has a total of 4 DCVs, with no hitch and no power beyond. How many total primary shuttle valves are required?
   A. Three.
   B. Four.
   C. Five.
   D. Six.
11. A tractor with an LSPC hydraulic system has a total of 4 DCVs, with no hitch and no power beyond. How many total secondary shuttle valves are required?
   A. Three.
   B. Four.
   C. Five.
   D. Six.
12. Technician A states that LSPC systems can use fixed-displacement pumps. Technician B states that LSPC systems can use variable-displacement pumps. Who is correct?
   A. Technician A.
   B. Technician B.
   C. Both A and B.
   D. Neither A nor B.
13. An LSPC hydraulic system with a plugged orifice in the signal network will exhibit what?
   A. Only low system pressure.
   B. Only moderate system pressure.
   C. Only high system pressure.
14. All of the following can be used to describe an unloading valve's spring, *EXCEPT*:
   A. margin.
   B. differential.
   C. standby.
   D. stall.
15. All of the following are required for a gear pump and two DCVs to be used in a load-sensing system, *EXCEPT*:
   A. closed-center DCVs.
   B. unloading valve.
   C. pressure-reducing valves.
   D. shuttle valves.
16. All of the following are required for a pressure-compensated DCV to operate a bidirectional motor, *EXCEPT*:
   A. closed-center DCV.
   B. unloading valve.
   C. pressure-reducing valve.
   D. shuttle valve.
17. An inline axial variable-displacement pump that is used in an LSPC torque-limiting hydraulic system has how many control spools (used for controlling the pump flow and pressure)?
   A. Zero.
   B. One.
   C. Two.
   D. Three.
18. What is a common system pressure in an LSPC hydraulic system when the control valves are in the neutral position?
   A. 300–550 psi.
   B. 550–1000 psi.
   C. 1000–1500 psi.
   D. 2000–2500 psi.

19. What type of pressure will an LSPC hydraulic system have in the stall mode?
    A. Zero pressure.
    B. Low pressure.
    C. Moderately average pressure.
    D. High pressure.
20. When an LSPC system is running a hydraulic motor, signal pressure plus margin pressure equals _____.
    A. pump outlet pressure
    B. working pressure
    C. regulated pressure
    D. relief pressure
21. Anytime pump outlet pressure is directly connected to signal pressure what will occur?
    A. Standby mode.
    B. Working mode.
    C. Stall mode.
    D. Carryover mode.
22. In a traditional LSPC variable-displacement pump equipped with two spools, what is the name of the small spool that has a hydraulic port for connecting a hydraulic line?
    A. Pressure compensator spool.
    B. Torque-limiting spool.
    C. Flow control spool.
23. When actuated, a signal valve (sometimes called a jammer valve) does what hydraulically in an LSPC system?
    A. Puts the system in a low-pressure standby.
    B. Connects case pressure to signal pressure.
    C. Connects pump outlet pressure to signal pressure.
    D. Lowers pump outlet pressure.
24. Which of the following systems has the ability to provide either high pressures or high flows for a given amount of horsepower?
    A. Open-center.
    B. Pressure compensating.
    C. Load sensing pressure compensating.
    D. LSPC and torque limiting.
25. Within a torque-limiting LSPC pump, which of the following components has the responsibility of conveying the angle of the swash plate?
    A. Pressure compensator spool.
    B. Flow control spool.
    C. Sensing piston.
    D. Bias piston.
26. Power beyond is being added to an LSPC hydraulic system. What else must be added?
    A. An extra shuttle valve.
    B. A relief valve.
    C. Torque-limiting controls.
    D. A variable-displacement pump.
27. Looking at a PQ graph, the corner horsepower point illustrates what system condition?
    A. Minimum pressure, minimum flow.
    B. Maximum pressure, maximum flow.
    C. Wasted horsepower.
    D. Usable horsepower.
28. A variable-displacement LSPC hydraulic system that uses a jammer solenoid for all of its functions can eliminate which of the following from the system?
    A. Signal network.
    B. Flow control valve.
    C. Pressure compensator valve.
    D. Hydraulic pump.
29. An LSPC DCV has a manual detent kickout. The DCV's primary shuttle will send oil to all of the following, *EXCEPT*:
    A. DCV detent kickout.
    B. pressure compensator valve.
    C. DCV secondary shuttle.
    D. main system relief valve.
30. Which one of the following systems will consume the least amount of horsepower when the DCVs are in a neutral position?
    A. Traditional open-center system.
    B. Pressure-compensating system.
    C. Fixed-displacement LSPC system.
    D. Variable-displacement LSPC system.

Manufacturers use downstream compensation in their machines if the operator commonly selects multiple hydraulic functions simultaneously. A motor grader is a common machine that has numerous different hydraulic controls operating at one time, which is more effective if the machine is equipped with a flow-sharing hydraulic system.

Chapter 19

Flow Sharing/ Downstream Compensation

Objectives

After studying this chapter, you will be able to:

- ✓ Describe the benefit of flow-sharing hydraulic systems.
- ✓ Explain how post-spool compensation works.
- ✓ Explain the reason a signal relief valve is required in a post-spool-compensated hydraulic system.
- ✓ Describe the order of oil flowing through a flow-sharing DCV that is controlling a double-acting cylinder.
- ✓ List the pressures found inside a flow-sharing DCV that is operating an actuator.
- ✓ List examples of Caterpillar machines that use PPPC.
- ✓ List examples of Case flow-sharing machines.
- ✓ List other descriptions of flow-sharing hydraulic systems.

Introduction to Flow Sharing/ Post-Spool Compensation

As explained at the end of **Chapter 18**, LSPC pre-spool compensation, has some disadvantages. The most notable disadvantage is that any time an operator requests more flow than the pump is capable of delivering, the actuator with the lowest working pressure will receive oil first and the circuits with higher working pressures will not receive oil until the demands of the weaker circuits have been met.

See **Figure 19-1**. If the operator requests 20 gpm (75 lpm) of flow for DCV 1 and 20 gpm (75 lpm) for DCV 2, and if the pump is only capable of delivering 20 gpm (75 lpm), the cylinder with the lowest working pressure will receive the pump's oil flow, in this case DCV 2. DCV 1 will receive oil only after cylinder 2 reaches its end of travel and system pressure builds.

This disadvantage is magnified in LSPC systems that have numerous hydraulic functions requested simultaneously. A motor grader is a good example of a machine that has numerous hydraulic functions. A typical grader has the following functions: left blade lift, right blade lift, blade side shift, circle turn, circle shift, blade tip, front wheel steer, articulation steer, wheel lean, scarifier lift, ripper lift, and snowplow or dozer lift. An operator could easily select multiple hydraulic functions simultaneously. If the machine is equipped with an LSPC pre-spool compensation hydraulic system, the circuit with the lowest operating pressure will receive the majority of the oil flow first, and prevent oil from flowing to the controls that require higher working pressures.

Consider a LSPC system that contains two hydraulic motors that require different quantities of flow at two different pressures. See Figure 19-2. DCV 1 requires 8 gpm (30 lpm) at 400 psi (28 bar), and DCV 2 requires only 4 gpm (15 lpm), but at twice the pressure, 800 psi (55 bar).

Looking at Figure 19-2, if the pump is operating at max rpm and is capable of delivering only 10 gpm (37 lpm), the sequence of results would be the following:

- Motor 1 would first rotate, because it has the lightest load.
- After motor 1 receives its full supply of oil, motor 2 would begin to spin, but it would rotate slower than the operator requested, because the pump is not capable of delivering the full 4 gpm (15 lpm) requested.

Next consider what would occur if the engine rpm slowed down to an idle speed. What would be the sequence of events?

- Motor 2 would come to a complete stop first.
- Then, motor 1 would slow down to the speed at which the pump is capable of delivering flow.

The system shown in Figure 19-2 has a few other characteristics that are different from those found in a flow-sharing post-spool compensation system.

- The compensator valves are sensing their own individual working pressures.

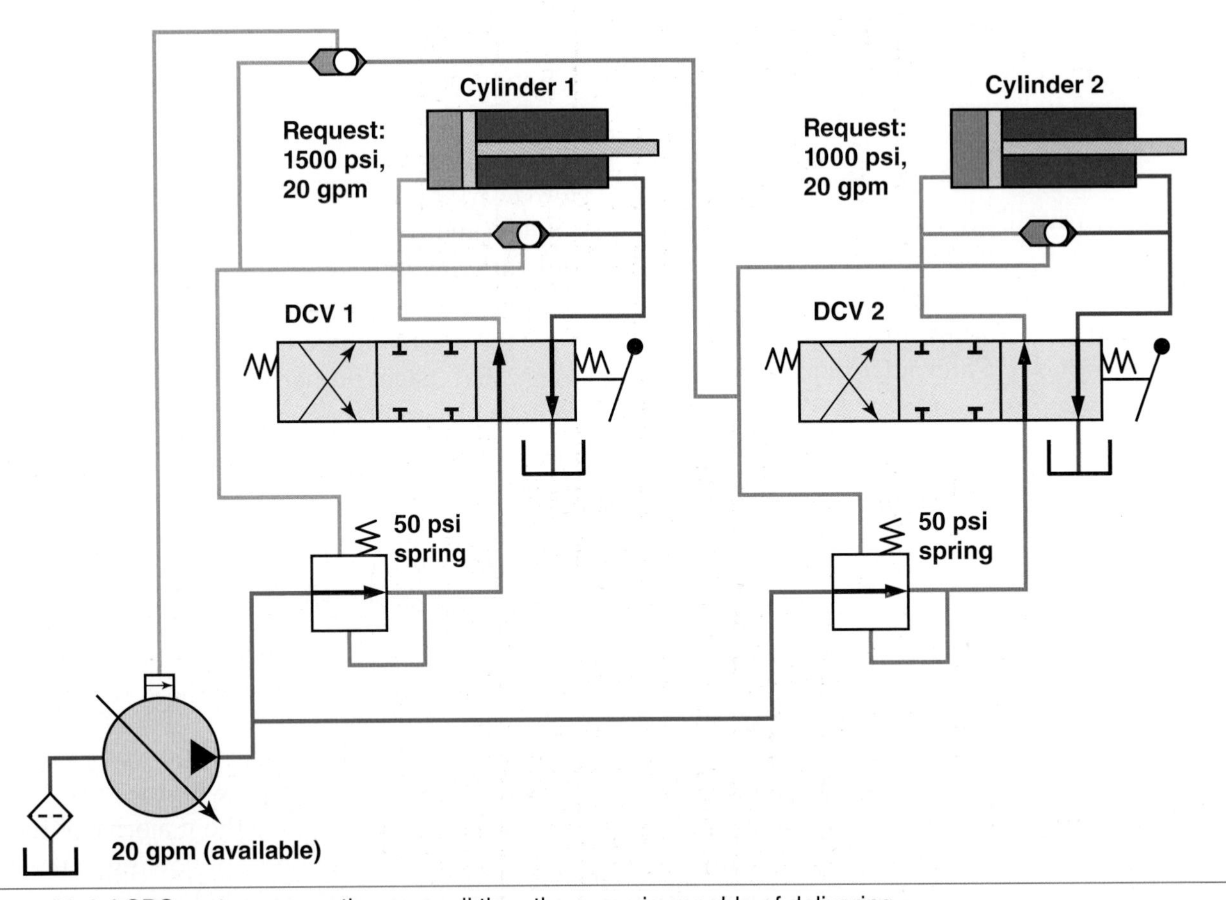

Figure 19-1. LSPC system requesting more oil than the pump is capable of delivering.

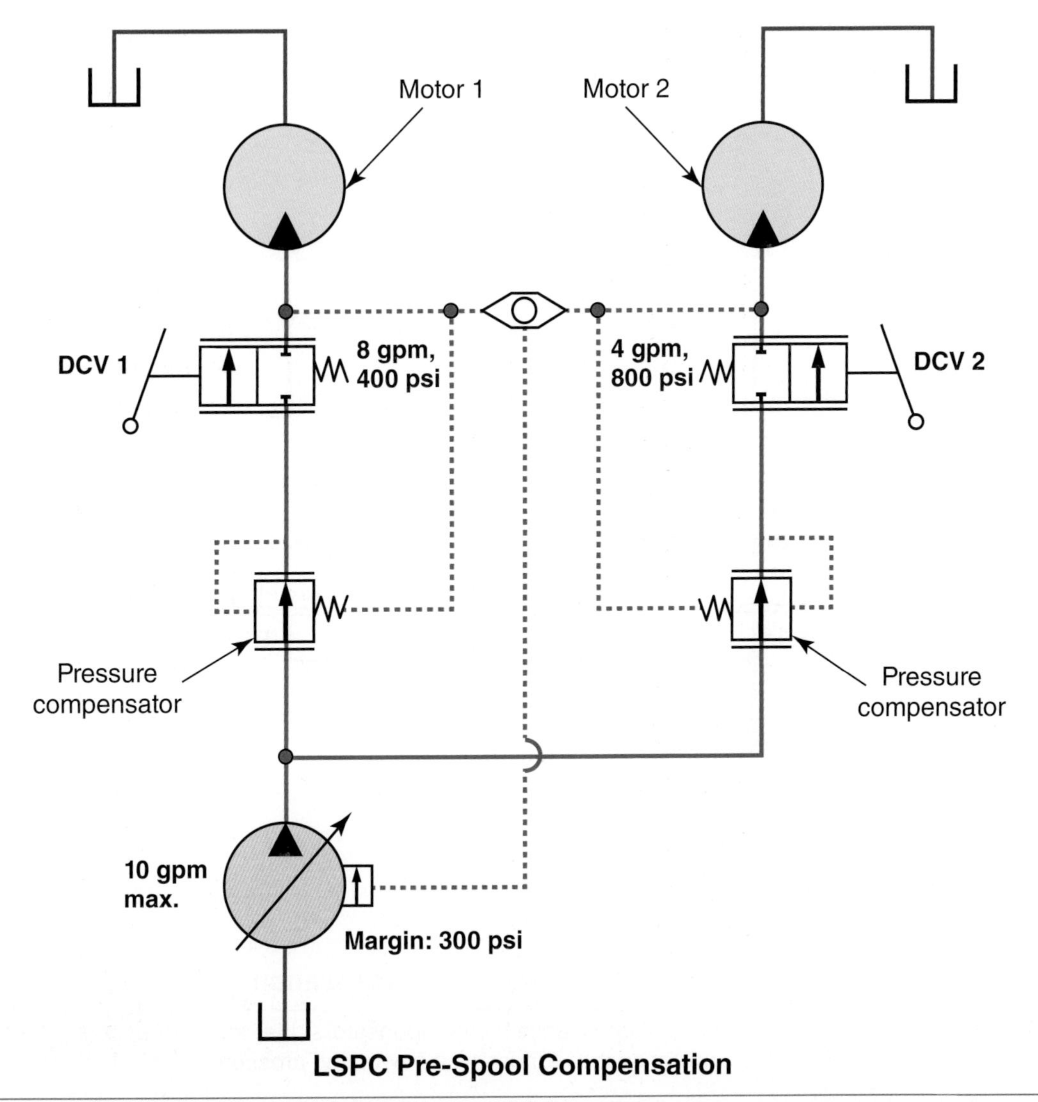

Figure 19-2. An LSPC pre-spool compensation example with two hydraulic motors requesting different flows at two different operating pressures.

- The shuttle valve chooses the highest system working pressure and sends that working pressure to the pump's compensator.

In order to convert a pre-spool LSPC hydraulic system to a flow-sharing system, a few changes must be made to the design:

- The pressure compensator valves must be placed after the DCV spools.
- The shuttle valve must send the highest system working pressure to the pump's compensator, and also to the post-spool compensators. This means that the post-spool compensators no longer sense their own individual working pressure, but instead sense the system's highest working pressure.
- The pressure compensators must be normally closed pressure-relief valves instead of normally open pressure-reducing valves.
- Later, it will be explained why a signal relief valve must also be added to the circuit for downstream, post-spool, compensation. See **Figure 19-3**.

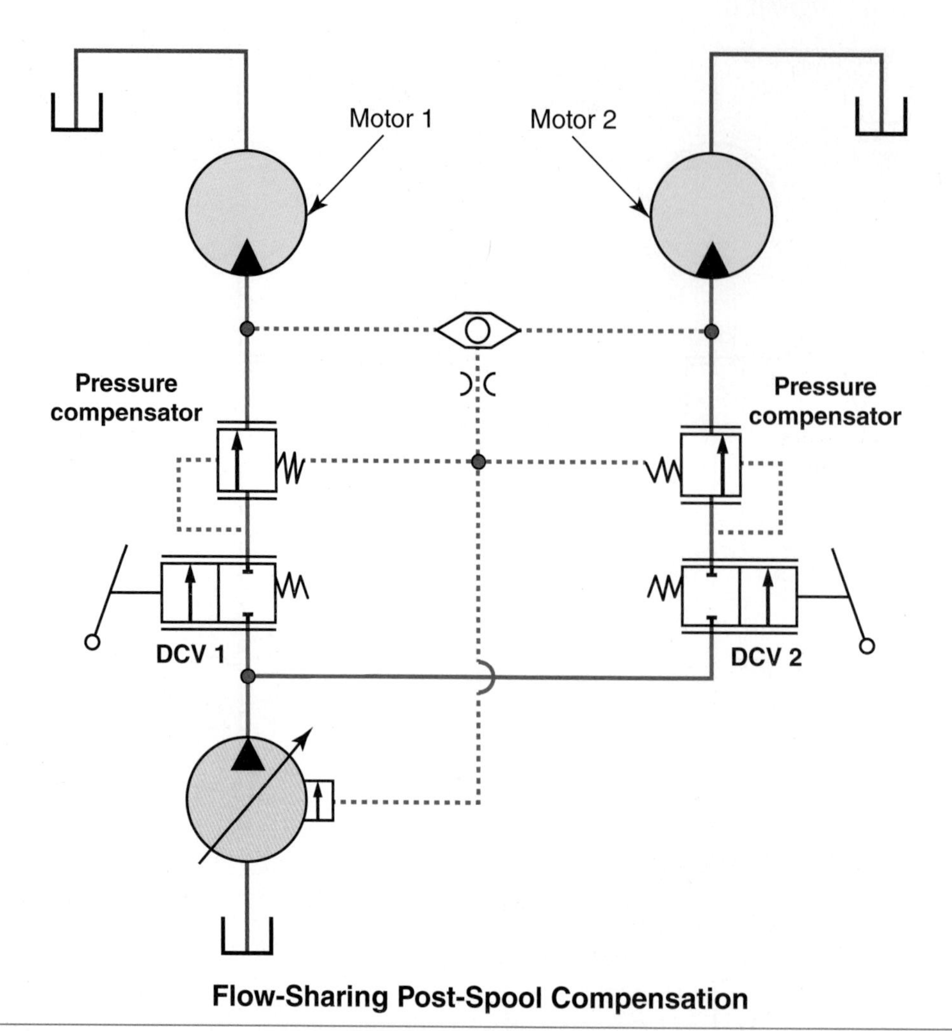

Figure 19-3. Flow-sharing post-spool compensation places the compensators after the DCV spools. The compensators are normally closed and sense the highest system working pressure.

A ***flow-sharing hydraulic system***, also known as ***post-spool-compensated*** or ***downstream compensation*** hydraulic system, will proportion the available flow to each of the DCVs based on the DCV's position. For example, if an operator shifted DCV 1 to 100% open position, DCV 2 to 50% open, and shifted DCV 3 to 25% open, the hydraulic system will proportion the system's flow to each of those DCVs based on the percentage requested by the operator. In this scenario, if the pump was capable of delivering 35 gpm (132 lpm), flow would be proportioned as follows:

- DCV 1 would receive 20 gpm (75 lpm).
- DCV 2 would receive 10 gpm (37 lpm) (half of the request of DCV 1).
- DCV 3 would receive 5 gpm (19 lpm) (half of the request of DCV 2).

How Post-Spool Compensation Works

In **Chapter 18**, it was explained that installing a pressure compensator valve prior to the DCV spool provided the benefit of having a very small pressure drop across the DCV spool, approximately 40 to 90 psi (3 to 6 bar).

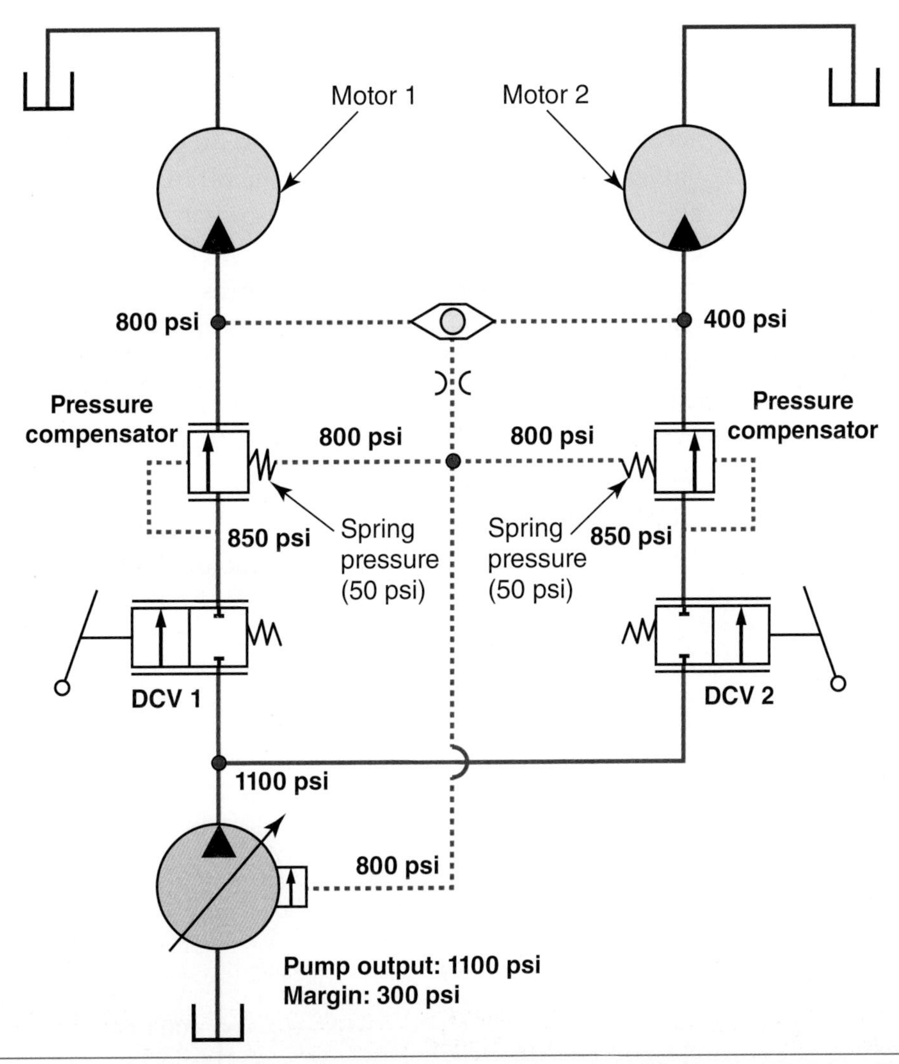

Figure 19-4. Post-spool compensation will proportion oil flow to the actuators in the percentages requested by the operator.

Refer to **Figure 19-1**. In LSPC systems, this pressure compensator senses its own working pressure, not the highest system working pressure. This design cannot proportion flow equally to all of the DCVs.

Two changes must be made to convert the LSPC system to flow sharing:

- The pressure compensator valves must be placed after the DCV spools.
- All of the pressure compensator valves must sense the highest system working pressure.

See **Figure 19-4**. Two key characteristics enable the post-spool compensation to proportion flow to the actuators:

- Margin pressure drops any time the requested flow is more than the pump can supply. Refer back to **Chapter 18** for a review of margin pressure. Why does margin pressure drop? When an operator requests too much flow, the pump cannot produce enough flow to maintain the pressure that the margin spool is requesting. When margin pressure drops, it causes a drop in pump outlet pressure, which effectively reduces the pressure drop across the DCV spool.

- Whenever the DCV spool experiences a reduced pressure drop, the working pressure remains the same, which enables the pressure compensator valves to further restrict and limit flow going to the actuators. This process occurs across all of the pressure compensator valves, which effectively divides the oil in proportion to the actuators, thus "flow sharing". Or said another way, when the combined flow request from all of the DCVs is more than the pump can maintain, the highest working pressure acts on all of the pressure compensator valves, and those valves simultaneously begin to limit flow to the actuators.

Considering Figure 19-4, notice that the pressure drop across the DCV spools is essentially margin pressure minus the pressure compensator valve's spring value:

300 psi–50 psi = 250 psi

It is this pressure that drops whenever too much flow is requested from the pump.

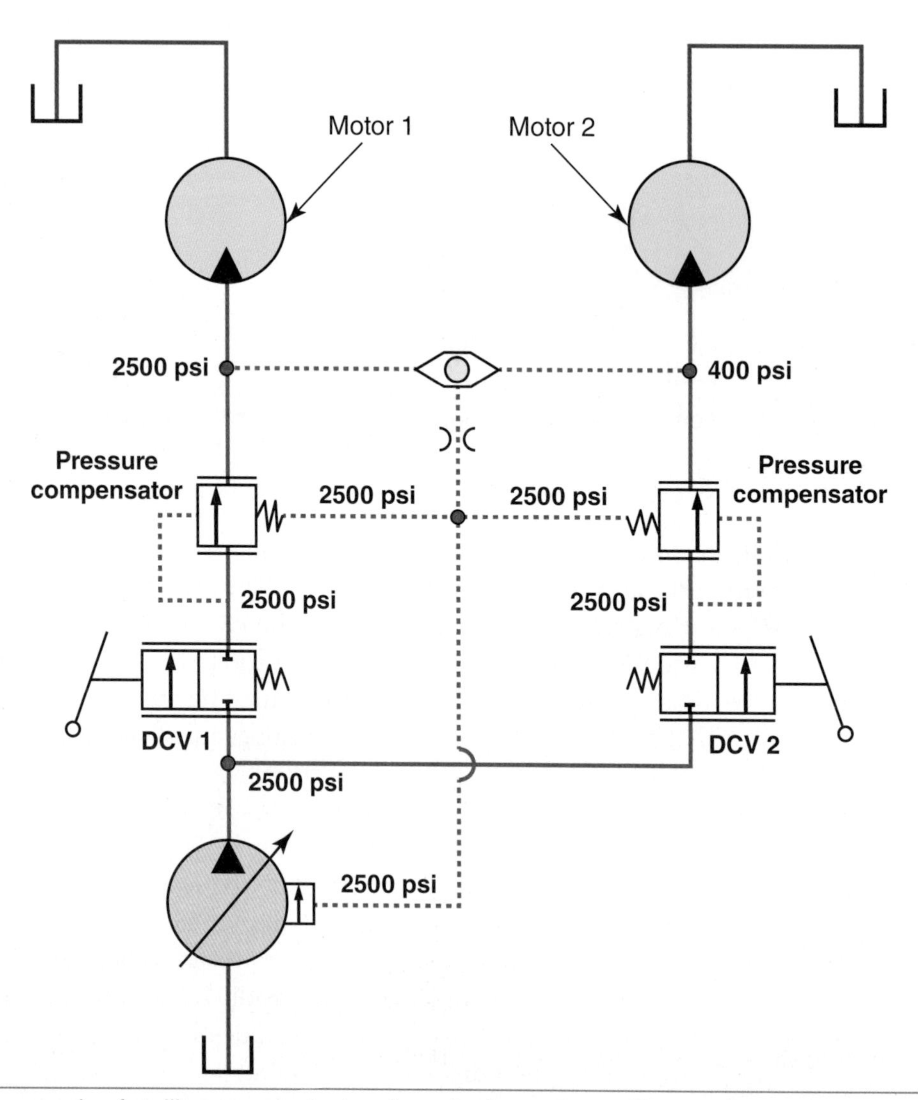

Figure 19-5. An example of stalling an actuator in a flow-sharing system without a signal relief valve.

Why Use a Signal Relief Valve?

The system shown in Figure 19-4 does have one drawback. If one of the actuators stalls at the pump's cutoff value, then all of the flow will be shut off to the actuators. How is this possible?

Consider that the pump's PC spool is set at 2500 psi (172 bar). Imagine the hydraulic motor is attempting to operate a winch, and the hydraulic motor stalls while attempting to winch a heavy load.

In Figure 19-5, motor 1 has reached the pump cutoff value of 2500 psi (172 bar). As a result, the signal pressure and the pump outlet pressure both will rise to a value of 2500 psi (172 bar). The pressure compensator valves now have 2500 psi (172 bar) being supplied to their inlet; however, the compensator valves also have 2500 psi (172 bar) signal pressure plus 50 psi (3 bar) spring pressure acting on the opposite side of the valve, and the combined 2550 psi (175 bar) will effectively cut off all flow to the actuators. A ***signal relief valve,*** also known as a ***signal limiter valve,*** will resolve this problem. See Figure 19-6.

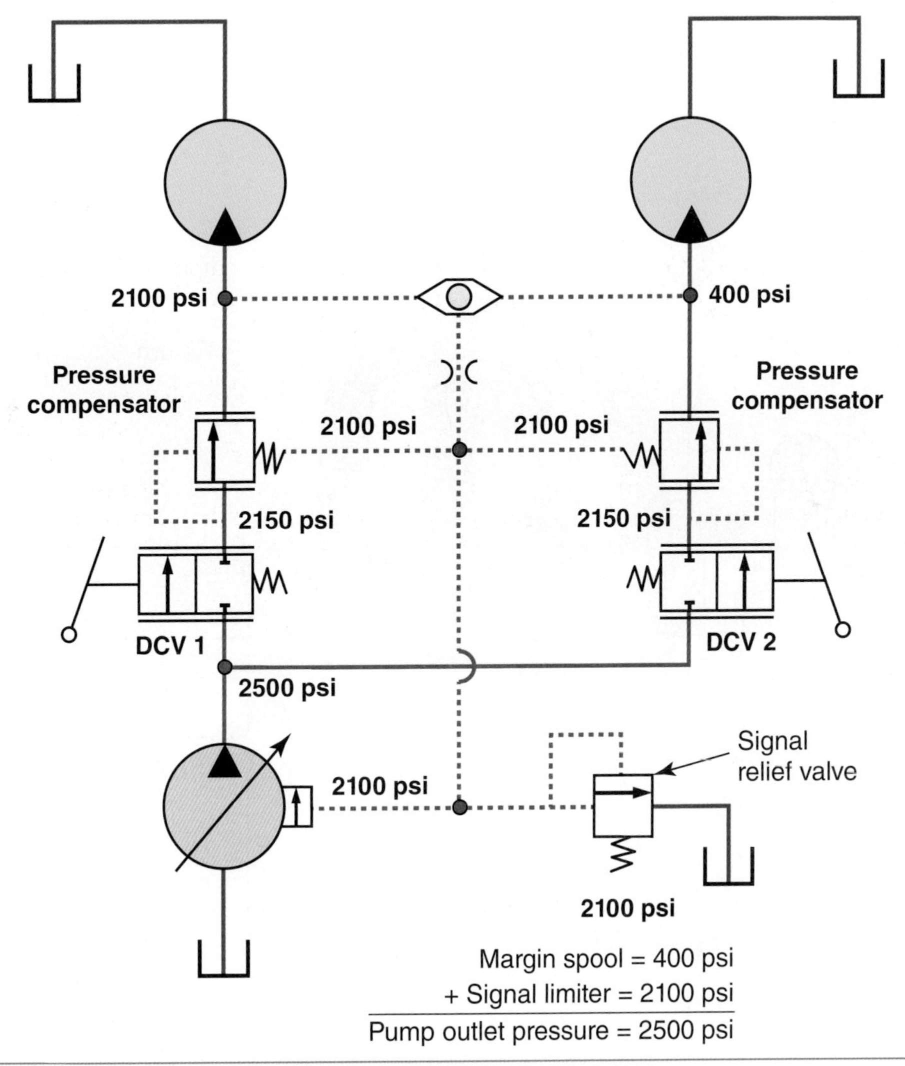

Figure 19-6. A signal relief valve in a flow-sharing system will allow the system to maintain a pressure drop across the DCV spool.

The value of the signal relief valve plus the value of margin pressure, instead of the pump's PC spool, now sets the pump's maximum output pressure. The use of a signal relief valve allows the hydraulic system to maintain margin pressure, which prevents the pressure compensator valves from blocking off all system flow. The signal relief valve enables the hydraulic system to maintain a pressure drop across the DCV spools, which allows flow to be metered to the actuators, even if one of the actuators stalls.

When using a signal relief valve, the pump may not have a PC spool valve. Sometimes the machine will continue to use both the external signal relief valve and a pump PC spool. Having both components in the system offers one advantage and one disadvantage. The advantage is the pump PC spool can act as a backup in case the signal relief valve fails. The disadvantage is that if a technician incorrectly adjusts the pump's PC spool to a value that is less than the combined values of the margin spool and the signal relief valve, then whenever an actuator stalls, all flow will be cutoff to the actuators. For this reason, whenever a machine uses both a pump PC spool and a signal relief valve, you must pay close attention to the valve adjustments and follow the procedures accurately.

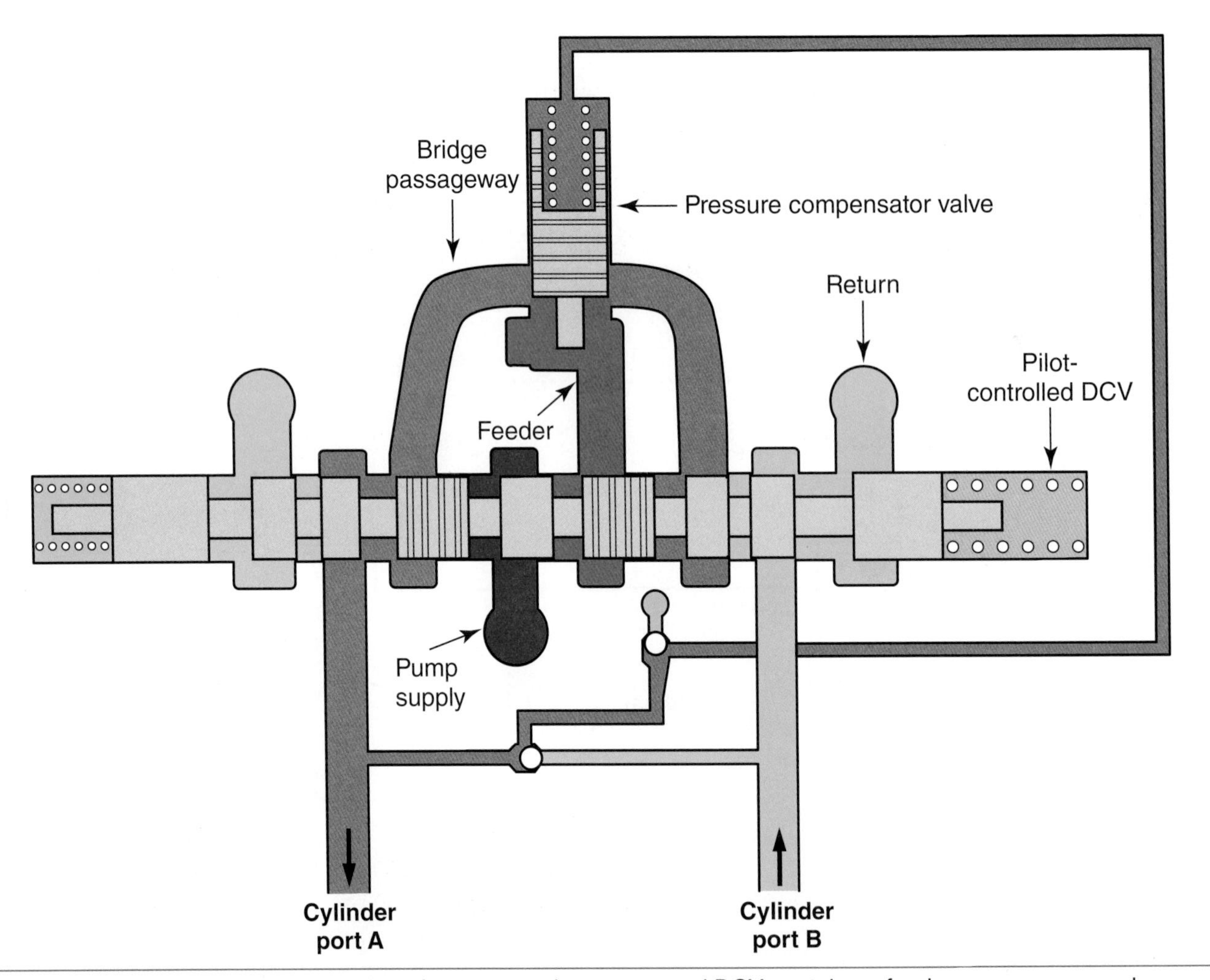

Figure 19-7. A cross-sectional drawing of a post-spool-compensated DCV contains a feeder passageway and a bridge passageway.

Flow Sharing with Double-Acting Actuators (Passageways)

The previous illustrations, Figures 19-2 through 19-6, oversimplify flow-sharing hydraulic systems because the actuator is a unidirectional/single-acting hydraulic motor. However, most flow-sharing hydraulic systems need to be able to control double-acting actuators. This adds more complexity to the system. For example, because the compensator valve must be located after the DCV spool and because the compensator must be able to compensate during cylinder extension and retraction, two additional passageways are added to a traditional DCV.

The first passageway connects the DCV spool to the compensator valve. Caterpillar calls it the ***feeder passageway***, John Deere's construction division calls it the ***compensator passageway*** and Case's construction division calls it the ***intermediate passageway***. The next passageway routes the oil from the compensator valve back to the DCV spool. This passageway is commonly labeled the ***bridge passageway***.

Figure 19-7 provides a cross-sectional drawing of a flow-sharing DCV. During cylinder operation, oil will flow through the DCV in the following order:

- Flows into the DCV.
- Past the DCV spool.
- Into the feeder passage/compensator passage/intermediate passage.
- To the compensator valve.
- Through the bridge passageway.
- Back to the DCV spool.
- Out to the hydraulic cylinder.

Figure 19-8 illustrates the order of oil flow that takes place in a post-spool-compensated DCV.

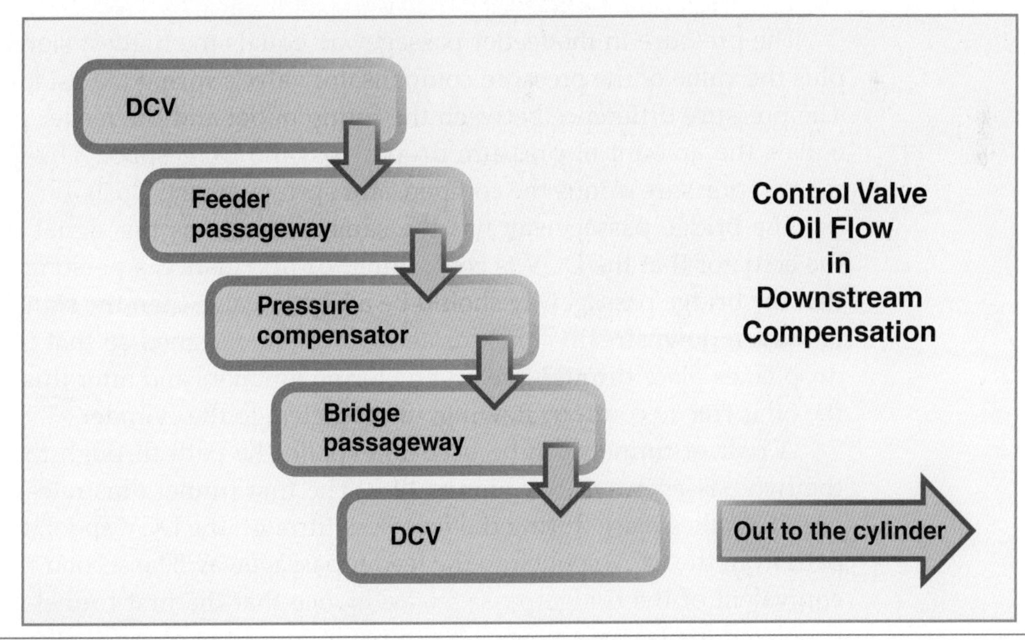

Figure 19-8. The order of oil flowing through a downstream post-spool-compensated DCV is more complex than the average DCV. It flows through the DCV spool twice, once before the compensator and once after the compensator.

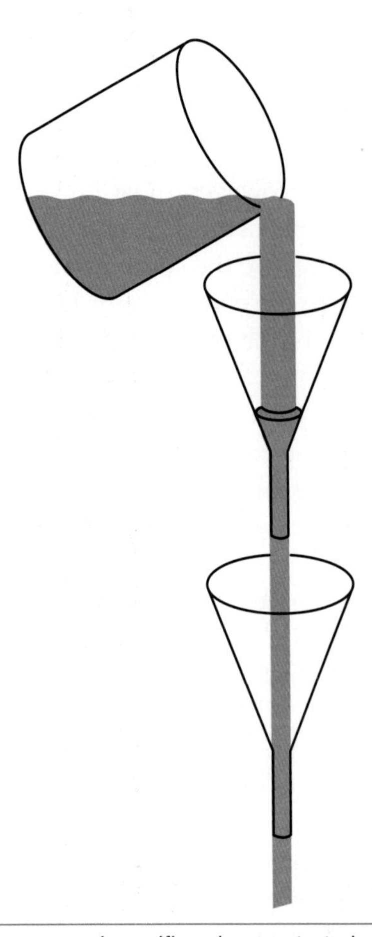

Figure 19-9. Two funnels with the same size orifice demonstrate how oil will flow uninhibited through the bridge passageway because the resistance occurred through the first pass of the DCV spool valve.

The pressure in the feeder passageway equals the highest signal pressure plus the value of the pressure compensator valve's spring (50 psi for example). The pressure difference between the pump outlet and the feeder passageway equals the amount of pressure drop across the DCV spool. This pressure is margin pressure minus the compensator spring value (50 psi).

The bridge passageway has the same pressure as the signal pressure of the actuator that the DCV is controlling. At first glance, a person might think that the bridge passageway should be a higher value than the signal pressure. However, downstream-compensated DCVs are designed so that the pressure drop takes place through the first metering junction, and after this restriction, the oil is free to continue flowing, unrestricted, to the cylinder.

A pair of funnels can be used to explain the path through the DCV and the two passageways. See **Figure 19-9**. The first funnel illustrates the restriction that takes place during the first pass through the DCV spool, which is the path from the DCV spool into the feeder passageway. The second funnel is the equivalent of the bridge passageway. Notice that the first funnel (junction of spool and feeder passageway) has already restricted oil flow, allowing the oil to flow freely through the second funnel (bridge passageway).

Operating Pressures in a Downstream-Compensated System with Demands Met

A downstream-compensated hydraulic system with two double-acting DCVs is illustrated in Figure 19-10. This illustration is based on the system needs being met, meaning that margin can be maintained and has not diminished. Notice that DCV 1 (on the left) is actuating a cylinder at 1800 psi (124 bar), and DCV 2 is actuating a cylinder at 1200 psi (83 bar). Note that the margin spool is set at 300 psi (21 bar). Based on the cylinder loads, the following pressures can be identified:

Pump output = 2100 psi (145 bar)
Feeder passages DCV 1 and DCV 2 = 1840 psi (127 bar)
Bridge passage DCV 1 = 1800 psi (124 bar)
Bridge passage DCV 2 = 1200 psi (83 bar)
Pressure drop across DCV 1 = 260 psi (18 bar)
Pressure drop across DCV 2 = 260 psi (18 bar)
Pressure drop across compensator DCV 1 = 40 psi (3 bar)
Pressure drop across compensator DCV 2 = 640 psi (39 bar)

Note: the compensator operating the lower pressure is more restricted than the compensator operating at the high pressure.

Caterpillar Proportional Priority Pressure Compensation (PPPC)

Caterpillar uses a flow-sharing load-sensing hydraulic system that they call ***proportional priority pressure compensation (PPPC).*** It was first introduced on the 5000 series shovels and H series motor graders. PPPC is commonly used on numerous Caterpillar machines. Some examples are: H series motor graders, M series motor graders, K series motor graders, 5000 series shovels, B and C large tracked excavators, M300 wheeled excavators, H series wheel loaders, G series midsized wheel loaders, TH 62/TH 63/TH 82/TH83 Telehandlers, B series Telehandlers, and E series backhoes.

Case Machines with Downstream Compensation

Case New Holland (CNH) manufactures both agricultural and construction equipment. Like other construction equipment manufacturers, they build machines with post-spool compensation. Some examples of CNH equipment with post-spool compensation are wheel loaders, dozers, backhoes, and telehandlers.

721F, 821F, and 921F Wheel Loaders

The Case 721F, 821F, 921F, and the W170C and W190C New Holland wheel loaders use downstream flow-sharing compensation and a variable-displacement pump. The DCVs on these machines are similar to the Caterpillar H series wheel loaders, E series backhoes, and D series track loaders. The shuttle valves have been eliminated. In these systems, the DCVs' pressure compensator

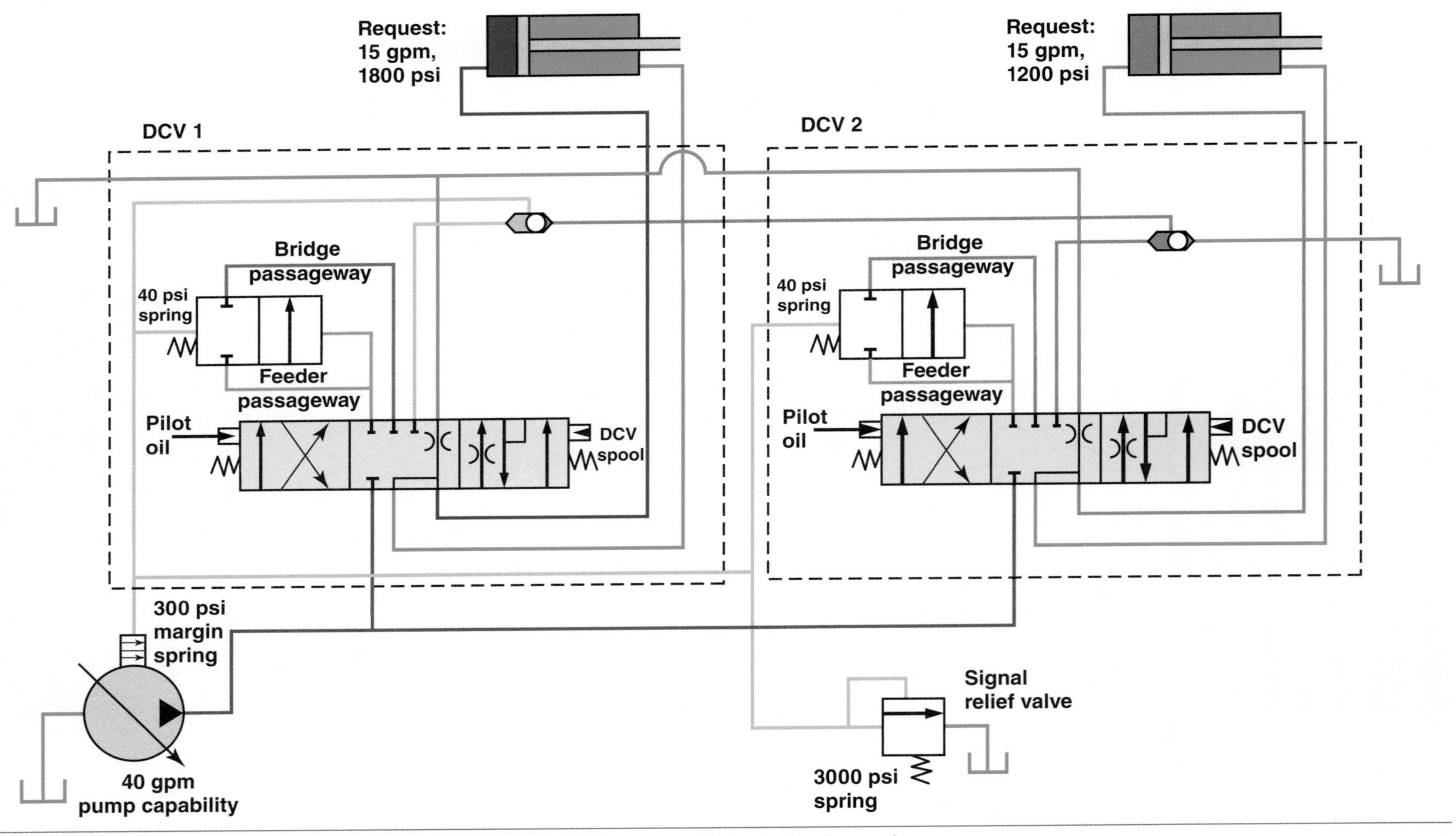

Figure 19-10. An example of a post-spool-compensated hydraulic system with the hydraulic needs met.

valves serve a second purpose. In addition to being pressure compensators, the valves also distribute the system's highest working pressure. All of the pressure compensator valves essentially block signal pressure from being sent, except for the pressure compensator valve operating at the highest signal pressure. That compensator valve will send the highest system signal pressure. As a result, the compensator valves effectively replace the shuttle valve network found in the systems described earlier.

1650K and 1850K Crawler's Use of an Unloading Valve

The Case 1650K and 1850K series I, II, and III large dozers all use a variable-displacement piston pump and are downstream post-spool compensated. The pump has three spools in the compensator assembly: a low-pressure standby spool (flow-control spool), high-pressure standby (PC) spool, and a torque-limiting spool. Even though the pump contains a low-pressure standby spool, the spool is not used.

This hydraulic system has an unloading valve in the DCV valve block. When the DCVs are in a neutral position, the unloading valve unloads the pump at 500 psi and flows approximately 5 gpm. This minimum 5 gpm flow at 500 psi provides the operator a hydraulic system that is more responsive and reactive.

The differential steer tractors use a hydraulic motor for the steering countershafts. Normal hydraulic functions operate at 3000 psi (207 bar) and the steering motors can request the pump to operate up to 5500 or 6000 psi (379 to 414 bar).

B-Series Motor Grader and 580/590N and SN Loader Backhoes

Other Case machines that employ downstream compensation are motor graders and loader backhoes. The B series Case motor graders use mechanically actuated DCVs. The pilot-controlled loader backhoes, for example 580 SN and 590 SN, use downstream compensation only for the following backhoe functions: stick, swing, bucket, boom, auxiliary, and extended-hoe. The stabilizers do not use any compensators. The steering and loader functions are controlled with an open-center gear pump.

Other Terminology for Flow Sharing

Many manufacturers in the industry incorporate post-spool compensation, but use different terminology. Hydraulic manufacturer Linde calls their system Linde Synchron Control (LSC). LSC was first introduced in 1984. Bosch Rexroth refers to their downstream-compensation systems as flow-matching systems. They also use the German nomenclature, Lastdruck Unabhangige Durchfluss Verteilung (LUVD).

Summary

- ✓ Post-spool compensation equally proportions oil flow to the actuators in the percentages requested by the DCV spool.
- ✓ Post-spool compensators sense the highest actuator working pressure.
- ✓ Post-spool compensators are normally closed.
- ✓ A signal relief valve is required in post-spool-compensated systems to maintain a pressure drop across the DCV spool when one of the actuators is stalled.
- ✓ The feeder passageway delivers oil from the DCV spool to the compensator.
- ✓ The bridge passageway delivers oil from the compensator back to the DCV spool.
- ✓ Caterpillar calls their flow-sharing hydraulic systems PPPC.

Technical Terms

bridge passageway
compensator passageway
downstream compensation
feeder passageway
flow-sharing hydraulic system
intermediate passageway
post-spool-compensated
proportional priority pressure compensation (PPPC)
signal limiter valve
signal relief valve

Review Questions

Answer the following questions using the information provided in this chapter.

1. Why are signal limiters used in flow-sharing hydraulic systems?
   A. To maintain margin pressure when an actuator is stalled.
   B. To magnify margin pressure when an actuator is stalled.
   C. To deplete margin pressure when an actuator is stalled.
   D. To eliminate margin pressure when an actuator is stalled.
2. Within a pre-spool LSPC system, what would the result be if the system capacity (pump capacity) was 20 gpm, and DCV 1 was requesting 15 gpm at 2000 psi, DCV 2 was requesting 10 gpm at 1000 psi, and DCV 3 was requesting 10 gpm at 500 psi?
   A. DCVs 1, 2, and 3 would all receive oil.
   B. DCVs 1 and 2 would receive oil.
   C. DCVs 2 and 3 would receive oil.
   D. DCVs 1 and 3 would receive oil.
3. A pre-spool LSPC system can be converted to a post-spool system by moving which of the following valves from in front of the DCV to the downstream side of the DCV?
   A. Dump valve.
   B. Relief valve.
   C. Compensator valve.
   D. Shuttle valve.
4. In a downstream post-spool-compensated flow-sharing DCV, the bridge passageway routes oil to the _____.
   A. cylinder
   B. DCV spool
   C. feeder passage
   D. pressure compensator

5. In a downstream post-spool-compensated flow-sharing DCV, when the oil first arrives at the DCV spool, where is the oil flow routed next?
   A. Bridge passageway.
   B. Cylinder.
   C. Feeder passageway.
   D. Pressure compensator.
6. In a downstream post-spool-compensated flow-sharing DCV, where does the feeder passage route the oil to next?
   A. Bridge passageway.
   B. Cylinder.
   C. DCV spool.
   D. Pressure compensator.
7. In a downstream post-spool-compensated flow-sharing DCV, where does oil go after the pressure compensator valve?
   A. Bridge passageway.
   B. Cylinder.
   C. DCV spool.
   D. Feeder passageway.
8. The feeder passageway can be called all of the following, *EXCEPT*:
   A. intermediate passageway.
   B. pressure compensator passageway.
   C. DCV spool valve passageway.
   D. None of the above.
9. In a downstream-compensated flow-sharing hydraulic system, the pressure compensator valves sense which of the following?
   A. Only their individual working pressures.
   B. The sum of all the working pressures.
   C. The highest system working pressure.
   D. Pump outlet pressure.
10. In an upstream-compensated LSPC hydraulic system, the pressure compensator valves sense which of the following?
   A. Only their individual working pressures.
   B. The sum of all the working pressures.
   C. The highest system working pressure.
   D. Pump outlet pressure.
11. Which of the following Caterpillar machines was the first to use PPPC hydraulics?
   A. H series motor graders.
   B. G series wheel loaders.
   C. B series telehandlers.
   D. E series backhoes.
12. In a flow-sharing downstream-compensation hydraulic system, if one DCV is stalled and if the signal relief pressure is set above the pump cutoff spring value, what will happen?
   A. The DCV that is stalled will be the only DCV to lose all flow.
   B. All DCVs will lose all flow.
   C. All DCVs will lose oil flow proportionally to pump maximum flow.
   D. None of the above.
13. Which of the following companies calls their flow-sharing system a flow-matching system?
   A. Bosch Rexroth.
   B. Eaton.
   C. Sauer Sundstrand.
   D. Linde.
14. Which of the following companies call their flow-sharing system "LSC"?
   A. Bosch Rexroth.
   B. Eaton.
   C. Sauer Sundstrand.
   D. Linde.
15. Two DCVs are operating two hydraulic motors with downstream compensation. One motor is operating at 1200 psi; the other motor is operating at 600 psi. Which pressure compensator valve is the most open?
   A. The pressure compensator valve operating the 1200 psi motor.
   B. The pressure compensator valve operating the 600 psi motor.
   C. Not enough information to know.

16. Two DCVs are actuating two hydraulic cylinders. One cylinder is extending at 700 psi, the other is extending at 1400 psi. Which pressure compensator valve is closed the most?
    A. The pressure compensator valve operating the 1400 psi cylinder.
    B. The pressure compensator valve operating the 700 psi cylinder.
    C. Not enough information to know.
17. In a downstream post-spool-compensated hydraulic system, which of the following is considered the same pressure as the cylinder's working pressure?
    A. Oil prior to the DCV.
    B. Feeder passageway.
    C. Bridge passageway.
    D. None of the above.
18. In a flow-sharing (downstream-compensated) hydraulic system, any time the operator exceeds the pump's capacity, what takes place?
    A. Margin pressure increases.
    B. Margin pressure drops.
    C. Margin pressure remains the same.
    D. None of the above.
19. Where is the pressure compensator located in flow-sharing, flow-matching, or Caterpillar PPPC hydraulic systems?
    A. Pre-spool.
    B. Post-spool.
    C. Both A and B.
    D. Neither A nor B.
20. In flow-sharing, flow-matching, or Caterpillar PPPC hydraulic systems, what does the pressure compensator valve sense?
    A. Its individual load-sensing pressure for that individual cylinder.
    B. The highest signal pressure.

Chapter 20 Excavator Pump Controls

Objectives

After studying this chapter, you will be able to:

- ✓ Describe the relationship between control pressure and pump flow for NFC and PFC systems.
- ✓ Explain minimum pump flow.
- ✓ Explain how an excavator pump regulator upstrokes and destrokes the pump.
- ✓ Explain the fundamental operation of an NFC hydraulic pump control.
- ✓ Explain the fundamental operation of a PFC hydraulic pump control.
- ✓ List the different sensing pressures used in NFC and PFC systems.
- ✓ Describe the operation of an EH PFC hydraulic system.

Introduction to Negative Flow Control (NFC) and Positive Flow Control (PFC) Systems

Traditional open-center systems use fixed-displacement pumps that generate constant flow at low pressures when the DCVs are in a neutral position. In traditional closed-center hydraulic systems, variable-displacement pumps use DCVs that block pump oil flow when they are in the neutral position. Negative flow control (NFC) and positive flow control (PFC) systems break both of those conventions. NFC and PFC systems use open-center DCVs and variable-displacement pumps. A ***negative flow control*** system, also known as negative control, is designed to upstroke the pump when a signal pressure (known as NFC pressure) drops. A ***positive flow control*** system, also known as positive control, is designed to upstroke the pump when a signal pressure (known as PFC pressure) increases. This chapter explains the fundamental operation of both systems.

NFC and PFC variable-displacement pumps produce ***minimum pump flow*** when their open-center DCVs are in a neutral position. This is different than traditional open-center hydraulic systems, which deliver maximum flow while the DCVs are in neutral. Also, the term "minimum" has a different meaning than it does when it is applied to traditional closed-center systems, which produce practically no flow when the DCVs are in a neutral position. As a general rule of thumb for both NFC and PFC, minimal displacement equals 15% to 25% of the pump's maximum displacement. This means that if an NFC or PFC pump has the ability to flow 100 gpm (378 lpm) when it is fully upstroked, the pump will flow approximately 15 to 25 gpm (57 to 95 lpm) when it is destroked.

The hydraulic excavator is one machine that regularly uses either the NFC or the PFC hydraulic system.

Note

Kawasaki is a hydraulic pump manufacturer that builds NFC and PFC pumps. Some Kawasaki pump controls, for example K3V and K5V, can easily be converted from NFC to PFC by moving the regulator's feedback linkage pivot point. This linkage will be explained later in the chapter.

Pump Regulators

Excavators frequently use two NFC pumps or two PFC pumps for providing oil for propulsion, boom, stick/arm, bucket, swing, and work tool controls. NFC and PFC systems commonly employ variable-displacement inline axial or bent-axis piston pumps.

Each pump is controlled by its own regulator. The ***pump regulator*** consists of a single, double-acting, servo piston that has a large end and a small end. See Figure 20-1. The pump regulator contains spool valve assemblies that have the responsibility to hydraulically actuate the pump's servo piston. As the servo piston actuates, it is attached to the pump swash plate and will increase or decrease pump flow depending on the movement of the servo piston.

NFC and PFC regulators are designed in such a way that the small servo piston end always has pilot oil pressure acting on it. The small servo piston end has the responsibility of moving the swash plate to a maximum angle to upstroke the pump to maximum flow. The small servo piston end upstrokes the pump when the regulator drains oil away from the large servo piston end.

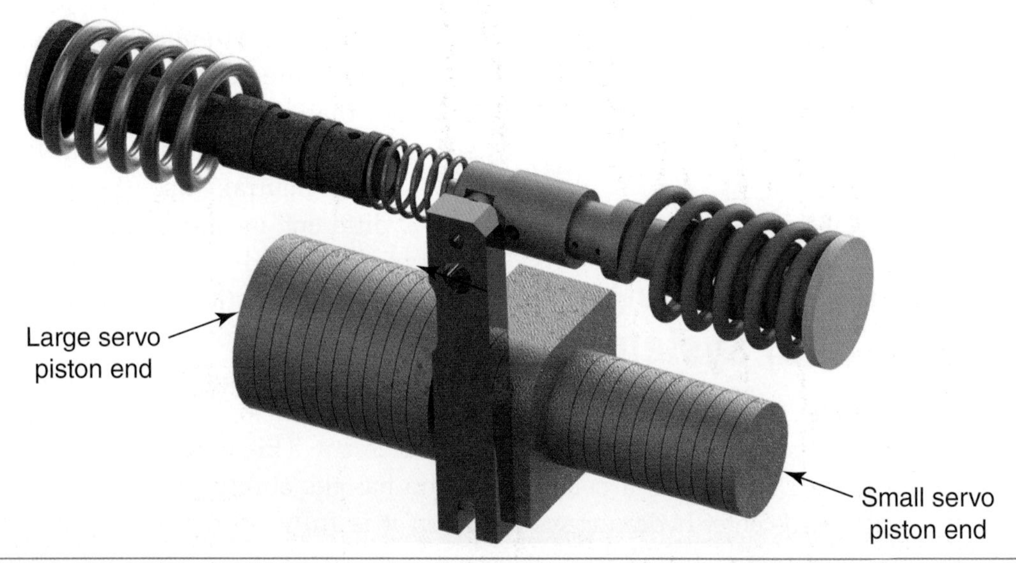

Figure 20-1. An example of an NFC pump regulator assembly. Notice the servo piston contains a large end and a small end.

The large servo piston end has the responsibility of moving the pump to a minimum angle, reducing the pump output to minimal flow. When the regulator directs oil to the larger servo piston, oil is acting on both the large servo piston end and the small servo piston end. The large servo piston end has the largest area and, therefore, the greatest force. As a result, the servo piston moves in the direction of the small end, destroking to the minimum flow. See **Figure 20-2**.

Figure 20-2 is an example of an NFC pump regulator similar to regulators found on Caterpillar and Case excavators. Kawasaki hydraulics labels this style of NFC regulator constant horsepower control. The ***constant horsepower control*** contains a regulator valve that senses the pump's outlet pressure. When that pressure increases too high, the regulator valve will destroke the pump back to a minimum position.

One hydraulic pressure that was intentionally left out of the schematic in **Figure 20-2** is ***companion pump pressure,*** which is the outlet pressure of the second piston pump. With the addition of the companion pump control pressure, Kawasaki labels the regulator ***total horsepower control,*** because each pump regulator can be destroked by sensing its own pump outlet pressure or the companion pump outlet pressure. See **Figure 20-3**.

Most excavators are configured in such a way that the operator can request more hydraulic horsepower than the engine is capable of delivering. For example, it is possible on some models that just one of the two pumps can require up to 85% of the engine's horsepower. For this reason, many excavator pump regulators, PFC or NFC, also sense their companion pump pressure to avoid stalling the engine. This will be discussed later in this chapter.

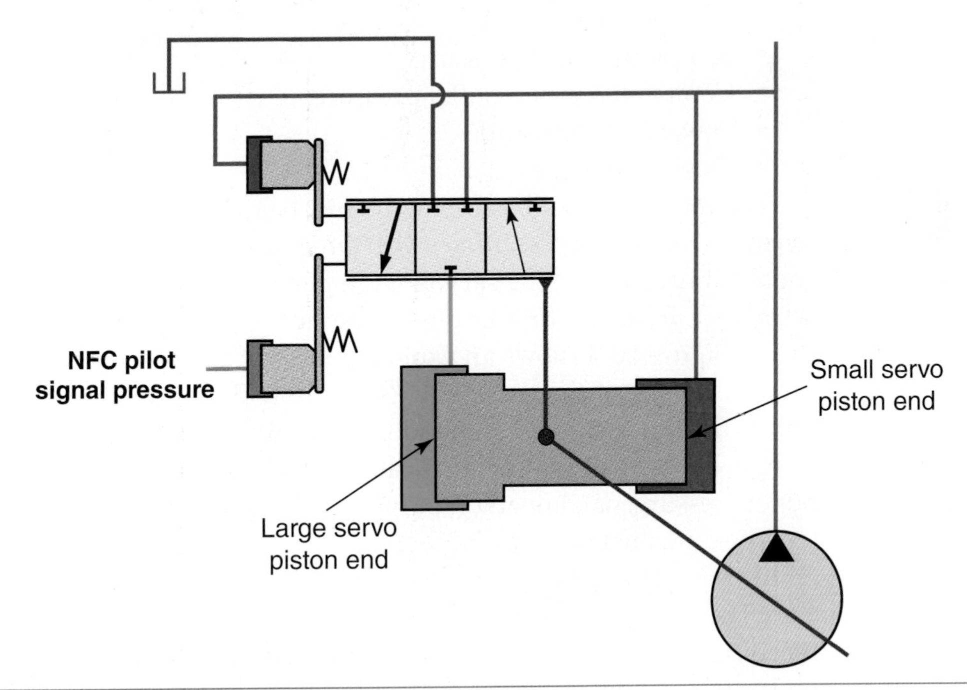

Figure 20-2. Example of an NFC pump regulator schematic that is used on NFC excavators, such as Case and Caterpillar. Notice that the NFC pilot pressure is acting on the regulator valve. When the NFC pilot pressure drops, the spring shifts, which effectively drains the oil from large servo piston end, causing the pump to upstroke.

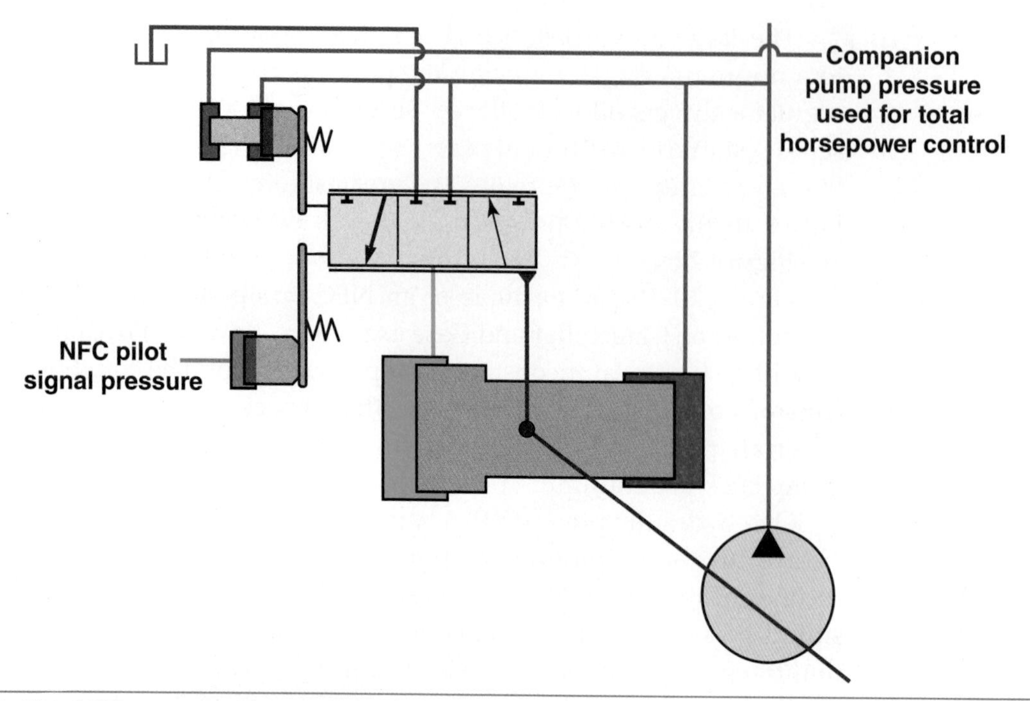

Figure 20-3. This NFC pump regulator senses both its own pump outlet pressure and the companion pump outlet pressure. This arrangement is called total horsepower control.

Figure 20-1 shows an NFC pump regulator that uses a single spool valve assembly for controlling pump flow. NFC and PFC pumps commonly use one to three spool valve assemblies. NFC pumps can have numerous control pressures acting on the pump's regulator spool valves. The following are examples of control pressures:

- NFC pressure.
- Self-pump outlet pressure.
- Companion pump outlet pressure.
- Power shift pressure.
- Flow limit pressure.

The regulator spool valves will shift based on the oil pressures they are sensing. Once the spool shifts, it will either direct oil to the servo's large piston end or drain oil from the servo's large piston end. As the servo piston moves, it causes a ***feedback link*** to pivot. The link is used to close the spool valve's sleeve. Figure 20-4 shows an exploded view of an NFC regulator.

The closing of the spool valve's sleeve will hold the servo piston in position of the last requested demand for flow. It will remain in this position until a change occurs in the input sensing pressures such as pilot pressure, pump outlet pressure, or companion pump pressure.

The difference between NFC and PFC is that NFC works on the principle of an inverse relationship, while PFC works on the principle of a positive relationship. The NFC system upstrokes the pump when pilot signal pressure decreases, meaning that a lower pilot signal pressure will result in greater pump flow. PFC, on the other hand, upstrokes the pump when pilot pressure increases, meaning that when the regulator receives increased pilot signal pressure, the pump flow increases.

Negative Flow Control (NFC) Hydraulic Systems

In an NFC system, the open-center DCVs are connected in series. When the DCVs are in the neutral position, they allow pump flow to return to the reservoir after dropping across an orifice. See **Figure 20-5**. An NFC orifice and an NFC relief valve are located inside a valve assembly. Case excavators label the valve assembly the ***destroke valve***. Whenever the DCVs are in a neutral position, the orifice has the responsibility of establishing the NFC back pressure.

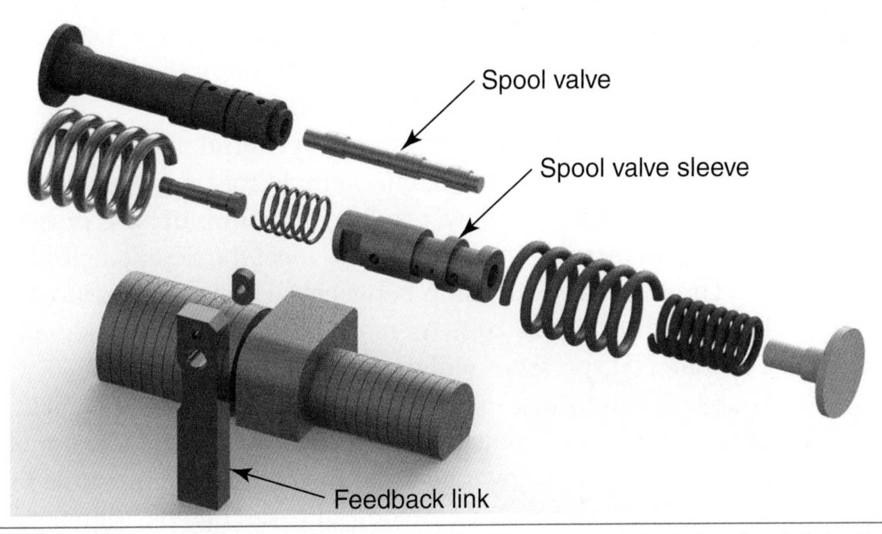

Figure 20-4. An exploded view of an NFC regulator assembly. The servo piston's feedback link attaches to the regulator's spool valve sleeve.

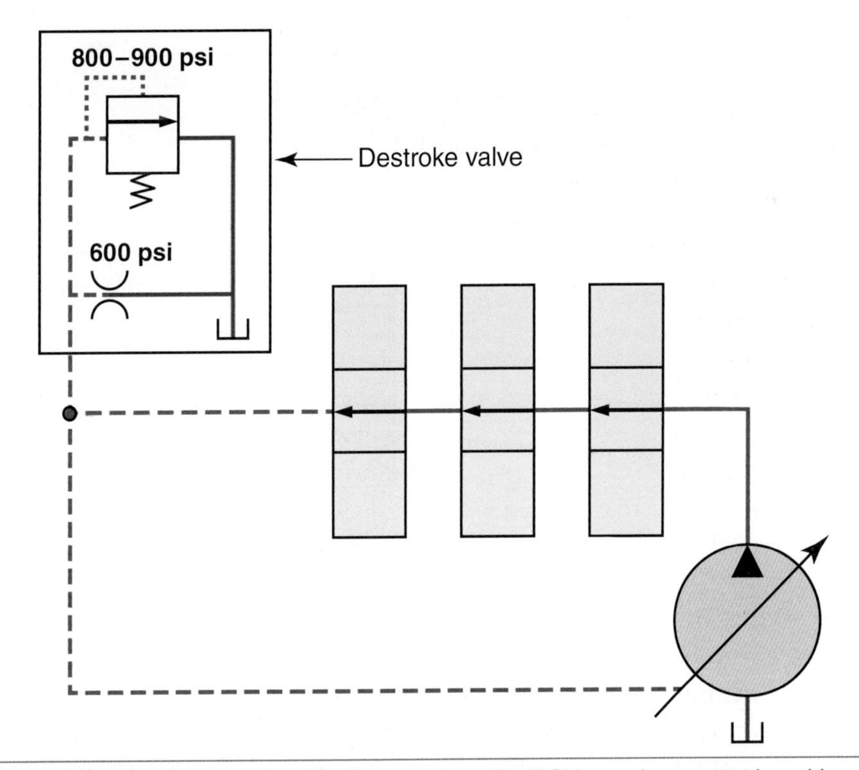

Figure 20-5. As this simplified NFC schematic illustrates, when the DCVs are in a neutral position, the orifice will establish the maximum NFC pressure. This causes the pump to destroke to its minimum flow.

A reasonable example of maximum NFC pressure when the DCVs are in a neutral position is 600 psi (41 bar). The variable-displacement pump works based on an inverse relationship, meaning that the pump will destroke when the NFC pressure is high. This is where the term negative flow control originates. If the NFC signal pressure reaches the maximum pressure, the pump will destroke to minimum pump flow.

If a DCV is actuated, for example to extend a cylinder, the NFC signal pressure drops, based on the operator-requested flow. A drop in NFC pressure will cause the pump to upstroke. For example, if the DCV spool is opened 30%, then 30% of the pump's flow will be directed to the cylinder while the remaining 70% of the pump flow will be routed through the DCV to the NFC orifice in destroking valve, and then to the reservoir.

Previously, when the DCVs were in neutral, 100% of pump flow was routed to the reservoir. With the current example, the pump flow to the reservoir would equal 70% of the total flow, resulting in a drop in NFC pressure, subsequently causing an increase in pump flow to meet the demand. See **Figure 20-6**.

The inverse relationship between NFC signal pressure and pump flow is illustrated in **Figure 20-7**. As more pump flow is requested to be sent to the cylinder, NFC pressure will drop.

On a Caterpillar excavator, after calibration, the NFC signal pressure is accurate enough that the machine's electronic control module (ECM) will deduce the pump's flow rate using the NFC signal pressure. **Figure 20-8** shows how an ECM estimates pump output based on NFC signal pressure. The ECM can then calculate hydraulic horsepower by measuring the NFC signal pressure (to estimate pump flow) and the pump outlet pressure.

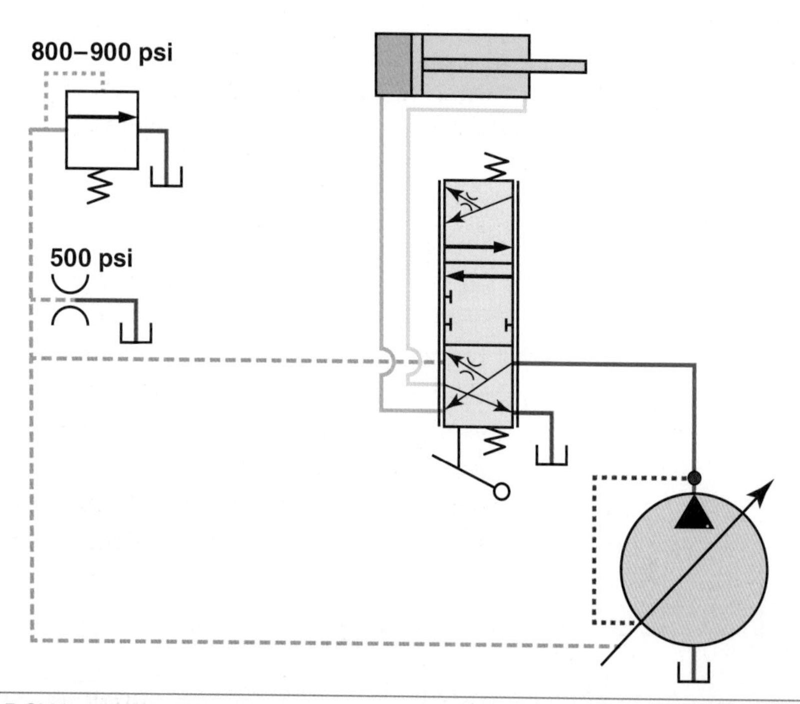

Figure 20-6. When a DCV is actuated to extend a cylinder, it causes a drop in NFC pressure, which causes the pump to upstroke. This schematic also illustrates how the pump's outlet pressure can be used to destroke the pump.

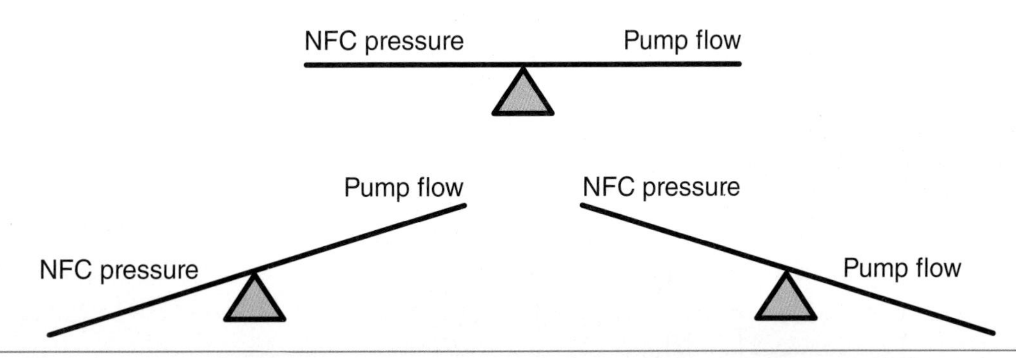

Figure 20-7. NFC pressure has an inverse relationship with the pump's flow. A decrease in NFC pressure causes the pump to upstroke. When NFC pressure peaks, the pump is destroked to minimum flow.

NFC Signal Pressure (psi)	Percentage of Pump Displacement	Example of Flow Rate (gpm)
600	15–25%	10
500	25–45%	20
400	45–65%	30
300	65–85%	40
200	85–100%	50

Figure 20-8. Some excavators use the NFC pressure to estimate the pump's flow rate. When the NFC pressure peaks, the pump will be destroked to minimum flow. When NFC pressure drops to the lowest pressure, then the pump will be upstroked to maximum flow.

Note

Late-model EH hybrid excavators use a swash plate angle sensor as a more direct method for determining pump flow.

Proportional Solenoids in NFC Circuits

It is possible for the operator to request more hydraulic horsepower than the engine is capable of delivering. For example, the operator might actuate multiple DCVs that require 300 hydraulic hp, but the engine might be capable of delivering only 200 hp. When this occurs, the ECM has the ability to destroke the pump to minimum pump flow by using a ***proportional solenoid valve (PSV)***. Both Case excavators and Caterpillar excavators use PSVs in their NFC excavators. These valves work in conjunction with the pump signal pressures for the purpose of destroking the pump to prevent the engine from stalling. See **Figure 20-9**.

Note

Case excavators label their PSV the ***horsepower control proportional solenoid***. Caterpillar excavators label their PSV the ***power shift solenoid***.

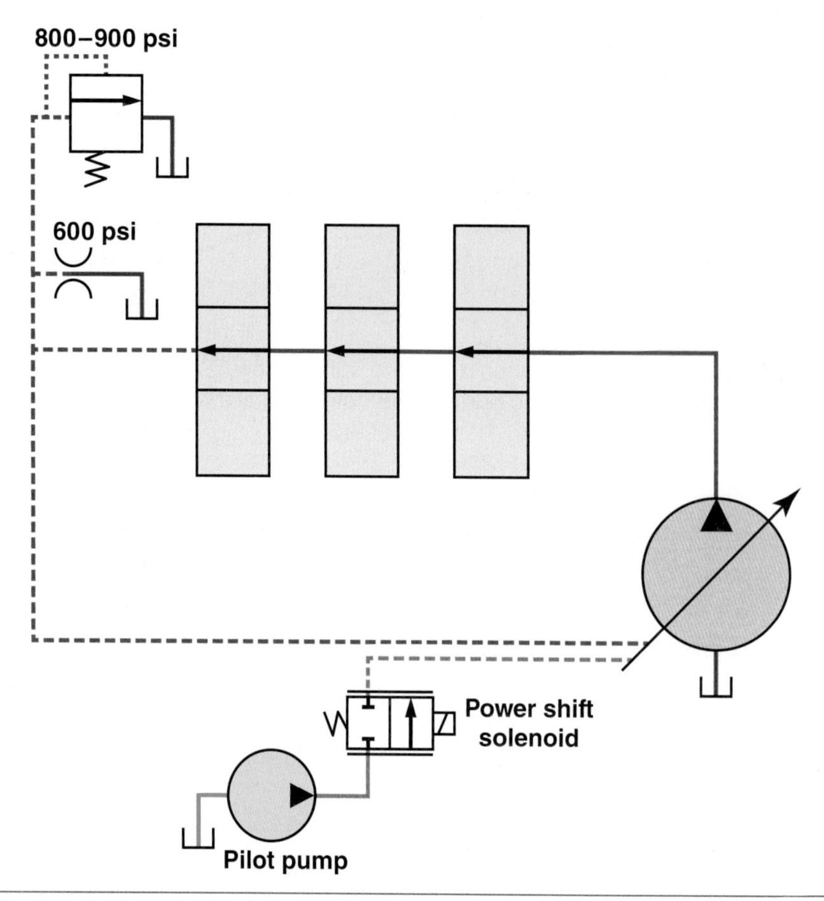

Figure 20-9. Caterpillar labels their PSV solenoid a power shift solenoid. It is used to destroke the pump to prevent the engine from lugging or stalling.

The ECM varies the current flow to the solenoid for the purpose of developing a signal pressure that will destroke the pump. This power shift signal pressure has the same inverse relationship to pump flow as NFC signal pressure. If power shift pressure is high, pump flow will be low. During pump standby, the ECM also energizes the power shift solenoid, generating as much as 350 psi (24 bar) of pilot pressure.

All 300 series Caterpillar excavators use a power shift solenoid. Most NFC Caterpillar excavators use a single power shift solenoid to destroke both NFC pumps. The solenoid is sometimes referred to as a PSPRV, ***power shift pressure reducing valve***. The power shift solenoid has the responsibility for preventing the engine from lugging or possibly stalling. It acts similarly to a torque limiter that is found in load-sensing hydraulic systems.

Pump Outlet Pressure

Pump outlet pressure also affects pump flow in the same relationship as NFC signal pressure and power shift pressure. Therefore, as pump pressure rises, it too can cause the pump to destroke to minimum flow. Power shift pressure is reactive. Therefore, if the system has a pressure spike, pump outlet pressure can destroke the pump faster than power shift pressure can.

As a review, the following three pressures have an inverse relationship on pump flow:

- NFC signal pressure.
- Power shift pressure.
- Pump outlet pressure.

A decrease in any of these three pressures will cause the pump flow to increase, and an increase in any of those three pressures will cause the pump flow to decrease. See Figure 20-10 and Figure 20-11.

Cross Sensing Pressure

Some Caterpillar NFC excavators use an average pump outlet pressure between both NFC pumps. This pressure is called the ***cross sensing pressure***. The sensing pressures of both pump outlets are connected through two orifices. It is the average of both pump pressures that acts on both pump regulators. See Figure 20-12. This pressure helps provide adequate engine horsepower to the NFC pumps. Caterpillar calls this constant horsepower control. Kawasaki refers to it as total horsepower control.

The pump regulator has spools that sense the three pressures: pump outlet pressure, NFC signal pressure, and power shift pressure. Examples of an NFC regulator are shown in Figures 20-1, 20-4, and 20-13.

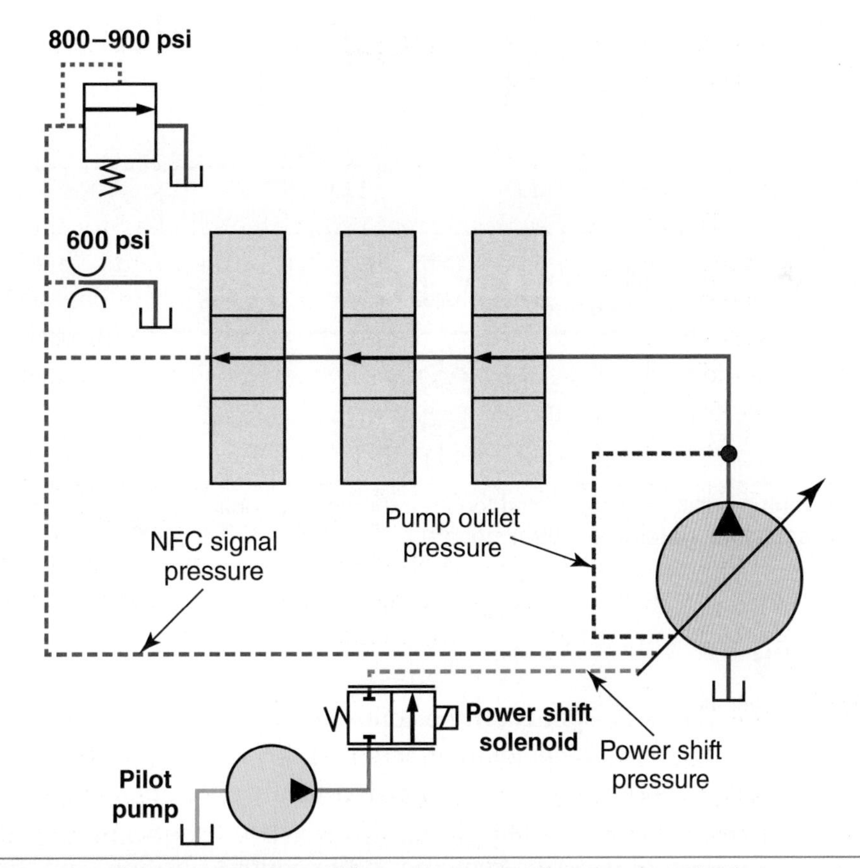

Figure 20-10. The pump regulator senses three different pressures that can destroke the pump to minimum flow: pump outlet pressure, NFC pressure, and power shift pressure.

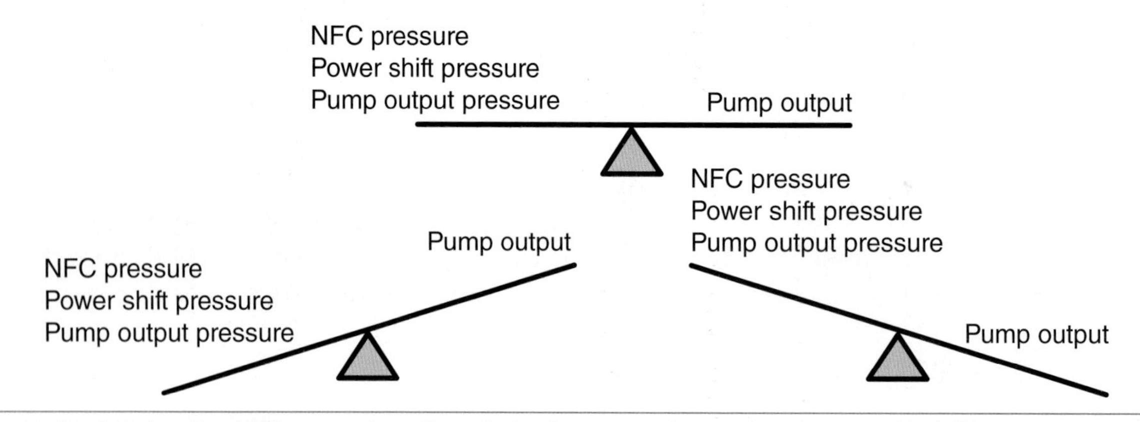

Figure 20-11. A Caterpillar NFC excavator will upstroke the pump when a drop is sensed in NFC pressure, power shift pressure, or pump outlet pressure.

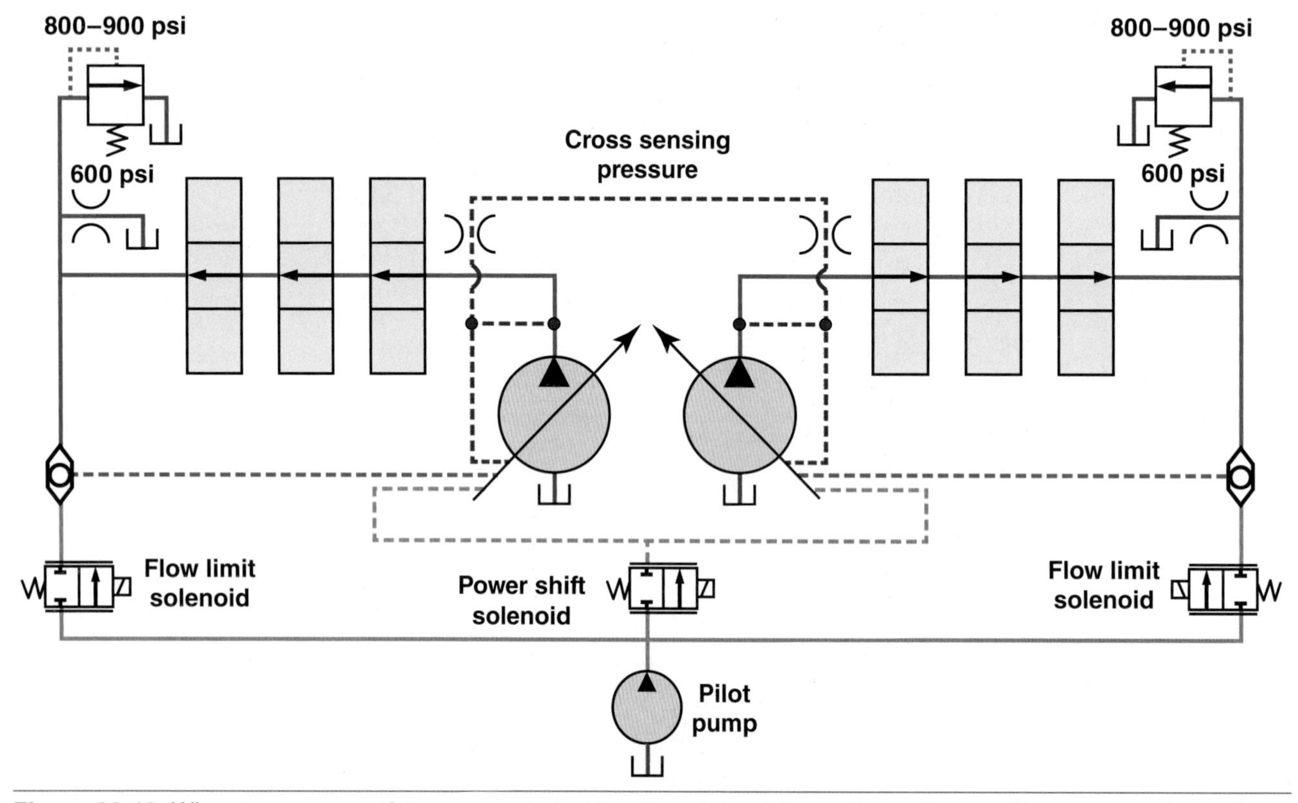

Figure 20-12. When two pump outlets are connected together through two orifices, the resulting pressure is used to help control both pump regulators. Caterpillar calls it cross sensing pressure.

Pilot-Controlled DCV

The DCV spools can be controlled hydraulically with pilot controls or by proportional solenoids that actuate pilot valves, which direct oil to the DCV spools. Figure 20-14 shows a hydraulically pilot-operated DCV spool valve that is controlled by a dual-solenoid DCV. When the solenoids are de-energized, the DCV spool is spring-centered to the neutral position, and the pump is in the standby mode. A flow limit solenoid is also shown in the illustration.

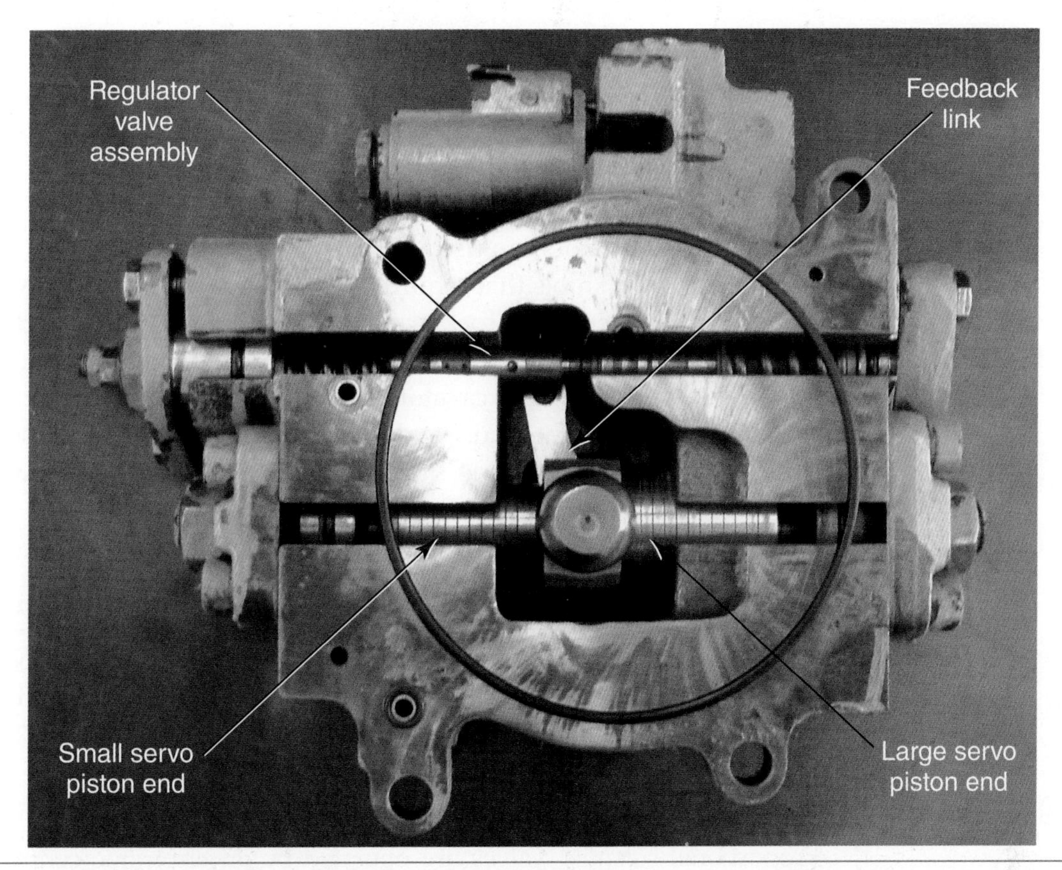

Figure 20-13. This pump regulator has been cut away to show the servo, feedback link, and regulator valve assembly. It is used on Caterpillar 320C, 320D, 320E NFC excavators.

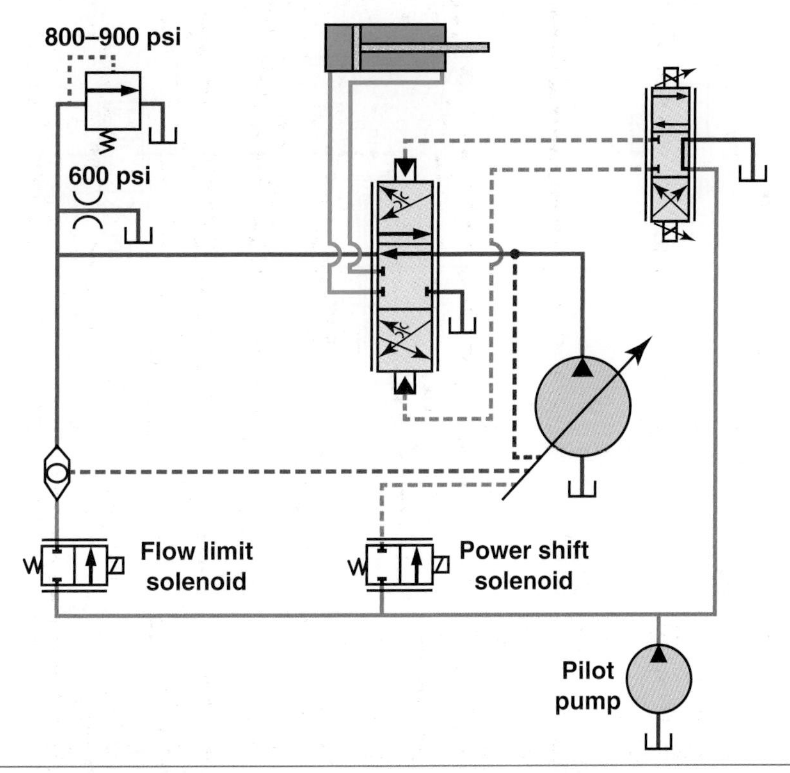

Figure 20-14. Pump output, NFC signal or flow limit, and power shift pressures act on the pump regulator. The flow limit solenoids are used in conjunction with work tools, such as a hammer. The DCV is pilot operated by a solenoid valve.

Flow Limit Solenoid

Two flow limit solenoids are used in some 300 series Caterpillar excavators, but not all. The ***flow limit solenoids*** have the responsibility to develop a false NFC signal pressure. The purpose of the solenoids is to ensure a constant amount of flow is available to the excavator's work tool circuit. The work tool circuit on some excavators can use as much as 80 or 90 gpm.

Examples of work tools that require a substantial amount of flow are hammers, tampers, mowers, and rock crushers. Shears and multi-handling tools can also have a momentary high demand for flow, but not continuous. On late-model Caterpillar NFC excavators, the flow limit solenoids are also used during boom down (regeneration) to destroke the pump when the boom is lowering. They are also used during the initial excavator swing to prevent pump flow from going over the relief.

The schematic in Figure 20-15 shows a cylinder extending. Notice the DCV solenoid is energized, causing the DCV spool to be pilot operated.

As mentioned, Caterpillar excavators commonly have two NFC variable-displacement pumps that generate two NFC signal pressures. The NFC signal pressures are always separate, meaning that NFC signal pressure 1 is used for pump 1, and NFC signal pressure 2 is used for pump 2. See Figure 20-16.

ECM Logic

Proportional solenoid valves are operated by an electronic control module (ECM). The ECM evaluates multiple electrical inputs, and then, using that information, energizes the solenoids to optimize the machine's hydraulic performance.

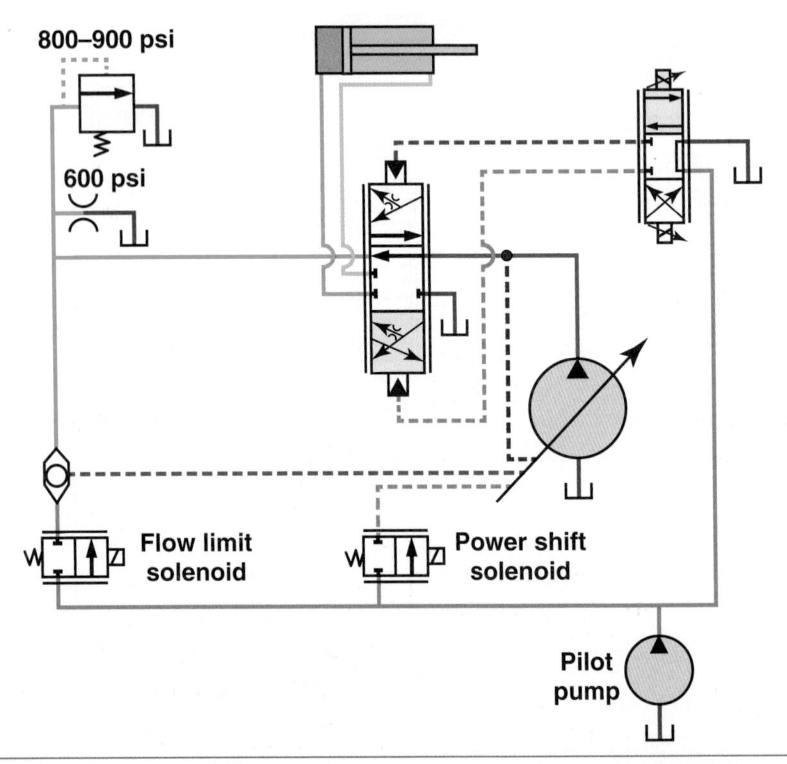

Figure 20-15. The DCV's solenoid pilot controls the DCV spool valve. The pump regulator senses NFC signal or flow limit, power shift, and pump outlet pressures to destroke the pump.

Inputs

Caterpillar NFC hydraulic systems monitor the following four electrical inputs:

- Engine rpm (actual speed).
- Pump output pressure.
- NFC signal pressure.
- Engine speed dial (requested speed).

The excavator's ECM will activate the outputs based on those four input signals. For example, when the engine speed dial is positioned between 1 and 7, a fixed voltage value is sent to the power shift solenoid. A slightly higher voltage value is sent as the dial approaches 1, which results in a higher power shift signal pressure and a decrease in pump flow.

Outputs

As previously mentioned, a Caterpillar excavator with NFC will be equipped with one power shift solenoid, and can be equipped with two flow limit solenoids. The ECM can indirectly control pump flow by energizing those solenoids.

The flow limit solenoid ensures the ECM's ability to provide constant flow to the work tool circuit. For example, when the excavator is operating a hammer and needs a substantial amount of flow, the flow limit solenoid would be activated. The power shift solenoid is used by the ECM to prevent the engine from lugging or potentially stalling the engine whenever the operator has requested too much hydraulic horsepower. Both the flow limit and the power shift solenoids are normally closed. When they are energized, they will begin sending a signal pressure to the pump's actuator. As that signal pressure increases, the pump output will decrease.

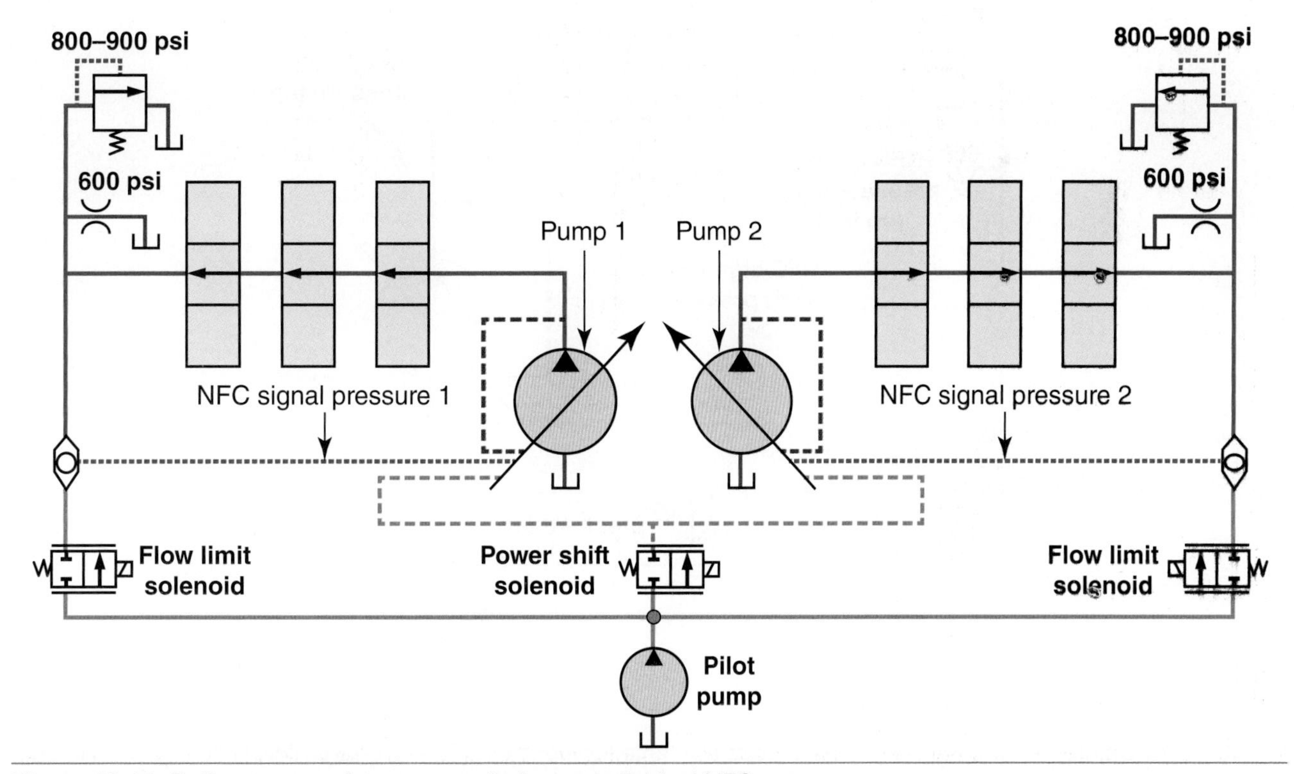

Figure 20-16. Both pump regulators sense their own individual NFC pressures.

If needed, technicians can directly measure the following pressures on the Caterpillar excavator:

- Pump outlet.
- NFC signal.
- Flow limit.
- Power shift.

See Figure 20-17. Note that the NFC plug is not needed because of the use of the dual NFC signal/flow limit pressure tap. Technicians can use this single tap to physically measure both NFC signal pressure and flow limit pressure.

Case Excavators

As previously mentioned, Case excavators also use a negative control open-center hydraulic system. Some examples of Case excavator models are CX 130, CX 160, CX 210, and CX 240. Case excavators are built by Sumitomo in Japan. The excavators have a hydraulic ***power save solenoid*** that reduces the engine load by 10% when the operator controls are at idle.

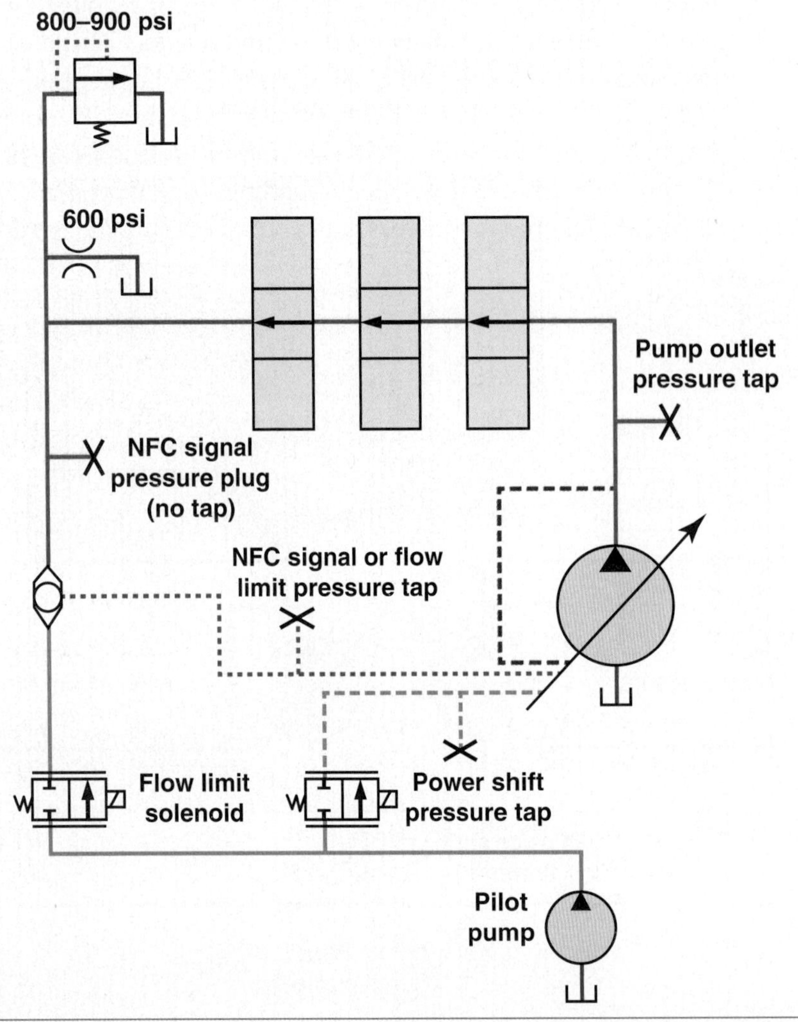

Figure 20-17. Caterpillar NFC excavators have ports available for measuring NFC signal or flow limit, power shift, and pump outlet pressure.

Kobelco Dynamic Acera SK 210–SK 290 Excavators

Kobelco SK 210 through SK 290 excavators can be equipped with positive control or negative control, depending on the serial number of the excavator. However, Kobelco excavators are equipped with positive control more often than negative control.

Both the negative and positive control Kobelco systems have a PSV for each pump regulator. With the negative control system, when the PSV is energized, it directs up to 465 psi (32 bar) of pilot pressure to the pump regulator to reduce the pump's displacement and minimum flow.

Positive Flow Control (PFC) Hydraulic Systems

Like negative flow control systems, positive flow control (PFC) systems also have open-center DCVs and variable-displacement pumps. However, PFC pump regulators must receive an increased pressure signal in order to upstroke the pump, which is opposite of NFC. See **Figure 20-18**.

Examples of machines that use PFC hydraulics include Hitachi excavators and John Deere excavators. Other brands of excavators, as well as other construction machines, have electro-hydraulic (EH) PFC hydraulic systems. Examples include Kobelco excavators, large Caterpillar wheel loaders, and large Caterpillar excavators. EH PFC systems are becoming more popular and will be explained later in this chapter.

PFC systems are highly responsive. Some operators prefer the feel of an NFC system over a reactive PFC system. Some proponents of PFC state that PFC systems provide improved metering control when attempting to feather a hydraulic function, such as slowly extending a hydraulic cylinder. The proponents also say that positive control systems provide a faster response. As mentioned earlier in the chapter, at least one style of pump can be converted from NFC to PFC by changing the pump regulator's pivot point.

A traditional PFC system, like the system used in John Deere G series and Hitachi Zaxis excavators, is more complex than a traditional NFC system. This style of PFC system requires the use of a hydraulic pilot computer, which will be explained later.

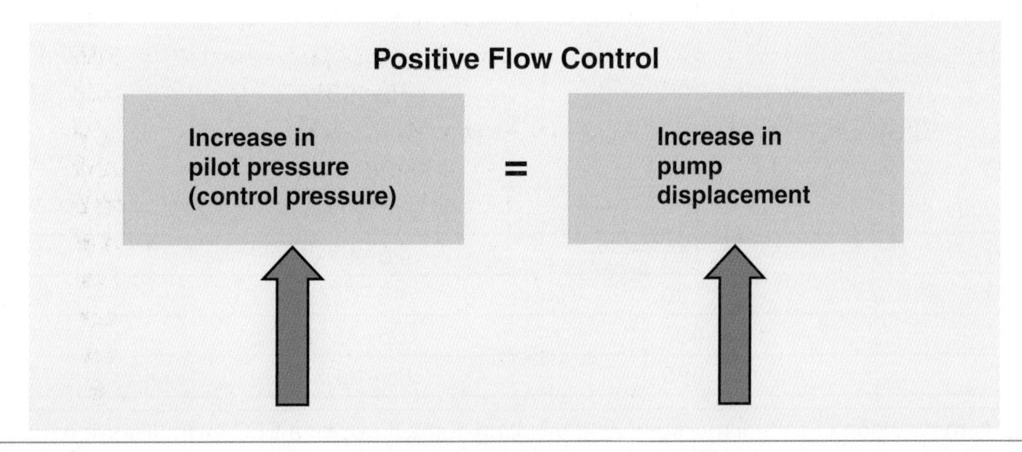

Figure 20-18. PFC pilot pressure is directly proportional to pump flow. When PFC pressure increases, so does pump flow.

Using the rule of thumb that NFC and PFC systems flow approximately 15% to 25% of the pump's displacement when the DCVs are in a neutral position, a comparison of horsepower consumption can be made. Note that most excavators use two variable-displacement piston pumps. Figure 20-19 provides a comparison of horsepower consumptions for a 50 gpm NFC pump and a 50 gpm PFC pump when the DCVs are in a neutral position.

Figure 20-20 shows the differences between PFC and NFC control pressures. As the NFC pressure increases, pump flow is decreased. As PFC pressure increases, pump flow also increases.

Pilot System

A traditional John Deere or Hitachi PFC hydraulic system is controlled by a complex pilot system. A gear pump commonly supplies the oil for the pilot circuit. However, the large Caterpillar wheel loaders use a variable-displacement pump for the pilot circuit on their EH PFC hydraulic system.

DCVs in Neutral				
NFC Signal Pressure		Percentage of Pump Displacement	Flow (gpm)	Horsepower
bar	psi			
40	580	15–25%	10	3.38
PFC Signal Pressure				
bar	psi			
2	30	15–25%	10	0.18

Figure 20-19. A comparison of horsepower consumptions between an NFC pump and a PFC pump when the DCVs are in a neutral position.

NFC Signal Pressure		Percentage of Pump Displacement	Flow (gpm)
bar	psi		
40	580	15–25%	10
35	500	25–45%	20
28	400	45–65%	30
21	300	65–85%	40
14	200	85–100%	50
PFC Signal Pressure		Percentage of Pump Displacement	Flow (gpm)
bar	psi		
2	30	15–25%	10
10	145	25–45%	20
20	290	45–65%	30
30	435	65–85%	40
40	580	85–100%	50

Figure 20-20. A comparison of control pressures and resulting pump flows for a PFC hydraulic system and an NFC hydraulic system.

Many construction machines have a ***hydraulic lockout*** control switch or lever. When the lockout switch or lever is actuated, it prevents the operator from actuating any of the machine's hydraulic controls. This prevents accidental movement of the machine's arms or implements. The lockout normally cuts off the machine's pilot oil, which makes the implement DCV spools ineffective.

On excavators, the lockout is commonly found on the left side of the operator seat in the form of a pilot lockout lever. Some pilot lockout levers engage the lockout when the lever is pushed down. Others engage the lockout when the lever is pulled up. See **Figure 20-21**.

When the operator disengages the lockout, a control valve (normally in the form of a solenoid) opens, which sends pilot oil to the hand- and foot-operated pilot control valves and to the pilot signal manifold. See **Figure 20-22**.

In the John Deere and Hitachi excavator PFC systems, as well as many other brands of excavators, the operator requests a hydraulic function by moving a simple joystick or foot-operated controller that consists of a pilot valve, sometimes called a ***pilot controller***. This controller directs oil to a ***pilot signal manifold***, also known as a ***hydraulic computer***. The computer contains shuttle valves and spool valves. The computer has the responsibility of directing oil to the appropriate components:

- Sends oil to the DCV spools to actuate the spool valves.
- Sends oil to the pump regulator to upstroke the PFC pump.
- Sends oil to additional valves as needed. For example, it might send oil to the swing brake to release it for excavator functions.

The pilot signal manifold contains two ***flow rate pilot valves***, one for each of the pumps. The flow rate pilot valve's purpose is precisely what its name implies, to control the pump's flow rate. The pilot signal manifold uses check valves to direct oil to each of the two flow rate valves. The valves are closed by spring pressure and opened by pilot pressure in proportion to the position of the pilot controllers. The flow rate valves then send oil to the corresponding pump regulator. See **Figure 20-23**. The amount of pilot pressure sent to the regulator is based on how far the operator moves the pilot controller. See **Figure 20-24**.

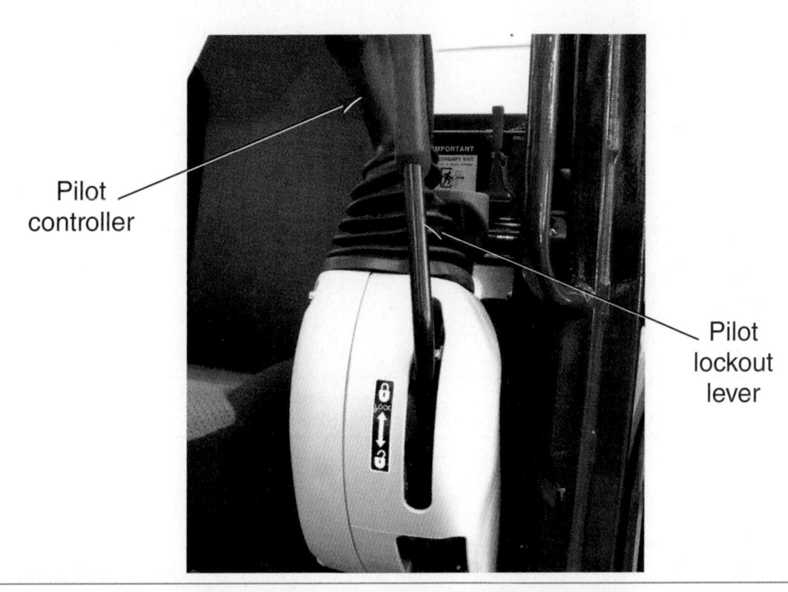

Figure 20-21. Many construction machines use a hydraulic lockout control switch or lever that, when actuated, will lock the hydraulic system.

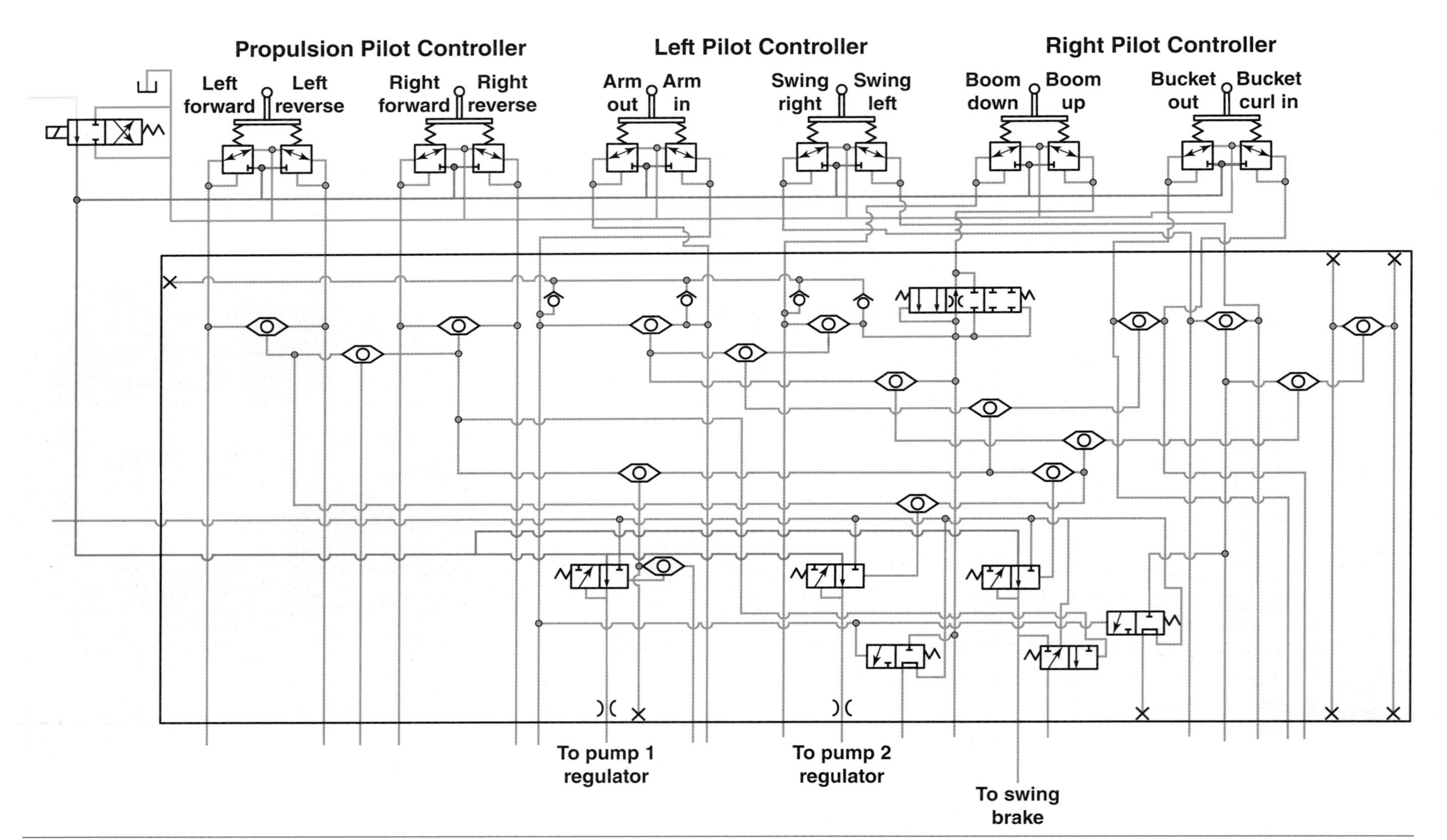

Figure 20-22. A pilot signal manifold is a complex hydraulic computer that controls the operation of the excavator. It receives pilot oil from the pilot controllers and directs pilot oil to the pump regulator valves and DCV spool valves.

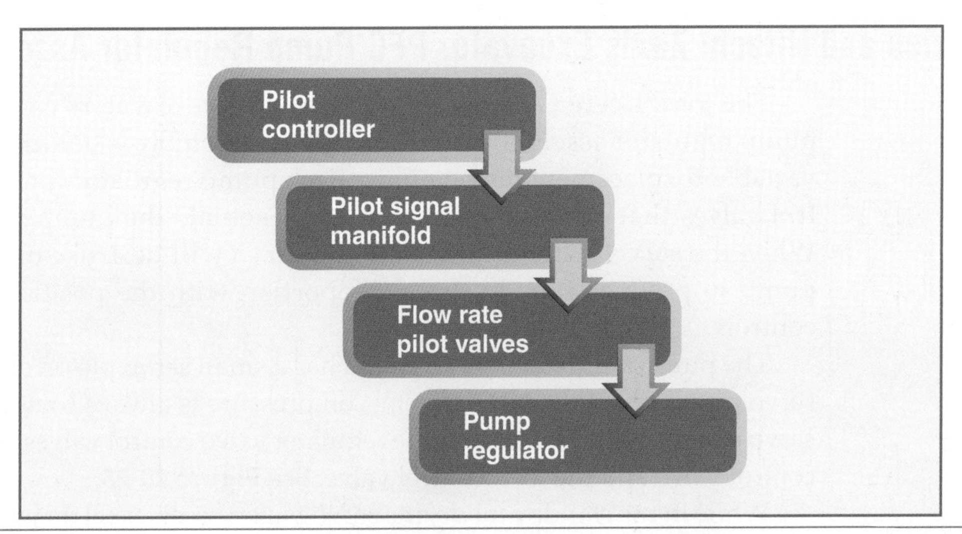

Figure 20-23. The pilot oil in a Deere and Hitachi PFC excavator flows from the pilot controller, to the pilot signal manifold, to the flow rate pilot valves, to the pump regulator, which upstrokes the pump.

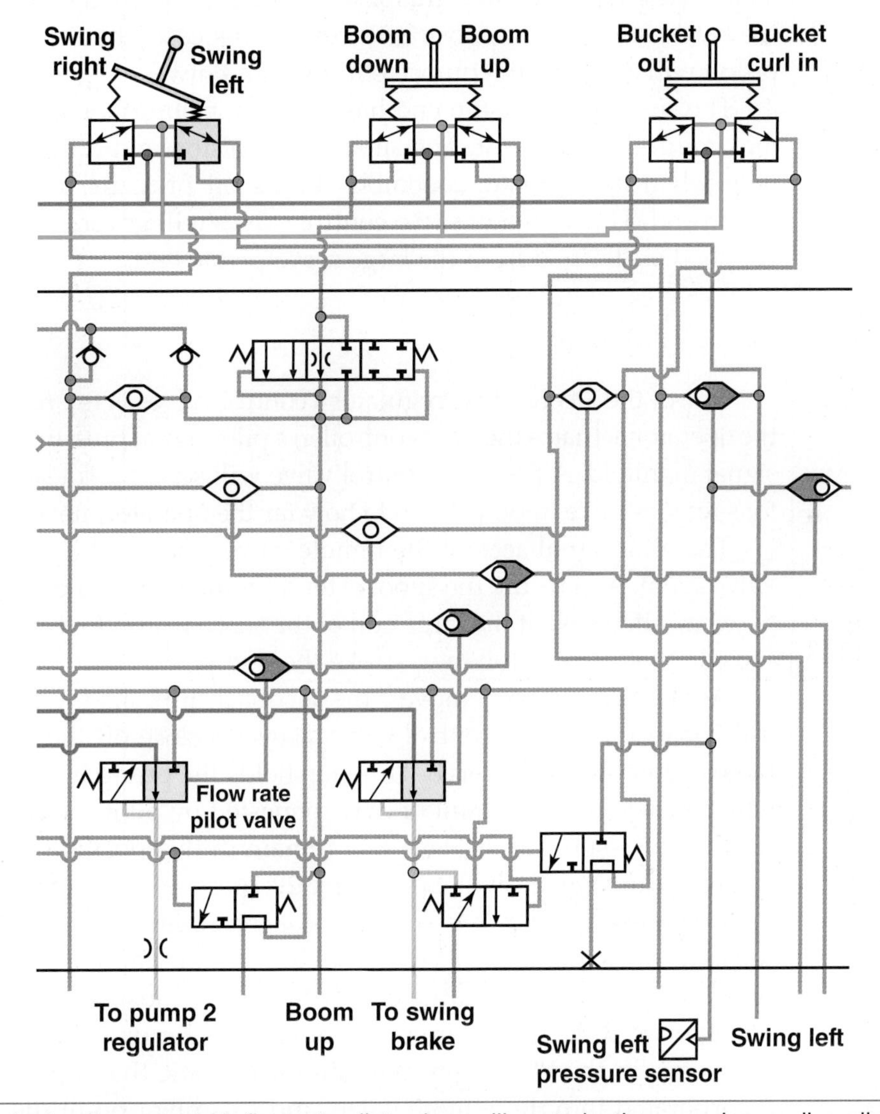

Figure 20-24. The hydraulic computer's flow rate pilot valves will upstroke the pump by sending pilot oil to the pump regulators in proportion to how far the operator moves the pilot controllers.

Deere G Series and Hitachi Zaxis Excavator PFC Pump Regulator Assemblies

The John Deere G series and Hitachi Zaxis excavators contain two PFC pump regulator assemblies. Each regulator assembly is dedicated to its own variable-displacement piston pump. Each pump regulator contains two control valves that are used to hydraulically actuate the pump's servo piston. When the servo piston is actuated, the servo will destroke or upstroke the pump to produce flow in direct proportion with the position of the pilot controls.

The pump regulator's servo piston has a small servo piston end and a large servo piston end. Like NFC systems, oil pressure is always routed to the small servo piston end. The names of the regulator's two control valves are the remote control valve and the load control valve. See Figure 20-25.

While in the standby mode, the servo piston receives oil on both sides of the piston, the small piston end and the large piston end. As a result, the pump is held at the minimum pump flow angle. In accordance with Pascal's law, when oil pressure is sent to an actuator with different areas, the larger area will win the battle of the two opposing forces. In this case, the large end of the servo piston will destroke the pump to minimum flow.

The small servo piston end has the responsibility of stroking the pump to maximum displacement. The small servo piston end can upstroke the pump only when its regulator assembly drains oil pressure away from the large piston end. The regulator's two control valves will govern whether or not oil is supplied or drained from the large servo piston end.

Regulator's Remote Valve

One of the Deere's PFC regulator's control valves is the ***remote valve***. When the operator actuates the pilot controller, a pilot signal pressure is sent to the pilot signal manifold. A flow rate control valve will send a pilot signal to the regulator's remote valve in proportion to how far the operator moved the controller.

The pilot signal acts on the remote valve piston, which pushes the remote valve spool. As a result, the spool extends, which opens a passageway. The passageway allows oil at the large end of the servo piston to drain to the reservoir, causing the servo piston to upstroke the pump.

As the servo piston moves, the feedback link also moves. The feedback link forces the remote control valve sleeve to close off the passageway to the large piston end of the servo. This step holds the pump in the current position, which provides the amount of flow requested by pilot controller position. The remote valve spool and sleeve will remain in this position until the regulator receives a change in input. See Figure 20-26.

When the operator returns the pilot controller to the neutral position, the remote valve's spring forces the spool back to the standby position. As a result, pilot oil is routed to the large end of the servo, causing the pump to destroke and operate at the minimum pump angle. See Figure 20-27.

Notice in the PFC pump regulator schematic that a pivot point has been incorporated into the remote valve, and this pivot point did not exist in the NFC pump regulator schematic. As the PFC pilot oil is sent to the pump regulator's remote valve (from the hydraulic computer's flow rate valve), the remote valve pivots. This pivoting action causes the remote valve to move the spool in the opposite direction than it would in an NFC regulator. Kawasaki NFC and PFC regulator schematics look practically identical, except the PFC regulator schematic has a pivot.

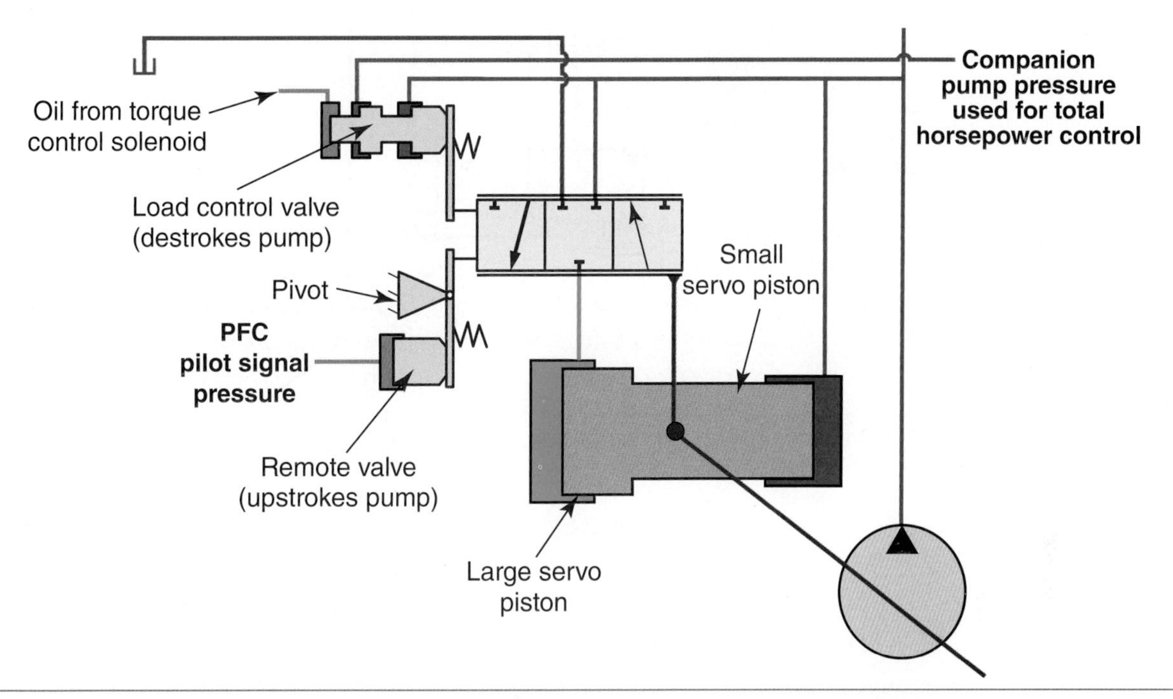

Figure 20-25. A schematic of the type of pump regulator used in Deere G series and Hitachi Zaxis excavators.

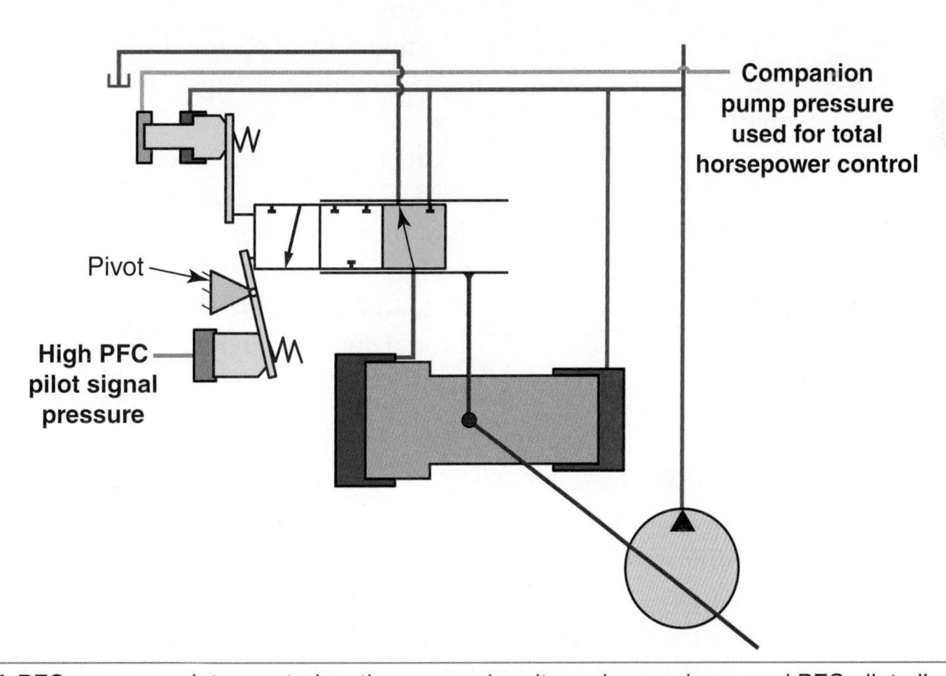

Figure 20-26. A PFC pump regulator upstrokes the pump when it receives an increased PFC pilot oil pressure. The remote valve will drain oil from the large end of the servo piston, causing the pump to upstroke.

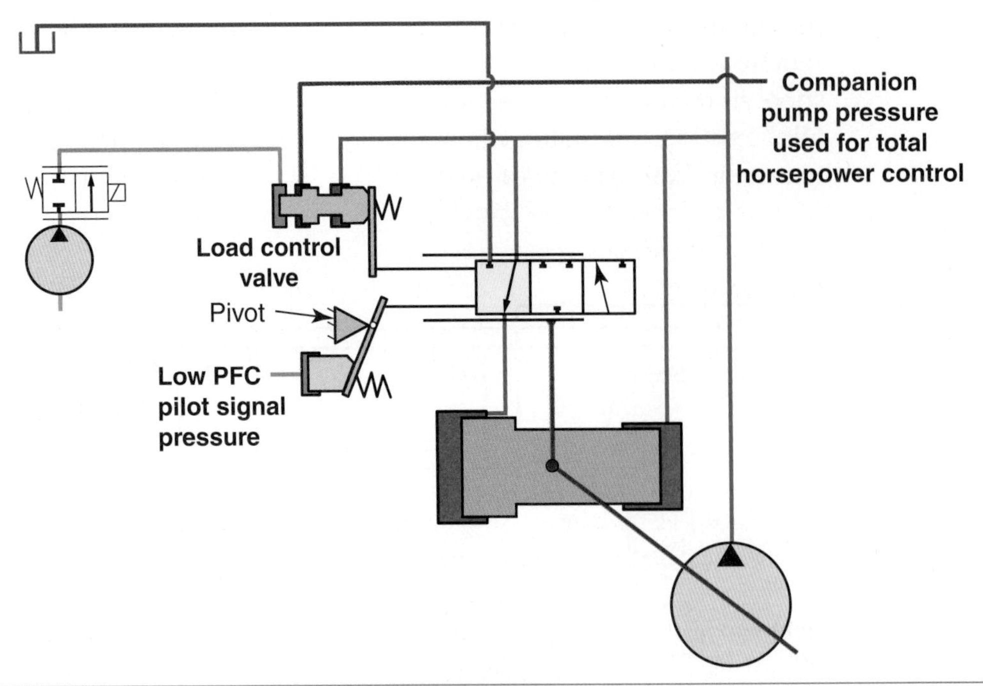

Figure 20-27. When the pilot controllers are in a neutral position, the PFC remote valve's spring will cause the regulator to deliver oil to the large end of the servo piston. This destrokes the pump to a minimum swash plate angle.

Load Control Valve

The pump regulator assembly also contains a ***load control valve***. This valve contains a piston that has three different effective piston areas. The following three oil pressures act on the load control valve piston:

- Self-pump pressure.
- Companion pump pressure.
- Oil pressure from the torque control solenoid.

If the combination of those pressures changes, it causes the load control valve to shift. If the load control valve actuates, pilot oil pressure is routed to the large end of the servo piston, causing the pump to destroke. See Figure 20-28.

In PFC excavators, like NFC excavators, just one pump by itself has the capacity to request up to 85% of the engine horsepower. The load control valve has the responsibility of destroking the pump when too much hydraulic horsepower has been requested. Note that the Deere G series excavator and Hitachi Zaxis excavator also use a proportional solenoid valve (PSV) called a ***torque control solenoid***. The solenoid sends pilot pressure to the load control valve in order to destroke the pump to the minimum flow angle.

Changing a Pilot Controller's Pattern

John Deere G series and Hitachi Zaxis excavators can have the pilot controllers' patterns switched from an excavator-style control pattern to a backhoe-style pattern. If the system is equipped with a ***pattern changer valve***, simply rotating the valve's control knob will change the pattern. The pattern changer valve contains three positions: excavator position, neutral middle position, and backhoe position. The neutral position prevents hydraulic operation and can be used to prevent vandalism or any other unwanted operation. The pattern changer can sometimes be added as an option during the purchase of the machine or added later.

On some machines, it is possible to switch the control pattern from excavator to backhoe style even if the excavator is not equipped with a pattern changer. On these machines, the pattern is switched by swapping the location of four hydraulic hoses. The four hoses are located between the pilot controllers and the signal manifold. The hoses are swapped at the pilot signal manifold. Be sure to follow the manufacturer's specific instructions.

The excavators are shipped from the factory in the ISO excavator control pattern. The two different types of control patterns are shown in Figure 20-29.

Electro-Hydraulic (EH) PFC Systems

Many PFC systems can be categorized as an ***electro-hydraulic (EH) PFC hydraulic system***. The machines require the use of an electronic control module (ECM) to energize a proportional solenoid valve (PSV). The PSV directs pilot oil pressure to the pump regulator valve to upstroke the pump. It is the fact that the signal pressure is delivered by a PSV that causes the system to be labeled an EH PFC. Each of the pump regulators commonly has its own dedicated PSV. See Figure 20-30.

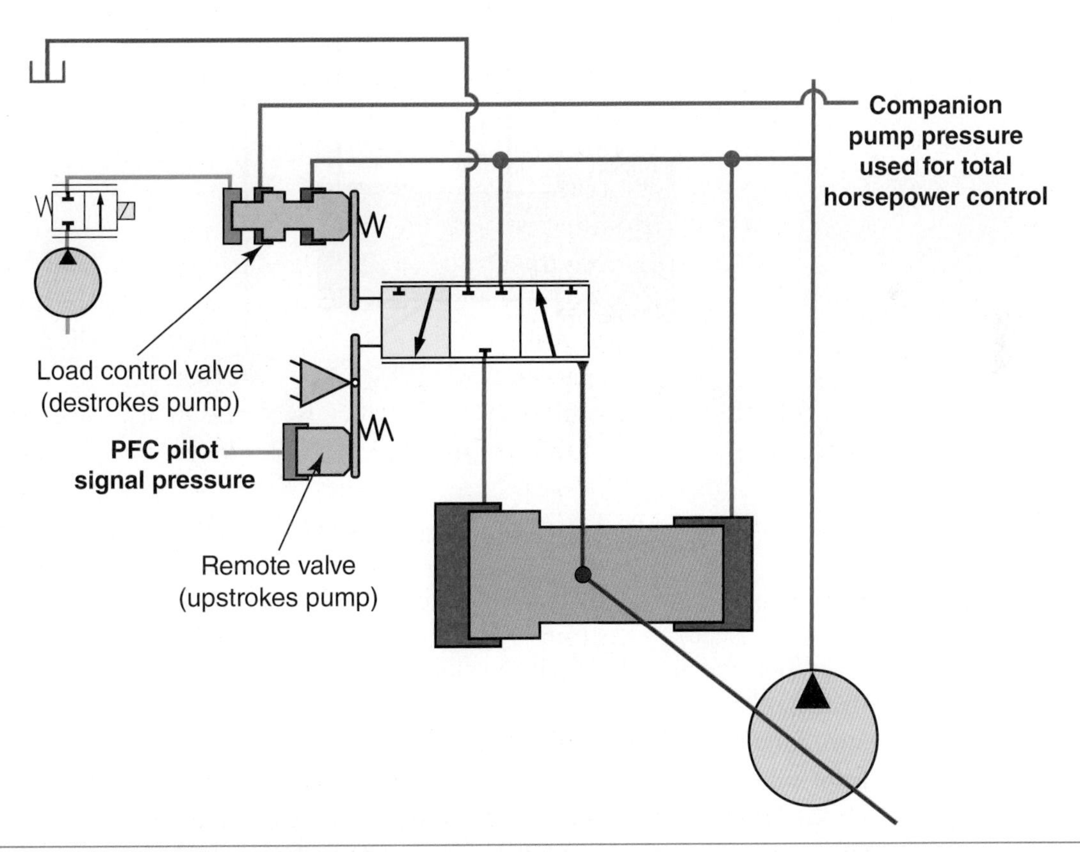

Figure 20-28. A John Deere and Hitachi PFC regulator can be destroked by receiving oil pressure from the self-pump outlet, companion pump outlet, or the torque control solenoid.

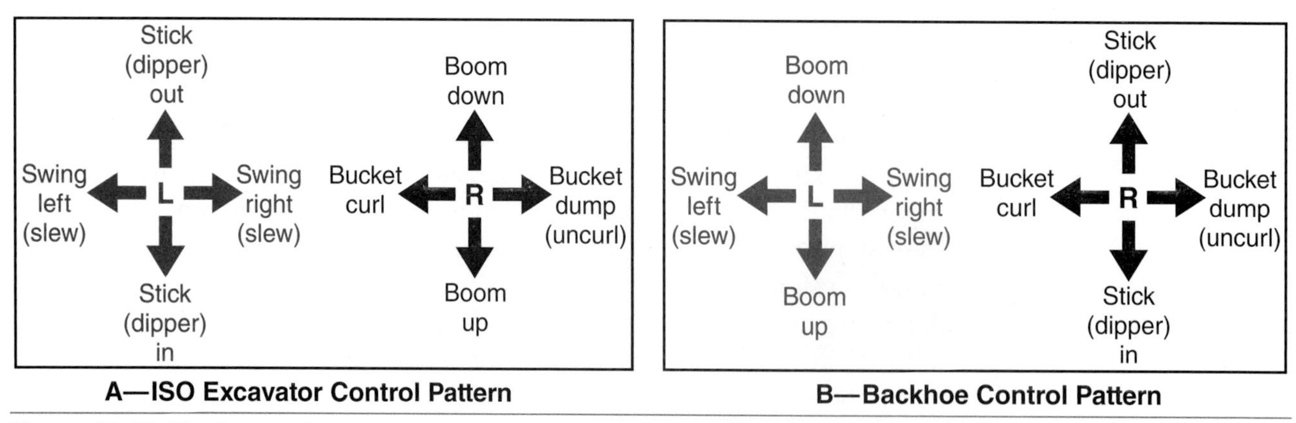

Figure 20-29. The image shows common controller patterns. A—Traditional ISO excavator control pattern. B—Backhoe control pattern.

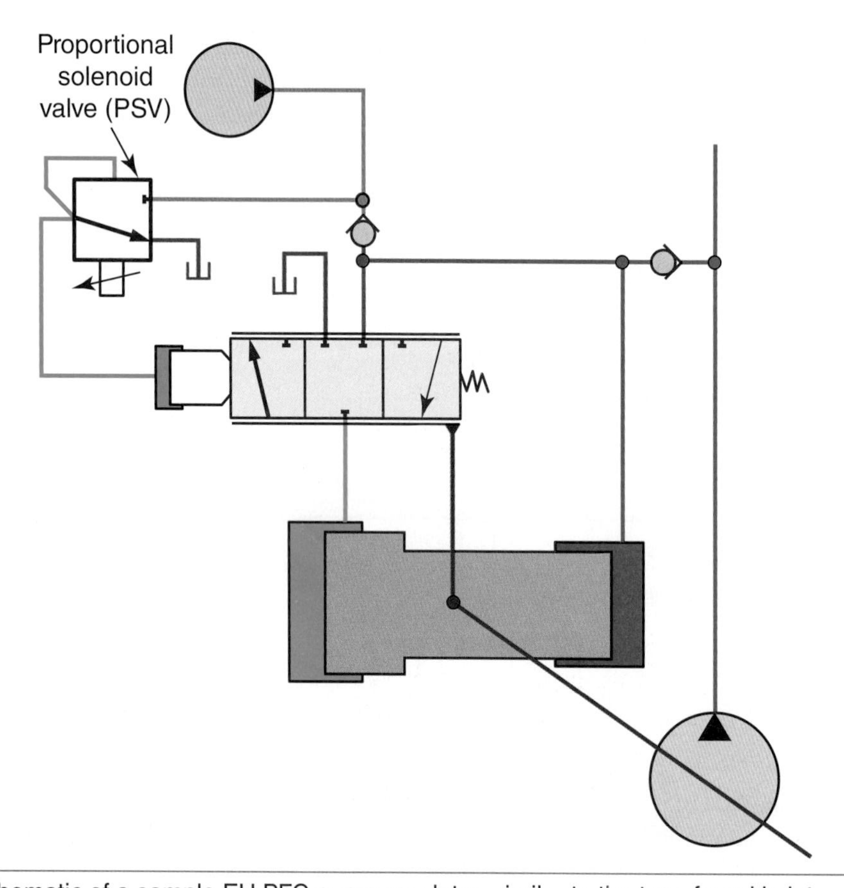

Figure 20-30. A schematic of a sample EH PFC pump regulator, similar to the type found in late-model excavators.

Standby Mode

When the machine is in the standby mode, the PSV is de-energized, which drains oil away from the regulator's spool valve. See **Figure 20-31**. Spring pressure forces the spool to shift to the left, resulting in the spool valve directing pilot oil to the large end of the servo piston. Because oil pressure is being applied to both the small end of the servo and the large end of the servo piston, the large end of the servo has the higher force, causing the pump to destroke to minimum displacement.

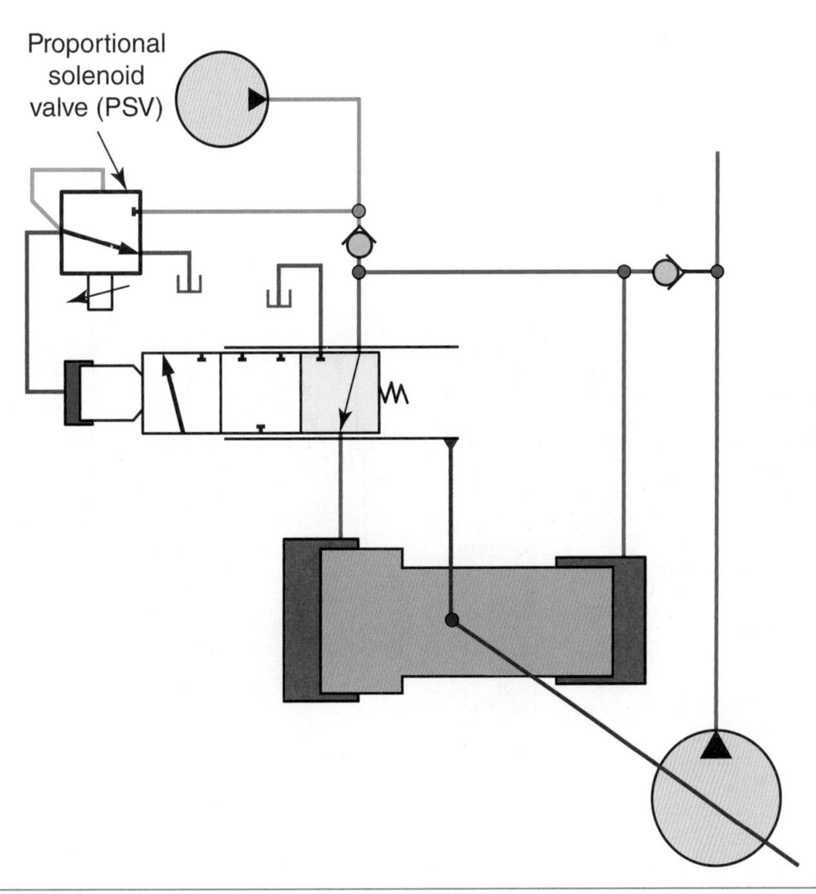

Figure 20-31. An EH PFC pump schematic in the standby mode. The PSV is de-energized, allowing the regulator spring to shift to the left, which causes oil pressure to be routed to the large end of the servo piston, causing the pump to destroke to the minimum pump flow.

Upstroked

When an operator requests a hydraulic function, the electronic joysticks will send an input to the ECM. Based on the inputs, the ECM will upstroke the pump as needed by energizing the PSV. The PSV directs pilot oil pressure to the pump regulator's valve. See Figure 20-32. The pilot pressure overcomes spring pressure, pushing the regulator valve to the right, which causes the spool valve to drain oil away from the large end of the servo piston. The small end of the servo piston receives oil pressure, causing the servo piston to shift to the left, which causes the pump to increase pump flow.

Controlling the Proportional Solenoid Valve (PSV)

The ECM controls pump flow by varying the electrical current flow to the PSV. An example PSV solenoid has an approximate resistance value of 17–25 ohms. The ECM varies the current, for example from 350 milliamps (mA) to 780mA, resulting in 80 to 500 psi of pilot oil pressure being directed to the pump regulator valve. As the PSV increases pressure on the pump regulator, the pump flow increases. See Figure 20-33. Notice that a graph depicts the variable PSV's electrical current flow and the pump's variable hydraulic flow.

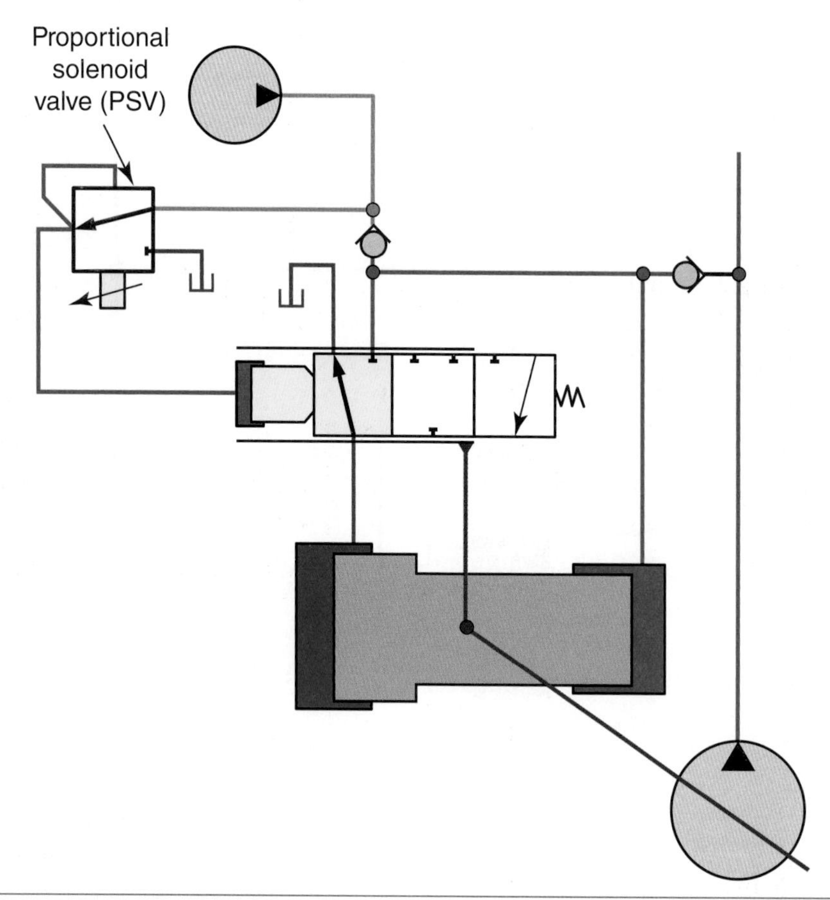

Figure 20-32. An EH PFC pump schematic in the upstroke mode. The PSV is energized, causing a positive pressure to act on the regulator valve, which will drain oil from the large end of the servo piston, causing the pump to upstroke.

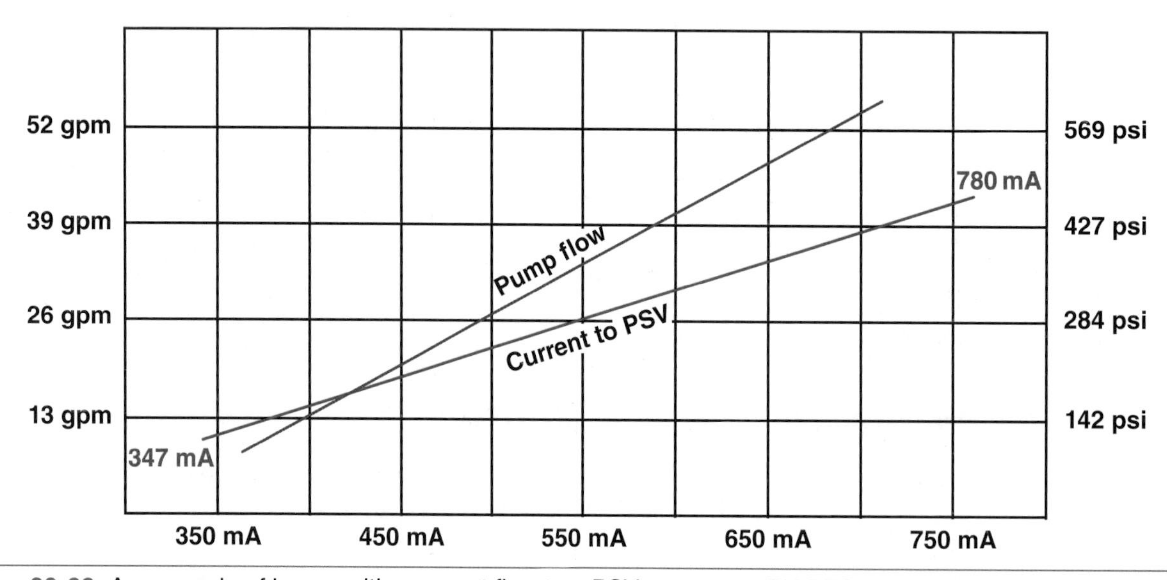

Figure 20-33. An example of how positive current flow to a PSV causes an EH PFC pump to increase pump flow.

Summary

- ✓ Excavators commonly use two variable-displacement pumps.
- ✓ Excavator pumps normally use either NFC or PFC controls.
- ✓ NFC pumps will upstroke when the control pressure drops.
- ✓ PFC pumps will upstroke when the control pressure increases.
- ✓ Excavator pump regulators commonly use a single servo piston with a large end and a small end. Sending oil to the small end upstrokes the pump. Sending oil to the large end destrokes the pump.
- ✓ NFC and PFC systems use open-center DCVs.
- ✓ NFC circuits use an orifice to build NFC pressure when the DCVs are in a neutral position.
- ✓ NFC regulators sense the following pressures: self-pump, companion pump, and PSV pressure, which Caterpillar calls power shift pressure and flow limit pressure.
- ✓ NFC pressure can be used by the ECM to estimate the pump's flow rate.
- ✓ NFC pressure and pump outlet pressure can be used to estimate hydraulic horsepower.
- ✓ Caterpillar and Case have manufactured NFC-controlled excavators.
- ✓ PFC hydraulic systems are found in John Deere and Hitachi excavators.
- ✓ PFC systems use a hydraulic computer for controlling pump regulators and DCV spool valves.
- ✓ The load control valve in PFC pump regulators destrokes the pump.
- ✓ The remote valve in PFC pump regulators upstrokes the pump.
- ✓ Many late-model excavators now use EH PFC hydraulic systems.
- ✓ EH PFC hydraulic systems use a solenoid that directly controls the pump regulator.

Technical Terms

companion pump pressure
constant horsepower control
cross sensing pressure
destroke valve
electro-hydraulic (EH) PFC hydraulic system
feedback link
flow limit solenoids
flow rate pilot valves
horsepower control proportional solenoid
hydraulic computer
hydraulic lockout
load control valve
minimum pump flow
negative flow control
pattern changer valve
pilot controller
pilot signal manifold
positive flow control
power save solenoid
power shift pressure reducing valve
power shift solenoid
proportional solenoid valve (PSV)
pump regulator
remote valve
torque control solenoid
total horsepower control

Review Questions

Answer the following questions using the information provided in this chapter.

1. As NFC signal pressure *increases*, pump output flow will _____.
   A. stay the same
   B. increase
   C. decrease
   D. increase or decrease, depending on load
2. As power shift pressure *increases*, pump output flow will _____.
   A. stay the same
   B. increase
   C. decrease
   D. increase or decrease, depending on load
3. In reference to a Caterpillar NFC excavator, Technician A states that the ECM determines hydraulic horsepower by measuring two different pressures. Technician B states that the ECM has no method for estimating hydraulic horsepower. Who is correct?
   A. Technician A.
   B. Technician B.
   C. Both A and B.
   D. Neither A nor B.
4. What is the purpose of flow limit solenoids?
   A. Develop a false power shift pressure.
   B. Develop a false NFC pressure.
   C. Develop a false pilot pressure.
   D. None of the above.
5. Why does the ECM control NFC or flow limit pressure?
   A. To maintain a constant flow to the work tool circuit.
   B. For the purpose of providing a main system relief.
   C. To operate the pump in standby.
6. NFC directional control valve spools are _____.
   A. open-center
   B. closed-center
   C. Neither A nor B.
7. Late-model Caterpillar hybrid (EH) NFC excavators determine pump flow based on which of the following?
   A. Swash plate angle sensor.
   B. Power shift pressure.
   C. NFC pressure.
   D. Pilot oil pressure.
8. Which of the following is a reasonable value for NFC signal pressure when the spools are in the neutral position?
   A. 200 psi.
   B. 400 psi.
   C. 600 psi.
   D. 800 psi.
9. When the DCV spools are in a neutral position, which of the following is responsible for establishing the NFC signal pressure?
   A. Relief valve.
   B. Orifice.
   C. Closed-center DCV spool.
   D. None of the above.
10. What happens to the NFC signal pressure when the DCV spool is actuated?
    A. Drops.
    B. Stays the same.
    C. Increases.
    D. Drops or increases, depending on the load.
11. The power shift solenoid receives oil from what supply?
    A. Main pump.
    B. Pilot pump.
    C. Secondary pump.
    D. None of the above.

12. If the operator requests more hydraulic horsepower than the engine is capable of delivering, which of the following has the responsibility of preventing excessive lugging or even stalling?
    A. Main system relief.
    B. Power shift solenoid.
    C. NFC pressure.
    D. None of the above.
13. On older Caterpillar excavators, what circuit requires the use of the flow limit solenoid?
    A. Start of boom swing.
    B. Boom down.
    C. Stick or bucket.
    D. Work tool.
14. Which of the following serves a purpose similar to that of torque limiting?
    A. Flow limit solenoid.
    B. Power shift solenoid.
    C. NFC pressure.
    D. None of the above.
15. Cross sensing oil pressure is an averaged _____.
    A. NFC signal pressure
    B. flow limit pressure
    C. power shift pressure
    D. pump outlet pressure
16. Which of the following OEMs does *not* manufacture an NFC excavator?
    A. Caterpillar.
    B. Case.
    C. Hitachi.
    D. All of the above.
17. How many servo pistons are used in a traditional NFC or PFC pump?
    A. One.
    B. Two.
    C. Three.
    D. Four.
18. In a traditional NFC or PFC pump, which of the following *always* has oil being supplied to it?
    A. Small end of the servo piston.
    B. Large end of the servo piston.
19. In a traditional NFC or PFC pump regulator, which of the following is responsible for stroking the pump to maximum displacement?
    A. Small end of the servo piston.
    B. Large end of the servo piston.
20. In a traditional NFC or PFC pump regulator, which of the following is responsible for destroking the pump to minimum displacement?
    A. Small side of the servo piston.
    B. Large side of the servo piston.
21. All of the following are common control pressures that act on an NFC pump regulator, *EXCEPT*:
    A. load sensing pressure.
    B. NFC.
    C. self-pump outlet pressure.
    D. companion pump outlet pressure.
22. If a pump regulator senses both self-pump outlet pressure and companion pump outlet pressure, the pump control can be labeled _____.
    A. NFC
    B. PFC
    C. constant horsepower
    D. total horsepower
23. What purpose does the NFC pump regulator feedback link serve?
    A. To actuate the servo piston in order to upstroke the pump.
    B. To actuate the servo piston in order to destroke the pump.
    C. To actuate the spool sleeve as the servo piston moves, to hold the pump in the last commanded position.
24. When an NFC pump is in the destroke mode, how much pump flow is being generated?
    A. Practically zero flow.
    B. 5–15% of the pump's maximum displacement.
    C. 15–25% of the pump's maximum displacement.
    D. 25–35% of the pump's maximum displacement.

25. Choose the correct order of oil flow for upstroking a pump in a traditional PFC excavator.
    A. Pilot controller, pilot signal manifold, remote valve.
    B. Remote valve, pilot signal manifold, pilot controller.
    C. Remote valve, pilot controller, pilot signal manifold.
    D. Pilot signal manifold, pilot controller, remote valve.
26. In a traditional PFC excavator, which of the following pump regulator valves has the responsibility for upstroking the pump?
    A. Load control valve.
    B. Remote valve.
    C. Torque control solenoid.
    D. None of the above.
27. What is another name for a pilot signal manifold?
    A. ECM.
    B. Hydraulic computer.
    C. Pump regulator.
    D. All of the above.
28. All of the following can be considered an output of the pilot signal manifold, *EXCEPT*:
    A. pump regulator.
    B. pilot controller.
    C. swing brake release valve.
    D. flow combiner valve.
29. All of the following are used to destroke a PFC pump, *EXCEPT*:
    A. torque control solenoid pressure.
    B. pump outlet pressure.
    C. companion pump pressure.
    D. PFC pilot pressure.
30. In a traditional PFC excavator, which of the following would produce the most pump flow?
    A. 80 psi of PFC pilot pressure.
    B. 130 psi of PFC pilot pressure.
    C. 240 psi of PFC pilot pressure.
    D. 350 psi of PFC pilot pressure.

Chapter 21

Hydraulic Test Equipment

Objectives

After studying this chapter, you will be able to:

- ✓ List the different attributes measured by hydraulic test equipment.
- ✓ Describe the different types of flowmeters and their features.
- ✓ List three different ways a flowmeter can harm a hydraulic system.
- ✓ Explain the advantages of using diagnostic pressure taps.
- ✓ List the different types of mechanical pressure gauges.
- ✓ Describe a common application for using a differential pressure gauge.
- ✓ Explain the benefits of a Quadrigage™.
- ✓ Explain the operational features of the OTC electronic digital pressure meter.
- ✓ Explain the difficulties of measuring pump inlet vacuum.
- ✓ List the causes for high pump inlet vacuum.
- ✓ List examples for using an infrared temperature gun.
- ✓ List examples for using a photo tachometer.
- ✓ Explain the need for a stop watch.
- ✓ Describe the features of a repair shop's hydraulic pump and motor test stand.
- ✓ List the features of a data logger.
- ✓ Describe the advantage of using a portable particle counter.

Introduction to Hydraulic Test Equipment

Hydraulic technicians have a variety of test equipment available to test and diagnose hydraulic systems. Test instruments are used to measure numerous hydraulic system parameters, including the following:

- Flow.
- Pressure.
- Temperature.
- Shaft speeds.
- Cycle times.

Test equipment is often installed in the system for the sole purpose of diagnostics. Technicians sometimes leave the test equipment installed on the machine to allow personnel to continually monitor the system. Many late-model electronically controlled tractors display different hydraulic parameters, such as fluid temperature or fluid pressure in real-time, using the machine's in-cab monitor or via the manufacturer's electronic service tool when the tool is connected to the system.

This chapter will explain the correct usage and operation of hydraulic test equipment. The advantages and disadvantages of using different hydraulic test tools will also be provided.

Caution

Gauges and flowmeters retain oil after they are used and are often outfitted with a hydraulic quick coupler. Care must be taken to avoid cross-contaminating hydraulic systems. As mentioned in **Chapter 10**, the chemical characteristics of different hydraulic fluids vary tremendously. Many are incompatible with certain hydraulic systems. In addition, after a pressure gauge or flowmeter is used to diagnose a possibly contaminated system, it is highly likely that the pressure gauge or flowmeter will be contaminated also. The test equipment must be flushed thoroughly before its next use.

Flowmeters

A hydraulic ***flowmeter***, or ***flow rater*** as it is sometimes called, is an instrument designed for measuring the volume of flow in a hydraulic circuit or system. The meter is normally installed only for diagnostic purposes and rarely left permanently connected to the machine. Finding the correct plumbing to connect a flowmeter to a machine can be a challenge but it is of crucial importance to produce an accurate reading. Flowmeters can come with many different features and are available in a variety of sizes and types. They measure flow either electronically or mechanically.

Mechanical Flowmeters

Mechanical flowmeters are the most common type used for measuring flow in mobile hydraulic systems. Mechanical flowmeters are normally designed to use either a tapered magnetic poppet or a spring-loaded piston and a tapered shaft.

Tapered Magnetic Poppet Flowmeter

Figure 21-1 illustrates the tapered magnetic poppet-style flowmeter. The poppet contains an internal magnet. A calibrated spring holds the tapered poppet against a sharp-edged orifice. As oil is directed through the inlet of the meter, the tapered poppet is pushed backward by pressure, compressing the spring. The magnet inside of the poppet forces a magnetic needle on the analog gauge to move based on the poppet's linear position. The meter's analog gauge displays the system's flow rate. This style of flowmeter is manufactured by Webtec.

An example of a Webtec analog flowmeter is shown in **Figure 21-2**. This particular meter is used for measuring low flow rates, up to 4 or 4.5 gpm (15 or 17 lpm), depending on the viscosity of the oil. Technicians can find flowmeters with different gauge ranges that they can use to measure any hydraulic system.

Tapered Shaft Magnetic Follower Flowmeter

Another style of mechanical flowmeter uses a tapered shaft, which is also called a tapered metering pin. See **Figure 21-3**. The tapered shaft is stationary within the flowmeter's internal body. A sharp-edged orifice and magnet assembly resembling a piston encircles the tapered shaft. The assembly is held by a calibrated spring that is fully extended when no oil is flowing through the

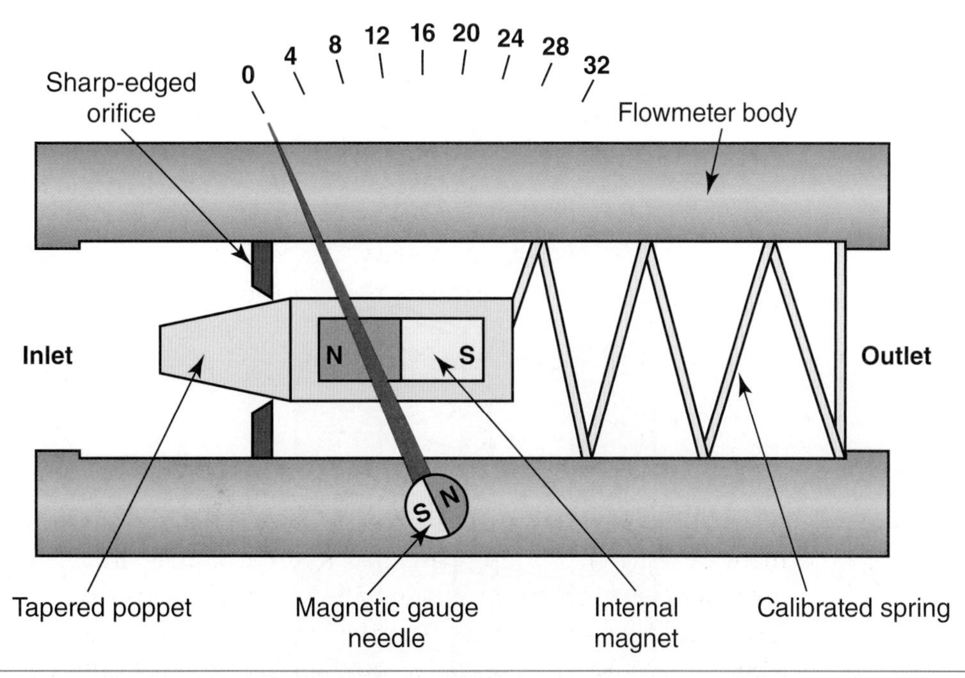

Figure 21-1. The tapered poppet inside this flowmeter contains an internal magnet. As fluid flows into the meter's inlet, the poppet compresses the spring. The poppet's magnet causes the meter's analog needle to move. The needle indicates the quantity of fluid flowing through the meter.

Figure 21-2. This meter uses an analog gauge for displaying the system's flow. The gauge displays a gallon per minute measurement for heavy oil on the top scale and a gpm measurement for light oil on the bottom scale.

meter. As oil enters the inlet of the meter, the orifice/magnet assembly compresses the spring because fluid flowing through the orifice and around the tapered pin creates a pressure difference across the assembly. A magnetic follower with an indicator line marked on it encircles the outside of the flowmeter's internal body. The follower is magnetically coupled to the orifice/magnet assembly for synchronized movement. As oil causes the assembly to compress the spring, the follower's indicator slides along a linear scale printed on the meter's exterior housing. Reading the position of the indicator line on the

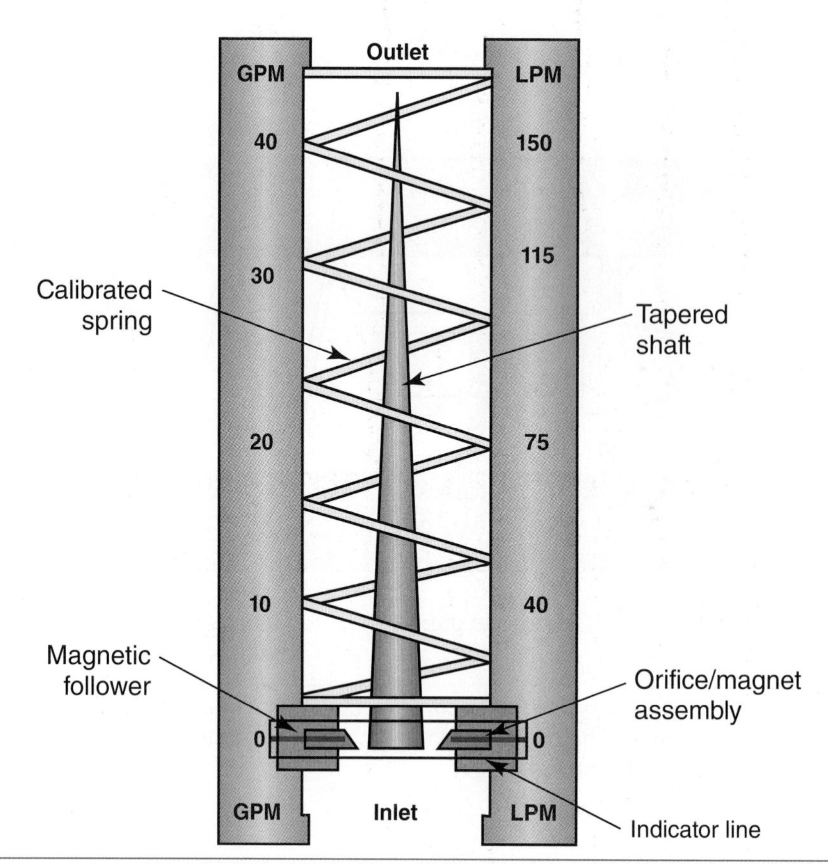

Figure 21-3. A tapered metering pin-type flowmeter uses a sharp-edged orifice that is held by a fully extended spring. The orifice contains a magnet that couples with a magnetic follower to slide along a linear scale on the face of the meter. As fluid flows into the meter's inlet, the orifice/magnet assembly compresses the spring, retracting the indicator up the scale to mark the flow rate measurement.

scale tells the amount of oil flowing through the meter. As a higher rate of oil flows through the flowmeter, it creates a greater pressure difference across the assembly. As a result, the spring compresses further. As the flow rate increases, the taper of the shaft shapes the enlargement of the orifice's area.

An example of a small-scale flowmeter that uses a tapered shaft is shown in **Figure 21-4**.

Electronic Flowmeters

Electronic flowmeters are categorized into three different types based on how the meter senses flow:

- Turbines.
- Differential pressure sensing meters.
- Positive-displacement meters.

Electronic Turbine Flowmeters

Electronic turbine flowmeters use a rotating turbine inside the meter housing. As fluid flows past the turbine, it will spin at a speed proportional to the fluid flow. The greater the flow, the faster the turbine will rotate. The meter uses an electronic pickup mounted on the outside of the meter to sense the turbine's rotational speed. The pickup generates a signal based on the rotor's

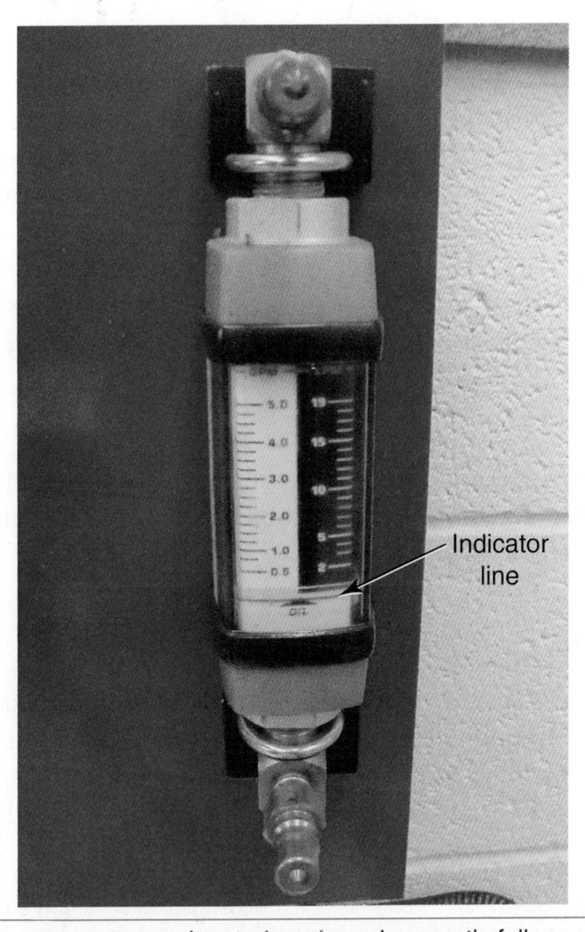

Figure 21-4. This small flowmeter uses a tapered metering pin and magnetic follower to measure flow rates up to 5 gpm (19 lpm). The pin, orifice/magnetic assembly, and calibrated spring are located inside the meter. Note the red indicator line marked on the magnetic follower near the base of the meter's scale.

speed, and sends that signal to the ECM. The turbine-style flowmeter is used in the aircraft industry. It can be used in various systems with different temperature ranges and fluid viscosities. The meter's precision and ability to work with remote-mounted electronics are its main advantages. The negatives are that the flowmeter relies on the smooth operation of precision bearings and requires electronics to read the system's flow. This means that an electronic turbine flowmeter is more susceptible to wear than a simple mechanical flowmeter and that it also must be supplied with battery voltage, at a minimum.

Electronic Positive-Displacement Flowmeters

An ***electronic positive-displacement flowmeter*** is similar to a hydraulic motor that rotates at a speed proportional to the system flow. It uses an electronic pulse generator to develop a signal based on the systems flow rate. Positive-displacement flowmeters vary in design, such as piston flowmeters, gear flowmeters and helical flowmeters. These meters are the only type that directly measures the volume of oil flowing through a circuit. The other flowmeters measure pressure and speed and estimate the volume of oil flow. These meters are not commonly used in mobile machinery, but are used in industrial settings, such as municipal water districts.

Electronic Differential Pressure-Sensing Flowmeters

Electronic differential pressure-sensing flowmeters use two pressure sensors that measure the pressure before the meter's restriction and after the meter's restriction. The meter works on the Bernoulli principle. The electronic pressure transducers transmit the pressure to the ECM. The ECM computes the flow based on the area of the restriction and the pressure drop across the restriction. The advantage of these meters is that they do not have moving, mechanical parts. The drawback is that the meter infers the flow rate based on other data, and as a result, the meter is less accurate at low flow ranges.

Flowmeter Features

When purchasing a flowmeter, technicians must consider what meter features, including the following, are needed for their type of work:

- Pressure gauge.
- Load valve.
- Thermometer.
- Burst disc protection.
- Capacity to allow oil to flow in both directions.

Although many flowmeters are equipped to only measure flow, that measurement is only half the hydraulic horsepower equation. To optimize the diagnostic capabilities of the flowmeter, it should contain a pressure gauge, like the flowmeter pictured in Figure 21-5.

Load Valve

By using a flowmeter equipped with a pressure gauge, a technician cannot magically cause the hydraulic system to build pressure. Adding a ***load valve***, which consists of a simple needle valve, enables the technician to place a load on the pump by restricting the pump's flow. This restriction will cause the system's pressure to rise. Figure 21-5 shows a tapered shaft-style flowmeter with a pressure gauge on the bottom and the load valve on the top.

Caution

Flow rating a pump has numerous risks and must be done according to procedures specified by the pump manufacturer to avoid component damage and personal injury. Operating a flowmeter will be discussed later in this chapter as well as in **Chapter 22**.

The flowmeter in Figure 21-6 contains a thermometer, a flow rate gauge, and a pressure gauge. The design of the meter allows all three gauges to be read simultaneously. The handwheel below the pressure gauge signifies that the meter has a load valve.

Flowmeter Burst Protection

Many flowmeters use ***burst discs***. A set of burst discs with installation instructions is shown in Figure 21-7. The discs act like fuses in an electrical circuit and rupture if the meter is over-pressurized. Once the discs rupture, the burst disc assembly will either bypass oil internally or exhaust oil externally.

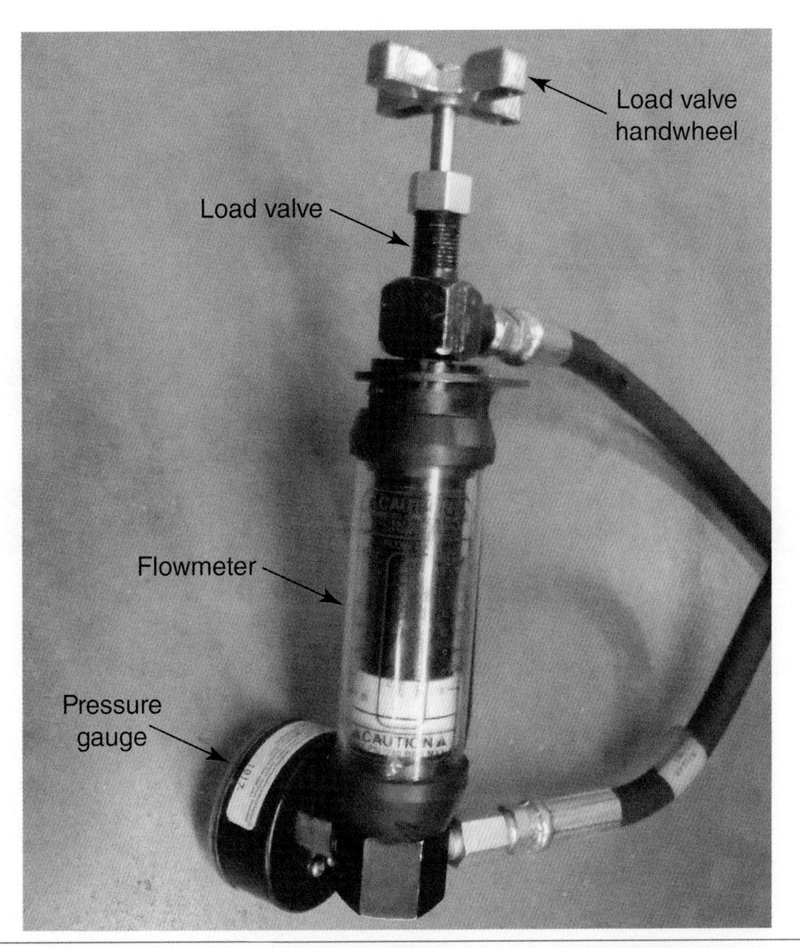

Figure 21-5. A flowmeter with a load valve on top and a pressure gauge on the bottom. The handwheel on the load valve changes the size of the restriction. This particular gauge was poorly designed because a technician cannot read the flow and the pressure gauge simultaneously. It also does not contain a thermometer.

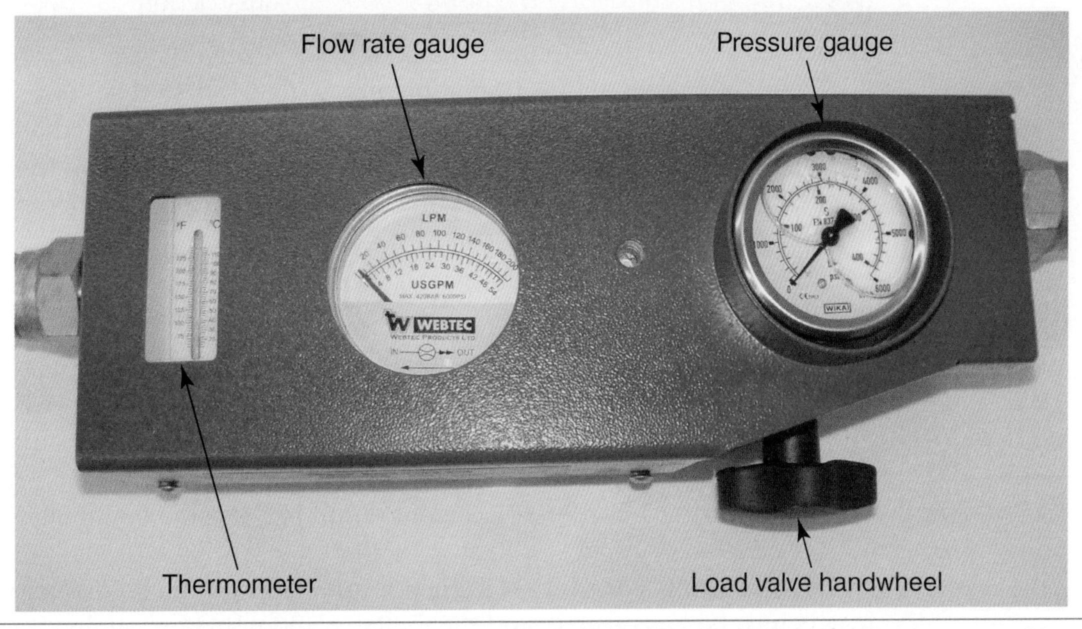

Figure 21-6. The thermometer, flow rate gauge, and pressure gauge can be quickly read and compared by a technician. The location of the load valve is marked by the handwheel below the pressure gauge. This flow meter can measure up to 54 gpm (210 lpm) and 6000 psi (414 bar).

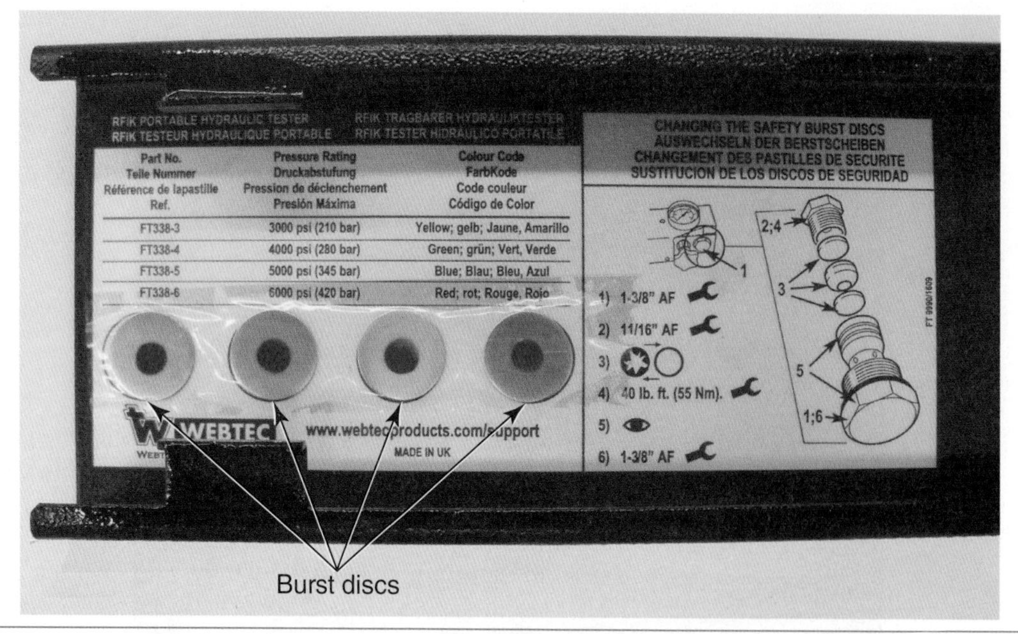

Figure 21-7. Four burst discs taped to the inside cover plate of a flowmeter. Manufacturer instructions for changing the burst discs are listed on the label.

Both burst protection assemblies have positive and negative aspects that may change based on the application. If a flowmeter contains an internal bypass assembly, a technician that unknowingly uses the meter after the discs have ruptured might incorrectly condemn a hydraulic pump, believing the pump cannot build pressure. The inability to quickly identify when the discs have ruptured is a downside to this assembly design. As mentioned in **Chapter 1**, some job sites are very sensitive to spilling any type of oil, and, if a spill does occur, it might be necessary to document the steps taken to restore the soil back to a normal condition. A burst disc assembly designed to bypass oil internally would be required at such job sites. In this situation, the design is a positive aspect.

Some hydraulic shops prefer meters with a burst disc assembly that exhausts oil externally because the resulting oil spray immediately alerts technicians of the need to replace the discs. Using this design on the environmentally sensitive job site, however, could create serious problems once the assembly expelled oil.

Installation of a Flowmeter

Mistakes, including the following, can easily be made by a technician during the installation or operation of a hydraulic flowmeter:

- Cavitating a pump because the flowmeter returned the oil to the wrong reservoir or location.
- Rupturing a fixed-displacement pump because the flowmeter's load valve was closed.
- Rupturing a fixed-displacement pump because the flowmeter was installed incorrectly and the meter would not allow oil flow in a reverse direction. If the flowmeter allows oil to flow in both directions, the meter normally only measures the flow in a single direction.

It is important to use the manufacturer's service literature when flow rating the hydraulic pump to avoid making mistakes. Without the specifications, a technician would have no method to validate the results. In addition to installation instructions, the service literature will specify the following parameters:

- Rated engine speed.
- Fluid operating temperature.
- System pressure.
- Pump's flow rate.

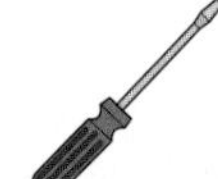

Note

Flowmeters do not remain in calibration and will lose accuracy after repeated use. It is important to periodically check the flowmeter's accuracy to determine if the meter needs to be recalibrated. Although cross-checking a flowmeter's accuracy against another flowmeter may be a common practice in the industry, the best and safest practice is to send flowmeters to a calibration lab on a regular schedule to ensure meter accuracy.

PQ Curve

As mentioned in **Chapter 18**, manufacturers sometimes require a technician to plot a PQ curve for diagnosing a hydraulic pump. A PQ curve lists a pump's flow rate on one axis, commonly the left axis, and the pump's outlet pressure on the other axis, commonly the bottom axis. **Figure 21-8** illustrates an example of a PQ curve that displays the pump manufacturer's high and low flow rate specifications. The machine's actual pump flow drops below the minimum specification at pressures above approximately 2000 psi. During a service training class, technicians are sometimes required to enter the pump's flows and pressures into a Microsoft Excel spreadsheet, and the software will automatically plot the PQ curve.

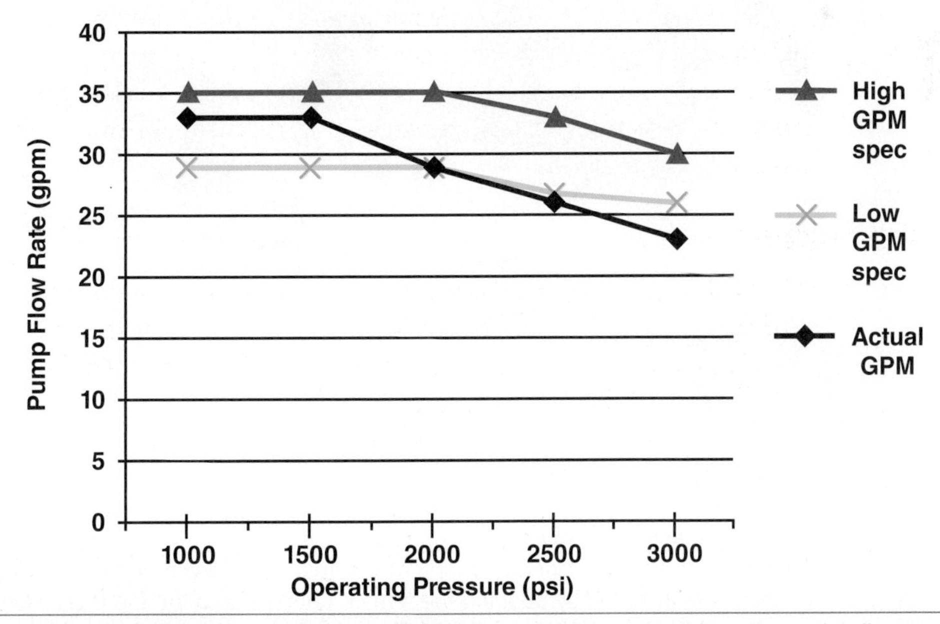

Figure 21-8. This PQ graph shows the maximum and minimum curves provided from the hydraulic pump manufacturer. The actual pump flow rate drops below minimum specifications when outlet pressure rises above 2000 psi.

Pressure Testing Equipment

Measuring fluid pressure is one of the most common tests performed by mobile hydraulics technicians. Manufacturers offer different types of testing equipment, including the following, that are used in the measuring or monitoring of pressures:

- Diagnostic couplers.
- Traditional pressure gauge.
- Differential pressure gauge.
- Quadrigage™.
- Electronic digital pressure meter.
- Snubbers.

Diagnostic Pressure Taps

Years ago, technicians looked for a plugged port in the system to measure system pressure. They removed the plug and threaded a gauge into the cavity. If the system did not have a plugged port, a T-fitting was installed in line, and the gauge was installed into the T-fitting. Fortunately, today many manufacturers place ***diagnostic pressure taps*** throughout a machine. The taps enable technicians to quickly connect to a hydraulic system, making it very easy to test pressures. Quick couplers attached to the pressure gauges allow a technician to quickly snap a gauge onto the male pressure tap. See **Figure 21-9**.

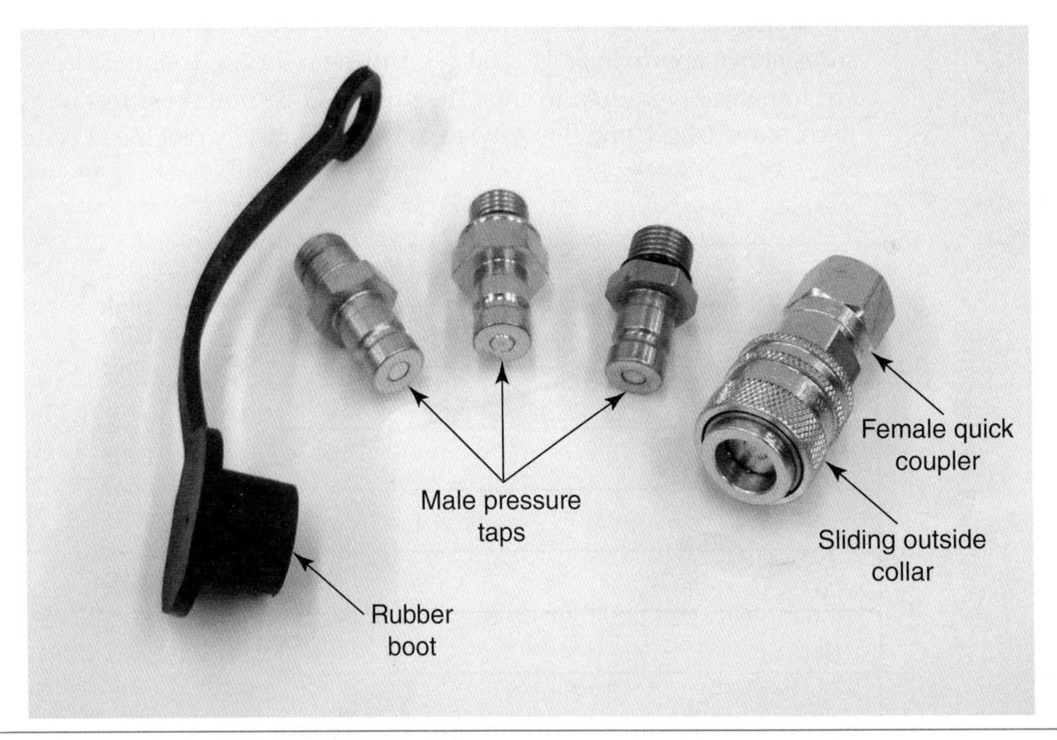

Figure 21-9. Male diagnostic pressure taps come in a variety of thread types and sizes. The taps are commonly installed in hydraulic systems to speed the process of measuring pressures. The female quick coupler is threaded onto the pressure gauge. The coupler is installed by pushing the female coupler onto the male pressure port until the collar automatically snaps in place. The coupler is removed from the pressure tap by sliding the coupler's outside collar rearward and pulling the female coupler off the male end.

The pressure taps are a standardized SAE J1502 interchange, and are found on numerous makes of machines. The quick couplers work similarly to air tool quick couplers and require the outside collar to be slid rearward in order for the coupler to be removed from the pressure tap. However, one difference between the air tool couplers and the pressure taps is that the pressure tap's collar does not have to be slid rearward during installation. The female coupler must simply be pushed onto the male tip for the assembly to snap together as a sealed connection.

Warning

Never remove a pressure gauge from a running hydraulic system or a system that is pressurized. Be sure to look at the gauge prior to removal. If it is registering pressure, do not remove the gauge until the system pressure has been depleted.

The advantage to using quick couplers is that the pressure taps can be left installed in the hydraulic system, preventing the unwanted contamination that occurs any time a gauge is threaded in or out of the system. The disadvantage is that the couplers can seep oil and have the potential to allow contamination to enter the hydraulic system. For that reason, the couplers frequently use a rubber boot or dust cap to cover the test port when not in use. John Deere calls the pressure taps "DR" for diagnostic receptacles.

Pressure taps can be purchased from the machine manufacturer or from a supplier. Some common pressure taps are 1/4" NPT and 1/8" NPT threads. They can also use ORB, ORF, and JIC threads. Sometimes it can be difficult to find a part number for a pressure tap. **Figure 21-10** provides part numbers for a variety of different pressure taps.

Caution

Always be sure to follow the OEM service literature when installing a pressure tap. Pressure taps can be designed for low-pressure applications, so be sure that the fitting is made of steel prior to installing it in a high-pressure system. Some OEMs also use a steel washer in conjunction with the O-ring.

The pressure taps provide another advantage. Most gauges are pipe threads and will easily leak with repeated removal and installation. The design of the pressure taps ensure a leak-free connection for the ports in the hydraulic system and for gauges equipped with a quick coupler.

Traditional Pressure Gauge

A traditional mechanical pressure gauge, **Figure 21-11**, is the most common pressure measuring device used by technicians due to its economical price. In fact, the gauge is sometimes cheaper than the female coupler that is attached to the gauge.

Gauges can be purchased at automotive stores, hydraulic shops, tractor supply stores, and dealerships. They can be designed to measure vacuum, low pressure, medium pressure, and high pressure. The physical size of the gauges can vary—a common size is approximately 2.5" in diameter.

SAE J1502 Interchange	CNH	Caterpillar	John Deere	Parker	Eaton/ Aeroquip/ Weatherhead
1/8″ –27 NPT male tap	H434164	8T-3613	XPD323	PD323	FD90-1012-02-04
1/4″ –18 NPT male tap	9845229	6V-3966	XPD343	PD343	FD90-1012-04-04
3/8″ –24 ORB male tap	—	—	XPD331	PD331	FD90-1044-03-04
7/16″ –20 ORB male tap	R55912	—	XPD345	PD341	FD90-1044-04-04
1/2″ –20 ORB male tap	—	—	XPD351	PD351	FD90-1044-05-04
9/16″ –18 ORB male tap	87026252	6V-3965	XPD361	PD361	FD90-1044-06-04
1/4″ –18 NPT Female tap	—	6V-3989	—	PD342	FD90-1034-04-04
1/8″ –27 NPT Female tap	—	6V-4142	—	PD322	FD90-1034-02-04
M14 × 1.5 ORB male tap	84320565	—	XPD367A	PD367A-6	FD90-1046-06-04
M18 × 1.5 ORB male tap	358968A1	—	—	PD3127-6	—
1/2″ Tube ORF male tap	—	—	—	PD38BTL	—
7/16″ –20 JIC (1/4″ tube) male tap	R54805	—	—	PD34BTX	—
9/16″ –18 JIC (3/8″ tube) male tap	—	—	—	PD36BTX	—
3/4″ –16 JIC (1/2″ tube) male tap	—	—	—	PD38BTX	—
9/16″ –18 ORF female tap	190117A1	—	—	PD34BTL-6	—
11/16″ –16 ORF female tap	190119A1	—	—	PD36BTL-5	—
13/16″ –16 ORF female tap	190316A1	—	—	PD38BTL-6	—
Dust cap	86502604	6V-0852	R77175	PDN-02C	FD90-1040-04
1/8″ –27 NPT female coupler	—	6V-4143	—	PD222	FD90-1021-02-04
1/4″ –18 NPT female coupler	73163493	6V-4144	RE219698	PD242	FD90-1021-04-04

Figure 21-10. Quick coupler part numbers.

Caution

If a technician installs a low-pressure gauge in a high-pressure hydraulic system, the gauge will be damaged.

Mechanical gauges can be dry or liquid filled with glycerin. The liquid-filled gauge reduces the tendency of the needle to bounce due to vibrations and pulsations, but it must be positioned so that the vent is in a vertical position. The glycerin can cause the gauge to discolor if it is left outside in the sunlight for long periods of time.

Figure 21-11. This mechanical pressure gauge is liquid filled and has a quick coupler attached to it. The gauge will measure up to 1500 psi.

Bourdon-Tube Pressure Gauge

The ***Bourdon-tube pressure gauge*** is the most common type of mechanical pressure gauge used by hydraulic technicians. Eugene Bourdon invented the gauge in 1849. Bourdon-tube gauges can have one of several different configurations:

- C-shaped tube.
- Spiral tube.
- Helical tube.

The C-shaped Bourdon pressure gauge has a C-shaped tube that will flex in relation to fluid pressure. As the tube flexes, it begins to straighten at the fixed end. This causes the other sealed tube end to move slightly. A linkage mechanism attached to the sealed tip of the tube transmits the movement through a gear setup to move the gauge needle. See **Figure 21-12A**.

In the spiral Bourdon-tube pressure gauge, pressure is applied to a long tube that is wrapped in multiple turns around a center point, like a tightly coiled snake. See **Figure 21-12B**. As pressure is applied to the gauge, the tube begins to uncoil slightly, causing movement of the tube tip, which is transmitted to the gauge needle.

The helical Bourbon-tube gauge uses a tube that has a small helical coil formed in the middle of the tube as shown in **Figure 21-13**. The tube is connected to mechanical linkage that transmits the amount of pressure to the gauge's needle.

Both the helical and spiral Bourdon gauges operate with the same principles as the C-shaped gauge. Pressure gauges used to measure low pressures use a bellows or a diaphragm, instead of a Bourdon tube.

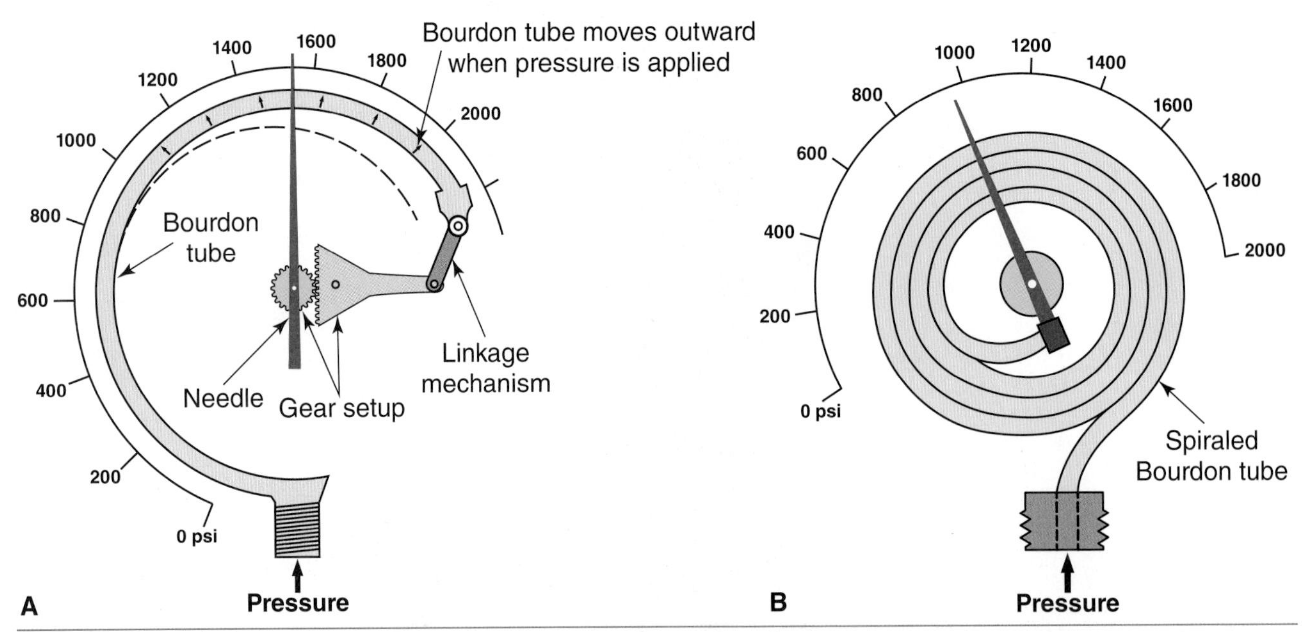

Figure 21-12. A—A C-shaped Bourdon-tube pressure gauge uses a C-shaped tube that will flex and straighten as hydraulic pressure is applied to the base of the gauge. Mechanical linkage attached to the sealed tip of the tube transfers tube deflection to the needle. B—A spiral Bourdon-tube pressure gauge contains a flattened tube spiraled around the center gauge needle.

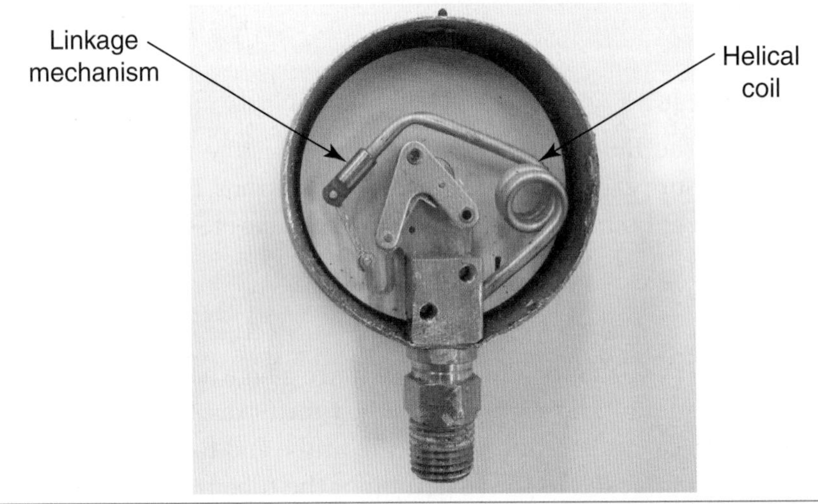

Figure 21-13. A helical Bourdon-tube pressure gauge disassembled to show its internal parts. Note the helical coil in the middle of the tube and the linkage mechanism attached to the tube's end.

Gauge Holder

Chapter 1 explained that pressure as low as 100 psi (7 bar) can penetrate the human skin, causing serious injury or potential death. It was also mentioned that reported hydraulic accidents involving pressures of 7000 psi (482 bar) and higher resulted in amputation nearly 100% of the time.

To reduce the risk of fluid injection injuries, it is a good idea to place pressure gauges in a hanger. This minimizes the need to handle live gauges. A semi-truck mud flap is the perfect material for constructing such a gauge hanger, Figure 21-14. The gauge is placed through the center of the holder and technicians can hang the gauge safely on the machine without having to hold a live hydraulic pressure gauge.

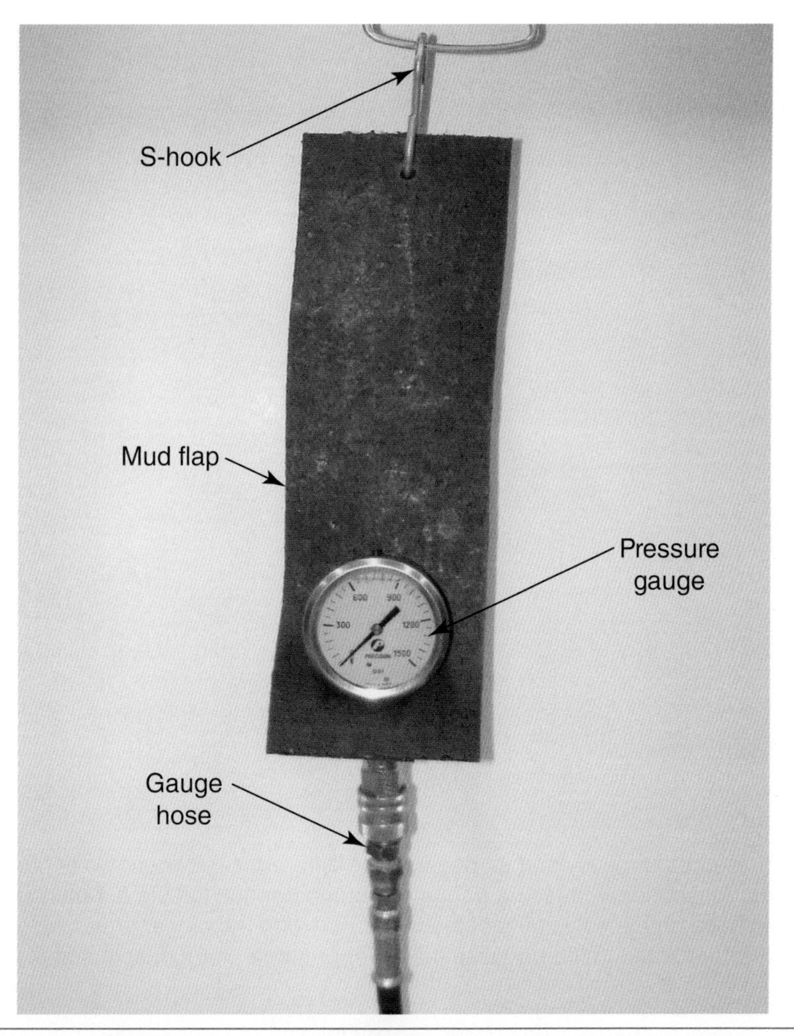

Figure 21-14. A piece of a truck's rubber mud flap can be used to make a pressure gauge holder. Simply cut or drill a large hole in the mud flap and insert the hose of the gauge through the hole. Connect the pressure gauge to the hose end protruding from the flap. Cut or drill a smaller hole in the opposite end of the flap. Slip one end of an S-hook through the small hole and use the other hook end to hang the gauge on the machine.

Differential Pressure Gauge

A differential pressure gauge allows a technician to measure the difference between two separate but simultaneous pressure values. As discussed in **Chapter 18**, a differential pressure gauge is helpful for measuring margin pressure. A technician connects the left pressure line to the higher pressure port and the differential pressure gauge. A second pressure line connects the right side of the differential pressure gauge to the lower pressure port. See **Figure 21-15**.

Quadrigage

Another style of pressure gauge is the MICO Quadrigage™, which is similar to the Caterpillar tetra gauge. The assembly contains three pressure gauges. Because the three gauges all have different ranges, the technician can use the gauge assembly to measure a very wide range of pressures.

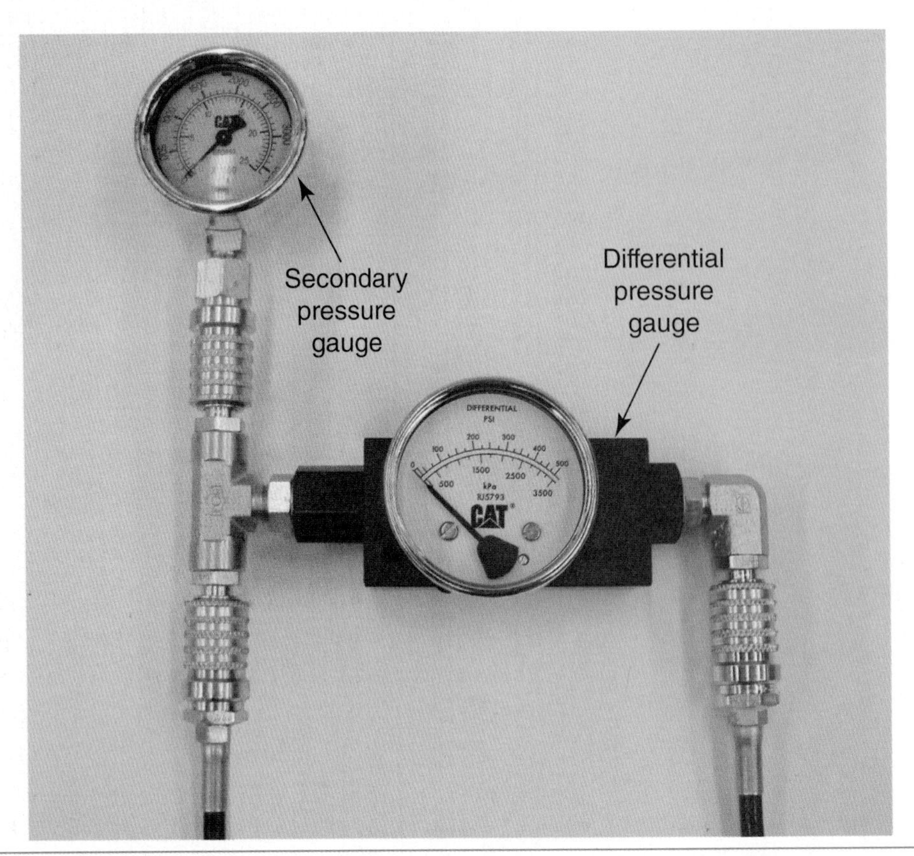

Figure 21-15. A differential pressure gauge is used to measure the difference between two separate but simultaneous pressure values. A secondary pressure gauge is mounted on the left via a T-fitting to monitor the highest system pressure. The hose on the left is connected to the port with the higher pressure, for example pump outlet. The hose on the right is connected to the port with the lower pressure, for example signal pressure.

The gauge's maximum pressure varies based on the model of the gauge. The gauge in Figure 21-16A can measure pressures ranging from a vacuum to as high as 6000 psi (414 bar). For example, if a technician needed to measure three different ports, but was unsure which one was the high pressure, medium pressure, or low pressure, this type of pressure gauge would be the tool of choice. Regardless of the port that the technician chose, the multipurpose gauge would provide an accurate reading with the possibility to cross four pressure scales:

- Gauge 1—A. Vacuum: 0 to 30 in Hg.
  B. Low pressure: 0 to 72 psi.
- Gauge 2—Medium pressure: 0 to 600 psi.
- Gauge 3—High pressure: 0 to 6000 psi.

The assembly contains an internal cutoff that prevents high pressure from damaging the two low-pressure gauges. For example, if the gauge was used to measure a pressure below 72 psi, only the low-pressure gauge would register a pressure. Once the pressure climbed above 72 psi, the reading would also register on the medium-pressure scale. If the pressure rose to over 600 psi, the reading would register on the high-pressure gauge. The technician would read the system pressure from only one gauge, which is the gauge reading the highest pressure.

Figure 21-16B shows a John Deere service tool used for measuring hydraulic pressures. It too has multiple gauges (160, 600, and 5000 psi) and operates in the same fashion as the Caterpillar Tetra Gauge and the MICO Quadrigage™. Notice that this gauge has a port and hand valve on the right and on the left. The gauge assembly is sometimes used to measure a differential pressure, but the gauge cannot display an active difference in pressures between port A and port B.

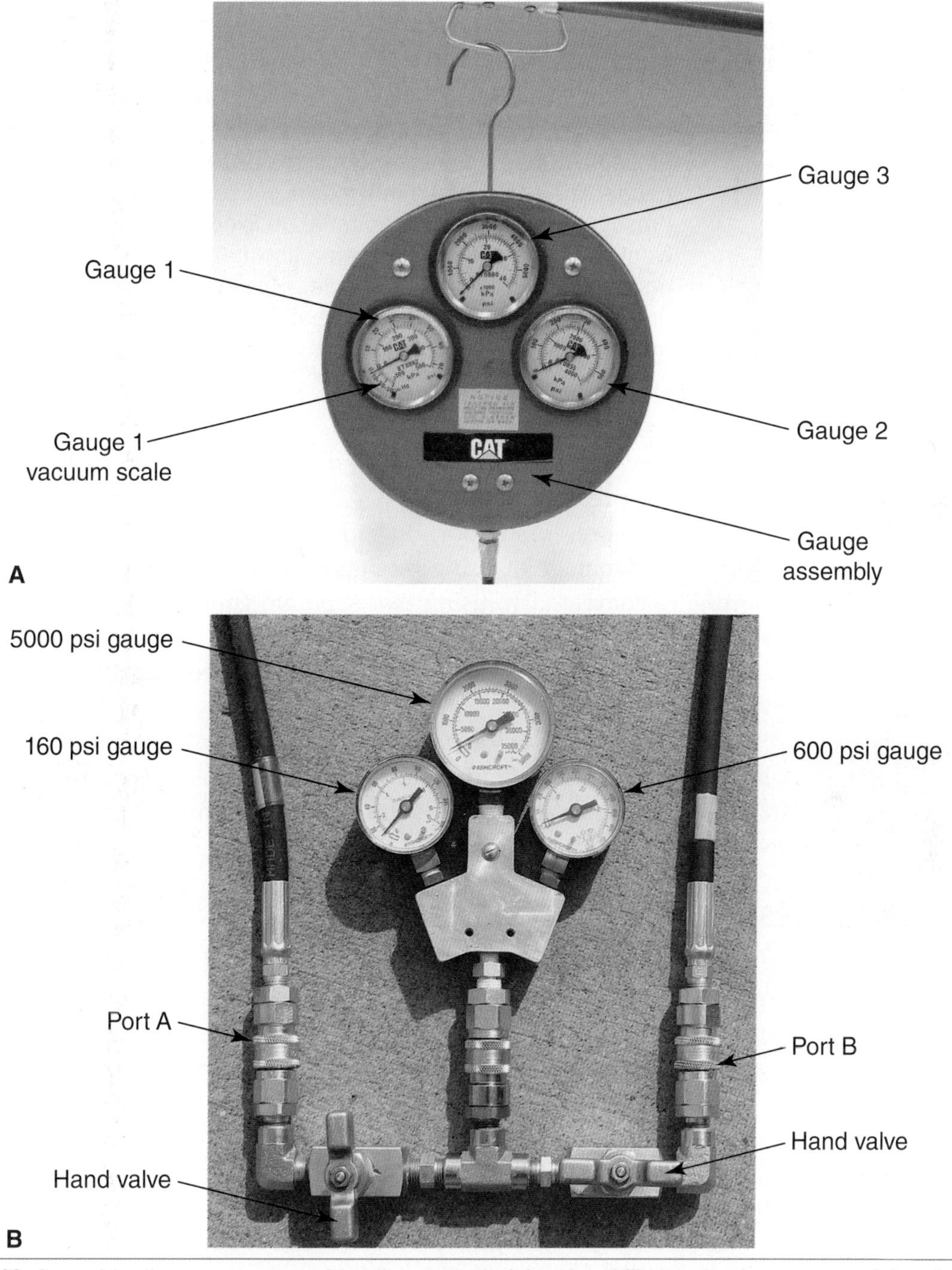

Figure 21-16. A combination gauge assembly allows a technician the ability to attach one gauge into a test port that can have a pressure value ranging from a vacuum to high pressure. A—This Caterpillar gauge can accurately measure pressure within a vacuum to 6000 psi range. Note the three gauges in the assembly. B—This John Deere gauge assembly contains a T-fitting and two shutoff valves that allow a technician to connect to two separate ports and measure each port one-at-a-time by shutting off the valve to the opposite port.

An example of operation for measuring differential pressure would be:

1. Connect the higher-pressure port to port A on the left (pump outlet pressure).
2. Connect the lower-pressure port to port B on the right (signal pressure).
3. Shut off the hand valve isolating port B from the gauge.
4. Open the hand valve on the right so that the gauge can display the higher system pressure.

Note

This gauge is not simultaneously measuring two pressures and displaying the difference (margin). It is not an active differential pressure gauge, but acts more like a passive differential pressure gauge. Therefore, when this gauge is used for measuring margin pressure, the hydraulic load must be fixed (constant). An example would be using a load valve on a flowmeter, instead of lifting an implement, like a dozer blade, which would cause the two pressures (pump outlet and signal) to vary as the implement moved.

5. Record the pressure of port A.
6. Shut off the hand valve to port A and open the hand valve to port B.
7. Record the pressure of port B.
8. Subtract the pressure of port B from the pressure of port A.

The one advantage of using this gauge assembly for measuring differential pressure compared to using two separate gauges is that the same gauge is being used to measure both pressures, ensuring that the readings are accurate relative to each other. If two separate gauges are used, there may be some deviation between the gauges.

Using the gauge assembly also has an advantage over using a single gauge. A single gauge would require shutting off the machine, depressurizing the circuits, and switching the single gauge from the higher-pressure port to the lower-pressure port.

Electronic Digital Pressure Meter

Pressures can also be measured electronically. An electronic pressure transducer is installed in place of mechanical pressure gauge. The electronic pressure transducer uses an internal sensing device such as a diaphragm, bellows, or Bourdon tube and converts its mechanical movement into an electrical output. The mechanical movement is converted into an electrical signal. The electronic signal is sent to a meter or ECM in the form of voltage, current or a frequency. The most common transducer electrical output varies either the voltage or current.

The OTC electronic digital pressure meter is a common electronic pressure meter used by hydraulic technicians. It offers numerous measurement options, many of which are also offered on other electronic digital pressure meters:

- Up to four different pressure ports.
- Pressure ranges from vacuum to 10,000 psi.

- Delta zero.
- Minimum/Maximum.
- Differential pressure.
- Temperature.

The greatest benefit of this type of pressure measuring tool is safety. A technician measuring a system with 10,000 psi (689 bar) can handle a low-voltage pressure transducer and a meter, rather than a live 10,000 psi (689 bar) pressure gauge.

The meter offers three different pressure transducers, each of which is designed to measure a certain pressure range. The low-pressure transducer is used for measuring pressures from a negative pressure (vacuum) up to 500 psi (34 bar). The medium-pressure transducer is used to measure pressures up to 5000 psi (345 bar). The high-pressure transducer measures pressure up to 10,000 psi (689 bar).

Caution

Pressure transducers on electronic pressure meters can cost four to five times the price of a traditional liquid-filled gauge. If a low-pressure transducer is placed in a high-pressure system, it will ruin the transducer. The meter has controls that must be set to the appropriate range. See **Figure 21-17**.

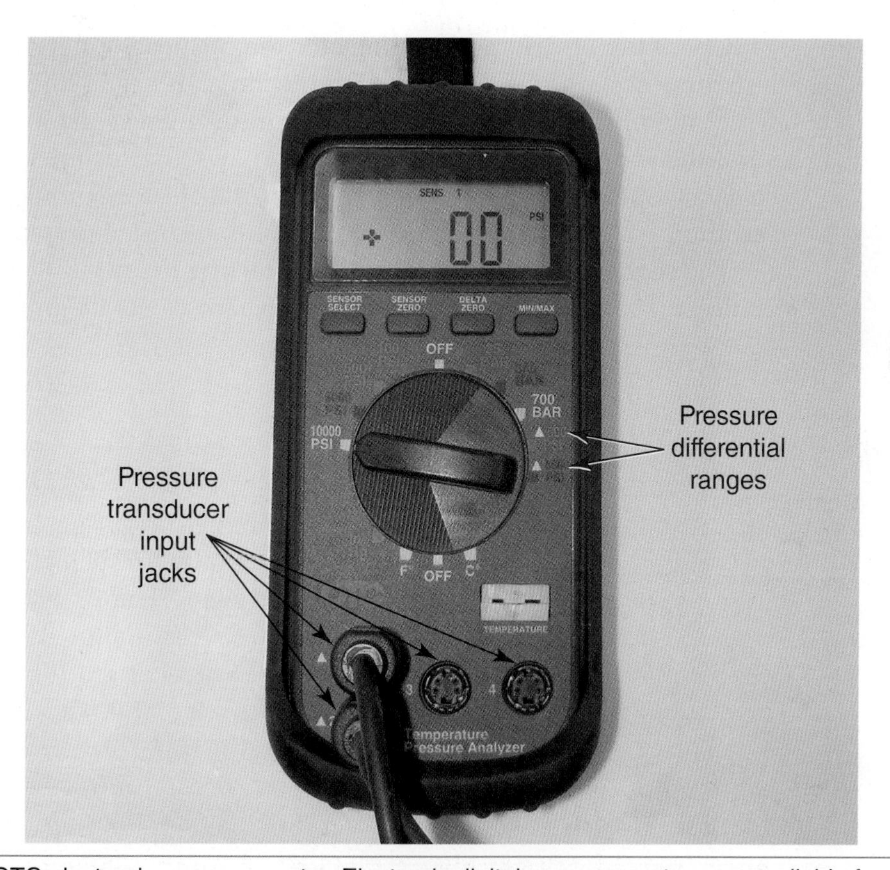

Figure 21-17. An OTC electronic pressure meter. Electronic digital pressure meters are available from multiple manufacturers. This particular electronic meter offers numerous features such as delta zero, differential pressure, minimum/maximum, and is a safe method for measuring very high system pressures.

A technician should install a quick coupler on the transducer to ensure a leak-free connection to a diagnostic tap. See **Figure 21-18**. The meter has electrical leads that connect the transducers to the meter.

Multiple Sensor Inputs

If a technician needs to measure multiple pressure ports, the meter is designed with multiple sensor input jacks and can be connected to up to four pressure ports at once. The technician does not have to handle four separate pressure gauges. However, the meter does not simultaneously display all four pressures onscreen. The technician must cycle through the four sensor readings, one-at-a-time, using the meter's controls.

Delta Zero

Delta zero is a feature that allows a technician to monitor changes for a given pressure in units of positive or negative numbers. Entering the meter's delta zero mode causes the existing system pressure reading to become the base zero for the meter's pressure scale.

As the system pressure changes, the meter will show a positive or negative reading. For example, if the system pressure is 500 psi (34 bar) and the delta zero mode is selected, a system pressure drop of 50 psi (3 bar) will make the meter read "–50 psi".

Minimum/Maximum

Operating or driving machines for the purpose of performing diagnostic tests imposes risks and dangers. If a technician needs to operate a machine while attempting to monitor the gauge for the highest and lowest system pressure, the ***minimum/maximum*** feature allows the technician to concentrate solely on operating the machine. The min/max feature provides the minimum recorded system pressure and the maximum recorded pressure.

The sequence of operation requires starting the machine so that the meter is measuring a fluid pressure. The technician selects the min/max feature on

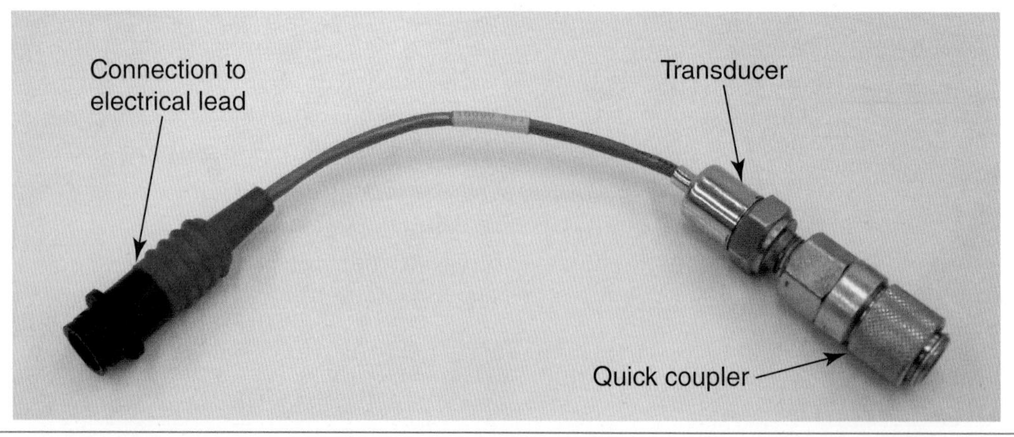

Figure 21-18. Pressure transducers use a pipe thread fitting to connect into hydraulic systems. This type of fitting is not designed for repeated use. A quick coupler installed onto the end of the transducer will lower the risk of leaks and contamination.

the meter's controls. After the machine operates for a period of time, the technician selects the min/max feature again to display the recorded minimum pressure. When the technician selects the min/max feature a third time, the meter displays the recorded maximum pressure.

Differential Pressure

Margin pressure was discussed in **Chapter 18** and **Chapter 19**. As mentioned, it is much easier to measure margin pressure using a differential pressure gauge than by comparing two different pressure gauges. The differential pressure option requires using the first and second sensor input jacks in the meter. The appropriate pressure range is chosen on the meter and sensor 1 is zeroed with the controls. The same steps are done for sensor 2. Then, select the proper pressure differential range on the meter. Pressure differential settings are often denoted with a triangle (Greek symbol delta—Δ). The delta symbol is commonly used to denote a change or difference in values.

The meter will display the pressure difference of sensor 2 in respect to sensor 1. For example, if sensor 2 is measuring pump outlet pressure of 2500 psi (172 bar) and sensor 1 is measuring a signal pressure of 2350 psi (162 bar), the meter would read "+150 psi". The pressure at sensor 2 is 150 psi (10 bar) greater than at sensor 1.

Snubbers

Hydraulic snubbers can be installed within pressure measuring gauges to eliminate pulsations that occur as a result of the hydraulic system flows and pressures. They are installed in the threaded fluid inlet of many traditional pressure gauges. Snubbers come in the following varieties:

- Sintered metal insert.
- Threaded plug.
- Porous element.
- Metering pin.
- Small Bourdon tube, such as a 1/8" coiled tube.

Selecting a Pressure Gauge

Prior to measuring a system pressure, a technician must select the proper pressure gauge. The technician should avoid the following mistakes when choosing a hydraulic pressure gauge:

- Selecting a gauge with too low of a pressure limit.
- Selecting a gauge with an inappropriate scale (too wide or narrow of a range).
- Selecting a gauge with poor accuracy.

If a gauge with too low of a pressure limit is accidentally installed on a hydraulic system, the pressure gauge will be damaged and will need to be replaced. This is a common mistake made by inexperienced technicians. The safer method is to assume no knowledge of a system's pressure.

Following the safer method requires selecting a pressure gauge with a large range, for example, up to 10,000 psi (689 bar). Once a rough pressure reading is obtained, a gauge with a lower graduated scale must be used to gain an accurate pressure reading. Before the appropriately scaled pressure gauge can be installed by the technician, the machine must be shut off again and the system has to be fully depressurized. A rule of thumb for selecting the appropriate scale range on a traditional Bourdon-tube pressure gauge is to choose a gauge with a range that is 40%–50% higher than the highest expected system pressure. Ideally, the gauge's needle should stay in the middle third of the scale.

Caution

If the pressure gauge is to be mounted on the machine permanently, professionals recommend installing a gauge with a peak scale value that is twice as great as the highest operating pressure of the system. If the system has a pressure spike with such a gauge installed in the system, the gauge will not fail.

Pressure Gauge Accuracy

Pressure gauges are offered in a variety of accuracies ranging from ± 0.5% to ± 3.0%. A progressive dealership will use the most accurate pressure gauges (± 0.5%) during the pre-delivery inspection of a machine to ensure that the customer receives a strong hydraulic machine with the correct system pressures. Technicians who use a less accurate gauge (one with a high tolerance percentage) may create additional problems for themselves. For example, suppose a technician has been tasked with adjusting the main system pressure, which has a specification of 3000 psi (207 bar) ± 50 psi (3 bar). If a technician is using a 6000 psi (414 bar) pressure gauge that has a ± 3.0% tolerance, the pressure reading could be 180 psi (12 bar) too high or too low. The problem could be further compounded if the pressures are checked with cold hydraulic oil. The customer might receive a machine with a lower than specified relief pressure, resulting in weaker breakout forces.

Calibration

Pressure gauges need to be checked periodically for accuracy. A rule of thumb, based on minimum standards, is to perform an annual accuracy check on pressure gauges. Many technicians/shops will perform calibration tests on a shorter timeline. Any time a gauge is suspected of producing incorrect readings, however, it should *not* be used again until it is tested.

The pressure gauge tester in **Figure 21-19** uses a series of weights. The weights are placed on the tester to verify the calibrated pressure. A calibrated electronic pressure gauge is often used to cross-check for correct calibration of the pressure tester. Once the tester's pressure has been authenticated, the pressure gauge in question can be tested.

Measuring Pump Inlet Pressure

Occasionally it may be necessary to measure pump inlet pressure on a machine that does not have any provision for easily installing a pressure gauge.

A

B

Figure 21-19. The accuracy of a pressure gauge needs to be periodically tested. A—A pressure gauge tester kit. B—The tester uses a series of weights to establish a known set pressure for correct calibration of the gauge.

In such cases, it might be easier to modify a current system fitting rather than attempt to install a large, modified T-fitting with an adapted pressure tap. The pump inlet fitting in **Figure 21-20** was removed, drilled, and tapped so that a pressure tap with NPT threads could be installed into the fitting.

A common pump inlet specification is no more than 7 in Hg. Less is best regarding pump inlet vacuum.

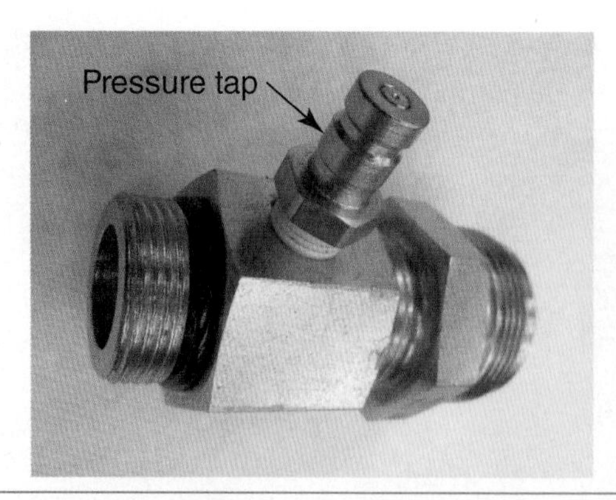

Figure 21-20. Many hydraulic systems are not configured for easy measurement of pump inlet vacuum. This pump inlet fitting was drilled and tapped so that a pressure tap could be threaded into the pump's fitting.

Note

In hydraulic systems, a positive head pressure at the pump's inlet is considered ideal compared to a vacuum. However, the hydraulic system designs required to achieve this (placing the reservoir high above pump inlet; using a large-diameter suction hose; adding a charge pump) are not always feasible and/or economical on mobile hydraulics applications.

As explained in **Chapter 2**, several factors can cause an increase in pump inlet vacuum: wrong oil, cold oil, plugged suction filter, plugged suction screen, wrong type of suction hose (collapsible), kinked suction hose, too small of diameter suction hose, or an internally failed suction hose.

Miscellaneous Service Equipment

Although pressure gauges and flowmeters are the two most common measuring devices used by hydraulic technicians, additional tools can be equally important to diagnosing and properly maintaining a hydraulic system. Examples of these tools include infrared temperature meters, photocell tachometers, stopwatches, pump test stands, and particle counters.

Infrared Temperature Meter

Fluid temperature can be an excellent indicator for diagnosing hydraulic systems. Infrared temperature guns can be purchased at local tool stores and are an invaluable diagnostic tool for a hydraulic technician. A technician simply points the gun's laser at the component and reads the temperature on the LCD screen. See Figure 21-21.

A temperature gun is often used to measure a hydraulic system's reservoir temperature. Agricultural technicians use temperature guns to locate a faulty cylinder on an implement. Large implements contain multiple hydraulic cylinders, making them difficult to diagnose. A technician can point the gun's laser at each cylinder until the faulty cylinder is identified by an abnormal temperature reading.

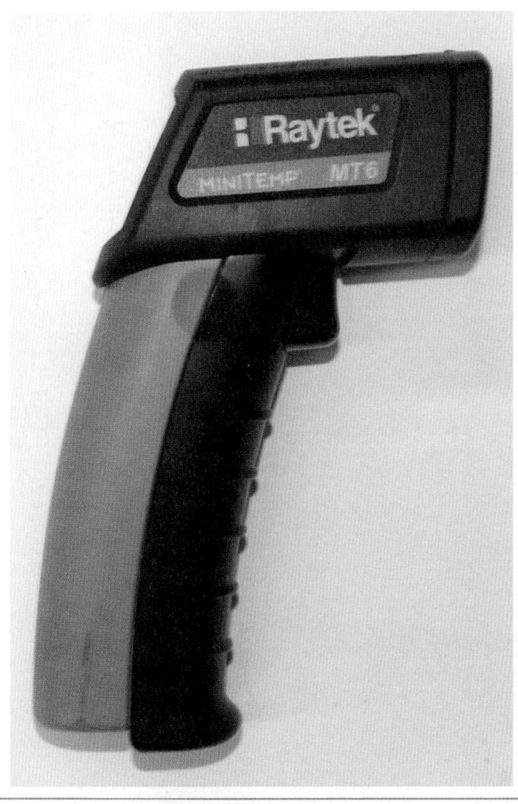

A

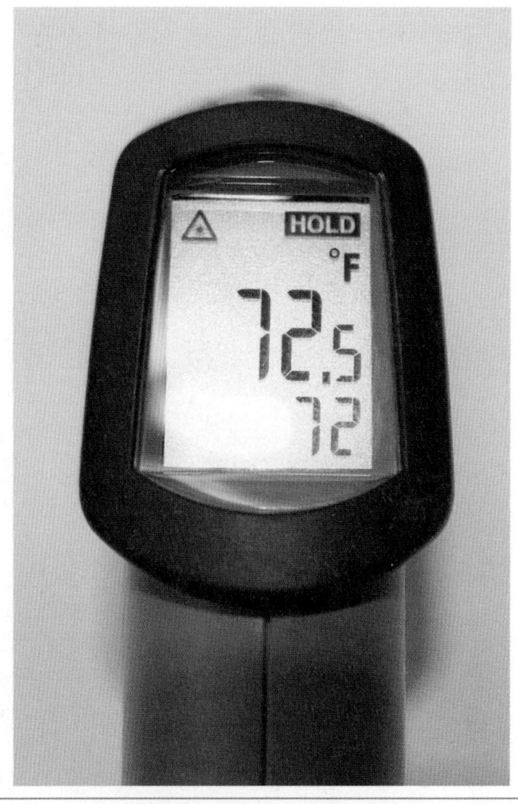

B

Figure 21-21. Infrared temperature guns are an excellent tool for measuring temperatures in hydraulic systems. The use of infrared light allows this type of thermometer to measure temperature without physical contact between itself and the matter. A—The profile view of an infrared thermometer gun. B—A close-up of the gun's LCD screen.

Photocell Tachometer

Some hydraulic symptoms result in a loss of shaft speed. A photocell tachometer, **Figure 21-22**, is a device that allows a technician to measure the actual shaft speed, which provides critical diagnostic information. Two examples of components that can experience shaft speed fluctuations are oil cooler fans and hydraulic motors.

Two steps are necessary to measure shaft speed with a photocell tachometer. A piece of reflective tape is placed on the shaft. The tachometer emits infrared light that is reflected by the tape as it rotates. The pulses of reflected light are received by the tachometer and used to count the number of times the reflective tape goes around during a set period of time. From this information, the tachometer calculates and displays the speed of the rotating shaft.

Stopwatch

A cycle time is the length of time it takes an actuator to extend or retract. Service manuals provide actuator cycle times, which are helpful for diagnosing sluggish hydraulic systems. Years ago, a technician needed to carry a handheld stopwatch. Fortunately, today's smart phones provide technicians with this helpful diagnostic tool.

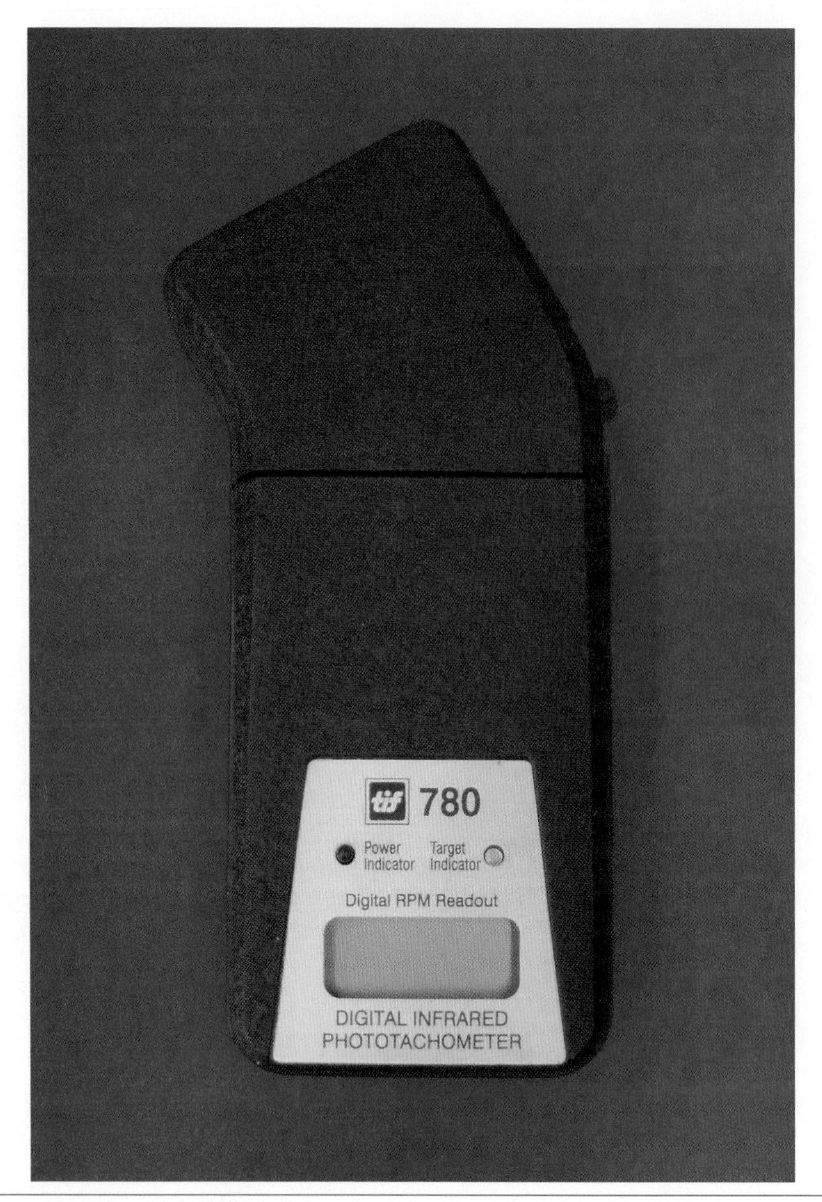

Figure 21-22. A photocell tachometer is used to measure the speed of a rotating shaft. The measurement is given in revolutions per minute (rpm).

Pump and Motor Test Stand

Many hydraulic repair shops have the ability to test a rebuilt pump or motor. The test stand has controls to choose the direction and speed of pump rotation, as well as the tools to measure pump output flow, output pressure, case drain flow, and case drain pressure. The stand can apply a load to the pump or motor to determine if the pump or motor meets the manufacturer's specified volumetric efficiency. See Figure 21-23.

Data Logger

Another hydraulic service tool is a data logger. Webtec offers several different types of data loggers. The instrument is used to monitor hydraulic pressure data.

Figure 21-23. This pump/motor test stand is used to load a pump or motor to test its volumetric efficiency.

It offers the advantage of analyzing and graphing multiple channels of data simultaneously. Channels can be switched on and off to display fluid pressure graphs as needed. When complete, the data report can\ be modified, printed, or saved in a file to be e-mailed.

Portable Particle Counter

Chapter 12 detailed the advantages of portable particle counters. Years ago, technicians had to send oil samples to an oil analysis lab several hundreds of miles away in order to determine the cleanliness of the oil. Today, technicians using portable particle counters have the luxury of analyzing particle counts immediately in the field instead of waiting for a lab's report.

Summary

- ✓ Hydraulic test equipment is used to measure flows, pressures, temperatures, shaft speeds, and cylinder cycle times.
- ✓ Diagnostic pressure taps are installed on most machines, which speeds the process for measuring pressures.
- ✓ Mechanical pressure gauges are economical and usually glycerin filled to dampen pulsations.
- ✓ A differential pressure gauge simplifies the process of measuring margin pressure.
- ✓ A Quadrigage allows a technician to tap into a hydraulic system with less worry of harming the gauge due to its ability to accurately measure a wide range of pressures.
- ✓ The OTC electronic digital pressure meter provides a safer method for measuring high system pressures.
- ✓ The OTC electronic digital pressure meter offers numerous features including measuring four different pressures, delta zero mode, minimum/maximum, and differential pressure measurement.
- ✓ Snubbers reduce gauge pulsations caused by vibration and fluid fluctuations.
- ✓ Selecting a pressure gauge with too low of a pressure limit will result in a damaged gauge. Select a gauge that will read in the middle third of the scale.
- ✓ Pressure gauge accuracies range from ± 0.5% to ± 3%. Using an inaccurate gauge can cause poor machine performance.
- ✓ Pressure gauges should be checked annually for accuracy. A pressure gauge tester uses a series of weights that allows shops to recalibrate electronic pressure gauges, which are used to cross-check traditional Bourdon tube pressure gauges.
- ✓ It can be difficult to measure pump inlet pressure. Sometimes it is easiest to install a pressure tap into the pump's inlet fitting.
- ✓ Infrared thermometers can be used to help find hydraulic inefficiencies, such as overheated hydraulic cylinders that are bypassing oil.
- ✓ A stopwatch is helpful for measuring a cylinder's cycle time.
- ✓ Hydraulic repair facilities use pump and motor stands for measuring the component's volumetric efficiency.
- ✓ Data loggers can measure and graph multiple circuit pressures simultaneously and provide the information in a customizable report.
- ✓ Portable particle counters enable a technician to determine system contamination levels on the job site, allowing the technician to take immediate action.

Technical Terms

Bourdon-tube pressure gauge
burst discs
delta zero
diagnostic pressure taps
electronic differential pressure-sensing flowmeters
electronic positive-displacement flowmeter
electronic turbine flowmeters
flowmeter
flow rater
load valve
minimum/maximum

Review Questions

Answer the following questions using the information provided in this chapter.

1. Technician A states that personnel who are flow rating a pump should use the service manual in order to obtain the flow specification. Technician B states that personnel who are flow rating a pump should be cautious and avoid blocking the flow of a fixed-displacement pump. Who is correct?
   A. Technician A.
   B. Technician B.
   C. Both A and B.
   D. Neither A nor B.
2. Which of the following is the most commonly used device for measuring pressure?
   A. Bellows pressure gauge.
   B. Bourdon-tube pressure gauge.
   C. Quadrigage.
   D. Electronic pressure meter.
3. What is the purpose of a load valve in a flowmeter?
   A. Creates a vacuum in the circuit.
   B. Ruptures the burst discs.
   C. Depressurizes the hydraulic system.
   D. Restricts the hydraulic pump's flow to increase pressure.
4. How many gauges are in the Caterpillar tetra gauge assembly?
   A. Two.
   B. Three.
   C. Four.
   D. Five.
5. What liquid is used to fill "liquid-filled" pressure gauges?
   A. Glycerin.
   B. Mineral oil.
   C. Kerosene.
   D. Hydraulic fluid.
6. Snubbers are used to do what in hydraulic systems?
   A. Lower flow.
   B. Lower pressure.
   C. Eliminate pulsations.
   D. Increase pressure.
7. What type of fitting does John Deere abbreviate as "DR"?
   A. Elbow.
   B. Diagnostic pressure tap.
   C. T-fitting.
   D. Shut-off valve.
8. What is the primary advantage of using an electronic pressure meter?
   A. Costs less than a comparable Bourdon-tube pressure gauge.
   B. Measures atmospheric pressure.
   C. Technician safety.
   D. Creates flow and pressure pulsations in a hydraulic system.
9. Which feature on the OTC electronic pressure meter is used to measure margin pressure?
   A. Min/Max.
   B. Delta zero mode.
   C. Differential pressure.
   D. Temperature.
10. Which feature on the OTC electronic pressure meter is used for monitoring the positive or negative change of a single point in a circuit?
    A. Min/Max.
    B. Delta zero mode.
    C. Differential pressure.
    D. Temperature.

11. Which feature on the OTC electronic pressure meter is used to record the highest and lowest system pressure for a given test port?
    A. Min/Max.
    B. Delta zero mode.
    C. Differential pressure.
    D. Temperature.
12. How many pressures will the OTC electronic pressure meter display at one time on its screen?
    A. One.
    B. Two.
    C. Three.
    D. Four.
13. Technician A states that Bourdon-tube pressure gauges suffer from an accuracy tolerance near ± 7% due to their mechanical design. Technician B states that Bourdon-tube pressure gauges are configured with a S-shaped tube. Who is correct?
    A. Technician A.
    B. Technician B.
    C. Both A and B.
    D. Neither A nor B.
14. Technician A states that diagnostic pressure taps can be designed with NPT, ORB, ORF, or JIC threads. Technician B states that diagnostic pressure taps are removed from the hydraulic system after each use. Who is correct?
    A. Technician A.
    B. Technician B.
    C. Both A and B.
    D. Neither A nor B.
15. Which of the following variables is *not* part of a PQ curve?
    A. Temperature.
    B. Flow rate.
    C. Pressure.
16. A technician needs to connect a pressure gauge to a hydraulic test port. The technician does not know which of the three test ports is the high-pressure port and which are the two low-pressure ports. In order to speed the process and avoid damaging equipment, which of the following pressure measuring devices is recommended?
    A. Bellows pressure gauge
    B. Bourdon-tube pressure gauge.
    C. Quadrigage.
    D. Electronic pressure meter.
17. Technician A states that pressure gauges will maintain correct calibration for the life of the gauge. Technician B states that the lower the accuracy tolerance percentage of a pressure gauge, the greater its measurement accuracy. Who is correct?
    A. Technician A.
    B. Technician B.
    C. Both A and B.
    D. Neither A nor B.
18. All of the following hydraulic pump specifications for flow rating a pump will be included in the manufacturer's service literature, *EXCEPT*:
    A. pump's flow rate.
    B. particle count contamination.
    C. fluid operating temperature.
    D. rated engine speed.
19. On an OTC electronic pressure meter, the blue electronic pressure transducer is capable of measuring pressure up to _____.
    A. 500 psi (34 bar)
    B. 5000 psi (345 bar)
    C 7500 psi (510 bar)
    D. 10,000 psi (689 bar)
20. Which of the following tools is used for measuring actuator cycle times?
    A. Photocell tachometer.
    B. Infrared thermometer.
    C. Stopwatch.
    D. OTC electronic pressure meter.

21. Which of the following tools requires the use of reflective tape?
    A. Photocell tachometer.
    B. Infrared thermometer.
    C. Stopwatch.
    D. OTC electronic pressure meter.
22. Which of the following tools is used for measuring the speed of a rotating shaft?
    A. Photocell tachometer.
    B. Infrared thermometer.
    C. Portable pressure counter.
    D. OTC electronic pressure meter.
23. Which of the following tools can be used by agricultural technicians to locate an implement's hydraulic cylinder that contains a bad piston seal?
    A. Photocell tachometer.
    B. Infrared thermometer.
    C. OTC electronic pressure meter.
    D. Portable particle counter.
24. Which of the following tools is used to check the cleanliness of a hydraulic system?
    A. Photocell tachometer.
    B. Infrared thermometer.
    C. Portable particle counter.
    D. OTC electronic pressure meter.
25. All of the following measurements can be gathered by a hydraulic pump and motor stand, *EXCEPT*:
    A. particle counts in the hydraulic fluid.
    B. pump output flow.
    C. case drain pressure.
    D. case drain flow.
26. Technician A states that flowmeters can have a burst disc assembly that bypasses oil internally or leaks oil externally. Technician B states that flowmeter burst discs act like fuses in an electrical circuit. Who is correct?
    A. Technician A.
    B. Technician B.
    C. Both A and B.
    D. Neither A nor B.
27. All of the following features are commonly found on flowmeters, *EXCEPT*:
    A. thermometer.
    B. pressure gauge.
    C. load valve.
    D. timer.
28. Technician A states that a mechanical flowmeter that allows oil flow in both directions will also measure the oil flow in both directions. Technician B states that all mechanical flowmeters allow the oil flow in only one direction. Who is correct?
    A. Technician A.
    B. Technician B.
    C. Both A and B.
    D. Neither A nor B.
29. When a flowmeter is installed, all of the following can cause damage to a hydraulic system, *EXCEPT*:
    A. leaving the load valve open.
    B. connecting the flowmeter backward.
    C. routing the flowmeter's return line to the wrong reservoir.
    D. closing the load valve completely.
30. All of the following can cause an increase in pump inlet vacuum, *EXCEPT*:
    A. kinked suction hose.
    B. cold oil.
    C. plugged suction screen.
    D. suction hose with too large of a diameter.

Hydraulic technicians must follow a routine—including examining and understanding the system and questioning the machine operator—to successfully diagnose and repair hydraulic systems.

Chapter 22

Hydraulic Troubleshooting Principles

Objectives

After studying this chapter, you will be able to:

- ✓ Explain the seven-step process for troubleshooting a hydraulic system.
- ✓ Explain how to isolate problems during the diagnostic process.
- ✓ List components that can cause a group of hydraulic actuators to malfunction.
- ✓ List examples of how heat and noise can guide the process of troubleshooting a hydraulic system.
- ✓ List the common parameters tied to flow rating a pump.
- ✓ List some hydraulic parameters to measure during a machine's regular preventative maintenance inspection.

Introduction to Hydraulic Troubleshooting

Troubleshooting a hydraulic system can be a difficult and overwhelming process. The machine's schematic might be quite large and intimidating. However, there are multiple steps and tips explained in this chapter that will help guide a technician through the process. Every technician's goal should be to find the solution to the hydraulic fault in the most efficient and safest manner.

One of the most important principles of any type of troubleshooting is to avoid the temptation to make a hurried repair without fully diagnosing the system for the sake of trying to act quickly. Working smart instead of working hard can be the most productive step in finding the solution.

John Deere specifies seven steps for troubleshooting a hydraulic system:

1. Understand the system.
2. Question the machine operator.
3. Operate the hydraulic system.
4. Examine the hydraulic system.
5. Identify the possible causes.
6. Isolate the true cause.
7. Test your conclusion.

This chapter will provide examples for all seven steps as well as discuss other important diagnostic principles.

Understand the System

Can a technician fix a hydraulic system if he or she does not understand the system? It may be possible for a technician who is unfamiliar with a system to stumble on a solution by following basic diagnostic procedures, but it is more likely the technician will do more harm than good.

What are the risks of attempting to fix a system without having the proper information and knowledge? For starters, the technician may not be able to recognize proper operation of the system. This could lead to an embarrassing situation in which the technician attempts to "repair" a system that is performing exactly as it was designed to. Even worse, the technician's attempts at repair could actually create a dangerous situation, undermine customer confidence, and tarnish his or her reputation.

For example, imagine a technician is sent on a service call for a machine with a two-speed hydrostatic drive motor that downshifts from high to low speed once the drive pressure reaches 3100 psi (214 bar). If the upset customer pushes to have the machine returned quickly, and the unknowledgeable technician chooses to order parts for a symptom that is normal, time will be wasted and good parts will be discarded. This poor decision can cause the customer to lose respect for the service department. The dealership could receive a lower ***customer satisfaction index (CSI)*** rating, which is generated through customer surveys. A low dealership CSI places the dealership's contract with the equipment manufacturer at risk.

In a worst-case scenario, a technician unfamiliar with a particular hydraulic system may attempt to re-engineer some aspect of the system in a manner that is not approved by the manufacturer. For example, incorrectly adding a hydraulic attachment could result in a complete machine loss when the return line overheats and catches fire, burning the machine to the ground. When the smoke subsides, the question will be, "Who is stuck with paying for a new machine?" The consequences would be even greater if someone is physically injured by the mistake.

Therefore, some of a technician's best spent time is learning as much as possible about the hydraulic system before commencing to fix the machine. Technicians should avoid rushing diagnosis and service, even if the customer is pushing them to act fast.

Question the Machine Operator

Machine operators can offer a tremendous amount of valuable information to a technician during the diagnostic process. Veteran troubleshooters have learned the art of asking the right questions. They are also able to pick up clues from the operator's ***nonverbal communication***, such as answering a question with a complete lack of confidence.

Technicians might be asked to diagnose systems over the phone, without being able to see, touch, or hear the machine. Communication skills are by far the most important factor in this scenario. Strong listening skills are a necessity as operators can provide critical pieces of information. For example, the operator might mention that the hydraulic system exhibits a certain type of noise at a specific location on the machine, or the machine malfunctions only after completing a certain sequence of steps, or the oil temperature light illuminates only after a particular series of events. Insight into recent machine repairs or service is extremely helpful also. An operator may state that the system worked fine until a leaky component was recently replaced.

Occasionally, troubleshooters are asked to resolve a problem for a person who is not forthcoming with the whole story. A skilled troubleshooter can sense when someone is withholding pertinent information in fear of embarrassment or blame. In such circumstances, it is helpful to calmly share with the individual that the story does not add up, and that the whole story needs to be explained. If a technician approaches the discussion with a careful and understanding ear, many times the person will realize that the technician is simply trying to do his or her job and does not intend to condemn or ridicule them.

Care should also be taken to avoid overemphasizing something that is most likely only a coincidence and cannot truly be related to the original symptom. After struggling with a problem for an extended period, stick to the proven diagnostic principles and knowledge gained through the work. Do not turn the focus to something that cannot be related, but is only coincidental.

Operate the Hydraulic System

The machine must be operated during the diagnostic process for a host of reasons. The most important reason is the necessity of ***duplicating the problem***, also known as reproducing the symptom. If an actuator intermittently malfunctions, it might be extremely difficult to resolve the issue because the problem is difficult to reproduce. A customer can become very frustrated when a problem does not occur in the presence of the technician.

If the problem cannot be duplicated, it is practically impossible to fix the machine. If the technician cannot reproduce the problem at the start of the job, it is impossible to be certain of the cause. It also makes it impossible to determine if any actions taken, such as replacing a suspect component, have truly fixed the problem.

A technician needs to operate the machine to identify the problem in the correct context. For example, if the customer states a cylinder leaks down too fast, operating the machine allows a technician to witness that although the cylinder drift appears relatively fast, it operates within the ***engineering specifications***. Service literature provides many different types of engineering specifications, including the following:

- Cycle time.
- Operating temperature.
- Pump flow rate.
- System pressure.

The problem can be compounded if the customer owns multiple identical machines. If only one of the machines has a cylinder leaking down but is deemed to be operating within engineering specifications, the problem might cause a rift between the customer and manufacturer/dealership. The dealership and manufacturer must make a decision to demonstrate ***goodwill*** to the multiple-machine owner. As a gesture of goodwill, the dealer might attempt to improve a machine's performance by replacing components or making adjustments, even though the machine operates within the acceptable tolerances or is no longer covered under a warranty.

Technicians should not be hesitant to operate machinery and check system pressures. For instance, consider a constant flow open-center system that is malfunctioning and running at high pressure, or a variable-displacement pressure-compensating system that is malfunctioning and running at low pressure. In each situation, the pressure is critical to gaining an accurate understanding of the system's symptom. Inexperienced technicians can waste valuable time before uncovering this critical data.

Some technicians allow their pride to slow the diagnostic process. A good hydraulic technician cannot possibly know how to operate every type of hydraulic machine on earth. A humble technician will ask the operator to demonstrate the hydraulic symptom by operating the machine. After using the machine for a period of time, operators tend to develop tremendous skill at its controls, and a wise technician will harness that expertise by requesting their help.

Examine the Hydraulic System

During a machine inspection, nothing is more important than the hydraulic fluid. Make no assumptions regarding even the basic parts of a system. At a minimum, a technician should physically check the condition of the reservoir and its fluid level. Even seasoned technicians make the mistake of overlooking essential machine specifications during an examination. Certain specifications could be difficult or time-consuming to check and a hurried technician might skip the step. In some of these cases, a simple check and correction to one of a system's basic elements was the solution to the hydraulic problem.

Case Study

The Need for Careful Observation

Over the course of two days, a hotline troubleshooter assisted a technician who was trying to resolve several hydraulic problems on one machine. The hotline troubleshooter approached the problems methodically, but the diagnostic process kept hitting a dead end. After going in circles with technician over the course of those two days, the hotline troubleshooter became convinced that the system lacked oil, despite the technician's assurances that it was full. The troubleshooter sent the technician back to the machine, insisting that he recheck the oil level. The technician called back shortly, apologizing that he had simply looked at the sight glass and it was yellowed over, giving him the impression that the system was full. A more careful check revealed that the system's oil level was, in fact, very low.

As mentioned in **Chapter 14**, the position of the machine's implements influences the oil level of the system. If a forklift's mast lift cylinders are fully raised when a technician is checking the reservoir, a large portion of the oil is allocated to those cylinders. The directions in the operator's manual for checking the oil level must be carefully followed to avoid arriving at a false reading.

While checking the oil level, look for air bubbles. Aerated fluid causes a host of problems, including a reduction in bulk modulus of the fluid. This leads to spongy actuator operation or loss of control, reduced machine efficiency, and additional noise. It can also increase oil temperature, oil oxidation, oil degradation, cavitation, and pump wear, while reducing viscosity and shortening oil life.

Investigate if the machine has the correct oil. Oils have different properties and viscosities that will cause hydraulic problems if used in an incompatible hydraulic system. If the oil is also used to apply a clutch mechanism, check to see if the oil is dark or smells burnt.

While inspecting the machine, look for brand-new components that might have been recently installed. Look at the paint on the components to determine if a component has been overheated.

A savvy technician will use the machine inspection period to also install test gauges. Determine the pressures when the system is in a neutral position as well as the main system relief pressure. If the machine has an easy location for installing a flowmeter (for example, auxiliary couplers), a technician should flow rate the hydraulic system for additional information.

Identify Possible Causes

During the diagnosis, it often helps technicians to write all of their thoughts on paper. List as many potential causes as possible so that the ideas can later be critically analyzed. After this stage, the next phase requires carefully breaking down these possible causes to reach the conclusion. Sources that can help generate possible causes are service manuals, schematics, technical service bulletins, operator's manuals, and service training manuals. Veteran technicians can be an invaluable information resource for any technician who is struggling with this step.

Isolate the True Cause

Experience pays dividends when attempting to determine the cause of the problem. There are a few techniques that experienced technicians use when trying to isolate the cause of the malfunction.

Functional or Instrumentation Problems

One of the simplest notions of isolating malfunctions is to first deem the problem as a functional problem or an instrumentation problem. An example of an ***instrumentation problem*** is a machine that has a fault code for a shaft speed problem; however, the shaft speed is operating correctly and the sensing circuit is at fault. If the malfunction was a ***functional problem***, the shaft speed would truly be too slow, too fast, or not rotating.

Consider a four-engine military aircraft that is flying at 40,000 feet when one of the engine's instruments starts to indicate a malfunction, such as loss of oil pressure. Before shutting off the engine, the crew might swap the electronic oil pressure gauge with a known accurate gauge and see if the problem remains with the same engine or if the problem moves to the adjacent engine. If the problem moves, the crew knows that they have an instrumentation problem (a bad gauge) rather than a functional engine problem.

Start where Energies Change Form

One trick that veteran troubleshooters often use is starting the search for causes at locations with ***changing energies***, such as where electrical energy changes to hydraulic energy. For example, if an actuator is controlled with a solenoid-operated DCV, start directly at the solenoid. If the cylinder will not actuate, it is helpful to know if the problem is on the electrical side or the hydraulic side. If the solenoid is not being energized, measuring hydraulic flow and pressures will not matter. Energies also change at the hydraulic pump. If a pump is belt-driven, check the condition of the belt before fabricating hoses to hook up a flowmeter.

Is It the Hydraulic Pump or the Relief Valve?

A technician must take special precautions when flow rating a fixed-displacement pump with an external relief valve. Flow rating a gear pump with the relief valve left in the circuit, as shown in Figure 22-1, provides additional safety in the event something goes wrong during the procedure. If there is a problem, the relief valve will still release the excess pressure from the circuit.

Chapter 21 listed two potential causes for rupturing a fixed-displacement pump while attempting to flow rate the pump:

- If the flowmeter is installed backwards, the system could become over pressurized, rupturing the pump or a hydraulic line.
- If a flowmeter's load valve is accidentally closed, the same result could occur.

The downside of flow rating a pump as illustrated in Figure 22-1 is that if the tests results are out of specification, the technician cannot definitively state whether the problem is caused by the pump or by a weak relief valve.

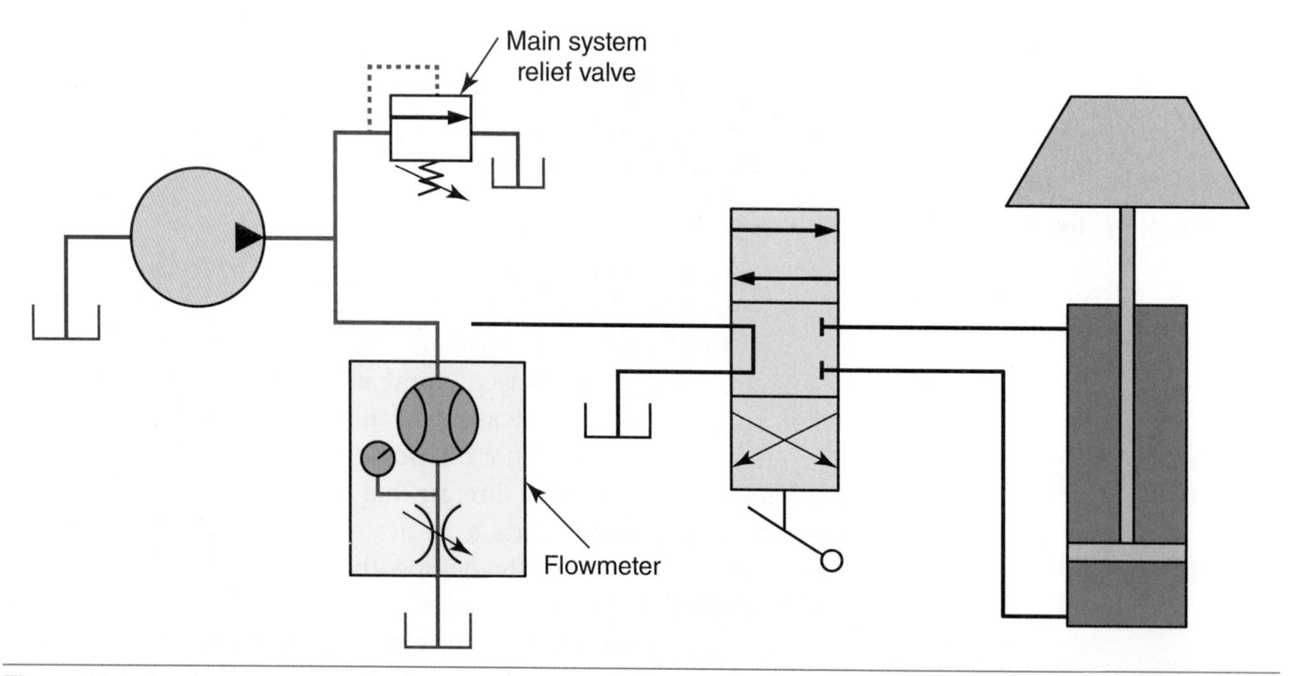

Figure 22-1. A schematic example of flow rating a fixed-displacement pump with the main system relief valve connected in the circuit.

The technician can install the flowmeter in the circuit as shown in **Figure 22-2** if test results prove unsatisfactory with the main relief valve connected to the test circuit. The advantage is that the flowmeter can now isolate a problem to determine if it is a weak pump or a weak relief valve. However, with fixed-displacement pumps, the flowmeter installation must be errorless or the pump or hose could rupture, because the system has no relief valve protection.

Caution

Before measuring a system's flow rate, a technician must check that the flowmeter is plumbed in the correct direction and its load valve is opened to avoid damage to the pump and test equipment.

Common Circuits

Unfortunately, customers and operators do not always provide technicians with a comprehensive list of what works and what does not work on a machine. After a major problem is noted, the operator often will not know that another common feature is also malfunctioning. It takes time to test the other functions to determine if they are also affected. For example, consider a customer has a combine with a system, like the one shown in **Figure 22-3**, that will not actuate the unloader swing auger. If the customer is harvesting corn (meaning that the header does not have a reel to actuate, move fore or aft, or lift), and if the combine is not equipped with the optional header tilt; anytime the jammer solenoid is malfunctioning, the customer will not know that the machine also has problems with the reel lift, reel fore, and reel aft. The customer would only know that the unloading auger does not swing in or swing out.

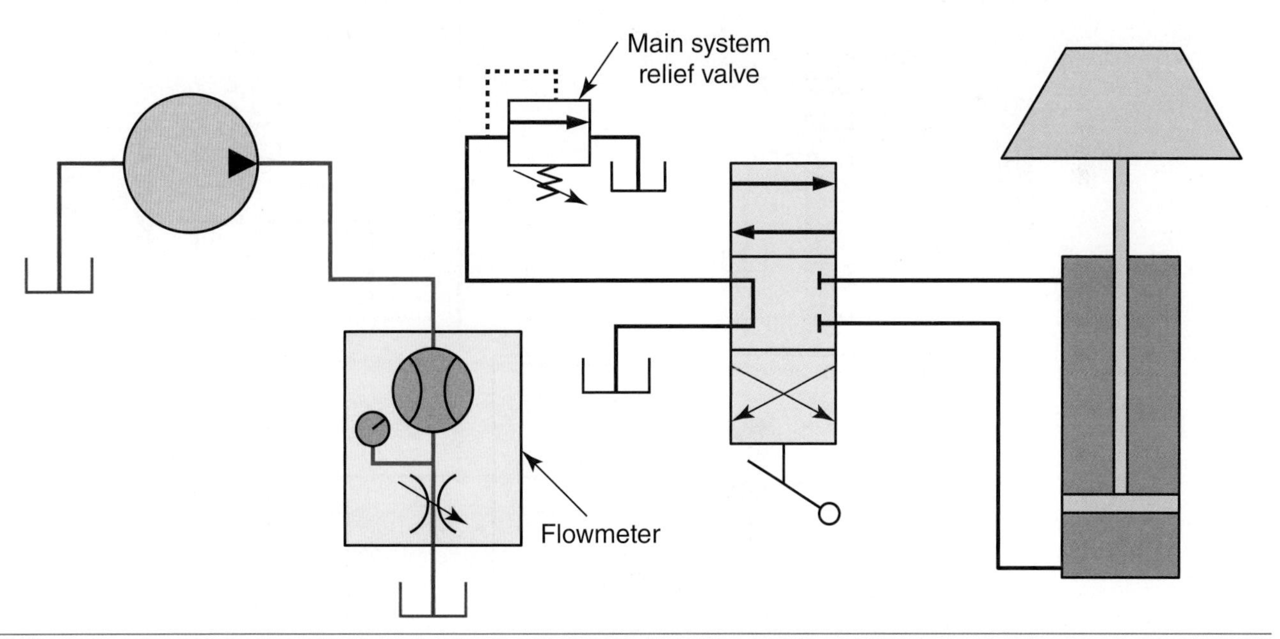

Figure 22-2. A schematic example of flow rating a fixed-displacement pump with no relief valve protection connected in the circuit. Although more dangerous in the event of a pressure buildup, this test circuit setup lets a technician determine if the hydraulic pump or the relief valve is the source of a problem.

While looking for commonalities, determine if other legs of the circuit are shared and, if so, determine if they are also malfunctioning. Some components that can cause problems for a common group of actuators are listed here:

- Return manifold.
- Relief valve.
- Hydraulic pump.
- Flow divider valve.
- Priority valve.
- Reservoir.

In Figure 22-3, two additional components, other than the jammer solenoid, could also cause a loss of reel fore, reel aft, reel lift, unloader swing, and header tilt. Those components are the secondary relief valve and the 1.5 gpm flow divider valve.

The one item rarely addressed by service literature is the hydraulic pump's inlet specifications. Due to the design of the trucks, concrete mixing truck manufacturers provide instructions for testing the pump inlet pressure on their

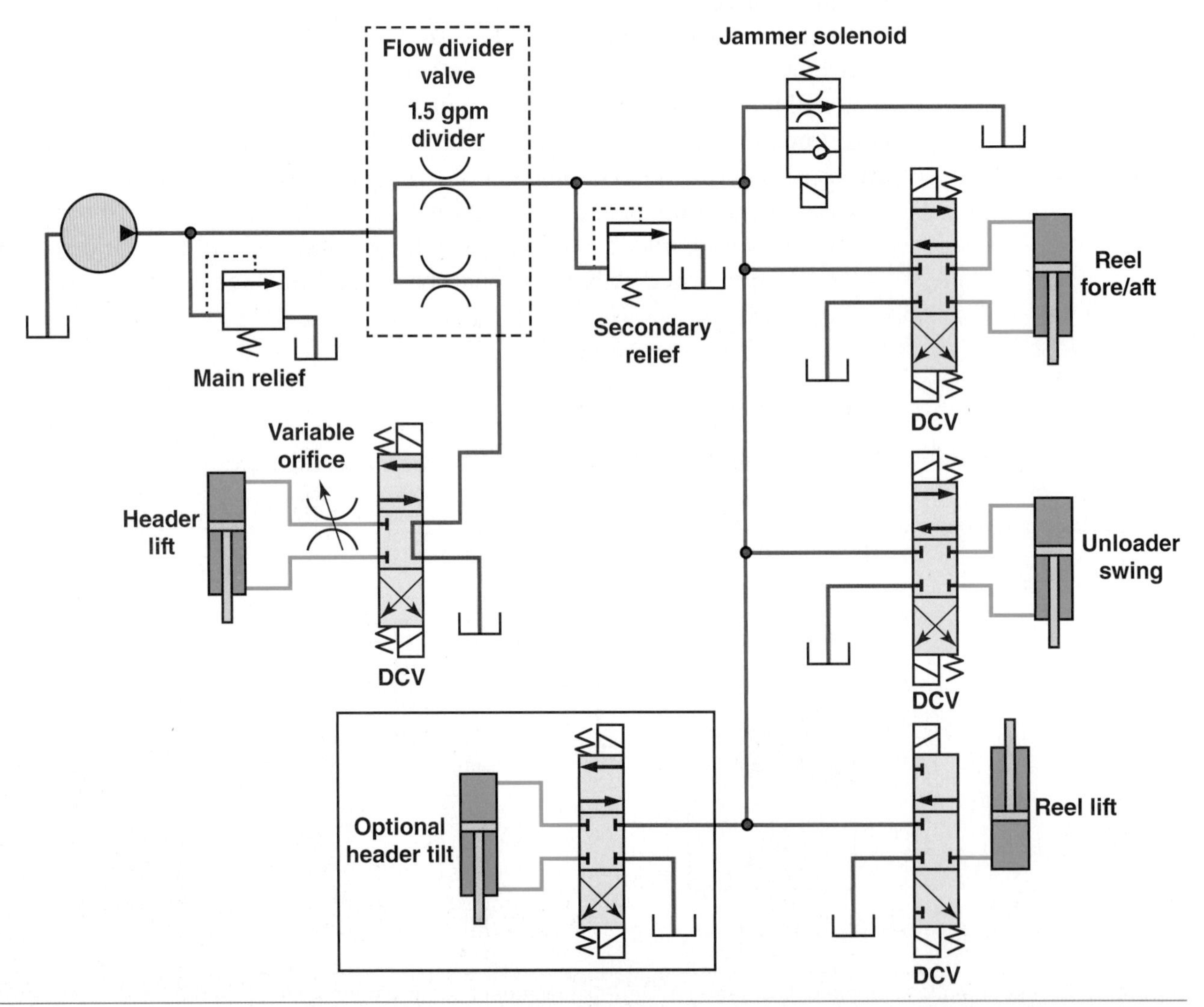

Figure 22-3. In this combine schematic, the 1.5 gpm flow divider, the secondary relief, and the jammer solenoid can all cause problems with the four DCVs: reel fore/aft, unloader swing, reel lift, and the optional header tilt.

systems. Some concrete trucks place the hydraulic pump for the mixing drum drive at the front of the truck with the oil reservoir located near the hydraulic motor, as shown in Figure 22-4. As a result, the charge pump has to draw oil through a long suction hose, and measuring the charge pump inlet pressure provides the necessary data to determine if the pump inlet has a restriction. Refer back to **Chapter 21** for the basics on measuring pump inlet vacuum and to see the causes of high pump inlet vacuum.

Test Your Conclusion

When it is time to test the conclusion, it is best to propose the solution on paper first, especially when the solution comes with an exorbitant price tag. A written justification for the planned repair helps both the technician's boss and the customer. Management can act as a second set of eyes to examine and scrutinize the proposal and the steps taken to reach the conclusion, especially before deciding to authorize an expensive repair. The customer can fully understand the need for the solution. If the repair is performed, the technician should log the solution in the machine's electronic file or paper folder for future reference.

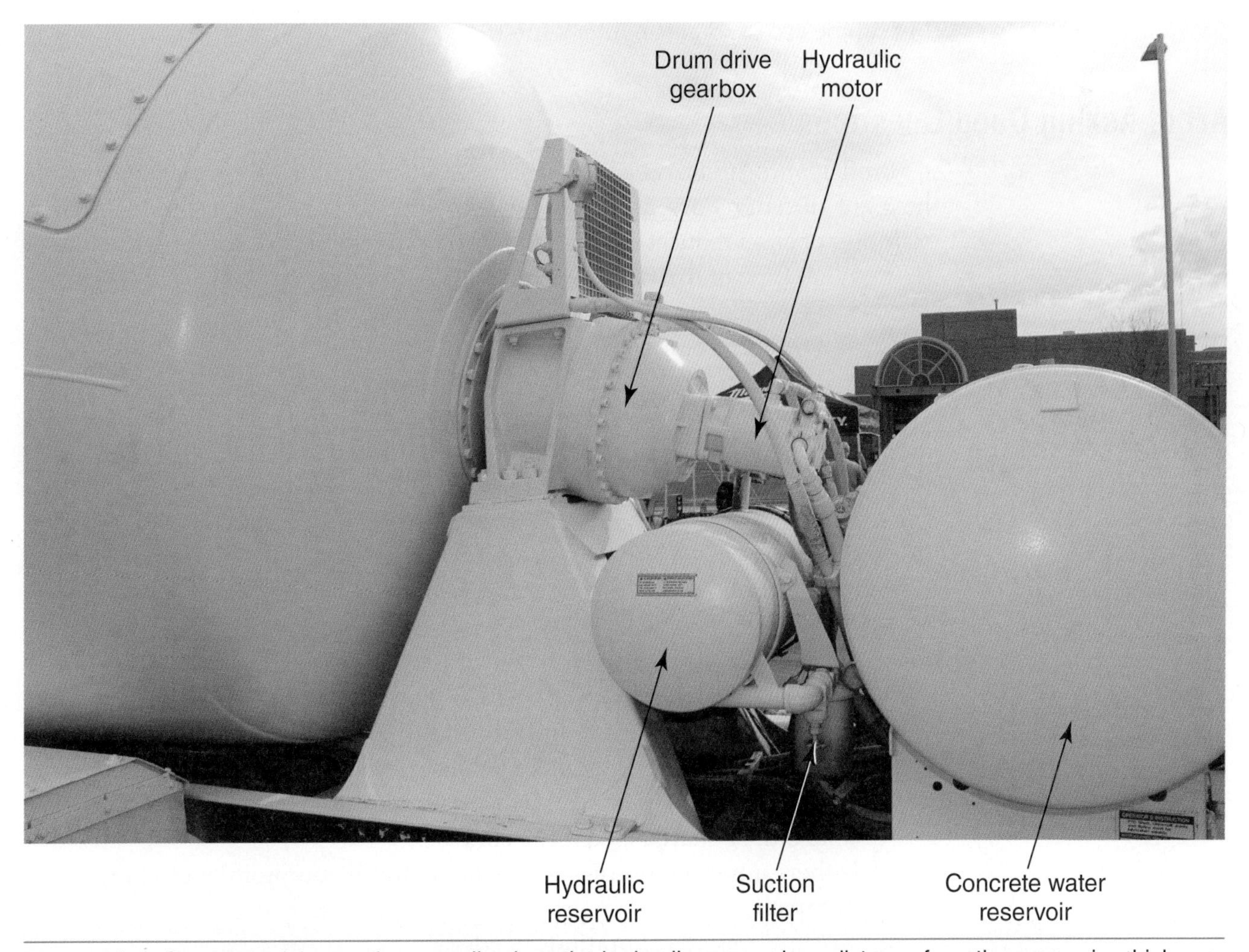

Figure 22-4. Concrete mixing trucks normally place the hydraulic pump a long distance from the reservoir, which can cause potential inlet problems because of the long suction hose.

Additional Troubleshooting Principles

Additional factors, such as the ones listed, must be considered when diagnosing hydraulic systems:

- Return circuit.
- Art of asking good questions.
- Noise.
- Heat.
- Pressure and flows.
- Cycle times.

Return Circuit

In the same fashion that the electrical ground can cause a myriad of electrical problems, so can the hydraulic return circuit also cause malfunctions. The return circuit is frequently overlooked during diagnosis. If the return pressure is too high or restricted, it can cause numerous hydraulic symptoms. Machines might not have a return circuit diagnostic pressure tap. In this case, a technician must add a T-fitting in order to measure pressure in the return circuit. If the system uses a return filter, the filter head might have a test port that can be accessed at the filter's inlet.

Art of Asking Good Questions

Proficient technicians have mastered the art of posing great questions. If the right questions are asked, it can expedite the diagnostic process. Some examples of crucial questions posed by veteran technicians are provided here:

- What type of hydraulic system is it?
  - Fixed displacement/open center?
  - Pressure compensating?
  - Load sensing pressure compensating?
  - Flow sharing?
  - Negative flow control?
  - Positive flow control?
- What types of pressures, flows, heat, and noise should be produced during normal operation?
- Can the problem be duplicated?
- Does the reservoir have oil? Or, was the reservoir level properly checked?
- Has any component been replaced or changed recently?
- How long has the symptom been occurring and how long does the symptom last?
- What are the commonalities, such as shared reservoir, shared return, shared pump, shared relief valve, shared flow divider, shared jammer solenoid, or shared electronic controls?
- Is there anything unique about the hydraulic pump inlet or charge pump circuit?
- Are any special calibrations needed?
- Are any special sequences required for the actuator to operate?

Noise

All hydraulic systems produce noise. However, if the operator is able to recognize a noticeable change in noise when the machine malfunctions, it can help pinpoint the problem or at least alert the operator that something is not normal.

Chapter 17 detailed a case study example on a PC system in a loader backhoe. One of the key factors that helped in diagnosing that system was the noise that the operator heard at the back of the machine.

Cavitation produces another noise that technicians frequently use to diagnose a hydraulic system. Service manuals sometimes instruct the technician to listen for cavitation, which occurs when the pump's inlet has too much restriction.

As mentioned in the **Chapter 16** case study, a technician was asked to fix a malfunctioning hydraulic system. The owner stated that the hydraulic system sounded like it was going to explode. This noise helped the operator recognize something was wrong and compelled him to seek help.

Heat

Heat can be a trait that aids in diagnosing hydraulic systems. Heat can also cause unforeseen problems. Hydraulic horsepower is a function of pressure and flow. Machines are efficient when the pressure and flow is performing useful work, such as extending a cylinder or rotating a hydraulic motor. However, if fluid is flowing over a relief valve, the pressure and flow is not performing useful work. The hydraulic horsepower is being wasted in the form of heat energy that is dissipating into the atmosphere.

Infrared temperature guns were mentioned in **Chapter 21** as a tool for locating hydraulic system inefficiencies. For example, an agricultural technician who was having trouble identifying which of the implement cylinders had a leaky piston seal could use a temperature gun to pinpoint the hotter cylinder, which is bypassing oil.

It is helpful to have a good benchmark of normal hydraulic operating temperatures, especially if working on similar machines. To maintain consistency, technicians can draw targets on the hydraulic components so that they are consistently measuring the same location on the components.

It is recommended that temperature measurements of marked components be recorded as part of a regular ***preventative maintenance (PM) inspection*** on the machine. The inspection is a comprehensive process that takes place on a regular interval, such as every 500 hours, and requires access to shielded and covered parts of the machine. The process involves a thorough inspection of hoses, seals, and components that are not normally checked during an operator's daily visual inspection. The benchmark temperatures are used as a reference standard for future measurements and are trended over a period of time.

Heat can also cause unforeseen problems. For example, if a hydraulic system was poorly designed and caused the oil to overheat, and if that oil was supplied to a cylinder, the cylinder's pressure would drop as the cylinder's oil cooled. The drop in cylinder pressure can cause technicians to think the cylinder's seals are leaking. The true cause of the problem in this scenario is the overheating of the oil.

Paint can affect the temperature of hydraulic systems. Painting the shiny metal surfaces of a hydraulic system can reduce its operating temperature by as much as 10°F. Flat, or non-glossy, white paints have great emissivity and therefore provide excellent heat reduction. White paints have the added benefit of making hydraulic leaks more evident.

Pressures and Flows

The first time an inexperienced technician is sent to flow rate a hydraulic system, he or she might be unsure of what to look for during the test. The service manual typically specifies a maximum pressure and maximum flow at a rated engine rpm and specified operating temperature. If the technician is able to achieve the rated flow and pressure at the rated rpm and temperature, the pump is performing normally. However, if the system does not meet specifications, the technician should examine the following factors:

- Determine if the pressures or flows are erratic or jumpy.
- Determine if the pressures or flows vary greatly with small changes of temperature.
- As mentioned in **Chapter 18**, a manufacturer's service manual might specify a PQ curve and have correction factors based on pump rpm. Most systems do not have a correction factor for pump rpm. Therefore, be sure the engine rpm is accurate before condemning a suspect pump.
- Do not overlook the pump's inlet. If pressure and flow are weak, it is possible that the hydraulic pump does not have a good supply. The supply may be comprised of a complex charge circuit, much more complex than the main piston pump circuit. In that case, the charge circuit might be the culprit for a poor performing main piston pump circuit.

Cycle Times

Most construction equipment has cycle times specified in either the service manual or a performance handbook. The times can be listed for an individual circuit, such as a backhoe's stick, or the times can be a combined portion of the boom, stick, and bucket cycles.

In general, the average operator will not notice when a machine is operating slightly below its optimal efficiency percentage. By the time a technician is called to fix the machine, the cycle times are usually noticeably below the specification. The times should be accurately checked and compared to the specification. Measuring the cycle time is essential to determining if the symptom is a legitimate problem. Cycle times should be another measurement recorded during a 500 hour preventative maintenance inspection.

Case Study

Slow to Rise

A skid steer's bucket is slow to rise. See **Figure 22-5**. What are some of the components/features that a technician needs to consider when searching for a solution?

- Reservoir.
- Cycle time.
- Hydraulic pump.
- Main system relief valve.
- Common actuators such as bucket curl or auxiliary.
- Cylinder.
- Cylinder port relief valve.
- Return circuit.
- Hydraulic pump inlet.

If the reservoir level was good, the pump was not audibly cavitating, and the cycle time was below the specification, a technician could perform the following steps.

- Check the other common actuators to determine their performance.
- Flow rate the hydraulic system.
- Check the temperature of the cylinder.
- Determine if the cylinder drifts downward when the DCV is in a neutral position.

Let's consider each of the possible steps. Checking the other common actuators would be a simple, fast step that could answer questions about the pump and the main system relief valve. If the bucket curl and the auxiliary hydraulic circuit perform well, the pump and main system relief valve would appear to be okay.

The only reason to flow rate the hydraulic system is if the other actuators were performing poorly. Common components that could affect all three DCVs are the reservoir, pump, main system relief valve, and the return circuit.

If the cylinder's cycle time was too slow and the other actuators were performing well, the cylinder's piston seal could be suspect. An infrared temperature gun could help determine if the piston is bypassing oil.

If it was determined that only the cylinder rise was slow, and the other actuators performed well, it would be helpful to determine if the lift cylinder drifted downward while the DCV is in a neutral position. If the cylinder did drift, the cylinder port relief valve could be weak causing it to drift and also causing it to be slow to rise. In this case study, the cylinder port relief valve was at fault.

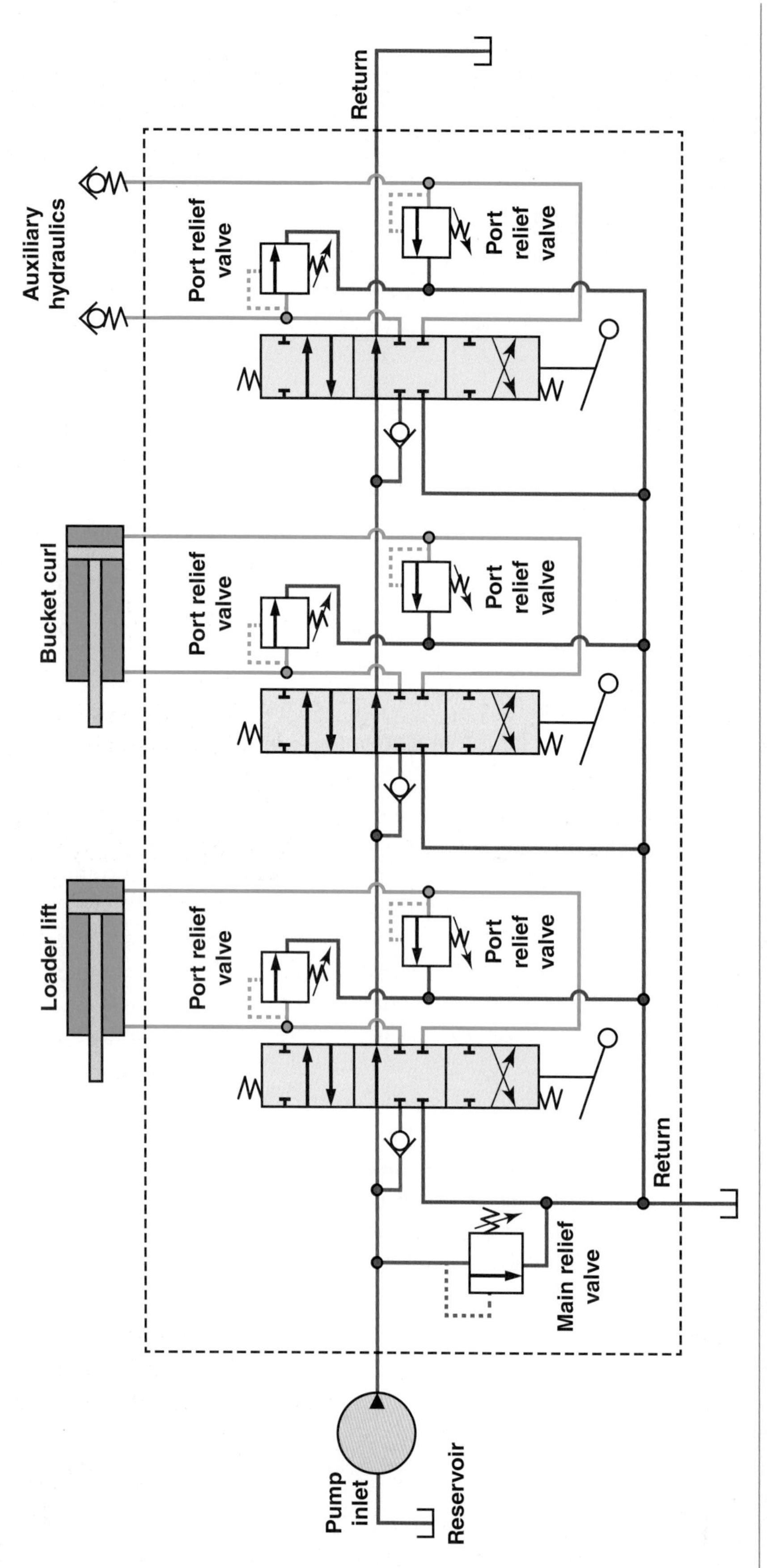

Figure 22-5. This simplified skid-steer schematic illustrates a fixed-displacement open-center hydraulic system protected with a main system relief valve. The three common hydraulic functions are loader lift and lower, bucket dump and curl, and auxiliary hydraulics.

Summary

- ✓ It is extremely important and relevant for a technician to first learn a hydraulic system before attempting to fix it.
- ✓ A hydraulic problem that cannot be duplicated by a technician is impossible to diagnose or successfully repair with any certainty.
- ✓ Questioning the machine operator can yield essential data that is helpful for diagnosing the system.
- ✓ Before attempting to fix a hydraulic system, be sure to identify the style/design of the hydraulic system and the machine's functions.
- ✓ When inspecting the machine, determine if the hydraulic pressures are normal or abnormal based on service literature specifications.
- ✓ A technician should pay extra attention to the condition, level, and type of oil used in the problematic hydraulic circuit.
- ✓ After inspecting a system, technicians should create and carefully analyze a list of possible causes of the problem.
- ✓ Before attempting to diagnose a system, first determine if the problem is a functional problem or an instrumentation problem.
- ✓ When diagnosing a system, start where energies change form.
- ✓ When installing a flowmeter in a fixed-displacement pump circuit with no relief valve protection, be sure that the flowmeter is not backwards and that its load control valve is open.
- ✓ Determine if there are other common hydraulic functions that are also malfunctioning in conjunction with the main problem.
- ✓ Complex repairs, especially expensive ones, should be proposed on paper for review by management and the customer before they are carried out.
- ✓ Do not overlook a restricted return line as a possible reason for the hydraulic malfunction.
- ✓ Veteran troubleshooters have learned the art of asking good questions as a means for skillfully solving hydraulic problems.
- ✓ Understanding abnormal hydraulic noises can help a technician find the malfunction efficiently.
- ✓ Temperature can be an attribute used for diagnosing malfunctioning systems.
- ✓ Determine if pressures and flows are erratic.
- ✓ Measuring cylinder cycle times is essential to determining if the symptom is an actual problem.
- ✓ Record temperatures, cycle times, and other pertinent information during the machine's regularly scheduled preventative maintenance inspection.

Technical Terms

changing energies
customer satisfaction index (CSI)
duplicating the problem
engineering specifications
functional problem
goodwill
instrumentation problem
nonverbal communication
preventative maintenance (PM) inspection

Review Questions

Answer the following questions using the information provided in this chapter.

1. Technician A states that attempting to fix a hydraulic system with no knowledge of the system can result in a low CSI for the dealership. Technician B states that this situation can lead to a potential safety hazard for the customer. Who is correct?
   A. Technician A.
   B. Technician B.
   C. Both A and B.
   D. Neither A nor B.
2. Technician A states that before attempting to diagnose a hydraulic system, a technician must know or learn about that specific system. Technician B states that every technician's goal should be to fix the hydraulic problem with minimal help or information from personnel. Who is correct?
   A. Technician A.
   B. Technician B.
   C. Both A and B.
   D. Neither A nor B.
3. When diagnosing a hydraulic system, Technician A states that the fluid level of the reservoir should be physically verified by the technician. Technician B states that asking the operator questions should be avoided. Who is correct?
   A. Technician A.
   B. Technician B.
   C. Both A and B.
   D. Neither A nor B.
4. All of the following are common troubleshooting mistakes, *EXCEPT*:
   A. hurrying to solve a problem on an unknown system to keep a customer happy.
   B. incorrectly checking the reservoir.
   C. duplicating the problem.
   D. not measuring pressures.
5. All of the following are correct statements regarding checking the fluid level in a hydraulic reservoir, *EXCEPT*:
   A. the service literature will specify the position of the machine implements and attachments for a correct fluid level reading.
   B. the sight glass should be used to expedite the process.
   C. check the fluid for air bubbles.
   D. confirm the correct type of fluid is being used.
6. All of the following techniques should be used for diagnosing hydraulic systems, *EXCEPT*:
   A. listen for abnormal noises.
   B. measure component temperature.
   C. use hands to locate an external leak.
   D. measure cycle time.
7. All of the following can cause a common group of actuators to malfunction, *EXCEPT*:
   A. lazy cylinder port relief valve.
   B. weak hydraulic pump.
   C. restricted return line.
   D. plugged priority valve.
8. What type of hydraulic system is most likely to have a pump inlet problem due to the length of the suction hose?
   A. Combine harvester.
   B. Concrete mixing truck.
   C. Excavator.
   D. Dozer.
9. What is the primary advantage of flow rating a fixed-displacement hydraulic pump with the relief valve connected to the test circuit?
   A. Safety.
   B. The ability to distinguish between a pump and a relief valve malfunction.
   C. Fast.
   D. Economical.

10. What is the primary advantage of flow rating a fixed-displacement pump without the relief valve connected to the test circuit?
    A. Safety.
    B. The ability to distinguish between a pump and a relief valve malfunction.
    C. Fast.
    D. Economical.
11. When asking a machine operator about a problem, if the story seems to not make sense, and it appears as if the operator might be hiding some relevant facts, what should a technician do?
    A. Stop asking questions.
    B. Ask the operator for the rest of the story.
    C. Assume the operator's mistake caused the problem.
    D. Check the fuses.
12. When troubleshooting a hydraulic circuit, a technician should _____.
    A. isolate functional problems from instrumentation problems
    B. hurry up and skip novice diagnostic steps
    C. avoid duplicating the problem for complex issues
    D. check system pressures but decline the chance to operate the machine
13. A technician is flow rating a hydraulic pump to determine if the pump is bad. All of the following must be done as part of the procedure, *EXCEPT*:
    A. measure both flow and pressure.
    B. operate the hydraulic pump at the specified speed.
    C. ensure fluid is at operating temperature.
    D. slow the hydraulic pump to a minimum speed.
14. Which of the following is the best guideline for diagnosing a hydraulic system?
    A. Begin with electrical components.
    B. Begin with the hydraulic piping.
    C. Begin with the mechanical components.
    D. Begin where energy changes forms.

For Questions 15 and 16, refer to the following schematic when answering the questions.

15. Which of the following choices would be a good place to begin troubleshooting if the cylinder will not operate, and the electrical motor is operating?
    A. Electric switch.
    B. Relief valve.
    C. Pump inlet vacuum.
    D. DCV solenoid.
16. In the schematic, the electric switch is used for which of the following purposes?
    A. Protects against electrical surges.
    B. Actuates either side of the DCV spool to move the cylinder.
    C. Controls the electric motor's speed.
    D. All of the above.

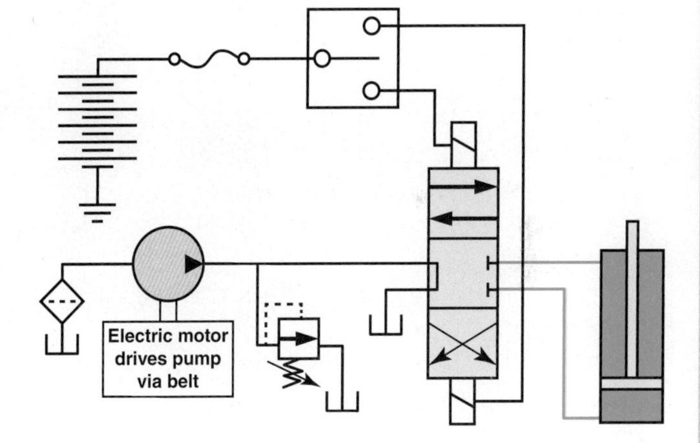

Skid steers use a dual-path hydrostatic transmission that allows the operator to propel and steer the machine. Hydrostatic drives are used on many mobile hydraulic machines in numerous industries.

Chapter 23

Hydrostatic Drives

Objectives

After studying this chapter, you will be able to:

- ✓ Explain the difference between a hydrodynamic drive and a hydrostatic drive.
- ✓ Describe open- and closed-loop HSTs.
- ✓ Explain the difference between single-path and dual-path HSTs.
- ✓ List examples of off-road HST applications.
- ✓ List HST advantages and disadvantages.
- ✓ Explain the purposes of an HST charge pump.
- ✓ Describe the different controls used to operate an HST pump.
- ✓ Explain the flow of oil in an HST circuit.
- ✓ List the sequence of events that must take place for charge pressure to drop 30 psi (2 bar).
- ✓ Explain the operation of variable-speed HST motors.
- ✓ List two reasons why an HST propulsion lever can operate backward.
- ✓ Explain how an HST increases speed with an infinitely variable pump and an infinitely variable motor.
- ✓ Describe the operation of inching valves, manual bypass valves, pressure release solenoids, IPOR valves, and anti-stall control valves.

Introduction to Fluid Drives

Off-highway machines use two different styles of fluid drives: hydrostatic and hydrodynamic. The two drives are sometimes confused with one another. Both require the use of fluid as a means of transmitting power. But the two fluid drives have significant operating differences.

Hydrodynamic Drive

The most common type of hydrodynamic drive in the off-highway industry is the ***torque converter***. The converter serves two purposes: it acts like an automatic clutch and it multiplies engine torque. The converter consists of three finned members: (1) impeller, (2) turbine, and (3) stator. See Figure 23-1. The engine drives the input member known as the impeller. The impeller acts like a pump, sending fluid energy to drive the turbine, which is the output member of the converter. The turbine most commonly drives a mechanical transmission's input shaft. The stator is a recycling mechanism that receives oil exhausting from the turbine and realigns the returning oil flow so that it helps drive the impeller. This causes the converter to multiple torque. Torque converters can also contain lockup clutches and impeller clutches, such as the converter used in Caterpillar's 994D wheel loader.

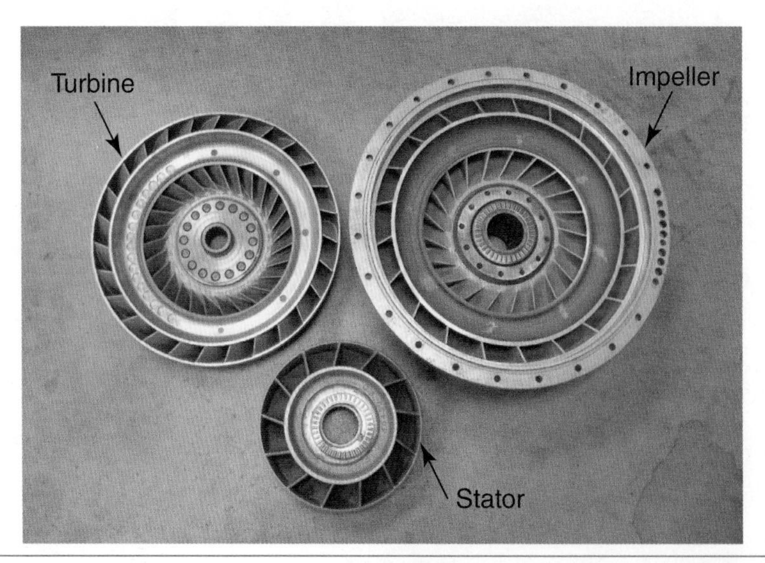

Figure 23-1. A torque converter is a hydrodynamic drive consisting of an impeller, a turbine, and a stator.

Hydrodynamic drives operate at relatively low fluid pressures. The fluid's mass and velocity is responsible for transmitting power. This fluid drive does not have tight sealing surfaces between the impeller and turbine, allowing the drive to slip if the engine is running and the service brakes are applied. Caterpillar has used ***torque dividers***, which consist of a torque converter and an internal planetary gear set. The planetary gear set provides additional torque multiplication.

Prior to torque converters, a hydrodynamic drive was a simple ***fluid coupling***, which consisted of just two straight-finned members: an impeller and a turbine. The fluid coupling did not contain a stator. For that reason, it could not multiply torque.

Basic Hydrostatic Drive

A ***hydrostatic drive***, also known as a hydrostatic transmission (HST), consists of a hydraulic pump and a hydraulic motor. See Figure 23-2. The fluid drive allows the operator to change the machine's speed, torque, and direction of travel. The HST is used in place of other styles of propulsion systems, for example PowerShift transmissions, automatic transmissions, or manual clutch synchronized transmissions.

Most HSTs use a variable-displacement reversible engine-driven hydraulic pump. The hydraulic motor can be fixed displacement or variable displacement. Examples of agricultural HST applications are combines, cotton pickers, tractors, swathers, sprayers, and lawn tractors. Examples of HST applications for construction equipment are skid steers, dozers, track loaders, wheel loaders, trenchers, excavators, and concrete mixing trucks. Note that differential steer tractors, such as Caterpillar dozers, Challenger rubber track tractors, and John Deere agricultural twin track tractors, use a hydraulic pump and motor that is very similar to a traditional HST. However, in these machines, the pump and motor are used only for hydraulic steering input, and not for propulsion.

Open-Loop and Closed-Loop HSTs

HSTs are classified as open loop or closed loop. An ***open-loop HST*** routes the motor's return oil directly back to the reservoir. In these applications, the pump must draw all of the inlet oil from the reservoir. See Figure 23-3. Excavators commonly use this style of hydraulic propulsion. See Figure 23-4.

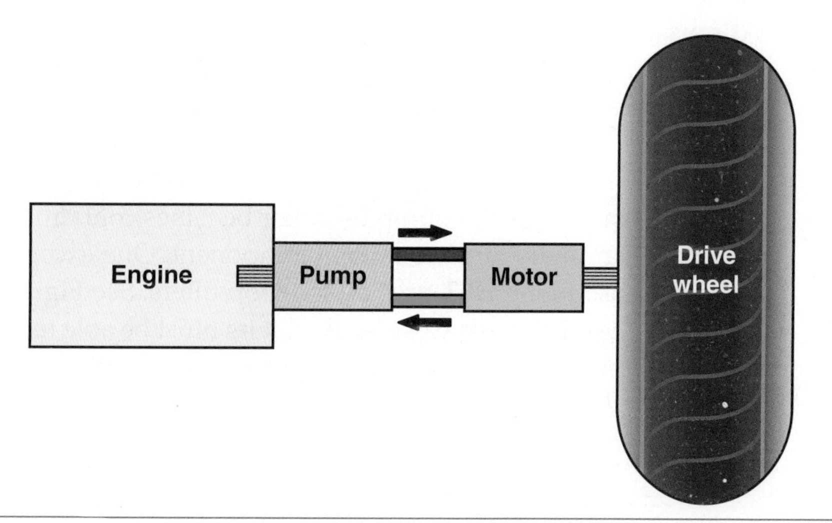

Figure 23-2. A hydrostatic transmission is simply a hydraulic pump and a hydraulic motor that provides forward and reverse propulsion and variable speed.

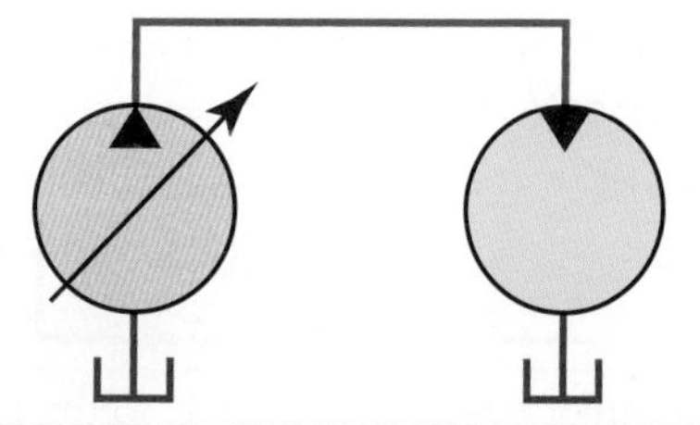

Figure 23-3. An open-loop HST draws new pump oil directly from the reservoir and does not have the aid of using the hydraulic motor's return oil.

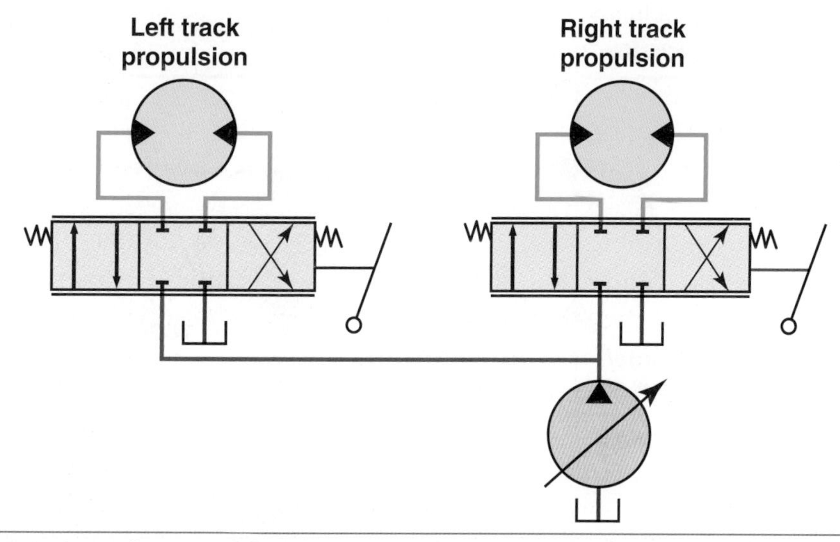

Figure 23-4. Excavators normally use unidirectional open-loop pumps to supply oil to DCVs for the purpose of propelling the machine.

Closed-loop HSTs are the most common style of propulsion systems. The hydraulic motor closes the loop by returning its oil back to the pump's inlet instead of the reservoir. The pump inlet no longer has to draw all of the oil from the reservoir. See **Figure 23-5**.

HSTs also use a charge pump to provide make-up oil to compensate for losses due to pump and motor inefficiencies. Charge pump circuits will be discussed later in this chapter.

When a closed-loop pump or motor fails or becomes contaminated, both the pump and the motor must be rebuilt. Many customers or technicians want to condemn the pump or the motor and simply replace the one part. However, in a closed-loop drive, the pump feeds the motor and the motor feeds the pump. Therefore, if the pump or motor becomes contaminated, it will directly inject contaminants into the other component. One exception to this scenario is if the closed-loop HST uses closed-loop filters. See **Figure 23-6**. These filters are rarely used, for two reasons: the filters must be able to withstand very high drive pressures, for example up to 7000 psi (483 bar), and the filters must allow oil flow in both forward and reverse directions.

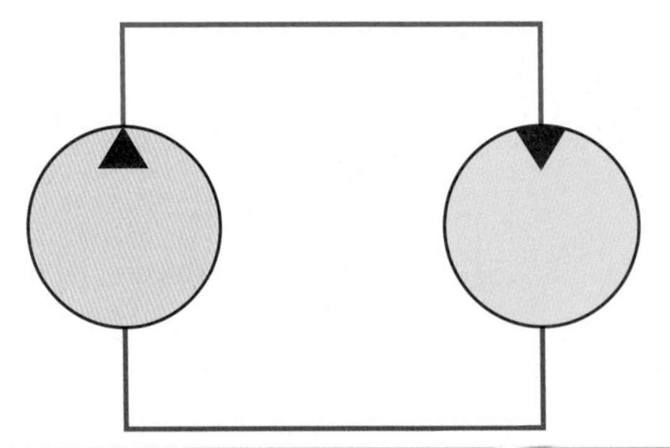

Figure 23-5. A closed-loop HST sends the motor's return oil directly to the pump's inlet.

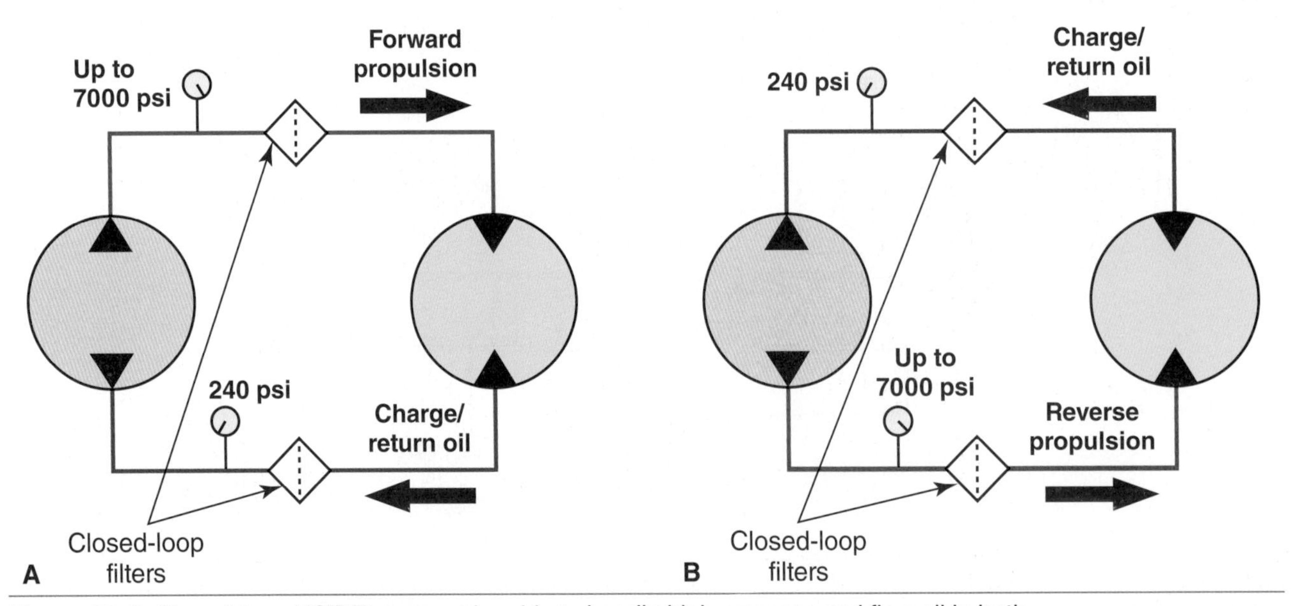

Figure 23-6. Closed-loop HST filters must be able to handle high pressure and flow oil in both forward (A) and reverse (B) directions.

Applications of Hydrostatic Drives

A variety of hydrostatic drives are available to meet the various demands of different applications. Several common designs used in various applications will be discussed in the following sections.

Single Path

Off-highway applications vary in their design. The simplest design is a ***single-path HST***, which consists of one variable-displacement reversible hydraulic pump and either a fixed-displacement or variable-displacement hydraulic motor. See **Figure 23-7**. The single-path design is used in compact utility tractors, combine harvesters, older cotton pickers, concrete mixing trucks, and lawn tractors, **Figure 23-8**.

Figure 23-7. This single-path HST consists of a variable-displacement pump that drives a fixed-displacement motor. This trainer contains the traditional components found in an older combine. The motor provides input into a three-speed mechanical transmission that contains a differential. The differential side gears deliver power out of the transmission's left and right output shafts, which can be held by the service brake and parking brake.

Figure 23-8. A common HST-propelled lawn tractor uses a belt-driven HST pump that drives a HST motor. The pump and motor are often part of a single transaxle assembly that also includes a differential. The HST motor delivers power to the differential, which drives the rear left and right wheels. The HST pump normally has a fan mounted to the input shaft to cool the transmission.

Dual Path

Dual-path HST drives use two separate hydrostatic transmissions for the purpose of propelling the machine and steering the machine. One pump and one motor will be used to drive the left side of the machine, and one pump and one motor will be used to drive the right side of the machine. Examples of dual-path applications are skid steers, track loaders, dozers, swathers, and zero turning radius (ZTR) lawn mowers. See **Figures 23-9** and **23-10**.

Dual-path HSTs will propel straight forward or reverse when both motors are being driven in the same direction at the same speed. When the motors are driven at different speeds or different directions, then the motor will steer to the left or to the right.

Swathers, also known as self-propelled windrowers, use dual-path HST drives. The dual-path HST propels the front wheels, while the back wheels are mounted on casters and swivel as needed. See **Figure 23-11**.

Swathers use two piston pumps that are often directly driven by the engine in tandem. One pump is responsible for directly controlling the left drive wheel while the other pump directly controls the right drive wheel. See **Figure 23-12**. This dual path HST will control both the direction of steering (left or right) and machine travel (forward or reverse).

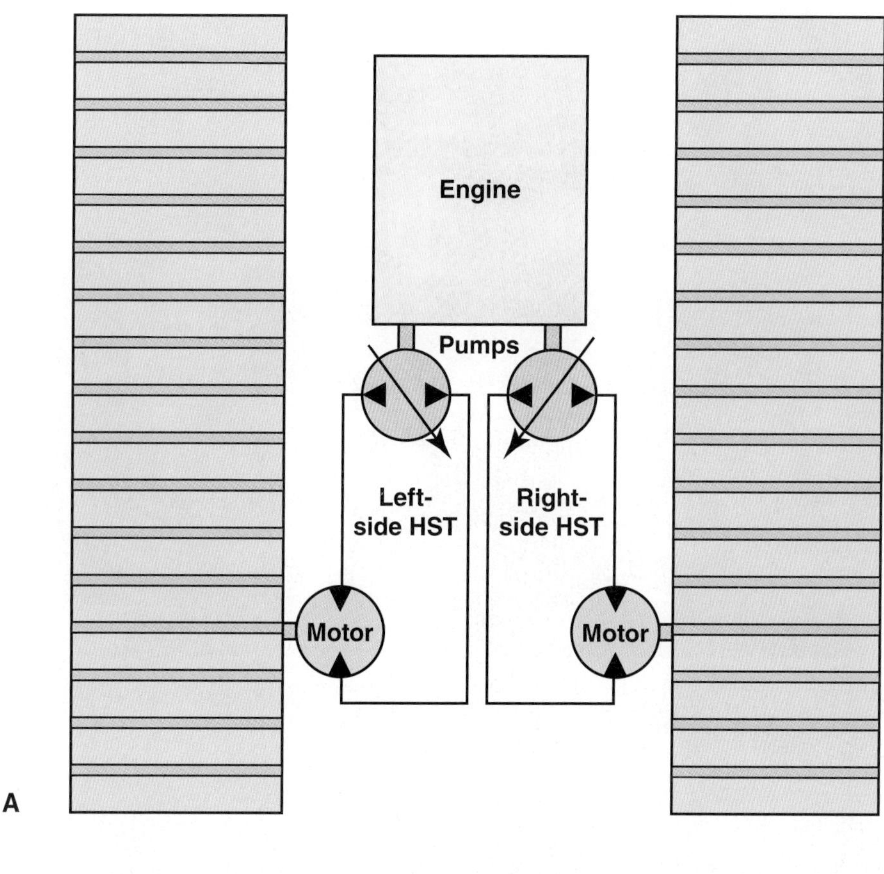

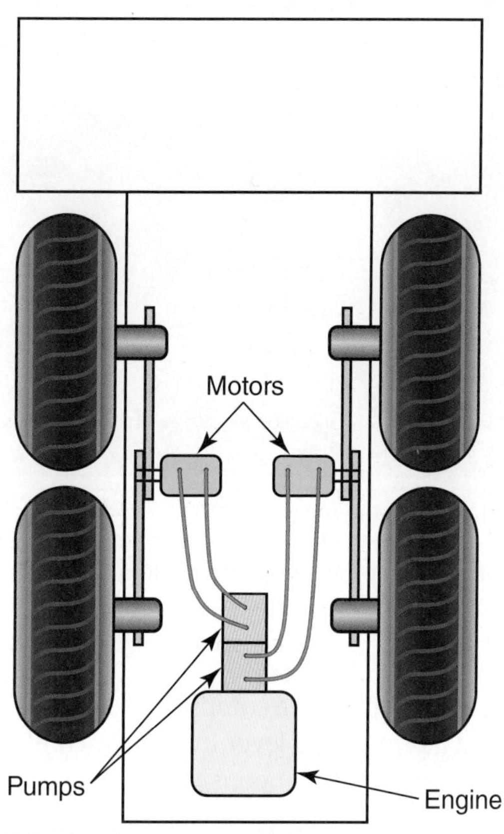

Figure 23-9. Dual-path systems. A—HST-propelled dozers and track loaders use a pump and motor for each track. B—On a skid steer, each side is driven by a separate pump and motor.

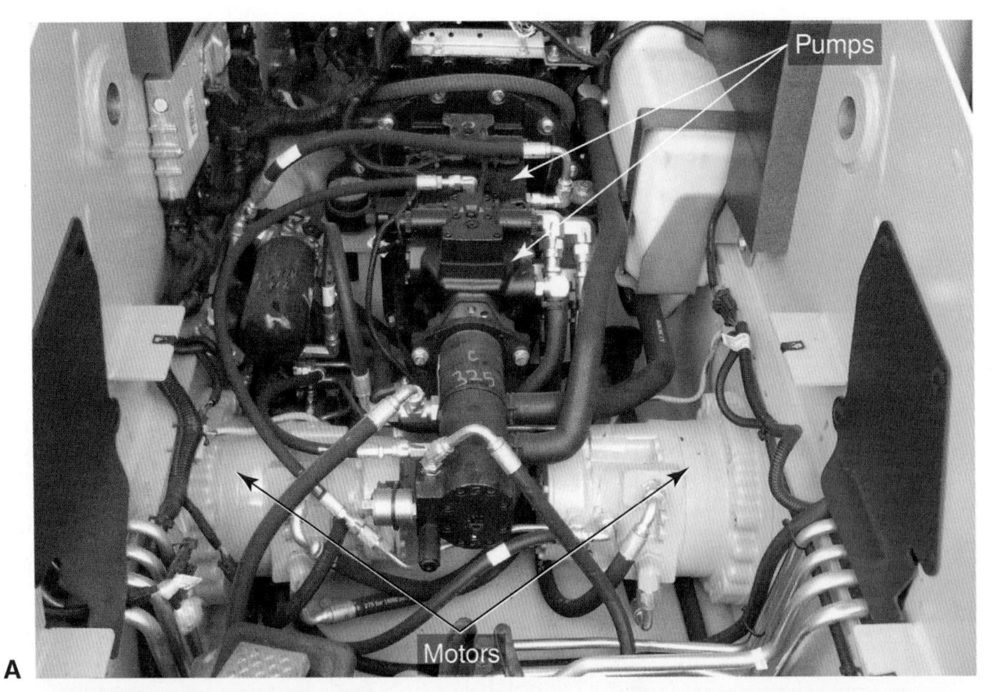

Figure 23-10. Dual-path system pump and motor arrangements. A—Skid steer pumps normally are driven in tandem. The two HST piston pumps are also coupled to the two implement pumps. Notice this dual path HST uses two cam lobe motors. B—A zero turning radius (ZTR) mower is equipped with a dual-path pump and motor used to drive each side of the mower. The left propulsion lever controls the left HST, and the right propulsion lever controls the right HST.

Older swathers use a complex mechanical linkage that consists of a steering input and a propulsion lever input. The combination of the two inputs operates the two pumps simultaneously, but independently from each other. When the parking brake is released and the steering wheel is steered, the steering shaft linkage strokes one pump in one direction and the other pump in the opposite direction. This causes the swather to turn sharply to the left or to the right.

Figure 23-11. A swather is a self-propelled windrower that is used to cut hay and forage crops. The swather will cut the crop, gather it to the center of the header, and deposit it in a single narrow row that can later be baled by a hay baler or harvested by a combine.

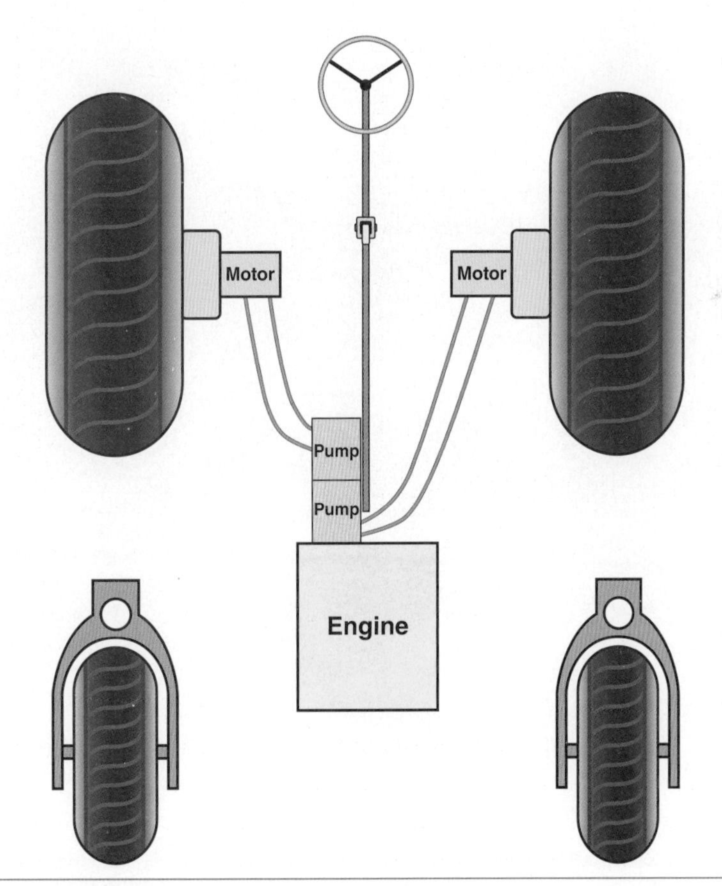

Figure 23-12. A swather commonly has two pumps driven in tandem. One pump controls the left drive wheel, and the other pump controls the right drive wheel.

The steering shaft must contain left-hand threads to operate one pump and right-hand threads to operate the other pump in the opposite direction. As the steering wheel is turned, the steering shaft coordinates the independent control of the two pumps, moving one pump's swash plate forward and the other pump's swash plate backward. When the operator is actuating only the propulsion lever and the steering wheel remains in a neutral position, the two pump swash plates stroke in the same direction at the same swash plate angle.

The swather in **Figure 23-13** uses mechanical linkage for controlling the HST pumps. The steering wheel turns the steering shaft. When the machine

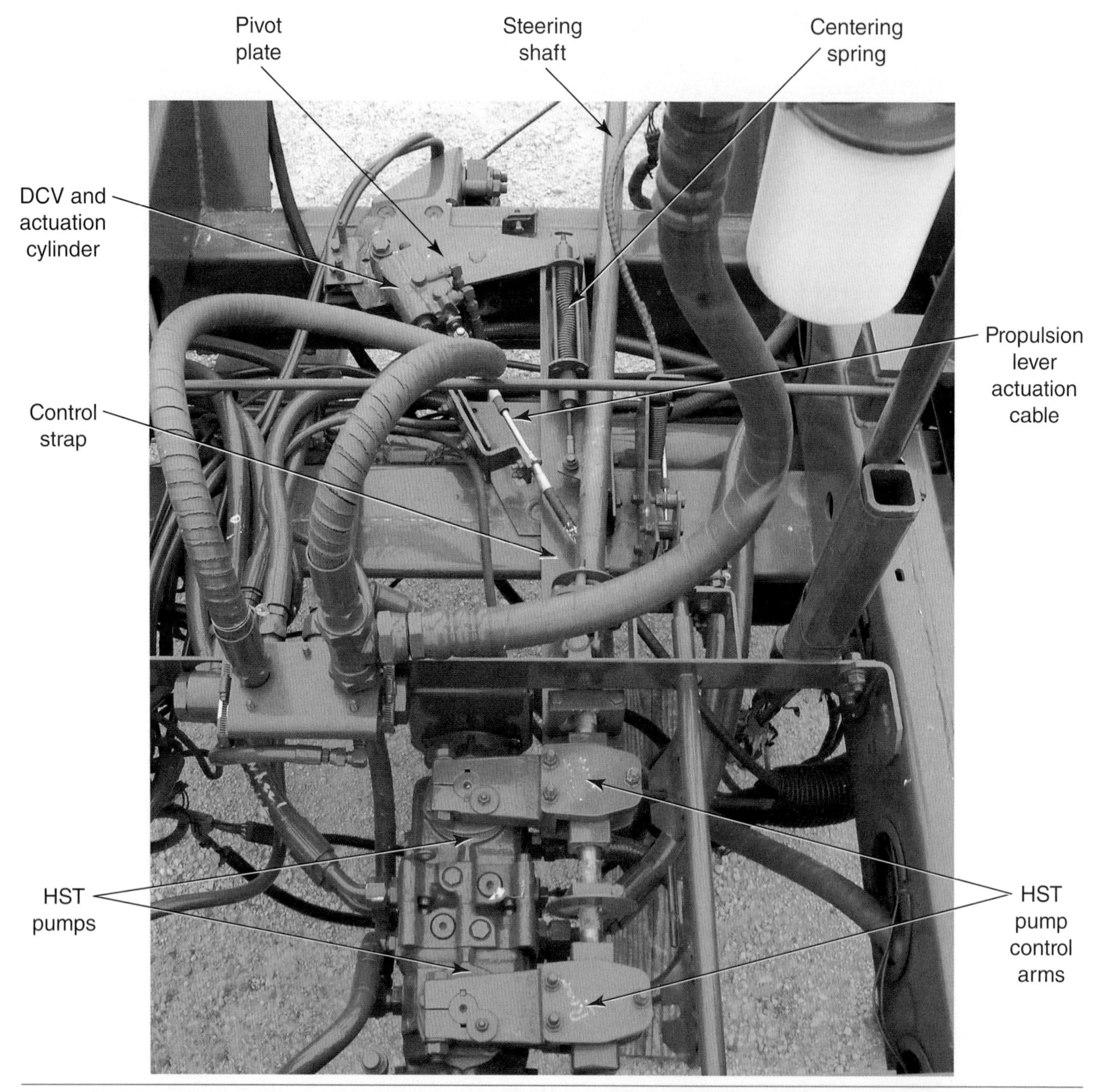

Figure 23-13. Swather HST controls consist of a steering shaft, a propulsion lever, DCV and actuation cylinder, pivot plate, centering spring, and a control strap.

is sitting still, if the steering shaft is rotated, the HST pump control arms both pivot toward each other (inward) or both pivot away from each other (outward), depending on the direction of steering wheel rotation.

When the propulsion lever is actuated, it operates the DCV valve, which causes the actuation cylinder to operate the pivot plate. The pivot plate strokes the control strap, which causes the HST pump control arms to operate in the same direction, either both forward or both rearward, depending on the direction the propulsion lever is moved. When the propulsion lever and the steering wheel are operated at the same time, the control arms operate independently of one another.

One Pump and Two Motor Applications

The previously explained single-path and dual-path HST configurations dedicate one pump to drive one motor. A few other applications can use one or more hydraulic pump(s) to drive two or more hydraulic motors. Common applications include excavators, agricultural sprayers, and Caterpillar wheel loaders.

Excavators

The HSTs used in excavators are similar to dual-path systems. Most excavators use two implement pumps rather than a dedicated hydrostatic pump for propulsion. The implement pumps control other hydraulic functions, such as boom, stick, bucket, and swing in addition to track propulsion. Excavators can be designed so that one pump drives one or both track motors. The hydraulic motor drives a planetary final drive, which is responsible for driving the track's drive sprocket. The implement pump is a unidirectional pump, which requires the use of a DCV for controlling the speed and direction of oil flowing to the motor. Refer back to **Figure 23-4**.

Agricultural Sprayer

Four-wheel sprayers commonly have one hydrostatic drive motor located at each wheel. The John Deere 4630 sprayer uses two hydrostatic pumps to supply oil to four hydrostatic motors. The pumps drive two motors located diagonally from each other. The front pump is responsible for supplying oil to the left-front drive motor and the rear-right drive motor. The rear pump is responsible for supplying oil to the right-front drive motor and the rear-left drive motor. If one of the front wheels loses traction, the other pump will still drive the other front wheel.

Agricultural sprayers use narrow tires that travel between rows of crop. The machine is elevated to drive over the standing crops. See **Figure 23-14**.

Fixed- and Variable-Displacement Hydrostatic Motors

Most off-highway HSTs use a variable-displacement pump. However, the motor can be fixed or variable displacement. The simplest application is a fixed-displacement motor, sometimes called a ***single-speed motor***. This application has a fixed swash plate angle and can be found in compact utility tractors, older combines and cotton pickers. See **Figure 23-15**.

Figure 23-14. This image shows the hydraulic hoses that are connected to the sprayer's drive wheels. The sprayer contains a drive motor at each of the four wheels.

Figure 23-15. A single-speed motor uses a fixed swash plate.

Other off-highway machines use some type of variable-displacement motor to provide the operator a wider range of machine speeds and torque. A larger degree of swash plate angle will equal a slower speed and increased torque because it takes more oil to complete one motor revolution. A smaller swash plate angle will equal a faster travel speed and reduced output torque. Some combines have the option of a ***two-speed motor***, which provides the customer two specific swash plate angles, such as a high-speed (15°) swash plate angle and a low-speed (18°) swash plate angle.

Wheel loaders are available with an infinitely variable hydrostatic drive motor. Caterpillar 924K, 930K, and 938K wheel loaders use a unique variable-displacement motor application. The loader uses one variable-displacement pump and two variable-displacement motors. Both motors provide power into a single gearbox that splits power flow to the front and rear axles. When the wheel loader requires a slower travel speed or higher torque, both hydraulic motors provide an input to the gearbox. When the loader requires a higher travel speed or lower torque, only the smaller displacement motor provides an input to the gearbox. The system operation will be explained later in this chapter.

Hydrostatic Transmission Advantages

HSTs provide numerous advantages. The output shaft speed can be maintained even as the engine speed varies. Output shaft speed and direction can be controlled remotely and accurately and can be infinitely variable. The machine speed can be varied without having to change other functional speeds. The machine direction can be quickly reversed without having to mechanically shift gears and without a shock load to the machine. The drive offers overload protection in the form of relief valves. The transmissions use relatively small components for the amount of power they are capable of transferring.

The transmissions do not coast or freewheel. Instead, they provide hydrostatic engine braking. The braking occurs when the operator reverses the propulsion lever, which causes the hydraulic motor to act like a pump and the pump to act like a motor, resulting in engine braking. This form of braking virtually eliminates wear on the machine's service brakes.

All of these advantages lead to the most common reason why engineers and customers choose HSTs over traditional mechanical drives, which is increased overall machine productivity. With no need to disengage a clutch or shift gears or brakes, the operator can make the most efficient use of the machine.

Hydrostatic Transmission Disadvantages

The HST however does have disadvantages. It is not as energy efficient as a mechanical transmission. A mechanical transmission can have an overall efficiency of 92% or higher. A HST will have an overall efficiency of 85% or lower. As a result, a machine equipped with an HST will use more fuel than a machine equipped with a traditional mechanical transmission. HSTs are also noisy and sensitive to contamination and heat.

Configurations

HSTs can be categorized into two configurations: split and integral. A ***split HST*** is one of the most common designs found in the off-highway industry. It allows the engine-driven pump to be located at a different location a considerable distance away from the HST motor. Examples of split configurations are used in combines, cotton pickers, and dozers. Notice in Figure 23-16 that the split HST configuration requires hydraulic hoses or tubing to route oil from the pump to the motor.

Integral HSTs eliminate the need for external hoses or tubing. The pump and the motor are directly connected to each other. Integral configurations can range from inline, to U-shaped, to S-shaped. See Figure 23-16.

Hydrostatic Charge Pumps and Main Piston Pumps

A HST commonly uses a charge pump. A fixed-displacement gear pump is normally used as the transmission's charge pump. It is typically driven in tandem off the back of the transmission's piston pump. See Figure 23-17.

The charge pump serves four purposes:

- Supercharges the transmission's piston pump.
- Provides make-up oil to the closed loop to prevent cavitation.
- Supplies oil to the transmission's directional control valve.
- Cools the transmission by replenishing case lubrication.

See the schematic in Figure 23-18. The charge pump flows oil into the HST's closed loop through two check valves that are also known as make-up valves. When the HST is operating, one check valve prevents high-pressure oil from flowing into the charge pump circuit, while the other check allows charge pump flow into the charge loop.

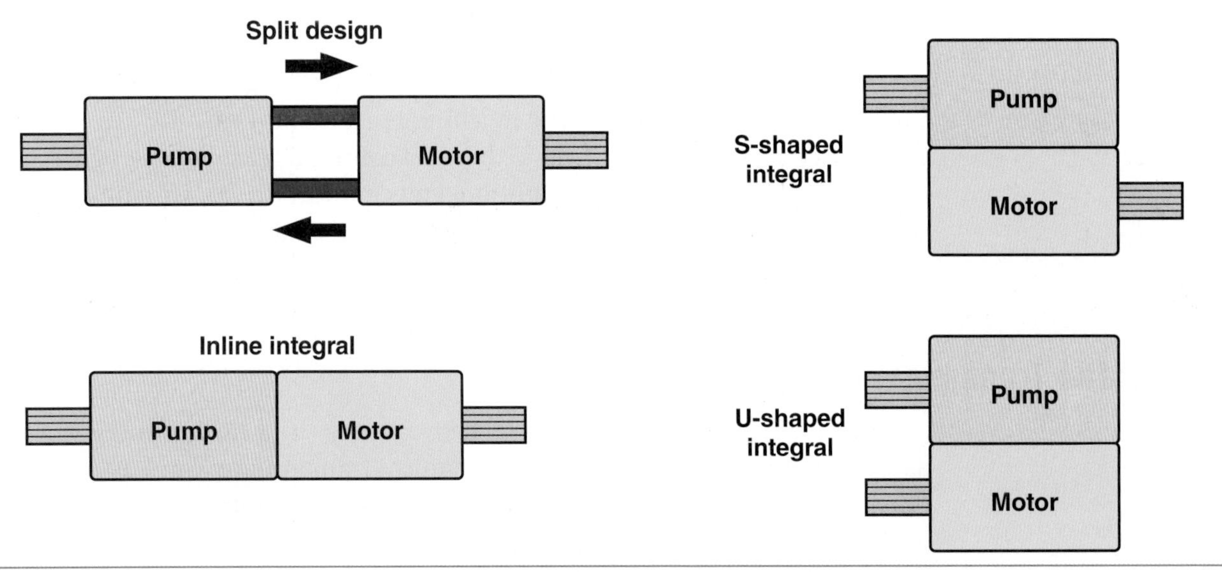

Figure 23-16. HSTs can be split or integral. Split HSTs require the use of hydraulic hoses or tubing. Integral HSTs can be S-shaped, inline, or U-shaped.

Figure 23-17. The charge pump has been removed from the Eaton HST piston pump. Notice the charge pump's driveshaft has a notch that fits into a matching notch on the piston pump's shaft.

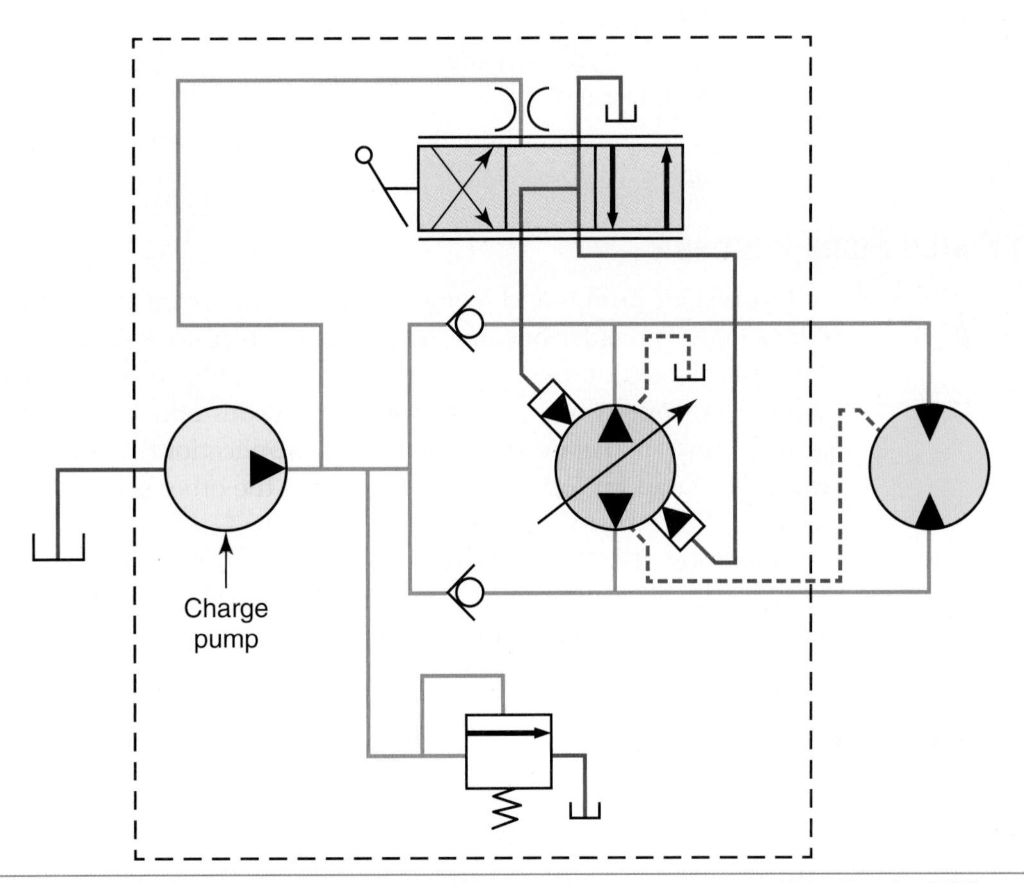

Figure 23-18. A HST charge pump will charge the piston pump and closed loop, supply oil to the HST control valve, and cool the HST by replenishing case drain oil.

Sizing a Charge Pump

Charge pumps are sized to prevent piston pump cavitation. Charge pump flow is most critical when the transmission is operating at high drive pressures. As a rule of thumb, charge pumps are sized to provide 19% of the piston pump's displacement. This value is used to enable the charge pump to provide cavitation-free operation if the pump and motor each fall to a 90% volumetric efficiency. For example:

0.90 (pump efficiency) × 0.90 (motor efficiency) = 0.81 (transmission volumetric efficiency)

1 – 0.81 = 19% charge pump displacement

If a piston pump was flowing 50 gpm at the rated engine rpm, then the charge pump would need to provide the following flow:

50 gpm × 0.19 = 9.5 charge gpm

The Eaton *Heavy Duty Hydrostatic Transmission Pump and Motor Sizing Guide* (No. 3-409) recommends assuming a maximum volumetric efficiency of 96% for a pump and 97% for a fixed-displacement motor. Based on the previous piston pump example of 50 gpm and a charge pump flow of 9.5 gpm, how much extra flow would the charge pump be flowing in a new hydrostatic transmission application?

0.96 (pump) × 0.97 (motor) = 0.9312 overall transmission volumetric efficiency

1 – 0.9312 = 0.0688 (necessary charge pump displacement for a new transmission)

50 gpm × 0.0688 = 3.44 gpm (necessary charge flow for a new transmission)

9.5 gpm (charge pump flow) – 3.44 gpm (necessary charge flow) = 6.06 gpm extra flow

Styles of Piston Pump Frames

Hydrostatic pumps and motors commonly have one of two different styles of housings. The older housing used two servo pistons for stroking a swash plate that pivoted on two trunnion-tapered roller bearings. See **Figure 23-19**. This style of pump has been around for decades and is still available from Eaton, for use in "heavy duty" hydrostatic applications. One servo is used to stroke the pump for forward propulsion and the other servo is used to stroke the pump for reverse operation.

The newer style of hydrostatic pump frame uses a single servo piston to rotate a cradle bearing swash plate. See **Figure 23-20**. The single servo piston is stroked in one direction for propelling the machine forward, and stroked in the opposite direction for propelling the machine in reverse.

Types of Hydrostatic Pump Controls

Hydrostatic pumps are commonly actuated by one or two servo pistons that are controlled with a manual servo control valve, electronic proportional solenoids, or an electronic servo valve. It is also possible to find HST pumps that use a mechanical lever to directly control the pump's swash plate. The mechanically controlled HST pumps are normally lower power applications or require some type of mechanical advantage, such as a long lever, to actuate the swash plate.

Manually Controlled Hydrostatic Pumps

The manually controlled servo valve, also called a HST DCV, has been around the longest. The actuation of the servo control valve will control the speed and the direction of the HST. A propulsion lever is located in the machine's cab or operator station, and a cable connects the propulsion lever to the pump's control valve. The propulsion lever is sometimes called a FNR lever for ***forward and reverse lever***.

As the propulsion lever is stroked, a cable or manual linkage moves a spool valve located inside the HST control valve. The charge pump supplies oil to the servo control valve via an orifice. See **Figure 23-21**.

In a dual-servo pump, the spool valve receives oil via the orifice and sends control oil to the appropriate servo piston to stroke the pump forward or reverse. See **Figure 23-22**.

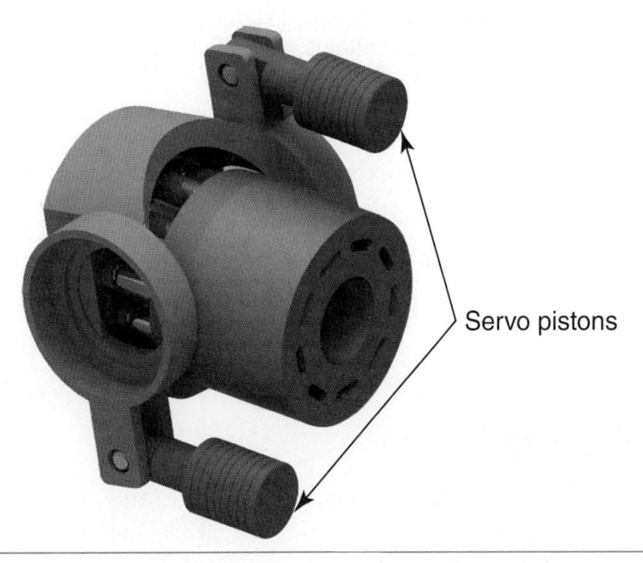

Figure 23-19. A dual-servo HST pump uses a reverse servo piston and a forward servo piston.

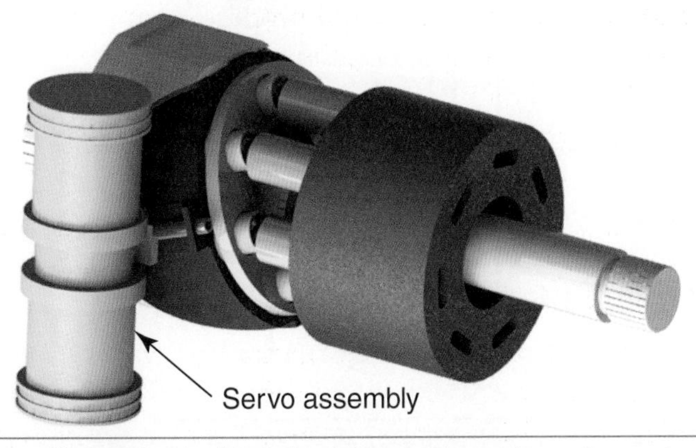

Figure 23-20. A single-servo cradle bearing hydrostatic piston pump achieves forward and reverse propulsion by moving the single piston. The piston moves up for one direction and moves down for the reverse direction.

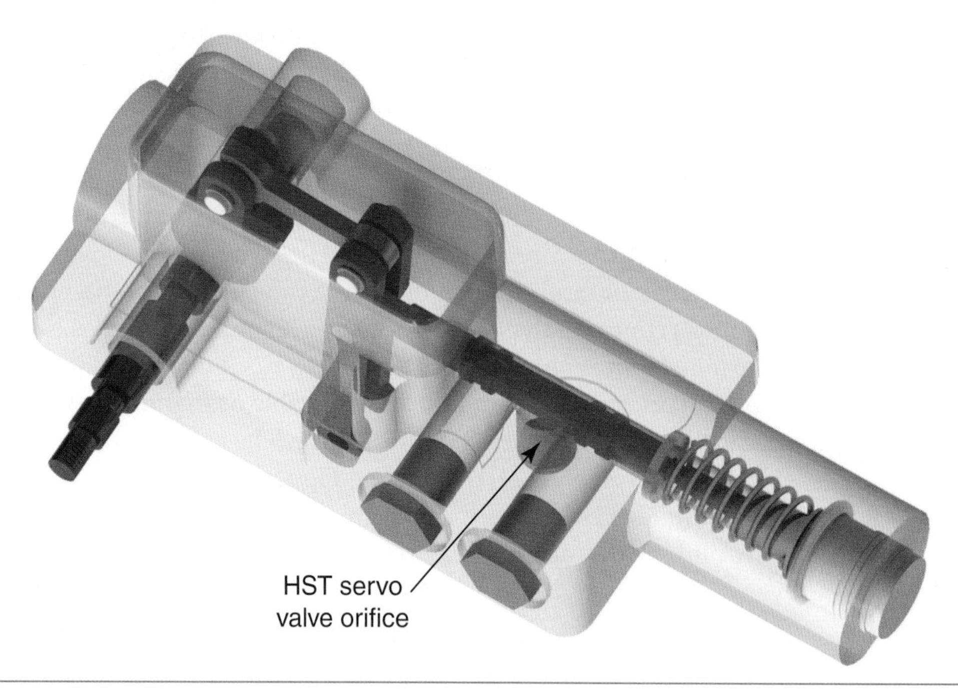

Figure 23-21. A manual HST servo valve contains a spool valve that is spring centered. An orifice meters the oil into the spool valve.

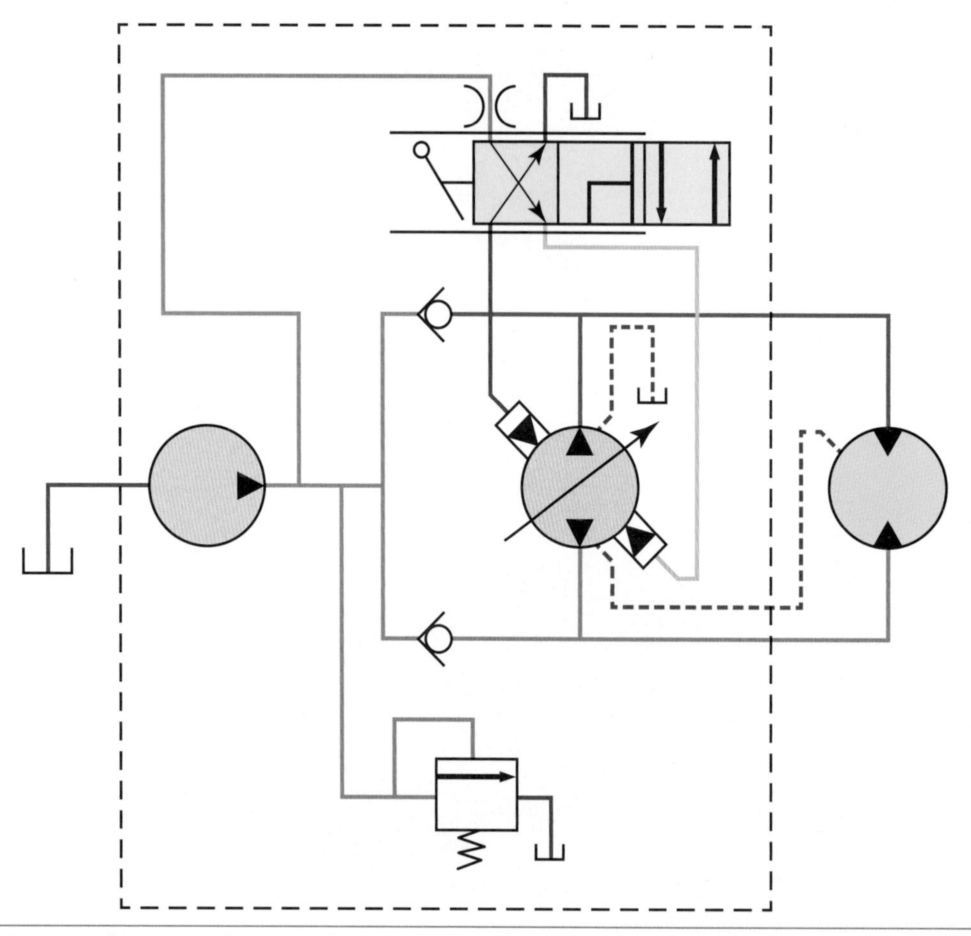

Figure 23-22. A manually operated servo valve receives oil through an orifice for the purpose of actuating the forward servo piston or the reverse servo piston.

Manually operated servo control valves can include a ***neutral safety switch***, which will block electrical current flow, preventing the engine from starting anytime the lever is outside of neutral. See Figure 23-23.

Warning

If the switch fails or the servo valve switch assembly becomes misadjusted, the machine can start even when the propulsion lever is in a forward or reverse position. Always be prepared in the event that a component fails or becomes misadjusted.

Electronically Controlled Hydrostatic Pumps

Many late-model HSTs are electronically controlled, with either a proportional solenoid or a servo motor. **Chapter 9** explained the three types of electronically controlled motors. Electric servo motors are more expensive and less common than proportional solenoids.

Solenoid-controlled HSTs require one solenoid for each direction of propulsion, for example, one reverse solenoid and one forward solenoid. See Figure 23-24. Each solenoid has a power wire and a ground. The electronic control module can control either the power or the ground. An example range of solenoid current flow is 350 milliamperes to 850 milliamperes.

The hydraulic schematic of a solenoid-controlled hydrostatic transmission looks similar to that of a manually controlled transmission. In Figure 23-25, the dual-servo hydrostatic pump schematic illustrates a pump that uses two proportional solenoids.

Figure 23-23. A HST control valve's neutral safety switch is designed to prevent the engine from starting when the lever is in a forward or reverse position.

Figure 23-24. This Rexroth HST pump uses two solenoids to control the pump.

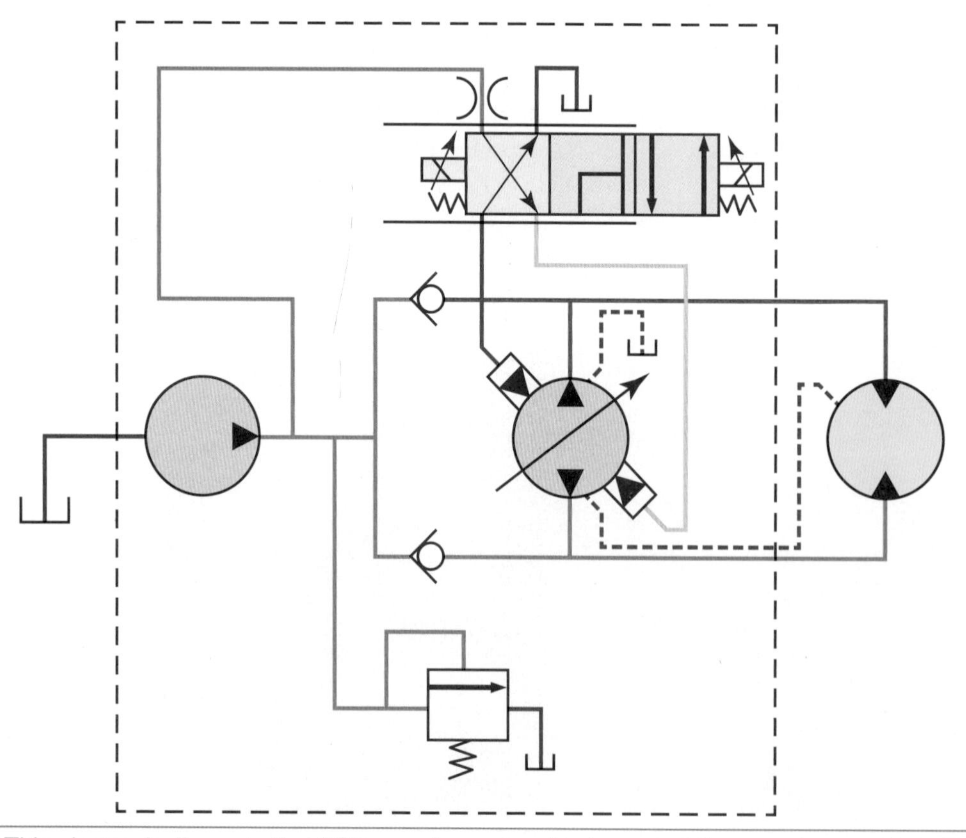

Figure 23-25. This electronically controlled HST uses two solenoids to control the pump's operation. The schematic shows the pump propelling the machine in a forward direction.

Feedback Link

Hydrostatic pumps use a *feedback link*. The link provides a direct mechanical feedback of the position of the swash plate to the pump's control valve. When the operator actuates the control valve, the spool directs servo oil to the servo piston, which causes the swash plate to pivot, resulting in the pump flowing oil and the machine moving forward or reverse. As the control valve directs oil to one servo piston, the other servo piston is ported to the tank. If the spool valve was allowed to continue to direct oil to the servo piston, the pump's swash plate would continue to pivot, resulting in increased machine travel. To limit an uncontrolled acceleration of machine travel, as the swash plate moves, the feedback link repositions the spool so that it can hold the pump swash plate at the exact angle that the operator requested. See **Figure 23-26**.

Hydrostatic Drive Operation and Oil Flow

A sequence of hydraulic events takes place in order to propel an HST. The charge pump draws oil from the reservoir into the charge pump's inlet. The charge pump delivers charge oil to the following three areas:

- The main piston pump's inlet.
- The control valve orifice.
- The transmission's closed loop.

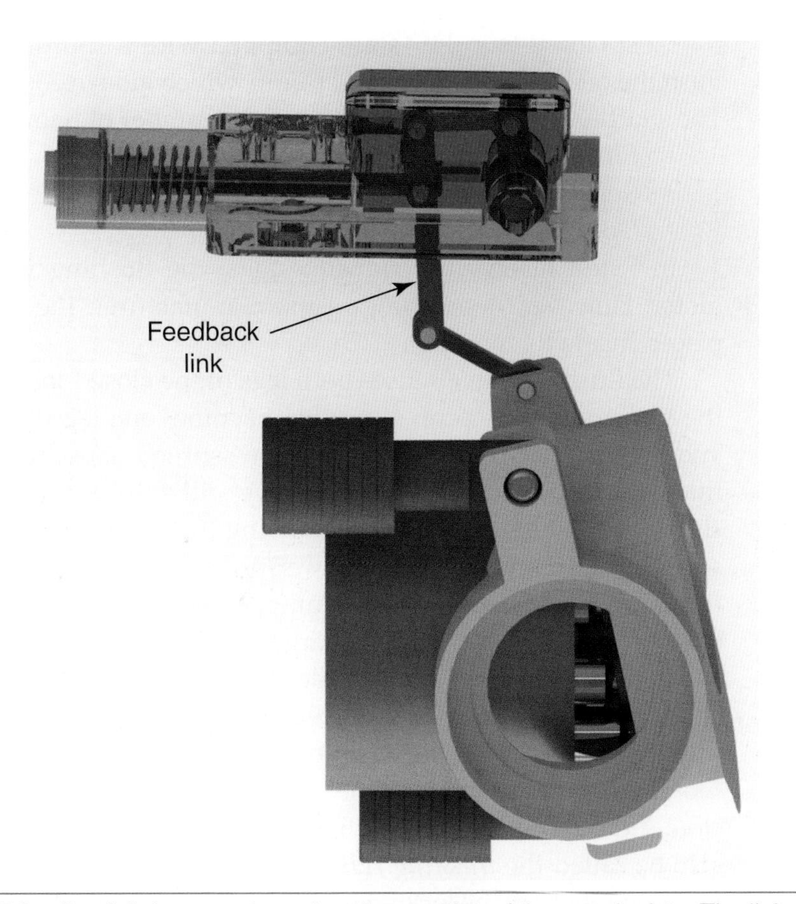

Figure 23-26. The HST feedback link moves based on the position of the swash plate. The link will hold the control valve's spool in the last commanded position.

Servo Oil

The control valve's orifice supplies oil to the spool valve. If the propulsion lever is in the neutral position, the spool valve blocks oil flow. When the operator strokes the propulsion lever, the spool valve is actuated and will route ***servo oil***, sometimes called control oil, to actuate a servo piston. As the servo piston moves, it causes the piston pump's swash plate to stroke the pump. See Figure 23-27.

Note that the piston pump is a reversible pump. The engine drives the pump in a specific direction, either clockwise or counterclockwise. The operator chooses which direction the oil will flow by moving the propulsion lever, either forward or reverse. HST pumps are the most common application of reversible pumps.

As the piston pump swash plate is actuated, it causes the piston pump to draw oil into its inlet, and pushes oil out of its outlet. The pump's inlet and outlet switch functions when the swash plate is reversed.

The drive oil leaving the pump is sent to the motor to propel the machine. As oil enters the hydrostatic motor, it exerts pressure against the pistons inside the motor's rotating group. Once the drive pressure reaches the threshold necessary to propel the machine, the pistons are pushed against the swash plate. This causes the rotating group to spin, which, in turn, rotates the motor's output shaft.

Flushing Valve

Most closed-loop HSTs are equipped with a shuttle valve that is used to flush the oil from the motor and pump, which aids in cooling the transmission. The shuttle valve can be referred to by a number of names:

- ***Flushing valve***.
- Replenishing valve.
- Hot-oil purge valve.

In Eaton heavy-duty transmissions, the flushing valve is placed in the motor. However, depending on the manufacturer, the flushing valve can be placed inside the pump.

The flushing valve senses both legs of the closed-loop transmission. When the transmission is propelling the machine, one leg becomes drive pressure and the other leg remains as charge pressure. Because drive pressure is higher than charge pressure, the shuttle valve will shift. As soon as the shuttle valve shifts, the shuttle valve opens, allowing the charge oil to act on a lower-pressure charge relief valve. This lower-pressure charge relief valve is similar to the charge relief located in the pump, except that it is set approximately 30 psi (2 bar) lower than the pump's charge relief.

Considering Figure 23-27, the charge relief located in the pump can be called the ***neutral relief valve***, because it is in command whenever the transmission is in neutral. The lower pressure charge relief valve located inside the motor can be associated as the charge relief for forward and reverse propulsion. Although some manufacturers call this relief a charge relief, the valve can also be called the ***flushing relief valve***. See Figure 23-28. The lower-pressure flushing relief will be located in the same housing as the flushing shuttle valve, both in the pump or both in the motor.

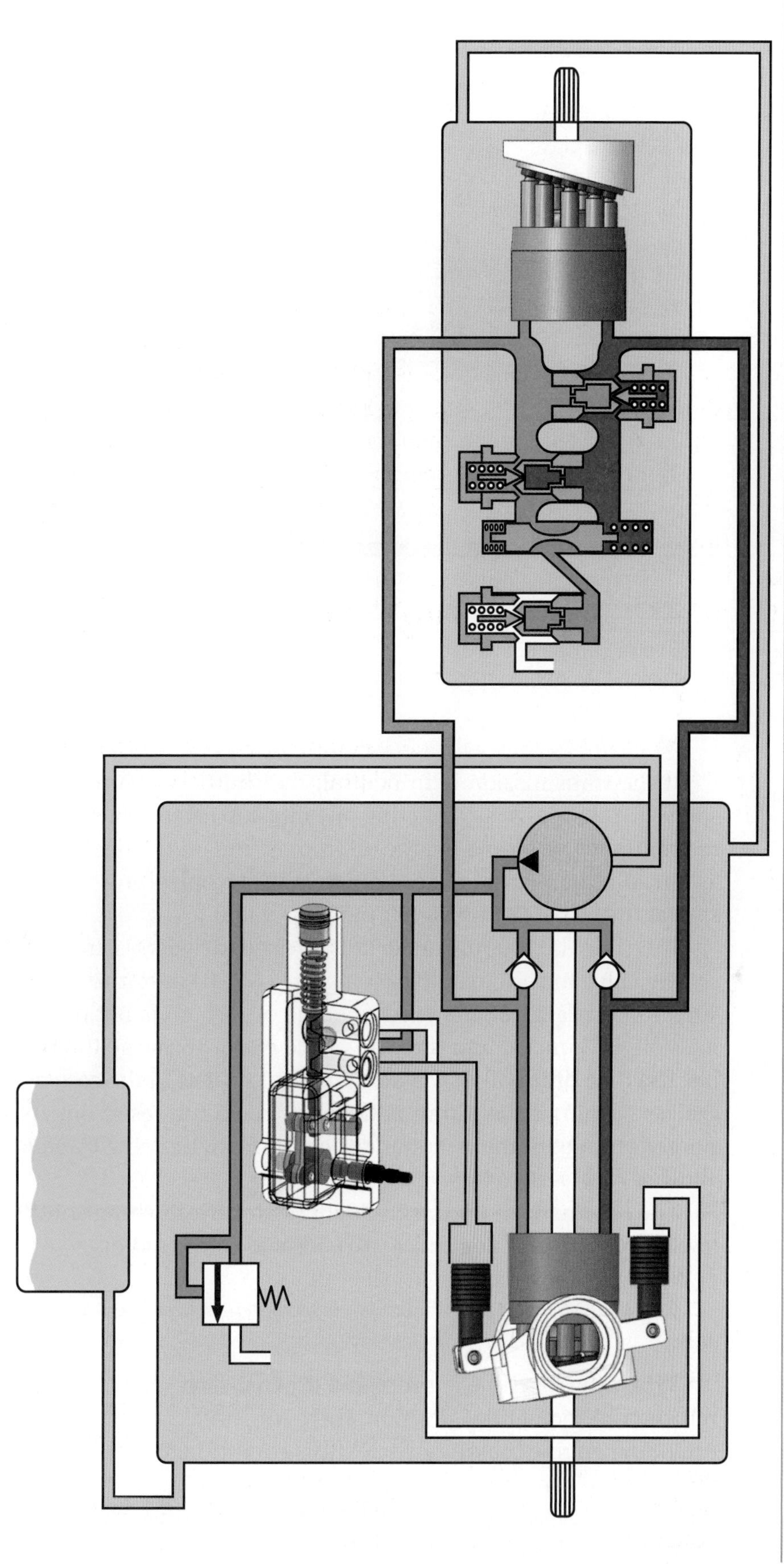

Figure 23-27. A dual-servo HST pump is operated by a control valve. The pump directs oil to the motor, causing it to rotate.

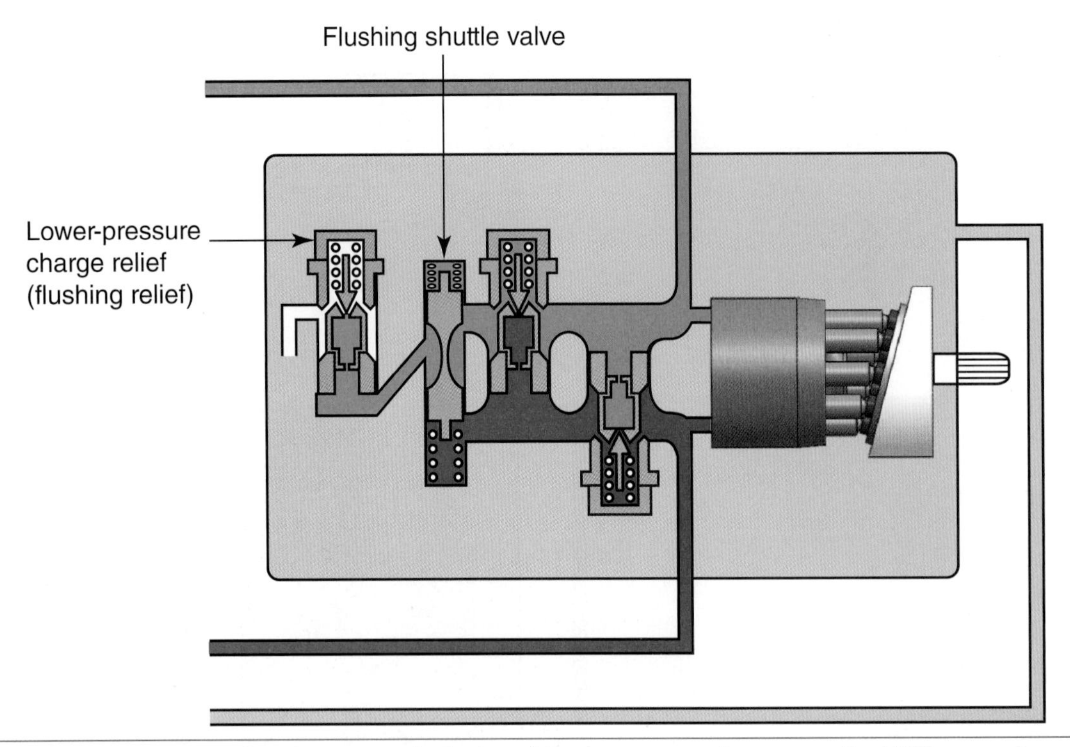

Figure 23-28. The flushing shuttle valve senses both closed loop pressures. As soon as the HST moves forward or reverse, the shuttle valve shifts, causing the charge pressure to be controlled by the lower-pressure charge relief valve.

If the transmission is in neutral, the shuttle valve remains in a balanced state with charge pressure acting on both sides of the shuttle valve. In this state, the shuttle valve blocks charge oil from acting on the flushing relief valve, and as a result, the neutral charge relief valve is controlling the HST charge pressure at the 30 psi (2 bar) higher pressure value.

Anytime the propulsion lever is stroked, drive pressure builds causing the shuttle valve to shift, resulting in a 30 psi (2 bar) pressure drop in charge pressure. In Figure 23-27, the excess charge oil, which is not needed by the transmission, is dumped into the case of the motor to purge the transmission's hot oil. The case drain oil in the motor is then routed to the case of the pump, and the case drain of the pump is then routed to the reservoir. As a result, oil is purged from both the motor and the pump anytime the transmission is in forward or reverse. See Figure 23-29.

The schematic in Figure 23-30 illustrates the flushing shuttle valve and the lower-pressure flushing relief valve located in the motor.

Deceleration

When the machine is propelled down a slope or if the propulsion lever is returned to the neutral position, the hydrostatic transmission will decelerate, providing engine braking. During this condition, the motor is being driven by the machine's momentum. The shuttle valve will shift in the opposite direction. After deceleration, once the pump begins driving the motor again, the shuttle valve shifts back to the original position.

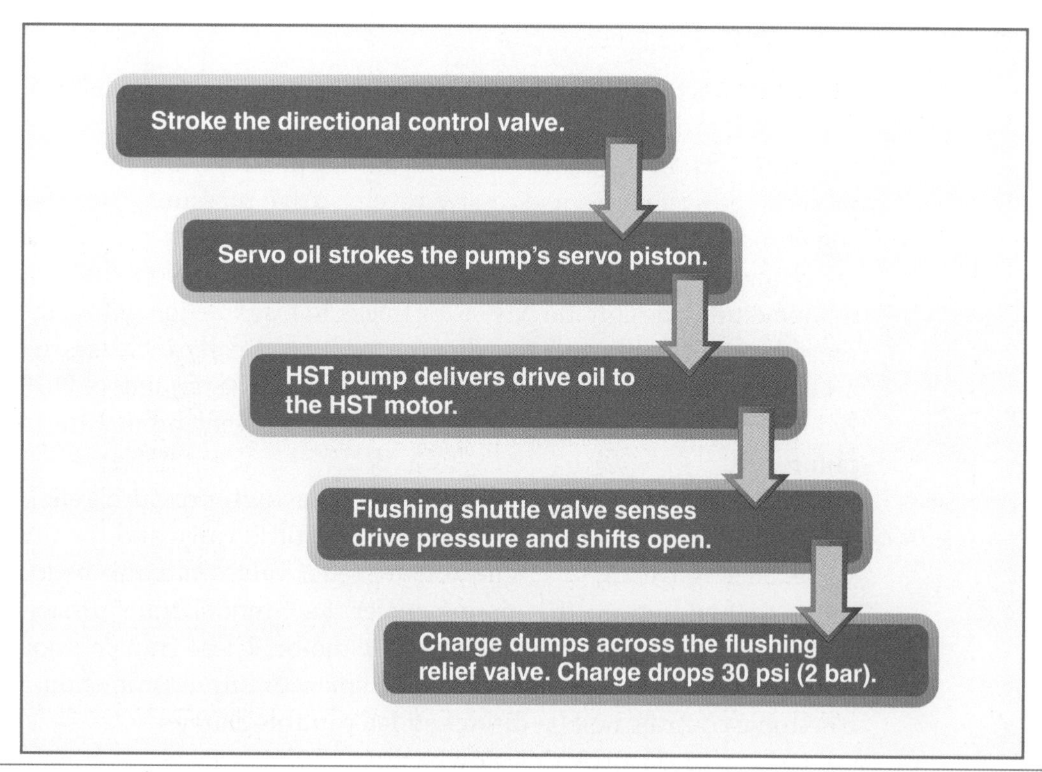

Figure 23-29. A sequence of actions is required in order for the HST charge pressure to drop 30 psi (2 bar).

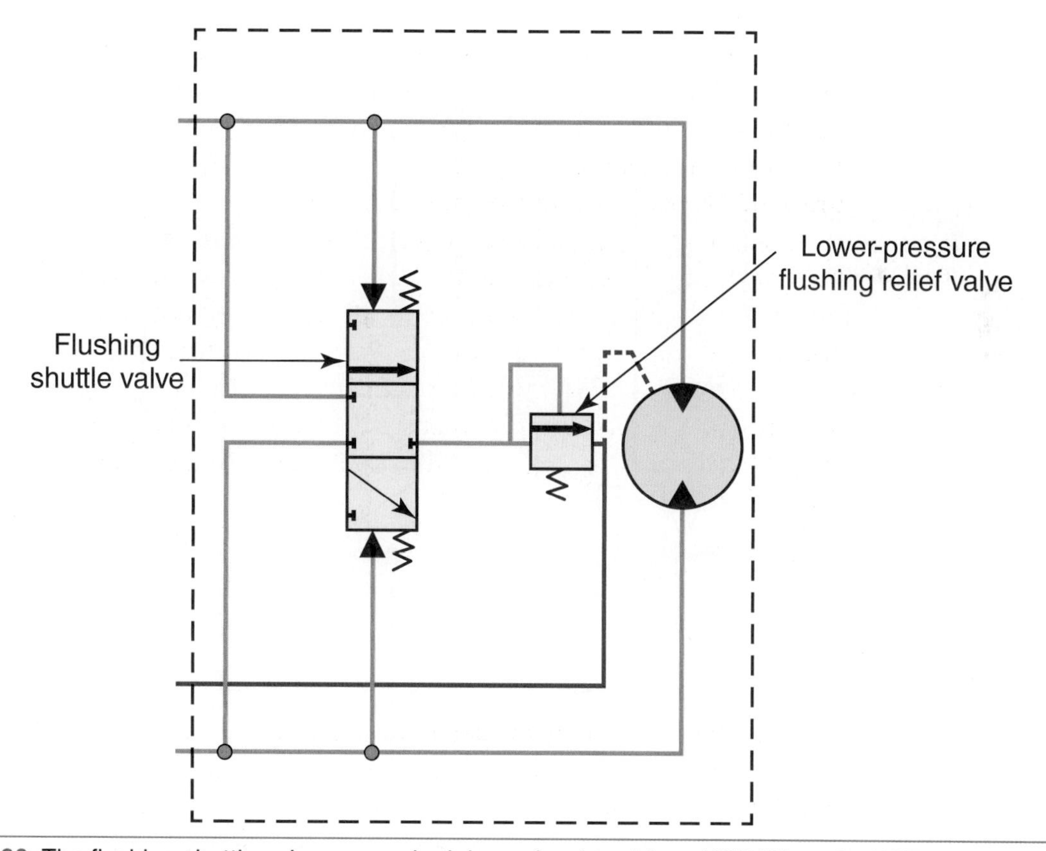

Figure 23-30. The flushing shuttle valve senses both legs of a closed-loop HST. When the HST is in forward or reverse, the shuttle valve will shift up or down, which will allow charge oil to be controlled by the lower-pressure flushing relief valve.

High-Pressure Relief Valves

If the machine's load becomes excessive or stalls, two high-pressure relief valves will provide circuit protection by relieving the high pressure. One relief valve protects the system from forward drive pressure, and the other relief valve protects the system from excessive reverse drive pressure. The relief valves are sometimes called cross-over reliefs because they dump directly into the opposite leg of the closed-loop transmission. See **Figure 23-31**. Note that some machine manufacturers design the wheels or tracks to lose traction before the HST drive pressure stalls. Although the slipping traction greatly increases wear on tires and tracks and should be avoided, it does reduce the number of HST stalls. This reduces excessive heat and wear, helping to prevent premature transmission failure.

In Eaton heavy-duty transmissions, the high-pressure relief valves are located in the motor block along with the shuttle valve and the flushing relief valve. See **Figure 23-32**. High-pressure relief valves can also be located in the pump, depending on the manufacturer. Late-model transmissions use electronic controls to prevent stalling of the motor. These transmissions only rely on the high-pressure relief valves for spike or surge protection. Late-model electronic controls will be discussed later in this chapter.

Figure 23-33 provides flowcharts for the path of oil flowing through a HST.

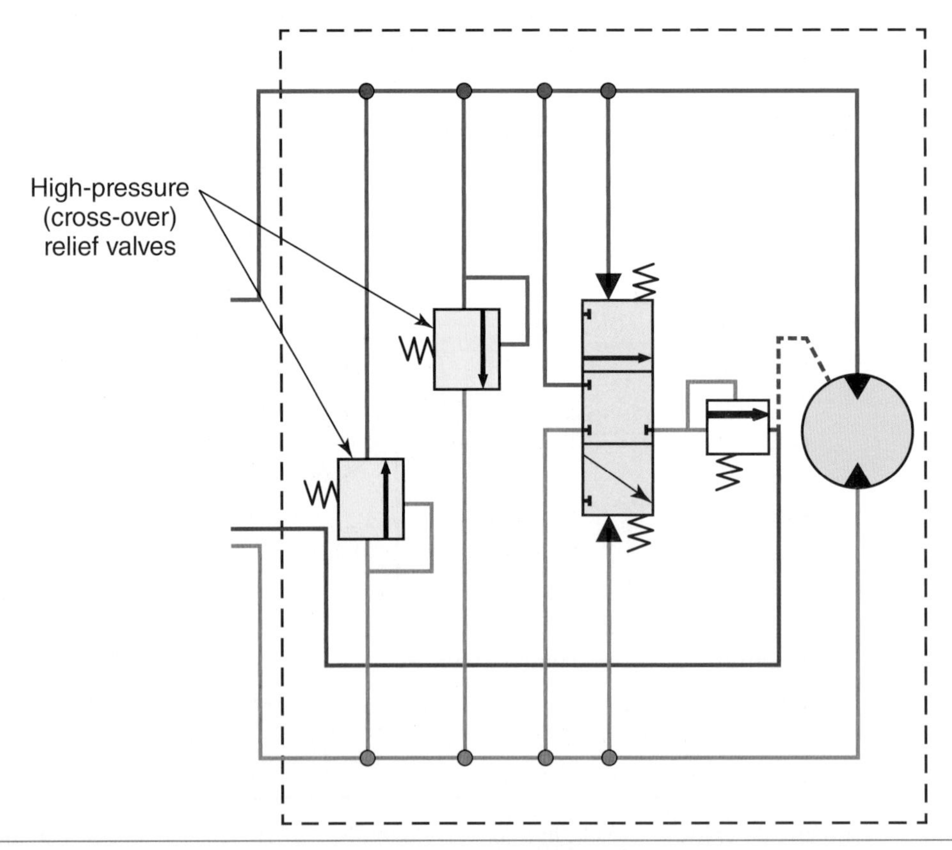

Figure 23-31. The yellow highlighted cross-over relief valves will dump the high-pressure drive oil into the charge loop if the drive pressure reaches the high-pressure relief valve setting.

Figure 23-32. A fixed-displacement HST motor is commonly used on older combines and cotton pickers. The rectangle block contains two high-pressure relief valves, a lower-pressure flushing relief valve, and a shuttle valve.

Reverse Propulsion

When the operator reverses the propulsion lever, the servo valve's spool sends oil to the opposite servo piston, causing it to reverse the pump's swash plate angle. The pump now reverses its oil flow, causing the motor to rotate in the opposite direction. See Figure 23-34.

Hydrostatic Transmission Filtration and Cooling

Earlier in the chapter, it was explained that most closed-loop HSTs do not use closed-loop filtration because the filters would have to filter oil in both directions and be able to withstand high pressures. On older machines, transmissions commonly had suction strainers placed in the reservoir and a suction filter prior to the charge pump. With those old practices, the pump inlet would initially be protected from contamination. However, it is difficult to determine if a suction strainer is plugged. Suction strainers are also difficult to replace because they require the reservoir to be drained. Plugged suction strainers and suction filters and poor maintenance practices can cause pump cavitation, resulting in catastrophic harm. If the transmission has suction filtration, a bypass valve should be used to reduce the risk of catastrophic failure. See Figure 23-35.

Many machines that incorporate filter bypass valves will use a warning system to alert the operator when the filter is bypassing the oil. Some suction filters that have a bypass valve will use a filter indicator. The indicator needs to be regularly checked to determine if the filter needs to be replaced.

If the transmission does not use closed-loop filtration or suction filtration, it can employ three other filtration methods:

- Case drain filtration.
- Charge pressure filtration.
- Off-line kidney-loop filtration.

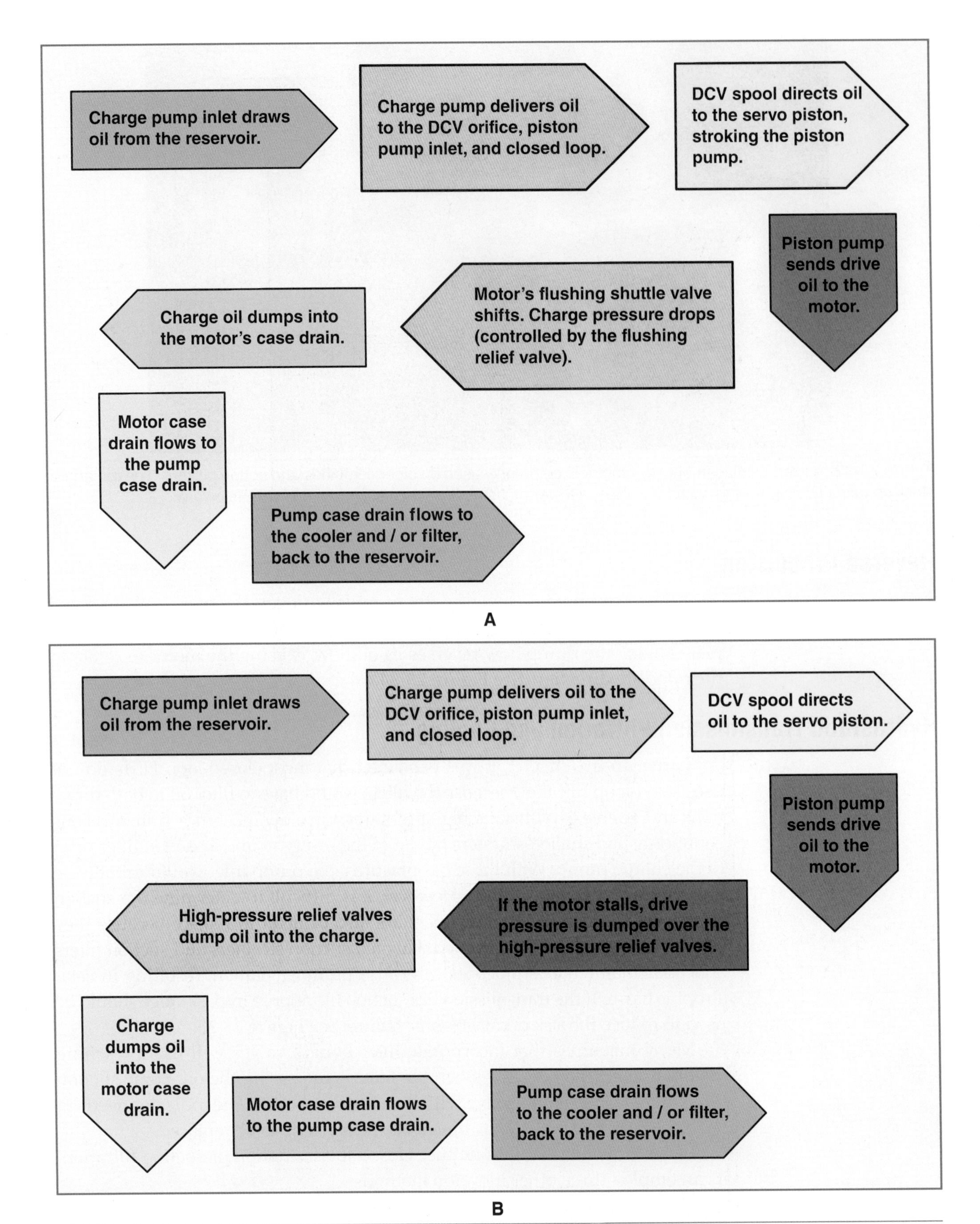

Figure 23-33. HST oil flow diagrams. A—This flowchart depicts the low-pressure oil path through an HST. B—This flowchart depicts oil flowing through an HST, including the high-pressure oil path.

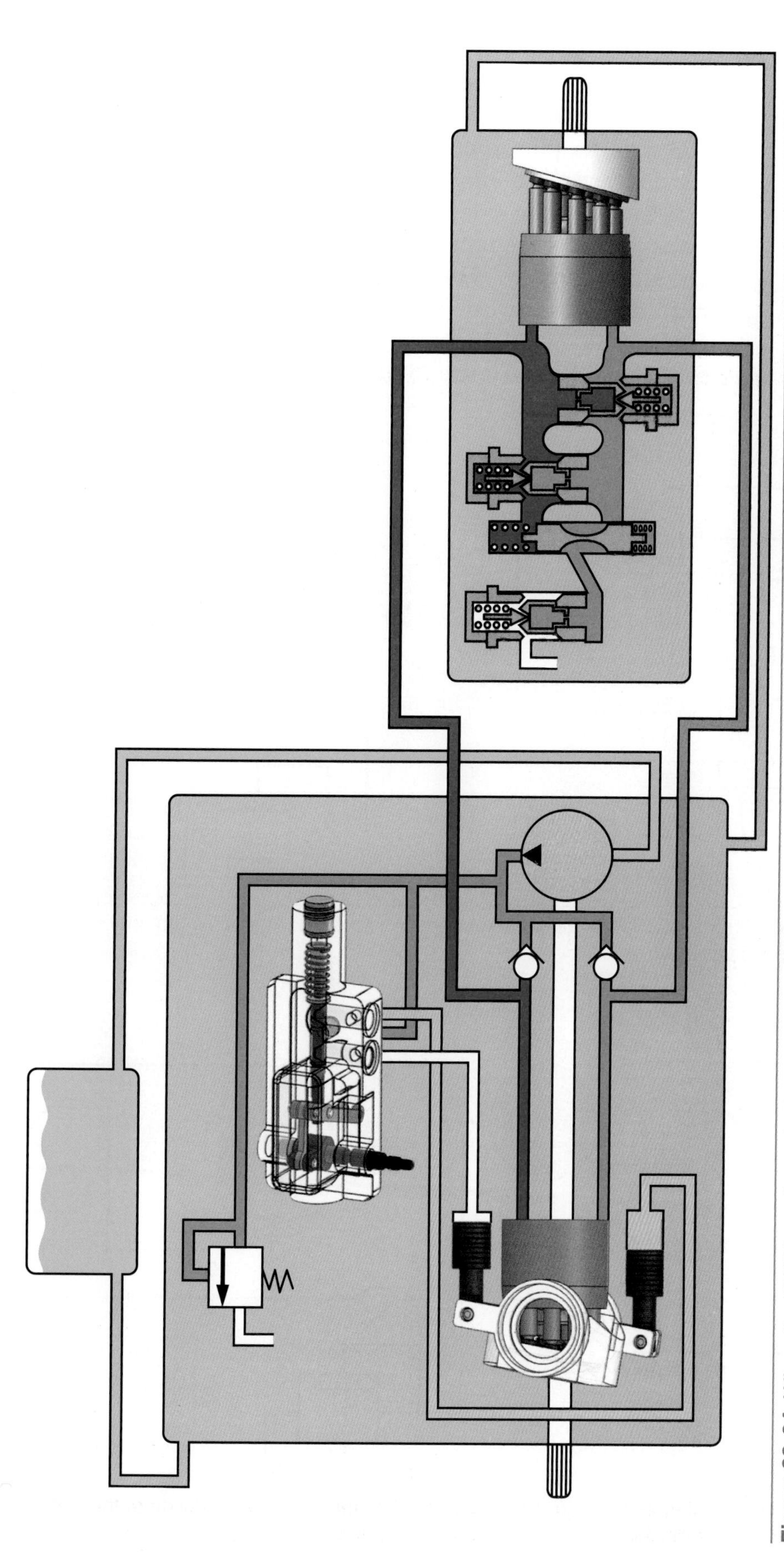

Figure 23-34. When the HST lever is stroked in a rearward direction, the reverse servo piston causes the pump to deliver reverse drive pressure to the motor. As a result, the tractor moves in a reverse direction.

Some manufacturers choose to filter case drain. There are two challenges with filtering case drain. The first is that case drain is also commonly used for cooling. The backpressure of filtering case drain plus the backpressure caused by the case drain oil cooler can cause the case pressure to rise. Most case drain pressures for pumps and motors are quite low, for example 15 psi (1 bar) or less. However, cooling and filtering case drain flow can cause case pressures to rise higher than 45 psi (3 bar). The rise in case pressure poses two problems. One concern is the potential for a motor or pump shaft seal to start leaking. The second concern is that high internal leakage results in high case pressure. High case pressure can cause a technician to incorrectly diagnose a transmission as faulty when, in reality, the filter or cooler is plugged, causing too much backpressure. See Figure 23-36.

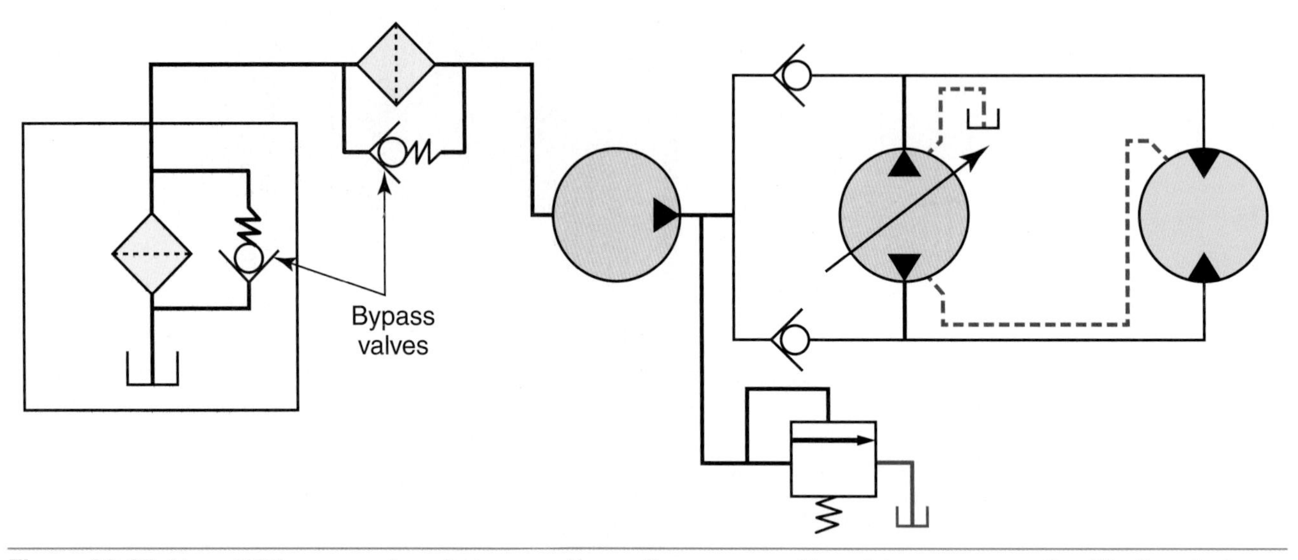

Figure 23-35. Many HSTs use suction strainers and/or suction screens. At a minimum, any suction filtration device should contain a bypass valve.

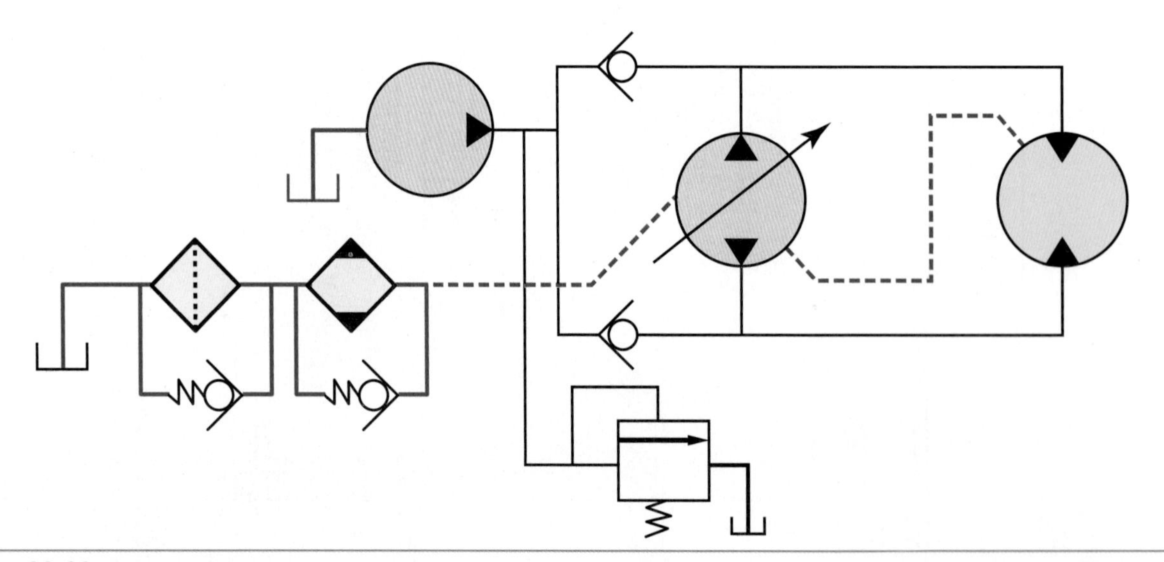

Figure 23-36. If a manufacturer chooses to cool and filter case drain oil, the restriction of the cooler and filter will elevate the case drain pressure. As a result, the shaft seal is more likely to leak.

Filtering the charge pump flow also creates two potential problems. The first is that the filter must be able to withstand a pressure as high as 300 psi (20 bar). The second challenge is that most charge pumps are driven in tandem directly off the back side of the piston pump, and the pump flow is usually routed internally through the housing. If the charge pump flow needs to be filtered prior to the pump's inlet, the charge pump flow must be routed outside of the tandem pump housing and then back into the pump housing. See **Figure 23-37**.

As mentioned in **Chapter 11**, many of today's machines use off-line kidney-loop filtration. Any time the machine needs an inspection, maintenance, or service, a filter cart can be connected to the hydraulic reservoir to filter the reservoir's oil. See **Figure 23-38**.

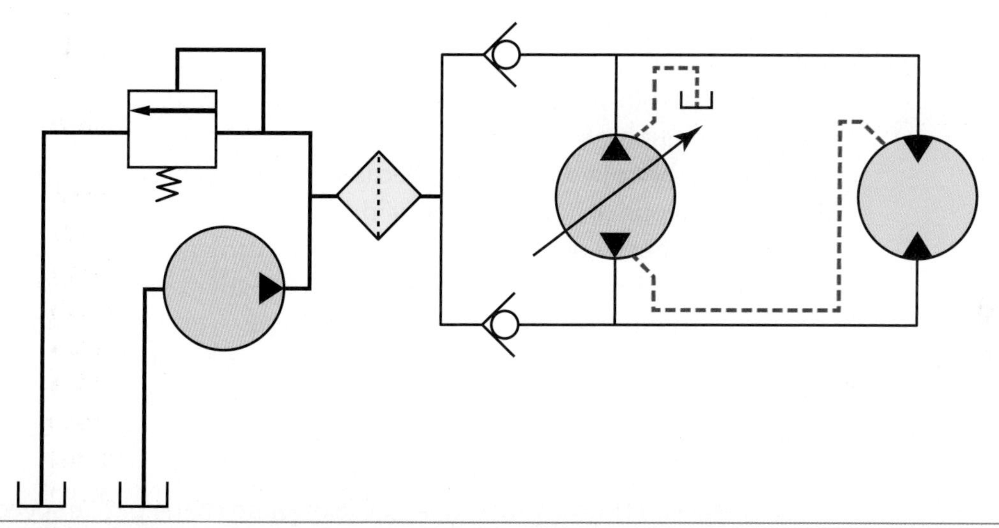

Figure 23-37. Filtering a HST charge pump requires the filter to be plumbed in series between the pump and the closed loop.

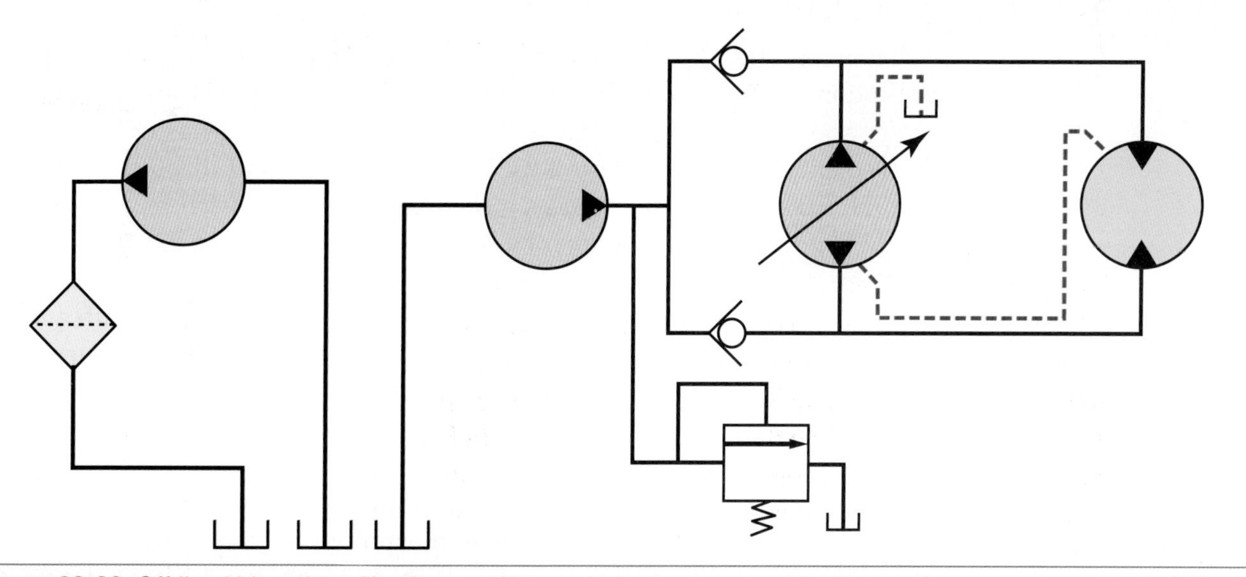

Figure 23-38. Off-line kidney-loop filtration can filter particles to a very small micron rating as compared to the other filtration options.

Variable Hydrostatic Drive Motor Applications

Practically all hydrostatic drives use a variable-displacement pump that allows the operator to change the machine's travel speed. However, sometimes the pump does not provide enough variation of travel speed. For that reason, some manufacturers offer a variable-displacement hydraulic motor.

Two-Speed Motor Applications

Many machines are equipped with a two-speed HST that uses a variable-displacement pump to drive a variable-displacement motor that contains two specific displacements. The motor is configured with a large displacement for low speed and a smaller displacement for high speed. The two specific displacements give the HST two operating ranges, low speed and high speed. Technically, the motor has variable displacement. However, the motor is not infinitely variable. It has only a low-speed position and a high-speed position. Sometimes the low-speed position is called *field speed*, as it is used during the normal operation of the machine while operating in the field. The high-speed position can be called *road speed*, because it is often used when the machine is driving from one field to another field.

Some examples of machines and manufacturers that currently use or have used two-speed motors are skid steers, track loaders, John Deere dozers (450J), Case IH combines (1600, 2100, and 2300 series), and excavators. Two-speed motors can have one of three different types of motor frames. Inline axial piston motors are commonly used on combines and dozers. Bent-axis piston motors are used in dozers. Radial piston motors are used in skid steers and combine rear-power drive axles.

Two-speed HST transmissions can also obtain two separate speeds by means of an ECM varying the pump to provide two specific pump displacements. However, most two-speed HSTs are achieved using a two-speed HST motor.

Eaton Two-Speed Axial Piston Motor

Eaton has offered axial two-speed piston motors for decades. One common application of these motors is agricultural combines. On the Case IH combine, the low speed swash plate angle is 18° and the high speed swash plate angle is 15°. A solenoid is used to control the motor's two different displacements. The Case IH 2300 series combine uses a separate oil source, with regulated pressure, to supply the oil to the motor's control valve. See **Figure 23-39**.

When the solenoid is de-energized, a regulated gear pump supplies 300 psi (21 bar) of pressure to act on the spool, causing the spool to shift against its spring. As the spool shifts, it directs servo oil to the low-speed servo (S1), which places the motor in the low-speed position.

When the operator selects the motor's high position, the motor's solenoid is energized. See **Figure 23-40**. The solenoid drains the control oil that was previously acting on the spool. The spool's spring shifts the valve, which causes the valve to direct higher-pressure oil to the motor's high-speed servo (S2). The solenoid shifts the swash plate to the high-speed position, which is a decreased swash plate angle.

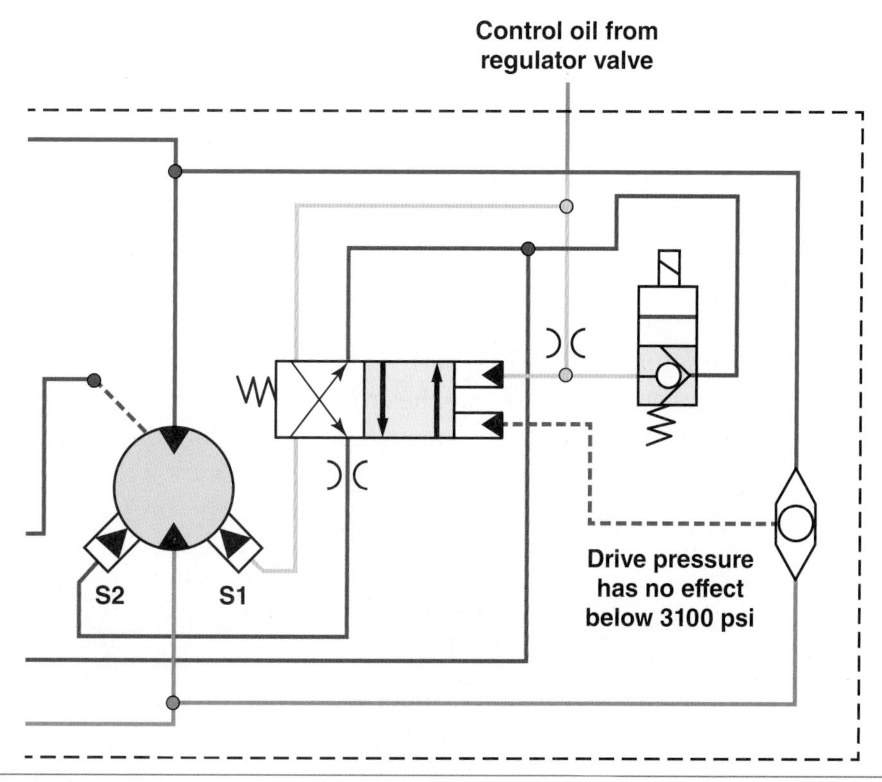

Figure 23-39. When the two-speed solenoid is de-energized, the control pressure causes the spool valve to shift. This results in servo oil actuating the low-speed servo (S1), which increases swash plate angle.

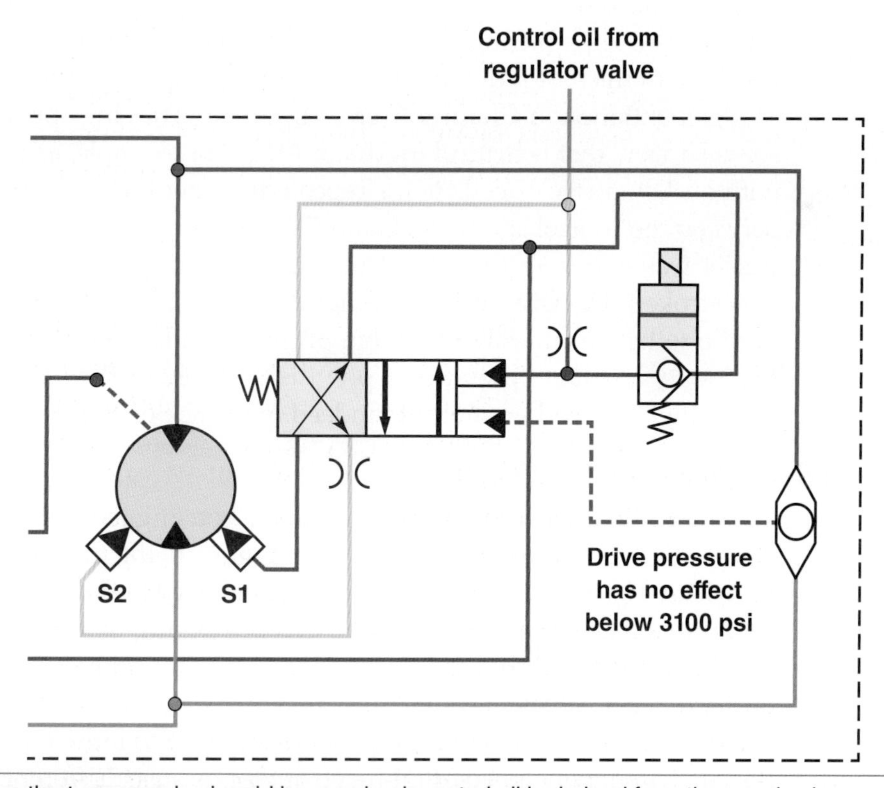

Figure 23-40. When the two-speed solenoid is energized, control oil is drained from the spool valve, causing the spring to shift the spool to the right. Control oil is then directed to the high-speed servo (S2), causing the motor's swash plate to shift to the decreased angle.

While in the high-speed position, a shuttle valve is used to sense reverse or forward drive pressure. When drive pressure reaches a threshold of 3100 psi (214 bar), the high pressure causes the valve to downshift the motor to the low-speed position. To summarize:

- 3100 psi (214 bar) of drive pressure will force the motor to shift to low.
- If the solenoid is de-energized, 300 psi (21 bar) of regulated pressure will cause the motor to shift to low.

Based on that summarization, a person might ask how is it possible that a pressure as low as 300 psi (21 bar) has the same effect as a pressure ten times that amount, 3100 psi (214 bar). The answer is the pressures are being applied to different surface areas. See the cross-sectional drawing in Figure 23-41.

When the motor is in the low-speed position, the solenoid is de-energized. As a result, 300 psi (21 bar) of control oil pressure will act on the control spool, which shifts the pressure response spool. The pressure response spool will then direct servo oil to the S1 low-speed servo piston.

The solenoid must be energized to shift to the high-speed position. See Figure 23-42. When the solenoid is energized, it connects the two passageways, allowing the control oil to drain to the reservoir. The pressure response spool spring will shift the spool to the right, which opens the passageway for servo oil to be sent to the high-speed servo (S2).

The cross-sectional drawing in Figure 23-42 shows that the control spool is shifted to the right when the motor is in the high-speed position. Notice a shuttle valve and needle roller are located on the right side of the control valve. The shuttle valve senses if the vehicle is moving forward or reverse. When drive pressure reaches 3100 psi (214 bar), the drive pressure forces the needle roller to the left, shifting the control spool and the pressure response spool, resulting in the motor downshifting to the low-speed position.

As mentioned in **Chapter 3**, cross-sectional drawings provide service personnel a view that is helpful for diagnostics. For example, a combine equipped with an Eaton two-speed motor repeatedly blew the gasket that was located between the control spool block and the needle roller block. Every time a new gasket was installed, the gasket would fail each time the propulsion handle was stroked. Using a cross-sectional drawing, it was determined that if the needle roller was missing, the high pressure could no longer be isolated from the lower, regulated pressure. Upon disassembly of the valve, the technician found that the needle roller was indeed missing. See Figure 23-43.

Identifying an Eaton Dual-Servo Pump and Eaton Dual-Servo Motor

The Eaton dual-servo two-speed motor housing looks similar to the Eaton dual-servo piston pump housing. At first glance, the average person will not be able to distinguish between the two housings. However, the pump housing contains the charge pump, and the motor housing contains the valve block. The valve block includes the flushing shuttle valve, flushing relief valve, high-pressure relief valves, and two pressure jumper hoses used to sense drive pressure for the purpose of downshifting the motor. See Figure 23-44.

An Eaton two-speed dual-servo motor is also different in that the S1 low-speed servo uses spacers and shims to limit the swash plate angle. See Figure 23-45. The dual-servo pump uses equal-length servo pistons.

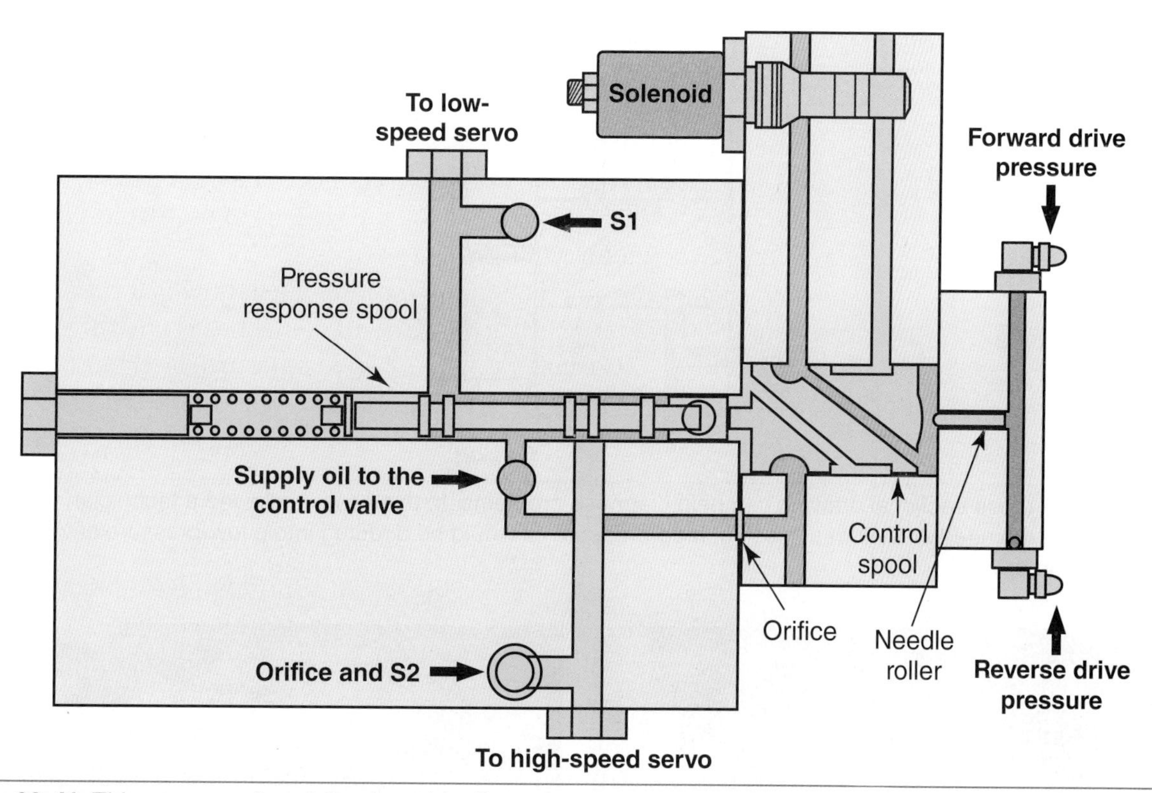

Figure 23-41. This cross-sectional drawing of an Eaton two-speed motor control valve illustrates that the solenoid is de-energized, which allows the solenoid to direct oil to act on the control spool. Notice that drive pressure acts on a needle roller, which has an area ten times smaller than the control spool.

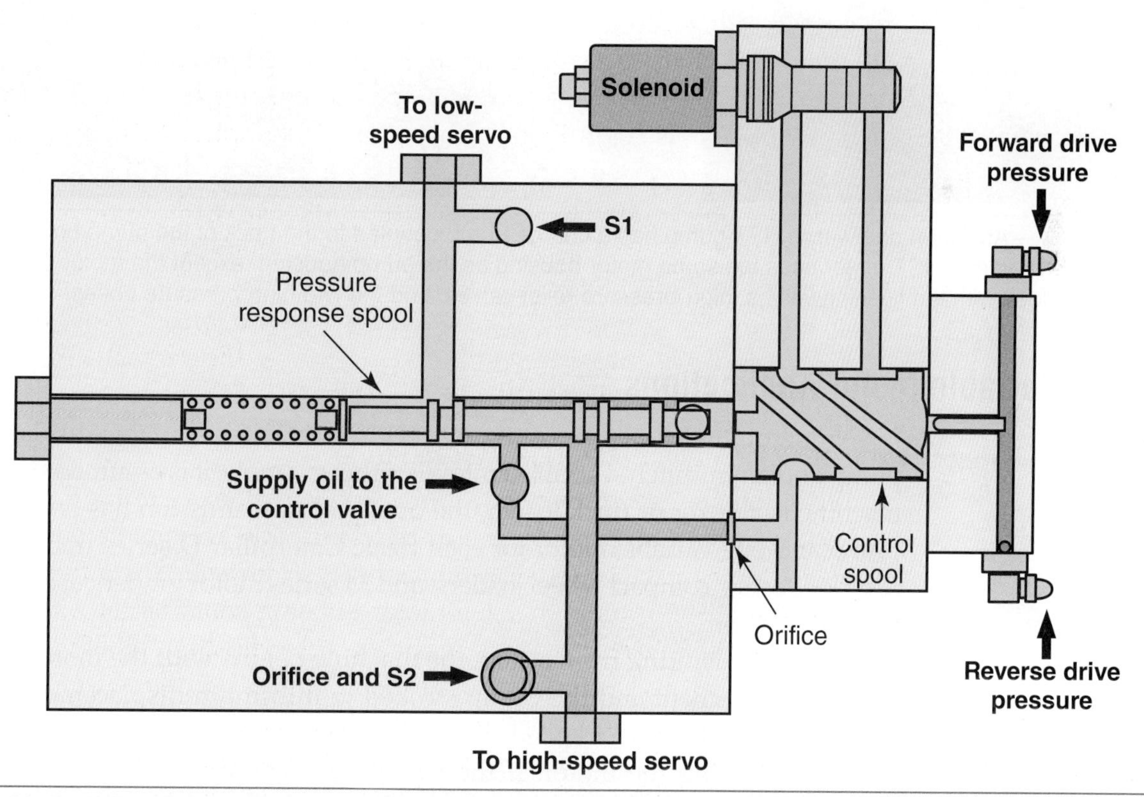

Figure 23-42. When the solenoid is energized, the control oil is drained to the reservoir. The spring shifts the spool, causing servo oil to be sent to the high-speed servo.

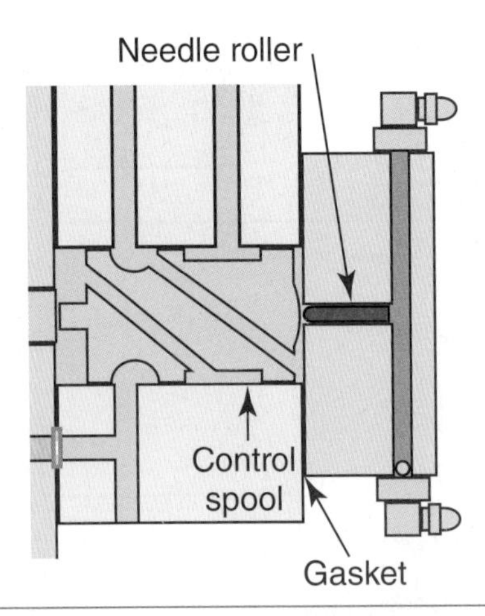

Figure 23-43. A cross-sectional drawing can help diagnose problems. In this case, it allowed a technician to determine that if the needle roller was missing, high-pressure oil would be directed into a low-pressure cavity, causing the gasket to rupture.

Figure 23-44. A—An Eaton dual-servo HST pump has a charge pump coupled to the back of the pump housing. B—An Eaton two-speed HST motor uses the same motor housing as the pump housing, except the motor contains a rectangular block with the flushing valves, high-pressure relief valves, and the two high-pressure hoses.

Infinitely-Variable Motor Applications

Numerous machines are configured with an infinitely-variable-displacement pump and an infinitely-variable motor. For this example, it is assumed that if the machine is a dozer or track loader, the dual-path transmission has one variable pump and one variable motor for each track. Caterpillar D series track loaders, K series dozers, compact wheel loaders, and M series motor graders use this type of configuration.

When accelerating from a stop, the machine begins with the motor at maximum displacement and the pump stroked to minimum displacement. While trying to increase travel speed, the ECM will first fully upstroke the pump before destroking the motor. If the pump has achieved maximum displacement and the operator has requested more travel speed, the ECM will begin to reduce the motor's displacement. See **Figure 23-46**.

Figure 23-45. An Eaton two-speed HST motor has spacers on one servo and no spacers on the opposite servo.

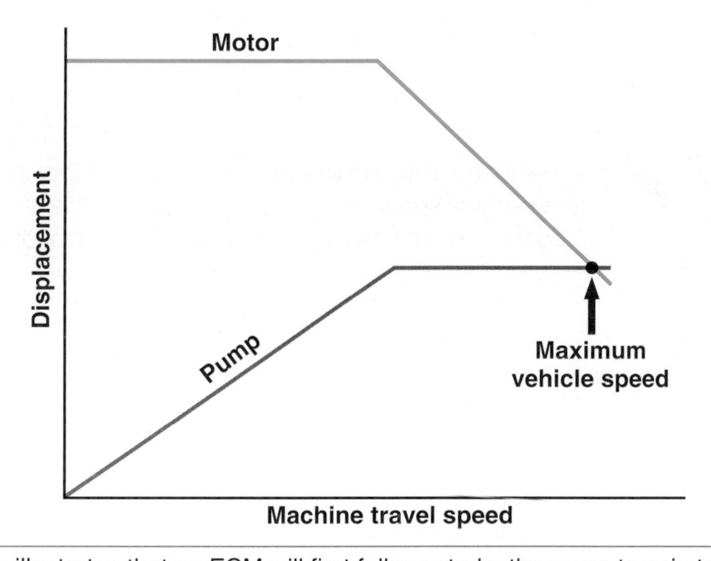

Figure 23-46. This graph illustrates that an ECM will first fully upstroke the pump to gain travel speed. If more travel speed is required after the pump has been fully upstroked, the motor displacement will be decreased.

Variability by Using One or Two Variable-Displacement Motors

Earlier in the chapter, it was explained that the Caterpillar 924K, 930K, and 938K wheel loader transmissions use one variable-displacement pump and two variable-displacement motors. Both motors provide power into a single gearbox that splits power flow to the front and rear axles. One motor has a small displacement and the other motor has a larger displacement.

Both motors are plumbed in parallel and always receive oil from the same hydrostatic pump. However, at high travel speeds, the ECM will nullify the large-displacement motor, making it ineffective. The ECM must perform two tasks in order to nullify the large-displacement motor. The large-displacement motor is coupled to the gearbox through a clutch mechanism. The ECM will disengage the clutch to prevent the larger motor from providing input into the gearbox. If the motor was allowed to simply freewheel, the pump's oil flow would take the path of least resistance. Therefore, the second task the ECM must perform is to destroke the motor's swash plate to a neutral angle. When an axial piston motor swash plate is placed in an exact neutral angle, it can no longer rotate even if it is receiving pump flow.

The small motor is always in use and will be rotating anytime the wheel loader is moving, regardless of the travel speed. When the wheel loader requires more torque, the pump drives both hydrostatic motors, which increases the total motor displacement. As a result of using both motors, torque is increased and travel speed is reduced. When the machine is traveling at high speed, only the small motor is used, and the large motor does not provide power into the gearbox. See the graph of the displacements of the pump and motors in **Figure 23-47**.

The Caterpillar 924K, 930K, and 938K wheel loaders also offer a ***creeper control***, which provides very slow travel speeds while allowing a large amount of implement-hydraulic flow to operate attachments such as brooms, brush cutters, or snowblowers. In range 1, the customer can limit the maximum travel speed as slow as 0.6 mph or as fast as 8 mph. The default maximum travel speed in range 1 is 4.4 mph. The operator can choose one of four travel speed ranges listed in **Figure 23-48**.

Note

The loader has two foot pedals that enable the operator to control the engine speed and travel speed independently. The left foot pedal controls the machine's travel speed. The right foot pedal controls the engine speed.

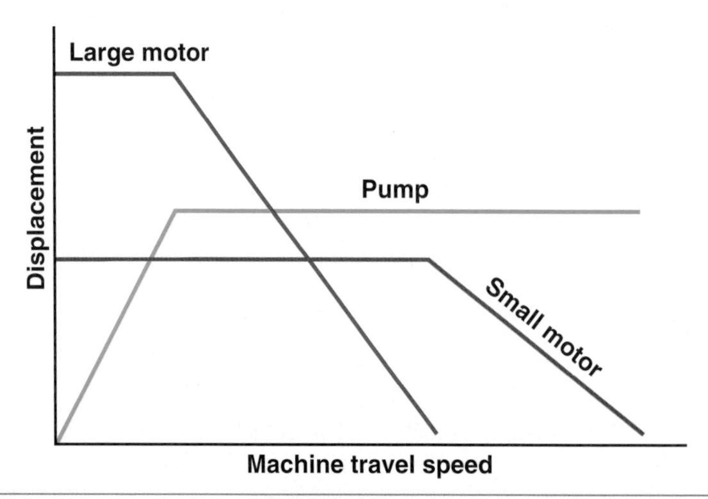

Figure 23-47. On the Caterpillar 924K, 930K, and 938K wheel loaders, the ECM will first fully upstroke the pump in order to increase travel speed. If the loader needs more travel speed, the ECM will begin reducing the large motor's displacement. If more travel speed is required, the ECM will finally reduce the small motor's displacement.

Single-Speed, Fixed-Displacement Hydrostatic Motor

Many hydrostatic transmissions use a simple fixed-displacement motor, which is also known as a single-speed motor. If the motor is an inline axial motor, Figure 23-49, it contains a fixed swash plate. In this application, travel speed is changed by adjusting the variable pump.

A fixed-displacement motor is the simplest design used in off-highway equipment. It does not require any controls, such as electrical current or control oil, because the pump is responsible for the change of speed. The motor simply rotates at the speed and direction commanded by the pump. A single-speed motor lends itself to one potential problem when it is being replaced. If the technician is not careful, the new motor can easily be installed upside down, which can cause a dangerous situation if not noticed.

Mode	Maximum Travel Speed (mph)	Maximum Travel Speed (km/h)	Application
Range 1	0–8.0 mph	0–13 km/h	Creeper—operations requiring high hydraulic flows, like brooms and snowblowers
Range 2	0–17 mph	0–27 km/h	Truck loading
Range 3	0–25 mph	0–40 km/h	Load and carry
Range 4	0–25+ mph	0–40+ km/h	Roading (transport)

Figure 23-48. Caterpillar 924K, 930K, and 938K wheel loaders offer four ranges of travel speeds. Each range allows an operator to vary the machine speed from a stop to the maximum travel speed for that range.

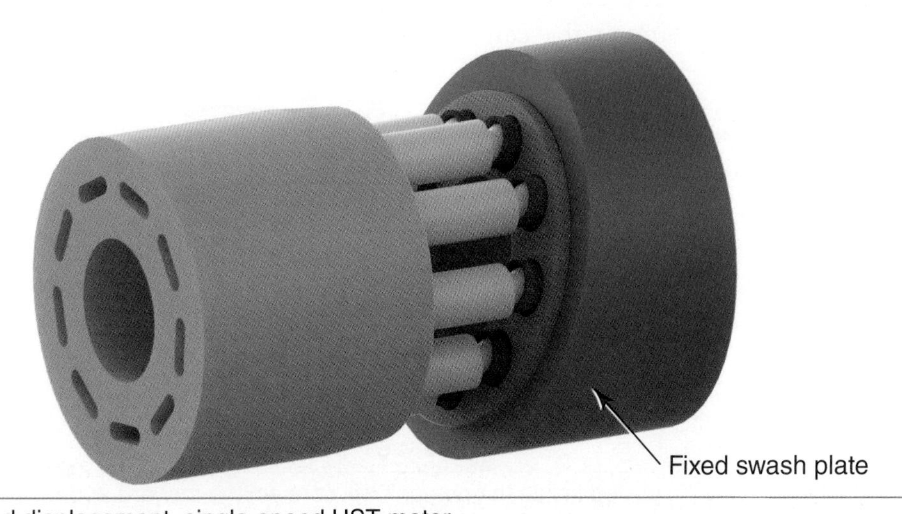

Figure 23-49. A fixed displacement, single-speed HST motor.

Warning

If a motor is installed upside down, the swash plate will also be upside down. This will cause the propulsion lever to work opposite of the expected way. It is possible, but less likely, for the same symptom to occur if a technician was physically able to install the closed-loop drive hoses to the wrong ports. To prevent this problem from occurring, the inlets and outlets of the motors are labeled, for example port "A" and port "B". Be sure to install the motor and drive lines in the original configuration to prevent the HST from operating backward.

Additional Hydrostatic Valving

HSTs can be equipped with additional types of control valves that offer different features. The following controls are commonly used in mobile HST applications.

Inching Valve

Some HSTs are equipped with an ***inching valve***. The valve is commonly operated by a foot pedal and acts similar to a transmission clutch. The valve can perform the following functions:

- Provide a method to control the steady acceleration from a stop, by slowly releasing the foot pedal after the propulsion lever has been actuated.
- Enable the operator to slowly inch up to an implement to ease the installation of the implement.
- Allow an operator to coast to a stop.
- Enable the operator to disengage the transmission and use the service brakes for stopping.

Some machines use the hydrostatic motor to drive a two-, three-, or four-speed gearbox that provides the machine additional operating ranges. On these machines, the gearing typically does not use synchronizers and the operator's manual will specify to change the ranges only when the machine is stopped. If the machine is equipped with a two-, three-, or four-speed gearbox, depressing the inching valve can help the operator change ranges.

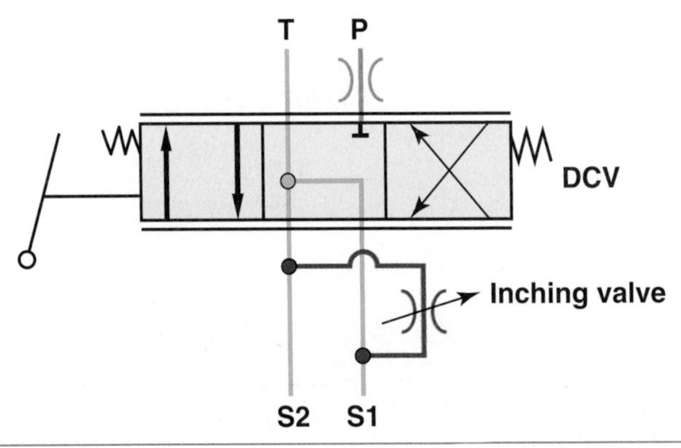

Figure 23-50. If the inching valve is placed inside the pump, it will connect the forward and reverse servo ports when actuated, neutralizing the pump's flow.

Inching Valve Located inside the Pump

The inching valve can be designed to work in conjunction with the pump or the motor. If the inching valve is incorporated with the pump, it is located inside the pump's servo control valve. See the schematic in **Figure 23-50**. As the inching valve is operated, it connects the forward servo port to the reverse servo port, which causes the servo pistons to return to the balanced neutral position.

Inching Valve Used in Conjunction with the Motor

The foot-and-inch valve on older Case IH combines, for example 1600, 2100, and 2300 series, was used in conjunction with the hydrostatic motor. For this style foot-and-inch valve system, the pedal valve worked in combination with the motor's two high-pressure relief valves. Together, all three valves formed a pilot relief system. See **Figure 23-51**.

The foot-and-inch valve sets the pilot pressure that is held against both high-pressure relief valves. When the pedal is pressed, the pilot oil is dumped. As a result, the high-pressure relief valves will drop to a very low value and the motor will no longer be able to develop enough torque to propel the machine.

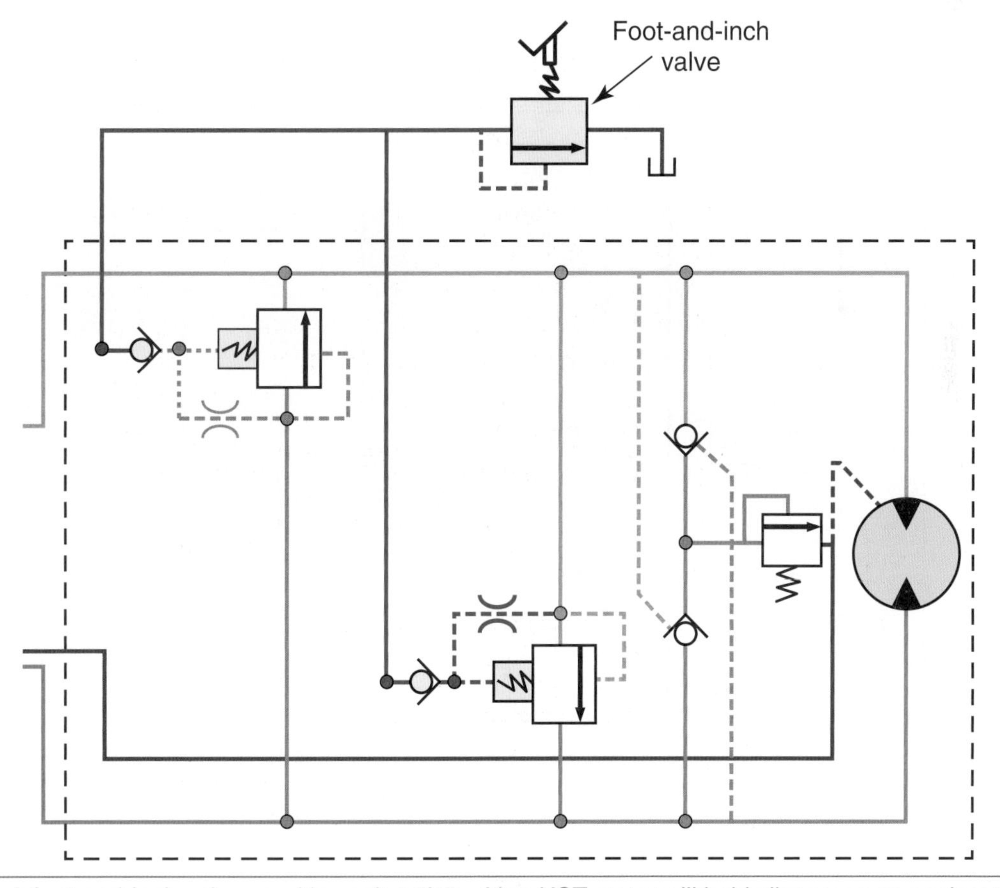

Figure 23-51. A foot-and-inch valve used in conjunction with a HST motor will hold pilot pressure against the high-pressure relief valves. When the foot-and-inch pedal is depressed, the high-pressure relief valve values fall to practically no pressure, which neutralizes the HST.

On the style of foot-and-inch valve system illustrated in **Figure 23-51**, technicians can adjust the transmission's high-pressure setting for both forward and reverse by adding or removing shims inside the foot-and-inch pedal valve. As a word of caution, HSTs have been damaged as a result of a technician adjusting the pressure too high. In one scenario, a customer capped off the line going to the foot-and-inch valve, which resulted in the customer destroying the transmission.

Manual Bypass Valve

A motor can also be equipped with a ***manual bypass valve***. See **Figure 23-52**. The valve allows the fluid from one leg of the drive loop to be bypassed into the other leg of the loop. The valve is used in several situations:

- To ease shifting of a two-, three-, or four-speed multi-gear transmission.
- To allow an inoperative machine to be towed.
- To deactivate the hydrostatic transmission for safety purposes.

The manual bypass valve is a hand-operated rotary valve. It is designed to be either fully open or fully closed. The valve is to only be used when the pump is in neutral and the machine is stationary. The valve will unlock the motor's output shaft, enabling it to be rotated. During normal hydrostatic transmission operation, the valve will be closed.

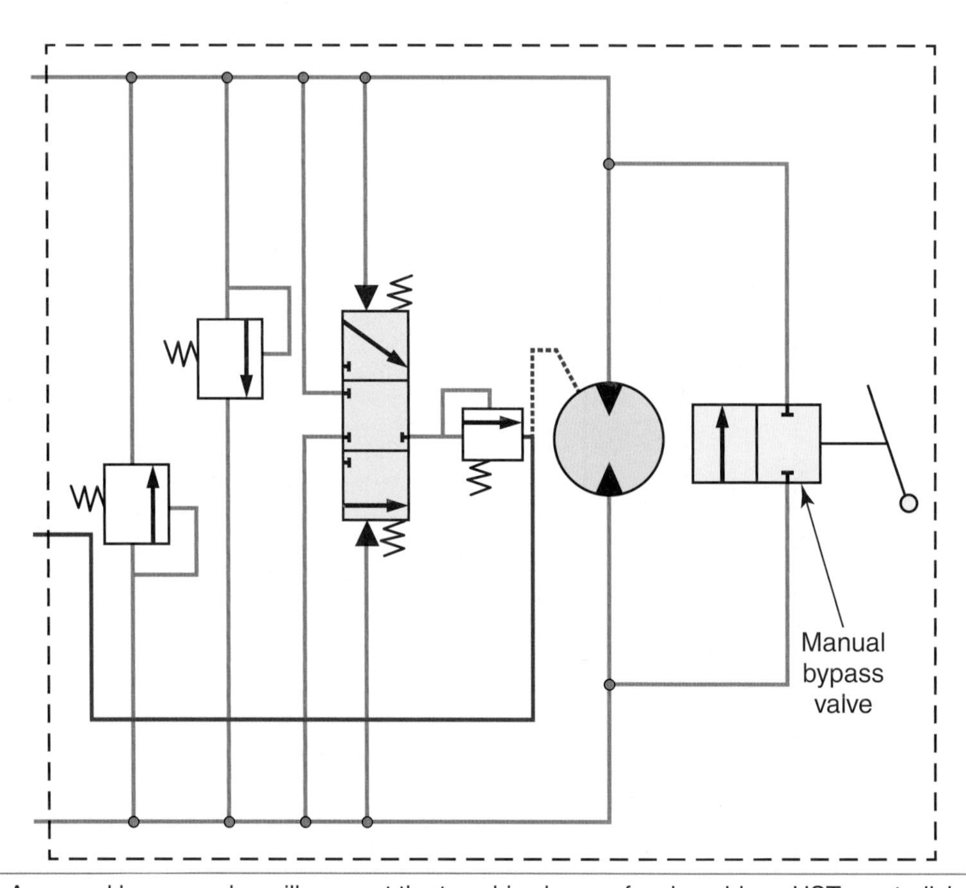

Figure 23-52. A manual bypass valve will connect the two drive loops of a closed-loop HST, neutralizing the HST.

Electronic Pressure-Release Solenoid

Late-model HSTs use a solenoid similar to the manual bypass valve. On a Case IH 7120-9120 series combine, the solenoid is called a ***pressure-release solenoid***. When energized, the solenoid connects the two legs of the closed loop, allowing the motor to freewheel, which makes it easier to shift the transmission gearbox. When the propulsion lever is in neutral and the operator requests a new speed range in the gearbox, the ECM will automatically energize the pressure-release solenoid to make it easier to shift the gearbox. See **Figure 23-53**.

Internal Pressure Override

HSTs can incorporate an ***internal pressure override (IPOR)***. The valve is placed in series between the charge pump and the inlet to the drive pump's servo control valve. The IPOR senses forward and reverse drive pressures. When the pressure reaches the IPOR valve's setting, the valve will dump the supply oil to the servo control valve, which causes the pump to return to a neutral state. The IPOR acts like a high-pressure limiter. See **Figure 23-54**.

The IPOR valve is used to protect the machine from high-pressure overloads for extended periods of time. Combines and cotton pickers are a couple of examples that have used IPOR valves. An operator unfamiliar with this style of control might state that the machine is malfunctioning because the transmission loses power at high-pressure settings. In this scenario, if the machine is equipped with a multispeed gearbox, the transmission should be shifted to a lower gear ratio.

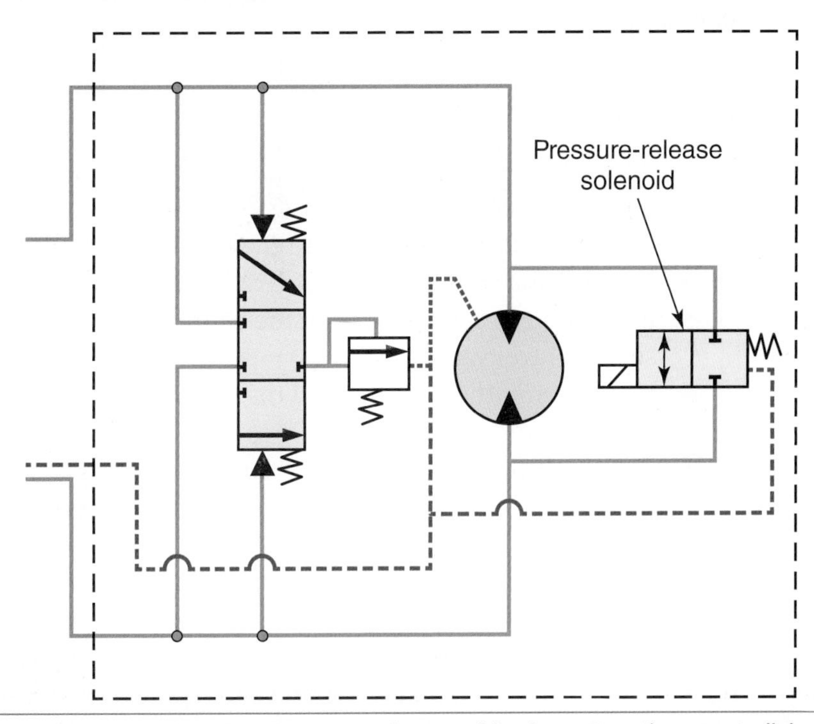

Figure 23-53. A pressure-release solenoid will connect the two drive loops together, neutralizing the HST.

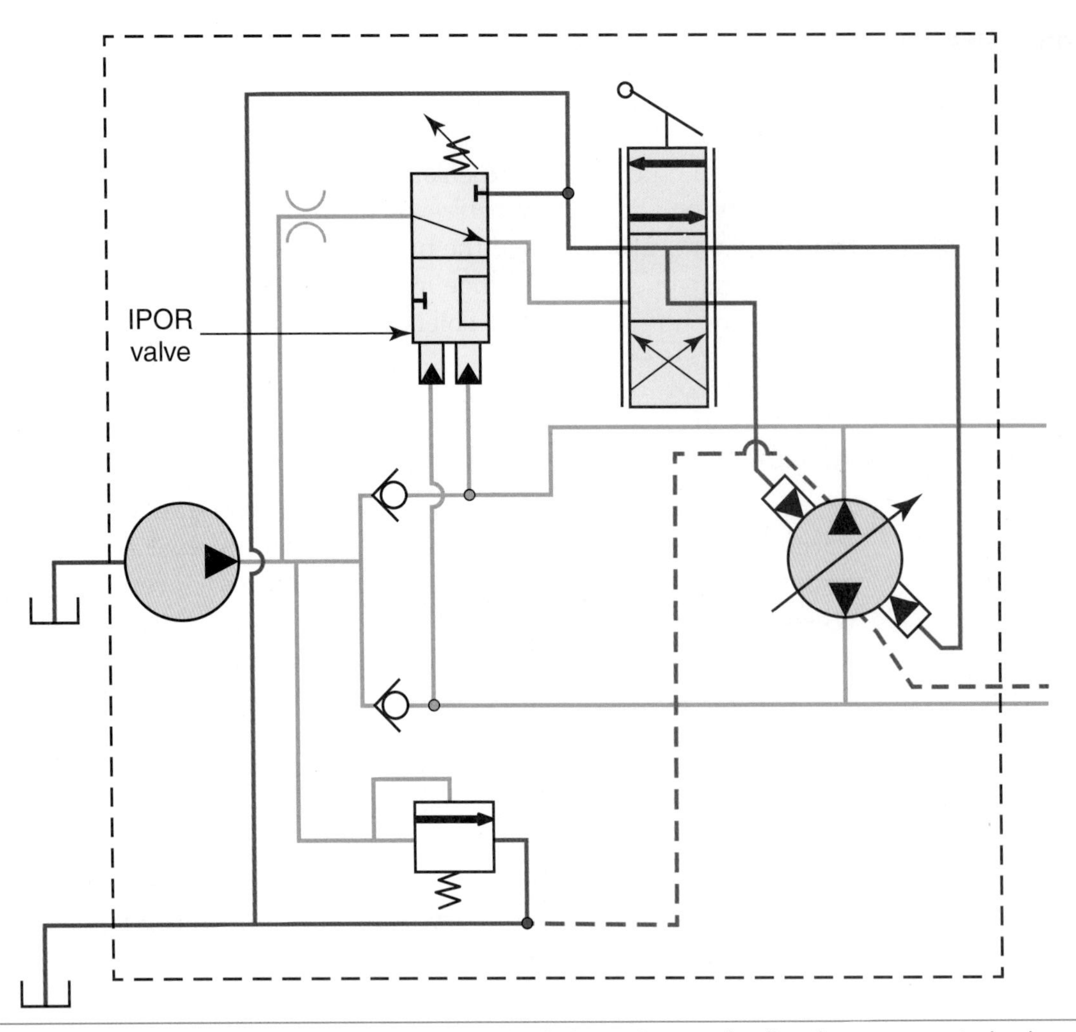

Figure 23-54. The IPOR valve senses high pressure and will dump the supply oil to the servo control valve when the high pressure reaches the IPOR setting.

Pressure Cutoff Valve

Late-model Case IH combines (7130–9130) use a similar valve called a ***pressure cutoff valve***. It is similar to the IPOR in the following ways:

- It senses forward and reverse drive pressure.
- It is placed in series between the charge pump and the supply to the servo control valve.
- It will dump the supply oil to the servo control valve if drive pressure reaches the cutoff limit.

In this application, the pressure cutoff valve controls the transmission's high pressure during gradual pressure buildup. The pump also contains high-pressure relief valves, known as cross-over relief valves, which protect the system from sudden or rapid pressure increases. See **Figure 23-55**.

In the Case IH 7130–9130 combines, the high-pressure relief valves are set 10% higher than the high-pressure cutoff valve setting. Once a high-pressure relief valve opens from a sudden pressure spike, it is possible to see the high pressure drop 10% to the high-pressure cutoff valve setting.

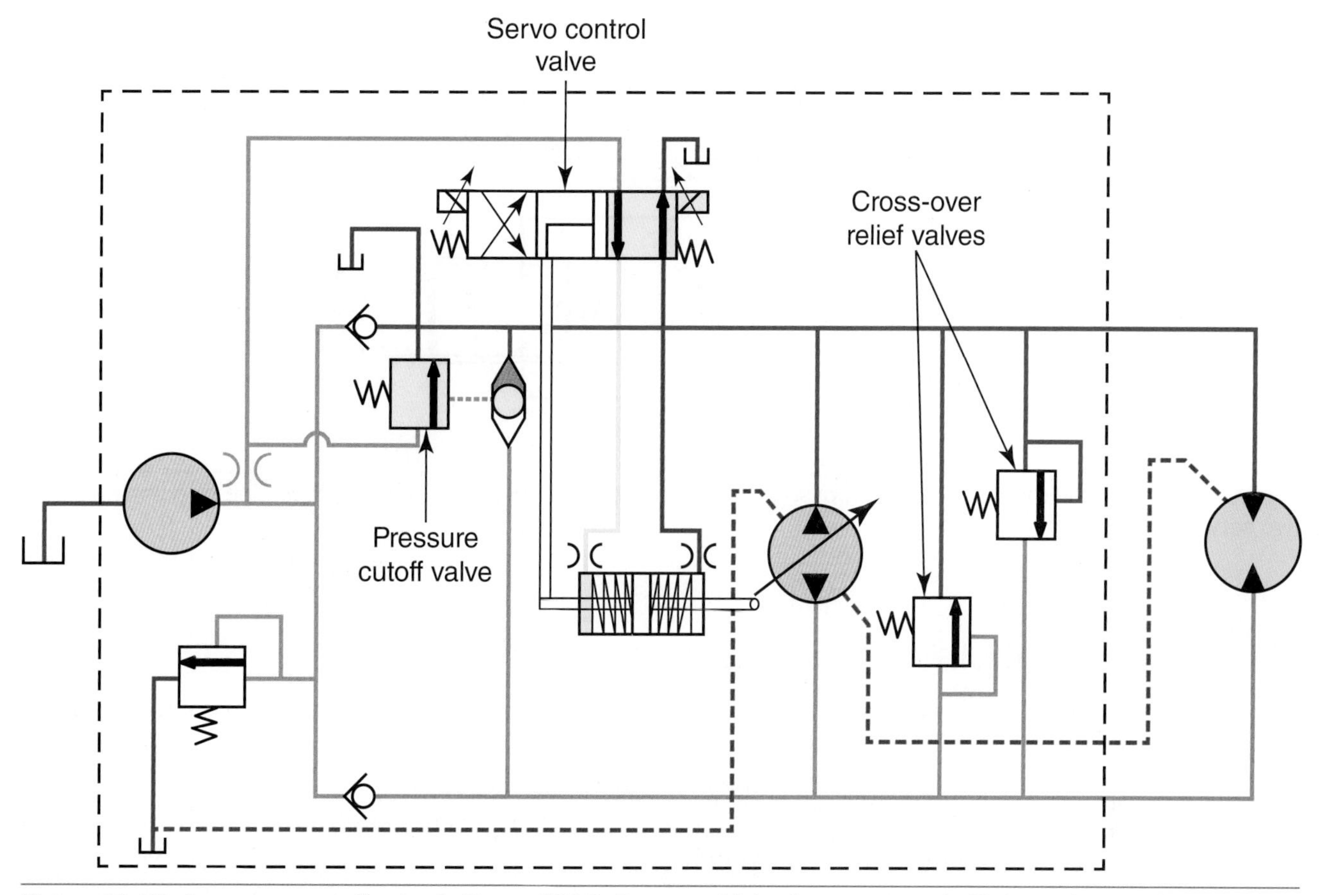

Figure 23-55. A pressure cutoff valve is like an IPOR. It senses high oil pressure and will dump the supply to the servo control valve when pressure reaches the relief setting.

One benefit of the system in Figure 23-55 is that it contains two separate high-pressure controls in the pump and not the motor. Therefore, if a blockage like a pinched high-pressure line were to occur between the pump and motor, the pump would still have high-pressure relief protection. If the system contained high-pressure relief valves only in the motor, any type of blockage between the pump and motor would result in overpressurization, damaging the pump and/or high-pressure drive plumbing.

Electronic Anti-Stall Control

Transmissions can be equipped with an ***electronic anti-stall control***. The two-position destroke solenoid is normally closed. An ECM will monitor the machine's engine speed. When engine speed drops, the ECM will modulate the solenoid. As the solenoid is energized, it hydraulically connects the pump's forward and reverse servo circuit, which reduces the pump's displacement. The control allows an ECM to electronically limit stalling of the pump. The larger the difference between the engine speed setting and the actual engine speed, the more the ECM will modulate the solenoid. The control enables the operator to operate the machine at full power and reduce the possibility of lugging the engine. See Figure 23-56.

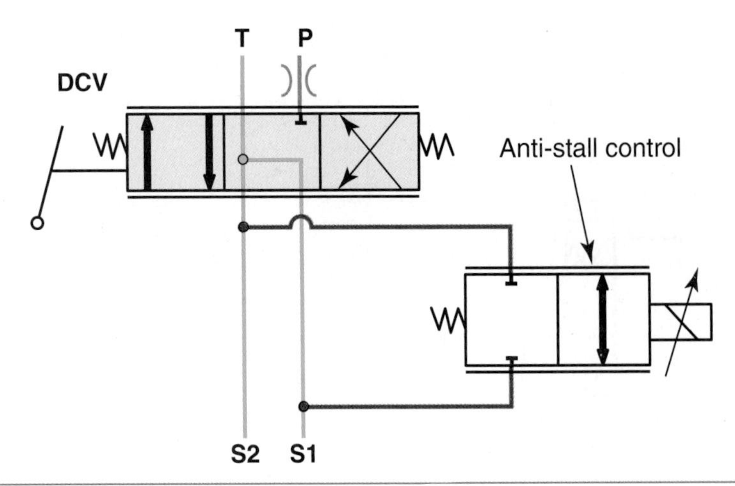

Figure 23-56. An electronic anti-stall control solenoid will prevent the HST from lugging the engine by combining the two servo pressures.

Electronically Controlled Displacement

The previous illustration exhibited an electronic anti-stall control solenoid used in conjunction with a manual directional control valve. When the HST pump is electronically actuated, the ECM can provide a similar benefit by reducing the current to the HST pump's solenoids, eliminating the need for an anti-stall control solenoid.

Accumulator Effect

This chapter has detailed numerous mechanisms used to control or limit high drive pressure, for example, IPOR, pressure cutoff valve, and anti-stall destroke solenoid. If a traditional hydrostatic transmission is allowed to stall, and if the closed loop uses hoses instead of steel tubing, it is possible for the pump to cavitate. The condition occurs when extremely high drive pressures are achieved, causing the hydraulic hoses to swell. **Chapter 15** details this effect, called *volumetric expansion*.

During normal operation, the motor is rotating and exhausting the return oil into the closed-loop low-pressure leg. The return loop helps supercharge the piston pump. However, if the motor stalls while building high pressure, the high pressure can cause the hydrostatic drive hose to swell, which can cause a momentary point of pump cavitation. During the brief moment of cavitation, the pump is not receiving any of the return oil while the motor is stalled. In some unique applications, accumulators are added to the charge circuit to prevent pump cavitation anytime a motor is stalled.

Summary

- ✓ Hydrodynamic and hydrostatic drives use fluid energy to propel a machine.
- ✓ Torque converters, fluid couplings, and torque dividers are hydrodynamic drives.
- ✓ Open-loop HSTs draw intake oil from only the reservoir.
- ✓ The intake of a closed-loop HST pump draws oil from the HST motor's return.
- ✓ A single-path HST uses one pump and one motor.
- ✓ Dual-path HSTs use one pump and one motor to drive each side of a machine.
- ✓ HSTs increase the machine's productivity.
- ✓ HSTs are noisy, sensitive to heat, and are mechanically inefficient.
- ✓ HST charge pumps supply oil to the servo control valve, supercharge the piston pump, provide make-up oil to the closed loop, and help cool the circuit by replenishing the closed loop's oil.
- ✓ The feedback link is actuated by the swash plate for the purpose of holding the servo control spool in the last commanded position.
- ✓ A flushing valve senses the two legs of a closed-loop HST.
- ✓ When the HST is operated, the drive pressure causes the flushing shuttle valve to shift, which allows charge oil to be controlled by the lower-pressure flushing relief valve.
- ✓ The flushing relief valve dumps oil into the motor's case. The oil is then routed to the pump's case for the purpose of purging and replacing the hot oil in the HST pump and motor.
- ✓ A HST's charge pressure is highest when the transmission is in neutral and the engine is at high idle.
- ✓ HST high-pressure relief valves dump the oil into the opposite leg of the closed loop.
- ✓ Closed-loop HST filtration is rare because the filter must be able to filter oil in two different directions and filter oil at pressures up to 7000 psi (483 bar).
- ✓ HST case drain is often cooled and sometimes filtered.
- ✓ Variable-speed motors can be inline axial, bent axis, or radial piston style.
- ✓ Variable-speed motors can have two different specific speeds or can be infinitely variable.
- ✓ A HST with a variable-displacement pump and a variable-speed motor will vary the pump from minimum displacement to maximum displacement, then vary the motor from maximum displacement to minimum displacement in order to increase the machine speed from slow to fast.
- ✓ A single-speed motor is sometimes accidentally installed 180° upside down, causing the HST to operate backward.
- ✓ Inching valves and manual bypass valves can be used to help shift a two-, three-, or four-speed manual gearbox while the machine is sitting still.
- ✓ Inching valves can be used in conjunction with the HST pump or the HST motor.
- ✓ IPOR valves sense high oil pressure and dump the supply oil to the servo control valve, neutralizing the HST.
- ✓ Many late-model HSTs use electronic controls to prevent the HST from stalling or lugging the engine.
- ✓ Hydraulic drive hoses can swell due to excessive drive pressure, causing the HST pump to cavitate.

Technical Terms

closed-loop HST
creeper control
dual-path HST
electronic anti-stall control
fluid coupling
flushing relief valve
flushing valve
forward and reverse lever
hydrodynamic drive
hydrostatic drive
inching valve
integral HST
internal pressure override (IPOR)
manual bypass valve
neutral relief valve
neutral safety switch
open-loop HST
pressure cutoff valve
pressure-release solenoid
servo oil
single-path HST
single-speed motor
split HST
torque converter
torque dividers
two-speed motor

Review Questions

Answer the following questions using the information provided in this chapter.

1. Which of the following is a hydrostatic drive advantage?
   A. Quiet.
   B. Component cost.
   C. Mechanical efficiency.
   D. Overall productivity.
2. In a closed-loop hydrostatic transmission, after the oil leaves the motor, it is sent to the _____.
   A. reservoir
   B. high-pressure relief valve
   C. pump inlet
   D. servo piston
3. Which type of machine uses a hydrostatic motor at each of the drive wheels and also uses a complex steering mechanism and rear caster tires?
   A. Combine.
   B. Swather.
   C. Sprayer.
   D. Dozer.
4. Which of the following machines is *not* considered a dual-path hydrostatic drive?
   A. Skid steer.
   B. Dozer.
   C. Track loader.
   D. Combine.
5. Which of the following uses high velocities and low pressures for transmitting power?
   A. Hydrostatic drive.
   B. Hydrodynamic drive.
   C. Collarshift drive.
   D. Synchroshift drive.
6. Which of the following machines commonly uses open-loop hydrostatic drives?
   A. Combines.
   B. Agricultural tractors.
   C. Dozers.
   D. Excavators.
7. A cradle bearing–style HST uses _____ servo piston(s).
   A. zero
   B. one
   C. two
   D. three
8. A trunnion bearing HST uses _____ servo piston(s).
   A. zero
   B. one
   C. two
   D. three

9. What are the two responsibilities of the HST control valve?
   A. Relieve case drain, and relieve charge pressure.
   B. Relieve drive pressure, and relieve charge pressure.
   C. Increase/decrease speed, and change forward or reverse directions.
   D. Isolate high pressure/charge pressure circuits, and cool the HST.
10. What has to happen in order to get charge pressure to drop 30 psi (2 bar)?
    A. The high-pressure shuttle valve must shift.
    B. The HST pump control valve must be actuated.
    C. High-pressure relief oil must act directly on top of the charge relief.
    D. Engine speed must be raised to high idle.
11. What is the reason the charge pressure drops 30 psi (2 bar)?
    A. To aid in cooling the HST.
    B. To save horsepower.
    C. To lower pump inlet vacuum.
    D. To seat the motor's high-pressure relief valves.
12. Whenever the HST is driven forward, which charge pressure relief valve is controlling charge pressure?
    A. Pump charge relief valve.
    B. Flushing relief valve.
    C. High-pressure relief valve.
    D. IPOR valve.
13. A hydrostatic charge pump generally should provide at least how much flow in relationship to the pump's flow?
    A. 19%.
    B. 42%.
    C. 85%.
    D. 100%.
14. All of the following are used to describe the hot oil purge valve, *EXCEPT*:
    A. flushing valve.
    B. replenishing valve.
    C. shuttle valve.
    D. case drain valve.
15. Which of the following pressures will *not* act on the hot oil purge valve?
    A. Drive pressure.
    B. Charge pressure.
    C. Servo pressure.
    D. All of the above.
16. A HST pump has an efficiency of 85%, and the HST motor has an efficiency of 85%. If the HST pump flow at high idle is 50 gpm, what would be the required charge pump flow rate?
    A. 13.875 gpm.
    B. 15 gpm.
    C. 27.75 gpm.
    D. 30 gpm.
17. How many charge relief valves are normally used in a closed-loop HST?
    A. Zero.
    B. One.
    C. Two.
    D. Three.
18. Technician A states the high-pressure relief valves can be located in the HST pump. Technician B states that the high-pressure relief valves can be located in the HST motor. Who is correct?
    A. Technician A.
    B. Technician B.
    C. Both A and B.
    D. Neither A nor B.
19. When high-pressure oil opens the high-pressure relief valve, where is the oil dumped?
    A. Directly to the case.
    B. Directly to charge.
    C. Directly to reservoir.
20. What conditions result in the highest charge pressure?
    A. Low idle, neutral.
    B. Low idle, forward or reverse.
    C. High idle, neutral.
    D. High idle, forward or reverse.
21. The IPOR valve senses _____ pressures.
    A. case
    B. charge
    C. servo
    D. drive

22. Once the pressure reaches the IPOR spring setting, which of the following pressures will the IPOR valve dump to the tank?
    A. Case pressure.
    B. Servo control valve supply oil.
    C. Drive pressure.
    D. None of the above.
23. How does the anti-stall solenoid hydraulically reduce drive pressure?
    A. It dumps charge pressure.
    B. It dumps high-pressure relief.
    C. It connects both legs of the closed loop.
    D. It connects both servo pressures.
24. The Case 2100–2300 series combine's two-speed HST motor will downshift from high speed to low speed when a specific drive pressure is reached. This several thousand psi of drive pressure will act on the control valve assembly and cause the control spool to shift back to low. When the control valve is in the neutral position, why is only 300 psi required to shift the control valve assembly to the low position?
    A. Pressure intensification.
    B. Length differences of the control spool and piston.
    C. Difference in areas of the roller needle and piston.
    D. Because of the sequencing valve.
25. A Caterpillar K series wheel loader uses which type of HST configuration?
    A. One pump and one motor.
    B. One pump and two motors.
    C. Two pumps and one motor.
    D. Two pumps and two motors.
26. When an inline hydrostatic piston motor's swash plate is in a neutral angle, what will be the result?
    A. The pump will drive the motor at a higher speed.
    B. The pump will drive the motor at a lower speed
    C. The pump will not drive the motor.
    D. Motor direction will reverse.
27. What is the name of the linkage that connects the servo control valve to the swash plate?
    A. Torque limiter link.
    B. Hi-low speed link.
    C. Feedback link.
    D. None of the above.
28. What is the purpose of the linkage that connects the control valve to the swash plate?
    A. To reposition the spool so that it can hold the pump swash plate at the exact angle requested by the operator.
    B. To tell the pump to generate flow.
    C. To relieve charge pressure.
    D. All of the above.
29. A combine's foot-and-inch pedal is similar to what type of pedal found on tractors?
    A. Service brake.
    B. Transmission clutch.
    C. Engine decelerator.
    D. None of the above.
30. A technician just replaced a single-speed hydrostatic motor. After installing the motor, he or she found that the propulsion lever works backward. What is wrong?
    A. The motor was installed 180° upside down.
    B. The wrong motor was installed.
    C. The linkage was connected backward.
    D. The wrong servo control valve was installed.

Chapter 24

Hydrostatic Drive Service and Diagnostics

Objectives

After studying this chapter, you will be able to:

- ✓ List the steps required to adjust the neutral position on two different styles of hydrostatic pumps.
- ✓ List the steps for starting a hydrostatic transmission after replacing a pump or motor.
- ✓ Describe the process for troubleshooting the following hydrostatic transmission symptoms:
  - Overheating.
  - Low power in both directions.
  - Low power in just one direction.
  - Machine will not move.

Centering Adjustments

As mentioned in **Chapter 23**, hydrostatic transmissions do not coast or freewheel by design. If a hydrostatic pump is misadjusted, the pump will not have a true neutral, but instead will creep either forward or rearward. Regardless of whether the pump is a single-servo cradle bearing style or a dual-servo trunnion bearing style, there are two types of adjustments for setting the pump to a neutral position. One adjustment is to set the control valve to neutral, and the other adjustment is to set the single-servo piston or dual-servo pistons to neutral.

Safety precautions must be taken prior to making any centering adjustments, which will cause the machine to move. Follow the manufacturer's safety procedures when making pump adjustments. Many manufacturers specify to place the machine on jack stands to prevent personnel or equipment from being accidentally run over. If the pump is electronically controlled, the manufacturer may specify to unplug one or more speed sensors to prevent error codes for uncommanded machine propulsion

In addition to a misadjusted pump, two other conditions can cause a pump to creep: misadjusted pump linkage or a shorted solenoid circuit. Both of these potential culprits must be considered as a possible source for a creeping transmission. In some manually controlled dual-path hydrostatic drives, such as those on older skid steers and older swathers, the process for adjusting the dual-pump linkage can be complex. The manufacturer service literature details the methods for centering both pumps.

For electronically controlled pumps, remove the pump control valve solenoid coils and see if the machine still creeps in neutral. If the machine still creeps, a shorted solenoid circuit is not at fault.

Single-Servo Cradle Bearing Pump

Newer HST pumps use a single-servo piston to control a cradle bearing design swash plate. The servo piston will actuate the pump's swash plate to provide forward and reverse propulsion. The pump requires two different adjustments for centering the pump to neutral:

- Centering the servo piston.
- DCV null adjustment.

Centering the Servo Piston

A single-servo pump includes a threaded screw that is used for placing the servo piston in an exact neutral position. See **Figure 24-1**. Note that a jam nut holds threaded rod in position. As the threaded rod is screwed in or out, the position of the servo piston shifts, changing the angle of the pump's swash plate. See **Figure 24-2**.

Prior to adjusting the servo piston, a technician must first hydraulically loop the servo piston control pressures together. A hydraulic jumper hose is attached to the forward servo test port and the reverse test port, S1 and S2.

Warning

Note that as this adjustment is made, the machine will move. For this reason, safety precautions must be taken. Follow the manufacturer's safety precautions.

Figure 24-3 shows a pump on a stand, not a running transmission. However, the photo shows the S1 and S2 pressure ports looped together. Gauges are installed to measure pressure at the two drive pressure ports. This setup is used when adjusting the servo piston's position to equalize the two drive pressures. The service literature will specify the appropriate size of pressure gauge for measuring the two drive pressures. Because most manufacturers recommend having the machine placed on jack stands, it normally takes only a small amount of drive pressure to propel the unloaded tires or tracks.

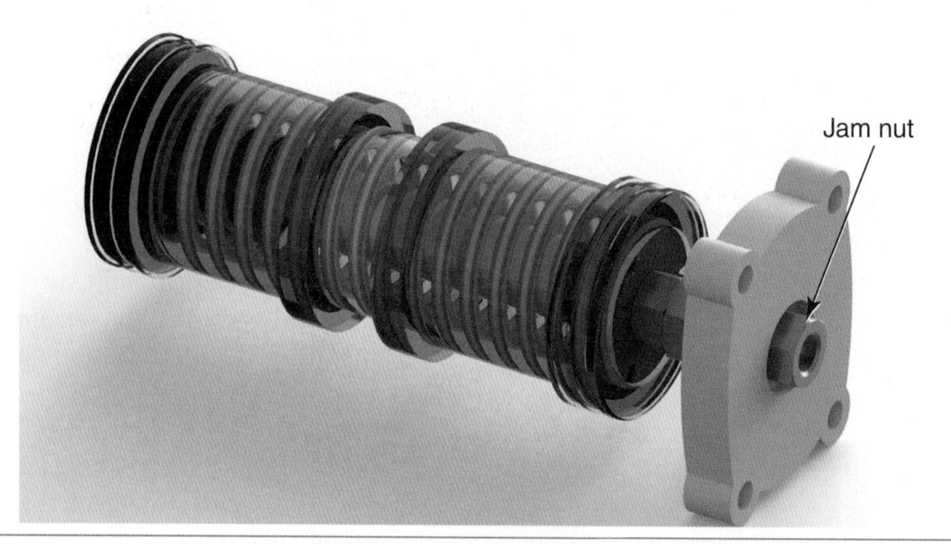

Figure 24-1. A single-servo cradle bearing hydrostatic pump contains an end plate with a threaded adjustment for centering the servo piston.

Figure 24-2. As the threaded adjustment is turned, it causes the swash plate to actuate. The threaded rod is adjusted until the swash plate is centered in a neutral position.

Figure 24-3. The servo is adjusted by first connecting the forward and reverse servo pressures together with a looped hydraulic hose. While monitoring the drive pressures, adjust the servo until the pressures are equal.

Caution

If a pressure gauge with a low pressure range is being used, be sure to avoid applying the brakes. Applying the brakes would cause the drive pressure to reach stall pressure, which would damage low-pressure gauges.

While adjusting the servo piston's threaded adjustment, technicians monitor pressures on both legs of the closed loop circuit. With the servo piston pressures hydraulically looped together, the technicians adjust the servo piston until the pressures of both legs of the closed loop are equal. The pump in **Figure 24-3** has the benefit of being able to simultaneously monitor both forward and reverse drive pressures.

Single-Servo Pumps with a Single Drive Pressure Test Port

Some single-servo pumps use a shuttle valve and a single drive pressure test port. See **Figure 24-4**. A technician cannot measure both forward and reverse drive pressures simultaneously with separate pressure gauges because the shuttle valve directs drive pressure to a single test port, limiting the technician to measuring only drive pressure rather than individual forward drive pressure or individual reverse drive pressure. To center a pump with this configuration, a couple of extra steps are required. As previously mentioned, any time a technician is making pump neutral-adjustments, the machine will literally drive forward or rearward. Precautions must be taken to prevent the machine from running over personnel or equipment. Place the machine on jack stands as directed by the manufacturer's service procedure.

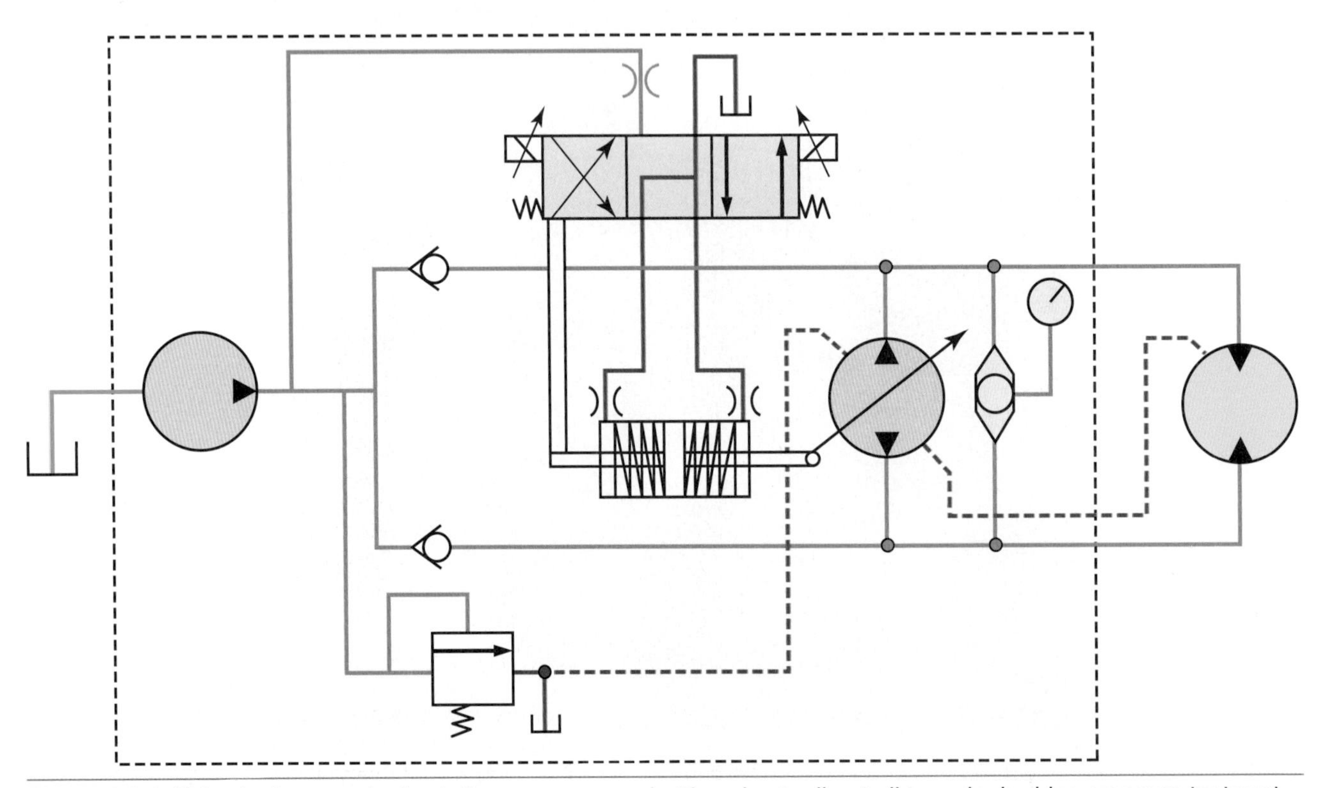

Figure 24-4. This single-servo hydrostatic pump uses a shuttle valve to direct oil to a single drive pressure test port.

Although this type of pump has only one drive pressure test port, it is still equipped with two servo pressure test ports. After connecting the hydraulic jumper hose to the two servo pressure test ports, loosen the jam nut to the servo piston's adjustment bolt. Most adjustment bolts are turned with an Allen wrench. Gradually turn the adjustment bolt in one direction until the pump reaches maximum drive pressure. As previously mentioned, drive pressure will be lower than normal because the machine is off the ground and has very little load. Stop moving the adjustment bolt when the machine reaches maximum drive pressure. Use a marker to mark the exact location of the Allen wrench. Next, move the adjustment bolt in the opposite direction while monitoring the drive pressure gauge. The goal is to gradually sweep the adjustment in the opposite direction until maximum drive pressure is reached. Mark the exact position of the Allen wrench. See **Figure 24-5**. Next, place the servo piston in the exact neutral position by moving the Allen wrench exactly in between the two marked locations. This procedure is used for centering the differential steering hydraulic pump on a MT Challenger tractor.

DCV Null Adjustment

The second neutral adjustment on a single-servo piston is called the null adjustment. The null adjustment centers the spool inside the solenoid valve assembly. The null adjustment commonly uses two Allen-head screws. One screw acts like a jam nut and locks the null adjustment in a neutral position. A solenoid valve assembly with the two Allen-head screws is located in the center of the housing. See **Figure 24-6**.

The null adjustment screw is technically not an adjustment screw, but an eccentric shaft. As the eccentric is rotated, the solenoid's spool valve is adjusted in and out of neutral. See **Figure 24-7**.

Figure 24-5. When the servo piston on a single-servo hydrostatic pump equipped with just one drive pressure test port is being centered, the position of the adjustment screw must be marked on the pump at the point when the pump initially begins to drive forward. It must also be marked when the pump initially begins to drive rearward. The adjustment screw is then centered between the two marked lines.

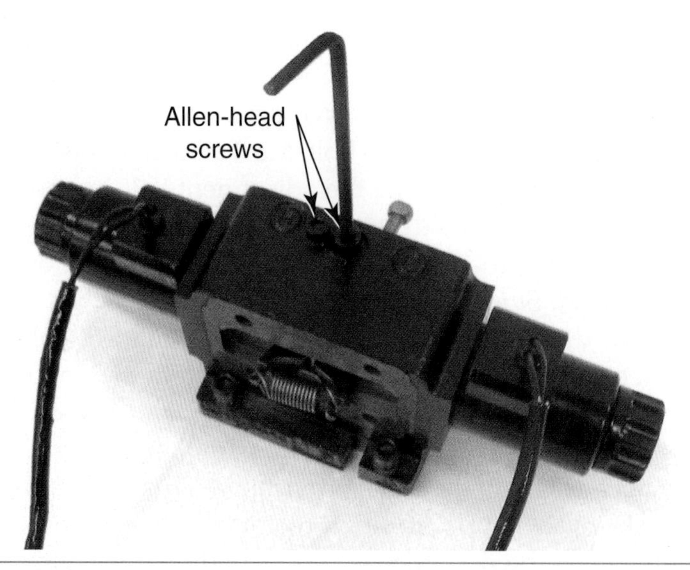

Figure 24-6. A single-servo hydrostatic piston pump contains an eccentric shaft that is centered with an Allen wrench. The adjustment process is called the null adjustment.

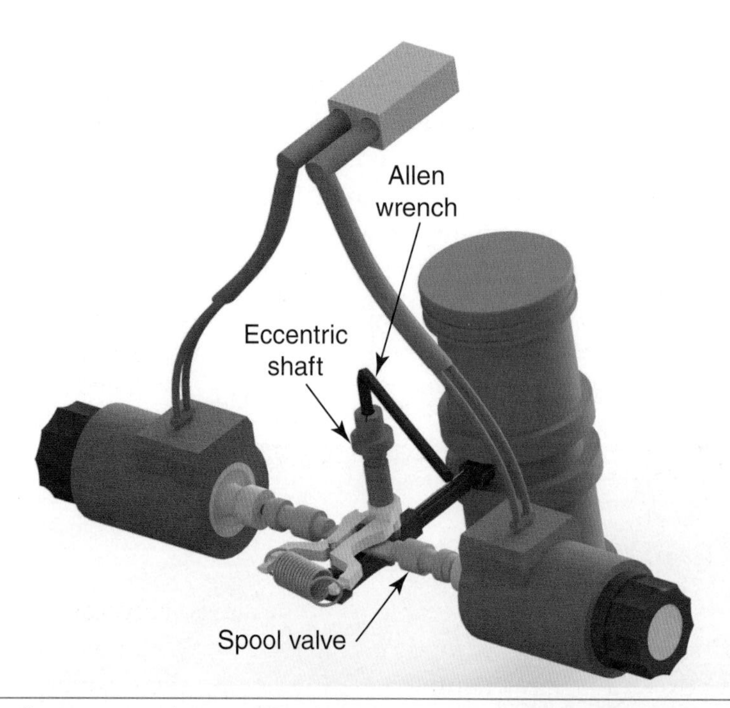

Figure 24-7. The null adjustment requires an Allen wrench to rotate an eccentric shaft for the purpose of centering the solenoid's spool. The adjustment does not have threads and can be rotated freely through 360° of rotation. The adjustment can be very sensitive, causing the servo pressures to build with very little movement to the eccentric shaft.

Two low-pressure gauges must be installed in the servo test ports (S1 and S2) before the adjustment is made. See **Figure 24-8**. The goal is to adjust the null adjuster until the two servo control pressures equalize. After the outside locking screw is loosened, an Allen wrench is used to adjust the center null eccentric. The null eccentric is very sensitive and requires slow and careful turning of the adjuster.

Figure 24-8. Adjusting the pump's null adjustment requires monitoring the forward servo pressure and reverse servo pressure. Note that this photo was taken with the pump off of the tractor. When the adjustment is made on a live machine, the eccentric is rotated until the pressures equalize, for example both gauges read 30 psi (2 bar). Any time the pressures are unequal, the pump is stroked in either a forward or reverse position and the drive tires will be moving.

Warning

If a technician is careless and backs the lock screw out too far, it is possible for the eccentric rod to become dislodged. If this happens, oil pressure can spew out of the housing and the machine will default to maximum forward or reverse propulsion. Therefore, be careful to follow the manufacturer's service and safety specifications.

Dual-Servo Trunnion Bearing Pump Adjustments

The Eaton dual-servo trunnion bearing hydrostatic pumps also have two different adjustments for centering the pump to a neutral position. The pump's control valve has a neutral adjustment, and the position of the two servo pistons affect the pump's swash plate angle. If the servo pistons are not centered when the control valve is in the neutral position, the pump's swash plate will not be at the 0° position.

Adjusting the Pump DCV Neutral Position

The pump's control valve uses a centering spring to hold the control spool in a neutral position. Similar to the electronic null adjustment, the control spool needs to be adjusted so that the servo control oil pressures are balanced. A cap commonly covers the DCV centering spring. The centering spring has an adjustment screw for placing the spool in a neutral position. See **Figure 24-9**.

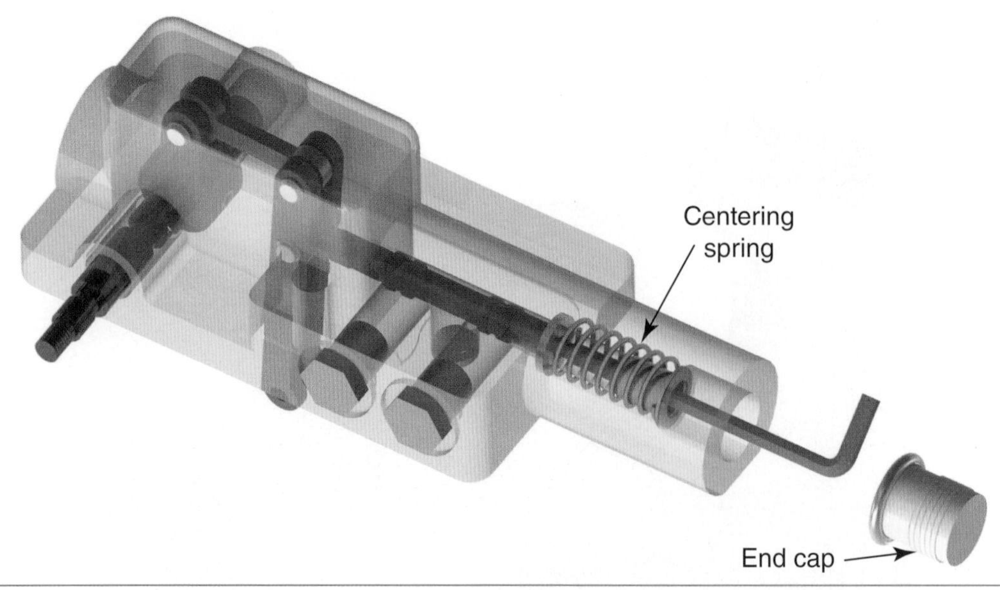

Figure 24-9. An Eaton dual-servo pump can contain a manual control valve that is centered with a spring. The end cap is removed to gain access to the centering spring adjustment.

Adjusting the Dual-Servo Pistons

Dual-servo piston pumps have threaded servo caps installed over both of the servo pistons. As the servo caps are installed, a depth micrometer or a dial indicator is used to measure the swash plate. Both sides of the swash plate are measured. Eaton states that the two measurements should be within 0.0005″ of each other. It is critical to mark the caps before removal to aid the assembly process. See **Figure 24-10**.

Shaft Run-Out

Hydrostatic pumps and motors are mounted in different configurations. If the mount or adapter is incorrectly machined or not flat, the shaft splines can wear prematurely. Some manufacturers provide detailed instructions for checking the pump mount surface and the shaft drive for run-out. The measuring tool of choice is a dial indicator. If the shaft's splines fail prematurely, the pump or motor can be disassembled and the shaft replaced. Be sure to fix the problem that caused the shaft to fail prematurely. Installing a new shaft is typically cheaper than purchasing a new pump or motor. If the failure is prematurely worn shaft splines, most technicians will simply replace the shaft, even if the hydrostatic transmission is a closed-loop drive, because the rotating group is usually unrelated to the worn shaft splines.

Note

With good maintenance, service, and contamination control, a pump can often last well over 10,000 hours.

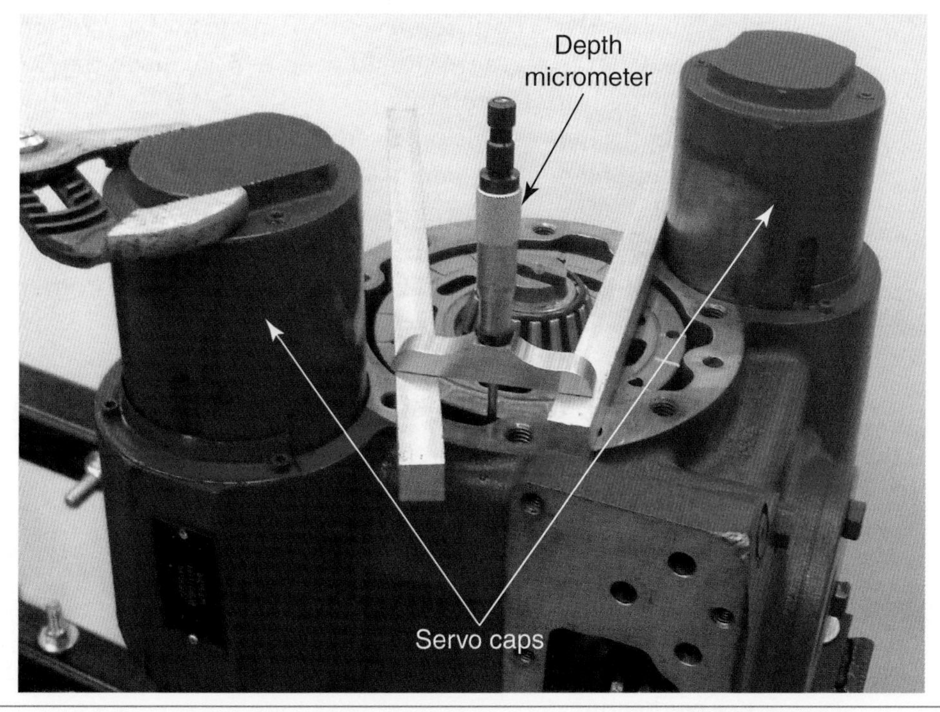

Figure 24-10. The installation of the servo caps adjusts the pump's swash plate angle. Prior to removing the caps, mark their location with a permanent marker. A depth micrometer or a dial indicator is used to center the swash plate while installing the servo caps.

Startup (Commissioning) of a Hydrostatic Drive

Many mistakes can be made during the installation of a new pump or motor. The National Fluid Power Association labels the process of returning a system to service following pump or motor replacement as ***commissioning*** the hydraulic system. If the component is not put back into service correctly, the life of the component can be drastically reduced. Be sure to follow the manufacturer's startup procedures when installing a new hydrostatic pump or motor, even if the component was only removed to have a new shaft installed.

The following is an example of one manufacturer's startup procedure:

1. Prevent the machine's engine from running by removing a key or pulling a fuse.
2. Fill the case (of the pump and or motor) with oil via the case drain plug.
3. Install a pressure gauge in the charge pressure test port.
4. Place the multispeed range gearbox in the neutral position (manufacturer's instructions differ regarding whether or not to also release the brake and chock the machine's wheels). The startup procedure should be designed so that the hydrostatic transmission has no load on the motor's shaft.
5. Start and run the engine at low idle while monitoring charge pressure. If pressure fails to build to specification in less than 15 seconds, shut off the machine, and diagnose the cause.
6. If charge pressure meets specification, increase the engine speed to 1500 rpm to purge air through the system.

7. Slowly move the propulsion lever halfway forward and operate 4 to 5 minutes. Charge pressure should drop a little due to the flushing relief when the hydrostat is rotating.
8. Slowly move the propulsion lever back through neutral and halfway reverse and allow it to operate 4 to 5 minutes.
9. Return the propulsion lever back to neutral and set the parking brake.
10. Repeat the startup procedure three times.
11. Operate all the control pressure circuits to purge air from the system.
12. Refill the reservoir as needed.
13. Check all pressures.
14. Replace the hydraulic filters after 10 to 15 hours of operation.

Hydrostatic Drive Diagnostics by Symptom

Hydrostatic transmissions can exhibit numerous types of symptoms, making it a challenge to accurately diagnose a problem. The last half of this chapter provides instruction for diagnosing HSTs based on the following symptoms and presents important tips relating to troubleshooting HSTs:

- Overheating.
- Transmission creeps forward or backward while in neutral.
- Low power or sluggish in forward and reverse.
- Low power or sluggish in just one direction.
- Coasting or freewheeling transmission.
- Transmission will not move in either direction.
- Charge pressure is too high.

Overheating

Many off-highway machines have a transmission overheating indicator light, monitor, and/or buzzer to alert the operator that the machine is overheating. When the transmission's overheating light is illuminated, a technician needs to first determine if the problem is with the indicator or an actual overheating transmission. The manufacturer service literature should provide the temperature threshold for illuminating the overheating light, as well as the location of the temperature sensor. One off-highway machine will illuminate the light once the oil reaches 200°F.

If manufacturer service literature lacks this information, a rule of thumb is that the oil should not be more than 100°F warmer than the ambient temperature. For example, if a combine is harvesting in July and it is 97°F outside, the rule of thumb states the oil should not exceed 197°F. Use an infrared thermometer to measure the temperature near the oil temperature sending unit. If the temperature is well below the manufacturer's specified threshold, diagnose it as an indication problem rather than true overheating. In **Figure 24-11**, the sending unit could be shorted, the relay could be shorted, and a short to ground could exist between the sending unit and the instrument panel.

If the transmission oil temperature is indeed too hot, it could be caused by a variety of factors: a worn pump or motor, a restricted oil cooler, a lazy oil cooler bypass, or a purge relief valve swapped with the neutral charge relief valve.

Many transmissions cool the case drain as previously discussed in **Chapter 23**. If the case drain oil cooler becomes internally restricted or externally plugged, the transmission will overheat. The transmission case drain pressure should be measured. If the case pressure is the same value as the oil cooler bypass valve, for example 60 psi (4 bar), the next step would be to diagnose the oil cooler. If possible, measure the case drain oil flow after the bypass and after the oil cooler. See **Figure 24-12**.

If the oil cooler is not plugged, inspect the oil cooler bypass valve. More than one technician has made the costly mistake of presuming the excessive leakage in the pump and motor are causing the system to overheat. They needlessly replaced the pump and motor, only to later find a malfunctioning oil cooler bypass valve. The oil will take the path of least resistance, and if it is easier for the oil to bypass the cooler and go straight to the reservoir, then it will.

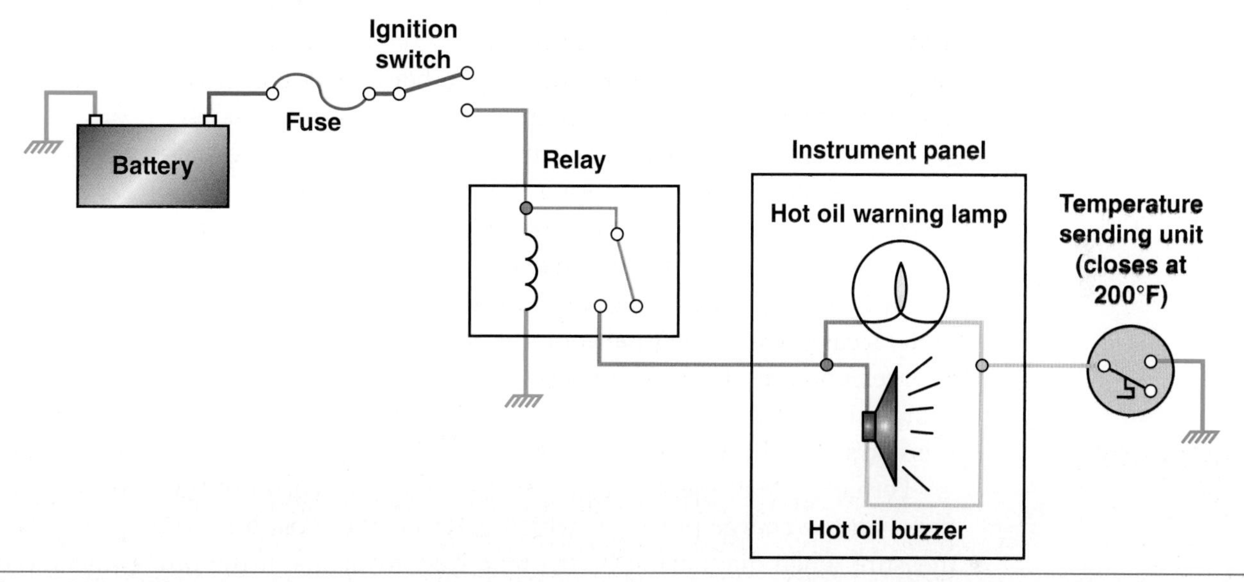

Figure 24-11. HSTs commonly employ instrumentation to warn operators when the oil temperature has exceeded the normal operating range.

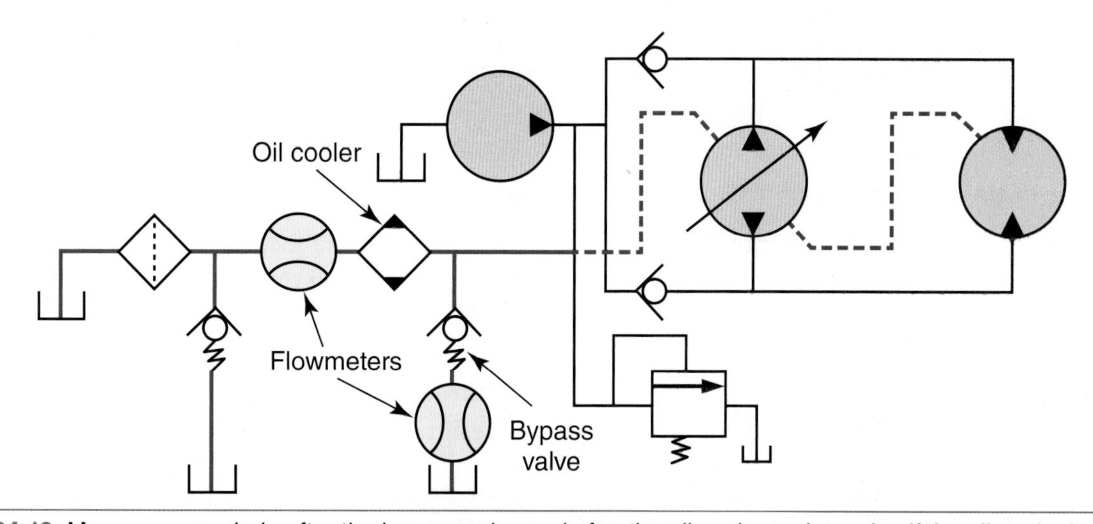

Figure 24-12. Measure case drain after the bypass valve and after the oil cooler to determine if the oil cooler is plugged.

Measuring Case Drain Flow

If the oil cooler and oil cooler bypass valve are good, the next focus is to determine if the transmission has excessive internal leakage. Flow rating case drain is a little tricky. Some experts state that, if the flushing relief valve is located in the motor, measuring only the motor's case drain will essentially equal the total charge pump flow. For example, when the machine is moving, the flushing relief valve normally dumps the charge oil into the motor's case. Therefore, if the flushing relief valve is located in the motor, it is only possible to isolate the pump's case drain flow and determine if the pump has excessive case drain. The motor's case drain will equal charge pump flow. Or, if the flushing relief valve is located in the pump, the case drain of the pump will equal charge pump flow, and it will only be possible to isolate the motor's case drain to determine the motor's case drain flow.

However, due to the complexity of some charge circuits, this is not always 100% accurate. The following are some design variations that affect measurement of case drain flow:

- Some charge pumps are used for other machine functions such as pilot controls.
- Some hydrostatic drives use a separate pump (shared with other systems) for supercharging the hydrostatic closed loop.
- Some hydrostatic drives direct the flushing relief valve's oil to a pressure compensator valve before dumping into the case.

At least one manufacturer has confirmed that measuring the hydrostatic motor's case drain flow (that contains a flushing relief valve) will indeed yield a larger volume of case drain flow. Due to the potential complexity of some charge circuits, it is critical to use solid OEM data prior to measuring any transmission's case drain flow.

Perhaps one of the best tests for diagnosing a worn-out transmission is to focus on the charge pressure, which is by far the most important pressure to measure when diagnosing hydrostatic transmissions. If the pump and motor have too much internal leakage, it can cause charge pressure to drop and case pressure to rise. However, it is possible that an abnormally high case pressure is due to a restriction in the case drain oil cooler or case drain filter, misleading a technician into thinking the pump and motor are excessively worn. Charge pressure is the key to diagnosing a suspect transmission. The role of charge pressure in diagnosis will be discussed in more depth later in this chapter.

Hot-Oil Purge Valve Mislocated

If the hot-oil purge valve was swapped in place of the neutral charge relief, the transmission would not cool. Anytime the propulsion lever is moved past neutral, the charge pressure should drop 30 psi (2 bar). If the charge pressure is 30 psi (2 bar) low in neutral, it is possible that the hot-oil purge relief valve and the neutral charge relief valve were swapped. Be sure to place the hot-oil purge relief in the correct location, as designed by the manufacturer, or the transmission will overheat.

Case Study

Overheating Hydraulic Oil When Driving under Light Loads

A skid steer OEM had experienced overheating hydraulic systems in skid steers any time the machine was being propelled under light loads. The machine manufacturer found that the HST motor's flushing valves were not opening because drive pressures were relatively low. To remedy the situation, the machine manufacturer recommended driving the machine in a fast figure-eight pattern to cool the oil.

The machines were equipped with dual-path HSTs (one pump and one motor for each side). If such a machine is propelled fast and steered in a tight figure-eight for 15 minutes, it causes both shuttle valves in the HST motors to open alternately, cooling the dual-path HST. Keep in mind that the symptom only occurred under light loads. Also note that the HST supplier was hesitant to use flushing shuttle valve springs with a weaker pressure setting because the weaker springs might prevent the HST from reaching proper operating temperature in certain climates. In addition, this was the only machine application that was experiencing the overheating problem under light loads.

Creeping Hydrostatic Transmission

A misadjusted transmission may continually creep forward or in reverse, indicating that the transmission will not remain in neutral. With the propulsion lever in the neutral position, both legs of the closed loop should have equal pressure, which equals charge pressure. If the propulsion lever is in neutral and one of the legs of the closed loop has a higher pressure than the other, the transmission is not truly in neutral. Details were provided earlier in this chapter for centering the hydrostatic pump. The focus should be on the pump and the directional control valve. It is very unlikely that a motor would cause a transmission to creep forward or rearward.

Low Power or Sluggish in Forward and Reverse

One of the most common and difficult HST symptoms to diagnose is an HST with low power. The operator may describe the transmission as sluggish or unable to pull under a load. The first step should be to determine if the low power occurs in both forward and reverse. The following seven areas must be investigated to determine the root cause of the lack of power.

Engine

Before suspecting that the transmission is at fault, a technician needs to be sure that the engine is not at fault. A poor-performing engine can be mistaken for a weak hydrostatic transmission. Therefore, the engine must be investigated before the transmission is diagnosed.

One manufacturer's service literature states to place the machine's gearbox in the high gear range. With the brakes applied, stroke the propulsion lever. If the engine speed drops due to heavy load, the machine is performing correctly. If not, determine if the engine is producing full power and rpm. If the engine is performing properly, proceed to diagnosing the HST.

Transmission Controls and Other Features and Options

Depending on the transmission's configuration, the transmission can respond differently to the test listed above. For example, **Chapter 23** explains that if the machine has an IPOR valve, the pump will be destroked once drive pressure reaches the IPOR valve's setting. If the pump is configured with an ECM anti-stall mode, the transmission would also appear to have low power when the ECM destrokes pump.

If the transmission did not respond properly by drawing the engine speed down, the next step would be to eliminate other hydrostatic drive functions. For example, if the machine is a combine equipped with a power guide axle (PGA), the PGA should be shut off and the transmission should be retested. If the transmission can then draw down engine speed, the PGA must be diagnosed. However, keep in mind that when the PGA is engaged, the pump is sending oil to the main hydrostatic motor, plus the two rear PGA motors. The additional two rear motors should provide much more torque with the PGA engaged, rather than when the PGA is disengaged. Engaging the PGA will also lower drive pressure due to the increased motor displacement.

If the engine produces adequate power and if no other control, such as a power guide axle, is affecting the lower power symptom, then the following areas need to be investigated:

1. Oil (check type and level).
2. Pump inlet (check for cavitation and/or aeration).
3. Charge (check for proper pressure).
4. Filtration (take an oil sample from case drain and inspect the oil filter).
5. High-pressure relief valves.

Reservoir

Before proceeding too far into diagnosing a poor performing hydrostatic transmission, it is first critical to check the reservoir. Look to see if the oil level is low or the oil is aerated. Determine if the machine is using the correct type of oil. For example, some manufacturers require an oil with a higher viscosity index number if the machine is equipped with a hydrostatic transmission. If the customer purchased the standard viscosity oil, that could be affecting the pump's performance.

Charge Pump Inlet

After ensuring the machine has the correct oil at the right level, check to see if the charge pump is receiving a good supply of oil. Some technicians will first listen for cavitation before installing a vacuum gauge. Some manufacturers might not list a pump inlet specification. A rule-of-thumb is *less* is best, but preferably the pump has a positive head pressure and no pump inlet vacuum. One pump manufacturer states the charge pump inlet vacuum should not exceed 7 inches of mercury. See **Figure 24-13**.

Manufacturers today are less likely to use suction strainers or suction filters due to their tendency to starve the pump of oil when clogged. However, some machines still have a lengthy suction hose, for example a concrete mixing truck's drum drive. **Chapter 21** provided instructions for tapping into a pump inlet to measure pump vacuum. If charge pump inlet vacuum is a concern, be sure to measure the vacuum directly at the charge pump inlet.

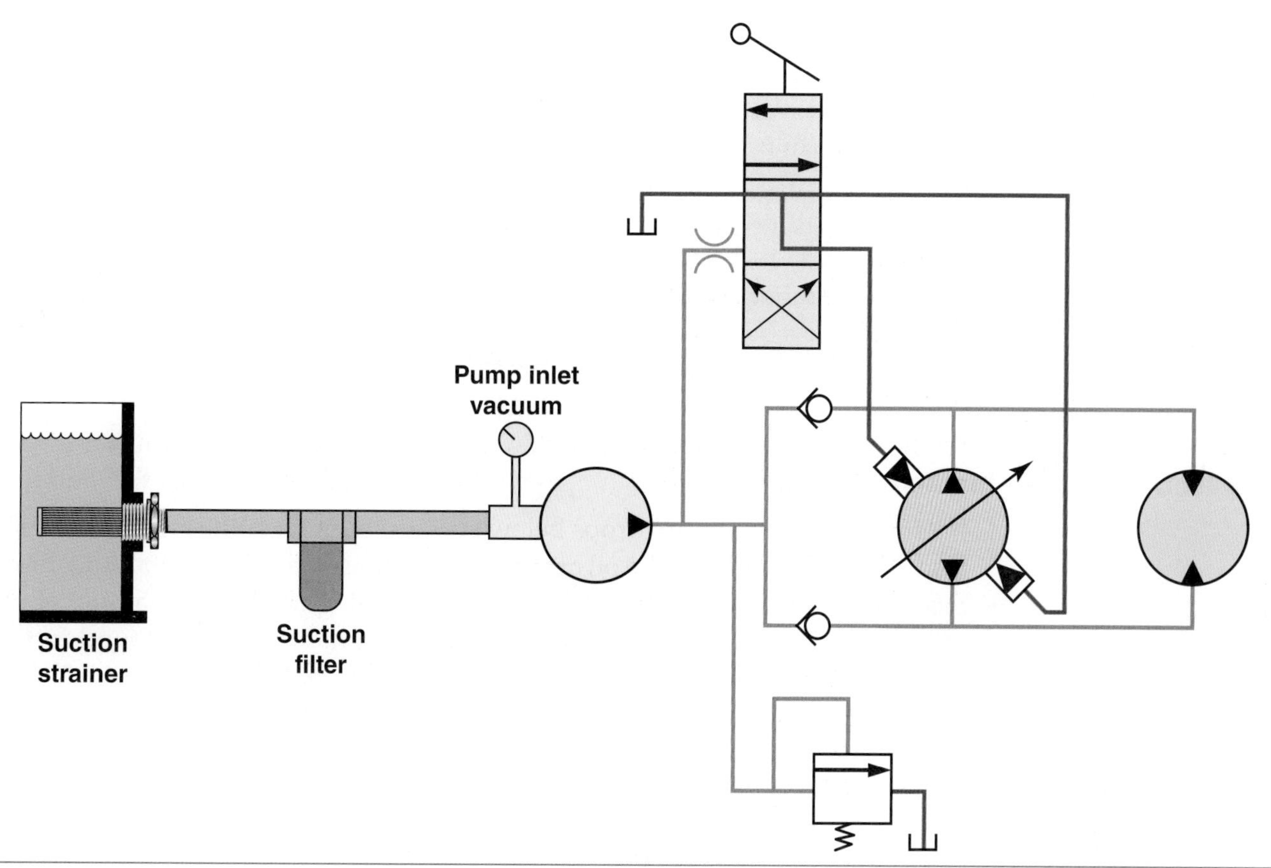

Figure 24-13. HST pump vacuum is measured at the inlet to the pump.

Charge Pressure

The next step is to investigate charge pressure. When the transmission's piston pump and motor become excessively worn, the charge pump volume will not be able to overcome the internal leakage losses. As a result of the worn hydrostatic transmission, charge pressure will drop. When charge pressure drops, the piston pump will cavitate, causing further damage to the hydrostatic transmission.

Low charge pressure can be a telltale sign that the transmission has excessive leakage, resulting in a low power complaint. When a low power complaint is being diagnosed on a hydrostatic transmission, charge pressure is arguably the most important pressure to monitor. For a refresher on the charge system and flushing valve, see **Chapter 23**.

With the oil at operating temperature, measure charge pressure when the transmission is under load and compare it to the manufacturer's specification. The charge pressure should drop approximately 30 psi (2 bar) when the propulsion lever is shifted from neutral to either forward or reverse. The manufacturer will normally specify an engine speed for the test, 1900 rpm for instance.

If the charge pump and the rest of the charge circuit are okay, the charge pressure measurement in neutral should be close to the manufacturer's specification. Explained another way, if the hydrostatic pump and motor are weak, it is possible for the neutral charge pressure to be within specification and the forward/reverse charge pressure to drop below specification. See **Figure 24-14** for an example of two transmissions, one good and one bad.

Good Transmission	Specification	Measured Pressure
Neutral Charge Pressure	300 psi (20 bar)	300 psi (20 bar)
Forward/Reverse Charge Pressure (under load)	270 psi (18 bar)	270 psi (18 bar)
Good Charge Pump, Bad Piston Pump and Motor	**Specification**	**Measured Pressure**
Neutral Charge Pressure	300 psi (20 bar)	300 psi (20 bar)
Forward/Reverse Charge Pressure (under load)	270 psi (18 bar)	240 psi (20 bar)

Figure 24-14. Examples of two charge pressure measurements, one good and one bad.

Inspect Oil and Filter for Contamination

If the charge pressure drops below specification when the system is under load, it is time to inspect the oil and filter for contamination. One manufacturer recommends removing the motor's case drain plug to obtain a sample of oil and cutting open the transmission oil filter. If contamination is found in the filter, the pump case, or the motor case, and if the transmission is a closed loop-drive, both the pump and the motor must be replaced. In addition, the entire system, including the hoses, reservoir, and cooler, must be flushed. Install new oil and a filter.

Drive Pressure

If the motor's oil sample and filter have no contamination, the next step is to measure the system's drive pressure. Notice that this book lists measuring drive pressure last. Many technicians, especially those unfamiliar with hydrostatic transmissions, will want to measure drive pressure first. If drive pressure is low, study the schematic and inspect any valve that can limit drive pressure. Examples include the internal pressure override (IPOR) valve, pressure cutoff valve, high-pressure relief (cross-over relief valves), foot-and-inch valve, anti-stall solenoid, manual bypass valve, and electronic pressure-release solenoid. Note that some shops have the capacity to individually test the high-pressure relief valves. See **Figure 24-15**.

After everything listed above (engine, transmission controls, reservoir, pump inlet, charge pressure, oil and filter, drive pressure, and other valves that can limit drive pressure) have been found to be operating properly, the solution is for the customer to shift the gearbox to a lower operating range. It is possible that the customer will desire the machine speed of second gear and the torque of first gear. The customer might have purchased a machine that is too small to meet his or her needs. For a fixed amount of horsepower, the customer can get an increase in torque or an increase in speed, but not both. The only way to increase both speed and torque is to increase engine horsepower. See **Figure 24-16**.

Low Power or Sluggish in Just One Direction

If a hydrostatic transmission is noticeably under powered in only one direction, one of the high-pressure relief valves may be at fault. Most transmissions use the same type of high-pressure relief valve for both forward and reverse propulsion. If the manufacturer indeed uses the same type of valve for forward and reverse, a technician can swap the forward and reverse high-pressure relief valves to see if the low power symptom changes to the opposite direction. If the transmission is then underpowered in the opposite direction, the high-pressure relief valve is at fault. As previously mentioned, the neutral charge relief should never be swapped with the flushing relief valve, otherwise the transmission will overheat, greatly reducing the life of the transmission.

It is also possible for a make-up valve to cause an HST to exhibit low power in only one direction. A leaky make-up valve will also cause charge pressure to be too high in one direction.

Figure 24-15. Relief valves can be bench tested to determine their relief pressure setting.

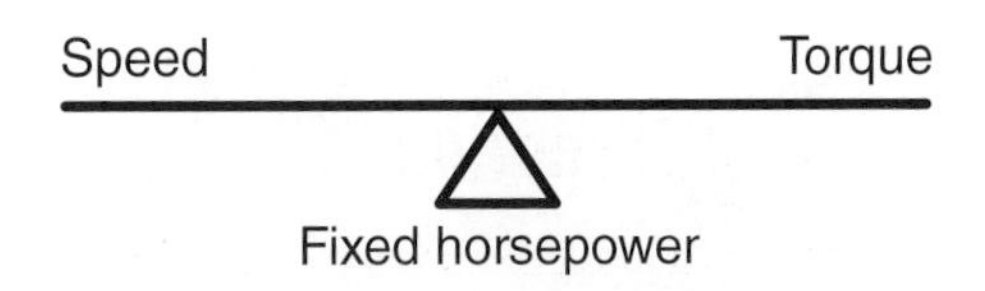

Figure 24-16. For a fixed amount of horsepower, speed is inversely proportional to torque.

Coasting or Freewheeling Transmission

Hydrostatic transmissions provide the benefit of hydrostatic braking. If an operator notices the transmission freewheeling or coasting, the machine should also lack power. A machine with these symptoms should be diagnosed. See the recommendations previously listed for diagnosing low power in both directions.

Transmission Will Not Move in Either Direction

If a transmission will not move in either direction and the pump is electronically controlled, check to see if the solenoids are being energized. If the problem is electrical, use the manufacturer's service information to diagnose the problem. If the solenoids are receiving electrical power or if the pump is manually controlled, the next step is check for proper charge pressure, because the charge is responsible for supplying oil to the directional control valve.

If charge pressure is good, see if the pump can build servo pressure. Manufacturers rarely publish any specifications for servo pressure. The process of balancing servo pressures to center a pump's null adjustment was explained earlier in this chapter.

When the pump's control valve is actuated, the DCV receives supply oil from the charge pump via the orifice. If the pump has good charge pressure but is unable to build any servo pressure, the supply orifice is likely plugged.

Figure 24-17 shows a plugged orifice a technician found on a hydrostatic transmission. Previously, the owner had multiple technicians attempt to diagnose why the machine would not propel itself. It was a pleasant surprise to find that the machine could not develop any servo pressure, because it is almost a given that DCV orifice is plugged. In this case, a portion of an O-ring had plugged the orifice.

Figure 24-17. The control valve's orifice was plugged with material from an O-ring.

On this machine with the plugged orifice, the charge pressure would not drop, and both legs of the closed loop had the same pressures. Once the orifice was fixed, servo pressure could be developed, causing the pump's swash plate to pivot. This resulted in the development of drive pressure, which caused the flushing shuttle valve to shift, and charge pressure drop.

Effects of the Directional Control Valve Orifice

Note that the DCV orifice controls the responsiveness of the transmission. The larger the orifice, the more responsive a transmission will be. A smaller orifice causes the transmission to be less responsive. A machine's service literature may provide several different part numbers for different orifice options.

Charge Pressure Is Too High

If a technician encounters a machine with charge pressure that is too high when the machine is operated in one direction, one of the closed-loop make-up valves could be allowing high-pressure oil to leak back into the charge circuit. In this scenario, the charge pressure would be considerably higher than specification. See **Figure 24-18**. If one of the make-up valves is leaking, it can also cause low power in one direction.

Do Not Attempt to Isolate a Hydrostatic Pump

If a technician is diagnosing a weak closed-loop hydrostatic transmission, he or she should never attempt to isolate the pump. There are two reasons for this. First, if the motor contains the high-pressure relief valves, the technician can rupture a pump or drive hose while trying to isolate the pump. This is likely to happen if the technician caps off the closed-loop drive lines. Second, it should be unnecessary to isolate a pump. If the transmission is weak and the pump is bad, the motor should be replaced as well, because the HST is a closed-loop transmission. This concept was explained in **Chapter 23**.

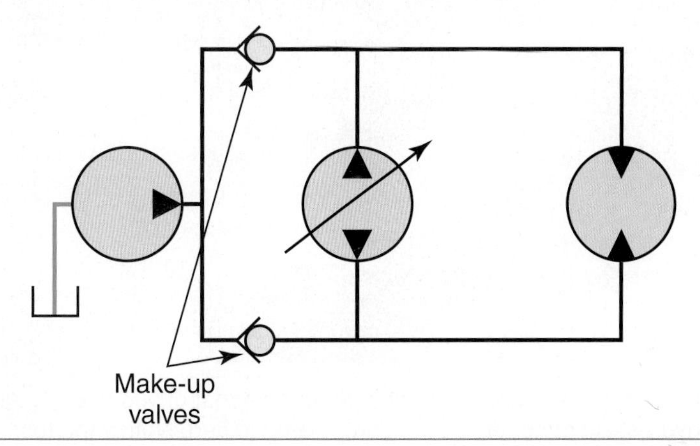

Figure 24-18. Make-up valves are check valves that allow the charge pump to supercharge the closed loop.

Using a Hydrostatic Transmission Test Stand

Some hydraulic shops have the ability to test a hydrostatic transmission's pump and motor. Some of those shops use a diesel engine to power the test stand. See Figure 24-19. Notice that the test stand uses a pump and motor to drive the pump that was rebuilt. The test stand uses another pump to load the transmission's motor that was rebuilt.

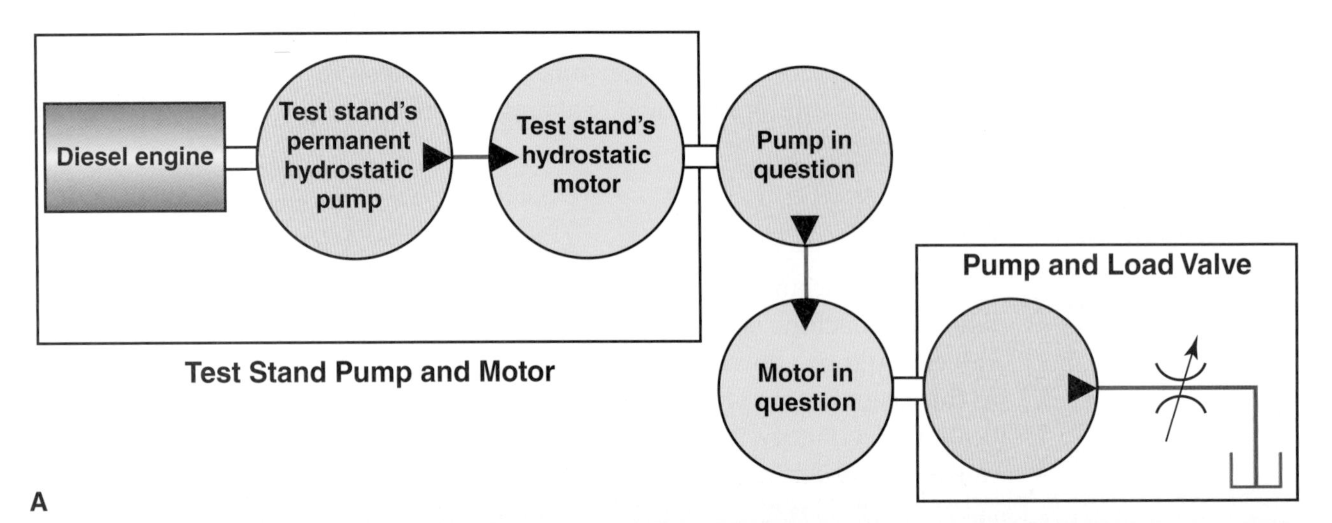

Figure 24-19. A—A hydrostatic drive test stand consists of an engine that drives a test stand pump and motor assembly. The test stand's hydrostatic motor drives the pump being tested. Notice the test stand also uses a pump and a load valve to load the transmission. B—A hydrostatic transmission test stand.

Summary

✓ Servo-operated hydrostatic pumps have at least two separate adjustments for setting the pump to neutral: DCV null and servo piston(s). The machine must be safely placed on jack stands prior to adjusting the pump, because the adjustments will cause the machine to move.

✓ Charge pressure is the most important pressure to measure when diagnosing a low power symptom. It will drop approximately 30 psi when the transmission is shifted from neutral to forward or reverse. A worn hydrostatic transmission typically results in a drop in charge pressure, and will contaminate the oil and the filter.

✓ If the charge pump is believed to be cavitating, measure the pump inlet's vacuum. The pump inlet vacuum should be no more than 7 inches of mercury.

✓ If the transmission is overheating, perform the following checks:

- Ensure the problem is not an indication problem.
- Check case drain and oil cooler flow.
- Measure charge pressure to see if the transmission is excessively worn.
- Determine if someone mistakenly swapped the neutral charge relief valve with the hot-oil purge relief valve.

Technical Term

commissioning

Review Questions

Answer the following questions using the information provided in this chapter.

1. All of the following can cause a hydrostatic transmission to creep, *EXCEPT*:
   A. a misadjusted pump DCV.
   B. a misadjusted pump servo piston.
   C. a shorted pump solenoid.
   D. a shorted two-speed motor solenoid.
2. When centering the servo piston on a hydrostatic pump that is equipped with only one servo piston, which of the following pressures must be measured?
   A. Drive.
   B. Charge.
   C. Servo.
   D. Case.
3. When making the DCV null adjustment on a hydrostatic transmission equipped with a single-servo piston pump, which of the following pressures must be measured?
   A. Drive.
   B. Charge.
   C. Servo.
   D. Case.
4. When centering the servo piston on a hydrostatic pump that is equipped with only one servo piston, which of the following pressures must the technician hydraulically loop together?
   A. Drive.
   B. Charge.
   C. Servo.
   D. Case.

5. When making the DCV null adjustment on a hydrostatic transmission equipped with a single-servo piston pump and electronic solenoids, what is the Allen wrench turning/adjusting?
   A. The solenoid's armature.
   B. An eccentric shaft.
   C. The servo piston.
   D. The solenoid's coil.
6. A hydraulic shop has rebuilt a hydrostatic pump that is equipped with a single-servo piston. The technician is adjusting the pump to ensure it is in neutral. The pump configuration has how many different adjustments for setting neutral?
   A. One.
   B. Two.
   C. Three.
   D. Four.
7. A technician is installing the servo piston caps on a dual-servo piston hydrostatic pump. The technician is measuring the swash plate to ensure the pump is neutrally centered. What is the maximum difference allowed from one side to the other?
   A. .010″
   B. .005″
   C. .001″
   D. .0005″
8. A technician has installed a new hydrostatic motor. All of the following are necessary steps for starting (commissioning) the transmission, *EXCEPT*:
   A. filling the motor's case with oil.
   B. monitoring charge pressure.
   C. starting the machine with the gearbox in the neutral position.
   D. monitoring case pressure.
9. A technician has installed a new hydrostatic pump. During the startup procedure, how should the transmission be operated?
   A. Under heavy load.
   B. Under moderate load.
   C. With no load.
   D. Load does not matter.
10. A customer asks a shop to investigate a lack of power in a hydrostatic transmission. The technician places the gearbox in high gear, applies the parking brake, service brake and strokes the propulsion lever, resulting in engine lugging. Technician A states that the hydrostatic transmission is responding correctly. Technician B states the suction filter might be plugged. Who is correct?
    A. Technician A.
    B. Technician B.
    C. Both A and B.
    D. Neither A nor B.
11. A hydrostatic transmission is overheating. Which of the following hydraulic oils should be investigated first?
    A. Case drain.
    B. Charge.
    C. Drive.
    D. Servo.
12. A hydrostatic transmission sounds as if it is cavitating. Which of the following pressures should be measured?
    A. Case drain.
    B. Charge.
    C. Servo.
    D. Suction.
13. A hydrostatic transmission lacks power. Which of the following is the most important pressure to measure during the diagnostic process?
    A. Case drain.
    B. Charge.
    C. Drive.
    D. Suction.
14. A hydrostatic transmission has good charge pressure, but will not move forward or backward. A technician strokes the propulsion lever, but cannot obtain any servo pressure. What is most likely at fault?
    A. A bad high-pressure relief valve.
    B. A plugged DCV orifice.
    C. A bad flushing shuttle valve.
    D. A stuck-open make-up valve.

15. Which of the following has a large effect on the responsiveness of the hydrostatic transmission?
    A. Make-up valve.
    B. DCV orifice.
    C. Flushing relief valve.
    D. Pressure rating of the drive loop hoses.
16. A hydrostatic transmission is creeping forward. Which of the following is *least* likely to be the cause?
    A. The position of pump's DCV.
    B. The position of pump's servo piston.
    C. The position of motor's swash plate.
    D. A shorted pump DCV solenoid.
17. A transmission is sluggish in forward, but operates strong in reverse. Which of the following steps is the most common procedure performed by veteran technicians to diagnose this problem?
    A. Swap the high-pressure relief valves.
    B. Swap the charge relief valves.
    C. Swap the make-up valves.
    D. Swap the drive hoses.
18. When is it okay to swap the charge relief valves?
    A. When charge pressure is low.
    B. When charge pressure is high.
    C. When the transmission lacks performance in one direction.
    D. Never.
19. When is it okay to isolate a hydrostatic pump and cap off its drive lines?
    A. When charge pressure is low.
    B. When charge pressure is high.
    C. When the transmission lacks performance in one direction.
    D. Never.
20. A technician has been investigating a hydrostatic transmission with a low power complaint in both directions. All of the following are diagnostic steps for investigating a *low* power symptom, *EXCEPT*:
    A. cutting open a filter.
    B. pulling a sample from the motor's case drain.
    C. measuring charge pressure.
    D. measuring servo pressure.
21. A technician has been investigating a hydrostatic transmission with a low power complaint in both directions. What is the recommended procedure if all of the test results are good?
    A. Increase the setting for high-pressure relief valves.
    B. Change the fluid and filters.
    C. Shift to a lower gear.
    D. Swap the charge relief valves.
22. All of the following are normal hydrostatic transmission characteristics, *EXCEPT*:
    A. it coasts or freewheels.
    B. it is loud during high drive pressures.
    C. the charge pressure drops 30 psi (2 bar) when moving.
    D. it provides hydrostatic braking.
23. A hydrostatic transmission is overheating. All of the following can cause overheating, *EXCEPT*:
    A. a lazy oil cooler bypass.
    B. a restricted oil cooler.
    C. a worn hydrostatic pump and motor.
    D. a faulty indicator lamp.
24. Charge pressure is too high when driving in one direction. Which of the following is most likely at fault?
    A. A leaking make-up valve.
    B. A leaking charge relief valve.
    C. A leaking drive relief valve.
    D. A leaking DCV spool.
25. A technician is adjusting the hydrostatic pump's neutral on a machine. Which of the following steps must be completed to prevent the machine from running over the technician?
    A. Removing the pump.
    B. Disconnecting the linkage.
    C. Removing the solenoid coils.
    D. Placing the machine on jack stands.

26. A technician is testing a hydrostatic transmission's charge pressure. At wide-open throttle in neutral, the charge pressure measures 300 psi (20 bar). When the transmission is loaded in forward and reverse, the charge pressure drops to 270 psi (18 bar). Technician A states that the transmission is bad, and the pump and motor must be rebuilt and the entire system flushed. Technician B states the test results are good. Who is correct?
    A. Technician A.
    B. Technician B.
    C. Both A and B.
    D. Neither A nor B.
27. A transmission will not drop charge pressure when the DCV is stroked. All of the following could be at fault, *EXCEPT*:
    A. a plugged DCV orifice.
    B. a stuck flushing shuttle valve.
    C. a high-pressure relief set too high.
    D. a flushing relief valve set too high.
28. A technician is testing a hydrostatic transmission's charge pressure. At wide-open throttle in neutral, the charge pressure measures 300 psi (20 bar). Which of the following charge pressure readings measured under load in forward or reverse would indicate a worn transmission?
    A. 330 psi.
    B. 300 psi.
    C. 270 psi.
    D. 240 psi.

Chapter 25

Steering

Objectives

After studying this chapter, you will be able to:

- ✓ List the advantages and disadvantages of an SCU.
- ✓ List common names used to describe SCUs.
- ✓ Describe how the SCU internal components are connected to one another.
- ✓ Explain the order of oil flow through an SCU.
- ✓ Explain how an SCU can be operated with a dead engine.
- ✓ List and explain the purpose of optional valves used in SCUs.
- ✓ Explain non-load reaction, load reaction, wide-angle steering, cylinder dampening, and Q-Amp steering.
- ✓ Explain the operation of a steering priority valve including CF, EF, and LS pressures.
- ✓ Explain the difference between static and dynamic signal lines.
- ✓ List the possible locations of steering relief valve.
- ✓ Briefly describe the hydraulic steering troubleshooting steps.
- ✓ List examples and advantages of electronic steering applications.

Steering Control Units

Mobile equipment manufacturers have used a unique hydraulic power steering system in their machines for decades. The system uses a manually operated ***steering control unit (SCU)*** that is rotated by a steering wheel and steering shaft. See Figure 25-1. The hydraulic steering system has no direct mechanical connection between the SCU, the hydraulic pump, or the steering cylinder, as compared to the mechanical linkages found in automotive power steering and manual steering systems.

Flexibility and adaptability are two advantages of using this style of hydraulic steering. The SCU is often located directly below the steering column, underneath the floor of the operator's cab. The hydraulic hoses provide the necessary flexibility to deliver oil to the steering cylinder, which may be located a great distance away from the SCU.

The downside of a traditional SCU is that it is not electronically controlled and is difficult to adapt to machines that use GPS-guided steering systems, also known as ***auto guidance steering systems***. However, kits are available for converting a traditional SCU system into an auto guidance steering system.

SCUs can be called different names:

- Hand metering unit (HMU).
- Hydrostatic steering.
- Steering hand-pump.
- Steering control valve.

Char Lynn®, which was purchased by Eaton Corporation in 1970, was one of the first manufacturers to build this style of hydraulic steering system.

SCU Components, Design, and Operation

The SCU is responsible for completing two tasks related to steering a tractor. The SCU must control the direction of oil flow (right steer and left steer) and the quantity of oil flow (dictates how fast the wheels are turned).

Figure 25-2 shows the primary components that comprise an SCU:

- Gerotor assembly.
- Drive shaft.
- Control sleeve.
- Centering pin.
- Control spool.

Figure 25-1. The internal components of an SCU are rotated by means of a steering wheel and steering shaft. The SCU provides true hydraulic steering with no mechanical link between the SCU and the steering cylinder.

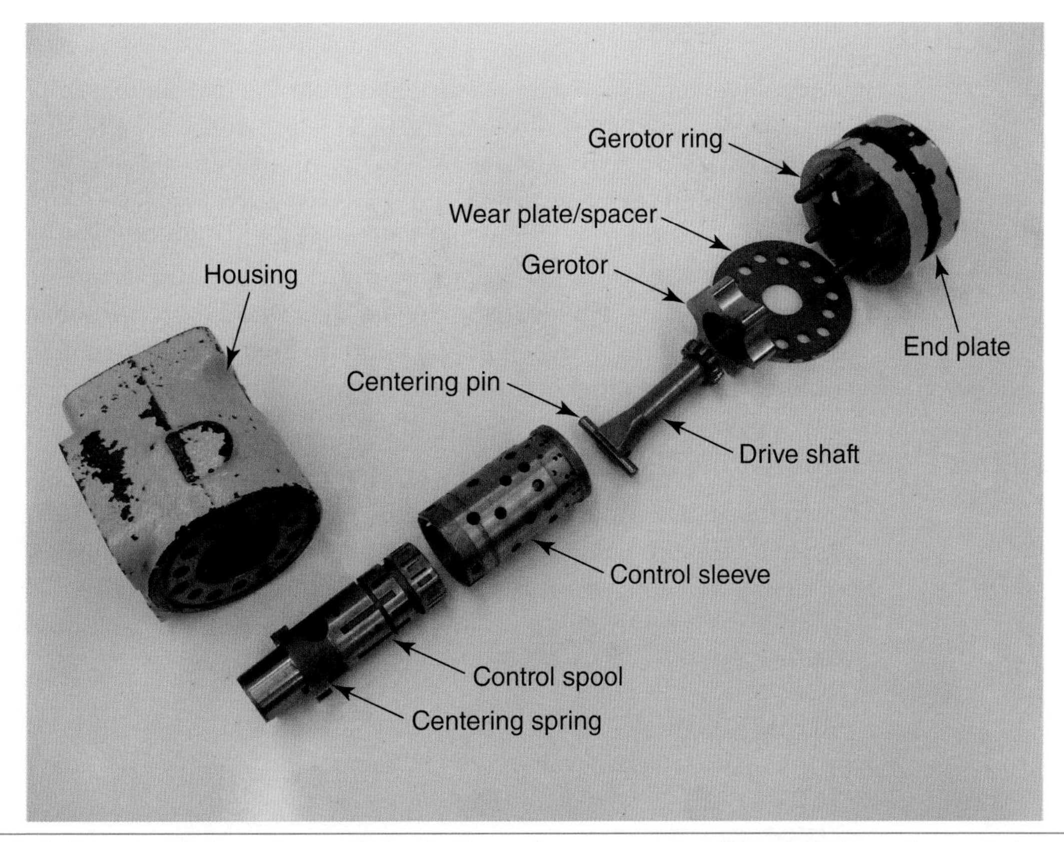

Figure 25-2. A disassembled steering control unit. The main components within an SCU are the control spool and control sleeve assembly and the gerotor assembly.

The SCU contains several components that are internally connected to one another. The way in which the coupled components operate is described in the following sequence:

- The steering wheel's drive shaft is splined to the control spool.
- The control spool is coupled to the control sleeve through a centering pin and centering springs.
- The centering pin fits loosely inside the control spool and has a tight fit inside the control sleeve.
- As the steering wheel turns the control spool, the control sleeve follows via the loose-fitting centering pin. When the steering wheel stops rotating, the centering spring will center the control spool/sleeve assembly.
- The drive shaft fits inside the control spool/sleeve assembly and has a yoke shape at the bottom of the shaft that slides over the centering pin.
- As the control spool/sleeve assembly spins, the centering pin causes the drive shaft to rotate.
- The drive shaft is splined to the internal gerotor that rotates inside the ring of the gerotor assembly.
- The SCU housing, control spool, and control sleeve work together to act as a rotary DCV.
- The stationary components within an SCU are the gerotor's ring gear, the housing, the spacer/wear plate and the end plate/cover plate.

Open-Center SCU

Open-center SCUs route oil supplied from the hydraulic pump back to the reservoir at a low pressure when the SCU is in neutral. The design, as shown in **Figure 25-5** through **Figure 25-7**, is not used with steering priority valves because steering priority valves must receive a load-sensing signal from the SCU. An open-center SCU is most likely used with either a fixed-displacement pump that is dedicated solely for steering, or a fixed-displacement pump and a proportional divider valve. The divider valve would proportion a percentage of pump flow to the open-center SCU, and the remaining percentage of oil flow would be used for other circuits, such as traditional hydraulic implements.

Closed-Center Non-Load-Sensing SCU

A closed-center non-load-sensing SCU, as shown in **Figure 25-3**, would be used in a pressure-compensating hydraulic system. Most tractors use a steering priority valve, which requires that the SCU be configured with a load-sensing signal line. This particular SCU would be rare because most SCUs are used in conjunction with a load-sensing steering priority valve.

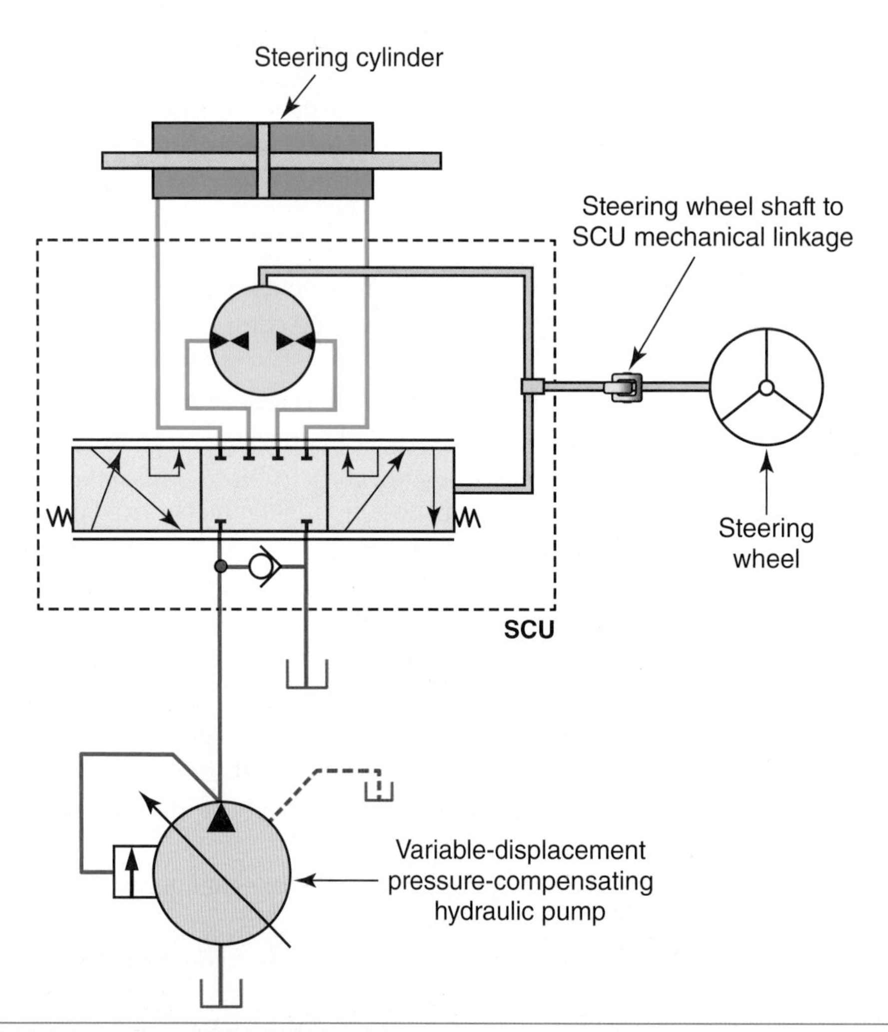

Figure 25-3. A schematic of a hydraulic steering system with a pressure-compensating hydraulic pump and a closed-center SCU without a load-sensing signal line.

Closed-Center Load-Sensing SCU

A traditional open-center or closed-center SCU is a four-way DCV that has four ports:

- Inlet.
- Outlet.
- Right steer.
- Left steer.

Most SCUs are a closed-center load-sensing design with a fifth hydraulic port, which is the load-sensing signal line. See Figure 25-4. Interestingly, this style of load-sensing SCU is used in multiple types of hydraulic systems:

- Open-center hydraulic system.
- Pressure-compensating hydraulic system.
- Load-sensing pressure-compensating hydraulic system.
- Flow-sharing hydraulic system.

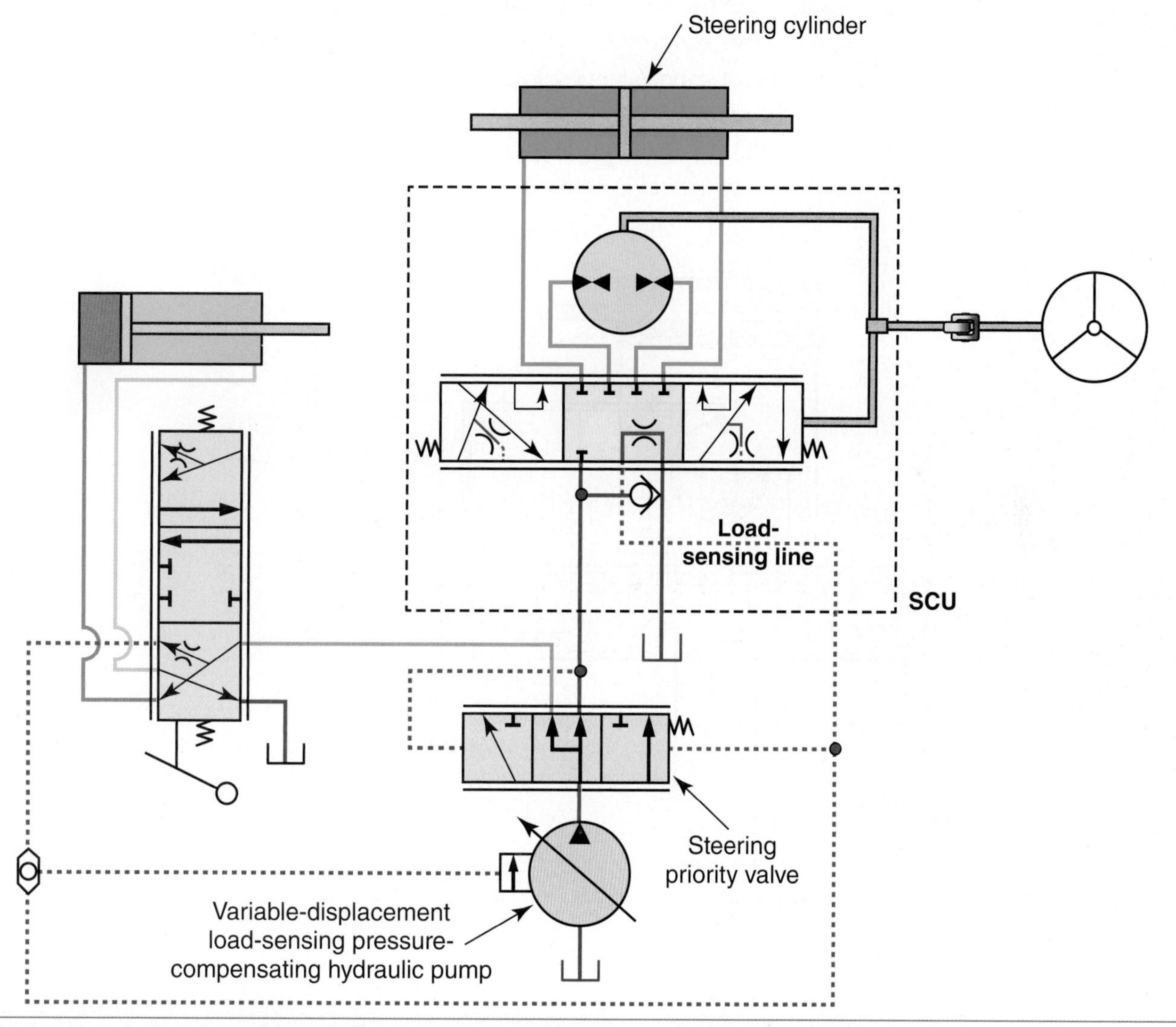

Figure 25-4. A closed-center load-sensing SCU is required if the machine is equipped with a steering priority valve. The signal line is also used for load-sensing hydraulic pumps. Note in this example that a load-sensing signal pressure is generated through the SCU, and does not use a primary shuttle valve.

In the traditional open-center hydraulic system and the traditional closed-center pressure-compensating hydraulic system, the SCU's load-sensing signal line is only used for the steering priority valve. However, in load-sensing hydraulic systems, whether it is pre-spool or post-spool, the SCU's signal line is used for the steering priority valve and the load-sensing hydraulic pump or unloading valve. Priority valves and load-sensing signal lines are detailed later in this chapter.

SCU Neutral Operation

The schematic in Figure 25-5 shows an SCU in a neutral operation, meaning the tractor is driving straight forward. In this example, the SCU is an open-center valve and the oil is routed back to the reservoir at a low-pressure.

SCU Left or Right Steering Mode of Operation

When the operator rotates the steering wheel to the left or to the right, the SCU will route the oil through the following path. See Figure 25-6.

1. The control spool/sleeve assembly receives oil from the hydraulic pump.
2. The control spool/sleeve assembly directs the oil to the gerotor assembly.
3. The gerotor assembly meters the quantity of oil that will eventually be sent to the steering cylinder. The quantity of oil is governed by how fast the steering wheel is rotated.

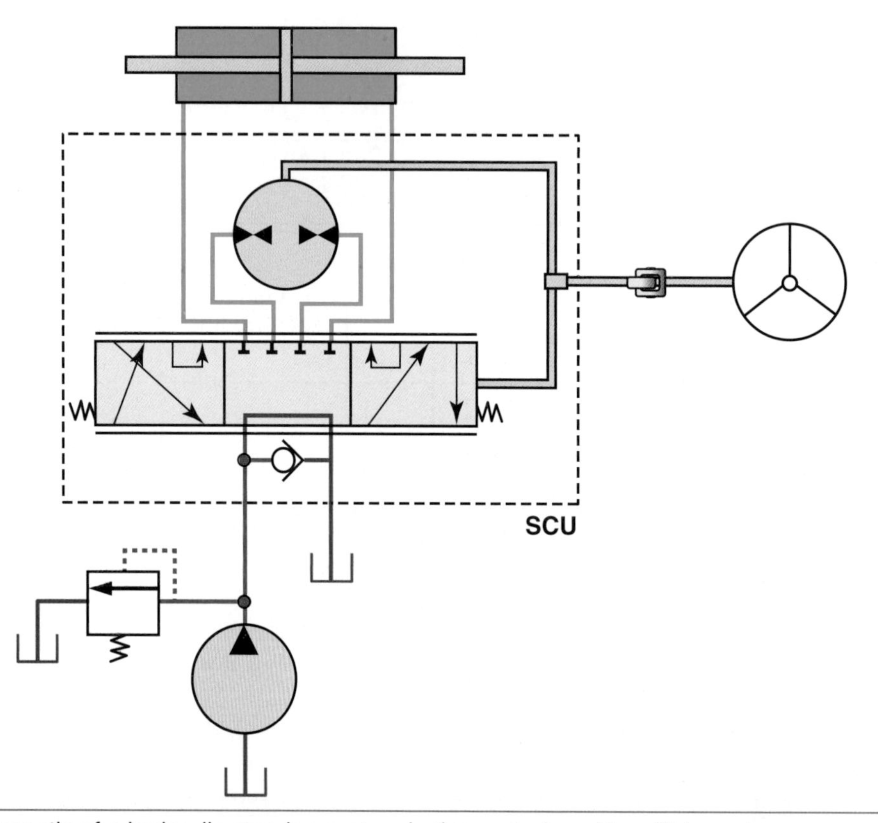

Figure 25-5. A schematic of a hydraulic steering system in the neutral position. This system uses an open-center SCU in its design. If the operator rotates the steering wheel, both the control spool/sleeve assembly and the gerotor assembly will all rotate at the same speed and in the same direction as the steering wheel.

4. The control spool/sleeve assembly receives the metered oil from the gerotor assembly and directs it to the steering cylinder to steer the tractor to the left or to the right, based on the operator's movement of the steering wheel.
5. The steering cylinder's return oil is routed back to the control spool/sleeve assembly.
6. The control spool/sleeve assembly directs the return oil to the reservoir.

SCU Dead Engine Steering Mode

As part of a truly 100% hydraulic steering system, an SCU has a special provision designed into the unit that enables the operator to steer the tractor in the event that the engine dies or the hydraulic pump fails. This feature is called ***emergency steering*** or ***manual steering mode***.

The ***manual steering check valve*** is located at the SCU's inlet and taps into the SCU inlet port and the SCU tank return port. If the operator attempts to steer when the hydraulic pump is unable to supply oil flow to the SCU, the manual steering check valve allows the SCU's gerotor to draw oil directly from the reservoir. The gerotor assembly uses this oil as the supply oil for emergency steering, **Figure 25-7**. As the operator manually rotates the steering wheel, the gerotor rotates within the gerotor ring, causing an expanding volume among the teeth near the internal gear's inlet port and a decreasing volume among the teeth near the internal gear's outlet port. The action of the gerotor during this mode causes it to act like a manual pump. For this reason, the SCU is sometimes called a ***steering hand-pump***.

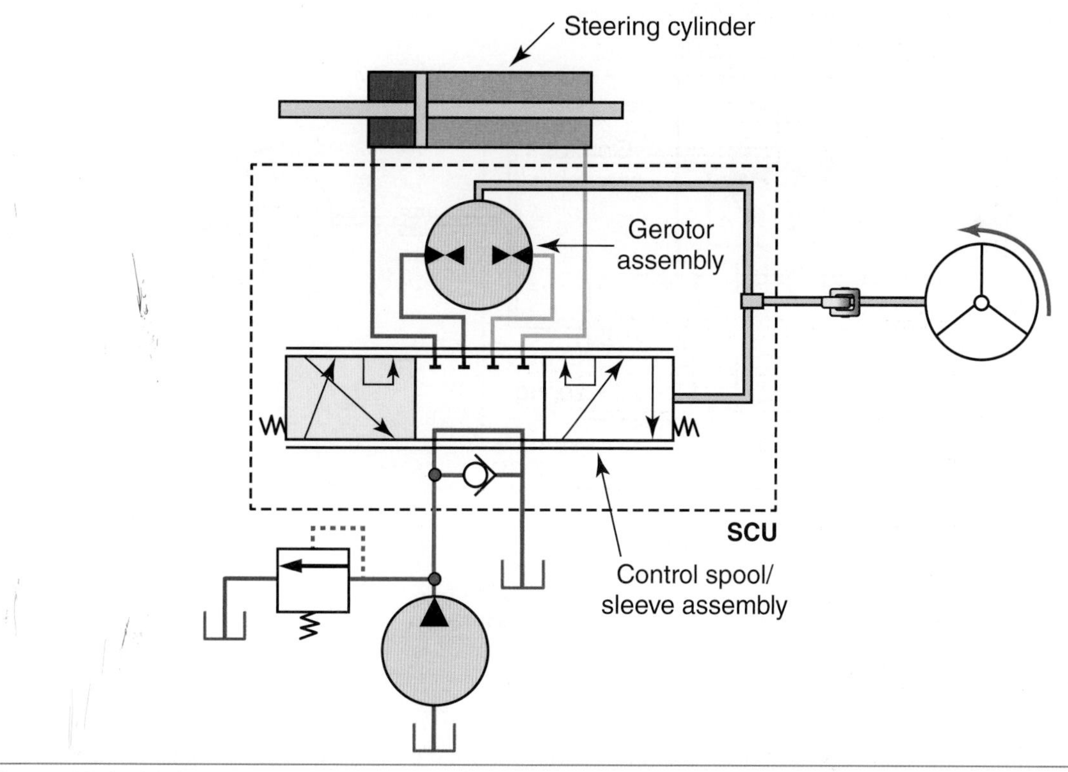

Figure 25-6. As the operator rotates the steering wheel to the left, the SCU control spool/sleeve assembly directs the oil in and out of the SCU to steer the tractor to the left. This schematic shows a left turn. Note that for a right turn, the control spool/sleeve shifts to the far right position and the oil follows the opposite flow path. Oil pressure forces the steering cylinder piston to the right end of its cylinder.

Optional SCU Internal Valves

In addition to emergency steering, SCU manufacturers offer machine manufacturers a variety of internal steering valve options:

- Inlet check valve.
- Inlet relief valve.
- Load-sensing relief valve.
- Cylinder port relief valves.
- Anti-cavitation check valve.

Inlet Check Valve

An inlet check valve is shown in Figure 25-8. If external loads exerted a force on the steering cylinder, it could cause the steering pressure to exceed the main system pressure. The steering oil would be able to bleed backwards through the SCU and cause the steering wheel to kick back, which is known as ***steering kickback***. The inlet check valve prevents steering kickback.

Inlet Relief Valve

Most hydraulic steering systems have a dedicated relief valve for the steering circuit. The relief valve is commonly placed either inside the SCU as shown in Figure 25-9 or inside the steering priority valve which is discussed later in the

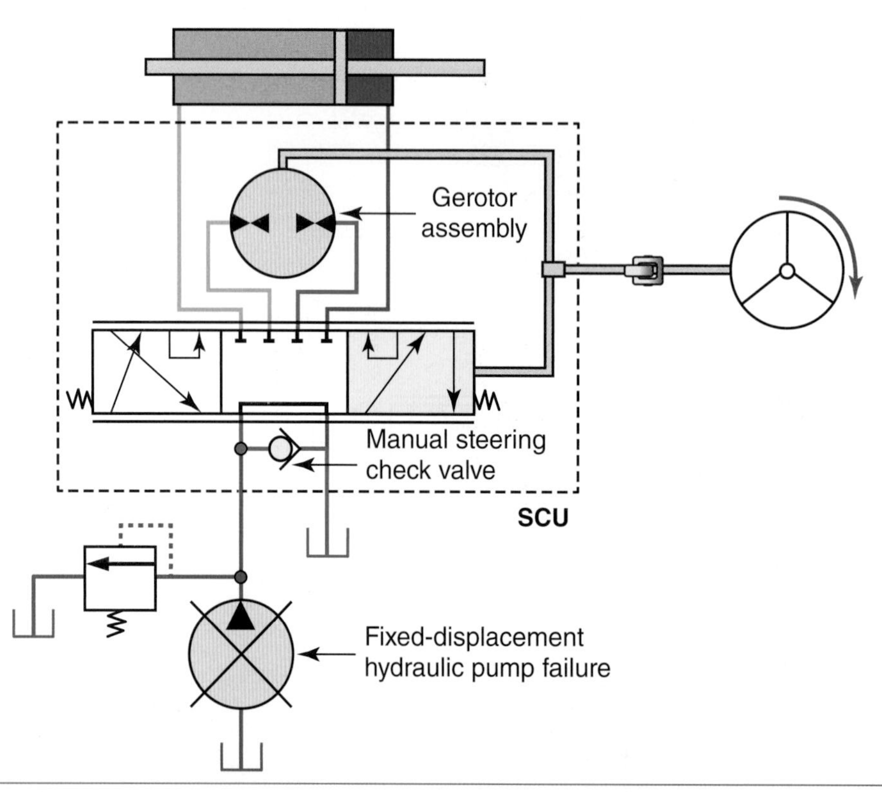

Figure 25-7. The manual steering check valve allows for emergency steering if the engine or hydraulic pump fails during operation. As the gerotor is spun, oil is drawn from the reservoir into the SCU for the purpose of providing manual steering. This example illustrates manual steering to the right. Trace the path of the oil as it is extracted from the reservoir and travels through the check valve, into the control spool/sleeve assembly, and through the gerotor assembly before reaching the steering cylinder.

chapter. Steering relief valve pressures are normally set below the main system pressure. For example, if the main system pressure is set at 3000 psi (207 bar), the steering relief valve pressure could be set as low as 2300 psi (159 bar).

If the hydraulic pump is used only for steering, the main system relief in Figure 25-9 would be redundant and unnecessary for a system equipped with an inlet relief. However, if the fixed-displacement pump's flow is divided by a proportional flow divider valve, the main system relief is required.

Load-Sensing Relief Valve

Load-sensing hydraulic steering systems may use a load-sensing relief valve that is combined into the SCU or the steering priority valve. The load-sensing relief valve establishes the steering relief pressure by dumping the steering signal pressure to the reservoir when the pressure exceeds its setting. See Figure 25-10.

Cylinder Port Relief Valves

As mentioned in **Chapter 7**, any time a DCV is in a neutral position, the circuitry located after the DCV has no protection from forces generated by external loads. SCUs can use cylinder port relief valves to protect the steering cylinder(s) and the steering hoses as shown in Figure 25-11.

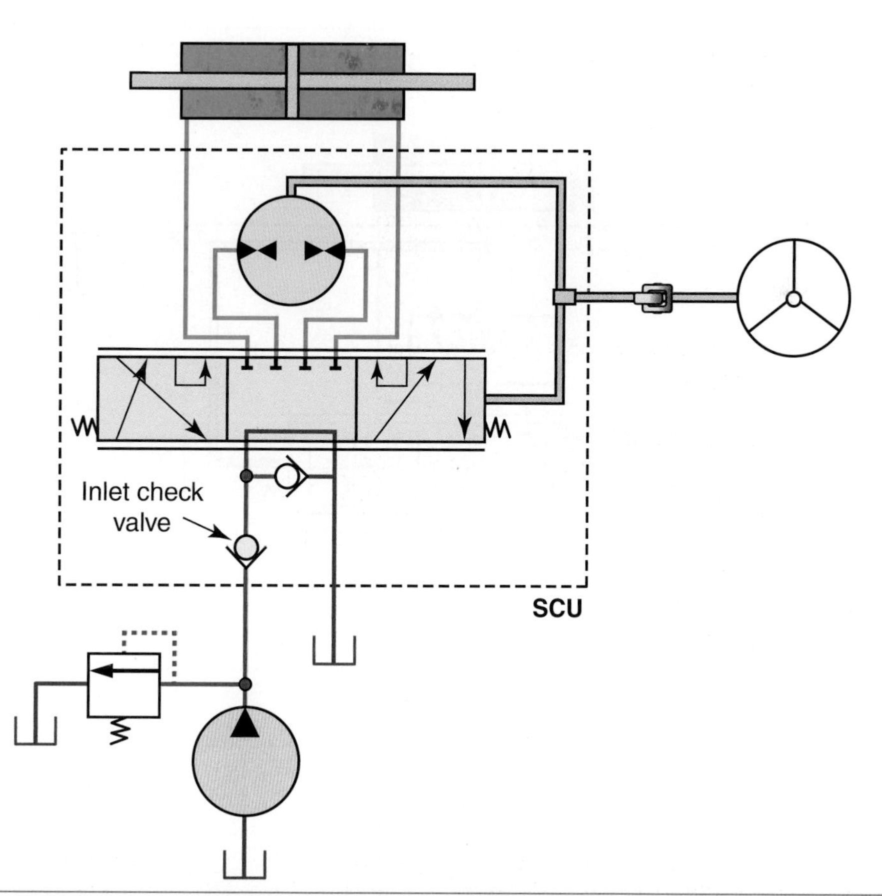

Figure 25-8. The inlet check valve prevents high-pressure steering oil from bleeding backwards through the SCU when the steering pressure exceeds system pressure. This prevents steering kickback.

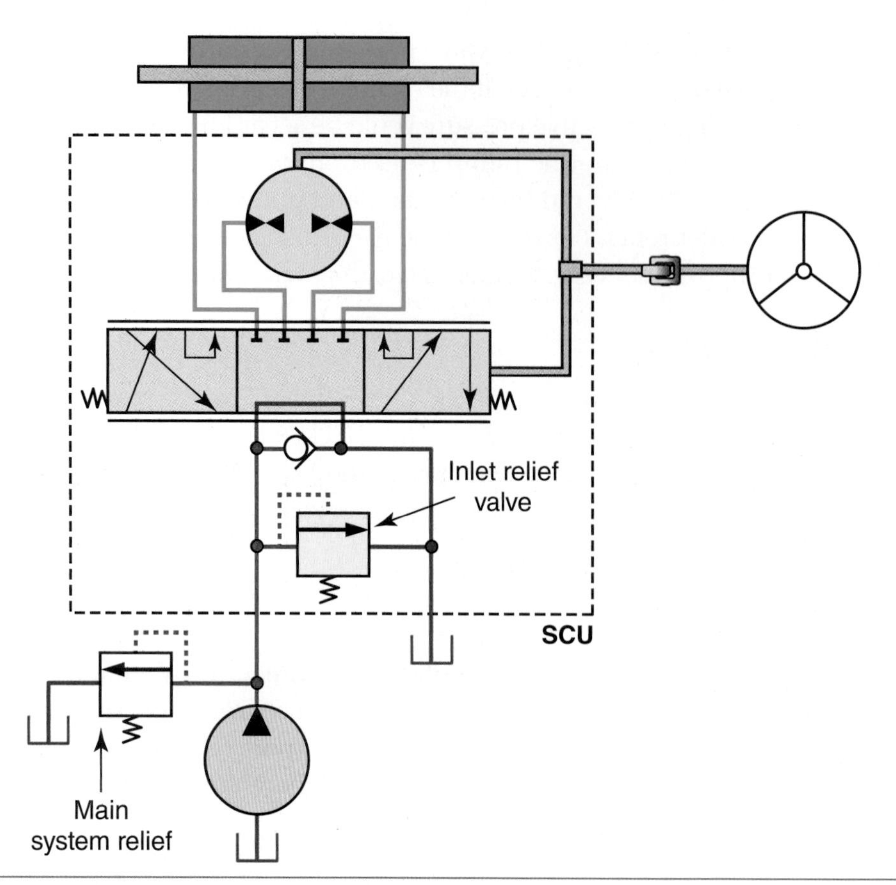

Figure 25-9. An inlet relief valve can be placed inside the SCU. It is used as a hydraulic steering system relief valve.

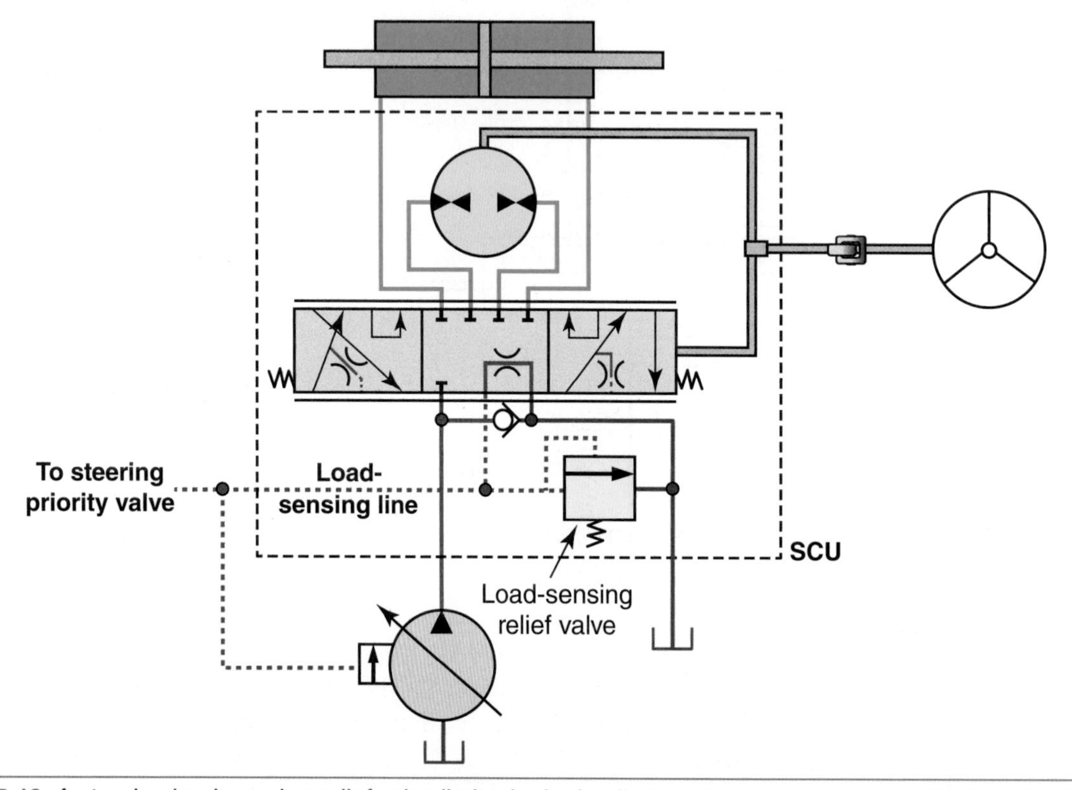

Figure 25-10. A steering load-sensing relief valve limits the hydraulic steering system pressure by dumping the steering signal pressure to the reservoir when the pressure exceeds its setting.

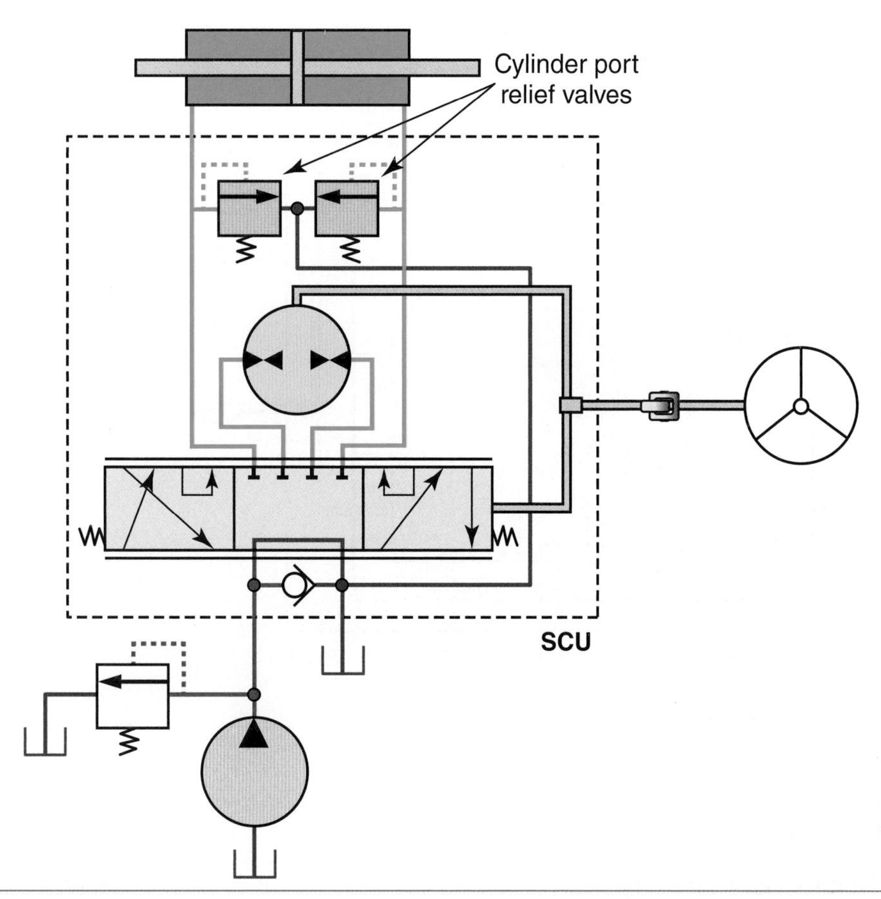

Figure 25-11. Cylinder port relief valves can be added inside the SCU. When the SCU is moved into a neutral position, they protect the steering working ports from pressure surges caused by external forces.

Anti-Cavitation Check Valves

Anti-cavitation valves, also known as make-up valves, can be installed inside an SCU. The valves are designed to prevent cylinder cavitation by drawing oil from the reservoir any time the cylinders actuate faster than what the hydraulic pump can supply the oil required. See Figure 25-12.

SCU Features

SCU manufacturers offer SCUs in several different configurations:

- Non-load reaction.
- Load reaction.
- Wide-angle steering.
- Cylinder dampening.
- Q-Amp steering.

Non-Load Reaction

In a ***non-load reaction SCU***, the steering cylinder's left and right work ports are blocked when the SCU is in a neutral position. This configuration will hold the steering cylinder(s) in a fixed position even when external loads are exerting forces on the steering cylinder(s). Figure 25-3 through Figure 25-12 all show non-load reaction SCUs.

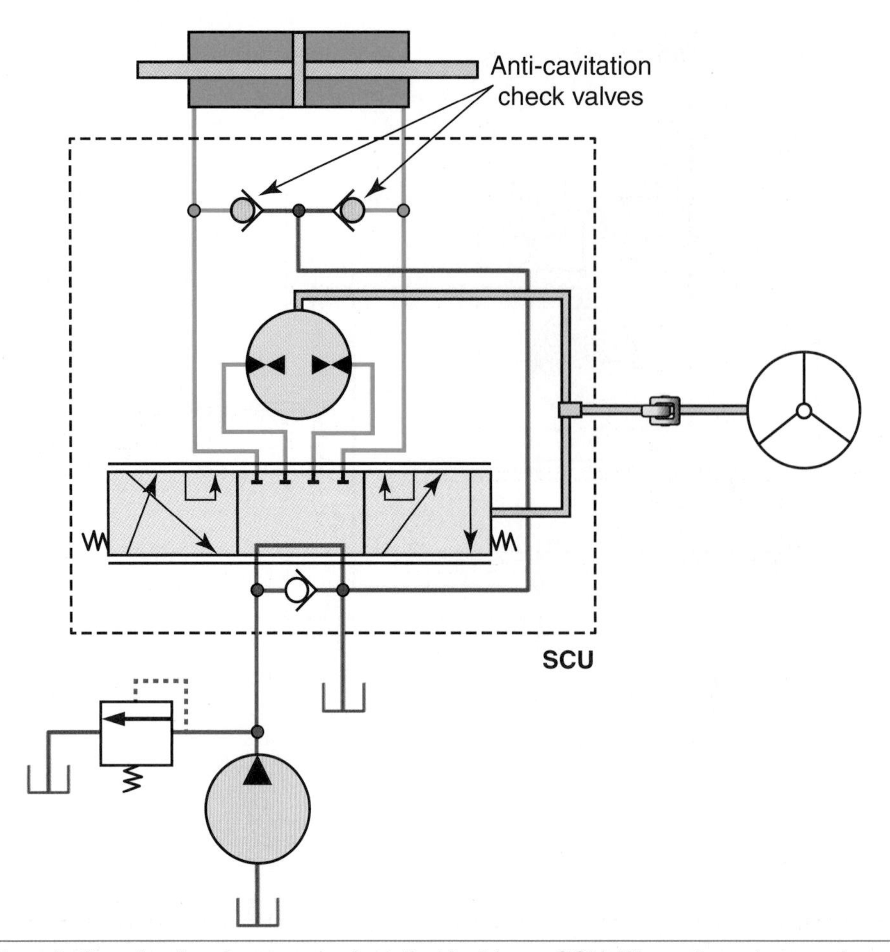

Figure 25-12. Anti-cavitation check valves can be installed inside an SCU. The valves draw make-up oil from the reservoir in the event that the steering cylinder(s) actuates fast enough that the pump alone is incapable of delivering all of the oil required.

Load Reaction

A ***load reaction SCU*** allows the external loads placed on the steering cylinder(s) to actuate the hydraulic steering system. The SCU's neutral position in Figure 25-13 has open passage between the gerotor assembly and the steering cylinder work ports. When the operator releases the steering wheel through the approximate midpoint of the turn, the forces being exerted on the tractor's axle will cause the steering cylinder(s) to return to the neutral position.

A single-acting differential cylinder cannot be used with a load reaction SCU. The steering cylinder(s) area must be the same for left and right steer. A double-rod cylinder will work, or two steering cylinders that have equal areas, plumbed in parallel will also work, as shown in Figure 25-13.

Wide-Angle Steering

A ***wide-angle SCU*** is used on large articulated tractors. It is specifically designed to smooth the bumpy steering motion characteristic of such vehicles. The SCU is able to smooth the steering motion by increasing the amount of deflection between the SCU's control spool and sleeve during a turn.

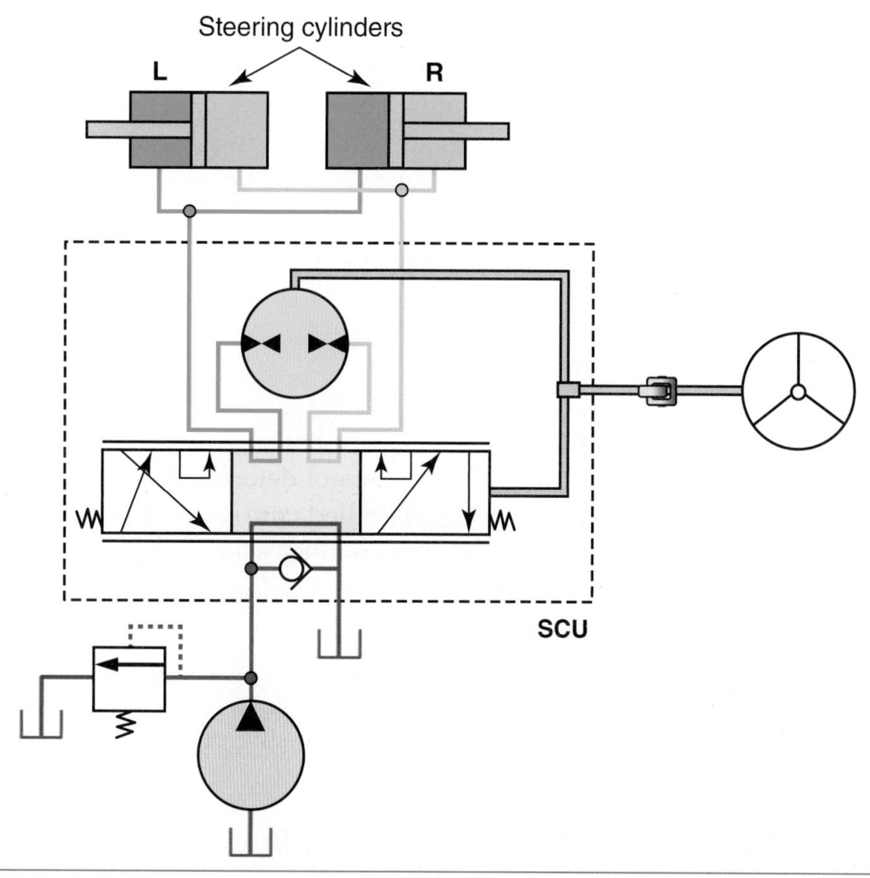

Figure 25-13. A load reaction SCU allows the external loads placed on the steering cylinder(s) to force the steering wheel back to a neutral position, which occurs when the operator releases the steering wheel midway through the turn.

Cylinder Dampening

A ***cylinder dampening SCU*** is designed to smooth the steering motion on large articulated tractors. The cylinder dampening is achieved by using adjustable orifices that divert a small amount of the fluid from the steering cylinder ports back to the reservoir whenever the steering wheel is turned.

Q-Amp Steering

The Eaton Q-Amp® SCU is designed to deliver a greater volume of steering oil than an equivalently sized traditional SCU. The increase in flow occurs when the steering shaft is rotated at a relatively high speed, such as 10 rpm. The SCU uses internal variable orifices to divert a portion of the oil straight to the steering cylinder rather than being metered through the gerotor assembly.

When the steering shaft is rotated at slower speeds—less than 10 rpm—the gerotor assembly meters all of the oil sent to the steering cylinders. The Q-Amp SCU can produce 60% more oil flow than the traditional gerotor is capable of delivering. The advantages of the Q-Amp SCU include the following:

- A smaller SCU can be used to deliver a larger amount of steering oil.
- When the vehicle is traveling at a high road speed, the SCU will not be overly reactive as the operator slowly rotates the steering wheel.
- It is capable of fast turns when the steering wheel is rotated quickly.

Traditional Steering Priority Valves

As mentioned in **Chapter 8**, the majority of mobile hydraulic systems use a steering priority valve to ensure that steering demands are met before sending the remainder of the oil to the secondary implement circuits. An example of a steering priority valve is shown in **Figure 25-14**.

A steering priority valve normally has five ports:

- Supply.
- Return.
- Load-sensing input.
- Excess flow (EF).
- Controlled flow (CF).

The priority valve's control spool determines whether the pump's supply oil is sent to the SCU, which is called ***controlled flow (CF)***, or if the oil is sent downstream to the secondary circuits, which is called ***excess flow (EF)***. The control spool is operated by three variables:

- Controlled flow pressure.
- Load-sensing pressure.
- Bias spring pressure.

CF pressure opposes both load-sensing pressure and spring pressure.

Steering Priority Valve Operation with the SCU in Neutral

If the closed-center SCU is in a neutral position (operator is not steering), the CF pressure will increase due to the blocked ports. The CF pressure is routed to the end of the control spool by means of an orifice and cross-drilled passageway through the center of the spool. With the SCU in a neutral position, there is no steering signal pressure.

The CF pressure builds until it can overcome the priority valve's bias spring. For this example, assume the bias spring equals 150 psi (10 bar) and the control spool's surface area equals one square inch. Once the CF pressure builds to 150 psi (10 bar), the control spool shifts, compressing the spring. All of the pump's supply oil is routed out of the EF port and sent to the forklift's mast DCV block. The forklift's mast DCVs are open centered and route the oil back to the reservoir at a low-pressure—such as 150 psi (10 bar).

Steering Priority Valve Operation with the SCU Steering Left or Right

As the operator rotates the steering wheel, the SCU develops a steering load-sensing signal pressure that is sent to the priority valve. The load-sensing pressure plus the bias spring pressure ensures that the control spool meets the oil demand requested by the SCU. After the SCU's oil demand is met, the CF pressure builds once again, assuming the fixed-displacement pump is producing excess oil. When the priority valve senses that the SCU's flow demand is met, the CF pressure overcomes the combined bias spring pressure and signal pressure, resulting in the control spool directing the remaining balance of oil flow to the secondary circuits (EF port).

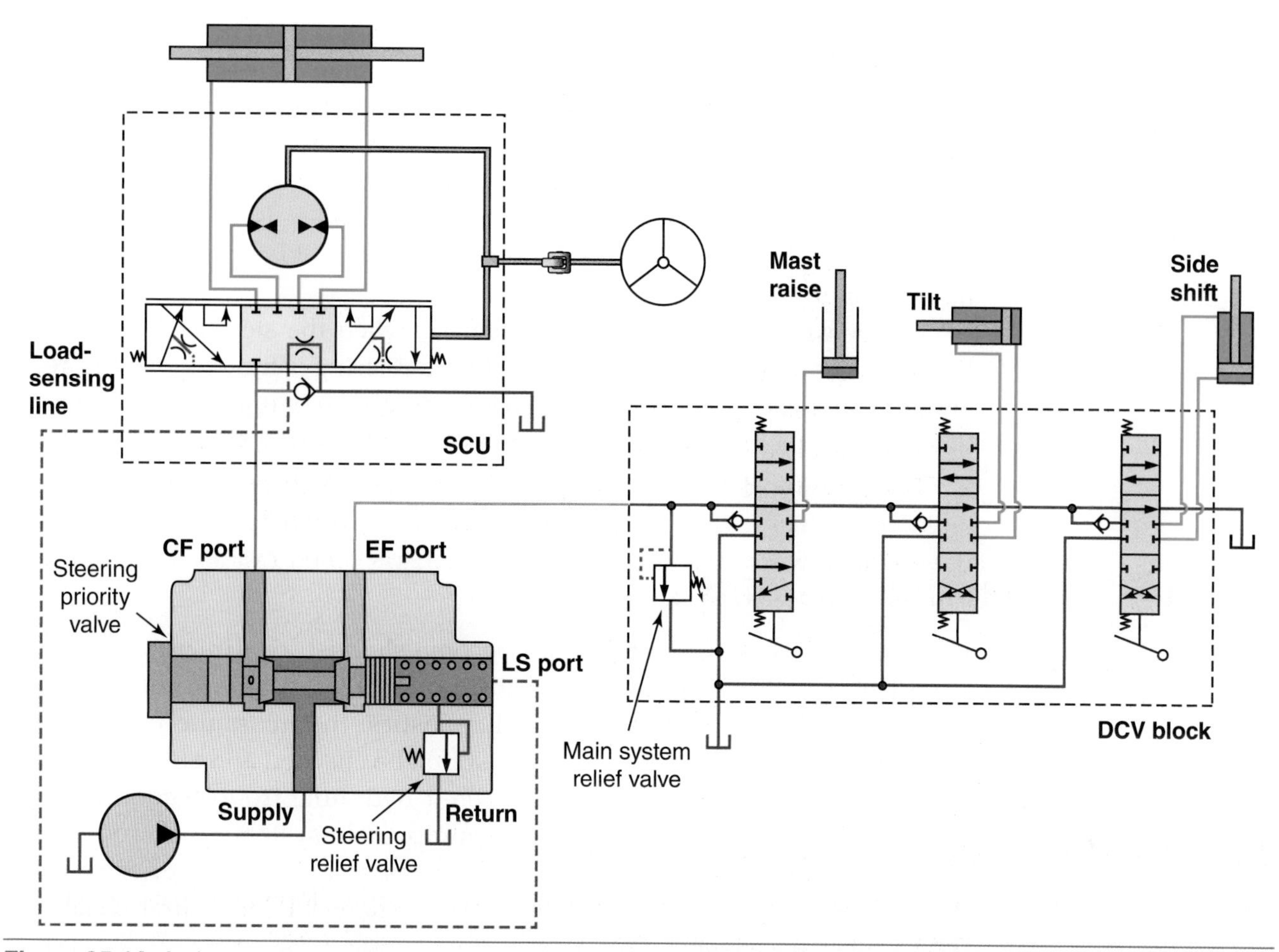

Figure 25-14. A closed-center load-sensing SCU is commonly used with a steering priority valve and a fixed-displacement pump. The schematic resembles a Case 586e forklift hydraulic system. Note that the main system relief valve is located downstream inside the forklift's DCV block and the steering relief is located in the priority valve, which dumps the steering LS pressure.

Steering Priority Valve Load-Sensing Relief

As mentioned earlier, the priority valve can contain a load-sensing relief valve, as depicted in the schematic in Figure 25-14. The load-sensing relief controls the steering's maximum system pressure by dumping the load-sensing signal line to the reservoir when the signal pressure reaches the LS relief valve's setting.

Load-Sensing Line

Steering priority circuits often use one of two types of signal line configurations, either a static load-sensing signal or a dynamic load-sensing signal. The ***static load-sensing signal*** is the standard type of signal configuration. The priority valve's control spool receives a static pressure signal, meaning that the oil in the signal circuit is not flowing while the SCU is in a neutral position. Static load-sensing lines are used when the steering system has sufficient steering performance, does not lack response, and the steering load-sensing signal lines are relatively short (less than six feet).

The ***dynamic load-sensing signal*** circuit is used if the machine's signal line is longer than six feet, if the machine's steering circuit is less than stable, or if the steering operation is inconsistent. When the SCU is in a neutral position, the priority valve is generating a small signal pressure. The dynamic steering signal pressure is the result of a small amount of oil flowing back to the reservoir through an adjustable orifice. The dynamic load-sensing signal improves hydraulic steering responsiveness and also helps in cold climates by warming up the circuit.

Engineers choose dynamic load sensing when the steering load-sensing line is longer than six feet or if the machine's steering is too sluggish (delayed) and needs to respond more quickly to the operator's commands. The dynamic flow of oil in the signal line generates a constant steering signal pressure, for example 50 to 150 psi, any time the tractor is running.

Diagnosing Steering Systems

Hydraulic steering systems can become sluggish or difficult to steer. A short troubleshooting process is used to diagnose such problems. Before a technician should proceed to SCU diagnostics, the hydraulic pump, main system relief valve, and return line must first be checked and ruled out as the cause of the problem. **Chapter 22** covered hydraulic troubleshooting principles. Presuming the steering system pump, main system relief, and return line are operating properly, the technician can also check the priority valve pressures (EF, CF, and LS). However, it is usually easier to first check the SCU's internal leakage.

Be sure to follow the manufacturer's service literature for testing an SCU's internal leakage. An example sequence for performing an ***SCU internal leakage test*** consists of the following steps:

- Turn the steering wheel clockwise or counterclockwise until the wheels reach their end of travel.
- Ensure that the system has reached the operating temperature specified in the testing procedure.
- Use a torque wrench to apply the specified amount of torque to the steering wheel retaining bolt. A typical torque specification is 60 inch-pounds. Apply the torque in the same direction that the steering wheel was turned.
- With the wheels steered fully against their stops, apply the specified amount of torque to the steering wheel in the same direction the steering wheel was turned. The steering wheel should rotate slowly. Internal leakage in the SCU allows the steering wheel to rotate even though the wheels are locked.
- Record the number of steering wheel revolutions completed in one minute. The manufacturer's literature will specify the proper torque to apply, and a maximum acceptable rotational speed for the steering wheel, such as 4 rpm with 60 inch-pounds applied.
- Repeat the test procedure, turning the steering wheel and applying torque in the opposite direction.

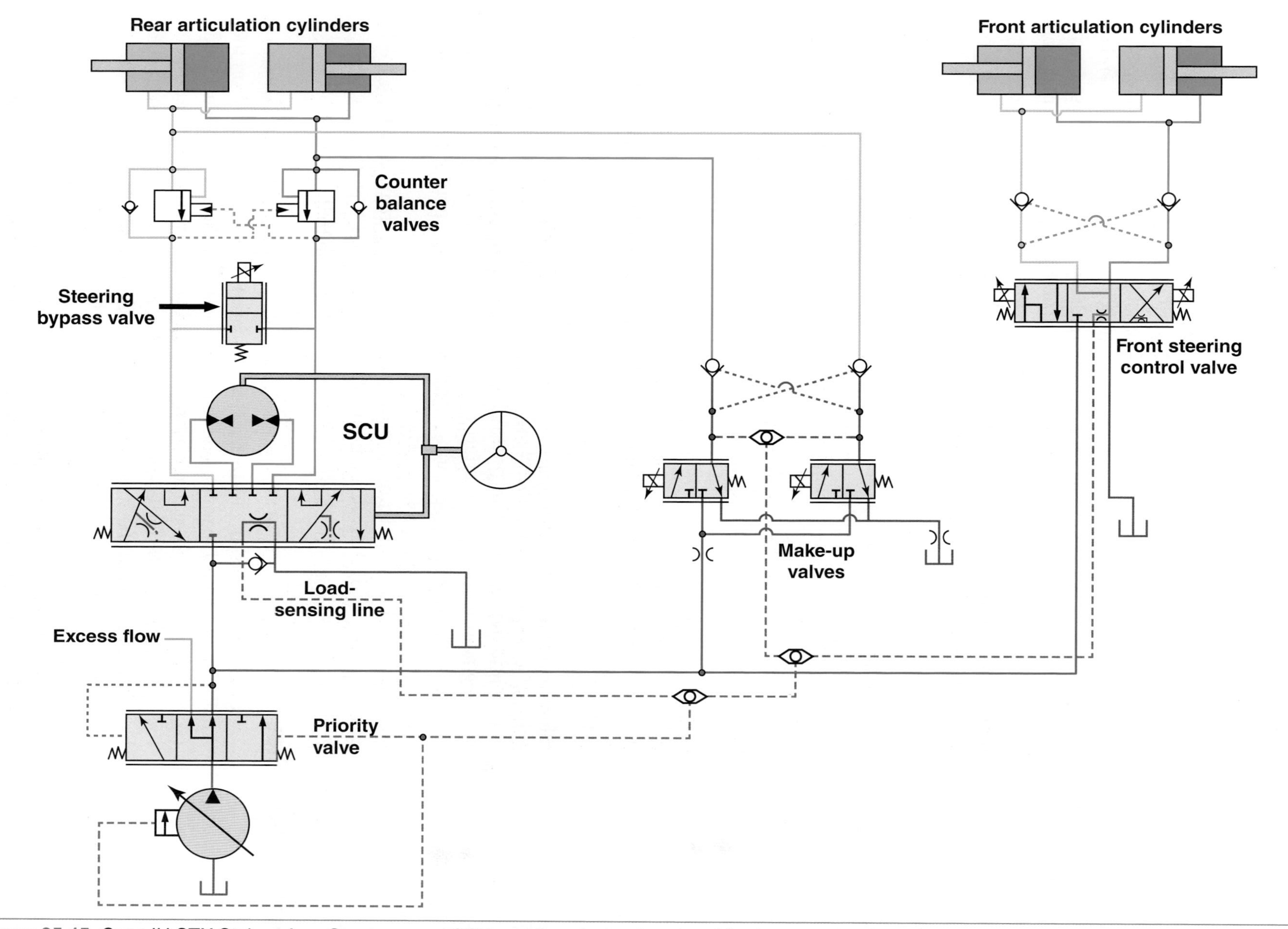

Figure 25-15. Case IH STX Steiger AccuSteer uses an SCU and five electronic solenoid valves to control the tractor's double-articulated steering. The SCU and make-up valves provide oil only to the rear articulation cylinders. The front articulation cylinders receive oil from the front steering solenoid valves.

If the rotational speed of the steering wheel exceeds the specification, the next step is to determine if the leakage is occurring in the steering cylinder or in the SCU. Most service literature recommends capping the steering cylinder's hydraulic hoses and repeating the tests. If the steering wheel's rotational speed is within specification with the cylinder hoses capped, the steering cylinder is internally leaking and needs to be rebuilt. If the speed is still too high with the steering cylinder hoses capped, the SCU has too much internal leakage and must be replaced.

If the SCU and steering cylinder(s) are found to be operating efficiently and the priority valve pressures meet requirements, the priority valve should be inspected to determine if the spring is broke, orifice is plugged, or if the spool is seized in its bore.

Electronic Steering Features

Many mobile machines no longer use SCUs, but instead use electronic controls to steer the tractor, also known as ***steer by wire***. The electronic controls coupled with global positioning satellite (GPS) systems allow manufacturers to offer automatic guidance steering systems. A tractor equipped with such a system will create less overlap of chemicals when spraying farm fields, which saves fuel and reduces input costs. The guidance systems also reduce operator fatigue.

Specific manufacturers offer other unique electronic steering features such as the John Deere ActiveCommand Steering and Caterpillar Quick Steer.

John Deere ActiveCommand Steering™

John Deere offers an ***ActiveCommand Steering system*** that uses electronic inputs, including a gyroscope to sense the tractor's balance, to aid the operator while steering. The steering system has multiple features, such as variable ratio steering and variable effort steering. ***Variable ratio steering*** optimizes the number of turns the steering wheel must rotate depending on the speed of the tractor and the degree of sharpness of the turn being performed. ***Variable effort steering*** adjusts the turning effort required by the operator to best control the tractor under changing operating conditions, such as more wheel resistance when turning at road speeds. The ActiveCommand Steering also eliminates steering wheel drift and prevents oversteering. An operator driving a tractor at road speed with ActiveCommand Steering will need to correct the steering wheel less than he or she would with a conventional steering system.

Caterpillar Quick Steer

Caterpillar offers an electronic steering mode, called ***Quick Steer***, on their small K series wheel loaders. The feature is best used when the wheel loader is cycling in short stops while loading trucks. With the machine in neutral and the parking brake off, the operator presses the appropriate button on the electronic monitor to put the steering system into Quick Steer mode.

On entering the Quick Steer mode, the machine's transmission is shifted into the first gear range. In this range, the loader has adjustable ground speed control and generates more flow for steering. Previously, the steering wheel needed to be turned three revolutions to fully steer from stop to stop when the loader was stationary. The Quick Steer mode dramatically decreases the operator's turns of the steering wheel. It allows the operator to fully steer the loader from left to right (or vice versa) by only rotating the steering wheel from the 10 o'clock position to the 2 o'clock position.

Case IH 4WD Combination Steering

For several decades, small- and mid-frame Case IH four-wheel-drive Steiger tractors have offered combination steering, which later became termed AccuSteer™. The feature allows a customer to choose between two options:

- Articulation only steering.
- Combination articulation and front axle steering.

In the combination steer mode, the front axle is steered first. After the front articulation is steered 10°, the rear articulation is pivoted an additional 42°, providing a total of 52° of articulated steering.

The early Steigers used a conventional SCU along with an electronically controlled double-selector valve. When the double-selector valve was de-energized, the valve directed oil to the articulation cylinders. When the steering electronic control module energized the double-selector valve, the valve directed oil to the front axle steering.

Late-model STX Steiger AccuSteer tractors incorporate the use of electronic steering controls in conjunction with a traditional SCU to enable GPS auto guidance. The STX Steiger's steering controller monitors the following input data:

- Rear articulation axle potentiometer.
- Front articulation axle potentiometer.
- Steering control pressure sensor.
- Steering wheel motion sensor.
- Transmission output speed sensor.

Based on data received from the inputs, the steering controller can energize the following outputs:

- Front steer left solenoid valve.
- Front steer right solenoid valve.
- Left make-up solenoid valve.
- Right make-up solenoid valve.
- Selector bypass solenoid valve.

The steering priority valve delivers oil to the three control valves in parallel, as shown in **Figure 25-15**:

- Front steering solenoid control valves.
- Make-up solenoid valves.
- SCU.

The SCU only controls the rear articulation steering cylinders. The bypass solenoid valve is a normally closed solenoid. When the bypass is de-energized, the SCU oil actuates the rear articulation cylinders. When the bypass valve is energized, a portion or all of the SCU oil can be bypassed and sent back to the reservoir instead of actuating the rear articulation cylinders. The make-up solenoids are used to maintain the position of the rear articulation cylinders.

The front steering solenoid valves directly control and maintain the position of the front articulation steering cylinders. The SCU, make-up solenoids, and bypass valve only affect the rear articulation cylinders.

Summary

- ✓ SCUs provide a true hydraulic steering system with no mechanical connection between the SCU and the steering cylinder.
- ✓ The purpose of the SCU's gerotor assembly is to meter the quantity of the steering oil.
- ✓ The purpose of the SCU's control spool/ sleeve assembly is to control the direction of oil flow to a tractor's steering cylinder(s), which steers the tractor to the left or to the right.
- ✓ SCUs can be called hand metering units, hydrostatic steering, steering hand-pumps, or steering control valves.
- ✓ A traditional open-center or closed-center SCU has only four hydraulic ports (supply, return, left steer, and right steer).
- ✓ Most SCUs are closed center and contain a fifth hydraulic port for the load-sensing signal line.
- ✓ A load-sensing SCU is required if the tractor contains a steering priority valve.
- ✓ An SCU contains a check valve at its inlet that connects the supply line to the return line, allowing the SCU's gerotor assembly to be used like a hand pump to supply emergency steering oil in the event of engine or hydraulic pump failure.
- ✓ A check valve located in series to the SCU's supply is used to prevent steering kickback.
- ✓ A steering relief valve can be located inside the SCU or inside the steering priority valve.
- ✓ The neutral position of a non-load reaction SCU blocks off the steering cylinder actuator ports.
- ✓ The neutral position of a load reaction SCU has open actuator work ports.
- ✓ Wide-angle steering and cylinder dampening are used in large articulated tractors to smooth the bumpy steering motion.
- ✓ Q-Amp steering provides conventional gerotor metering if the steering wheel is turned slowly and increased flow (up to 60% more) if the steering wheel is turned quickly.
- ✓ Controlled flow is the priority valve port that supplies oil to the SCU.
- ✓ Excess flow is the priority valve port that supplies oil to the secondary circuits.
- ✓ A dynamic load-sensing signal line generates a low pressure steering signal by constantly flowing oil in the steering signal line. They are used when the steering signal line is six feet or longer, or if the steering response is slow, less stable, or inconsistent.
- ✓ SCUs have a specified internal leakage specified in steering wheel rpms.
- ✓ Modern mobile hydraulic machines are steered by a network of electronic controls, called a steer-by-wire system, instead of an SCU.
- ✓ John Deere's ActiveCommand steering system provides variable ratio steering and variable effort steering. It eliminates steering wheel drift and prevents oversteering.
- ✓ Caterpillar Quick Steer provides an increase in steering oil flow when the loader is operating at a slow speed, enabling the operator to fully steer the loader with minimal movement of the steering wheel.
- ✓ Case IH 4WD combination steering (AccuSteer) provides both center articulation and front articulation steering.

Technical Terms

ActiveCommand Steering system
auto guidance steering systems
controlled flow (CF)
cylinder dampening SCU
dynamic load-sensing signal
emergency steering
excess flow (EF)
load reaction SCU
manual steering check valve
manual steering mode
non-load reaction SCU
Quick Steer
SCU internal leakage test
static load-sensing signal
steer by wire
steering control unit (SCU)
steering hand-pump
steering kickback
variable effort steering
variable ratio steering
wide-angle SCU

Review Questions

Answer the following questions using the information provided in this chapter.

1. The steering drive shaft splines to which SCU component?
   A. Control sleeve.
   B. Gerotor.
   C. Control spool.
   D. Centering pin.
2. Which SCU component is directly responsible for rotating the gerotor?
   A. Control sleeve.
   B. Drive shaft.
   C. Control spool.
   D. Centering springs.
3. Which SCU component is responsible for metering the quantity of steering oil that is sent to the steering cylinder?
   A. Drive shaft.
   B. Inlet check valve.
   C. Control spool/sleeve assembly.
   D. Gerotor assembly.
4. Which SCU component is responsible for directing the oil in and out of the SCU?
   A. Drive shaft.
   B. Inlet check valve.
   C. Control spool/sleeve assembly.
   D. Gerotor assembly.
5. Technician A states an SCU capable of working in an auto guidance steering system does not need to be electronically controlled. Technician B states that SCUs have no mechanical linkage between the SCU and the steering cylinder. Who is correct?
   A. Technician A.
   B. Technician B.
   C. Both A and B.
   D. Neither A nor B.
6. A SCU can be called by all of the following terms, *EXCEPT*:
   A. hand metering unit.
   B. steering hand-pump.
   C. steering control valve.
   D. selective control valve.
7. What two SCU components are coupled with a centering pin and centering springs?
   A. Gerotor and wear plate/spacer.
   B. Control spool and control sleeve.
   C. Drive shaft and gerotor.
   D. Control spool and end plate.

8. All of the following parts of an SCU are stationary, *EXCEPT*:
   A. the housing.
   B. the gerotor ring gear.
   C. the control spool.
   D. the wear plate/spacer.
9. In the dead engine steering mode, which of the following components allows the SCU to be operated as a small hand-pump to provide oil for steering?
   A. Manual steering check valve.
   B. Main system relief valve.
   C. Steering priority valve.
   D. Make-up valve.
10. The inlet check valve located in series to the SCU's supply is responsible for _____.
    A. preventing cavitation
    B. preventing steering kickback
    C. protecting the steering cylinder when the SCU is in a neutral position
    D. minimizing hydraulic steering pressure
11. Which feature should be added to the trapped oil lines after the control spool/ sleeve assembly in a closed-center SCU to prevent pressure surges that may damage steering cylinder(s) or hoses?
    A. Fuses.
    B. Accumulator.
    C. Anti-cavitation check valve.
    D. Cylinder report relief valves.
12. What task does an SCU's inlet relief valve perform?
    A. Prevents cavitation.
    B. Prevents steering kickback.
    C. Protects the steering cylinder when the SCU is in a neutral position.
    D. Limits maximum hydraulic steering pressure.
13. Technician A states that a load-sensing SCU features five hydraulic ports. Technician B states that open-center SCUs are used with priority valves in load-sensing pressure-compensating hydraulic systems. Who is correct?
    A. Technician A.
    B. Technician B.
    C. Both A and B.
    D. Neither A nor B.
14. Which SCU component is responsible for keeping the control spool/sleeve assembly aligned while the SCU is in a neutral position?
    A. Centering pin.
    B. Centering springs.
    C. Drive shaft.
    D. Gerotor.
15. Which of the following choices lists the correct route of oil flow through an SCU during an active steering condition?
    A. Oil enters SCU, directed through the gerotor, routed to the control spool/ sleeve, directed to the steering cylinder, back to SCU through the spool/sleeve, and routed back to the tank.
    B. Oil enters SCU, directed through control spool/sleeve, routed through gerotor, directed through control spool/sleeve, directed to steering cylinders, return oil routed back to SCU through the control spool/sleeve, directed out of SCU back to the tank.
    C. Oil enters SCU, directed through the control spool/sleeve, routed to the steering cylinders, routed back to control spool/sleeve, directed through the gerotor, back to control spool/sleeve, and routed to the tank.
    D. Oil enters SCU, directed through the control spool/sleeve, routed through the gerotor, flow sent to the main system relief valve, and directed to the tank.

16. Which of the following steering priority valve ports supplies inlet flow to the SCU?
    A. Controlled flow.
    B. Load-sensing.
    C. Excess flow.
    D. Hydraulic pump supply.
17. Within a steering priority valve, if the SCU is in a neutral position, where is the majority of the oil exiting the priority valve?
    A. Excess flow.
    B. Controlled flow.
    C. Load-sensing.
    D. Return.
18. A tractor is using a fixed-displacement hydraulic pump, a steering priority valve, and a load-sensing SCU in its hydraulic steering system. Where is the signal pressure being sent?
    A. Hydraulic pump.
    B. Steering priority valve.
    C. Both A and B.
    D. Neither A nor B.
19. A tractor is using a LSPC hydraulic pump, a steering priority valve, and a load-sensing SCU in its hydraulic steering system. Where is the signal pressure being sent?
    A. Hydraulic pump.
    B. Steering priority valve.
    C. Both A and B.
    D. Neither A nor B.
20. What is a typical amount of torque applied to the steering wheel for an SCU internal leakage test?
    A. 5 inch pounds
    B. 60 inch pounds
    C. 120 inch pounds
    D. 220 inch pounds
21. The SCU leakage test results exceeded specifications. The technician capped the hydraulic lines going to the steering cylinder and repeated the test. The second test results were within specifications. What is the problem?
    A. SCU.
    B. Steering cylinder.
    C. Both A and B.
22. A technician is reading a service manual to diagnose a possible leaking SCU. The service literature specification for the SCU is 4 revolutions per minute. If the SCU is good, what type of results will the technician obtain during the test?
    A. 4 or fewer revolutions per minute.
    B. 4 or more revolutions per minute.
23. Which SCU configuration is designed to block both left and right steering cylinder ports when the SCU is in a neutral position?
    A. Load reaction.
    B. Load sensing.
    C. Non-load reaction.
    D. Q-Amp steering.
24. Which SCU configuration is designed to allow the external forces placed on a steering cylinder to return the SCU to a neutral position once the operator releases the steering wheel at mid-turn?
    A. Load reaction.
    B. Load sensing.
    C. Non-load reaction.
    D. Q-Amp steer.
25. Which SCU configuration smoothes the steering operation of an articulated tractor by increasing the amount of deflection between the SCU's control spool and sleeve during a turn?
    A. Wide-angle steering.
    B. Cylinder dampening.
    C. Load reaction.
    D. Non-load reaction.
26. Which SCU configuration smoothes the steering operation of an articulated tractor by means of adjustable orifices?
    A. Wide-angle steering.
    B. Cylinder dampening.
    C. Load reaction.
    D. Non-load reaction.
27. Which type of load sensing signal is used when the signal line is 6′ or longer?
    A. Static load sensing signal.
    B. Dynamic load sensing signal.

28. Which electronic steering option provides variable ratio steering and variable effort steering?
    A. John Deere ActiveCommand Steering.
    B. Caterpillar Quick Steer.
    C. Auto guidance steering system.
    D. Case IH AccuSteer.
29. Which electronic steering option provides a full stop-to-stop steer by rotating the wheel from the 10 o'clock position to the 2 o'clock position?
    A. John Deere ActiveCommand Steering.
    B. Caterpillar Quick Steer.
    C. Auto guidance steering system.
    D. Case IH AccuSteer.
30. Which electronic steering option provides both articulation and front axle steering?
    A. John Deere ActiveCommand Steering.
    B. Caterpillar Quick Steer.
    C. Auto guidance steering system.
    D. Case IH AccuSteer.

Image Credits

Chapter Openers. tdhster/Shutterstock.com

Chapter 1

Figure 1-5. ten43/Shutterstock.com; Kenneth Sponsler/ Shutterstock.com

Figure 1-6. Rob Byron/Shutterstock.com

Chapter 2

Figure 2-1. Pecold/Shutterstock.com

Chapter 3

Figure 3-5. AridOcean/Shutterstock.com

Chapter 4

Figure 4-6. Peter Ivanov Ishmiriev/Shutterstock.com

Figure 4-38. Lukas Beran/Shutterstock.com

Chapter 5

Figure 5-26B. CNH Industrial

Chapter 6

Figure 6-6. TRIG/Shutterstock.com

Figure 6-10. Brian McEntire/Shutterstock.com

Figure 6-20. L Barnwell/Shutterstock.com

Figure 6-24. Bryan Bell

Chapter 10

Figure 10-3B. Alexey Stiop/Shutterstock.com

Chapter 11

Figure 11-9. Scrugglegreen/Shutterstock.com

Chapter 14

Figure 14-19. Artic Fox® a brand of Phillips & Temro Industries

Additional Art

Page 124. safakcakir/Shutterstock.com

Page 292. Vereshchagin Dmitry/Shutterstock.com

Page 318. Dima Fadeev/Shutterstock.com

Page 378. mihalec/Shutterstock.com

Page 398. Taina Sohlman/Shutterstock.com

Page 416. Taina Sohlman/Shutterstock.com

Page 456. Naypong/Shutterstock.com

Page 534. Corepics VOF/Shutterstock.com

Page 552. Dmitry Kalinovsky/Shutterstock.com

Glossary

A

absolute rating. Largest pore size of a filter's element, measured in microns. (11)

accumulator: Vessel used to store fluid pressure. (2)

acid number (AN). Measure of an oil's acidic contamination and additive depletion. It is also an indirect measure of the fluid's rate of oxidation. (12)

ActiveCommand Steering™ system. Electronic steering system designed by John Deere that incorporates the inputs from a variety of on-board sensors to automatically assist the operator's steering. (25)

actuator filters. High-pressure filters that must allow and provide filtration for oil flow in both directions. (11)

air cooler. Oil cooler that transfers heat from the oil to the surrounding air. (14)

air separation ability (ASA). Ability of a hydraulic fluid to separate entrained air. (10)

attenuator. Noise suppression device used in hydraulic systems. It consists of a heavy steel shell and tubes and is designed to reduce the fluid pulsations that normally cause noise to resonate. (2)

auto guidance steering systems. Steering systems that use electronic controls coupled with a GPS system to program and direct the machine's path of movement. (25)

B

baffle. Divider plate located inside a reservoir to prevent return oil from immediately cycling back into the pump's inlet. (14)

bearing plate. Plate that is pinned to the end of the barrel and rotates along with the barrel assembly. (4)

bent-axis piston motors. Hydraulic piston motors that do not contain a swash plate. The motor gets its fixed angle from the difference in angle between the shaft and the motor's barrel. (5)

Beta ratio. Ratio of the number of particles injected upstream compared to the number of particles that were able to pass through the filter and make it downstream during the test. (11)

bias piston. Piston that attempts to push the swash plate to a maximum angle, increasing displacement. It works in opposition to the control piston. (4)

bias spring. Spring that acts to push the swash plate to a maximum angle, increasing displacement. It works in opposition to the control piston. (4)

biodegradable fluids. Hydraulic fluids that are less toxic, less harmful to the environment, and degrades much faster than traditional hydraulic fluid. (10)

bladder accumulator. Type of gas accumulator in which the gas charge is contained in a synthetic rubber bag, called a bladder. (13)

bleeder valve. Valve that is used to depressurize a sealed reservoir. (14)

boundary lubrication. Film developed between two sliding components that come into close contact with each other, to the point that tiny edges on their surfaces, known as asperities, can break-off and cause friction and heat. (10)

Bourdon-tube pressure gauge. Instrument that measures pressure by means of mechanically transferring the pressure-generated deflection of the metal Bourdon tube to movement of the gauge's needle. It is the most common style of mechanical pressure gauge in use. (21)

breather. Device that prevents contaminants, and in some cases moisture, from entering a vented reservoir. (11)

bridge passageway. DCV fluid duct that routes oil from the pressure compensator valve to the DCV spool in a flow-sharing hydraulic system. (19)

bulk modulus. Measure of a fluid's resistance to compression. (2)

burst discs. Thin, circular pieces of rated metal placed inside a flowmeter and manufactured to rupture if the meter is over-pressurized. They act in the same manner as a fuse in an electrical circuit. (21)

burst pressure. Minimum pressure that will rupture a hose. (15)

C

cam lobe motor. Most popular type of radial piston motor used in the mobile hydraulic industry. It is commonly used in hydrostatic transmission applications. (5)

cap end. End of a hydraulic cylinder opposite the rod end. (6)

case drain. Oil leakage inside a pump or motor's case. (4)

case drain line. Unpressurized return line that carries oil from the pump or actuator case to the reservoir. (14)

centistokes (cSt). International System of Units (SI) unit for measuring viscosity. (10)

changing energies. Transforming one type of force into a different form of power. (22)

charge pressure filter. Filter that protects components that are located downstream from the charge pump. (11)

charge pump. Auxiliary pump used to supercharge the main pump's inlet with oil flow. (4)

clean-out plate. Removable plate that provides access for cleaning a reservoir. (14)

closed-loop HSTs. Hydrostatic transmissions in which the motor returns its oil to the pump. As a result, the pump does not need to draw all of its oil from the reservoir. (24)

closed-loop pump. Pump that relies on an external source to help supply a positive charge of oil into the pump's inlet. The external source may be return oil from the actuator or a charge pump. (4)

combination drawings. Drawings that are some combination of cutaway, graphic, pictorial, and exploded component drawings. (3)

commissioning. Term used by the National Fluid Power Association referring to the process of returning a hydraulic system to service following pump or motor replacement. (24)

companion pump pressure. Outlet pressure of a secondary piston pump. (20)

compensator passageway. DCV fluid duct that routes oil from the DCV spool to the pressure compensator valve in a flow-sharing hydraulic system. John Deere uses this term in their construction equipment service literature. See also *feeder passageway* or *intermediate passageway*. (19)

compressibility. Ease with which a certain mass of liquid can be reduced in volume. Mathematically, it is the reciprocal of bulk modulus. (10)

constant horsepower control. Name Kawasaki Hydraulics gives to a negative flow control regulator that senses its own pump outlet pressure. (20)

controlled flow (CF). Oil that is routed through a priority valve's outlet port. (8)

controlled flow (CF). Hydraulic pump oil flow that is delivered to the hydraulic steering control unit via the steering priority valve. (25)

control piston. Piston that attempts to push the swash plate back to a neutral angle, thereby shutting off the pump flow. Works in opposition to the bias spring or bias piston. (4)

corner horsepower. Point on a PQ curve that shows the total amount of horsepower required to deliver both maximum system hydraulic pressure and maximum system hydraulic flow simultaneously. (18)

counterbalance valves. Valves used to control loads when the cylinder is operating an overrunning load. The counterbalance valve prevents cavitation, provides load-holding ability, and prevents a load from extending a cylinder too fast. (7)

cover. Outer layer of a hose that protects it from the elements, such as chemicals, weather, and abrasion. (15)

cracking pressure. Pressure at which a relief valve begins to separate from its seat, allowing oil to flow. (7)

cradle bearing axial piston pump. Style of pump uses two bearings or bushings in a cradle-shape form, similar to the main bearings in an engine. The cradle bearings allow the swash plate to pivot. (4)

crankcase outlet valve. Valve installed as part of the pump compensator in a radial piston variable-displacement hydraulic pump. It upstrokes the pump by draining oil out of the pump's crankcase, which causes piston reciprocation due to the low case pressure. (17)

creeper control. Feature found in off-highway equipment. It enables very slow travel speeds while allowing a large amount of implement-hydraulic flow to operate attachments such as brooms, brush cutters, or snow blowers. (24)

crescent internal-toothed gear pump. Internal-toothed gear that uses a crescent spacer to separate the external-toothed drive gear and the internal-toothed driven gear. (4)

crimping. Process of forcing hose ends through dies with finger-like protrusions that create pleats in the fitting to permanently affix it to the hose. (15)

cross sensing pressure. Average pump outlet pressure between both NFC pumps on Caterpillar NFC excavators. (20)

cross-over relief valves. Pressure relief valves that dump relief oil back into the motor's inlet. (5)

cushion. Internal plunger that blocks off a large portion of the cylinder's flow to slow the cylinder as it reaches the end of its travel. (6)

customer satisfaction index (CSI). Rating system based on client surveys that evaluates a company's product/service, reputation, public image, and customer happiness. (22)

cutaway drawing. Drawing that depicts an assembly that has been virtually sliced in half, or cutaway, for the purpose of showing its internal parts. (3)

cycle time. Amount of time it takes an actuator to either fully extend, fully retract, or both fully extend and fully retract. (8)

cylinder block. Barrel containing the pump cylinders that is splined to the piston pump's input shaft. (4)

cylinder dampening SCU. Steering control unit configured to smooth steering motion by using adjustable orifices that divert some of the fluid from the steering cylinder ports back to the reservoir whenever the steering wheel is turned. (25)

cylinder port relief valves. Reliefs that are located between a DCV spool and actuator. They are normally direct-acting valves. Also known as *line-relief valves* or *actuator-relief valves*. (7)

D

dash size. Size designation for hydraulic hoses and tubes. For hoses, the dash size equals the hose's inside diameter in sixteenths of an inch. For tubes, the dash size equals the tube's outside diameter in sixteenths of an inch. (15)

dead-headed. Refers to a cylinder that is not moving because it has stalled or reached the end of its travel. (7)

delta zero. Electronic digital pressure meter feature that sets the existing pressure reading as the base zero number and shows changes from this pressure in units of positive or negative numbers. (21)

depth filter. Filter element consisting of multiple layers of filtering media or a single thick layer of media. (11)

destroked. Adjusted to have reduced piston movement, which results in reduced flow. Also used to describe a variable-displacement pump that is producing very little flow. (17)

destroke valve. Name of a valve assembly in Case excavators that consists of an NFC orifice and an NFC relief valve inside a single valve assembly. (20)

destroking solenoid valve. Electronically actuated valve that reduces hydraulic pump displacement to cutoff pump outlet flow. (17)

diagnostic pressure taps. Test ports in a hydraulic system that are used to measure system pressure. Their design speeds the process of measuring pressures and also reduces the risk of system contamination during testing. (21)

diaphragm accumulator. Small vessel that contains a flexible metal diaphragm with a synthetic rubber diaphragm that separates the gas charge and the hydraulic oil. (13)

differential cylinder. Double-acting hydraulic cylinder that is equipped with a single rod. (6)

differential pressure. Difference in pressure between the motor's inlet and the motor's outlet. (5)

differential pressure. Pressure variation between hydraulic pump outlet pressure and the highest system working pressure. See also *margin pressure*. (18)

diffuser. Perforated screen or a perforated housing inside a reservoir that reduces fluid velocity (14)

dime valves. Term used for shuttle valves in John Deere agricultural equipment service literature because of the part's visual resemblance to the US ten cent coin. (18)

direct-acting relief valve. Relief valve that contains a single spring. Also known as a *simple relief valve*. (7)

dithering. Constant slight valve motion or agitation intended to prevent stiction and minimize hysteresis. (9)

double-acting. Type of hydraulic cylinder that uses fluid pressure to extend the cylinder and fluid pressure to retract the cylinder. (6)

downstream compensation. Pressure compensation design that places the pressure compensator valve after a DCV spool within the hydraulic system. See also *post-spool compensated*. (19)

drain valve. Type of valve installed in reservoirs to aid in draining and kidney loop filtration. It contains a built-in check valve that is opened when the mating fitting is threaded onto the drain valve and closes when the fitting is loosened. See also *ecology valve*. (12)

dual-path HST. Hydrostatic drive system consisting of two separate hydrostatic transmissions for the purpose of propelling the machine and steering the machine. One pump and one motor will be used to drive the left side of the machine and one pump and one motor will be used to drive the right side of the machine. (24)

dump valve. Valve used to drain an accumulator by routing the charged oil to the reservoir. (13)

dump valve. A valve used in open-center hydraulics systems as a means to build maximum system pressure. Also known as *jammer solenoid valve*. (16)

duplicating the problem. Creating the exact conditions and symptoms of a machine's flaw for the purpose of accurately diagnosing and correcting the issue. Also known as *reproducing the symptom*. (22)

duty cycle. Percentage of time that a solenoid is energized during a given time span. (9)

dynamic-load sensing signal. Steering signal that has constant flow and pressure. It is a more responsive signal system than a static steering signal system. (25)

E

ecology valve. Type of valve installed in reservoirs to aid in draining and kidney loop filtration. It contains a built-in check valve that is opened when the mating fitting is threaded onto the drain valve and closes when the fitting is loosened. See also *drain valve*. (12)

electro-hydraulic (EH) PFC hydraulic system. System in which an electronic control module (ECM) is used to energize a proportional solenoid valve (PSV), which, in turn, directs pilot oil pressure to the pump regulator valve to upstroke the pump. (20)

electronic anti-stall control. Normally closed two-position destroke solenoid that is modulated by an ECM in order to limit stalling of the pump. (24)

electronic differential pressure-sensing flowmeters. Flowmeters designed with two pressure sensors that measure the pressure before the meter's internal restriction and after the meter's restriction to calculate flow rate. The meters operate on the Bernoulli principle. (21)

electronic positive-displacement flowmeter. Style of flowmeter that uses an electronic pulse generator to develop a signal based on the volume of oil passed through the meter during each revolution of its internal components. It is the only style of flowmeter that directly measures the volume of oil flowing through the circuit. (21)

electronic turbine flowmeters. Flowmeters designed with an electronic pickup that senses the rotational speed of the in-meter turbine as it spins at a speed proportional to fluid flow. (21)

element collapse pressure. Pressure at which a filter's internal media collapses. (11)

elemental analysis. Laboratory procedure used to identify traces of elements within the oil. Also known as *elemental wear analysis*. (12)

emergency steering. Controlling the actuation of a hydraulic steering system manually in the event of an engine or hydraulic failure. See also *manual steering mode*. (25)

engineering specifications. Exact measurements, criteria, and tolerances of a machine's part, assembly, or system that serve as benchmarks for efficient machine operation. (22)

Environmental Protection Agency (EPA). United States' federal agency that is responsible for protecting the nation's land and bodies of water. (1)

excess flow (EF). Flow that is dumped through a valve's bypass port. (8)

excess flow (EF). Hydraulic pump oil flow that is delivered to the secondary circuit(s) via the steering priority valve after the steering control unit's oil demand is met. (25)

exploded component drawing. Drawing that shows each of the individual parts that are located inside of a component or assembly. (3)

external-toothed gear motors. Hydraulic motors that have two spur gears. One gear is called the drive gear, which is coupled to the output shaft. The other external-toothed gear is the idler. As fluid is supplied to the motor's inlet, the pressure causes the gears to rotate. (5)

external-toothed gear pump. Gear pump with two external-toothed gears, a drive gear and a driven gear. The displaced oil travels around the outside perimeter of the pump. (4)

F

feedback link. Link that provides a direct mechanical feedback of the position of the swash plate to a pump's control valve. (20)

feeder passageway. DCV fluid duct that routes oil from the DCV spool to the pressure compensator valve in a flow-sharing hydraulic system. Caterpillar uses this term in their service literature. See also *compensator passageway* or *intermediate passageway*. (19)

field attachable. Able to be attached by a technician in the field, without the need of a crimping machine or a swaging machine. (15)

filter bypass valve. Valve that opens due to the pressure differential through a clogged filter element. It allows oil to bypass the filter if it becomes excessively restrictive. (11)

filter indicators. Devices that measure a filter's incoming pressure and the outgoing pressure. As the pressure difference approaches and exceeds the design limit, the indicator provides a visual indication that the filter must be replaced. (11)

filters. Filtration devices designed to remove particles that are finer than those removed by a strainer. (11)

fire point. Temperature at which a fluid can sustain a continued fire after the ignition source as been removed. (10)

fire-resistant hydraulic fluids. Fluids that have elevated flash and fire points. These fluids are denoted by the ISO with an HF prefix. (10)

fixed-displacement motor. Hydraulic motor that operates at a constant speed for a given amount of input flow. (5)

fixed-displacement pump. Pump in which the chambers have a set volume and produce a fixed amount of flow for a given pump speed. (2) (4)

flare fitting. Type of fitting that mates a cone-shaped flare on the male fitting to a tube that has a matching angled seat. Also known as a *flared tube fitting*. (15)

flaretite seal. Metal, cone-shaped seal with Loctite™ sealant baked onto it. It is designed specifically for flare fittings. (15)

flash point. Temperature at which a fluid will temporarily ignite when exposed to an ignition source. (10)

flats from finger tight (FFFT). Procedure for tightening a fitting by first tightening the fitting's hex nut finger tight and then continuing to tighten it with a wrench until a specified number of flats on the hex nut have passed a reference point on the fitting body. (15)

float. Function of a hydraulic system that allows an implement such as a loader bucket, dozer blade, or motor grader blade to glide along a hard surface. (9)

flooded pump inlet. Pump inlet that is located below the reservoir, which ensures that gravity helps feed fluid into the inlet. (10)

flow. Quantity of fluid moving past a given point during a specified time period. (2)

flow control spool. Term used to describe the pump unloading control device that is part of an LSPC variable-displacement hydraulic pump. Also known as a *margin spool valve* or *low-pressure standby spool valve*. (18)

flow limit solenoids. Two valves used on some 300 series Caterpillar excavators. The valves develop a false NFC signal pressure to ensure a constant amount of flow is available to the excavator's work tool circuit. (20)

flowmeter. Instrument designed for measuring the volume of oil flow in a hydraulic circuit or system. See also *flow rater*. (21)

flow rate pilot valves. Two valves in a pilot signal manifold that control the pump's flow rate in a PFC hydraulic system. (20)

flow rater. Volumetric flow measurement tool. (21)

flow-sharing hydraulic system. Hydraulic system design that when the combined flow request from all of the DCVs is more than the hydraulic pump can maintain, the highest working pressure acts on all of the pressure compensator valves, and those valves simultaneously begin to proportion flow to the actuators based on the percentage of flow requested from each of the DCVs. It requires *downstream compensation*. (19)

fluid coupling. Hydrodynamic drive that transfers power and acts like an automatic clutch, but is not capable of multiplying torque due to its design. (24)

fluid injection. Injury that occurs when pressurized fluid penetrates the skin. (1)

flushing relief valve. Lower-pressure charge relief valve inside a closed-loop HST. It provides charge pressure relief when the transmission is in forward or reverse. (24)

flushing valve. Shuttle valve that is used to flush the oil from the motor and pump, which aids in cooling the transmission. Also known as a *replenishing valve* or *hot-oil purge valve*. (24)

force. Push or pull applied to an object. Measured in Newtons or pounds. (2)

force motor. Reversible linear electric motor that is used to actuate a pilot valve. (9)

forward and reverse lever. Lever that is connected to the pump's control valve and is used to control the speed and the direction of the HST. Also known as a *propulsion lever*. (24)

full film lubrication. Oil film that completely separates the surfaces of two moving components. (10)

full flow filtration. Type of filtration system that filters the entire system's oil flow. Full flow filters can be placed before the pump or in the return circuit. (11)

full-flow pressure. Pressure at which a relief valve completely opens and allows the system's maximum flow. (7)

functional problem. Legitimate machine malfunction originating from a defect or failure of a component, assembly, or system. (22)

G

galling. Material lifting off of one component and adhering to another component due to friction and lack of lubrication. (10)

gallons per minute (gpm). US Customary unit for measuring flow. (2)

gas accumulators. Accumulators that store energy by compressing a gas charge. Also known as *pneumatic accumulators* and *hydro-pneumatic accumulators*. (13)

gate valve. Adjustable flow-control valve that allows oil to flow straight through its passageway when the gate is opened. (8)

gear pumps. Fixed-displacement pumps in which gears provide the pumping action. (4)

gear track. Path cut into the bore of an aluminum pump housing by the steel gears. (4)

geroller motors. Popular type of internal-gear motor found in mobile equipment. They typically contain an inner rotor, a fixed outer ring, several rollers, an inner drive shaft coupled to the motor's output shaft, and some type of valving device to direct fluid to the motor's expanding oil chambers. (5)

gerotor internal-toothed gear pump. Internal-toothed pump that has a smaller external-toothed drive gear placed inside a larger internal-toothed driven gear. Teeth on the external-toothed gear have lobes rather than spur-shaped teeth. As the smaller external-toothed gear is driven, the outer gear with internal teeth orbits around the external-toothed gear. (4)

gerotor motor. Motor that has an internal-toothed ring gear and a star-shaped external-toothed gear. Both the internal- and external-toothed gears have lobe-shaped teeth. This type of motor is sometimes called an orbital motor, because the small star-shaped gear orbits inside of the ring gear. (5)

gland. Removable component that supports and guides the cylinder rod and acts like a cylinder rod bearing or bushing. (6)

globe valve. Valve similar to a gate valve, but in which oil must make a right-angle turn through the block when the globe is opened. (8)

goodwill. Charitable gesture, commonly opposing company guidelines, but done for a customer to show appreciation and to create a favorable public image for the business. (22)

graphic drawings. Drawing that uses symbols to depict a complete circuit or system. These drawings show all of the system components and how they are connected or related to the other components within the system. (3)

H

hand pump. Lever-operated hydraulic pump consisting of a piston, reservoir, two check valves, and an actuating lever. (4)

head end. Term that can describe either the rod end of a hydraulic cylinder (as defined by National Fluid Power Society) or the end opposite end of the rod (as defined by some construction equipment manufacturers). (6)

heat exchanger. Device used to cool the hydraulic oil in a system. See also *oil cooler*. (14)

high-pressure filters. Filters that are designed to withstand high system pressures and are installed in a heavy housing. (11)

horsepower control proportional solenoid. Name that Case assigns to their proportional solenoid valves (PSV). (20)

hose cutoff factor. Amount of hose that protrudes into a hose end and must be accounted for when cutting a hose to length. (15)

hydraulic computer. Assembly of shuttle and spool valves in a PFC hydraulic system that direct oil to the appropriate components, such as DCV spools, pump regulator, and additional valves. See also *pilot signal manifold*. (20)

hydraulic heaters. Devices that use heat from engine coolant or an electrical heating element to warm the hydraulic oil. The heating element is commonly placed in the reservoir. See also *hydraulic warmers*. (14)

hydraulic horsepower. Unit used to describe power in hydraulic systems. It is calculated as the product of flow and pressure. (2)

hydraulic hose. Flexible conductor for hydraulic oil. (15)

hydraulic hybrids. Hydraulic equipment that use a pump and motor coupled with an accumulator for the purpose of recovering hydraulic energy that would otherwise be transformed into heat energy and dissipated into the atmosphere. (13)

hydraulic lockout. Switch or lever that, when engaged, prevents the operator from actuating any of the machine's hydraulic controls. (20)

hydraulic motor. Positive-displacement actuator, containing several oil chambers that receive oil flow for the purpose of driving an output shaft. (5)

hydraulics. Use of a liquid to perform work, such as extending a cylinder to lift a heavy load. (2)

hydraulic warmers. Devices that use heat from engine coolant or an electrical heating element to warm the hydraulic oil. The heating element is commonly placed in the reservoir. See also *hydraulic heaters*. (14)

hydrodynamic. Having the ability to transfer power via fluids in motion. (2)

hydrodynamic drives. Fluid drive system that operates at a relatively low fluid pressure and relies on the fluid's mass and velocity for transmitting power. (24)

hydrolytic stability. A fluid's resistance to reaction with water. (10)

hydrostatic. Having the ability to transfer power via fluids at rest or under pressure. (2)

hydrostatic drive. Hydraulic drive system that uses fluids under pressure for the purpose of driving a machine and changing speed and torque. See also *hydrostatic transmission (HST)*. (24)

hydrostatic transmission (HST). Hydraulic drive system, consisting of a pump and motor, that allows the operator to change the machine's speed, torque, and direction. See also *hydrostatic drive*. (2)

hysteresis. Lag that occurs when the ECM commands the solenoid or torque motor to actuate, but that command is delayed due to stiction forces placed on the valve. (9)

I

inching valve. Foot-operated valve used on some HSTs to perform functions similar to a transmission clutch. (24)

inline axial piston motors. Hydraulic motors that contain a rotating group consisting of a barrel, pistons, a slipper retaining ring, and a bearing plate. In addition to the rotating group, the motors contain a swash plate and valve plate. (5)

inline axial pump. Piston pump that has a rotating cylinder block, also known as a barrel, that is splined to the pump's input shaft. As the input shaft rotates the cylinder, the pistons reciprocate in and out of their cylinders. (4)

inner tube. Inner part of a hose, made from oil-resistant materials and in direct contact with the oil. (15)

instrumentation problem. Malfunction that is caused by a fault in the sensing circuit and/or gauge and not in the operation of the component, assembly, or system. (22)

integral HSTs. Hydrostatic drive systems in which the pump and motor are directly connected to each other. (24)

integral reservoir. Reservoir that also functions as the machine's transmission housing or axle housing. (14)

intermediate passageway. DCV fluid duct that routes oil from the DCV spool to the pressure compensator valve in a flow-sharing hydraulic system. Case uses this term in their construction equipment service literature. See also *feeder passageway* or *compensator passageway*. (19)

internal pressure override (IPOR). Valve placed in series between a charge pump and the inlet to the drive pump's servo control valve. It senses drive pressure and protects the machine from high pressure overloads for extended periods of time. (24)

internal-toothed gear motors. Hydraulic motors that contain an internal-toothed gear. They are commonly configured with a large internal-toothed ring gear that surrounds a smaller external-toothed gear. (5)

intra-vane pump. Vane pump in which a small vane is contained inside a larger vane. The two vanes form a smaller oil chamber used for extending the vanes. (4)

ISO range code. Most common standard for evaluating system cleanliness in the mobile equipment industry. (12)

ISO viscosity grade (ISO VG). Scale developed by the International Standards Organization (ISO) for indicating a fluid's viscosity. A grade is assigned to a fluid based on its viscosity range at a temperature of 104°F (40°C). (10)

isolation valve. Valve used to isolate a component, such as an accumulator, from the hydraulic system. (13)
isolators. Term used for shuttle valves in John Deere construction equipment. (18)

J

jammer solenoid valve. A valve used in both open-center hydraulics systems and closed-center LSPC hydraulic systems, as a means to build maximum system pressure. Also known as a *dump valve* in open-center systems. (16)

K

kidney loop filter. Bypass filtration system that is separate from the main hydraulic system. It is designed to slowly filter the fluid inside the reservoir, and allows a very efficient filter to be used without interfering with system operation. (11)

L

laminar oil flow. Smooth, uniform, turbulence-free flow of oil. (10)
lapping machine. Machine with a flat, rotating surface to which an abrasive slurry is applied. It is used to machine the flat surfaces of the rotating group. (4)
law of conservation of energy. Law of physics that states that energy cannot be created or destroyed, but can only be transformed from one form to another. (2)
lay line. Line printed along the length of a hose, which a technician can use as a visual indicator for determining if the hose is twisted. (15)
lift check valves. Valves designed to hold the cylinder's position when the DCV spool is first actuated. Once system pressure is higher than the load's pressure, the lift check opens and causes the cylinder to move. (7)
lip seal. Piston seal that is designed to hold fluid in one direction. Fluid pressure forces the lip to seal against the barrel and piston, providing a snug connection. (6)
liters per minute (lpm). Metric unit for measuring flow. (2)
load control valve. Valve used in a PFC hydraulic system that has the responsibility of destroking the pump when too much hydraulic horsepower has been requested. (20)
load-holding valve. Valve that prevents an actuator from moving or drifting when the DCV is in a neutral position. (7)
load reaction SCU. Steering control unit configured to allow the external loads placed on the steering cylinder(s) during a turn to actuate the hydraulic steering system. (25)
load-sensing (LS) hydraulic system. Hydraulic system design designed to control pump outlet flow rate based on the highest working pressure delivered to the pump via the signal network. (18)
load-sensing pressure-compensating (LSPC) hydraulic system. Hydraulic system that uses shuttle valves to sense the actuator's working pressure for the purpose of sending signal pressure to the pump's flow control spool, or to an unloading valve, and to the DCV compensators. The compensators can be pre-spool or post-spool. (9)
load-sensing (LS) pumps. Pump that has the ability to vary its flow based on the actuators' working pressures. (4)
load valve. Adjustable needle valve added to a flowmeter to build pressure in the system by restricting the hydraulic pump's flow. (21)
lubricity. Ability to provide lubrication. (10)

M

magnetostrictive cylinder sensor. Internal linear-displacement transducer (LDT) that directly measures a cylinder's position. (6)
make-up valves. Low-pressure check valves that prevent cavitation. (5)
manual bypass valve. Valve that allows the fluid from one leg of an HST drive loop to be bypassed into the other leg of the loop. (24)
manually controlled DCVs. Directional control valve that is directly controlled by the machine operator, typically using a lever. (9)
manual steering check valve. Valve connecting the steering control unit's inlet and tank return ports to allow the SCU's gerotor assembly to draw oil directly from the reservoir. This oil is used for steering in the event of a hydraulic pump or engine failure. (25)
manual steering mode. Steering control unit feature that enables the operator to steer the tractor in the event of an engine or hydraulic failure. See also *emergency steering*. (25)
margin pressure. Difference between hydraulic pump outlet pressure and signal pressure. See also *differential pressure*. (18)
mechanical efficiency. Measure of the motor's efficiency that accounts for energy losses that occur due to friction and drag. It is determined by dividing the actual motor's torque by the theoretical torque and multiplying that product times 100. (5)
mechanical horsepower. Unit used to describe power in mechanical systems. Equal to 33,000 ft-lb/min (550 ft-lb/second). (2)
metal-to-metal fittings. Fittings that create a seal by mating the metal surfaces of the two fittings. No O-ring is required to provide the seal. (15)
metering-in. Controlling the speed of an actuator by limiting the flow of oil into a cylinder. (6)
metering-out. Controlling the speed of an actuator by limiting the flow of a cylinder's return oil. (6)
micron (µm). One millionth of a meter. Also known as a *micrometer*. (11)
Mine Safety and Health Administration (MSHA). United States federal agency responsible for ensuring mine site safety. (1)

minimum bend radius. Tightest bend that is allowable for a given hose or tube. (15)

minimum/maximum. Electronic digital pressure meter feature that provides the minimum and the maximum recorded pressure in a hydraulic system during a chosen length of operation. (21)

minimum pump flow. Flow produced by NFC and PFC variable-displacement pumps when their DCVs are in a neutral position, typically 15–25% of the pump's maximum displacement. (20)

N

National Aerospace Standard (NAS). Standard occasionally used for evaluating system cleanliness in the mobile equipment industry. (12)

National Pipe Tapered (NPT). Course thread type used on pipe fittings. The thread is designed so that the male component's tapered exterior threads are compressed into the mated port's or pipe's tapered interior threads during tightening. (15)

National Pipe Tapered Fuel (NPTF). Fine thread type used on pipe fittings. The thread is designed so that the male component's tapered exterior threads are compressed into the mated port's or pipe's tapered interior threads during tightening. (15)

National Response Center (NRC). Federal communication center to which the spill of any reportable quantity of oil must be reported. The center will evaluate the spill and coordinate the response. (1)

needle valve. Valve having fine threads on the valve stem, which allow small, precise adjustments to the valve's orifice size and the resulting pressure and flow. (8)

negative flow control. Type of pump control system that is designed to upstroke the pump when a signal pressure (known as NFC pressure) drops. (20)

neutral relief valve. Charge relief located in a closed-loop HST. It is so named because it provides charge pressure relief when the transmission is in neutral. (24)

neutral safety switch. Electrical switch that prevents the engine from starting anytime the propulsion lever is outside of neutral. (24)

nitrogen charging pumps. Device that draws nitrogen from the atmosphere and filters out the oxygen and other gases. The pump can then supply the nitrogen at pressures up to 5000 psi (350 bar) to charge accumulators. (13)

nitrogen gas. Inert gas preferred for use in gas accumulators. (13)

nominal rating. Somewhat arbitrary number that indicates the smallest diameter of particles that the filtration device is expected to capture with relative consistency. (11)

nomograph. Graph that can be used to find a third variable when any two variables are known. (2)

non-conductive. Resistant to the flow of electricity. (15)

non-load reaction SCU. Steering control unit configured to hold the steering cylinder(s) in a fixed position even when external loads are exerting forces on the steering cylinder(s). (25)

non-positive-displacement pump. Pump in which the impeller does not have a tight sealing surface within the pump body. As the impeller is driven, fluid is flung outward by centrifugal force, providing flow. (4)

non-pressure-compensated. Having an output that varies with changes in input pressure or flow. (8)

nonverbal communication. Interaction among people based on visual cues and not with spoken words. (22)

O

Occupational Safety and Health Administration (OSHA). United States federal agency that is responsible for ensuring that employees have a safe work environment. (1)

oil cooler. Device used to cool the hydraulic oil in a system. See also *heat exchanger*. (14)

oil gallery. Internal passage or area designed within a hydraulic pump that temporarily holds hydraulic fluid before it is distributed to internal pump components. (17)

one-way check valves. Valves that allow flow in only one direction and block flow in the other direction. (9)

open-center hydraulic system. Hydraulic system that uses a fixed-displacement pump that is matched with open-center DCVs. When the DCVs are in a neutral position, an open passageway allows the pump's flow to be routed to the reservoir at a low pressure. (9)

open-loop HST. Hydrostatic transmission in which the return oil is routed directly to the reservoir, and the pump must draw all of its oil from the reservoir. (24)

open-loop pump. Pump that draws all of its inlet oil directly from the reservoir. (4)

orifice. Non-pressure-compensated flow-control valve that consists of a small passageway that limits the flow of oil. (8)

O-ring boss (ORB) fitting. Fitting that has a groove machined into the sealing surface of the male fitting to hold an O-ring. The male fitting compresses the O-ring into a machined seat located in the female port of the housing. (15)

O-ring face seal (ORFS) fitting. Fitting with a groove machined into the face of the thread body of the male fitting to hold the O-ring. The female fitting has a flat face. The male fitting compresses the O-ring into the female fitting's flat face when assembled. (15)

overrunning load. Load placed on an actuator that causes the actuator to move before oil can actuate the cylinder or motor. (5)

oxidation stability. Fluid's resistance to reaction with oxygen. (10)

P

parallel hydraulic hybrid. Hydraulic hybrid system in which the hydraulic motor is not the sole source responsible for driving the vehicle. The hydraulic motor can be decoupled from power train, allowing the machine to be propelled solely by the engine and power train, with no aid from the hydraulic motor. (13)

particle counter. Machine that counts the number of contaminants based on the size of the contaminants. (12)

Pascal's law. Law of physics that dictates that a force acting upon a liquid in a container will pressurize the liquid, and that the pressure will act equally in all directions. (2)

pattern changer valve: Valve used to change an excavator's control pattern. It allows any of three control configuration to be selected: excavator position, neutral middle position, and backhoe position. The neutral position prevents hydraulic operation and can be used to prevent vandalism or any other unwanted operation. (20)

personal protective equipment (PPE). Equipment and clothing that is designed to protect the employees from potential injuries or illnesses. (1)

petroleum fluids. Fluids obtained by refining crude oil. These are the fluids most commonly used in mobile hydraulics. (10)

pictorial drawings. Drawings that depict the actual appearance of hydraulic components. (3)

pilot-controlled DCV. Directional control valve that is remotely controlled by a low-pressure hydraulic or pneumatic signal. (9)

pilot controller. Pilot valve that directs oil to a DCV. (20)

pilot-operated relief valve. Relief valve that contains two poppets, a large main poppet and a smaller pilot poppet, that work in unison. Also known as a *compound relief valve* or a *balanced relief valve*. (7)

pilot signal manifold. Assembly of shuttle and spool valves in a PFC hydraulic system that direct oil to the appropriate components, such as DCV spools, pump regulator, and additional valves. See also *hydraulic computer*. (20)

pipe. Thick-wall, rigid hydraulic fluid conductor that is *not* designed to be bent. (15)

pipe union. Three-piece component designed to speed the process of connecting and disconnecting two pieces of pipe. (15)

piston accumulator. Type of gas accumulator in which the gas charge is sealed in the cylinder and acts on one side of a piston. System oil acts on the other side of the piston. As system pressure increases, the piston compresses the gas charge, storing energy. (13)

pneumatics. Use of a gas to perform work, such as extending a cylinder to lift a heavy load. (2)

positive displacement pump. Pump with tight sealing surfaces. This type of pump ejects practically all of the oil that it takes in during an operational cycle. (4)

positive flow control. Type of pump control system that is designed to upstroke the pump when a signal pressure (known as PFC pressure) increases. (20)

post-spool-compensated. Pressure compensation design that places the pressure compensator valve after a DCV spool within the hydraulic system. See also *downstream compensation*. (19)

post-spool compensation. Method of pressure compensation in which the pressure-compensating valve is located after the DCV spool. (8)

potentiometer. Three-wire sensor that provides a variable resistance based on the location of the signal wiper. (6)

pour point. Lowest temperature at which the specified fluid will flow. (10)

power beyond. Optional hydraulic system feature that enables additional hydraulic control valves to be added to a tractor's hydraulic system. The additional DCVs are usually located on the implement. (9)

power save solenoid. Valve, used on Case excavators built by Sumitomo, that reduces the engine load by 10% when the operator controls are at idle. (20)

power shift pressure reducing valve. Single power shift solenoid valve used to destroke both NFC pumps on Caterpillar NFC excavators. (20)

power shift solenoid. Name that Caterpillar assigns to their proportional solenoid valves (PSV) used on NFC excavators to prevent the engine from lugging or stalling. (20)

PQ curve. Graph depicting a variable-displacement hydraulic pump's flow rate (Q) at different operating pressures (P). The flow rate is commonly placed on the Y axis and the operating pressure on the X axis. (18)

precharge pressure. Pressure of the gas in a gas accumulator when no oil is present. (13)

pre-spool compensation. Pressure compensation design that places the pressure compensator valve before a DCV spool within the hydraulic system. See also *upstream compensation*. (8)

pressure. Force distributed over a given unit of area. It is measured in pounds per square inch (psi), bar, or kilopascals (kPa). (2)

pressure-compensating (PC) hydraulic system. Hydraulic system that uses a variable-displacement pump that is matched with closed-center DCVs. It can alter hydraulic pump displacement to control pump outlet flow rate based on a predetermined maximum system pressure limit. See also *pressure-limiting hydraulic system*. (9)

pressure compensation. Ability to maintain a constant actuator speed (cylinder or hydraulic motor) based on a fixed position (opening) of the DCV spool. (8)

pressure cutoff spool. Term used to describe the pressure compensator control device that is part of an LSPC variable-displacement hydraulic pump. Also known as a *high-pressure standby spool*. (18)

pressure cutoff valve. Valve that provides protection from high-pressure overloads in an HST on late-model Case IH combines. This valve control's the system's high pressure during gradual pressure buildup rather than from pressure spikes, which are managed by the pump's high-pressure relief valves. (24)

pressure-limiting hydraulic system. Hydraulic system design that can alter hydraulic pump displacement to control pump outlet flow rate based on a predetermined maximum system pressure limit. See also *pressure-compensating hydraulic system.* (17)

pressure override. Pressure range over which a relief valve provides protection. It is calculated by subtracting the valve's cracking pressure from the full-flow pressure. (7)

pressure-reducing valve (PRV). Normally open valve that limits its outlet pressure by sensing downstream oil pressure. (7)

pressure-release solenoid. Solenoid valve, used in the HST on a Case IH 7120-9120 series combine, that performs the same basic function as a manual bypass valve. (24)

pressure-relief valve. Normally closed valve that opens to allow fluid to pass once the pressure reaches the valve's setting. (7)

pressure-sequence valve. Pressure-control valve that directs flow to a secondary circuit after system pressure has risen enough in the primary circuit to overcome the closing pressure of the spring within the valve. (7)

pressurized return lines. Main return lines that send low-pressure oil back to the reservoir. (14)

preventative maintenance (PM) inspection. Comprehensive assessment of a machine's operation performed on a scheduled interval to inspect components, measure machine parameters, monitor efficiencies, and avoid major problems during its service life. (22)

primary shuttle valve. Designation of shuttle valve that determines an individual actuator's higher working pressure. (18)

priority valves. Valves that ensure the demand of the primary circuit is met before any remaining oil is sent to secondary circuits. (8)

programmable kick-outs. Electronic controls that are used to reduce cylinder shock loads by slowing cylinder rod travel as it approaches its designated position. (6)

proportional flow divider valves. Valves that are designed to divide a quantity of oil flow in a fixed ratio, supplying flow to two or more branches. (8)

proportional flow filtration. Filtration system that filters only a partial amount of the hydraulic system's oil flow. Also known as *bypass filtration.* (11)

proportional priority pressure compensation (PPPC). Flow-sharing load-sensing hydraulic system developed and used by Caterpillar. (19)

proportional solenoid valve. Valve that is opened and closed by a pulse-width modulated solenoid. Also known as a *rapid on-off solenoid valve.* (9)

proportional solenoid valve (PSV). Valve, used in NFC hydraulic systems, that works in conjunction with pump signal pressures for the purpose of destroking the pump to prevent the engine from stalling. (20)

pulse-width modulated (PWM) solenoid. Solenoid that varies output based on the duty cycle at which the coil is energized and de-energized. See also *proportional solenoid valve.* (9)

pump compensator. Mechanism designed to apply pressurized fluid to a control piston within a variable-displacement pump. It destrokes the pump when the pump outlet pressure reaches the pump compensator's spring value. See also *pump cutoff valve.* (17)

pump cutoff valve. Valve used in a variable-displacement hydraulic pump to shut off the pump's flow by destroking the pump when system pressure overcomes compensator spring pressure. Also known as a *high-pressure standby spool.* (17)

pump regulator. Single, double-acting, servo piston, used in an NFC or PFC pump, that adjusts the pump's swash plate angle to increase or decrease pump flow. (20)

Q

quantity fuse. Device that protects a hydraulic system from excessive flow. It is similar to a velocity fuse, but is used in low-flow circuits with a fixed flow rate. (8)

Quick Steer. Electronic steering mode developed by Caterpillar that decreases the operator's turns of the steering wheel to complete a full steer from left to right or right to left by roughly 90%. (25)

R

radial piston motors. Hydraulic motors in which the pistons are positioned perpendicular to the motor's output shaft. The motor can be a rotating cam design, which uses a stationary cylinder block, or a rotating piston block design. (5)

ram. Single-acting cylinder that uses a cylinder rod that is the same diameter as the cylinder's piston. They are used in applications that require long stroke and rod rigidity. (6)

reference oil. Sample of new oil, used to establish a baseline so changes in the oil's condition can be properly evaluated. (12)

regeneration. Re-routing of return oil to the inlet of an unloaded differential cylinder as it is extending. Its purpose is to increase the speed at which an unloaded cylinder is extended. (6)

reinforcement. Center structure of a hose that gives the hose its strength. Three styles are commonly used: braided, spiral, and helical. (15)

remote pressure compensation. Type of system in which an external pressure compensator valve is located in a remote location, away from the pump. (17)

remote valve. One of two PFC regulator control valves in Deere systems. It receives a signal from a flow rate control valve, which varies in proportion based on how far the operator moved the controller. (20)

reportable quantity (RQ). Quantity of spilled oil that requires someone to file a report documenting the spill with a local, state, or federal agency. (1)

reservoir. Containment device that houses a machine's hydraulic fluid. Also known as a *hydraulic tank* or *sump*. (14)

resolvers. Term used for shuttle valves in Caterpillar equipment. (18)

return filter. Filter that is installed in the system's return line and is designed to remove contaminants before the oil re-enters the reservoir. (11)

reusable hose ends. Fittings that are designed to be removed from an old hose and reused on a replacement hose. (15)

reversible pump. Pump that provides proper pumping action regardless of the direction the shaft rotates, or a pump that is driven in one direction, either CCW or CW, but has the capacity to internally reverse the direction of the pump's flow. (4)

Reynolds number (Re). Measure of the smoothness of fluid flow. (10)

rod seal. Dynamic-type seal installed in the cylinder gland. It is designed to hold pressure as the rod moves in and out of the cylinder. (6)

rotary actuators. Hydraulic device that transforms fluid energy into rotational mechanical energy. (5)

rotating-block radial piston pumps. Pumps that have a block assembly that is driven by the pump's input shaft. As the block rotates, the pistons move in and out of their cylinders, providing the pumping action. (4)

rotating cam piston motor. Type of hydraulic piston motor in which the pistons force a cam output shaft assembly to rotate. (5)

rotating cam radial piston pump. Pump in which the pump housing is stationary and the pistons reciprocate in and out of their individual bores, creating an expanding volume and a decreasing volume. (4)

rotating cylinder block piston motor. Hydraulic piston motor design in which several pistons are located inside a rotating cylinder block. Fluid pressure is directed to the pistons causing them to slide out against a ring, which causes the cylinder block to rotate. The cylinder block is splined to the motor's output shaft. (5)

rotating group. Assembly of parts that rotate together in a piston pump. It consists of the pump's barrel, piston slipper assemblies, slipper retaining ring, and bearing plate. (4)

rust and oxidation inhibitors. Additives in hydraulic fluid that help prevent rust and oxidation. (10)

S

safety block. Term used to describe a dump valve and an isolation valve that are placed inside one valve block assembly. (13)

safety data sheets (SDS). Information sheets that provide end users important information about a product. The 16 categories of information included in the sheets include safety and hazards, physical and chemical properties, and transport and disposal, among others. (1)

safety factor. Ratio between the working pressure and burst pressure. For a hydraulic hose, the safety factor is four, meaning that the burst pressure of the hose is four times greater than its working pressure. (15)

safety fuse. Device that protects accumulator circuits from excessive pressure or excessive heat (which leads to excessive pressure). (13)

sampling probe. Fitting installed on the end of tubing so the tubing can be used to draw an oil sample from the system's sampling valve. (12)

sampling valve. Valve used exclusively for drawing oil samples and strategically placed in the system so it will provide a representative sample using the system's turbulent oil flow. (12)

Saybolt Universal Seconds (SUS). Unit of measure for kinematic viscosity. This unit is commonly used in the United States. Also known as *Saybolt Seconds Universal (SSU)*. (10)

scavenge pump. Pump that is designed to draw oil from a sump and return the oil back to a reservoir. (14)

schedule number. Code that indicates the wall thickness of pipe with a given outside diameter. (15)

SCU internal leakage test. Procedure to assess the amount of tolerated oil seepage between the parts operating inside of a steering control unit. (25)

scheduled oil sampling (SOS). Phrase coined by Caterpillar to identify periodic oil sampling for the purpose of monitoring oil condition over time. (12)

seat and poppet valve. Valve consisting of a poppet and a matching seat. When the poppet moves away from the seat, a passage for oil flow is created. Flow is blocked when the poppet reseats. (9)

secondary shuttle valve. Designation of shuttle valve that determines the higher working pressure between two different DCVs' working pressures. (18)

Selective Control Valve (SCV). Term used by John Deere to describe a directional control valve. (8)

series hydraulic hybrid. Hydraulic hybrid system in which the engine is designed to drive the pump, and the pump is designed to drive a motor. The motor is responsible for driving the wheels through a final drive and differential. (13)

service bulletins. Documents used by manufacturers to distribute technical-update information to dealerships after a product has been released. (3)

service letters. Documents sent only to the manufacturer's field representatives to make them aware of short-term solutions or fixes that are not ready to be distributed throughout the dealership network. (3)

servo oil. Oil flow that is used to actuate a servo piston. Also known as *control oil*. (24)

servo valve. Valve that is operated by a servo. (9)

shuttle valve. Valve that can sense two different working pressures and send the higher pressure to the next destination in the circuit. (9)

signal limiter valve. Valve that enables a flow-sharing hydraulic system to maintain a pressure drop across the DCV spools, which allows oil flow to be metered to the actuators, even if one of the actuators stalls. See also *signal relief valve*. (19)

signal network. Group of primary and secondary shuttle valves that sense actuator working pressure in a load-sensing hydraulic system. (18)

signal pressure. Hydraulic fluid force developed by and based on the load placed on a cylinder. The pressure is sent through the load-sensing line when the cylinder is actuated. (18)

signal relief valve. Valve that enables a flow-sharing hydraulic system to maintain a pressure drop across the DCV spools, which allows oil flow to be metered to the actuators, even if one of the actuators stalls. See also *signal limiter valve*. (19)

single-acting cylinder. Hydraulic cylinder that is hydraulically actuated in just one direction. The cylinder uses an outside force to return the cylinder back to its original state. (6)

single-path HST. Hydrostatic drive system consisting of one variable-displacement reversible hydraulic pump and one hydraulic motor. (24)

single-speed motor. Fixed-displacement motor. (24)

skiving. Process of removing the outer covering from the end of a hose that will be inserted into a fitting. (15)

sliding-vane pump. Vane pump that consists of a rotor that is driven by the pump's input shaft, has vanes that slide in and out of the rotor, and a cam-shaped housing that forms the outer edge of the pump chamber. (4)

slipper pad. Shoe that is attached to the piston and rides against the swash plate's surface as the pump's cylinder block rotates. (4)

slipper retaining ring. Ring that holds the slipper pads as they ride along the swash plate. (4)

solenoid. Electrical coil with a plunger that extends or retracts when the coil is energized. (9)

spanner wrench. Tool designed to grip a cylinder gland by attaching to the gland's key slots or dowel holes. (6)

specific gravity (Sg). Ratio of the fluid's weight in relation to the weight of water. (2)

spectrometer. Instrument that measures the different frequencies of light emitted from a heated oil sample in order to identify the chemical makeup of the oil and any contaminants it contains. (12)

Spill Prevention, Control, and Countermeasure (SPCC) Act. Regulation that defines the requirements for oil spill prevention, preparedness, and response in the event of
a spill. (1)

split-flange O-ring fitting. Fitting with a groove machined into the face of the thread body of the male fitting to hold the O-ring. The female fitting has a flat face. The male fitting is held to the flat face on the female fitting by two clamps and four bolts. (15)

split HST. Hydrostatic drive system in which an engine-pump is separate from the hydraulic motor and connected to it by hoses or tubing. (24)

spool valve. Valve that contains a core with lands separated by interland spaces, referred to as a spool. The valve housing contains a bore for the spool. As the spool moves in its bore, it opens and seals off different ports in the housing, changing or blocking the flow of oil through the valve. (9)

stall mode. Hydraulic pump condition that occurs anytime system pressure deadheads and the pump cannot build enough pressure to overcome both the signal pressure and the margin spring value. The pump is destroked and system pressure builds. (18)

standby mode. Hydraulic pump condition that occurs when the DCVs are in a neutral position. The pump is destroked and pump outlet pressure equals margin pressure. (18)

standby pressure. High-pressure setting that a pressure-compensating hydraulic pump maintains when the DCVs are in a neutral position and the pump is waiting to perform work. (17)

static load-sensing signal. Steering signal that has a pressure based on the steering cylinder's working pressure. Unlike the dynamic load-sensing signal, it does not have flow when the SCU is in a neutral position. (25)

steer by wire. Steering system that uses an electronic control module to transform the steering actions of the machine operator into electrical signals. These signals are transmitted to electronic actuators that steer the wheels of the machine. (25)

steering control unit (SCU). Manually operated group of hydraulic components rotated by a steering wheel and used to control the actuation of a hydraulic steering system. See also *steering hand-pump*. (25)

steering hand-pump. Another term for a steering control unit (SCU) based on the way the gerotor assembly can serve as a manually actuated pump if required. (25)

steering kickback. Jerkiness or recoil in a steering wheel caused by steering oil bleeding backwards through the steering control unit. (25)

stepper motor. Motor that precisely controls shaft rotation by pulsing electromagnets that surround the motor's armature. This on/off pulsing of the electromagnets divides the movement of the armature into discrete, controlled segments. (9)

strainers. Filtration device, typically made of wire mesh, designed to remove only large particles. (11)
stroke control valve. Valve installed as part of the pump compensator in a radial piston variable-displacement hydraulic pump. It destrokes the pump by delivering pressurized oil into the crankcase to hold the pistons retracted. (17)
suction filter. Filter installed upstream from a pump's inlet. (11)
suction hose. Hose designed to provide the hydraulic pump with a supply of oil. The hose is reinforced to prevent it from collapsing under vacuum. (15)
surface-type filtering devices. Class of filtration device that consists of a single thin surface layer through which oil passes. Strainers are the most common surface-type filtering devices. (11)
swaging. Process of forcing a hose end through a set of smooth-tapered dies. (15)
synthetic fluids. Fluids that are designed by engineers, rather than simply refined from crude oil. (10)

T

telescoping cylinder. Linear actuator that contains up to six cylinder tubes that extend to provide a very long cylinder stroke. During retraction, the tubes collapse into a compact cylinder. (6)
temperature controller. Heat exchanger that uses engine coolant to cool or warm the hydraulic oil. (14)
tension load. Load that occurs when a cylinder rod is subjected to a pulling force. Also known as *shear load*. (6)
thermal relief valve. Valve designed to prevent the cylinder and hoses from rupturing due to this heated oil expansion by dumping the overpressurized oil to the tank. (7)
three-way pressure-compensated flow-control valve. Flow-control valve that has three hydraulic ports: an inlet port, an outlet port, and a bypass port. Also known as a *bypass flow-control valve*. (8)
thrust load. Load that results from a cylinder rod applying a pushing force. Also known as *compression load*. (6)
tie bolt cylinder. Hydraulic cylinder with a housing that contains four long bolts that mate the rod end with the cap end. The design simplifies the repair and rebuilding of the cylinder. (6)
torque. Force being applied through an arcing motion, such as a rotating lever. (5)
torque control solenoid. Proportional solenoid valve used in Deere and Hitachi excavators to send pilot pressure to the load control valve in order to destroke the pump to the minimum flow angle. (20)
torque converter. Hydrodynamic drive that transfers power, acts like an automatic clutch, and multiplies torque. (24)
torque dividers. Drive assembly consisting of a torque converter and an internal planetary gear set. The planetary gear set provides additional torque multiplication. (24)
torque-limiting control. Mechanism added to a LSPC variable-displacement pump that destrokes the pump when the operator requests too much hydraulic horsepower. (18)
torque motor. Electronically controlled electric motor that uses a "T-shaped" armature that pivots when activated. (9)
total horsepower control. Name Kawasaki Hydraulics applies to a pump control system in which each pump regulator can be destroked by sensing its own pump outlet pressure or the companion pump outlet pressure. (20)
trended. Evaluated based on incremental changes over a period of time. (12)
trunnion bearing axial piston pump. Pump in which the swash plate pivots on two trunnions that are located inside two bearings. (4)
tubing. Thin-walled, non-flexible steel hydraulic oil conductor designed to be bent and shaped as required by its application within the system. (15)
two-speed motor. Motor that can be stroked between two fixed swash plate angles, a high-speed angle and a low-speed angle. (24)
two-way pressure-compensated flow-control valve. Flow control valve having two ports and in which differential pressure causes the compensator spool to move away from or toward the inlet port, restricting or increasing the flow of oil to the outlet port. Also known as a *restrictor*. (8)

U

unidirectional. Configured to rotate in only one direction, either clockwise or counterclockwise. (5)
unidirectional pump. Pump that is designed to be rotated in only one direction, either clockwise or counterclockwise, and that will produce flow in only one direction. (4)
unloading valve. Pressure-control valve that is used to unload a hydraulic pump, allowing the system to operate at a low pressure value. (7)
upstream compensation. Pressure compensation design that places the pressure compensator valve before a DCV spool within the hydraulic system. See also *pre-spool compensation*. (18)
upstroked. Adjusted to provide increased pump oil flow. Also used to describe a variable-displacement pump that has been adjusted to provide maximum oil flow. (17)

V

vacuum pump. Manually actuated pump used to draw oil from the hydraulic reservoir. It is used to take an oil sample. (12)
valve overlap. Distance a DCV spool must move from its centered position before it initially opens the valve's port(s). (9)

valve plate. Stationary plate containing one or two metering slots, which determine the direction of rotation for the pump. It directs oil in and out of the pump's rotating group. (4)
vane motors. Hydraulic motor in which the rotor is equipped with sliding vanes and is splined to the motor's output shaft. (5)
vane pump cartridge. Preassembled unit containing the cam ring, rotor, vanes, and side plates. It is used to rebuild a vane pump. (4)
variable adjustment. Changes made by the operator to vary the speeds and positions of component while the machine is in operation. (2)
variable-displacement pumps. Pump in which the chamber's effective volumes can be changed to increase or decrease flow for a given engine speed. (4)
variable effort steering. Electronic steering feature that adjusts the steering force required by the operator to best control the tractor under changing operating conditions. (25)
variable ratio steering. Electronic steering feature that optimizes the number of turns the steering wheel must rotate depending on tractor speed and the degree of sharpness of the turn being performed. (25)
variable-volume reservoir (VVR). Reservoir in which the oil is contained in a synthetic bladder and pressured by a coiled spring acting on that bladder. The sealed design prevents outside air from entering the reservoir. (14)
velocity fuse. Device that protects the hydraulic system from excessive flow. It works on the principle of a pressure drop across an orifice. Also known as a *flow fuse*. (8)
viscosity. Liquid's resistance to flow based on its thickness. (8)
viscosity index (VI). Rating assigned to an oil to indicate how much or how little the oil's viscosity changes across a wide temperature spectrum. (10)
volumetric efficiency. Ratio between the amount of fluid actually displaced by a pump and the theoretical maximum amount of fluid the pump could displace. (4)
volumetric expansion (VE). Swelling of a hydraulic hose due to pressurization of fluid within the hose. (15)

W

water-type oil coolers. Class of oil coolers that use water or engine coolant to cool the oil. (14)
wear ring. Ring installed on a cylinder's piston to prevent the piston from rubbing the barrel. (6)
whole goods kits. Attachments that are purchased by a customer and installed as a kit after the machine has been delivered. (3)
wide-angle SCU. Steering control unit configured to smooth steering motion by increasing the amount of deflection between the SCU's control spool and sleeve during a turn. (25)
wiper. External seal that protrudes from a cylinder gland and removes contamination from the surface of the cylinder rod as it retracts. (6)
wire mesh number. Number that indicates the fineness of a wire mesh screen. The number is equal to one half of the total number of wires in each square inch of the strainer. (11)
work. Force applied through a distance, which occurs anytime a force is used to move an object. (2)
working mode. Hydraulic pump condition anytime a machine operator moves a DCV to route oil to operate an actuator. The pump is upstroked and pump outlet pressure equals the signal pressure plus the margin pressure. (18)
working pressure. Maximum pressure at which a hose can safely be used. (15)

Index

C

D

E

G

H

I

J

K

L

M

Q

R

S